Profiles in Mathematics

Brief historical sketches and vignettes present the stories of people who have advanced the discipline of mathematics.

PROFILE IN MATHEMATICS

John Napier

During the seventeenth century, the growth of scientific fields such as astronomy required the ability to perform often unwieldy calculations. Scottish mathematician John Napier (1550–1617) made great contributions toward solving the problem of computing these... clude... and a... plicat... Napie... oped...

Mathematics Today

This feature relates mathematics to everyday life, helping students to recognize the need for math and gain an appreciation for math in their lives.

MATHEMATICS TODAY

The New Old Math

Sometimes the *new* way to solve a problem is to use an *old* method. Such is the case in some elementary schools with regard to multiplying whole numbers. In addition to teaching the traditional multiplication algorithm, some schools are also teaching the lattice method of multiplication. Lattice multiplication, along with the Hindu–Arabic numerals, was introduced to Europe in the thirteenth century by Fibonacci. (See the *Profile in Mathematics?* on page 298.) Lattice multiplication is recommended by the National Council of Teachers of Mathematics as a way to help students increase their conceptual understanding as they are developing computational ability. This method gained in popularity when it was included in several elementary school textbooks as a way to help students understand how the traditional multiplication algorithm works.

Did You Know?

These colorful and engaging boxed features highlight the connections between mathematics and a wide variety of other disciplines, including history, the arts and sciences, and technology.

DID YOU KNOW?

Speaking to Machines

Pit Land

On digital video discs (DVDs) and compact discs (CDs), video and sound... digitally encoded on the underside... the disc in a binary system of pits... "lands" (nonpits). To play the dis... laser beam tracks along the spira... is reflected when it hits a land (si... sent = 1), but it is not reflected... pits (no signal = 0). The binary s... quence is then converted into vi... images and music.

For the past 600 year... have used the Hindu–A... system of numeration w... change. Our base 10 nume... system seems so obvious t... perhaps because of our 10 f... and 10 toes, but it would b... to think that numbers in... bases are not useful. In fact,... the most significant nume... systems is the binary syste... base 2. This system, with it... mental simplicity, is what is u... computers to process inform... and "talk" to one another. W... computer receives a comma... data, every character in the... mand or data must first be... verted into a binary numeral f... computer to understand and...

RECREATIONAL MATHEMATICS

74. Determine b, by trial and error, if $1304_b = 204$.

75. In a base 4 system, each of the four numerals is represented by one of the following colors:

Determine the value of each color if the following addition is true in base 4.

INTERNET/RESEARCH ACTIVITIES

76. Write a report on how computers use the binary (base 2), octal (base 8), and hexadecimal (base 16) numeration systems.

Exercise Sets

The diverse and numerous exercise sets include exercise types such as Concept/Writing, Practice the Skills, Problem Solving, Challenge Problem/ Group Activity, Recreational Mathematics, and Internet/ Research Activities.

A Survey of

MATHEMATICS

with Applications

EIGHTH EDITION

EIGHTH EDITION

A Survey of
MATHEMATICS
with Applications

ALLEN R. ANGEL

Monroe Community College

CHRISTINE D. ABBOTT

Monroe Community College

DENNIS C. RUNDE

Manatee Community College

PEARSON

Addison
Wesley

Boston   San Francisco   New York
London   Toronto   Sydney   Tokyo   Singapore   Madrid
Mexico City   Munich   Paris   Cape Town   Hong Kong   Montreal

Publisher: Greg Tobin

Executive Editor: Anne Kelly

Acquisitions Editor: Marnie Greenhut

Senior Project Editor: Rachel Reeve

Associate Project Editor: Elizabeth Bernardi

Editorial Assistant: Leah Goldberg

Senior Managing Editor: Karen Wernholm

Senior Production Supervisor: Peggy McMahon

Design Supervisor: Barbara T. Atkinson

Cover Designer: Barbara T. Atkinson

Cover Image: Rows of speakers, © Getty Images, Inc.

Cover Photo Photographer: Chad Baker

Photo Researcher: Beth Anderson

Media Producer: Ashley O'Shaughnessy

Software Development: Eric Gregg and Eileen Morse

Executive Marketing Manager: Becky Anderson

Marketing Coordinator: Bonnie Gill

Senior Author Support/Technology Specialist: Joe Vetere

Rights and Permissions Advisor: Dana Weightman

Manufacturing Manager: Evelyn Beaton

Text Design, Production Coordination, Illustrations, and Composition: Nesbitt Graphics, Inc.

For permission to use copyrighted material, grateful acknowledgment has been made to the copyright holders on page C-1 in the back of the book, which is hereby made part of this copyright page.

Many of the designations used by manufacturers and sellers to distinguish their products are claimed as trademarks. Where those designations appear in this book, and Pearson Education was aware of a trademark claim, the designations have been printed in initial caps or all caps.

Library of Congress Cataloging-in-Publication Data

Angel, Allen R., 1942–
        A survey of mathematics with applications.—8th ed. / Allen R. Angel,
    Christine D. Abbott, Dennis C. Runde.
            p. cm.
        Includes index.
        ISBN 0-321-50107-1–ISBN 0-321-50132-2
        1. Mathematics–Textbooks. I. Abbott, Christine D. II. Runde, Dennis C.
QA39.3.A54 2009
510—dc22   2007060062

ISBN-13: 978-0-321-50107-3   ISBN-10: 0-321-50107-1
3 4 5 6 7 8 9 10—VHP—11 10 09

C O N T E N T S

TO THE STUDENT

Mathematics is an exciting, living study. Its applications shape the world around you and influence your everyday life. We hope that as you read this book you will realize just how important mathematics is and gain an appreciation of both its usefulness and its beauty. We also hope to teach you some practical mathematics that you can use every day and that will prepare you for further mathematics courses.

The primary purpose of this text is to provide material that you can read, understand, and enjoy. To this end, we have used straightforward language and tried to relate the mathematical concepts to everyday experiences. We have also provided many detailed examples for you to follow.

The concepts, definitions, and formulas that deserve special attention are in boxes or are set in boldface type. Within each category, the exercises are graded, with more difficult problems appearing at the end. Exercises with numbers set in color are writing exercises. At the end of most exercise sets are Challenge Problem/Group Activity Exercises that contain challenging or exploratory exercises. At the end of each chapter are Group Projects that reinforce the material learned or provide related material.

Each chapter has a summary, review exercises, and a chapter test. Be sure to read the chapter summary, work the review exercises, and take the chapter test. The answers to the odd-numbered exercises, all review exercises, all chapter test exercises, and selected recreational mathematics exercises appear in the answer section in the back of the text. You should, however, use the answers only to check your work.

It is difficult to learn mathematics without becoming involved. To be successful, we suggest that you read the text carefully *and work each exercise in each assignment in detail*. Check with your instructor to determine which supplements are available for your use.

We welcome your suggestions and your comments. You may contact us at the following address:

Allen Angel
c/o Marketing
Mathematics and Statistics
Addison-Wesley
75 Arlington St., Suite 300
Boston, MA 02116

Or by email at:

math@awl.com
Subject: for Allen Angel

Good luck in your adventure in mathematics!

Allen R. Angel
Christine D. Abbott
Dennis C. Runde

(M)(A)T(H) We Use It Every Day!

We present *A Survey of Mathematics with Applications,* eighth edition, with the vision in mind that we use math every day. In this edition, we stress how mathematics is used in our daily lives whenever possible. Our primary goal is to give students a text they can read, understand, and enjoy while learning how mathematics affects the world around them. Numerous real-life applied examples motivate topics. A variety of interesting exercises demonstrate the real-life nature of mathematics and its importance in students' lives.

The text is intended for students who require a broad-based general overview of mathematics, especially those majoring in the liberal arts, elementary education, the social sciences, business, nursing, and allied health fields. It is particularly suitable for those courses that satisfy the minimum competency requirement in mathematics for graduation or transfer.

The expanded version of *A Survey of Mathematics with Applications,* eighth edition, contains all the material covered in the basic text, with additional chapters on graph theory and on voting and apportionment. Although many wished inclusion of these topics, others did not. Therefore, we have two versions of the book.

CONTENT REVISION

In this edition, we have revised and expanded certain topics to introduce new material and to increase understanding.

Chapter 1 "Critical Thinking Skills," has been updated with exciting and current examples and exercises.

Chapter 2 "Sets," now includes new material on the difference of two sets and the Cartesian product of two sets.

Chapter 3 "Logic," has more exercises and a greater variety of exercises, such as in Section 3.1. Certain material has been rewritten for greater clarity. A new section (3.7) on switching circuits has been added.

Chapter 4 "Systems of Numeration," has increased use of the bases 2, 8, and 16, which have applications to modern computers and electronics. See Section 4.3.

Chapter 5 "Number Theory and the Real Number System," has the most current number theory information (largest prime number, most accurate value of pi). Updated examples and exercises include the most current economic numbers, such as federal debt and gross domestic product.

Chapter 6 "Algebra, Graphs, and Functions," has more and a greater variety of examples and exercises dealing with real-life situations. Additional exercises involving exponential equations have been included.

Chapter 7   "Systems of Linear Equations and Inequalities," now includes annotations and more detailed explanations of certain topics.

Chapter 8   "The Metric System," has many examples and interesting photographs of real-life (metric) situations taken from around the world.

Chapter 9   "Geometry," now includes information on finding surface area. Some information on non-Euclidean geometry has been rewritten.

Chapter 10   "Mathematical Systems," has additional exercises and examples. We have tied this material to real-life situations more.

Chapter 11   "Consumer Mathematics," contains current interest rates and updated information on items that may be of interest to students, including material on credit and mutual funds. Additional information on stocks, bonds, and mutual funds is also included. Section 11.6, "Ordinary Annuities, Sinking Funds, and Retirement Investments" has been added.

Chapter 12   "Probability," has a greater variety of examples and exercises, and many that deal with real-life situations have been added. Certain material has been rewritten for greater clarity. Calculator keystrokes for solving permutation and combination exercises are included.

Chapter 13   "Statistics," now includes calculator keystrokes and commands for using Microsoft Excel to determine several statistical measures (see the Technology Tip on page 906). Material related to standard scores has been revised.

NEW AND REVISED FEATURES

Several important improvements in presentation have also been made.

- Each section opening contains interesting and motivational applications to introduce each section and illustrate the real-world nature of the material in that section.
- Technology Tips that explain how a graphing calculator or a computer spreadsheet may be used to work certain problems have been added in selected sections.
- The number of examples has been increased throughout the text to promote student understanding.
- Sources have been added; up-to-date tables, graphs, and charts make the material more relevant and encourage students to read graphs and analyze data.
- The feature previously called Mathematics Everywhere is now called Mathematics Today. Current mathematical topics related to material presented in this book are included in the Mathematics Today boxes. The information provided in these boxes relates mathematics to students' everyday lives. This material will help students see the need for mathematics and gain appreciation for it.
- Additional Did You Know? and Profiles in Mathematics features have been added throughout the book.
- In various exercise sets, the number and variety of exercises has been increased. Approximately 40% of the exercises have been revised or updated to reflect current data, new material in the text, and the needs and interests of today's students.

CONTINUING FEATURES

Several features appear throughout the book, adding interest and provoking thought.

- *Chapter openings* Interesting and motivational applications introduce each chapter and illustrate the real-world nature of the chapter topics.
- *Problem solving* Beginning in Chapter 1, students are introduced to problem solving and critical thinking. The theme of problem solving is then continued throughout the text, and special problem-solving exercises are presented in the exercise sets.
- *Critical thinking skills* In addition to a focus on problem solving, this book also features sections on inductive and deductive reasoning, estimation, and dimensional analysis.
- *Profiles in Mathematics* Brief historical sketches and vignettes present the stories of people who have advanced the discipline of mathematics.
- *Did You Know?* The colorful, engaging, and lively Did You Know? boxes highlight the connection of mathematics to history, the arts and sciences, technology, and a broad variety of disciplines.
- *Group Projects* At the end of each chapter are suggested projects that can be used to have students work together. These projects can also be assigned to individual students if desired.
- *Chapter Summaries, Review Exercises, and Chapter Tests* End-of-chapter summaries, exercises, and tests help students review material and prepare for exams.

TECHNOLOGY RESOURCES

MyMathLab®

MyMathLab is a series of text-specific, easily customizable online courses for Pearson Education textbooks in mathematics and statistics. Powered by CourseCompass™ (Pearson Education's online teaching and learning environment) and MathXL® (our online homework, tutorial, and assessment system), MyMathLab gives you the tools you need to deliver all or a portion of your course online, whether your students are in a lab setting or working from home. MyMathLab provides a rich and flexible set of course materials, featuring free-response exercises that are algorithmically generated for unlimited practice and mastery. Students can also use online tools, such as video lectures, animations, and a multimedia textbook, to improve their understanding and performance independently. Instructors can use MyMathLab's homework and test managers to select and assign online exercises correlated directly to the textbook, and they can also create and assign their own online exercises and import TestGen tests for added flexibility. MyMathLab's online gradebook—designed specifically for mathematics and statistics—automatically tracks students' homework and test results and gives the instructor control over how to calculate final grades. Instructors can also add offline (paper-and-pencil) grades to the gradebook. MyMathLab is available to qualified adopters. For more information, visit our website at www.mymathlab.com or contact your Pearson Education sales representative.

MathXL®, www.mathxl.com

MathXL® is a powerful online homework, tutorial, and assessment system that accompanies Pearson Education textbooks in mathematics or statistics. With MathXL, instructors can create, edit, and assign online homework and tests using algorithmically generated exercises correlated at the objective level to the textbook. They can also create and assign their own online exercises and import TestGen tests for added flexibility. All student work is tracked in MathXL's online gradebook. Students can take chapter tests in MathXL and receive personalized study plans based on their test results. The study plan diagnoses weaknesses and links students directly to tutorial exercises for the objectives they need to study and retest. Students can also access supplemental animations and video clips directly from selected exercises. MathXL is available to qualified adopters. For more information, visit our website at www.mathxl.com or contact your sales representative.

InterAct Math Tutorial Website: www.interactmath.com

Get practice and tutorial help online! The interactive InterAct Math tutorial website provides algorithmically generated practice exercises that correlate directly to the exercises in the book. Students can retry an exercise as many times as they like with new values each time for unlimited practice and mastery. Each exercise is accompanied by an interactive guided solution that provides helpful feedback for incorrect answers. Students can also view a worked-out sample problem that guides them through an exercise similar to the one they're working on.

Video Lectures on CD with Optional Captioning

The video lectures for this text are also available on CD-ROM, making it easy and convenient for students to watch the videos from a computer at home or on campus. The complete digitized video set, affordable and portable for students, is ideal for distance learning or supplemental instruction (ISBN 13: 978-0-321-51091-4; ISBN 10: 0-321-51091-7).

S U P P L E M E N T S for Students

Student's Solutions Manual
ISBN 13: 978-0-321-51089-1; ISBN 10: 0-321-51089-5

This for-sale manual contains solutions to all odd-numbered exercises and to all review and chapter test exercises.

Addison-Wesley Math Tutor Center

The Addison-Wesley Math Tutor Center is staffed by qualified mathematics instructors who provide students with tutoring on examples and odd-numbered exercises from the textbook. Tutoring is available via toll-free telephone, toll-free fax, e-mail, or the Internet. White Board technology allows tutors and students to actually see problems worked while they "talk" in real time over the Internet during tutoring sessions. www.aw-bc/tutorcenter

S U P P L E M E N T S for Instructors

Annotated Instructor's Edition
ISBN 13: 978-0-321-50132-5; ISBN 10: 0-321-50132-2

This special edition of the text includes answers next to the exercises, when they fit. Answers that do not fit next to the exercise are placed in a separate section in the back of the book. Answers to all text exercises are included.

Instructor's Solutions Manual
ISBN 13: 978-0-321-51092-1; ISBN 10: 0-321-51092-5

This manual contains solutions to all exercises in the text and answers to Group Projects.

Instructor's Testing Manual
ISBN 13: 978-0-321-51093-8; ISBN 10: 0-321-51093-3

This manual includes three alternate tests per chapter.

New! Insider's Guide
ISBN 13: 978-0-321-52833-9; ISBN 10: 0-321-52833-6

This manual includes resources and helpful teaching tips designed to help both new and adjunct faculty with course preparation and classroom management.

PowerPoint Lecture Presentation

Available through www.aw-bc.com/irc or at MyMathLab. This classroom presentation software covers all-important topics from sections in the text.

TestGen®

TestGen enables instructors to build, edit, print, and administer tests using a computerized bank of questions developed to cover all the objectives of the text. TestGen is algorithmically based, allowing instructors to create multiple but equivalent versions of the same question or test with the click of a button. Instructors can also modify test bank questions or add new questions. Tests can be printed or administered online. The software and test-bank are available for download from Pearson Education's online catalog.

A C K N O W L E D G M E N T S

We thank our spouses, Kathy Angel, Jason Abbott, and Kris Runde, for their support and encouragement throughout the project. They helped us in a great many ways, including proofreading, typing, and offering valuable suggestions. We are grateful for their wonderful support and understanding while we worked on the book.

We also thank our children: Robert and Steven Angel; Matthew and Jake Abbott; and Alex, Nicholas, and Max Runde. They also gave us support and encouragement and were very understanding when we could not spend as much time with them as we wished because of book deadlines. Without the support and understanding of our families, this book would not be a reality.

We thank Patricia Nelson and Paul Lorczak for their conscientious job of checking the text and answers for accuracy. We also thank Sherry Tornwall of the University of Florida for reading the manuscript and suggesting improvements to its content. And thanks to Becky Troutman for preparing the Index of Applications.

Many people at Addison-Wesley deserve thanks, including all those listed on the copyright page. In particular, we thank Anne Kelly, executive editor; Marnie Greenhut, acquisitions editor; Rachel Reeve, senior project editor; Elizabeth Bernardi, associate project editor; Leah Goldberg, editorial assistant; Peggy McMahon, senior production supervisor; Becky Anderson, executive marketing manager; Bonnie Gill, marketing assistant; Ashley O'Shaughnessy, media producer; Barbara Atkinson, senior designer; Beth Anderson, photo researcher; and Karen Wernholm, managing editor. We also thank Maria McColligan of Nesbitt Graphics, Inc., for her assistance as project manager for this project.

Elka Block and Frank Purcell also deserve our thanks for the excellent work they did on the *Student's Solutions Manual* and the *Instructor's Solutions Manual* and Debra McGivney for her work on the *Instructor's Testing Manual*.

Finally, we thank the reviewers from all editions of the book and all the students who have offered suggestions for improving it. A list of reviewers for all editions of this book follows. Thanks to you all for helping make *A Survey of Mathematics with Applications* the most successful liberal arts mathematics textbook in the country.

Allen R. Angel

Christine D. Abbott

Dennis C. Runde

REVIEWERS FOR THIS AND PREVIOUS EDITIONS

*Mary Anne Anthony-Smith, Santa Ana College

Frank Asta, College of DuPage, IL

Robin L. Ayers, Western Kentucky University

Hughette Bach, California State University–Sacramento

Madeline Bates, Bronx Community College, NY

Rebecca Baum, Lincoln Land Community College, IL

Vivian Baxter, Fort Hays State University, KS

Una Bray, Skidmore College, NY

David H. Buckley, Polk Community College, FL

Robert C. Bueker, Western Kentucky University

Carl Carlson, Moorhead State University, MN

Kent Carlson, St. Cloud State University, MN

Donald Catheart, Salisbury State College, MD

Yungchen Cheng, Southwest Missouri State University

Joseph Cleary, Massasoit Community College, MA

Donald Cohen, SUNY Ag & Tech College at Cobleskill, NY

David Dean, Santa Fe Community College, FL

*John Diamantopoulos, Northeastern State University, OK

*Greg Dietrich, Florida Community College at Jacksonville

Charles Downey, University of Nebraska

Annie Droullard, Polk Community College, FL

Ruth Ediden, Morgan State University, MD

Lee Erker, Tri-County Community College, NC

*Nancy Eschen, Florida Community College at Jacksonville

Karen Estes, St. Petersburg College, FL

Teklay Fessanaye, Santa Fe Community College, FL

Kurtis Fink, Northwest Missouri State University

Raymond Flagg, McPherson College, KS

Penelope Fowler, Tennessee Wesleyan College

Gilberto Garza, El Paso Community College, TX

Judith L. Gersting, Indiana University–Purdue University at Indianapolis

Lucille Groenke, Mesa Community College, AZ

John Hornsby, University of New Orleans, LA

Nancy Johnson, Broward Community College, FL

*Phyllis H. Jore, Valencia Community College, FL

*Heidi Kiley, Suffolk County Community College, NY

Daniel Kimborowicz, Massasoit Community College, MA

Mary Lois King, Tallahassee Community College, FL

David Lehmann, Southwest Missouri State University

Peter Lindstrom, North Lake College, TX

James Magliano, Union College, NJ

Yash Manchanda, East Los Angeles College & Fullerton College, CA

Don Marsian, Hillsborough Community College, FL

Marilyn Mays, North Lake College, TX

Robert McGuigan, Westfield State College, MA

Wallace H. Memmer, Brookdale Community College, NJ

Maurice Monahan, South Dakota State University

Julie Monte, Daytona Beach Community College, FL

Karen Mosely, Alabama Southern Community College

*Kathleen Offenholley, Brookdale Community College, NY

Edwin Owens, Pennsylvania College of Technology

Wing Park, College of Lake County, IL

Bettye Parnham, Daytona Beach Community College, FL

Joanne Peeples, El Paso Community College, TX

Nelson Rich, Nazareth College, NY

Kenneth Ross, University of Oregon

Ronald Ruemmler, Middlesex County College, NJ

Rosa Rusinek, Queensborough Community College, NY

Len Ruth, Sinclair Community College, OH

John Samoylo, Delaware County Community College, PA

Sandra Savage, Orange Coast College, CA

Gerald Schultz, Southern Connecticut State University

Richard Schwartz, College of Staten Island, NY

Kara Shavo, Mercer County Community College, NJ

Minnie Shuler, Chipola Junior College, FL

Paula R. Stickles, University of Southern Indiana

Kristin Stoley, Blinn College–Bryan, TX

Steve Sworder, Saddleback College, CA

Shirley Thompson, Moorhead College, GA

Alvin D. Tinsley, Central Missouri State University

*Sherry Tornwall, University of Florida

William Trotter, University of South Carolina

Zia Uddin, Lock Haven University of Pennsylvania

Sandra Welch, Stephen F. Austin State University, TX

Joyce Wellington, Southeastern Community College, NC

Sue Welsch, Sierra Nevada College

Robert F. Wheeler, Northern Illinois University

Susan Wirth, Indian River Community College, FL

James Wooland, Florida State University

*Judith B. Wood, Central Florida Community College

Jean Woody, Tulsa Community College, OK

Michael A. Zwick, Monroe Community College, NY

*Denotes reviewers for eighth edition.

2.1 SET CONCEPTS

Can you think of a few different categories or groups to which you belong? One way you could categorize yourself is by your gender. Another way is by your academic major. A third way is by your state of residence. In this section, we will discuss ways to sort or classify objects. We will also discuss different methods that can be used to indicate collections of objects.

▲ Horses may be classified in many ways.

We encounter sets in many different ways every day of our lives. A *set* is a collection of objects, which are called *elements* or *members* of the set. For example, the United States is a collection or set of 50 states plus the District of Columbia. The 50 individual states plus the District of Columbia are the members or elements of the set that is called the United States.

A set is *well defined* if its contents can be clearly determined. The set of U.S. presidents is a well-defined set because its contents, the presidents, can be named. The set of the three best movies is not a well-defined set because the word *best* is interpreted differently by different people. In this text, we use only well-defined sets.

Three methods are commonly used to indicate a set: (1) description, (2) roster form, and (3) set-builder notation.

The method of indicating a set by *description* is illustrated in Example 1.

EXAMPLE ❶ *Description of Sets*

Write a description of the set containing the elements Monday, Tuesday, Wednesday, Thursday, Friday, Saturday, Sunday.

SOLUTION The set is the days of the week. ●

Listing the elements of a set inside a pair of *braces*, { }, is called *roster form*. The braces are an essential part of the notation because they identify the contents as a set. For example, $\{1, 2, 3\}$ is notation for the set whose elements are 1, 2, and 3, but $(1, 2, 3)$ and $[1, 2, 3]$ are not sets because parentheses and brackets do not indicate a set. For a set written in roster form, commas separate the elements of the set. The order in which the elements are listed is not important.

Sets are generally named with capital letters. For example, the name commonly selected for the set of *natural numbers* or *counting numbers* is N.

NATURAL NUMBERS
$N = \{1, 2, 3, 4, 5, \dots\}$

The three dots after the 5, called an *ellipsis*, indicate that the elements in the set continue in the same manner. An ellipsis followed by a last element indicates that the elements continue in the same manner up to and including the last element. This notation is illustrated in Example 2(b).

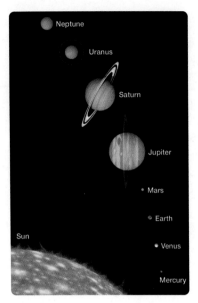

▲ The planets of Earth's solar system.

EXAMPLE ❷ *Roster Form of Sets*

Express the following in roster form.

a) Set A is the set of natural numbers less than 5.

b) Set B is the set of natural numbers less than or equal to 75.

c) Set P is the set of planets in Earth's solar system.

SOLUTION

a) The natural numbers less than 5 are 1, 2, 3, and 4. Thus, set A in roster form is
$A = \{1, 2, 3, 4\}$.

b) $B = \{1, 2, 3, 4, \ldots, 75\}$. The 75 after the ellipsis indicates that the elements continue in the same manner up to and including the number 75.

c) $P = \{\text{Mercury, Venus, Earth, Mars, Jupiter, Saturn, Uranus, Neptune}\}$*   ●

EXAMPLE ❸ *The Word* Inclusive

Express the following in roster form.

a) The set of natural numbers between 4 and 9.

b) The set of natural numbers between 4 and 9, inclusive.

SOLUTION

a) $A = \{5, 6, 7, 8\}$

b) $B = \{4, 5, 6, 7, 8, 9\}$. Note that the word *inclusive* indicates that the values of 4 and 9 are included in the set.   ●

The symbol $\in$, read, "is an element of," is used to indicate membership in a set. In Example 3, because 6 is an element of set A, we write $6 \in A$. This may also be written $6 \in \{5, 6, 7, 8\}$. We may also write $9 \notin A$, meaning that 9 is not an element of set A.

Set-builder notation (sometimes called *set-generator notation*) may be used to symbolize a set. Set-builder notation is frequently used in algebra. The following example illustrates its form.

$$D \quad = \quad \{ \quad x \quad | \quad \text{Condition(s)} \}$$

↑	↑	↑	↑	↑	↑
Set D	is	the set of	all elements x	such that	the condition(s) x must meet in order to be a member of the set.

Consider $E = \{x \mid x \in N \text{ and } x > 10\}$. The statement is read: "Set E is the set of all the elements x such that x is a natural number and x is greater than 10." The conditions that x must meet to be a member of the set are $x \in N$, which means that x must be a natural number, and $x > 10$, which means that x must be greater than 10. The numbers that meet both conditions are the set of natural numbers greater than 10. Set E in roster form is

$$E = \{11, 12, 13, 14, \ldots\}$$

---

*In August 2006, Pluto was reclassified as a dwarf planet.

EXAMPLE ❹ *Using Set-Builder Notation*

a) Write set $B = \{1, 2, 3, 4, 5\}$ in set-builder notation.

b) Write, in words, how you would read set B in set-builder notation.

SOLUTION

a) Because set B consists of the natural numbers less than 6, we write

$$B = \{x \mid x \in N \text{ and } x < 6\}$$

Another acceptable answer is $B = \{x \mid x \in N \text{ and } x \leq 5\}$.

b) Set B is the set of all elements x such that x is a natural number and x is less than 6.

EXAMPLE ❺ *Roster Form to Set-Builder Notation*

a) Write set $C = \{$North America, South America, Europe, Asia, Australia, Africa, Antarctica$\}$ in set-builder notation.

b) Write in words how you would read set C in set-builder notation.

SOLUTION

a) $C = \{x \mid x \text{ is a continent}\}$.

b) Set C is the set of all elements x such that x is a continent.

EXAMPLE ❻ *Set-Builder Notation to Roster Form*

Write set $A = \{x \mid x \in N \text{ and } 2 \leq x < 8\}$ in roster form.

SOLUTION $A = \{2, 3, 4, 5, 6, 7\}$

EXAMPLE ❼ *Largest Cities*

The chart shows the 10 most populated U.S. cities in 2005. Let set C be the set of cities in California that are among the 10 most populated U.S. cities in 2005. Write set C in roster form.

Ten Most Populated Cities in the U.S., 2005	Population
New York, New York	8,143,197
Los Angeles, California	3,844,829
Chicago, Illinois	2,842,518
Houston, Texas	2,016,582
Philadelphia, Pennsylvania	1,463,281
Phoenix, Arizona	1,461,575
San Antonio, Texas	1,256,509
San Diego, California	1,255,540
Dallas, Texas	1,213,825
San Jose, California	912,332

Source: U.S. Census Bureau

We learn to group objects according to what we see as the relevant distinguishing characteristics. One way used by educators to measure this ability is through visual cues. An example can be seen in this test, called "Creature Cards," offered by the Education Development Center. How would you describe membership in the set of Jexums?

SOLUTION By examining the chart we find that three California cities appear in the chart. They are Los Angeles, San Diego, and San Jose. Thus, set $C = \{$Los Angeles, San Diego, San Jose$\}$. ●

A set is said to be *finite* if it either contains no elements or the number of elements in the set is a natural number. The set $B = \{2, 4, 6, 8, 10\}$ is a finite set because the number of elements in the set is 5, and 5 is a natural number. A set that is not finite is said to be *infinite*. The set of counting numbers is one example of an infinite set. Infinite sets are discussed in more detail in Section 2.6.

Another important concept is equality of sets.

> Set A is **equal** to set B, symbolized by $A = B$, if and only if set A and set B contain exactly the same elements.

For example, if set $A = \{1, 2, 3\}$ and set $B = \{3, 1, 2\}$, then $A = B$ because they contain exactly the same elements. The order of the elements in the set is not important. If two sets are equal, both must contain the same number of elements. The number of elements in a set is called its *cardinal number.*

> The **cardinal number** of set A, symbolized by $n(A)$, is the number of elements in set A.

Both set $A = \{1, 2, 3\}$ and set $B = \{$England, Brazil, Japan$\}$ have a cardinal number of 3; that is, $n(A) = 3$, and $n(B) = 3$. We can say that set A and set B both have a cardinality of 3.

Two sets are said to be *equivalent* if they contain the same number of elements.

> Set A is **equivalent** to set B if and only if $n(A) = n(B)$.

Any sets that are equal must also be equivalent. Not all sets that are equivalent are equal, however. The sets $D = \{$a, b, c$\}$ and $E = \{$apple, orange, pear$\}$ are equivalent because both have the same cardinal number, 3. Because the elements differ, however, the sets are not equal.

Two sets that are equivalent or have the same cardinality can be placed in *one-to-one correspondence*. Set A and set B can be placed in one-to-one correspondence if every element of set A can be matched with exactly one element of set B and every element of set B can be matched with exactly one element of set A. For example, there is a one-to-one correspondence between the student names on a class list and the student identification numbers because we can match each student with a student identification number.

Consider set S, states, and set C, state capitals.

$$S = \{\text{North Carolina, Georgia, South Carolina, Florida}\}$$
$$C = \{\text{Columbia, Raleigh, Tallahassee, Atlanta}\}$$

Two different one-to-one correspondences for sets S and C follow.

$$S = \{\text{North Carolina, Georgia, South Carolina, Florida}\}$$
$$C = \{\text{Columbia, Raleigh, Tallahassee, Atlanta}\}$$

$$S = \{\text{North Carolina, Georgia, South Carolina, Florida}\}$$
$$C = \{\text{Columbia, Raleigh, Tallahassee, Atlanta}\}$$

Other one-to-one correspondences between sets S and C are possible. Do you know which capital goes with which state?

Null or Empty Set

Some sets do not contain any elements, such as the set of zebras that live in your house.

> The set that contains no elements is called the **empty set** or **null set** and is symbolized by { } or ∅.

Note that $\{\emptyset\}$ is not the empty set. This set contains the element ∅ and has a cardinality of 1. The set $\{0\}$ is also not the empty set because it contains the element 0. It also has a cardinality of 1.

EXAMPLE ❽ *Natural Number Solutions*

Indicate the set of natural numbers that satisfies the equation $x + 2 = 0$.

SOLUTION The values that satisfy the equation are those that make the equation a true statement. Only the number -2 satisfies this equation. Because -2 is not a natural number, the solution set of this equation is { } or ∅. ●

Universal Set

Another important set is a *universal set*.

> A **universal set,** symbolized by U, is a set that contains all the elements for any specific discussion.

When a universal set is given, only the elements in the universal set may be considered when working the problem. If, for example, the universal set for a particular problem is defined as $U = \{1, 2, 3, 4, \ldots, 10\}$, then only the natural numbers 1 through 10 may be used in that problem.

SECTION 2.1   EXERCISES

CONCEPT/WRITING EXERCISES

In Exercises 1–12, answer each question with a complete sentence.

1. What is a set?

2. What is an ellipsis, and how is it used?

3. What are the three ways that a set can be written? Give an example of each.

4. What is an infinite set?

5. What is a finite set?

6. What are equal sets?

7. What are equivalent sets?

8. What is the cardinal number of a set?

9. What is the empty set?

10. What are the two ways to indicate the empty set?

11. What is a universal set?

12. What does a one-to-one correspondence of two sets mean?

PRACTICE THE SKILLS

In Exercises 13–18, determine whether each set is well defined.

13. The set of the best books

14. The set of the easiest courses at your school

15. The set of states that have a common border with Kansas

16. The set of the four states in the United States having the largest areas

17. The set of astronauts who walked on the moon

18. The set of the most interesting teachers at your school

In Exercises 19–24, determine whether each set is finite or infinite.

19. $\{2, 4, 6, 8, \dots\}$

20. The set of multiples of 6 between 0 and 90

21. The set of odd numbers greater than 25

22. The set of fractions between 1 and 2

23. The set of odd numbers greater than 15

24. The set of apple trees in Gro-More Farms Orchards

In Exercises 25–34, express each set in roster form. You may need to use a world almanac or some other reference source.

25. The set of states in the United States whose names begin with the letter M

26. The set of oceans in the world

27. The set of natural numbers between 10 and 178

28. $C = \{x \mid x + 6 = 10\}$

29. $B = \{x \mid x \in N \text{ and } x \text{ is even}\}$

30. The set of states west of the Mississippi River that have a common border with the state of Florida

31. The set of football players over the age of 70 who are still playing in the National Football League

32. The set of states in the United States that have no common border with any other state

33. $E = \{x \mid x \in N \text{ and } 14 \leq x < 85\}$

34. The set of states in the United States that are not in the contiguous 48 states

In Exercises 35–38, use the following table, which shows the average price for a Big Mac in selected countries as of March 20, 2005. Let the 15 selected countries represent the universal set.

Average Price for a Big Mac

Country	Price
Switzerland	$5.46
Denmark	$4.97
Sweden	$4.46
United Kingdom	$3.61
Germany	$3.58
New Zealand	$3.16
United States*	$3.00
Turkey	$2.80
Peru	$2.74
Canada	$2.60
Chile	$2.56
Japan	$2.50
Australia	$2.46
Czech Republic	$2.45
China	$1.26

*Average price in four selected cities.
Source: *Economist*

Use the list to represent each set in roster form.

35. The set of countries in which the average price for a Big Mac is more than $3.00

36. The set of countries in which the average price for a Big Mac is less than $2.50

37. The set of countries in which the average price for a Big Mac is between $3.25 and $5.50

38. The set of countries in which the average price for a Big Mac is between $2.00 and $2.99

In Exercises 39–42, use the following graph, which shows the digital camera sales, in millions, for the years 1996–2005.

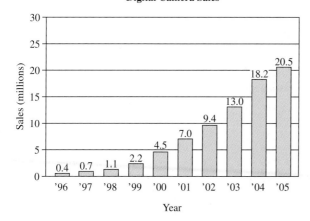

Digital Camera Sales

Source: PMA Marketing Research

Use the graph to represent each set in roster form.

39. The set of years included in the graph in which digital camera sales were more than 13 million

40. The set of years included in the graph in which digital camera sales were less than 6 million

41. The set of years included in the graph in which digital camera sales were between 1 million and 7 million

42. The set of years included in the graph in which digital camera sales were more than 25 million

In Exercises 43–50, express each set in set-builder notation.

43. $B = \{5, 6, 7, 8, 9, 10, 11, 12\}$

44. $A = \{1, 2, 3, 4, 5, 6, 7, 8, 9\}$

45. $C = \{3, 6, 9, 12, \ldots\}$

46. $D = \{5, 10, 15, 20, \ldots\}$

47. E is the set of odd natural numbers

48. A is the set of national holidays in the United States in July

49. C is the set of months that contain less than 30 days

50. $F = \{15, 16, 17, \ldots, 100\}$

In Exercises 51–58, write a description of each set.

51. $A = \{1, 2, 3, 4, 5, 6, 7\}$

52. $D = \{3, 6, 9, 12, 15, 18, \ldots\}$

53. $V = \{a, e, i, o, u\}$

54. $S = \{$Bashful, Doc, Dopey, Grumpy, Happy, Sleepy, Sneezy$\}$

55. $T = \{$oak, maple, elm, pine, $\ldots\}$

56. $E = \{x \mid x \in N \text{ and } 4 \leq x < 11\}$

57. $S = \{$Spring, Summer, Fall, Winter$\}$

58. B = {John Lennon, Ringo Starr, Paul McCartney, George Harrison}

▲ The Beatles

In Exercises 59–62, use the following list, which shows the 10 corporations with the highest favorable consumer opinion ratings as of December 2005. Let the 10 corporations in the list represent the universal set.

Corporation	Percent of Consumers with a Favorable Opinion
1. Johnson & Johnson	91
2. Google	91
3. Home Depot	90
4. Target	85
5. Coca-Cola	85
6. Toyota	84
7. Microsoft	83
8. Southwest Airlines	83
9. United Airlines	78
10. McDonald's	74

Source: Pew Research Center

Use the list to represent each set in roster form.

59. {x | x is a corporation in which at least 90 percent of consumers reported having a favorable opinion}

60. {x | x is a corporation in which fewer than 80 percent of consumers reported having a favorable opinion}

61. {x | x is a corporation in which between 75 and 80 percent of consumers reported having a favorable opinion}

62. {x | x is a corporation in which between 80 and 95 percent of consumers reported having a favorable opinion}

In Exercises 63–66, use the following graph which shows the number of Internal Revenue Service (IRS) audits of individual taxpayers, in millions, for the years 1996–2004.

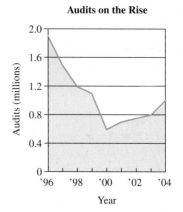

Audits on the Rise

Source: Internal Revenue Service

Use the graph to represent each set in roster form.

63. The set of years in which the number of IRS audits of individual taxpayers exceeded 1 million

64. The set of years in which the number of IRS audits of individual taxpayers were between 0.5 million and 2 million

65. The set of years in which the number of IRS audits of individual taxpayers were fewer than 1.4 million

66. The set of years in which the number of IRS audits of individual taxpayers exceeded 2 million

In Exercises 67–74, state whether each statement is true or false. If false, give the reason.

67. {e} $\in$ {a, e, i, o, u}

68. $b \in$ {a, b, c, d, e, f}

69. $h \in$ {a, b, c, d, e, f}

70. Mickey Mouse $\in$ {characters created by Walt Disney}

71. $3 \notin$ {x | $x \in N$ and x is odd}

72. Maui $\in$ {capital cities in the United States}

73. *Titanic* $\in$ {top 10 motion pictures with the greatest revenues}

74. $2 \in$ {x | x is an odd natural number}

*In Exercises 75–78, for the sets A = {2, 4, 6, 8}, B = {1, 3, 7, 9, 13, 21}, C = { }, and D = {#, &, %, □, *}.*

75. Determine $n(A)$.   76. Determine $n(B)$.

77. Determine $n(C)$.   78. Determine $n(D)$.

In Exercises 79–84, determine whether the pairs of sets are equal, equivalent, both, or neither.

79. $A = \{\text{algebra, geometry, trigonometry}\}$,
$B = \{\text{geometry, trigonometry, algebra}\}$

80. $A = \{7, 9, 10\}$, $B = \{a, b, c\}$

81. $A = \{\text{grapes, apples, oranges}\}$,
$B = \{\text{grapes, peaches, apples, oranges}\}$

82. A is the set of Siamese cats.
B is the set of cats.

83. A is the set of letters in the word *tap*.
B is the set of letters in the word *ant*.

84. A is the set of states.
B is the set of state capitals.

PROBLEM SOLVING

85. Set-builder notation is often more versatile and efficient than listing a set in roster form. This versatility is illustrated with the two sets.

$$A = \{x \mid x \in N \text{ and } x > 2\}$$
$$B = \{x \mid x > 2\}$$

a) Write a description of set A and set B.

b) Explain the difference between set A and set B.
(*Hint:* Is $4\frac{1}{2} \in A$? Is $4\frac{1}{2} \in B$?)

c) Write set A in roster form.

d) Can set B be written in roster form? Explain your answer.

86. Start with sets

$$A = \{x \mid 2 < x \le 5 \text{ and } x \in N\}$$

and

$$B = \{x \mid 2 < x \le 5\}$$

a) Write a description of set A and set B.

b) Explain the difference between set A and set B.

c) Write set A in roster form.

d) Can set B be written in roster form? Explain your answer.

*A cardinal number answers the question "How many?" An **ordinal number** describes the relative position that an element occupies. For example, Molly's desk is the third desk from the aisle.*

In Exercises 87–90, determine whether the number used is a cardinal number or an ordinal number.

87. J. K. Rowling has written 7 Harry Potter books.

▲ J. K. Rowling

88. Study the chart on page 25 in the book.

89. Lincoln was the sixteenth president of the United States.

90. Emily paid $35 for her new blouse.

91. Describe three sets of which you are a member.

92. Describe three sets that have no members.

93. Write a short paragraph explaining why the universal set and the empty set are necessary in the study of sets.

CHALLENGE PROBLEM/GROUP ACTIVITY

94. a) In a given exercise, a universal set is not specified, but we know that actor Orlando Bloom is a member of the universal set. Describe five different possible universal sets of which Orlando Bloom is a member.

b) Write a description of one set that includes all the universal sets in part (a).

INTERNET/RESEARCH ACTIVITY

95. Georg Cantor is recognized as the founder and a leader in the development of set theory. Do research and write a paper on his life and his contributions to set theory and to the field of mathematics. References include history of mathematics books, encyclopedias, and the Internet.

2.2   SUBSETS

▲ The set of occupations contains firefighters.

Consider the following sets. Set A = {architect, firefighter, mail carrier}. Set B = {architect, engineer, firefighter, mail carrier, teacher}. Note that each element of set A is also an element of set B. In this section, we will discuss how to illustrate the relationship between two sets, A and B, such that each element of set A is also an element of set B.

In our complex world, we often break larger sets into smaller more manageable sets, called *subsets*. For example, consider the set of people in your class. Suppose we categorize the set of people in your class according to the first letter of their last name (the A's, B's, C's, etc.). When we do so, each of these sets may be considered a subset of the original set. Each of these subsets can be separated further. For example, the set of people whose last name begins with the letter A can be categorized as either male or female or by their age. Each of these collections of people is also a subset. A given set may have many different subsets.

> Set A is a **subset** of set B, symbolized by $A \subseteq B$, if and only if all the elements of set A are also elements of set B.

The symbol $A \subseteq B$ indicates that "set A is a subset of set B." The symbol $\not\subseteq$ is used to indicate "is not a subset." Thus, $A \not\subseteq B$ indicates that set A is not a subset of set B. *To show that set A is not a subset of set B, we must find at least one element of set A that is not an element of set B.*

┌─ **EXAMPLE ❶** *A Subset?*

Determine whether set A is a subset of set B.
a) A = {marigold, pansy, geranium}
    B = {marigold, pansy, begonia, geranium}
b) A = {2, 3, 4, 5}
    B = {2, 3}
c) A = {$x \mid x$ is a yellow fruit}
    B = {$x \mid x$ is a red fruit}
d) A = {vanilla, chocolate, rocky road}
    B = {chocolate, vanilla, rocky road}

SOLUTION
a) All the elements of set A are contained in set B, so $A \subseteq B$.
b) The elements 4 and 5 are in set A but not in set B, so $A \not\subseteq B$ (A is not a subset of B). In this example, however, all the elements of set B are contained in set A; therefore, $B \subseteq A$.
c) There are fruits, such as bananas, that are in set A that are not in set B, so $A \not\subseteq B$.
d) All the elements of set A are contained in set B, so $A \subseteq B$. Note also that $B \subseteq A$. In fact, set A = set B.   ●

Proper Subsets

Set A is a **proper subset** of set B, symbolized by $A \subset B$, if and only if all the elements of set A are elements of set B and set $A \neq$ set B (that is, set B must contain at least one element not in set A).

Consider the sets $A = \{\text{red, blue, yellow}\}$ and $B = \{\text{red, orange, yellow, green, blue, violet}\}$. Set A is a *subset* of set B, $A \subseteq B$, because every element of set A is also an element of set B. Set A is also a *proper subset* of set B, $A \subset B$, because set A and set B are not equal. Now consider $C = \{\text{car, bus, train}\}$ and $D = \{\text{train, car, bus}\}$. Set C is a subset of set D, $C \subseteq D$, because every element of set C is also an element of set D. Set C, however, is not a proper subset of set D, $C \not\subset D$, because set C and set D are equal sets.

EXAMPLE ❷ *A Proper Subset?*

Determine whether set A is a proper subset of set B.
a) $A = \{\text{jazz, pop, hip hop}\}$
    $B = \{\text{classical, jazz, pop, rap, hip hop}\}$
b) $A = \{a, b, c, d\}$ $B = \{a, c, b, d\}$

SOLUTION

a) All the elements of set A are contained in set B, and sets A and B are not equal; thus, $A \subset B$.
b) Set $A =$ set B, so $A \not\subset B$. (However, $A \subseteq B$.)   ●

Every set is a subset of itself, but no set is a proper subset of itself. For all sets A, $A \subseteq A$, but $A \not\subset A$. For example, if $A = \{1, 2, 3\}$, then $A \subseteq A$ because every element of set A is contained in set A, but $A \not\subset A$ because set $A =$ set A.

Let $A = \{\ \}$ and $B = \{1, 2, 3, 4\}$. Is $A \subseteq B$? To show $A \not\subseteq B$, you must find at least one element of set A that is not an element of set B. Because this cannot be done, $A \subseteq B$ must be true. Using the same reasoning, we can show that *the empty set is a subset of every set, including itself.*

EXAMPLE ❸ *Element or Subset?*

Determine whether the following are true or false.
a) $3 \in \{3, 4, 5\}$
b) $\{3\} \in \{3, 4, 5\}$
c) $\{3\} \in \{\{3\}, \{4\}, \{5\}\}$
d) $\{3\} \subseteq \{3, 4, 5\}$
e) $3 \subseteq \{3, 4, 5\}$
f) $\{\ \} \subseteq \{3, 4, 5\}$

SOLUTION

a) $3 \in \{3, 4, 5\}$ is a true statement because 3 is a member of the set $\{3, 4, 5\}$.
b) $\{3\} \in \{3, 4, 5\}$ is a false statement because $\{3\}$ is a set, and the set $\{3\}$ is not an element of the set $\{3, 4, 5\}$.

c) $\{3\} \in \{\{3\}, \{4\}, \{5\}\}$ is a true statement because $\{3\}$ is an element in the set. The elements of the set $\{\{3\}, \{4\}, \{5\}\}$ are themselves sets.

d) $\{3\} \subseteq \{3, 4, 5\}$ is a true statement because every element of the first set is an element of the second set.

e) $3 \subseteq \{3, 4, 5\}$ is a false statement because the 3 is not in braces, so it is not a set and thus cannot be a subset. The 3 is an element of the set as indicated in part (a).

f) $\{\ \} \subseteq \{3, 4, 5\}$ is a true statement because the empty set is a subset of every set.   ●

Number of Subsets

How many distinct subsets can be made from a given set? The empty set has no elements and has exactly one subset, the empty set. A set with one element has two subsets. A set with two elements has four subsets. A set with three elements has eight subsets. This information is illustrated in Table 2.1. How many subsets will a set with four elements contain?

> The **number of distinct subsets** of a finite set A is 2^n, where n is the number of elements in set A.

Table 2.1 Number of Subsets

Set	Subsets	Number of Subsets
$\{\ \}$	$\{\ \}$	$1 = 2^0$
$\{a\}$	$\{a\}$ $\{\ \}$	$2 = 2^1$
$\{a, b\}$	$\{a, b\}$ $\{a\}, \{b\}$ $\{\ \}$	$4 = 2 \times 2 = 2^2$
$\{a, b, c\}$	$\{a, b, c\}$ $\{a, b\}, \{a, c\}, \{b, c\}$ $\{a\}, \{b\}, \{c\}$ $\{\ \}$	$8 = 2 \times 2 \times 2 = 2^3$

By continuing this table with larger and larger sets, we can develop a general formula for finding the number of distinct subsets that can be made from any given set.

EXAMPLE ❹ *Distinct Subsets*

a) Determine the number of distinct subsets for the set $\{S, L, E, D\}$.

b) List all the distinct subsets for the set $\{S, L, E, D\}$.

c) How many of the distinct subsets are proper subsets?

SOLUTION

a) Since the number of elements in the set is 4, the number of distinct subsets is
$2^4 = 2 \times 2 \times 2 \times 2 = 16$.

b) {S, L, E, D}   {S, L, E}   {S, L}   {S}   { }
   {S, L, D}   {S, E}   {L}
   {S, E, D}   {S, D}   {E}
   {L, E, D}   {L, E}   {D}
   {L, D}
   {E, D}

c) There are 15 proper subsets. Every subset except {S, L, E, D} is a proper subset.

EXAMPLE ⑤ *Variations of Pizza*

Sharon Bhatt is going to purchase a pizza at Pizza Hut. To her cheese pizza, she can add any of the following toppings: pepperoni, sausage, onions, mushrooms, anchovies, and ham. How many different variations of the pizza and toppings can be made?

SOLUTION Sharon can order the cheese pizza with no extra toppings, any one topping, any two toppings, any three toppings, and so on, up to six toppings. One technique used in problem solving is to consider similar problems that you have solved previously. If you think about this problem, you will realize that it is the same as asking how many distinct subsets can be made from a set with six elements. The number of different variations of the pizza is the same as the number of possible subsets of a set that has six elements. There are 2^6 or 64 possible subsets of a set with six elements. Therefore, there are 64 possible variations of the pizza and toppings.

DID YOU KNOW?

The Ladder of Life

Scientists use sets to classify and categorize knowledge. In biology, the science of classifying all living things is called *taxonomy*. More than 2000 years ago, Aristotle formalized animal classification with his "ladder of life": higher animals, lower animals, higher plants, lower plants. Today, living organisms are classified into six kingdoms (or sets) called animalia, plantae, archaea, eubacteria, fungi, and protista. Even more general groupings of living things are made according to shared characteristics. The groupings, from most general to most specific, are kingdom, phylum, class, order, family, genus, and species. For example, a zebra, *Equus burchelli*, is a member of the genus *Equus*, as is the horse, *Equus caballus*. Both the zebra and the horse are members of the universal set called the kingdom of animals and the same family, Equidae; they are members of different species (*E. burchelli* and *E. caballus*), however.

SECTION 2.2   EXERCISES

CONCEPT/WRITING EXERCISES

In Exercises 1–6, answer each question with a complete sentence.

1. What is a subset?

2. What is a proper subset?

3. Explain the difference between a subset and a proper subset.

4. Write the formula for determining the number of distinct subsets for a set with n distinct elements.

5. Write the formula for determining the number of distinct proper subsets for a set with n distinct elements.

6. Can any set be a proper subset of itself? Explain.

PRACTICE THE SKILLS

In Exercises 7–24, answer true or false. If false, give the reason.

7. Spanish $\subseteq$ {Spanish, French, Latin, German, Italian}

8. { } $\in$ {salt, pepper, basil, garlic powder}

9. { } $\subseteq$ {table, chair, sofa}

10. red $\subset$ {red, green, blue}

11. $5 \notin \{2, 4, 6\}$

12. {engineer, social worker} $\subseteq$ {physician, attorney, engineer, teacher}

13. { } $= \{\emptyset\}$

14. {motorboat, kayak} $\subseteq$ {kayak, fishing boat, sailboat, motorboat}

15. $\emptyset = \{ \ \}$

16. $0 = \{ \ \}$

17. $\{0\} = \emptyset$

18. $\{3, 8, 11\} \subseteq \{3, 8, 11\}$

19. {swimming} $\in$ {sailing, waterskiing, swimming}

20. $\{3, 5, 9\} \not\subset \{3, 9, 5\}$

21. { } $\subseteq$ { }

22. $\{1\} \in \{\{1\}, \{2\}, \{3\}\}$

23. {Panasonic, Sharp, Pioneer} $\subset$ {Sharp, Panasonic, Pioneer}

24. $\{b, a, t\} \subseteq \{t, a, b\}$

In Exercises 25–32, determine whether $A = B$, $A \subseteq B$, $B \subseteq A$, $A \subset B$, $B \subset A$ or if none of these answers applies. (There may be more than one answer.)

25. $A =$ {Chevrolet, Pontiac, Honda, Lexus}
    $B =$ {Chevrolet, Lexus}

26. $A = \{x \mid x \in N \text{ and } x < 6\}$
    $B = \{x \mid x \in N \text{ and } 1 \leq x \leq 5\}$

27. Set A is the set of states that border the Atlantic Ocean. Set B is the set of states east of the Mississippi River.

28. $A = \{1, 3, 5, 7, 9\}$
    $B = \{3, 9, 5, 7, 6\}$

29. $A = \{x \mid x \text{ is a brand of ice cream}\}$
    $B =$ {Breyers, Ben & Jerry's, Häagen-Dazs}

30. $A = \{x \mid x \text{ is a sport that uses a ball}\}$
    $B =$ {basketball, soccer, tennis}

31. Set A is the set of natural numbers between 2 and 7. Set B is the set of natural numbers greater than 2 and less than 7.

32. Set A is the set of all exercise classes offered at Gold's Gym. Set B is the set of aerobic exercise classes offered at Gold's Gym.

In Exercises 33–38, list all the subsets of the sets given.

33. $D = \varnothing$

34. $A = \{\bigcirc\}$

35. $B = \{\text{pen, pencil}\}$

36. $C = \{\text{steak, pork, chicken}\}$

PROBLEM SOLVING

37. For set $A = \{a, b, c, d\}$,

    a) list all the subsets of set A.

    b) state which of the subsets in part (a) are not proper subsets of set A.

38. A set contains nine elements.

    a) How many subsets does it have?

    b) How many proper subsets does it have?

In Exercises 39–50, if the statement is true for all sets A and B, write "true." If it is not true for all sets A and B, write "false." Assume that $A \neq \varnothing, U \neq \varnothing$, and $A \subset U$.

39. If $A \subseteq B$, then $A \subset B$. 　　40. If $A \subset B$, then $A \subseteq B$.

41. $A \subseteq A$ 　　　　　　　　42. $A \subset A$

43. $\varnothing \subset A$ 　　　　　　　　44. $\varnothing \subseteq A$

45. $A \subseteq U$ 　　　　　　　　46. $\varnothing \subset \varnothing$

47. $\varnothing \subset U$ 　　　　　　　　48. $U \subseteq \varnothing$

49. $\varnothing \subseteq \varnothing$ 　　　　　　　　50. $U \subset \varnothing$

51. *Building a House* The Jacobsens are planning to build a house in a new development. They can either build the base model offered by the builder or add any of the following options: deck, hot tub, security system, hardwood flooring. How many different variations of the house are possible?

52. *Salad Toppings* Donald Wheeler is ordering a salad at a Ruby Tuesday restaurant. He can purchase a salad consisting of just lettuce, or he can add any of the following items: cucumber, onion, tomato, carrot, green pepper, olive, mushroom. How many different variations of a salad are possible?

53. *Telephone Features* A customer with Verizon can order telephone service with some, all, or none of the following features: call waiting, call forwarding, caller identification,

three-way calling, voice mail, fax line. How many different variations of the set of features are possible?

54. *Hamburger Variations* Customers ordering hamburgers at Vic and Irv's Hamburger stand are always asked, "What do you want on it?" The choices are ketchup, mustard, relish, hot sauce, onions, lettuce, tomato. How many different variations are there for ordering a hamburger?

55. If $E \subseteq F$ and $F \subseteq E$, what other relationship exists between E and F? Explain.

56. How can you determine whether the set of boys is equivalent to the set of girls at a roller-skating rink?

57. For the set $D = \{a, b, c\}$

    a) is a an element of set D? Explain.

    b) is c a subset of set D? Explain.

    c) is $\{a, b\}$ a subset of set D? Explain.

CHALLENGE PROBLEM/GROUP ACTIVITY

58. *Hospital Expansion* A hospital has four members on the board of directors: Arnold, Benitez, Cathy, and Dominique.

    a) When the members vote on whether to add a wing to the hospital, how many different ways can they vote (abstentions are not allowed)? For example, Arnold—yes, Benitez—no, Cathy—no, and Dominique—yes is one of the many possibilities.

    b) Make a listing of all the possible outcomes of the vote. For example, the vote described in part (a) could be represented as (YNNY).

    c) How many of the outcomes given in part (b) would result in a majority supporting the addition of a wing to the hospital? That is, how many of the outcomes have three or more Y's?

RECREATIONAL MATHEMATICS

59. How many elements must a set have if the number of proper subsets of the set is $\frac{1}{2}$ of the total number of subsets of the set?

60. If $A \subset B$ and $B \subset C$, must $A \subset C$?

61. If $A \subset B$ and $B \subseteq C$, must $A \subset C$?

62. If $A \subseteq B$ and $B \subseteq C$, must $A \subset C$?

INTERNET/RESEARCH ACTIVITY

63. On page 57, we discussed the ladder of life. Do research and indicate all the different classifications in the Linnaean system, from most general to the most specific, in which a koala belongs.

2.3   VENN DIAGRAMS AND SET OPERATIONS

▲ Some soft drinks contain caffeine, some soft drinks contain sugar, and some soft drinks contain caffeine *and* sugar.

Suppose your boss sends you to the store to purchase a soft drink that contains caffeine *and* sugar. You were a bit distracted when he made his request and thought he said to purchase a soft drink that contains caffeine *or* sugar. Which soft drinks are contained in the set of soft drinks that contain caffeine *and* sugar? Which soft drinks are contained in the set of soft drinks that contain caffeine *or* sugar? These two questions are quite different. The first involves soft drinks joined by the word *and*. The second involves soft drinks joined by the word *or*. In this section, you will learn how to illustrate these and other set relationships.

Figure 2.1

A useful technique for illustrating set relationships is the Venn diagram, named for English mathematician John Venn (1834–1923). Venn invented the diagrams and used them to illustrate ideas in his text on symbolic logic, published in 1881.

In a Venn diagram, a rectangle usually represents the universal set, U. The items inside the rectangle may be divided into subsets of the universal set. The subsets are usually represented by circles. In Fig. 2.1, the circle labeled A represents set A, which is a subset of the universal set.

Two sets may be represented in a Venn diagram in any of four different ways, as shown in Fig. 2.2. Two sets A and B are *disjoint* when they have no elements in common. Two disjoint sets A and B are illustrated in Fig. 2.2(a). If set A is a proper subset of set B, $A \subset B$, the two sets may be illustrated as in Fig. 2.2(b). If set A contains exactly the same elements as set B, that is, $A = B$, the two sets may be illustrated as in Fig. 2.2(c). Two sets A and B with some elements in common are shown in Fig. 2.2(d), which is regarded as the most general form of a Venn diagram.

If we label the regions of the diagram in Fig. 2.2(d) using I, II, III, and IV, we can illustrate the four possible cases with this one diagram, Fig. 2.3, on page 61.

CASE 1: DISJOINT SETS When sets A and B are disjoint, they have no elements in common. Therefore, region II of Fig. 2.3 is empty.

CASE 2: SUBSETS When $A \subseteq B$, every element of set A is also an element of set B. Thus, there can be no elements in region I of Fig. 2.3. If $B \subseteq A$, however, then region III of Fig. 2.3 is empty.

Figure 2.2

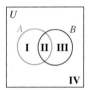

Figure 2.3

CASE 3: EQUAL SETS When set A = set B, all the elements of set A are elements of set B and all the elements of set B are elements of set A. Thus, regions I and III of Fig. 2.3 are empty.

CASE 4: OVERLAPPING SETS When sets A and B have elements in common, those elements are in region II of Fig. 2.3. The elements that belong to set A but not to set B are in region I. The elements that belong to set B but not to set A are in region III.

In each of the four cases, any element not belonging to set A or set B is placed in region IV.

Venn diagrams will be helpful in understanding set operations. The basic operations of arithmetic are $+, -, \times,$ and $\div$. When we see these symbols, we know what procedure to follow to determine the answer. Some of the operations in set theory are $'$, $\cap$, and $\cup$. They represent complement, intersection, and union, respectively.

Complement

Figure 2.4

> The **complement** of set A, symbolized by A', is the set of all the elements in the universal set that are not in set A.

In Fig. 2.4, the shaded region outside of set A within the universal set represents the complement of set A, or A'.

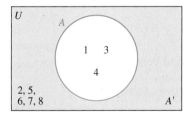

Figure 2.5

EXAMPLE ❶ *A Set and Its Complement*

Given

$$U = \{1, 2, 3, 4, 5, 6, 7, 8\} \text{ and } A = \{1, 3, 4\}$$

find A' and illustrate the relationship among sets U, A, and A' in a Venn diagram.

SOLUTION The elements in U that are not in set A are 2, 5, 6, 7, 8. Thus, $A' = \{2, 5, 6, 7, 8\}$. The Venn diagram is illustrated in Fig. 2.5. ●

Intersection

The word *intersection* brings to mind the area common to two crossing streets. The red car in the figure is in the intersection of the two streets. The set operation intersection is defined as follows.

> The **intersection** of sets A and B, symbolized by $A \cap B$, is the set containing all the elements that are common to both set A and set B.

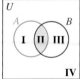

Figure 2.6

The shaded region, region II, in Fig. 2.6 represents the intersection of sets A and B.

EXAMPLE ❷ *Sets with Overlapping Regions*

Let the universal set, U, represent the 50 states in the United States. Let set A represent the set of states with a population of more than 10 million people as of 2005. Let set B represent the set of states that have at least one city with a population of more than 1 million people, as of 2005 (see the table). Draw a Venn diagram illustrating the relationship between set A and set B.

States With a Population of More Than 10 Million People	States with at Least One City with a Population of More Than 1 Million People
California	California
Texas	Texas
New York	New York
Florida	Illinois
Illinois	Pennsylvania
Pennsylvania	Arizona
Ohio	
Michigan	

Source: Bureau of the U.S. Census, U.S. Dept. of Commerce

SOLUTION First determine the intersection of sets A and B. The states common to both sets are California, Texas, New York, Illinois, and Pennsylvania. Therefore,

$$A \cap B = \{\text{California, Texas, New York, Illinois, Pennsylvania}\}$$

Place these elements in region II of Fig. 2.7. Complete region I by determining the elements in set A that have not been placed in region II. Therefore, Ohio, Michigan, and Florida are placed in region I. Complete region III by determining the elements in set B that have not been placed in region II. Thus, Arizona is placed in region III. Finally, place those elements in U that are not in either set within the rectangle but are outside both circles. This group includes the remaining 41 states, which are placed in region IV.   ●

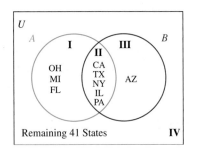

Figure 2.7

EXAMPLE ❸ *The Intersection of Sets*

Given

$$U = \{1, 2, 3, 4, 5, 6, 7, 8, 9, 10\}$$
$$A = \{1, 2, 3, 8\}$$
$$B = \{1, 3, 6, 7, 9\}$$
$$C = \{\ \}$$

find

a) $A \cap B$.   b) $A \cap C$.   c) $A' \cap B$.   d) $(A \cap B)'$.

SOLUTION

a) $A \cap B = \{1, 2, 3, 8\} \cap \{1, 3, 6, 7, 9\} = \{1, 3\}$. The elements common to both set A and set B are 1 and 3.

b) $A \cap C = \{1, 2, 3, 8\} \cap \{\ \} = \{\ \}$. There are no elements common to both set A and set C.

c) $A' = \{4, 5, 6, 7, 9, 10\}$
      $A' \cap B = \{4, 5, 6, 7, 9, 10\} \cap \{1, 3, 6, 7, 9\}$
      $\qquad = \{6, 7, 9\}$

d) To find $(A \cap B)'$, first determine $A \cap B$.
      $A \cap B = \{1, 3\}$ from part (a)
      $(A \cap B)' = \{1, 3\}' = \{2, 4, 5, 6, 7, 8, 9, 10\}$ ●

Union

The word *union* means to unite or join together, as in marriage, and that is exactly what is done when we perform the operation of union.

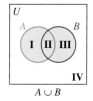

$A \cup B$

Figure 2.8

> The **union** of set A and set B, symbolized by $A \cup B$, is the set containing all the elements that are members of set A or of set B (or of both sets).

The three shaded regions of Fig. 2.8, regions I, II, and III, together represent the union of sets A and B. If an element is common to both sets, it is listed only once in the union of the sets.

EXAMPLE ❹ *Determining Sets from a Venn Diagram*

Use the Venn diagram in Fig. 2.9 to determine the following sets.

a) U b) A c) B' d) $A \cap B$
e) $A \cup B$ f) $(A \cup B)'$ g) $n(A \cup B)$

SOLUTION

a) The universal set consists of all the elements within the rectangle, that is, the elements in regions I, II, III, and IV. Thus, $U = \{9, \triangle, \square, \bigcirc, 3, 7, ?, \#, 8\}$.

b) Set A consists of the elements in regions I and II. Thus, $A = \{9, \triangle, \square, \bigcirc\}$.

c) B' consists of the elements outside set B, or the elements in regions I and IV. Thus, $B' = \{9, \triangle, \#, 8\}$.

d) $A \cap B$ consists of the elements that belong to both set A and set B (region II). Thus, $A \cap B = \{\square, \bigcirc\}$.

e) $A \cup B$ consists of the elements that belong to set A or set B (regions I, II, or III). Thus, $A \cup B = \{9, \triangle, \square, \bigcirc, 3, 7, ?\}$.

f) $(A \cup B)'$ consists of the elements in U that are not in $A \cup B$. Thus, $(A \cup B)' = \{\#, 8\}$.

g) $n(A \cup B)$ represents the *number of elements* in the union of sets A and B. Thus, $n(A \cup B) = 7$, as there are seven elements in the union of sets A and B.   ●

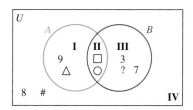

Figure 2.9

EXAMPLE ❺ *The Union of Sets*

Given

$$U = \{1, 2, 3, 4, 5, 6, 7, 8, 9, 10\}$$
$$A = \{1, 2, 4, 6\}$$
$$B = \{1, 3, 6, 7, 9\}$$
$$C = \{\ \}$$

determine each of the following.

a) $A \cup B$ b) $A \cup C$ c) $A' \cup B$ d) $(A \cup B)'$

SOLUTION

a) $A \cup B = \{1, 2, 4, 6\} \cup \{1, 3, 6, 7, 9\} = \{1, 2, 3, 4, 6, 7, 9\}$
b) $A \cup C = \{1, 2, 4, 6\} \cup \{\ \} = \{1, 2, 4, 6\}$. Note that $A \cup C = A$.
c) To determine $A' \cup B$, we must determine A'.

$$A' = \{3, 5, 7, 8, 9, 10\}$$
$$A' \cup B = \{3, 5, 7, 8, 9, 10\} \cup \{1, 3, 6, 7, 9\}$$
$$= \{1, 3, 5, 6, 7, 8, 9, 10\}$$

d) Find $(A \cup B)'$ by first determining $A \cup B$, and then find the complement of $A \cup B$.

$$A \cup B = \{1, 2, 3, 4, 6, 7, 9\} \text{ from part (a)}$$
$$(A \cup B)' = \{1, 2, 3, 4, 6, 7, 9\}' = \{5, 8, 10\}$$

●

EXAMPLE ❻ *Union and Intersection*

Given

$$U = \{a, b, c, d, e, f, g\}$$
$$A = \{a, b, e, g\}$$
$$B = \{a, c, d, e\}$$
$$C = \{b, e, f\}$$

determine each of the following.

a) $(A \cup B) \cap (A \cup C)$ b) $(A \cup B) \cap C'$ c) $A' \cap B'$

SOLUTION

a) $(A \cup B) \cap (A \cup C) = \{a, b, c, d, e, g\} \cap \{a, b, e, f, g\}$
$\qquad\qquad = \{a, b, e, g\}$
b) $(A \cup B) \cap C' = \{a, b, c, d, e, g\} \cap \{a, c, d, g\}$
$\qquad\qquad = \{a, c, d, g\}$
c) $A' \cap B' = \{c, d, f\} \cap \{b, f, g\}$
$\qquad\quad = \{f\}$

●

The Meaning of *and* and *or*

The words *and* and *or* are very important in many areas of mathematics. We use these words in several chapters in this book, including the probability chapter. The word *and* is generally interpreted to mean *intersection*, whereas *or* is generally interpreted to mean *union*. Suppose $A = \{1, 2, 3, 5, 6, 8\}$ and $B = \{1, 3, 4, 7, 9, 10\}$. The elements that belong to set A *and* set B are 1 and 3. These are the elements in the intersection of the sets. The elements that belong to set A *or* set B are 1, 2, 3, 4, 5, 6, 7, 8, 9, and 10. These are the elements in the union of the sets.

The Relationship Between $n(A \cup B)$, $n(A)$, $n(B)$, and $n(A \cap B)$

Having looked at unions and intersections, we can now determine a relationship between $n(A \cup B)$, $n(A)$, $n(B)$, and $n(A \cap B)$. Suppose set A has eight elements, set B has five elements, and $A \cap B$ has two elements. How many elements are in $A \cup B$? Let's make up some arbitrary sets that meet the criteria specified and draw a Venn diagram. If we let set $A = \{a, b, c, d, e, f, g, h\}$, then set B must contain five elements, two of which are also in set A. Let set $B = \{g, h, i, j, k\}$. We construct a Venn diagram by filling in the intersection first, as shown in Fig. 2.10. The number of elements in $A \cup B$ is 11. The elements g and h are in both sets, and if we add $n(A) + n(B)$, we are counting these elements twice.

To find the number of elements in the union of sets A and B, we can add the number of elements in sets A and B and then subtract the number of elements common to both sets.

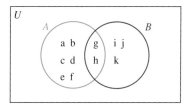

Figure 2.10

FOR ANY FINITE SETS A AND B,

$$n(A \cup B) = n(A) + n(B) - n(A \cap B)$$

EXAMPLE ❼ *How Many Visitors Speak Spanish or French?*

The results of a survey of visitors at the Grand Canyon showed that 25 speak Spanish, 14 speak French, and 4 speak both Spanish and French. How many speak Spanish or French?

SOLUTION If we let set A be the set of visitors who speak Spanish and let set B be the set of visitors who speak French, then we need to determine $n(A \cup B)$. We can use the above formula to find $n(A \cup B)$.

$$n(A \cup B) = n(A) + n(B) - n(A \cap B)$$
$$n(A \cup B) = 25 + 14 - 4$$
$$= 35$$

Thus, 35 of the visitors surveyed speak either Spanish or French.   ●

EXAMPLE ❽ *The Number of Elements in Set A*

Of the homes listed for sale with RE/MAX, 39 have either a three-car garage or a fireplace, 31 have a fireplace, and 18 have both a three-car garage and a fireplace. How many of these homes have a three-car garage?

SOLUTION If we let set A be the set of homes with a three-car garage and set B be the set of homes with a fireplace, we need to determine $n(A)$. We are given the number of homes with either a three-car garage or a fireplace, which is $n(A \cup B)$. We are also given the number of homes with a fireplace, $n(B)$, and the number of homes that have both a three-car garage and a fireplace, $n(A \cap B)$. We can use the formula $n(A \cup B) = n(A) + n(B) - n(A \cap B)$ to solve for $n(A)$.

$$n(A \cup B) = n(A) + n(B) - n(A \cap B)$$
$$39 = n(A) + 31 - 18$$
$$39 = n(A) + 13$$
$$39 - 13 = n(A) + 13 - 13$$
$$26 = n(A)$$

Thus, the number of homes listed for sale that have a three-car garage is 26.   ●

Two other set operations are the difference of two sets and the Cartesian product. We will first discuss the difference of two sets.

Difference of Two Sets

> The **difference** of two sets A and B, symbolized $A - B$, is the set of elements that belong to set A but not to set B.

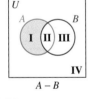

$A - B$

Figure 2.11

Using set-builder notation, the difference of two sets A and B is indicated by $A - B = \{x \mid x \in A \text{ and } x \notin B\}$. The shaded region, region I, in Fig. 2.11 represents the difference of two sets A and B, or $A - B$.

EXAMPLE ❾ *The Difference of Two Sets*

Given

$$U = \{a, b, c, d, e, f, g, h, i, j, k\}$$
$$A = \{b, c, e, f, g, h\}$$
$$B = \{a, b, c, g, i\}$$
$$C = \{h, e, g\}$$

determine

a) $A - B$ b) $A - C$ c) $A' - B$ d) $A - C'$

SOLUTION

a) $A - B$ is the set of elements that are in set A but not set B. The elements that are in set A but not set B are $e, f,$ and h. Therefore, $A - B = \{e, f, h\}$.

b) $A - C$ is the set of elements that are in set A but not set C. The elements that are in set A but not set C are $c, f,$ and h. Therefore, $A - C = \{c, f, h\}$.

c) To determine $A' - B$, we must first determine A'.

$$A' = \{a, d, i, j, k\}$$

$A' - B$ is the set of elements that are in set A' but not set B. The elements that are in set A' but not set B are d, j, and k. Therefore, $A' - B = \{d, j, k\}$.

d) To determine $A - C'$, we must first determine C'.

$$C' = \{a, c, d, f, h, i, j, k\}$$

$A - C'$ is the set of elements that are in set A but not set C'. The elements that are in set A but not set C' are b, e, and g. Therefore, $A - C' = \{b, e, g\}$.   ●

Next we discuss the Cartesian product.

Cartesian Product

> The **Cartesian product** of set A and set B, symbolized by $A \times B$ and read "A cross B," is the set of all possible *ordered pairs* of the form (a, b), where $a \in A$ and $b \in B$.

To determine the ordered pairs in a Cartesian product, select the first element of set A and form an ordered pair with each element of set B. Then select the second element of set A and form an ordered pair with each element of set B. Continue in this manner until you have used each element of set A.

┌ **EXAMPLE ❿** *The Cartesian Product of Two Sets*

Given $A = \{$orange, banana, apple$\}$ and $B = \{1, 2\}$, determine the following.

a) $A \times B$ b) $B \times A$ c) $A \times A$ d) $B \times B$

SOLUTION

a) $A \times B = \{($orange, 1$)$, (orange, 2) (banana, 1), (banana, 2), (apple, 1), (apple, 2)$\}$

b) $B \times A = \{(1,$ orange$)$, (1, banana), (1, apple), (2, orange), (2, banana), (2, apple)$\}$

c) $A \times A = \{($orange, orange$)$, (orange, banana), (orange, apple), (banana, orange), (banana, banana), (banana, apple), (apple, orange), (apple, banana), (apple, apple)$\}$

d) $B \times B = \{(1, 1), (1, 2), (2, 1), (2, 2)\}$ ●

We can see from Example 10 that, in general, $A \times B \neq B \times A$. The ordered pairs in $A \times B$ are not the same as the ordered pairs in $B \times A$ because (orange, 1) $\neq$ (1, orange).

In general, if a set A has m elements and a set B has n elements, then the number of ordered pairs in $A \times B$ will be $m \times n$. In Example 10, set A contains 3 elements and set B contains 2 elements. Notice that $A \times B$ contains 3×2 or 6 ordered pairs.

SECTION 2.3 EXERCISES

CONCEPT/WRITING EXERCISES

In Exercises 1–5, use Fig. 2.2 as a guide to draw a Venn diagram that illustrates the situation described.

1. Set A and set B are disjoint sets.

2. $A \subset B$

3. $B \subset A$

4. $A = B$

5. Set A and set B are overlapping sets.

6. How do we obtain the union of two sets A and B, $A \cup B$?

7. If we are given set A, how do we obtain A complement, A'?

8. In a Venn diagram with two overlapping sets, which region(s) represents $A \cup B$?

9. How do we obtain the intersection of two sets A and B, $A \cap B$?

10. In a Venn diagram with two overlapping sets, which region(s) represents $A \cap B$?

11. a) Which set operation is the word *or* generally interpreted to mean?

    b) Which set operation is the word *and* generally interpreted to mean?

12. Give the relationship between $n(A \cup B)$, $n(A)$, $n(B)$, and $n(A \cap B)$.

13. What is the difference of two sets A and B?

14. a) What is the Cartesian product of set A and set B?

    b) If set A has m elements and set B has n elements, how many elements will be in $A \times B$?

PRACTICE THE SKILLS/PROBLEM SOLVING

15. *National Parks* For the sets U, A, and B, construct a Venn diagram and place the elements in the proper regions.

    $U = \{$Badlands, Death Valley, Glacier, Grand Teton, Mammoth Cave, Mount Rainier, North Cascades, Shenandoah, Yellowstone, Yosemite$\}$

    $A = \{$Badlands, Glacier, Grand Teton, Mount Rainier, Yellowstone$\}$

    $B = \{$Death Valley, Glacier, Mammoth Cave, Mount Rainier, Yosemite$\}$

▲ Yosemite National Park

16. *Appliances and Electronics* For the sets U, A, and B, construct a Venn diagram and place the elements in the proper regions.

    $U = \{$microwave oven, washing machine, dryer, refrigerator, dishwasher, compact disc player, videocassette recorder, computer, camcorder, television$\}$

    $A = \{$microwave oven, washing machine, dishwasher, computer, television$\}$

    $B = \{$washing machine, dryer, refrigerator, compact disc player, computer, television$\}$

17. *Occupations* The following table shows the fastest-growing occupations for college graduates based on employment in 2002 and the estimated employment in 2012. Let the occupations in the table represent the universal set.

Fastest-Growing Occupations for College Graduates, 2002–2012

Occupation	Employment (in thousands of jobs) 2002	2012
Systems analysts	186	292
Physician assistants	63	94
Medical records technicians	147	216
Software engineers, applications	394	573
Software engineers, software	281	409
Physical therapist assistants	50	73
Fitness and aerobics instructors	183	264
Database administrators	110	159
Veterinary technicians	53	76
Dental hygienists	148	212

Source: U.S. Bureau of Labor Statistics

Let A = the set of fastest-growing occupations for college graduates whose 2002 employment was at least 150,000.

Let B = the set of fastest-growing occupations for college graduates whose estimated employment in 2012 is at least 400,000.

Construct a Venn diagram illustrating the sets.

18. *Golf Statistics* The following table shows the number of wins at the Masters Golf Tournament and the U.S. Open for selected golfers as of March 1, 2007. Let these golfers represent the universal set.

Golfer	Masters Golf Tournament	U.S. Open
Tiger Woods	4	2
Phil Mickelson	2	0
Payne Stewart	0	2
Fred Couples	1	0
Ray Floyd	1	1
Jack Nicklaus	6	4
Arnold Palmer	4	1
Ben Hogan	2	4
V. J. Singh	1	0
Retief Goosen	0	2

Let A = the set of golfers who won the Masters Golf Tournament at least two times.

Let B = the set of golfers who won the U.S. Open at least two times.

Construct a Venn diagram illustrating the sets.

19. Let U represent the set of animals in U.S. zoos. Let A represent the set of animals in the San Diego zoo. Describe A'.

▲ San Diego Zoo

20. Let U represent the set of U.S. colleges and universities. Let A represent the set of U.S. colleges and universities in the state of Mississippi. Describe A'.

In Exercises 21–26,

U is the set of insurance companies in the U.S.

A is the set of insurance companies that offer life insurance.

B is the set of insurance companies that offer car insurance.

Describe each of the following sets in words.

21. A'

22. B'

23. $A \cup B$

24. $A \cap B$

25. $A \cap B'$

26. $A \cup B'$

In Exercises 27–32,

U is the set of furniture stores in the U.S.

A is the set of furniture stores that sell mattresses.

B is the set of furniture stores that sell outdoor furniture.

C is the set of furniture stores that sell leather furniture.

Describe the following sets.

27. $A \cap B$

28. $A \cup C$

29. $B' \cap C$

30. $A \cap B \cap C$

31. $A \cup B \cup C$

32. $A' \cup C'$

In Exercises 33–40, use the Venn diagram in Fig. 2.12 to list the set of elements in roster form.

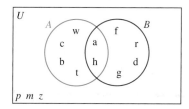

Figure 2.12

33. A

34. B

35. $A \cap B$

36. U

37. $A \cup B$

38. $(A \cup B)'$

39. $A' \cap B'$

40. $(A \cap B)'$

In Exercises 41–48, use the Venn diagram in Fig. 2.13 to list the set of elements in roster form.

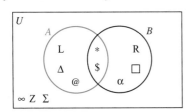

Figure 2.13

41. A

42. B

43. U

44. $A \cap B$

45. $A' \cup B$

46. $A \cup B'$

47. $A' \cap B$

48. $(A \cup B)'$

In Exercises 49–58, let

$$U = \{1, 2, 3, 4, 5, 6, 7, 8\}$$
$$A = \{1, 2, 4, 5, 8\}$$
$$B = \{2, 3, 4, 6\}$$

Determine the following.

49. $A \cup B$

50. $A \cap B$

51. B'

52. $A \cup B'$

53. $(A \cup B)'$

54. $A' \cap B'$

55. $(A \cup B)' \cap B$

56. $(A \cup B) \cap (A \cup B)'$

57. $(B \cup A)' \cap (B' \cup A')$

58. $A' \cup (A \cap B)$

In Exercises 59–68, let

$$U = \{a, b, c, d, e, f, g, h, i, j, k\}$$
$$A = \{a, c, d, f, g, i\}$$
$$B = \{b, c, d, f, g\}$$
$$C = \{a, b, f, i, j\}$$

Determine the following.

59. B'

60. $B \cup C$

61. $A \cap C$

62. $A' \cup B'$

63. $(A \cap C)'$

64. $(A \cap B) \cup C$

65. $A \cup (C \cap B)'$

66. $A \cup (C' \cup B')$

67. $(A' \cup C) \cup (A \cap B)$

68. $(C \cap B) \cap (A' \cap B)$

In Exercises 69–76, let

$$U = \{1, 2, 3, 4, 5, 6, 7, 8, 9, 10\}$$
$$A = \{1, 2, 4, 6, 9\}$$
$$B = \{1, 3, 4, 5, 8\}$$
$$C = \{4, 5, 9\}$$

Determine the following.

69. $A - B$

70. $A - C$

71. $A - B'$

72. $A' - C$

73. $(A - B)'$

74. $(A - B)' - C$

75. $C - A'$

76. $(C - A)' - B$

In Exercises 77–82, let

$$A = \{a, b, c\}$$
$$B = \{1, 2\}$$

77. Determine $A \times B$.

78. Determine $B \times A$.

79. Does $A \times B = B \times A$?

80. Determine $n(A \times B)$.

81. Determine $n(B \times A)$.

82. Does $n(A \times B) = n(B \times A)$?

PROBLEM SOLVING

In Exercises 83–96, let

$$U = \{x \mid x \in N \text{ and } x < 10\}$$
$$A = \{x \mid x \in N \text{ and } x \text{ is odd and } x < 10\}$$
$$B = \{x \mid x \in N \text{ and } x \text{ is even and } x < 10\}$$
$$C = \{x \mid x \in N \text{ and } x < 6\}$$

Determine the following.

83. $A \cap B$

84. $A \cup B$

85. $A' \cup B$

86. $(B \cup C)'$

87. $A \cap C'$

88. $A \cap B'$

89. $(B \cap C)'$

90. $(A \cup C) \cap B$

91. $(C' \cup A) \cap B$

92. $(C \cap B) \cup A$

93. $(A \cap B)' \cup C$

94. $(A' \cup C) \cap B$

95. $(A' \cup B') \cap C$

96. $(A' \cap C) \cup (A \cap B)$

97. When will a set and its complement be disjoint? Explain and give an example.

98. When will $n(A \cap B) = 0$? Explain and give an example.

99. *Pet Ownership* The results of a survey of customers at PetSmart showed that 27 owned dogs, 38 owned cats, and 16 owned both dogs and cats. How many people owned either a dog or a cat?

100. *Chorus and Band* At Henniger High School, 46 students sang in the chorus or played in the stage band, 30 students played in the stage band, and 4 students sang in the chorus and played in the stage band. How many students sang in the chorus?

101. Consider the formula

$$n(A \cup B) = n(A) + n(B) - n(A \cap B)$$

a) Show that this relation holds for $A = \{a, b, c, d\}$ and $B = \{b, d, e, f, g, h\}$.

b) Make up your own sets A and B, each consisting of at least six elements. Using these sets, show that the relation holds.

c) Use a Venn diagram and explain why the relation holds for any two sets A and B.

102. The Venn diagram in Fig. 2.14 shows a technique of labeling the regions to indicate membership of elements in a particular region. Define each of the four regions with a set statement. (*Hint:* $A \cap B'$ defines region I.)

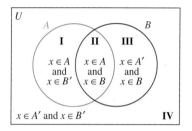

Figure 2.14

In Exercises 103–112, let $U = \{0, 1, 2, 3, 4, 5, \ldots\}$, $A = \{1, 2, 3, 4, \ldots\}$, $B = \{4, 8, 12, 16, \ldots\}$, *and* $C = \{2, 4, 6, 8, \ldots\}$. *Determine the following.*

103. $A \cup B$

104. $A \cap B$

105. $B \cap C$

106. $B \cup C$

107. $A \cap C$

108. $A' \cap C$

109. $B' \cap C$

110. $(B \cup C)' \cup C$

111. $(A \cap C) \cap B'$

112. $U' \cap (A \cup B)$

CHALLENGE EXERCISES/GROUP ACTIVITIES

In Exercises 113–120, determine whether the answer is $\emptyset$, A, *or* U. *(Assume* $A \neq \emptyset$, $A \neq U$.)

113. $A \cup A'$

114. $A \cap A'$

115. $A \cup \emptyset$

116. $A \cap \emptyset$

117. $A' \cup U$

118. $A \cap U$

119. $A \cup U$

120. $A \cup U'$

In Exercises 121–126, determine the relationship between set A and set B if

121. $A \cap B = B$.

122. $A \cup B = B$.

123. $A \cap B = \emptyset$.

124. $A \cup B = A$.

125. $A \cap B = A$.

126. $A \cup B = \emptyset$.

2.4   VENN DIAGRAMS WITH THREE SETS AND VERIFICATION OF EQUALITY OF SETS

Consider the set of the 10 highest-rated hospitals in the field of neurology. Next consider the set of the 10 highest-rated hospitals in the field of cardiology. Also, consider the set of the 10 highest-rated hospitals in the field of orthopedics. Are there any hospitals that are among the 10 highest rated in the fields of neurology, cardiology, and orthopedics? Are there any hospitals that are among the 10 highest rated in the fields of neurology and cardiology, but not orthopedics? In this section, we will learn how to use Venn diagrams to answer questions like these.

▲ Some of the highest-rated hospitals in the field of neurology are also among the highest-rated hospitals in the field of orthopedics.

In Section 2.3, we learned how to use Venn diagrams to illustrate two sets. Venn diagrams can also be used to illustrate three sets.

For three sets, *A*, *B*, and *C*, the diagram is drawn so the three sets overlap (Fig. 2.15), creating eight regions. The diagrams in Fig. 2.16 emphasize selected regions of three intersecting sets. *When constructing Venn diagrams with three sets, we generally start with region V and work outward* as explained in the following procedure.

Figure 2.15

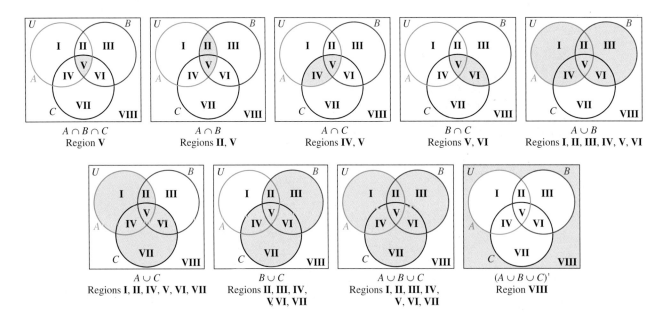

Figure 2.16

GENERAL PROCEDURE FOR CONSTRUCTING VENN DIAGRAMS WITH THREE SETS, A, B, AND C

1. Determine the elements to be placed in region V by finding the elements that are common to all three sets, $A \cap B \cap C$.

2. Determine the elements to be placed in region II. Find the elements in $A \cap B$. The elements in this set belong in regions II and V. Place the elements in the set $A \cap B$ that are not listed in region V in region II. The elements in regions IV and VI are found in a similar manner.

3. Determine the elements to be placed in region I by determining the elements in set A that are not in regions II, IV, and V. The elements in regions III and VII are found in a similar manner.

4. Determine the elements to be placed in region VIII by finding the elements in the universal set that are not in regions I through VII.

Example 1 illustrates the general procedure.

EXAMPLE ❶ *Constructing a Venn Diagram for Three Sets*

Construct a Venn diagram illustrating the following sets.

$$U = \{1, 2, 3, 4, 5, 6, 7, 8, 9, 10, 11, 12, 13, 14\}$$
$$A = \{1, 5, 8, 9, 10, 12\}$$
$$B = \{2, 4, 5, 9, 10, 13\}$$
$$C = \{1, 3, 5, 8, 9, 11\}$$

SOLUTION First find the intersection of all three sets. Because the elements 5 and 9 are in all three sets, $A \cap B \cap C = \{5, 9\}$. The elements 5 and 9 are placed in region V in Fig. 2.17. Next complete region II by determining the intersection of sets A and B.

$$A \cap B = \{5, 9, 10\}$$

$A \cap B$ consists of regions II and V. The elements 5 and 9 have already been placed in region V, so 10 must be placed in region II.

Now determine what numbers go in region IV.

$$A \cap C = \{1, 5, 8, 9\}$$

Since 5 and 9 have already been placed in region V, place the 1 and 8 in region IV. Now determine the numbers to go in region VI.

$$B \cap C = \{5, 9\}$$

Since both the 5 and 9 have been placed in region V, there are no numbers to be placed in region VI. Now complete set A. The only element of set A that has not previously been placed in regions II, IV, or V is 12. Therefore, place the element 12 in region I. The element 12 that is placed in region I is only in set A and not in set B or set C. Using set B, complete region III using the same general procedure used to determine the numbers in region I. Using set C, complete region VII by using the same procedure used to complete regions I and III. To determine the elements in region VIII, find the elements in U that have not been placed in regions I–VII. The elements 6, 7, and 14 have not been placed in regions I–VII, so place them in region VIII. ●

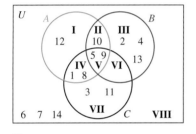

Figure 2.17

Venn diagrams can be used to illustrate and analyze many everyday problems. One example follows.

EXAMPLE ❷ *Blood Types*

Human blood is classified (typed) according to the presence or absence of the specific antigens A, B, and Rh in the red blood cells. Antigens are highly specified proteins and carbohydrates that will trigger the production of antibodies in the blood to fight infection. Blood containing the Rh antigen is labeled positive, $+$, while blood lacking the Rh antigen is labeled negative, $-$. Blood lacking both A and B antigens is called type O. Sketch a Venn diagram with three sets A, B, and Rh and place each type of blood listed in the proper region. A person has only one type of blood.

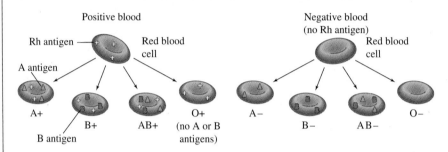

SOLUTION As illustrated in Chapter 1, the first thing to do is to read the question carefully and make sure you understand what is given and what you are asked to find. There are three antigens A, B, and Rh. Therefore, begin by naming the three circles in a Venn diagram with the three antigens; see Fig. 2.18.

Any blood containing the Rh antigen is positive, and any blood not containing the Rh antigen is negative. Therefore, all blood in the Rh circle is positive, and all blood outside the Rh circle is negative. The intersection of all three sets, region V, is AB+. Region II contains only antigens A and B and is therefore AB−. Region I is A− because it contains only antigen A. Region III is B−, region IV is A+, and region VI is B+. Region VII is O+, containing only the Rh antigen. Region VIII, which lacks all three antigens, is O−. ●

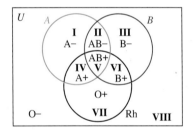

Figure 2.18

Verification of Equality of Sets

In this chapter, for clarity we may refer to operations on sets, such as $A \cup B'$ or $A \cap B \cap C$, as *statements involving sets* or simply as *statements*. Now we discuss how to determine if two statements involving sets are equal.

Consider the question: Is $A' \cup B = A' \cap B$ *for all sets A and B*? For the specific sets $U = \{1, 2, 3, 4, 5\}$, $A = \{1, 3\}$, and $B = \{2, 4, 5\}$, is $A' \cup B = A' \cap B$? To answer the question, we do the following.

Find $A' \cup B$	**Find $A' \cap B$**
$A' = \{2, 4, 5\}$	$A' = \{2, 4, 5\}$
$A' \cup B = \{2, 4, 5\}$	$A' \cap B = \{2, 4, 5\}$

For these sets, $A' \cup B = A' \cap B$, because both set statements are equal to $\{2, 4, 5\}$. At this point you may believe that $A' \cup B = A' \cap B$ for all sets A and B.

If we select the sets $U = \{1, 2, 3, 4, 5\}$, $A = \{1, 3, 5\}$, and $B = \{2, 3\}$, we see that $A' \cup B = \{2, 3, 4\}$ and $A' \cap B = \{2\}$. For this case, $A' \cup B \neq A' \cap B$. Thus, we have proved that $A' \cup B \neq A' \cap B$ for all sets A and B by using a *counterexample*. A counterexample, as explained in Chapter 1, is an example that shows a statement is not true.

In Chapter 1, we explained that proofs involve the use of deductive reasoning. Recall that deductive reasoning begins with a general statement and works to a specific conclusion. To verify, or determine whether set statements are equal for any two sets selected, we use deductive reasoning with Venn diagrams. Venn diagrams are used because they can illustrate general cases. To determine if statements that contain sets, such as $(A \cup B)'$ and $A' \cap B'$, are equal for all sets A and B, we use the regions of Venn diagrams. If both statements represent the same regions of the Venn diagram, then the statements are equal for all sets A and B. See Example 3.

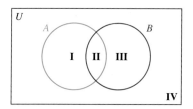

Figure 2.19

EXAMPLE ❸ *Equality of Sets*

Determine whether $(A \cap B)' = A' \cup B'$ for all sets A and B.

SOLUTION Draw a Venn diagram with two sets A and B, as in Fig. 2.19. Label the regions as indicated.

Find $(A \cap B)'$

Set	Corresponding Regions
A	I, II
B	II, III
$A \cap B$	II
$(A \cap B)'$	I, III, IV

Find $A' \cup B'$

Set	Corresponding Regions
A'	III, IV
B'	I, IV
$A' \cup B'$	I, III, IV

Both statements are represented by the same regions, I, III, and IV, of the Venn diagram. Thus, $(A \cap B)' = A' \cup B'$ for all sets A and B. ●

In Example 3, when we proved that $(A \cap B)' = A' \cup B'$, we started with two general sets and worked to the specific conclusion that both statements represented the same regions of the Venn diagram. We showed that $(A \cap B)' = A' \cup B'$ *for all sets A and B*. No matter what sets we choose for A and B, this statement will be true. For example, let $U = \{1, 2, 3, 4, 5, 6, 7, 8, 9, 10\}$, $A = \{3, 4, 6, 10\}$, and $B = \{1, 2, 4, 5, 6, 8\}$.

$$(A \cap B)' = A' \cup B'$$
$$\{4, 6\}' = \{3, 4, 6, 10\}' \cup \{1, 2, 4, 5, 6, 8\}'$$
$$\{1, 2, 3, 5, 7, 8, 9, 10\} = \{1, 2, 5, 7, 8, 9\} \cup \{3, 7, 9, 10\}$$
$$\{1, 2, 3, 5, 7, 8, 9, 10\} = \{1, 2, 3, 5, 7, 8, 9, 10\}$$

We can also use Venn diagrams to prove statements involving three sets.

Figure 2.20

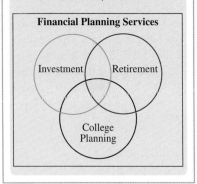
EXAMPLE ❹ *Equality of Sets*

Determine whether $A \cup (B \cap C) = (A \cup B) \cap (A \cup C)$ for all sets, A, B, and C.

SOLUTION Because the statements include three sets, A, B, and C, three circles must be used. The Venn diagram illustrating the eight regions is shown in Fig. 2.20.

First we will find the regions that correspond to $A \cup (B \cap C)$, and then we will find the regions that correspond to $(A \cup B) \cap (A \cup C)$. If both answers are the same, the statements are equal.

Find $A \cup (B \cap C)$

Set	Corresponding Regions
A	I, II, IV, V
$B \cap C$	V, VI
$A \cup (B \cap C)$	I, II, IV, V, VI

Find $(A \cup B) \cap (A \cup C)$

Set	Corresponding Regions
$A \cup B$	I, II, III, IV, V, VI
$A \cup C$	I, II, IV, V, VI, VII
$(A \cup B) \cap (A \cup C)$	I, II, IV, V, VI

The regions that correspond to $A \cup (B \cap C)$ are I, II, IV, V, and VI, and the regions that correspond to $(A \cup B) \cap (A \cup C)$ are also I, II, IV, V, and VI. The results show that both statements are represented by the same regions, namely, I, II, IV, V, and VI, and therefore $A \cup (B \cap C) = (A \cup B) \cap (A \cup C)$ for all sets A, B, and C.   ●

In Example 4, we proved that $A \cup (B \cap C) = (A \cup B) \cap (A \cup C)$ for all sets A, B, and C. Show that this statement is true for the specific sets $U = \{1, 2, 3, 4, 5, 6, 7, 8, 9, 10\}$, $A = \{1, 2, 3, 7\}$, $B = \{2, 3, 4, 5, 7, 9\}$, and $C = \{1, 4, 7, 8, 10\}$.

De Morgan's Laws

In set theory, logic, and other branches of mathematics, a pair of related theorems known as De Morgan's laws make it possible to transform statements and formulas into alternative and often more convenient forms. In set theory, *De Morgan's laws* are symbolized as follows.

> **DE MORGAN'S LAWS**
>
> 1. $(A \cap B)' = A' \cup B'$
> 2. $(A \cup B)' = A' \cap B'$

Law 1 was verified in Example 3. We suggest that you verify law 2 at this time. The laws were expressed verbally by William of Ockham in the fourteenth century. In the nineteenth century, Augustus De Morgan expressed them mathematically. De Morgan's laws will be discussed more thoroughly in Chapter 3, Logic.

SECTION 2.4   EXERCISES

CONCEPT/WRITING EXERCISES

1. How many regions are created when constructing a Venn diagram with three overlapping sets?

2. When constructing a Venn diagram with three overlapping sets, which region do you generally complete first?

3. When constructing a Venn diagram with three overlapping sets, after completing region V, which regions do you generally complete next?

4. A Venn diagram contains three sets, A, B, and C, as in Fig. 2.15 on page 72. If region V contains 4 elements and there are 12 elements in $B \cap C$, how many elements belong in region VI? Explain.

5. A Venn diagram contains three sets, A, B, and C, as in Fig. 2.15 on page 72. If region V contains 4 elements and there are 9 elements in $A \cap B$, how many elements belong in region II? Explain.

6. Give De Morgan's laws.

7. a) For $U = \{1, 2, 3, 4, 5\}$, $A = \{1, 4, 5\}$, and $B = \{1, 4, 5\}$, does $A \cup B = A \cap B$?

   b) By observing the answer to part (a), can we conclude that $A \cup B = A \cap B$ for all sets A and B? Explain.

   c) Using a Venn diagram, determine if $A \cup B = A \cap B$ for all sets A and B.

8. What type of reasoning do we use when using Venn diagrams to verify or determine whether set statements are equal?

PRACTICE THE SKILLS

9. Construct a Venn diagram illustrating the following sets.

$$U = \{a, b, c, d, e, f, g, h, i, j\}$$
$$A = \{c, d, e, g, h, i\}$$
$$B = \{a, c, d, g\}$$
$$C = \{c, f, i, j\}$$

10. Construct a Venn diagram illustrating the following sets.

$U = \{$Delaware, Pennsylvania, New Jersey, Georgia, Connecticut, Massachusetts, Maryland, South Carolina, New Hampshire, Virginia, New York, North Carolina, Rhode Island$\}$

$A = \{$New York, New Jersey, Pennsylvania, Massachusetts, New Hampshire$\}$

$B = \{$Delaware, Connecticut, Georgia, Maryland, New York, Rhode Island$\}$

$C = \{$New York, South Carolina, Rhode Island, Massachusetts$\}$

11. Construct a Venn diagram illustrating the following sets.

$U = \{$Circuit City, Best Buy, Wal-Mart, Kmart, Target, Sears, JCPenney, Costco, Kohl's, Gap, Gap Kids, Foot Locker, Old Navy, Macy's$\}$

$A = \{$Circuit City, Wal-Mart, Target, JCPenney, Old Navy$\}$

$B = \{$Best Buy, Target, Costco, Old Navy, Macy's$\}$

$C = \{$Target, Sears, Kohl's, Gap, JCPenney$\}$

12. Construct a Venn diagram illustrating the following sets.

$U = \{$*The Lion King, Aladdin, Cinderella, Beauty and the Beast, Snow White and the Seven Dwarfs, Toy Story, 101 Dalmatians, The Little Mermaid, The Incredibles*$\}$

$A = \{$*Aladdin, Toy Story, The Lion King, Snow White and the Seven Dwarfs*$\}$

$B = \{$*Snow White and the Seven Dwarfs, Toy Story, The Lion King, Beauty and the Beast*$\}$

$C = \{$*Snow White and the Seven Dwarfs, Toy Story, Beauty and the Beast, Cinderella, 101 Dalmatians*$\}$

13. Construct a Venn diagram illustrating the following sets.

$U = \{$peach, pear, banana, apple, grape, melon, carrot, corn, orange, spinach$\}$

$A = \{$pear, grape, melon, carrot$\}$

$B = \{$peach, pear, banana, spinach, corn$\}$

$C = \{$pear, banana, apple, grape, melon, spinach$\}$

14. Construct a Venn diagram illustrating the following sets.

$U = \{$Louis Armstrong, Glenn Miller, Stan Kenton, Charlie Parker, Duke Ellington, Benny Goodman, Count Basie, John Coltrane, Dizzy Gillespie, Miles Davis, Thelonius Monk$\}$

$A = \{$Stan Kenton, Count Basie, Dizzy Gillespie, Duke Ellington, Thelonius Monk$\}$

B = {Louis Armstrong, Glenn Miller, Count Basie, Duke Ellington, Miles Davis}

C = {Count Basie, Miles Davis, Stan Kenton, Charlie Parker, Duke Ellington}

15. *Olympic Medals* Consider the following table, which shows countries that won at least 22 medals in the 2004 Summer Olympics. Let the countries shown in the table represent the universal set.

Country	Gold Medals	Silver Medals	Bronze Medals	Total Medals
United States	35	39	29	103
Russia	27	27	38	92
China	32	17	14	63
Australia	17	16	16	49
Germany	14	16	18	48
Japan	16	9	12	37
France	11	9	13	33
Italy	10	11	11	32
South Korea	9	12	9	30
Great Britain	9	9	12	30
Cuba	9	7	11	27
Ukraine	9	5	9	23
Netherlands	4	9	9	22

Source: *2006 Time Almanac*

Let A = set of teams that won at least 48 medals.

Let B = set of teams that won at least 20 gold medals.

Let C = set of teams that won at least 10 bronze medals.

Construct a Venn diagram that illustrates the sets A, B, and C.

16. *Popular TV Shows* Let U = {*60 Minutes, American Idol–Tues., American Idol–Wed., CSI, CSI:Miami, Desperate Housewives, Friends, Grey's Anatomy, Law & Order, The Apprentice*}. Sets A, B, and C that follow show the five most watched television shows for the years 2004, 2005, and 2006, respectively (according to Nielsen Media Research). Then

A = {*CSI, American Idol–Tues., American Idol–Wed., Friends, The Apprentice*}

B = {*CSI, American Idol–Tues., American Idol–Wed., Desperate Housewives, CSI:Miami*}

C = {*American Idol–Tues., American Idol–Wed., CSI, Desperate Housewives, Grey's Anatomy*}

Construct a Venn diagram illustrating the sets.

Best Hospitals For Exercises 17–22, use the chart below which shows the *U.S. News and World Report* top 10 rankings of hospitals in 2005 in the fields of neurology, cardiology, and orthopedics. The universal set is the set of all U.S. hospitals.

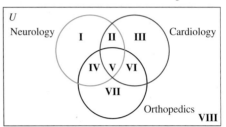

Figure 2.21

Top-Rated Hospitals

Neurology	Cardiology	Orthopedics
1. Mayo Clinic	1. Cleveland Clinic	1. Mayo Clinic
2. Johns Hopkins	2. Mayo Clinic	2. Hospital for Special Surgery
3. New York Presbyterian University Hospital	3. Johns Hopkins	3. Massachusetts General
4. Massachusetts General	4. Duke University Medical Center	4. Johns Hopkins
5. University of California, San Francisco	5. Massachusetts General	5. Cleveland Clinic
6. Cleveland Clinic	6. Brigham and Women's Hospital	6. UCLA Medical Center
7. St. Joseph's Hospital and Medical Center	7. New York–Presbyterian University Hospital	7. University of Iowa Hospitals and Clinics
8. Barnes-Jewish	8. Texas Heart Institute at St. Luke's Episcopal Hospital	8. Rush University Medical Center
9. UCLA Medical Center	9. Barnes-Jewish	9. University of Washington Medical Center
10. Methodist Hospital	10. University of Alabama	10. Duke University Medical Center

Source: *U.S. News and World Report*

Indicate in which region, I–VIII, in Fig. 2.21 each of the following hospitals belongs.

17. Mayo Clinic

18. UCLA Medical Center

19. Methodist Hospital

20. Barnes-Jewish

21. Brigham and Women's Hospital

22. Yale–New Haven Hospital

Rankings of Government Agencies For Exercises 23–28, use the following table, which shows the top 10 government agencies based on employee satisfaction in the areas of effective leadership, family-friendly culture, and pay and benefits. The universal set is the set of all government agencies.

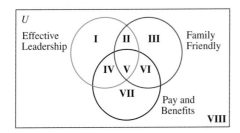

Figure 2.22

8. Department of Defense	8. Equal Employment Opportunity Commission	8. Agency for International Development
8. Agency for International Development	9. Department of Agriculture	9. Office of Personnel Management
10. Securities and Exchange Commission	9. Department of Education	9. Small Business Administration

Source: Partnership for Public Service

Indicate in which region, I–VIII, in Fig. 2.22 each of the following agencies belongs.

23. Department of Energy

24. National Aeronautics and Space Administration

25. Department of Justice

26. Small Business Administration

27. Department of Agriculture

28. Office of Personnel Management

Figures In Exercises 29–40, indicate in Fig. 2.23 the region in which each of the figures would be placed.

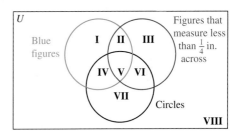

Figure 2.23

Rankings of Government Agencies

Effective Leadership	Family-Friendly Culture	Pay and Benefits
1. National Aeronautics and Space Administration	1. Federal Deposit Insurance Corp.	1. Federal Deposit Insurance Corp.
2. Nuclear Regulatory Commission	2. General Services Administration	2. Office of Management and Budget
3. Office of Management and Budget	3. Environmental Protection Agency	3. Securities and Exchange Commission
4. General Services Administration	4. Nuclear Regulatory Commission	4. National Aeronautics and Space Administration
5. National Science Foundation	5. National Aeronautics and Space Administration	5. Nuclear Regulatory Commission
6. Department of State	6. Office of Personnel Management	6. General Services Administration
7. Department of Energy	7. Department of Energy	7. Department of Commerce

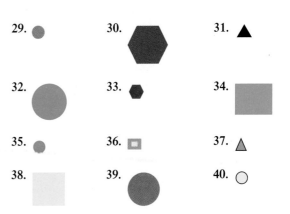

Senate Bills *During a session of the U.S. Senate, three bills were voted on. The votes of six senators are shown below the figure. Determine in which region of Fig. 2.24 each senator would be placed. The set labeled Bill 1 represents the set of senators who voted yes on Bill 1, and so on.*

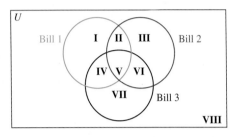

Figure 2.24

SENATOR	BILL 1	BILL 2	BILL 3
41. Feingold	yes	no	no
42. Lott	no	no	yes
43. McCain	no	no	no
44. Mikulski	yes	yes	yes
45. Obama	no	yes	yes
46. Specter	no	yes	no

In Exercises 47–60, use the Venn diagram in Fig. 2.25 to list the sets in roster form.

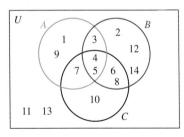

Figure 2.25

47. A

48. U

49. B

50. C

51. $A \cap B$

52. $A \cap C$

53. $(B \cap C)'$

54. $A \cap B \cap C$

55. $A \cup B$

56. $B \cup C$

57. $(A \cup C)'$

58. $A \cap (B \cup C)$

59. A'

60. $(A \cup B \cup C)'$

In Exercises 61–68, use Venn diagrams to determine whether the following statements are equal for all sets A and B.

61. $(A \cup B)'$, $A' \cap B'$

62. $(A \cap B)'$, $A \cup B'$

63. $A' \cup B'$, $A \cap B$

64. $(A \cup B)'$, $(A \cap B)'$

65. $A' \cap B'$, $(A \cap B)'$

66. $A' \cap B'$, $A \cup B'$

67. $(A' \cap B)'$, $A \cup B'$

68. $A' \cap B'$, $(A' \cap B')'$

In Exercises 69–78, use Venn diagrams to determine whether the following statements are equal for all sets A, B, and C.

69. $A \cap (B \cup C)$, $(A \cap B) \cup C$

70. $A \cup (B \cap C)$, $(B \cap C) \cup A$

71. $A \cap (B \cup C)$, $(B \cup C) \cap A$

72. $A \cup (B \cap C)'$, $A' \cap (B \cup C)$

73. $A \cap (B \cup C)$, $(A \cap B) \cup (A \cap C)$

74. $A \cup (B \cap C)$, $(A \cup B) \cap (A \cup C)$

75. $A \cap (B \cup C)'$, $A \cap (B' \cap C')$

76. $(A \cup B) \cap (B \cup C)$, $B \cup (A \cap C)$

77. $(A \cup B)' \cap C$, $(A' \cup C') \cap (B' \cup C)$

78. $(C \cap B)' \cup (A \cap B)'$, $A \cap (B \cap C)$

In Exercises 79–82, use set statements to write a description of the shaded area. Use union, intersection and complement as necessary. More than one answer may be possible.

79.

80.

81.

82.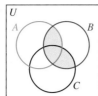

PROBLEM SOLVING

83. Let

$$U = \{1, 2, 3, 4, 5, 6, 7, 8, 9, 10\}$$
$$A = \{1, 2, 3, 4\}$$
$$B = \{3, 6, 7\}$$
$$C = \{6, 7, 9\}$$

a) Show that $(A \cup B) \cap C = (A \cap C) \cup (B \cap C)$ for these sets.

b) Make up your own sets A, B, and C. Verify that $(A \cup B) \cap C = (A \cap C) \cup (B \cap C)$ for your sets A, B, and C.

c) Use Venn diagrams to verify that $(A \cup B) \cap C = (A \cap C) \cup (B \cap C)$ for all sets A, B, and C.

84. Let

$$U = \{a, b, c, d, e, f, g, h, i\}$$
$$A = \{a, c, d, e, f\}$$
$$B = \{c, d\}$$
$$C = \{a, b, c, d, e\}$$

a) Determine whether $(A \cup C)' \cap B = (A \cap C)' \cap B$ for these sets.

b) Make up your own sets, A, B, and C. Determine whether $(A \cup C)' \cap B = (A \cap C)' \cap B$ for your sets.

c) Determine whether $(A \cup C)' \cap B = (A \cap C)' \cap B$ for all sets A, B, and C.

85. *Blood Types* A hematology text gives the following information on percentages of the different types of blood worldwide.

Type	Positive Blood, %	Negative Blood, %
A	37	6
O	32	6.5
B	11	2
AB	5	0.5

Construct a Venn diagram similar to the one in Example 2 and place the correct percent in each of the eight regions.

86. Define each of the eight regions in Fig. 2.26 using sets A, B, and C and a set operation. (*Hint:* $A \cap B' \cap C'$ defines region I.)

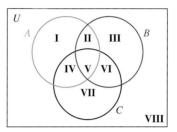

Figure 2.26

87. *Categorizing Contracts* J & C Mechanical Contractors wants to classify its projects. The contractors categorize set A as construction projects, set B as plumbing projects, and set C as projects with a budget greater than \$300,000.

a) Draw a Venn diagram that can be used to categorize the company projects according to the listed criteria.

b) Determine the region of the diagram that contains construction projects and plumbing projects with a budget greater than \$300,000. Describe the region using sets A, B, and C with set operations. Use union, intersection, and complement as necessary.

c) Determine the region of the diagram that contains plumbing projects with a budget greater than \$300,000 that are not construction projects. Describe the region using sets A, B and C with set operations. Use union, intersection, and complement as necessary.

d) Determine the region of the diagram that contains construction projects and nonplumbing projects whose budget is less than or equal to \$300,000. Describe the region using sets A, B, and C with set operations. Use union, intersection, and complement as necessary.

CHALLENGE PROBLEM/GROUP ACTIVITY

88. We were able to determine the number of elements in the union of two sets with the formula

$$n(A \cup B) = n(A) + n(B) - n(A \cap B).$$

Can you determine a formula for finding the number of elements in the union of three sets? In other words, write a formula to determine $n(A \cup B \cup C)$. [*Hint*: The formula will contain each of the following: $n(A)$, $n(B)$, $n(C)$, $n(A \cap B \cap C')$, $n(A \cap B' \cap C)$, $n(A' \cap B \cap C)$, and $2n(A \cap B \cap C)$.]

RECREATIONAL MATHEMATICS

89. **a)** Construct a Venn diagram illustrating four sets, *A*, *B*, *C*, and *D*. (*Hint*: Four circles cannot be used, and you should end up with 16 *distinct* regions.) Have fun!

   b) Label each region with a set statement (see Exercise 86). Check all 16 regions to make sure that *each is distinct*.

INTERNET/RESEARCH ACTIVITY

90. The two Venn diagrams on the right illustrate what happens when colors are added or subtracted. Do research in an art text, an encyclopedia, the Internet, or another source and write a report explaining the creation of the colors in the Venn diagrams, using such terms as union of colors and subtraction (or difference) of colors.

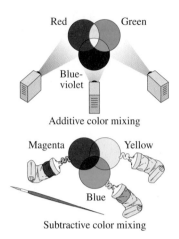

2.5   APPLICATIONS OF SETS

▲ We can use a Venn diagram to determine how many members of a health club use a particular piece of equipment.

The members of a health club were surveyed about their use of the equipment at the club. Suppose the results of the survey show how many members use the treadmill, how many members use the StairMaster, and how many members use both pieces of equipment. How can the manager of the club use this information to determine how many members use only the treadmill? In this section, we will learn how to use Venn diagrams to answer this type of question.

We can solve practical problems involving sets by using the problem-solving process discussed in Chapter 1: Understand the problem, devise a plan, carry out the plan, and then examine and check the results. First determine: What is the problem? or What am I looking for? To devise the plan, list all the facts that are given and how they are related. *Look for key words or phrases* such as "only set *A*," "set *A* and set *B*," "set *A* or set *B*," "set *A* and set *B* and not set *C*." Remember that *and* means intersection, *or* means union, and *not* means complement. The problems we solve in this section contain two or three sets of elements, which can be represented in a Venn diagram. Our plan will generally include drawing a Venn diagram, labeling the diagram, and filling in the regions of the diagram.

Whenever possible, follow the procedure in Section 2.4 for completing the Venn diagram and then answer the questions. *Remember, when drawing Venn diagrams, we generally start with the intersection of the sets and work outward.*

EXAMPLE ❶ *Fitness Equipment*

Fitness for Life health club is considering adding additional cardiovascular equipment. It is considering two types of equipment, treadmills (T) and Stair-Masters (S). The health club surveyed a sample of members and asked which equipment they had used in the previous month. Of 150 members surveyed, it was determined that

> 102 used the treadmills.
>
> 71 used the StairMasters.
>
> 40 used both types.

Of those surveyed,

a) how many did not use either the treadmill or the StairMaster?

b) how many used the treadmill but not the StairMaster?

c) how many used the StairMaster but not the treadmill?

d) how many used either the treadmill or StairMaster?

SOLUTION The problem provides the following information.

The number of members surveyed is 150: $n(U)$ is 150.

The number of members surveyed who used the treadmill is 102: $n(T) = 102$.

The number of members surveyed who used the StairMaster is 71: $n(S) = 71$.

The number of members surveyed who used both the treadmill and the Stair-Master is 40: $n(T \cap S) = 40$.

Figure 2.27

We illustrate this information on the Venn diagram shown in Fig. 2.27. We already know that $T \cap S$ corresponds to region II. As $n(T \cap S) = 40$, we write 40 in region II. Set T consists of regions I and II. We know that set T, the members who used the treadmill, contains 102 members. Therefore, region I contains $102 - 40$, or 62 members. We write the number 62 in region I. Set S consists of regions II and III. As $n(S) = 71$, the total in these two regions must be 71. Region II contains 40, leaving $71 - 40$ or 31 for region III. We write 31 in region III.

The total number of members surveyed who used the treadmill or the StairMaster is found by adding the numbers in regions I, II, and III. Therefore, $n(T \cup S) = 62 + 40 + 31 = 133$. The number in region IV is the difference between $n(U)$ and $n(T \cup S)$. There are $150 - 133$, or 17 members in region IV.

a) The members surveyed who did not use either the treadmill or the StairMaster are those members of the universal set who are not contained in set T or set S. The 17 members in region IV did not use the treadmill or StairMaster.

b) The 62 members in region I are those members surveyed who used the treadmill but not the StairMaster.

c) The 31 members in region III are those members surveyed who used the StairMaster but not the treadmill.

d) The members in regions I, II, or III are those members surveyed who used either the treadmill or the StairMaster. Thus, $62 + 40 + 31$ or 133 members surveyed used either the treadmill or the StairMaster. Notice that the 40 members in region II who use both types of equipment are included in those members surveyed who used either the treadmill or the StairMaster. ●

Similar problems involving three sets can be solved, as illustrated in Example 2.

EXAMPLE ❷ *Software Purchases*

CompUSA has recorded recent sales for three types of computer software: games, educational software, and utility programs. The following information regarding software purchases was obtained from a survey of 893 customers.

545 purchased games.

497 purchased educational software.

290 purchased utility programs.

297 purchased games and educational software.

196 purchased educational software and utility programs.

205 purchased games and utility programs.

157 purchased all three types of software.

Use a Venn diagram to answer the following questions. How many customers purchased

a) none of these types of software?

b) only games?

c) at least one of these types of software?

d) exactly two of these types of software?

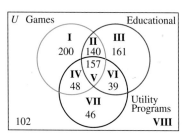

Figure 2.28

SOLUTION Begin by constructing a Venn diagram with three overlapping circles. One circle represents games, another educational software, and the third utilities. See Fig 2.28. Label the eight regions.

Whenever possible, work from the center of the diagram outwards. First fill in region V. Since 157 customers purchased all three types of software, we place 157 in region V. Next determine the number to be placed in region II. Regions II and V together represent the customers who purchased both games and educational software. Since 297 customers purchased both of these types of software, the sum of the numbers in these regions must be 297. Since 157 have already been placed in region V, $297 - 157 = 140$ must be placed in region II. Now we determine the number to be placed in region IV. Since 205 customers purchased both games and utility programs, the sum of the numbers in regions IV and V must be 205. Since 157 have already been placed in region V, $205 - 157 = 48$ must be placed in region IV. Now determine the number to be placed in region VI. A total of 196 customers purchased educational software and utility programs. The numbers in regions V and VI must total 196. Since 157 have already been placed in region V, the number to be placed in region VI is $196 - 157 = 39$.

Now that we have determined the numbers for regions V, II, IV, and VI, we can determine the numbers to be placed in regions I, III, and VII. We are given that 545 customers purchased games. The sum of the numbers in regions I, II, IV, and V must be 545. To determine the number to be placed in region I, subtract the amounts in regions II, IV, and V from 545. There must be $545 - 140 - 48 - 157 = 200$ in region I. Determine the numbers to be placed in regions III and VII in a similar manner.

$$\text{Region III} = 497 - 140 - 157 - 39 = 161$$
$$\text{Region VII} = 290 - 48 - 157 - 39 = 46$$

Now that we have determined the numbers in regions I through VII, we can determine the number to be placed in region VIII. Adding the numbers in regions I through VII yields a sum of 791. The difference between the total number of customers surveyed, 893, and the sum of the numbers in regions I through VII must be placed in region VIII.

$$\text{Region VIII} = 893 - 791 = 102$$

Now that we have completed the Venn diagram, we can answer the questions.

a) One hundred two customers did not purchase any of these types of software. These customers are indicated in region VIII.

b) Region I represents those customers who purchased only games. Thus, 200 customers purchased only games.

c) The words *at least one* mean "one or more." All those in regions I through VII purchased at least one of the types of software. The sum of the numbers in regions I through VII is 791, so 791 customers purchased at least one of the types of software.

d) The customers in regions II, IV, and VI purchased exactly two of the types of software. Summing the numbers in these regions $140 + 48 + 39$ we find that 227 customers purchased exactly two of these types of software. Notice that we did not include the customers in region V. Those customers purchased all three types of software.   ●

The procedure to work problems like those given in Example 2 is generally the same. Start by completing region V. Next complete regions II, IV, and VI. Then complete regions I, III, and VII. Finally, complete region VIII. When you are constructing Venn diagrams, be sure to check your work carefully.

TIMELY TIP When constructing a Venn diagram, the most common mistake made by students is forgetting to subtract the number in region V from the respective values in determining the numbers to be placed in regions II, IV, and VI.

EXAMPLE ❸ *Travel Packages*

Liberty Travel surveyed 125 potential customers. The following information was obtained.

68 wished to travel to Hawaii.

53 wished to travel to Las Vegas.

47 wished to travel to Disney World.

34 wished to travel to Hawaii and Las Vegas.

26 wished to travel to Las Vegas and Disney World.

23 wished to travel to Hawaii and Disney World.

18 wished to travel to all three destinations.

Use a Venn diagram to answer the following questions. How many of those surveyed

a) did not wish to travel to any of these destinations?

b) wished to travel only to Hawaii?

c) wished to travel to Disney World *and* Las Vegas, but not to Hawaii?

d) wished to travel to Disney World *or* Las Vegas, but not to Hawaii?

e) wished to travel to exactly one of these destinations?

▲ Hawaii

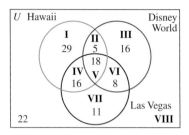

Figure 2.29

SOLUTION The Venn diagram is constructed using the procedures we outlined in Example 2. The diagram is illustrated in Fig. 2.29. We suggest you construct the diagram by yourself now and check your diagram with Fig. 2.29.

a) Twenty-two potential customers did not wish to travel to any of these destinations (see region VIII).

b) Twenty-nine potential customers wished to travel only to Hawaii (see region I).

c) Those potential customers in region VI wished to travel to Disney World *and* Las Vegas, but not to Hawaii. Therefore, eight customers satisfied the criteria.

d) The word *or* in this type of problem means one or the other or both. All the potential customers in regions II, III, IV, V, VI, and VII wished to travel to Disney World or Las Vegas. Those in regions II, IV, and V also wished to travel to Hawaii. The potential customers that wished to travel to Disney World or Las Vegas, but not to Hawaii, are found by adding the numbers in regions III, VI, and VII. There are $16 + 8 + 11 = 35$ potential customers who satisfy the criteria.

e) Those potential customers in regions I, III, and VII wished to travel to exactly one of the destinations. Therefore, $29 + 16 + 11 = 56$ customers wished to travel to exactly one of these destinations.

SECTION 2.5   EXERCISES

PRACTICE THE SKILLS/PROBLEM SOLVING

In Exercises 1–15, draw a Venn diagram to obtain the answers.

1. *Campus Activities* At a local college, a survey was taken to determine participation in different campus activities. Of 150 students surveyed, it was determined that
   32 participated in clubs
   27 participated in intramural sports
   18 participated in both clubs and intramural sports

   Of those surveyed,

   a) how many participated only in clubs?

   b) how many participated only in intramural sports?

   c) how many did not participate in either clubs or intramural sports?

2. *Landscape Purchases* Agway Lawn and Garden collected the following information regarding purchases from 130 of its customers.
   74 purchased shrubs.
   70 purchased trees.
   41 purchased both shrubs and trees.

   Of those surveyed,

   a) how many purchased only shrubs?

   b) how many purchased only trees?

   c) how many did not purchase either of these items?

3. *Real Estate* The Maiello's are moving to Wilmington, Delaware. Their real estate agent located 83 houses listed for sale, in the Wilmington area, in their price range. Of these houses listed for sale,
   47 had a family room.
   42 had a deck.
   30 had a family room and a deck.

   How many had

   a) a family room but not a deck?

   b) a deck but not a family room?

   c) either a family room or a deck?

4. *Soft Drink Taste Test* A soft drink company wishes to introduce a new soft drink. The company is considering two flavors, lemon lime and cherry cola. In a sample of 125 people, it was found that

80 liked lemon lime.
68 liked cherry cola.
42 liked both types.

a) How many liked only lemon lime?

b) How many liked only cherry cola?

c) How many liked either one or the other or both?

d) How many did not like either flavor?

5. *Cultural Activities* Thirty-three U.S. cities were researched to determine whether they had a professional sports team, a symphony, or a children's museum. The following information was determined.

    16 had a professional sports team.
    17 had a symphony.
    15 had a children's museum.
    11 had a professional sports team and a symphony.
     7 had a professional sports team and a children's museum.
     9 had a symphony and children's museum.
     5 had all three activities.

How many of the cities surveyed had

a) only a professional sports team?

b) a professional sports team and a symphony, but not a children's museum?

c) a professional sports team or a symphony?

d) a professional sports team or a symphony, but not a children's museum?

e) exactly two of the activities?

6. *Amusement Parks* In a survey of 85 amusement parks, it was found that

    24 had a hotel on site.
    55 had water slides.
    38 had a wave pool.
    13 had a hotel on site and water slides.
    10 had a hotel on site and a wave pool.
    19 had water slides and a wave pool.
     7 had all three features.

How many of the amusement parks surveyed had

a) only water slides?

b) exactly one of these features?

c) at least one of these features?

d) exactly two of these features?

e) none of these features?

▲ See Exercise 6

7. *Book Purchases* A survey of 85 customers was taken at Barnes & Noble regarding the types of books purchased. The survey found that

    44 purchased mysteries.
    33 purchased science fiction.
    29 purchased romance novels.
    13 purchased mysteries and science fiction.
     5 purchased science fiction and romance novels.
    11 purchased mysteries and romance novels.
     2 purchased all three types of books.

How many of the customers surveyed purchased

a) only mysteries?

b) mysteries and science fiction, but not romance novels?

c) mysteries or science fiction?

d) mysteries or science fiction, but not romance novels?

e) exactly two types?

8. *Movies* A survey of 350 customers was taken at Regal Cinemas in Austin, Texas, regarding the type of movies customers liked. The following information was determined.

    196 liked dramas.
    153 liked comedies.
     88 liked science fiction.
     59 liked dramas and comedies.
     37 liked dramas and science fiction.
     32 liked comedies and science fiction.
     21 liked all three types of movies.

Of the customers surveyed, how many liked

a) none of these types of movies?

b) only dramas?

c) exactly one of these types of movies?

d) exactly two of these types of movies?

e) dramas or comedies?

9. *Transportation* Seventy businesspeople in Sacramento, California, were asked how they traveled to work during the previous month. The following information was determined.

> 28 used public transportation.
> 18 rode in a car pool.
> 37 drove alone.
> 5 used public transportation and rode in a car pool.
> 9 used public transportation and drove alone.
> 4 rode in a car pool and drove alone.
> 3 used all three forms of transportation.

How many of those surveyed

a) only used public transportation?

b) only drove alone?

c) used public transportation and rode in a car pool, but did not drive alone?

d) used public transportation or rode in a car pool, but did not drive alone?

e) used none of these forms of transportation?

10. *Colleges and Universities* In a survey of four-year colleges and universities, it was found that

> 356 offered a liberal arts degree.
> 293 offered a computer engineering degree.
> 285 offered a nursing degree.
> 193 offered a liberal arts degree and a computer engineering degree.
> 200 offered a liberal arts degree and a nursing degree.
> 139 offered a computer engineering degree and a nursing degree.
> 68 offered a liberal arts degree, a computer engineering degree, and a nursing degree.
> 26 offered none of these degrees.

a) How many four-year colleges and universities were surveyed?

Of the four-year colleges and universities surveyed, how many offered

b) a liberal arts degree and a nursing degree, but not a computer engineering degree?

c) a computer engineering degree, but neither a liberal arts degree nor a nursing degree?

d) exactly two of these degrees?

e) at least one of these degrees?

11. *Homeowners' Insurance Policies* A committee of the Florida legislature decided to analyze 350 homeowners' insurance policies to determine if the consumers' homes were covered for damage due to sinkholes, mold, and floods. The following results of the analysis were determined.

> 170 homes were covered for damage due to sinkholes.
> 172 homes were covered for damage due to mold.
> 234 homes were covered for damage due to floods.
> 105 homes were covered for damage due to sinkholes and mold.
> 115 homes were covered for damage due to mold and floods.
> 109 homes were covered for damage due to sinkholes and floods.
> 78 homes were covered for damage due to all three conditions.

How many of the homes

a) were covered for damage due to mold but were not covered for damage due to sinkholes?

b) were covered for damage due to sinkholes or mold?

c) were covered for damage due to mold and floods, but were not covered for damage due to sinkholes?

d) were not covered for damage due to any of the three conditions?

12. *Appetizers Survey* Da Tulio's Restaurant hired Dennis Goldstein to determine what kind of appetizers customers liked. He surveyed 100 people, with the following results: 78 liked shrimp cocktail, 56 liked mozzarella sticks, and 35 liked both shrimp cocktail and mozzarella sticks. Every person interviewed liked one or the other or both kinds of appetizers. Does this result seem correct? Explain your answer.

13. **Discovering an Error** An immigration agent sampled cars going from the United States into Canada. In his report, he indicated that of the 85 cars sampled,

> 35 cars were driven by women.
> 53 cars were driven by U.S. citizens.
> 43 cars had two or more passengers.
> 27 cars were driven by women who are U.S. citizens.
> 25 cars were driven by women and had two or more passengers.
> 20 cars were driven by U.S. citizens and had two or more passengers.
> 15 cars were driven by women who are U.S. citizens and had two or more passengers.

After his supervisor reads the report, she explains to the agent that he made a mistake. Explain how his supervisor knew that the agent's report contained an error.

CHALLENGE PROBLEMS/GROUP ACTIVITIES

14. **Parks** A survey of 300 parks showed the following.

> 15 had only camping.
> 20 had only hiking trails.
> 35 had only picnicking.
> 185 had camping.
> 140 had camping and hiking trails.
> 125 had camping and picnicking.
> 210 had hiking trails.

Find the number of parks that

a) had at least one of these features.

b) had all three features.

c) did not have any of these features.

d) had exactly two of these features.

15. **Surveying Farmers** A survey of 500 farmers in a midwestern state showed the following.

> 125 grew only wheat.
> 110 grew only corn.
> 90 grew only oats.
> 200 grew wheat.
> 60 grew wheat and corn.
> 50 grew wheat and oats.
> 180 grew corn.

Find the number of farmers who

a) grew at least one of the three.

b) grew all three.

c) did not grow any of the three.

d) grew exactly two of the three.

RECREATIONAL MATHEMATICS

16. **Number of Elements** A universal set U consists of 12 elements. If sets A, B, and C are proper subsets of U and $n(U) = 12$, $n(A \cap B) = n(A \cap C) = n(B \cap C) = 6$, $n(A \cap B \cap C) = 4$, and $n(A \cup B \cup C) = 10$, determine

a) $n(A \cup B)$ b) $n(A' \cup C)$ c) $n(A \cap B)'$

2.6   INFINITE SETS

▲ Georg Cantor, founder of set theory

Which set is larger, the set of integers or the set of even integers? One might argue that because the set of even integers is a subset of the set of integers, the set of integers must be larger than the set of even integers. Yet both sets are infinite sets, so how can we determine which set is larger? This question puzzled mathematicians for centuries until 1874, when Georg Cantor developed a method of determining the cardinal number of an infinite set. In this section, we will discuss infinite sets and how to determine the number of elements in an infinite set

On page 48, we state that a finite set is a set in which the number of elements is zero or the number of elements can be expressed as a natural number. On page 48, we define a one-to-one correspondence. To determine the number of elements in a finite set, we can place the set in a one-to-one correspondence with a subset of the set of counting numbers. For example, the set $A = \{\#, ?, \$\}$ can be placed in one-to-one correspondence with set $B = \{1, 2, 3\}$, a subset of the set of counting numbers.

$$A = \{\#, ?, \$\}$$
$$\downarrow \downarrow \downarrow$$
$$B = \{1, 2, 3\}$$

Because the cardinal number of set B is 3, the cardinal number of set A is also 3. Any two sets such as set A and set B that can be placed in a one-to-one correspondence must have the same number of elements (therefore the same cardinality) and must be equivalent sets. Note that $n(A)$ and $n(B)$ both equal 3.

German mathematician Georg Cantor (1845–1918), known as the father of set theory, thought about sets that were not bounded. He called an unbounded set an *infinite set* and provided the following definition.

An **infinite set** is a set that can be placed in a one-to-one correspondence with a proper subset of itself.

In Example 1, we use Cantor's definition of an infinite set to show that the set of counting numbers is infinite.

EXAMPLE ❶ *The Set of Natural Numbers*

Show that $N = \{1, 2, 3, 4, 5, \ldots, n, \ldots\}$ is an infinite set.

SOLUTION To show that the set N is infinite, we establish a one-to-one correspondence between the counting numbers and a proper subset of itself. By removing the first element from the set of counting numbers, we get the set $\{2, 3, 4, 5, \ldots\}$, which is a proper subset of the set of counting numbers. Now we establish the one-to-one correspondence.

Counting numbers $= \{1, 2, 3, 4, 5, \ldots, \quad n \quad, \ldots\}$
$$\downarrow \downarrow \downarrow \downarrow \downarrow \qquad \downarrow$$
Proper subset $\quad = \{2, 3, 4, 5, 6, \ldots, n + 1, \ldots\}$

Note that for any number, n, in the set of counting numbers, its corresponding number in the proper subset is one greater, or $n + 1$. We have now shown the desired one-to-one correspondence, and thus the set of counting numbers is infinite. ●

Note in Example 1 that we showed the pairing of the general terms $n \rightarrow (n + 1)$. Showing a one-to-one correspondence of infinite sets requires showing the pairing of the general terms in the two infinite sets.

In the set of counting numbers, n represents the general term. For any other set of numbers, the general term will be different. The general term in any set should be written in terms of n such that when 1 is substituted for n in the general term, we get the first number in the set; when 2 is substituted for n in the general term, we get the second number in the set; when 6 is substituted for n in the general term, we get the sixth number in the set; and so on.

Consider the set $\{4, 9, 14, 19, \dots\}$. Suppose we want to write the general term for this set (or sequence) of numbers. What would the general term be? The numbers differ by 5, so the general term will be of the form $5n$ plus or minus some number. Substituting 1 for n yields $5(1)$, or 5. Because the first number in the set is 4, we need to subtract 1 from the 5. Thus, the general term is $5n - 1$. Note that when $n = 1$, the value is $5(1) - 1$ or 4; when $n = 2$, the value is $5(2) - 1$ or 9; when $n = 3$, the value is $5(3) - 1$ or 14; and so on. Therefore, we write the set of numbers with the general term as

$$\{4, 9, 14, 19, \dots, 5n - 1, \dots\}$$

Now that you are aware of how to determine the general term of a set of numbers, we can do some more problems involving sets.

EXAMPLE ❷ *The Set of Even Numbers*

Show that the set of even counting numbers $\{2, 4, 6, 8, \dots, 2n, \dots\}$ is an infinite set.

SOLUTION First create a proper subset of the set of even counting numbers by removing the first number from the set. Then establish a one-to-one correspondence.

Even counting numbers: $\{2, 4, 6, 8, \quad \dots, \quad 2n \quad, \dots\}$
$$\downarrow \downarrow \downarrow \downarrow \qquad\quad \downarrow$$
Proper subset: $\{4, 6, 8, 10, \dots, 2n + 2, \dots\}$

A one-to-one correspondence exists between the two sets, so the set of even counting numbers is infinite. ●

EXAMPLE ❸ *The Set of Multiples of Four*

Show that the set $\{4, 8, 12, 16, \dots, 4n, \dots\}$ is an infinite set.

SOLUTION

Given set: $\{4, \ 8, \ 12, \ 16, \ 20, \dots, \quad 4n \quad, \dots\}$
$$\downarrow \ \downarrow \ \downarrow \ \downarrow \ \downarrow \qquad\quad \downarrow$$
Proper subset: $\{8, 12, 16, 20, 24, \dots, 4n + 4, \dots\}$

Therefore, the given set is an infinite set. ●

Countable Sets

In his work with infinite sets, Cantor developed ideas on how to determine the cardinal number of an infinite set. He called the cardinal number of infinite sets "transfinite cardinal numbers" or "transfinite powers." He defined a set as *countable* if it is finite or if it can be placed in a one-to-one correspondence with the set of counting numbers. All infinite sets that can be placed in a one-to-one correspondence with the set of counting numbers have cardinal number *aleph-null*, symbolized $\aleph_0$ (the first Hebrew letter, aleph, with a zero subscript, read "null").

EXAMPLE ❹ *The Cardinal Number of the Set of Even Numbers*

Show that the set of even counting numbers has cardinal number $\aleph_0$.

SOLUTION In Example 2, we showed that a set of even counting numbers is infinite by setting up a one-to-one correspondence between the set and a proper subset of itself.

Now we will show that it is countable and has cardinality $\aleph_0$ by setting up a one-to-one correspondence between the set of counting numbers and the set of even counting numbers.

Counting numbers: $N = \{1, 2, 3, 4, \ldots, n, \ldots\}$

Even counting numbers: $E = \{2, 4, 6, 8, \ldots, 2n, \ldots\}$

For each number n in the set of counting numbers, its corresponding number is $2n$. Since we found a one-to-one correspondence between the set of counting numbers and the set of even counting numbers, the set of even counting numbers is countable. Thus, the cardinal number of the set of even counting numbers is $\aleph_0$; that is, $n(E) = \aleph_0$. As we mentioned earlier, the set of even counting numbers is an infinite set since it can be placed in a one-to-one correspondence with a proper subset of itself. Therefore, the set of even counting numbers is both infinite and countable.   ●

> Any set that can be placed in a one-to-one correspondence with the set of counting numbers has cardinality $\aleph_0$ and is infinite and is countable.

EXAMPLE ❺ *The Cardinal Number of the Set of Odd Numbers*

Show that the set of odd counting numbers has cardinality $\aleph_0$.

SOLUTION To show that the set of odd counting numbers has cardinality $\aleph_0$, we need to show a one-to-one correspondence between the set of counting numbers and the set of odd counting numbers.

Counting numbers: $N = \{1, 2, 3, 4, 5, \ldots, n, \ldots\}$

Odd counting numbers: $O = \{1, 3, 5, 7, 9, \ldots, 2n - 1, \ldots\}$

Since there is a one-to-one correspondence, the odd counting numbers have cardinality $\aleph_0$; that is, $n(O) = \aleph_0$.   ●

We have shown that both the odd and even counting numbers have cardinality $\aleph_0$. Merging the odd counting numbers with the even counting numbers gives the set of counting numbers, and we may reason that

$$\aleph_0 + \aleph_0 = \aleph_0$$

This result may seem strange, but it is true. What could such a statement mean? Well, consider a hotel with infinitely many rooms. If all the rooms are occupied, the hotel is, of course, full. If more guests appear wanting accommodations, will they be turned away? The answer is *no,* for if the room clerk were to reassign each guest to a new room

Welcome to
HOTEL INFINITY

▲ . . . where there's always room for one more. . .

with a room number twice that of the present room, all the odd-numbered rooms would become unoccupied and there would be space for more guests!

Cantor showed that there are different orders of infinity. Sets that are countable and have cardinal number $\aleph_0$ are the lowest order of infinity. Cantor showed that the set of integers and the set of rational numbers (fractions of the form p/q, where $q \neq 0$) are infinite sets with cardinality $\aleph_0$. He also showed that the set of real numbers (discussed in Chapter 5) could not be placed in a one-to-one correspondence with the set of counting numbers and that they have a higher order of infinity.

SECTION 2.6 EXERCISES

CONCEPT/WRITING EXERCISES

1. What is an infinite set as defined in this section?

2. a) What is a countable set?

   b) How can we determine if a given set has cardinality $\aleph_0$?

PRACTICE THE SKILLS

In Exercises 3–12, show that the set is infinite by placing it in a one-to-one correspondence with a proper subset of itself. Be sure to show the pairing of the general terms in the sets.

3. $\{5, 6, 7, 8, 9, \dots\}$

4. $\{20, 21, 22, 23, 24, \dots\}$

5. $\{3, 5, 7, 9, 11, \dots\}$

6. $\{20, 22, 24, 26, 28, \dots\}$

7. $\{3, 7, 11, 15, 19, \dots\}$

8. $\{4, 8, 12, 16, 20, \dots\}$

9. $\{6, 11, 16, 21, 26, \dots\}$

10. $\{1, \frac{1}{2}, \frac{1}{3}, \frac{1}{4}, \frac{1}{5}, \dots\}$

11. $\{\frac{1}{2}, \frac{1}{4}, \frac{1}{6}, \frac{1}{8}, \frac{1}{10}, \dots\}$

12. $\{\frac{6}{13}, \frac{7}{13}, \frac{8}{13}, \frac{9}{13}, \frac{10}{13}, \dots\}$

In Exercises 13–22, show that the set has cardinal number $\aleph_0$ by establishing a one-to-one correspondence between the set of counting numbers and the given set. Be sure to show the pairing of the general terms in the sets.

13. $\{3, 6, 9, 12, 15, \dots\}$

14. $\{50, 51, 52, 53, 54, \dots\}$

15. $\{4, 6, 8, 10, 12, \dots\}$

16. $\{0, 2, 4, 6, 8, \dots\}$

17. $\{2, 5, 8, 11, 14, \dots\}$

18. $\{6, 11, 16, 21, 26, \dots\}$

19. $\{5, 9, 13, 17, 21, \dots\}$

20. $\{\frac{1}{2}, \frac{1}{4}, \frac{1}{6}, \frac{1}{8}, \dots\}$

21. $\{\frac{1}{3}, \frac{1}{4}, \frac{1}{5}, \frac{1}{6}, \frac{1}{7}, \dots\}$

22. $\{\frac{1}{2}, \frac{2}{3}, \frac{3}{4}, \frac{4}{5}, \frac{5}{6}, \dots\}$

CHALLENGE PROBLEMS/GROUP ACTIVITIES

In Exercises 23–26, show that the set has cardinality $\aleph_0$ by establishing a one-to-one correspondence between the set of counting numbers and the given set.

23. $\{1, 4, 9, 16, 25, \dots\}$ 24. $\{2, 4, 8, 16, 32, \dots\}$

25. $\{3, 9, 27, 81, 243, \dots\}$ 26. $\{\frac{1}{3}, \frac{1}{6}, \frac{1}{12}, \frac{1}{24}, \frac{1}{48}, \dots\}$

RECREATIONAL MATHEMATICS

In Exercises 27–31, insert the symbol $<$, $>$, or $=$ in the shaded area to make a true statement.

27. $\aleph_0$ ▊ $\aleph_0 + \aleph_0$

28. $2\aleph_0$ ▊ $\aleph_0 + \aleph_0$

29. $2\aleph_0$ ▊ $\aleph_0$

30. $\aleph_0 + 5$ ▇ $\aleph_0 - 3$

31. $n(N)$ ▇ $\aleph_0$

32. There are a number of paradoxes (a statement that appears to be true and false at the same time) associated with infinite sets and the concept of infinity. One of these, called *Zeno's Paradox,* is named after the mathematician Zeno, born about 496 B.C. in Italy. According to Zeno's paradox, suppose Achelles starts out 1 meter behind a tortoise. Also, suppose Achelles walks 10 times as fast as the tortoise crawls. When Achelles reaches the point where the tortoise started, the tortoise is 1/10 of a meter ahead of Achelles; when Achelles reaches the point where the tortoise was 1/10 of a meter ahead, the tortoise is now 1/100 of a meter ahead; and

so on. According to Zeno's Paradox, Achelles gets closer and closer to the tortoise but never catches up to the tortoise.

a) Do you believe the reasoning process is sound? If not, explain why not.

b) In actuality, if this situation were real, would Achelles ever pass the tortoise?

INTERNET/RESEARCH ACTIVITIES

33. Do research to explain how Cantor proved that the set of rational numbers has cardinal number $\aleph_0$.

34. Do research to explain how it can be shown that the real numbers do not have cardinal number $\aleph_0$.

CHAPTER ② SUMMARY

IMPORTANT FACTS

And is generally interpreted to mean *intersection.*
Or is generally interpreted to mean *union.*

DE MORGAN'S LAWS

$$(A \cap B)' = A' \cup B'$$
$$(A \cup B)' = A' \cap B'$$

For any sets A and B,

$$n(A \cup B) = n(A) + n(B) - n(A \cap B).$$

Number of distinct subsets of a finite set with n elements is 2^n.

Symbol	Meaning
$\in$	is an element of
$\notin$	is not an element of
$n(A)$	number of elements in set A
$\emptyset$ or { }	the empty set
U	the universal set
$\subseteq$	is a subset of
$\not\subseteq$	is not a subset of
$\subset$	is a proper subset of
$\not\subset$	is not a proper subset of
$'$	complement
$\cap$	intersection
$\cup$	union
$-$	difference of two sets
$\times$	cartesian product
$\aleph_0$	aleph-null

CHAPTER ② REVIEW EXERCISES

2.1, 2.2, 2.3, 2.4, 2.6

In Exercises 1–14, state whether each is true or false. If false, give a reason.

1. The set of colleges located in the state of Oregon is a well-defined set.

2. The set of the three best cities in the United States is a well-defined set.

3. maple $\in$ {oak, elm, maple, sycamore}

4. { } $\subset \emptyset$

5. {3, 6, 9, 12, . . . } and {2, 4, 6, 8, . . . } are disjoint sets.

6. {$a, b, c, 1, 2$} is an example of a set in roster form.

7. {plane, train, automobile} = {train, plane, motorcycle}

8. {apple, orange, banana, pear} is equivalent to {tomato, corn, spinach, radish}.

9. If $A = $ {a, e, i, o, u}, then $n(A) = 5$.

10. $A = \{1, 4, 9, 16, \dots\}$ is a countable set.

11. $A = \{1, 4, 7, 10, \dots, 31\}$ is a finite set.

12. $\{2, 5, 7\} \subseteq \{2, 5, 7, 10\}$.

13. $\{x \mid x \in N \text{ and } 3 < x \leq 9\}$ is a set in set-builder notation.

14. $\{x \mid x \in N \text{ and } 2 < x \leq 12\} \subseteq \{1, 2, 3, 4, 5, \dots, 20\}$

In Exercises 15–18, express each set in roster form.

15. Set A is the set of odd natural numbers between 5 and 16.

16. Set B is the set of states that border Kansas.

17. $C = \{x \mid x \in N \text{ and } x < 162\}$

18. $D = \{x \mid x \in N \text{ and } 8 < x \leq 96\}$

In Exercises 19–22, express each set in set-builder notation.

19. Set A is the set of natural numbers between 52 and 100.

20. Set B is the set of natural numbers greater than 42.

21. Set C is the set of natural numbers less than 5.

22. Set D is the set of natural numbers between 27 and 51, inclusive.

In Exercises 23–26, express each set with a written description.

23. $A = \{x \mid x \text{ is a capital letter of the English alphabet from E through M inclusive}\}$

24. $B = \{\text{penny, nickel, dime, quarter, half-dollar}\}$

25. $C = \{x, y, z\}$

26. $D = \{x \mid 3 \leq x < 9\}$

In Exercises 27–36, let

$$U = \{1, 2, 3, 4, \dots, 10\}$$
$$A = \{1, 3, 5, 7\}$$
$$B = \{5, 7, 9, 10\}$$
$$C = \{1, 7, 10\}$$

Determine the following.

27. $A \cap B$

28. $A \cup B'$

29. $A' \cap B$

30. $(A \cup B)' \cup C$

31. $A - B$

32. $A - C'$

33. $A \times C$

34. $B \times A$

35. The number of subsets of set B

36. The number of proper subsets of set A

37. For the following sets, construct a Venn diagram and place the elements in the proper region.

$U = \{\text{lion, tiger, leopard, cheetah, puma, lynx, panther, jaguar}\}$

$A = \{\text{tiger, puma, lynx}\}$

$B = \{\text{lion, tiger, jaguar, panther}\}$

$C = \{\text{tiger, lynx, cheetah, panther}\}$

In Exercises 38–43, use Fig. 2.30 to determine the sets.

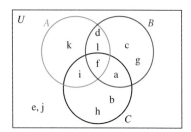

Figure 2.30

38. $A \cup B$ **39.** $A \cap B'$

40. $A \cup B \cup C$ **41.** $A \cap B \cap C$

42. $(A \cup B) \cap C$ **43.** $(A \cap B) \cup C$

Construct a Venn diagram to determine whether the following statements are true for all sets A, B, and C.

44. $(A' \cup B')' = A \cap B$

45. $(A \cup B') \cup (A \cup C') = A \cup (B \cap C)'$

In Exercises 46–51, use the following table, which shows the amount of sugar, in grams (g) and caffeine, in milligrams (mg), in an 8-oz serving of selected beverages. Let the beverages listed represent the universal set.

Beverage	Sugar (grams, g)	Caffeine (milligrams, mg)
Mountain Dew	31	37
Coca-Cola	27	23
Pepsi	27	25
Sprite	25	0
Brewed Coffee	0	85
Brewed tea	0	40
Orange juice	24	0
Grape juice	40	0
Gatorade	14	0
Water	0	0

Source: International Food Information Council

Let A be the set of beverages that contain at least 20 g of sugar.
Let B be the set of beverages that contain at least 20 mg of caffeine.
Indicate in Fig 2.31 in which region, I–IV, each of the following beverages belongs.

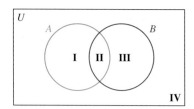

Figure 2.31

46. Pepsi

47. Brewed coffee

48. Orange juice

49. Water

50. Gatorade

51. Mountain Dew

2.5

52. *Pizza Survey* A pizza chain was willing to pay $1 to each person interviewed about his or her likes and dislikes of types of pizza crust. Of the people interviewed, 200 liked thin crust, 270 liked thick crust, 70 liked both, and 50 did not like pizza at all. What was the total cost of the survey?

53. *Cookie Preferences* The Cookie Shoppe conducted a survey to determine its customers' preferences.

    200 people liked chocolate chip cookies.
    190 people liked peanut butter cookies.
    210 people liked sugar cookies.
    100 people liked chocolate chip cookies and peanut butter cookies.
    150 people liked peanut butter cookies and sugar cookies.
    110 people liked chocolate chip cookies and sugar cookies.
    70 people liked all three.
    5 people did not like any of these cookies.

Draw a Venn diagram and then determine how many people

a) completed the survey.

b) liked only peanut butter cookies.

c) liked peanut butter cookies and chocolate chip cookies, but not sugar cookies.

d) liked peanut butter cookies or sugar cookies, but not chocolate chip cookies.

54. *TV Choices TV Guide* surveyed 510 subscribers asking which of the following three crime investigation shows they watched on a regular basis: *CSI:Crime Scene Investigation, CSI:Miami,* and *CSI:NY.* The results of the 510 questionnaires that were returned showed that

175 watched *CSI:NY.*
227 watched *CSI:Miami.*
285 watched *CSI:Crime Scene Investigation.*
100 watched *CSI:NY* and *CSI:Miami*
96 watched *CSI:NY* and *CSI:Crime Scene Investigation.*
87 watched *CSI:Miami* and *CSI:Crime Scene Investigation.*
59 watched all three shows.

Construct a Venn diagram and determine how many people

a) watched only *CSI:NY.*

b) watched exactly one of these shows.

c) watched *CSI:Miami* and *CSI:Crime Scene Investigation,* but not *CSI:NY.*

d) watched *CSI:NY* or *CSI:Crime Scene Investigation,* but not *CSI:Miami.*

e) watched exactly two of these shows.

▲ Actors from *CSI:NY*

2.6

In Exercises 55 and 56, show that the sets are infinite by placing each set in a one-to-one correspondence with a proper subset of itself.

55. $\{2, 4, 6, 8, 10, \dots\}$ **56.** $\{3, 5, 7, 9, 11, \dots\}$

In Exercises 57 and 58, show that each set has cardinal number $\aleph_0$ by setting up a one-to-one correspondence between the set of counting numbers and the given set.

57. $\{5, 8, 11, 14, 17, \dots\}$ **58.** $\{4, 9, 14, 19, 24, \dots\}$

CHAPTER ② TEST

In Exercises 1–9, state whether each is true or false. If the statement is false, explain why.

1. $\{1, y, \triangle, \$\}$ is equivalent to $\{p, \#, 5, \square\}$.

2. $\{3, 5, 9, h\} = \{9, 5, 3, j\}$

3. $\{\text{star, moon, sun}\} \subset \{\text{star, moon, sun, planet}\}$

4. $\{7\} \subseteq \{x \mid x \in N \text{ and } x < 7\}$

5. $\{\ \} \not\subseteq \{0\}$

6. $\{p, q, r, s\}$ has 15 subsets.

7. If $A \cap B = \{\ \}$, then A and B are disjoint sets.

8. For any set A, $A \cup A' = \{\ \}$.

9. For any set A, $A \cap U = A$.

In Exercises 10 and 11, use set

$$A = \{x \mid x \in N \text{ and } x < 9\}$$

10. Write set A in roster form.

11. Write a description of set A.

In Exercises 12–17, use the following information.

$$U = \{3, 5, 7, 9, 11, 13, 15\}$$
$$A = \{3, 5, 7, 9\}$$
$$B = \{7, 9, 11, 13\}$$
$$C = \{3, 11, 15\}$$

Determine the following.

12. $A \cap B$ **13.** $A \cup C'$

14. $A \cap (B \cap C)'$ **15.** $n(A \cap B')$

16. $A - B$ 17. $A \times C$

18. Using the sets provided for Exercises 12–17, draw a Venn diagram illustrating the relationship among the sets.

19. Use a Venn diagram to determine whether

$$A \cap (B \cup C') = (A \cap B) \cup (A \cap C')$$

for all sets A, B, and C. Show your work.

20. *Snacks at a Circus* Of the 155 people who purchased snacks at a circus,
76 purchased cotton candy.
90 purchased peanuts.
107 purchased popcorn.
52 purchased cotton candy and peanuts.
54 purchased cotton candy and popcorn.
57 purchased peanuts and popcorn.
35 purchased all three snacks.

Construct a Venn diagram and then determine how many people purchased

a) exactly one of these snacks.

b) none of these snacks.

c) at least two of these snacks.

d) cotton candy and peanuts, but not popcorn.

e) cotton candy or peanuts, but not popcorn.

f) only popcorn.

21. Show that the following set is infinite by setting up a one-to-one correspondence between the set and a proper subset of itself.

$$\{7, 8, 9, 10, \ldots\}$$

22. Show that the following set has cardinal number $\aleph_0$ by setting up a one-to-one correspondence between the set of counting numbers and the set.

$$\{1, 3, 5, 7, \ldots\}$$

G R O U P   P R O J E C T S

SELECTING A FAMILY PET

1. The Wilcox family is considering buying a dog. They have established several criteria for the family dog: It must be one of the breeds listed in the table, must not shed, must be less than 16 in. tall, and must be good with children.

   a) Using the information in the table,* construct a Venn diagram in which the universal set is the dogs listed. Indicate the set of dogs to be placed in each region of the Venn diagram.

   b) From the Venn diagram constructed in part (a), determine which dogs will meet the criteria set by the Wilcox family. Explain.

Breed	Sheds	Less than 16 in.	Good with children
Airedale	no	no	no
Basset hound	yes	yes	yes
Beagle	yes	yes	yes
Border terrier	no	yes	yes
Cairn terrier	no	yes	no
Cocker spaniel	yes	yes	yes
Collie	yes	no	yes
Dachshund	yes	yes	no
Poodle, miniature	no	yes	no
Schnauzer, miniature	no	yes	no
Scottish terrier	no	yes	no
Wirehaired fox terrier	no	yes	no

*The information is a collection of the opinions of an animal psychologist, Dr. Daniel Tortora, and a group of veterinarians.

CLASSIFICATION OF THE DOMESTIC CAT

2. Read the Did You Know? feature on page 57. Do research and indicate the name of the following groupings to which the domestic cat belongs.

a) Kingdom

b) Phylum

c) Class

d) Order

e) Family

f) Genus

g) Species

WHO LIVES WHERE

3. On Diplomat Row, an area of Washington, D.C., there are five houses. Each owner is a different nationality, each has a different pet, each has a different favorite food, each has a different favorite drink, and each house is painted a different color.

    The green house is directly to the right of the ivory house.

    The Senegalese has the red house.

    The dog belongs to the Spaniard.

    The Afghanistani drinks tea.

    The person who eats cheese lives next door to the fox.

    The Japanese eats fish.

    Milk is drunk in the middle house.

    Apples are eaten in the house next to the horse.

    Ale is drunk in the green house.

    The Norwegian lives in the first house.

    The peach eater drinks whiskey.

    Apples are eaten in the yellow house.

    The banana eater owns a snail.

    The Norwegian lives next door to the blue house.

For each house find

a) the color.

b) the nationality of the occupant.

c) the owner's favorite food.

d) the owner's favorite drink.

e) the owner's pet.

f) Finally, the crucial question is: Does the zebra's owner drink vodka or ale?

CHAPTER 3

Logic

▲ Logic is used to communicate effectively and make convincing arguments.

WHAT YOU WILL LEARN

- Statements, quantifiers, and compound statements
- Statements involving the words *not, and, or, if . . . then . . .,* and *if and only if*
- Truth tables for negations, conjunctions, disjunctions, conditional statements, and biconditional statements
- Self-contradictions, tautologies, and implications
- Equivalent statements, De Morgan's laws, and variations of conditional statements
- Symbolic arguments and standard forms of arguments
- Euler diagrams and syllogistic arguments
- Using logic to analyze switching circuits

WHY IT IS IMPORTANT

The study of logic enables us to communicate effectively, make more convincing arguments, and develop patterns of reasoning for decision making. The study of logic also prepares an individual to better understand the thought processes involved in learning other areas of mathematics and other subjects.

3.1 STATEMENTS AND LOGICAL CONNECTIVES

Almost everyday we see or hear advertisements that attempt to influence our buying habits. Advertisements often rely on spoken or written statements that are used to favorably portray the advertised product and form a convincing argument that will persuade us to purchase the product. Some familiar advertising statements are: *Don't leave home without it; It takes a licking and keeps on ticking; Sometimes you feel like a nut, sometimes you don't;* and *When it rains it pours.* In this section, we will learn how to represent statements using logic symbols that may help us better understand the nature of the statement. We will use these symbols throughout the chapter to analyze more complicated statements. Such statements appear in everyday life in, for example, legal documents, product instructions, and game rules in addition to advertising.

▲ "When it rains it pours." Logic symbols can be used to help us analyze statements made by advertisers.

History

The ancient Greeks were the first people to systematically analyze the way humans think and arrive at conclusions. Aristotle (384–322 B.C.) organized the study of logic for the first time in a work called *Organon.* As a result of his work, Aristotle is called the father of logic. The logic from this period, called *Aristotelian logic,* has been taught and studied for more than 2000 years.

Since Aristotle's time, the study of logic has been continued by other great philosophers and mathematicians. Gottfried Wilhelm Leibniz (1646–1716) had a deep conviction that all mathematical and scientific concepts could be derived from logic. As a result, he became the first serious student of *symbolic logic.* One difference between symbolic logic and Aristotelian logic is that in symbolic logic, as its name implies, symbols (usually letters) represent written statements. A self-educated English mathematician, George Boole (1815–1864), is considered to be the founder of symbolic logic because of his impressive work in this area. Among Boole's publications are *The Mathematical Analysis of Logic* (1847) and *An Investigation of the Law of Thought* (1854). Mathematician Charles Dodgson, better known as Lewis Carroll, incorporated many interesting ideas from logic into his books *Alice's Adventures in Wonderland* and *Through the Looking Glass* and his other children's stories.

Logic has been studied through the ages to exercise the mind's ability to reason. Understanding logic will enable you to think clearly, communicate effectively, make more convincing arguments, and develop patterns of reasoning that will help you in making decisions. It will also help you to detect the fallacies in the reasoning or arguments of others such as advertisers and politicians. Studying logic has other practical applications, such as helping you to understand wills, contracts, and other legal documents.

The study of logic is also good preparation for other areas of mathematics. If you preview Chapter 12, on probability, you will see formulas for the probability of A or B and the probability of A and B, symbolized as $P(A \text{ or } B)$ and $P(A \text{ and } B)$, respectively. Special meanings of common words such as *or* and *and* apply to all areas of mathematics. The meaning of these and other special words is discussed in this chapter.

Logic and the English Language

In reading, writing, and speaking, we use many words such as *and, or,* and *if . . . then . . .* to connect thoughts. In logic we call these words *connectives*. How are these words interpreted in daily communication? A judge announces to a convicted offender, "I hereby sentence you to five months of community service *and* a fine of $100." In this case, we normally interpret the word *and* to indicate that *both* events will take place. That is, the person must perform community service and must also pay a fine.

Now suppose a judge states, "I sentence you to six months in prison *or* 10 months of community service." In this case, we interpret the connective *or* as meaning the convicted person must either spend the time in jail or perform community service, but not both. The word *or* in this case is the *exclusive or*. When the exclusive *or* is used, one or the other of the events can take place, but *not both*.

In a restaurant, a waiter asks, "May I interest you in a cup of soup or a sandwich?" This question offers three possibilities: You may order soup, you may order a sandwich, or you may order both soup and a sandwich. The *or* in this case is the *inclusive or*. When the inclusive *or* is used, one or the other, *or both* events can take place. *In this chapter, when we use the word* or *in a logic statement, it will mean the* inclusive or *unless stated otherwise.*

If–then statements are often used to relate two ideas, as in the bank policy statement "If the average daily balance is greater than $500, then there will be no service charge." If–then statements are also used to emphasize a point or add humor, as in the statement "If the Cubs win, then I will be a monkey's uncle."

Now let's look at logic from a mathematical point of view.

Statements and Logical Connectives

A sentence that can be judged either true or false is called a *statement*. Labeling a statement true or false is called *assigning a truth value* to the statement. Here are some examples of statements.

1. The Brooklyn Bridge goes over San Francisco Bay.
2. Disney World is in Idaho.
3. The Mississippi River is the longest river in the United States.

In each case, we can say that the sentence is either true or false. Statement 1 is false because the Brooklyn Bridge does not go over San Francisco Bay. Statement 2 is false because Disney World is in Florida. By looking at a map or reading an almanac, we can determine that the Mississippi River is the longest river in the United States; therefore, statement 3 is true.

The three sentences discussed above are examples of *simple statements* because they convey one idea. Sentences combining two or more ideas that can be assigned a truth value are called *compound statements*. Compound statements are discussed shortly.

Quantifiers

Sometimes it is necessary to change a statement to its opposite meaning. To do so, we use the *negation* of a statement. For example, the negation of the statement "Emily is at home" is "Emily is not at home." The negation of a true statement is always a false

▲ The Brooklyn Bridge in New York City

statement, and the negation of a false statement is always a true statement. We must use special caution when negating statements containing the words *all*, *none* (or *no*), and *some*. These words are referred to as *quantifiers*.

Consider the statement "All lakes contain fresh water." We know this statement is false because the Great Salt Lake in Utah contains salt water. Its negation must therefore be true. We may be tempted to write its negation as "No lake contains fresh water," but this statement is also false because Lake Superior contains fresh water. Therefore, "No lakes contain fresh water" is not the negation of "All lakes contain fresh water." The correct negation of "All lakes contain fresh water" is "Not all lakes contain fresh water" or "At least one lake does not contain fresh water" or "Some lakes do not contain fresh water." These statements all imply that at least one lake does not contain fresh water, which is a true statement.

Now consider the statement "No birds can swim." This statement is false because at least one bird, the penguin, can swim. Therefore, the negation of this statement must be true. We may be tempted to write the negation as "All birds can swim," but because this statement is also false it cannot be the negation. The correct negation of the statement is "Some birds can swim" or "At least one bird can swim," each of which is a true statement.

Now let's consider statements involving the quantifier *some*, as in "Some students have a driver's license." This statement is true, meaning that at least one student has a driver's license. The negation of this statement must therefore be false. The negation is "No student has a driver's license," which is a false statement.

Consider the statement "Some students do not ride motorcycles." This statement is true because it means "At least one student does not ride a motorcycle." The negation of this statement must therefore be false. The negation is "All students ride motorcycles," which is a false statement.

The negation of quantified statements is summarized as follows:

Form of statement	Form of negation
All are.	Some are not.
None are.	Some are.
Some are.	None are.
Some are not.	All are.

The following diagram might help you to remember the statements and their negations:

The quantifiers diagonally opposite each other are the negations of each other.

EXAMPLE ❶ *Write Negations*

Write the negation of each statement.

a) Some snakes are poisonous.

b) All swimming pools are rectangular.

SOLUTION

a) Since *some* means "at least one," the statement "Some snakes are poisonous" is the same as "At least one snake is poisonous." Because it is a true statement, its negation must be false. The negation is "No snakes are poisonous," which is a false statement.

b) The statement "All swimming pools are rectangular" is a false statement since some pools are circular, some are oval, and some have other shapes. Its negation must therefore be true. The negation may be written as "Some swimming pools are not rectangular" or "Not all swimming pools are rectangular" or "At least one swimming pool is not rectangular." Each of these statements is true.

Compound Statements

Statements consisting of two or more simple statements are called **compound statements**. The connectives often used to join two simple statements are

and,   or,   if,...then...,   if and only if

In addition, we consider a simple statement that has been negated to be a compound statement. The word *not* is generally used to negate a statement.

To reduce the amount of writing in logic, it is common to represent each simple statement with a lowercase letter. For example, suppose we are discussing the simple statement "Leland is a farmer." Instead of writing "Leland is a farmer" over and over again, we can let p represent the statement "Leland is a farmer." Thereafter we can simply refer to the statement with the letter p. It is customary to use the letters p, q, r, and s to represent simple statements, but other letters may be used instead. Let's now look at the connectives used to make compound statements.

Not Statements

The negation is symbolized by $\sim$ and read "not." For example, the negation of the statement "Steve is a college student" is "Steve is not a college student." If p represents the simple statement "Steve is a college student," then $\sim p$ represents the compound statement "Steve is not a college student." For any statement p, $\sim(\sim p) = p$. For example, the negation of the statement "Steve is not a college student" is "Steve is a college student."

Consider the statement "Inga is not at home." This statement contains the word *not*, which indicates that it is a negation. To write this statement symbolically, we let p represent "Inga *is* at home." Then $\sim p$ would be "Inga is not at home." *We will use this convention of letting letters such as p, q, or r represent statements that are not negated. We will represent negated statements with the negation symbol, $\sim$.*

And Statements

The *conjunction* is symbolized by $\wedge$ and read "and." The $\wedge$ looks like an A (for And) with the bar missing. Let p and q represent the simple statements.

p: You will perform 5 months of community service.
q: You will pay a $100 fine.

Then the following is the conjunction written in symbolic form.

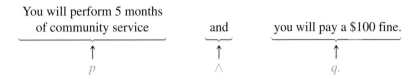

The conjunction is generally expressed as *and*. Other words sometimes used to express a conjunction are *but, however,* and *nevertheless*.

▲ Billie Joe Armstrong of Green Day

EXAMPLE ② *Write a Conjunction*

Write the following conjunction in symbolic form.

Green Day is not on tour, but Green Day is recording a new CD.

SOLUTION Let *t* and *r* represent the simple statements.

t: Green Day is on tour.
r: Green Day is recording a new CD.

In symbolic form, the compound statement is $\sim t \wedge r$. ●

In Example 2, the compound statement is "Green Day is not on tour, but Green Day is recording a new CD." This statement could also be represented as "Green Day is not on tour, but *they* are recording a new CD." In this problem, it should be clear the word *they* means *Green Day*. Therefore, the statement, "Green Day is not on tour, but they are recording a new CD" would also be symbolized as $\sim t \wedge r$.

Or Statements

The *disjunction* is symbolized by $\vee$ and read "or." The *or* we use in this book (except where indicated in the exercise sets) is the *inclusive or* described on page 102.

EXAMPLE ③ *Write a Disjunction*

Let

p: Maria will go to the circus.
q: Maria will go to the zoo.

Write the following statements in symbolic form.
a) Maria will go to the circus or Maria will go to the zoo.
b) Maria will go to the zoo or Maria will not go to the circus.
c) Maria will not go to the circus or Maria will not go to the zoo.

SOLUTION
a) $p \vee q$ b) $q \vee \sim p$ c) $\sim p \vee \sim q$ ●

Because *or* represents the *inclusive or,* the statement "Maria will go to the circus or Maria will go to the zoo" in Example 3(a) may mean that Maria will go to the circus, or that Maria will go to the zoo, or that Maria will go to both the circus *and* the zoo. The statement in Example 3(a) could also be written as "Maria will go to the circus or the zoo."

When a compound statement contains more than one connective, a comma can be used to indicate which simple statements are to be grouped together. When we write the compound statement symbolically, *the simple statements on the same side of the comma are to be grouped together within parentheses.*

For example, "Pink is a singer (p) or Geena Davis is an actress (g), and Dallas is in Texas (d)" is written $(p \lor g) \land d$. Note that the p and g are both on the same side of the comma in the written statement. They are therefore grouped together within parentheses. The statement "Pink is a singer, or Geena Davis is an actress and Dallas is in Texas" is written $p \lor (g \land d)$. In this case, g and d are on the same side of the comma and are therefore grouped together within parentheses.

EXAMPLE ❹ *Understand How Commas Are Used to Group Statements*

Let

p: Dinner includes soup.
q: Dinner includes salad.
r: Dinner includes the vegetable of the day.

Write the following statements in symbolic form.
a) Dinner includes soup, and salad or the vegetable of the day.
b) Dinner includes soup and salad, or the vegetable of the day.

SOLUTION

a) The comma tells us to group the statement "Dinner includes salad" with the statement "Dinner includes the vegetable of the day." Note that both statements are on the same side of the comma. The statement in symbolic form is $p \land (q \lor r)$.

   In mathematics, we always evaluate the information within the parentheses first. Since the conjunction, $\land$, is outside the parentheses and is evaluated *last,* this statement is considered a *conjunction*.

b) The comma tells us to group the statement "Dinner includes soup" with the statement "Dinner includes salad." Note that both statements are on the same side of the comma. The statement in symbolic form is $(p \land q) \lor r$. Since the disjunction, $\lor$, is outside the parentheses and is evaluated *last,* this statement is considered a *disjunction*.   ●

The information provided in Example 4 is summarized below.

Statement	Symbolic representation	Type of statement
Dinner includes soup, and salad or the vegetable of the day.	$p \land (q \lor r)$	conjunction
Dinner includes soup and salad, or the vegetable of the day.	$(p \land q) \lor r$	disjunction

A negation symbol has the effect of negating only the statement that directly follows it. To negate a compound statement, we must use parentheses. When a negation symbol is placed in front of a statement in parentheses, it negates the entire statement in parentheses. The negation symbol in this case is read, "It is not true that . . . " or "It is false that . . . "

EXAMPLE 5 *Change Symbolic Statements to Words*

Let

$$p: \quad \text{Jozsef is making breakfast.}$$
$$q: \quad \text{Arum is setting the table.}$$

Write the following symbolic statements in words.

a) $p \wedge {\sim}q$ b) ${\sim}p \vee {\sim}q$ c) ${\sim}(p \wedge q)$

SOLUTION

a) Jozsef is making breakfast and Arum is not setting the table.

b) Jozsef is not making breakfast or Arum is not setting the table.

c) It is false that Jozsef is making breakfast and Arum is setting the table.   ●

Recall that the word *but* may also be used in a conjunction. Therefore, Example 5(a) could also be written "Jozsef is making breakfast, *but* Arum is not setting the table."

Part (b) of Example 5 is a disjunction, since it can be written $({\sim}p) \vee ({\sim}q)$. Part (c), which is ${\sim}(p \wedge q)$, is a negation since the negation symbol negates the entire statement within parentheses. The similarity of these two statements is discussed in Section 3.4.

Occasionally, we come across a *neither–nor* statement, such as "John is neither handsome nor rich." This statement means that John is not handsome *and* John is not rich. If p represents "John is handsome" and q represents "John is rich," this statement is symbolized by ${\sim}p \wedge {\sim}q$.

If–Then Statements

The *conditional* is symbolized by $\rightarrow$ and is read "if–then." The statement $p \rightarrow q$ is read "If p, then q."* The conditional statement consists of two parts: the part that precedes the arrow is the *antecedent*, and the part that follows the arrow is the *consequent*.† In the conditional statement $p \rightarrow q$, the p is the antecedent and the q is the consequent.

In the conditional statement ${\sim}(p \vee q) \rightarrow (p \wedge q)$, the antecedent is ${\sim}(p \vee q)$ and the consequent is $(p \wedge q)$. An example of a conditional statement is "If you drink your milk, then you will grow up to be healthy." A conditional symbol may be placed between any two statements even if the statements are not related.

---

*Some books indicate that $p \rightarrow q$ may also be read "p implies q." Many higher-level mathematics books, however, indicate that $p \rightarrow q$ may be read "p implies q" only under certain conditions. Implications are discussed in Section 3.3.

† Some books refer to the antecedent as the hypothesis or premise and the consequent as the conclusion.

Sometimes the word *then* in a conditional statement is not explicitly stated. For example, the statement "If you get an A, I will buy you a car" is a conditional statement because it actually means "If you get an A, then I will buy you a car."

EXAMPLE ❻ *Write Conditional Statements*

Let

p: The portrait is a pastel.

q: The portrait is by Beth Anderson.

Write the following statements symbolically.

a) If the portrait is a pastel, then the portrait is by Beth Anderson.

b) If the portrait is by Beth Anderson, then the portrait is not a pastel.

c) It is false that if the portrait is by Beth Anderson then the portrait is a pastel.

SOLUTION

a) $p \to q$ b) $q \to \sim p$ c) $\sim(q \to p)$

▲ *Zoe* by Beth Anderson

EXAMPLE ❼ *Use Commas When Writing a Symbolic Statement in Words*

Let

p: Jorge is enrolled in calculus.

q: Jorge's major is criminal justice.

r: Jorge's major is engineering.

Write the following symbolic statements in words and indicate whether the statement is a negation, conjunction, disjunction, or conditional.

a) $(q \to \sim p) \lor r$ b) $q \to (\sim p \lor r)$

SOLUTION The parentheses indicate where to place the commas in the sentences.

a) "If Jorge's major is criminal justice then Jorge is not enrolled in calculus, or Jorge's major is engineering." This statement is a disjunction because $\lor$ is outside the parentheses.

b) "If Jorge's major is criminal justice, then Jorge is not enrolled in calculus or Jorge's major is engineering." This statement is a conditional because $\to$ is outside the parentheses.

If and Only if Statements

The *biconditional* is symbolized by $\leftrightarrow$ and is read "if and only if." The phrase *if and only if* is sometimes abbreviated as "iff." The statement $p \leftrightarrow q$ is read "p if and only if q."

▲ Martin St. Louis

EXAMPLE ⑧ *Write Statements Using the Biconditional*

Let

p: The Tampa Bay Lightning win the Stanley Cup.

q: Martin St. Louis is the MVP.

Write the following symbolic statements in words.

a) $p \leftrightarrow \sim q$ b) $\sim(q \leftrightarrow \sim p)$

SOLUTION

a) The Tampa Bay Lightning win the Stanley Cup if and only if Martin St. Louis is not the MVP.

b) It is false that Martin St. Louis is the MVP if and only if the Tampa Bay Lightning do not win the Stanley Cup.

You will learn later that $p \leftrightarrow q$ means the same as $(p \rightarrow q) \wedge (q \rightarrow p)$. Therefore, the statement "I will go to college if and only if I can pay the tuition" has the same logical meaning as "If I go to college then I can pay the tuition, and if I can pay the tuition then I will go to college."

A summary of the connectives discussed in this section is given in Table 3.1.

Table 3.1 Logical Connectives

Formal Name	Symbol	Read	Symbolic Form
Negation	$\sim$	"Not"	$\sim p$
Conjunction	$\wedge$	"And"	$p \wedge q$
Disjunction	$\vee$	"Or"	$p \vee q$
Conditional	$\rightarrow$	"If-then"	$p \rightarrow q$
Biconditional	$\leftrightarrow$	"If and only if"	$p \leftrightarrow q$

SECTION ③.① EXERCISES

CONCEPT/WRITING EXERCISES

1. **a)** What is a statement?

   b) What is a simple statement?

   c) What is a compound statement?

2. Fill in the blanks to make the following statements true.

   a) The negation of a true statement is always a _____ statement.

   b) The negation of a _____ statement is always a true statement.

3. Give three words that can be used as a quantifier in a statement.

4. Represent the statement, "The telephone does not have caller ID" symbolically. Explain your answer.

5. Write the general form of the negation for statements of the form

   a) none are.

   b) some are not.

   c) all are.

   d) some are.

6. Draw the symbol used to represent the

   a) conditional.

   b) disjunction.

   c) conjunction.

   d) negation.

   e) biconditional.

7. a) When the *exclusive or* is used as a connective between two events, can both events take place? Explain.

   b) When the *inclusive or* is used as a connective between two events, can both events take place? Explain.

   c) Which *or*, the *inclusive or* or the *exclusive or*, is used in this chapter?

8. Explain how a comma is used to indicate the grouping of simple statements.

PRACTICE THE SKILLS/PROBLEM SOLVING

In Exercises 9–22, indicate whether the statement is a simple statement or a compound statement. If it is a compound statement, indicate whether it is a negation, conjunction, disjunction, conditional, or biconditional by using both the word and its appropriate symbol (for example, "a negation," ~).

9. Jana Bryant is teaching statistics or she is teaching calculus.

10. If you burn the chicken wings, then you can feed them to the dog.

11. Time will go backwards if and only if you travel faster than the speed of light.

12. Louis Armstrong did not play the drums.

▲ Louis Armstrong, see Exercise 12

13. Bobby Glewen joined the Army and he got married.

14. The book was neither a novel nor an autobiography.

15. The hurricane did $400,000 worth of damage to DeSoto County.

16. Inhibor Melendez will be admitted to law school if and only if he earns his bachelor's degree.

17. It is false that Jeffery Hilt is a high school teacher and a grade school teacher.

18. If Cathy Smith walks 4 miles today, then she will be sore tomorrow.

19. Mary Jo Woo ran 4 miles today and she lifted weights for 30 minutes.

20. Nancy Wallin went to the game, but she did not eat a hot dog.

21. It is false that if John Wubben fixes your car then you will need to pay him in cash.

22. If Buddy and Evelyn Cordova are residents of Budville, then they must vote for mayor on Tuesday.

In Exercises 23–34, write the negation of the statement.

23. All butterflies are insects.

24. All houses are wired using parallel circuits.

25. No aldermen are running for mayor.

26. Some diet sodas contain saccharin.

27. Some turtles do not have claws.

28. No teachers made the roster.

29. No bicycles have three wheels.

30. All horses have manes.

31. Some pine trees do not produce pinecones.

32. No one likes asparagus.

33. Some pedestrians are in the crosswalk.

34. Some dogs with long hair do not get cold.

In Exercises 35–40, write the statement in symbolic form. Let

    p: The tent is pitched.
    q: The bonfire is burning.

35. The tent is not pitched.

36. The tent is pitched and the bonfire is burning.

37. The bonfire is not burning or the tent is not pitched.

38. The bonfire is not burning if and only if the tent is not pitched.

39. If the tent is not pitched, then the bonfire is not burning.

40. The bonfire is not burning, however the tent is pitched.

In Exercises 41–46, write the statement in symbolic form. Let

    p: The chili is spicy.
    q: The sour cream is cold.

41. The chili is not spicy, but the sour cream is cold.

42. Neither is the chili spicy nor is the sour cream cold.

43. The sour cream is not cold if and only if the chili is spicy.

44. If the chili is spicy, then the sour cream is not cold.

45. It is false that the chili is spicy or the sour cream is cold.

46. It is false that if the sour cream is not cold then the chili is spicy.

In Exercises 47–56, write the compound statement in words.
Let

    p: Ken Jennings won 74 games of *Jeopardy!*
    q: Ken Jennings won more than $3 million.

▲ Alex Trebek and Ken Jennings

47. $\sim q$ **48.** $\sim p$

49. $p \wedge q$ **50.** $q \vee p$

51. $\sim p \rightarrow q$ **52.** $\sim p \leftrightarrow \sim q$

53. $\sim p \vee \sim q$ **54.** $\sim(q \vee p)$

55. $\sim(p \wedge q)$ **56.** $\sim p \wedge \sim q$

In Exercises 57–66, write the statements in symbolic form.
Let

    p: The temperature is 90°.
    q: The air conditioner is working.
    r: The apartment is hot.

57. The temperature is 90° and the air conditioner is not working, and the apartment is hot.

58. The temperature is not 90° and the air conditioner is working, but the apartment is hot.

59. The temperature is 90° and the air conditioner is working, or the apartment is hot.

60. If the apartment is hot and the air conditioner is working, then the temperature is 90°.

61. If the temperature is 90°, then the air conditioner is working or the apartment is not hot.

62. The temperature is not 90° if and only if the air conditioner is not working, or the apartment is not hot.

63. The apartment is hot if and only if the air conditioner is working, and the temperature is 90°.

64. It is false that if the apartment is hot then the air conditioner is not working.

65. If the air conditioner is working, then the temperature is 90° if and only if the apartment is hot.

66. The apartment is hot or the air conditioner is not working, if and only if the temperature is 90°.

In Exercises 67–76, write each symbolic statement in words. Let

p: The water is 70°.

q: The sun is shining.

r: We go swimming.

67. $(p \lor q) \land \sim r$ **68.** $(p \land q) \lor r$

69. $\sim p \land (q \lor r)$ **70.** $(q \to p) \lor r$

71. $\sim r \to (q \land p)$ **72.** $(q \land r) \to p$

73. $(q \to r) \land p$ **74.** $\sim p \to (q \lor r)$

75. $(q \leftrightarrow p) \land r$ **76.** $q \to (p \leftrightarrow r)$

***Dinner Menu** In Exercises 77–80, use the following information to arrive at your answers. Many restaurant dinner menus include statements such as the following. All dinners are served with a choice of: Soup or Salad, and Potatoes or Pasta, and Carrots or Peas. Which of the following selections are permissible? If a selection is not permissible, explain why. See the discussion of the* exclusive or *on page 102.*

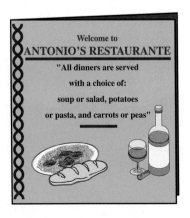

Welcome to
ANTONIO'S RESTAURANTE

"All dinners are served

with a choice of:

soup or salad, potatoes

or pasta, and carrots or peas"

77. Soup, salad, and peas

78. Salad, pasta, and carrots

79. Soup, potatoes, pasta, and peas

80. Soup, pasta, and potatoes

In Exercises 81–89, (a) select letters to represent the simple statements and write each statement symbolically by using parentheses and (b) indicate whether the statement is a negation, conjunction, disjunction, conditional, or biconditional.

81. I bought the watch in Tijuana and I did not pay $100.

82. If the conference is in Las Vegas, then we can see Wayne Newton or we can play poker.

83. It is false that if your speed is below the speed limit then you will not get pulled over.

84. If dinner is ready then we can eat, or we cannot go to the restaurant.

85. If the food has fiber or the food has vitamins, then you will be healthy.

86. If Corliss is teaching then Faye is in the math lab, if and only if it is not a weekend.

87. You may take this course, if and only if you did not fail the previous course or you passed the placement test.

88. If the car has gas and the battery is charged, then the car will start.

89. The classroom is empty if and only if it is the weekend, or it is 7 A.M.

CHALLENGE PROBLEMS/GROUP ACTIVITIES

90. *An Ancient Question* If Zeus could do anything, could he build a wall that he could not jump over? Explain your answer.

91. a) Make up three simple statements and label them p, q, and r. Then write compound statements to represent $(p \lor q) \land r$ and $p \lor (q \land r)$.

   b) Do you think that the statements for $(p \lor q) \land r$ and $p \lor (q \land r)$ mean the same thing? Explain.

RECREATIONAL MATHEMATICS

92. **Sudoku** Sudoku is a logic-based puzzle that originated in Japan. The goal is to enter numerical digits in each of the empty squares so that every digit from 1 to 9 appears exactly one time in each of the rows, in each of the columns, and in each of the nine 3 by 3 boxes. Completing the puzzle requires patience and logical ability. For more information, see www.websudoku.com. Complete the following Sudoku puzzle.

	1		4	8		5	6	
5					9	8		
	3				1	4		7
8	2			9		1		
6			1		4			9
		3		6			4	5
9		1	5				2	
		7	2					4
	5	2		7	8		3	

INTERNET/RESEARCH ACTIVITIES

93. **Legal Documents** Obtain a legal document such as a will or rental agreement and copy one page of the document. Circle every connective used. Then list the number of times each connective appeared. Be sure to include conditional statements from which the word *then* was omitted from the sentence. Give the page and your listing to your instructor.

94. Write a report on the life and accomplishments of George Boole, who was an important contributor to the development of logic. In your report, indicate how his work eventually led to the development of the computer. References include encyclopedias, history of mathematics books, and the Internet.

3.2 TRUTH TABLES FOR NEGATION, CONJUNCTION, AND DISJUNCTION

Consider the following statement: Idaho grows the most potatoes or Iowa grows the most corn. Under what conditions can the statement be considered true? Under what conditions can the statement be considered false? In this section, we will introduce a tool used to help us analyze such statements.

▲ Under what conditions is the statement *Idaho grows the most potatoes or Iowa grows the most corn* true?

A *truth table* is a device used to determine when a compound statement is true or false. Five basic truth tables are used in constructing other truth tables. Three are discussed in this section (Tables 3.2, 3.4, and 3.7), and two are discussed in the next section. Section 3.5 uses truth tables in determining whether a logical argument is valid or invalid.

Negation

The first truth table is for *negation*. If p is a true statement, then the negation of p, "not p," is a false statement. If p is a false statement, then "not p" is a true statement. For

Table 3.2 Negation

	p	*~p*
Case 1	T	F
Case 2	F	T

Table 3.3

	p	*q*
Case 1	T	T
Case 2	T	F
Case 3	F	T
Case 4	F	F

Table 3.4 Conjunction

	p	*q*	*p* ∧ *q*
Case 1	T	T	T
Case 2	T	F	F
Case 3	F	T	F
Case 4	F	F	F

example, if the statement "The shirt is blue" is true, then the statement "The shirt is not blue" is false. These relationships are summarized in Table 3.2. For a simple statement, there are exactly two true–false cases, as shown.

If a compound statement consists of two simple statements p and q, there are four possible cases, as illustrated in Table 3.3. Consider the statement "The test is today and the test covers Chapter 5." The simple statement "The test is today" has two possible truth values, true or false. The simple statement "The test covers Chapter 5" also has two truth values, true or false. Thus, for these two simple statements there are four distinct possible true–false arrangements. Whenever we construct a truth table for a compound statement that consists of two simple statements, we begin by listing the four true–false cases shown in Table 3.3.

Conjunction

To illustrate the conjunction, consider the following situation. You have recently purchased a new house. To decorate it, you ordered a new carpet and new furniture from the same store. You explain to the salesperson that the carpet must be delivered before the furniture. He promises that the carpet will be delivered on Thursday and that the furniture will be delivered on Friday.

To help determine whether the salesperson kept his promise, we assign letters to each simple statement. Let p be "The carpet will be delivered on Thursday" and q be "The furniture will be delivered on Friday." The salesperson's statement written in symbolic form is $p \wedge q$. There are four possible true–false situations to be considered (Table 3.4).

CASE 1: p is true and q is true. The carpet is delivered on Thursday and the furniture is delivered on Friday. The salesperson has kept his promise and the compound statement is true. Thus, we put a T in the $p \wedge q$ column.

CASE 2: p is true and q is false. The carpet is delivered on Thursday but the furniture is not delivered on Friday. Since the furniture was not delivered as promised, the compound statement is false. Thus, we put an F in the $p \wedge q$ column.

CASE 3: p is false and q is true. The carpet is not delivered on Thursday but the furniture is delivered on Friday. Since the carpet was not delivered on Thursday as promised, the compound statement is false. Thus, we put an F in the $p \wedge q$ column.

CASE 4: p is false and q is false. The carpet is not delivered on Thursday and the furniture is not delivered on Friday. Since the carpet and furniture were not delivered as promised, the compound statement is false. Thus, we put an F in the $p \wedge q$ column.

Examining the four cases, we see that in only one case did the salesperson keep his promise: in case 1. Therefore, case 1 (T, T) is true. In cases 2, 3, and 4, the salesperson did not keep his promise and the compound statement is false. The results are summarized in Table 3.4, the truth table for the conjunction.

The **conjunction** $p \wedge q$ is true only when both p and q are true.

EXAMPLE ❶ *Construct a Truth Table*

Construct a truth table for $p \wedge \sim q$.

SOLUTION Because there are two statements, p and q, construct a truth table with four cases; see Table 3.5(a). Then write the truth values under the p in the compound statement and label this column 1, as in Table 3.5(b). Copy these truth values directly from the p column on the left. Write the corresponding truth values under the q in the compound statement and call this column 2, as in Table 3.5(c). Copy the truth values for column 2 directly from the q column on the left. Now find the truth values of $\sim q$ by negating the truth values in column 2 and call this column 3,

Table 3.5

(a)

	p	q	$p \wedge \sim q$
Case 1	T	T	
Case 2	T	F	
Case 3	F	T	
Case 4	F	F	

(b)

p	q	$p \wedge \sim q$
T	T	T
T	F	T
F	T	F
F	F	F
		1

(c)

p	q	p	$\wedge$	$\sim$	q
T	T	T			T
T	F	T			F
F	T	F			T
F	F	F			F
		1			2

(d)

p	q	p	$\wedge$	$\sim$	q
T	T	T		F	T
T	F	T		T	F
F	T	F		F	T
F	F	F		T	F
		1		3	2

(e)

p	q	p	$\wedge$	$\sim$	q
T	T	T	F	F	T
T	F	T	T	T	F
F	T	F	F	F	T
F	F	F	F	T	F
		1	4	3	2

as in Table 3.5(d). Use the conjunction table, Table 3.4, and the entries in the columns labeled 1 and 3 to complete the column labeled 4, as in Table 3.5(e). The results in column 4 are obtained as follows:

Row 1: $T \wedge F$ is F.   Row 2: $T \wedge T$ is T.

Row 3: $F \wedge F$ is F.   Row 4: $F \wedge T$ is F.

The answer is always the last column completed. The columns labeled 1, 2, and 3 are only aids in arriving at the answer labeled column 4.   ●

The statement $p \wedge \sim q$ in Example 1 actually means $p \wedge (\sim q)$. In the future, instead of listing a column for q and a separate column for its negation, we will make one column for $\sim q$, which will have the opposite values of those in the q column on the left. Similarly, when we evaluate $\sim p$, we will use the opposite values of those in the p column on the left. This procedure is illustrated in Example 2.

In Example 1, we spoke about *cases* and also *columns*. Consider Table 3.5(e). This table has four cases indicated by the four different rows of the two left-hand (unnumbered) columns. The four *cases* are TT, TF, FT, and FF. In every truth table with two letters, we list the four cases (the first two columns) first. Then we complete the remaining columns in the truth table. In Table 3.5(e), after completing the two left-hand columns, we complete the remaining columns in the order indicated by the numbers below the columns. We will continue to place numbers below the columns to show the order in which the columns are completed.

In discussion of the truth table in Example 2, and all following truth tables, if we say column 1, it means the column labeled 1. Column 2 will mean the column labeled 2, and so on.

> **TIMELY TIP** When constructing truth tables it is very important to keep your entries in neat columns and rows. If you are using lined paper, put only one row of the table on each line. If you are not using lined paper, using a straightedge may help you correctly enter the information into the truth table's rows and columns.

EXAMPLE ❷ *Construct and Interpret a Truth Table*

a) Construct a truth table for the following statement: Jose is not an artist and Jose is not a musician.

b) Under which conditions will the compound statement be true?

c) Suppose "Jose is an artist" is a false statement and "Jose is a musician" is a true statement. Is the compound statement given in part (a) true or false?

SOLUTION

a) First write the simple statements in symbolic form by using simple nonnegated statements.

Let

p: Jose is an artist.

q: Jose is a musician.

Therefore, the compound statement may be written $\sim p \wedge \sim q$. Now construct a truth table with four cases, as shown in Table 3.6.

Fill in the column labeled 1 by negating the truth values under p on the far left. Fill in the column labeled 2 by negating the values under q in the second column from the left. Fill in the column labeled 3 by using the columns labeled 1 and 2 and the definition of conjunction.

In the first row, to determine the entry for column 3, we use false for $\sim p$ and false for $\sim q$. Since false $\wedge$ false is false (see case 4 of Table 3.4), we place an F in column 3, row 1. In the second row, we use false for $\sim p$ and true for $\sim q$. Since false $\wedge$ true is false (see case 3 of Table 3.4), we place an F in column 3, row 2. In the third row, we use true for $\sim p$ and false for $\sim q$. Since true $\wedge$ false is false (see case 2 of Table 3.4), we place an F in column 3, row 3. In the fourth row, we use true for $\sim p$ and true for $\sim q$. Since true $\wedge$ true is true (see case 1 of Table 3.4), we place a T in column 3, row 4.

Table 3.6

p	q	$\sim p$	$\wedge$	$\sim q$
T	T	F	F	F
T	F	F	F	T
F	T	T	(F)	F
F	F	T	(T)	T
		1	3	2

b) The compound statement in part (a) will be true only in case 4 (circled in blue) when both simple statements, p and q, are false, that is, when Jose is not an artist and Jose is not a musician.

c) We are told that p, "Jose is an artist," is a false statement and that q, "Jose is a musician," is a true statement. From the truth table (Table 3.6), we can determine that when p is false and q is true, the compound statement, case 3, (circled in red) is false.   ●

Disjunction

Consider the job description in the margin that describes several job requirements. Who qualifies for the job? To help analyze the statement, translate it into symbolic form. Let p be "A requirement for the job is a two-year college degree in civil technology" and q be "A requirement for the job is five years of related experience." The statement in symbolic form is $p \vee q$. For the two simple statements, there are four distinct cases (see Table 3.7).

CASE 1: p is true and q is true. A candidate has a two-year college degree in civil technology and five years of related experience. The candidate has both requirements and qualifies for the job. Consider qualifying for the job as a true statement and not qualifying as a false statement. Since the candidate qualifies for the job, we put a T in the $p \vee q$ column.

CASE 2: p is true and q is false. A candidate has a two-year college degree in civil technology but does not have five years of related experience. The candidate still qualifies for the job with the two-year college degree. Thus, we put a T in the $p \vee q$ column.

CASE 3: p is false and q is true. The candidate does not have a two-year college degree in civil technology but does have five years of related experience. The candidate qualifies for the job with the five years of related experience. Thus, we put a T in the $p \vee q$ column.

CASE 4: p is false and q is false. The candidate does not have a two-year college degree in civil technology and does not have five years of related experience. The candidate does not meet either of the two requirements and therefore does not qualify for the job. Thus, we put an F in the $p \vee q$ column.

In examining the four cases, we see that there is only one case in which the candidate does not qualify for the job: case 4. As this example indicates, an *or* statement will be true in every case, except when both simple statements are false. The results are summarized in Table 3.7, the truth table for the disjunction.

Civil Technician

Municipal program for redevelopment seeks on-site technician. **The applicant must have a two-year college degree in civil technology or five years of related experience.** Interested candidates please call 555-1234.

Table 3.7 Disjunction

p	q	$p \vee q$
T	T	T
T	F	T
F	T	T
F	F	F

The **disjunction**, $p \vee q$, is true when either p is true, q is true, or both p and q are true.

The disjunction $p \vee q$ is false only when p and q are both false.

Table 3.8

p	q	~	(~q	∧	p)
T	T	T	F	F	T
T	F	F	T	T	T
F	T	T	F	F	F
F	F	T	T	F	F
		4	1	3	2

EXAMPLE ❸ Truth Table with a Negation

Construct a truth table for ~(~q ∧ p).

SOLUTION First construct the standard truth table listing the four cases. Then work within parentheses. The order to be followed is indicated by the numbers below the columns (see Table 3.8). Under ~q, column 1, write the negation of the q column. Then, in column 2, copy the values from the p column. Next, complete the *and* column, column 3, using columns 1 and 2 and the truth table for the conjunction. The *and* column is true only when both statements are true, as in case 2. Finally, negate the values in the *and* column, column 3, and place these negated values in column 4. By examining the truth table you can see that the compound statement ~(~q ∧ p) is false only in case 2, that is, when p is true and q is false.   ●

A GENERAL PROCEDURE FOR CONSTRUCTING TRUTH TABLES

1. Study the compound statement and determine whether it is a negation, conjunction, disjunction, conditional, or biconditional statement, as was done in Section 3.1. The answer to the truth table will appear under ~ if the statement is a negation, under ∧ if the statement is a conjunction, under ∨ if the statement is a disjunction, under → if the statement is a conditional, and under ↔ if the statement is a biconditional.

2. Complete the columns under the simple statements, p, q, r, and their negations, ~p, ~q, ~r, within parentheses, if present. If there are nested parentheses (one pair of parentheses within another pair), work with the innermost pair first.

3. Complete the column under the connective within the parentheses, if present. You will use the truth values of the connective in determining the final answer in step 5.

4. Complete the column under any remaining statements and their negations.

5. Complete the column under any remaining connectives. Recall that the answer will appear under the column determined in step 1. If the statement is a conjunction, disjunction, conditional, or biconditional, you will obtain the truth values for the connective by using the last column completed on the left side and on the right side of the connective. If the statement is a negation, you will obtain the truth values by negating the truth values of the last column completed within the grouping symbols on the right side of the negation. Be sure to circle or highlight your answer column or number the columns in the order they were completed.

Table 3.9

p	q	(~p	∨	q)	∧	~p
T	T	F	T	T	F	F
T	F	F	F	F	F	F
F	T	T	T	T	T	T
F	F	T	T	F	T	T
		1	3	2	5	4

EXAMPLE ❹ Use the General Procedure to Construct a Truth Table

Construct a truth table for the statement (~p ∨ q) ∧ ~p.

SOLUTION We will follow the general procedure outlined in the box. This statement is a conjunction, so the answer will be under the conjunction symbol. Complete columns under ~p and q within the parentheses and call these columns 1 and 2, respectively (see Table 3.9). Complete the column under the disjunction, ∨, using the truth values in columns 1 and 2, and call this column 3. Next complete the column under ~p, and call this column 4. The answer, column 5, is determined from the definition of the conjunction and the truth values in column 3, the last column completed on the left side of the conjunction, and column 4.   ●

Table 3.10

	p	*q*	*r*
Case 1	T	T	T
Case 2	T	T	F
Case 3	T	F	T
Case 4	T	F	F
Case 5	F	T	T
Case 6	F	T	F
Case 7	F	F	T
Case 8	F	F	F

So far, all the truth tables we have constructed have contained at most two simple statements. Now we will explain how to construct a truth table that consists of three simple statements, such as $(p \wedge q) \wedge r$. When a compound statement consists of three simple statements, there are eight different true–false possibilities, as illustrated in Table 3.10. To begin such a truth table, write four Ts and four Fs in the column under *p*. Under the second statement, *q*, pairs of Ts alternate with pairs of Fs. Under the third statement, *r*, T alternates with F. This technique is not the only way of listing the cases, but it ensures that each case is unique and that no cases are omitted.

EXAMPLE ➎ Construct a Truth Table with Eight Cases

a) Construct a truth table for the statement "Santana is home and he is not at his desk, or he is sleeping."

b) Suppose that "Santana is home" is a false statement, that "Santana is at his desk" is a true statement, and that "Santana is sleeping" is a true statement. Is the compound statement in part (a) true or false?

SOLUTION

a) First we will translate the statement into symbolic form.
Let

> *p*: Santana is home.
>
> *q*: Santana is at his desk.
>
> *r*: Santana is sleeping.

In symbolic form, the statement is $(p \wedge \sim q) \vee r$.

Since the statement is composed of three simple statements, there are eight cases. Begin by listing the eight cases in the three left-hand columns; see Table 3.11. By examining the statement, you can see that it is a disjunction. Therefore, the answer will be in the $\vee$ column. Fill out the truth table by working in parentheses first. Place values under *p*, column 1, and $\sim q$, column 2. Then find the conjunctions of columns 1 and 2 to obtain column 3. Place the values of *r* in column 4. To obtain the answer, column 5, use columns 3 and 4 and the information for the disjunction contained in Table 3.7 on page 117.

Table 3.11

p	*q*	*r*	(*p*	$\wedge$	$\sim q$)	$\vee$	*r*
T	T	T	T	F	F	T	T
T	T	F	T	F	F	F	F
T	F	T	T	T	T	T	T
T	F	F	T	T	T	T	F
F	T	T	F	F	F	(T)	T
F	T	F	F	F	F	F	F
F	F	T	F	F	T	T	T
F	F	F	F	F	T	F	F
			1	3	2	5	4

b) We are given the following:

> p: Santana is home—false.
>
> q: Santana is at his desk—true.
>
> r: Santana is sleeping—true.

We need to find the truth value of the following case: false, true, true. In case 5 of the truth table, p, q, and r are F, T, and T, respectively. Therefore, under these conditions, the original compound statement is true (as circled in the table).   ●

We have learned that a truth table with one simple statement has two cases, a truth table with two simple statements has four cases, and a truth table with three simple statements has eight cases. In general, *the number of distinct cases in a truth table with n distinct simple statements is 2^n.* The compound statement $(p \vee q) \vee (r \wedge \sim s)$ has four simple statements, p, q, r, s. Thus, a truth table for this compound statement would have 2^4, or 16, distinct cases.

When we construct a truth table, we determine the truth values of a compound statement for every possible case. If we want to find the truth value of the compound statement for any specific case when we know the truth values of the simple statements, we do not have to develop the entire table. For example, to determine the truth value for the statement

$$2 + 3 = 5 \quad \text{and} \quad 1 + 1 = 3$$

we let p be $2 + 3 = 5$ and q be $1 + 1 = 3$. Now we can write the compound statement as $p \wedge q$. We know that p is a true statement and q is a false statement. Thus, we can substitute T for p and F for q and evaluate the statement:

$$p \wedge q$$
$$T \wedge F$$
$$F$$

Therefore, the compound statement $2 + 3 = 5$ and $1 + 1 = 3$ is a false statement.

EXAMPLE ❻ *Determine the Truth Value of a Compound Statement*

Determine the truth value for each simple statement. Then, using these truth values, determine the truth value of the compound statement.

a) 15 is less than or equal to 9.

b) George Washington was the first U.S. president or Abraham Lincoln was the second U.S. president, but there has not been a U.S. president born in Antarctica.

SOLUTION

a) Let

> p: 15 is less than 9.
>
> q: 15 is equal to 9.

The statement "15 is less than or equal to 9" means that 15 is less than 9 or 15 is equal to 9. The compound statement can be expressed as $p \vee q$. We know that both p and q are false statements since 15 is greater than 9, so we substitute F for p and F for q and evaluate the statement:

$$p \vee q$$
$$\text{F} \vee \text{F}$$
$$\text{F}$$

Therefore, the compound statement "15 is less than or equal to 9" is a false statement.

b) Let

  p: George Washington was the first U.S. president.

  q: Abraham Lincoln was the second U.S. president.

  r: There has been a U.S. president who was born in Antarctica .

The compound statement can be written in symbolic form as $(p \vee q) \wedge \sim r$. Recall that *but* is used to express a conjunction. We know that p is a true statement and that q is a false statement. We also know that r is a false statement since all U.S. presidents must be born in the United States. Thus, since r is a false statement, the negation, $\sim r$, is a true statement. So we will substitute T for p, F for q, and T for $\sim r$ and then evaluate the statement:

$$(p \vee q) \wedge \sim r$$
$$(\text{T} \vee \text{F}) \wedge \text{T}$$
$$\text{T} \wedge \text{T}$$
$$\text{T}$$

Therefore, the original compound statement is a true statement.   ●

EXAMPLE ❼ *Pet Ownership in the United States*

The number of pets owned in the United States in 2005 is shown in Fig. 3.1. Use this graph to determine the truth value of the following statement: There are more dogs owned than cats and there are fewer reptiles owned than birds, or the most numerous pets owned are not fish.

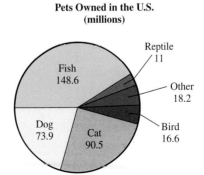

Pets Owned in the U.S.
(millions)

Reptile 11

Fish 148.6

Other 18.2

Dog 73.9

Cat 90.5

Bird 16.6

Source: American Pet Products Manufacturing Association

Figure 3.1

SOLUTION Let

p: There are more dogs owned than cats.

q: There are fewer reptiles owned than birds.

r: The most numerous pets owned are fish.

The given compound statement can be written in symbolic form as $(p \wedge q) \vee \sim r$. From Fig. 3.1, we see that statement p is false: there are actually more cats owned than dogs. We also see that statement q is true: there are fewer reptiles owned than birds. We also see that statement r is true: the most numerous pets owned are fish. Since r is true, its negation, $\sim r$, is false. Therefore, we substitute F for p, T for q, and F for $\sim r$, which gives

$$(p \wedge q) \vee \sim r$$
$$(\text{F} \wedge \text{T}) \vee \text{F}$$
$$\text{F} \vee \text{F}$$
$$\text{F}$$

Thus, the original compound statement is a false statement.

SECTION 3.2 EXERCISES

CONCEPT/WRITING EXERCISES

1. a) How many distinct cases must be listed in a truth table that contains two simple statements?

   b) List all the cases.

2. a) How many distinct cases must be listed in a truth table that contains three simple statements?

   b) List all the cases.

3. a) Construct the truth table for the conjunction $p \wedge q$.

   b) Under what circumstances is the conjunction true?

   c) Construct the truth table for the disjunction $p \vee q$.

   d) Under what circumstances is the disjunction false?

4. Suppose a compound statement contains five simple statements. How many distinct cases would the truth table for this compound statement have?

PRACTICE THE SKILLS/PROBLEM SOLVING

In Exercises 5–20, construct a truth table for the statement.

5. $p \wedge \sim p$

6. $p \vee \sim p$

7. $q \vee \sim p$

8. $p \wedge \sim q$

9. $\sim p \vee \sim q$

10. $\sim (p \vee \sim q)$

11. $\sim (p \wedge \sim q)$

12. $\sim (\sim p \wedge \sim q)$

13. $\sim q \vee (p \wedge r)$

14. $(p \vee \sim q) \wedge r$

15. $r \vee (p \wedge \sim q)$

16. $(r \wedge q) \wedge \sim p$

17. $(r \vee \sim p) \wedge \sim q$

18. $\sim p \wedge (q \vee r)$

19. $(\sim q \wedge r) \vee p$

20. $\sim r \vee (\sim p \wedge q)$

In Exercises 21–30, write the statement in symbolic form and construct a truth table.

21. The cookies are warm and the milk is cold.

22. The zoo is open, but it is not a nice day.

23. I have a new cell phone, but I do not have a new battery.

24. It is false that Wanda Garner is the president or that Judy Ackerman is the treasurer.

25. It is false that Jasper Adams is a tutor and Mark Russo is a secretary.

26. Mike made pizza and Dennis made a chef salad, but Gil burned the lemon squares.

27. The copier is out of toner, or the lens is dirty or the corona wires are broken.

28. I am hungry, and I want to eat a healthy lunch and I want to eat in a hurry.

29. The Congress must act on the bill, and the president must sign the bill or not sign the bill.

30. Gordon Langeneger likes the Mac Pro and he likes the MacBook Pro, but he does not like the Pentium IV.

In Exercises 31–42, determine the truth value of the statement if

a) p is true, q is false, and r is true.

b) p is false, q is true, and r is true.

31. $(\sim p \wedge r) \wedge q$

32. $\sim p \vee (q \wedge r)$

33. $(\sim p \vee \sim q) \vee \sim r$

34. $(\sim q \wedge \sim p) \vee \sim r$

35. $(p \vee \sim q) \wedge \sim (p \wedge \sim r)$

36. $(p \wedge \sim q) \vee r$

37. $(\sim r \wedge p) \vee q$

38. $\sim q \vee (r \wedge p)$

39. $(\sim q \vee \sim p) \wedge r$

40. $(\sim r \vee \sim p) \vee \sim q$

41. $(\sim p \vee \sim q) \vee (\sim r \vee q)$

42. $(\sim r \wedge \sim q) \wedge (\sim r \vee \sim p)$

In Exercises 43–50, determine the truth value for each simple statement. Then use these truth values to determine the truth value of the compound statement. (You may have to use a reference source such as the Internet or an encyclopedia.)

43. $18 \div 3 = 9$ or $56 \div 8 = 7$

44. $17 \geq 17$ and $-3 > -2$

45. Virginia borders the Atlantic Ocean or California borders the Indian Ocean.

46. Hawaii is the 50th state or Alaska lies on the equator, and Birmingham is the capital of Missouri.

47. Steven Spielberg is a movie director and Tom Hanks is an actor, but John Madden is not a sports announcer.

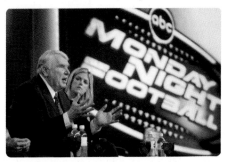

▲ John Madden and Melissa Stark

48. Quebec is in Texas or Toronto is in California, and Cedar Rapids is in Iowa.

49. Iraq is in Africa or Iran is in South America, and Syria is in the Middle East.

50. Holstein is a breed of cattle and collie is a breed of dogs, or beagle is not a breed of cats.

Carbon Dioxide Emissions In Exercises 51–54, use the table to determine the truth value of each simple statement. Then determine the truth value of the compound statement.

Countries with the Highest Carbon Dioxide (CO_2) Emissions in 2004

Country	CO_2 Emissions (in megatons)	Per Capita CO_2 Emissions (in tons)
United States	5652	19.66
China	3271	2.55
Russia	1503	10.43
Japan	1207	9.47
India	1016	0.97

Source: International Energy Agency

51. The United States had the lowest per capita CO_2 emissions and China had the highest per capita CO_2 emissions.

52. The United States had more CO_2 emissions than China and Russia had combined or the United States had more CO_2 emissions than Russia, Japan, and India had combined.

53. India had lower per capita CO_2 emissions than Japan had or China had lower per capita CO_2 emissions than India had.

54. Russia had lower per capita CO_2 emissions than the United States had, but Russia had higher per capita CO_2 emissions than Japan had.

Sleep Time In Exercises 55–58, use the graph, which shows the number of hours Americans sleep, to determine the truth value of each simple statement. Then determine the truth value of the compound statement.

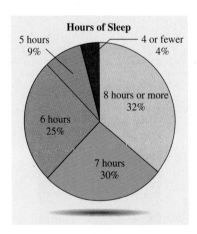

Hours of Sleep

- 5 hours 9%
- 4 or fewer 4%
- 8 hours or more 32%
- 6 hours 25%
- 7 hours 30%

55. It is false that 30% of Americans get 6 hours of sleep each night and 9% get 5 hours of sleep each night.

56. Twenty-five percent of Americans get 6 hours of sleep each night, and 30% get 7 hours of sleep each night or 9% do not get 5 hours of sleep each night.

57. Thirteen percent of Americans get 5 or fewer hours of sleep each night or 32% get 6 or more hours of sleep each night, and 30% get 8 or more hours of sleep each night.

58. Over one-half of all Americans get 7 or fewer hours of sleep each night, and over one-quarter get 6 or fewer hours of sleep each night.

In Exercises 59–62, let

p: Tanisha owns a convertible.
q: Joan owns a Volvo.

Translate each statement into symbols. Then construct a truth table for each and indicate under what conditions the compound statement is true.

59. Tanisha owns a convertible and Joan does not own a Volvo.

60. Tanisha does not own a convertible, but Joan owns a Volvo.

61. Tanisha owns a convertible or Joan does not own a Volvo.

62. Tanisha does not own a convertible or Joan does not own a Volvo.

In Exercises 63–66, let

p: The house is owned by an engineer.
q: The heat is solar generated.
r: The car is run by electric power.

Translate each statement into symbols. Then construct a truth table for each and indicate under what conditions the compound statement is true.

63. The car is run by electric power or the heat is solar generated, but the house is owned by an engineer.

64. The house is owned by an engineer and the heat is solar generated, or the car is run by electric power.

65. The heat is solar generated, or the house is owned by an engineer and the car is not run by electric power.

66. The house is not owned by an engineer, and the car is not run by electric power and the heat is solar generated.

Obtaining a Loan In Exercises 67 and 68, read the requirements and each applicant's qualifications for obtaining a loan.

a) Identify which of the applicants would qualify for the loan.

b) For the applicants who do not qualify for the loan, explain why.

67. To qualify for a loan of $40,000, an applicant must have a gross income of $28,000 if single, $46,000 combined income if married, and assets of at least $6,000.

   Mrs. Rusinek, married with three children, earns $42,000. Mr. Rusinek does not have an income. The Rusineks have assets of $42,000.

   Mr. Duncan is not married, works in sales, and earns $31,000. He has assets of $9000.

   Mrs. Tuttle and her husband have total assets of $43,000. One earns $35,000, and the other earns $23,500.

68. To qualify for a loan of $45,000, an applicant must have a gross income of $30,000 if single, $50,000 combined income if married, and assets of at least $10,000.

   Mr. Argento, married with two children, earns $37,000. Mrs. Argento earns $15,000 at a part-time job. The Argentos have assets of $25,000.

   Ms. McVey, single, has assets of $19,000. She works in a store and earns $25,000.

   Mr. Siewert earns $24,000 and Ms. Fox, his wife, earns $28,000. Their assets total $8000.

69. *Airline Special Fares* An airline advertisement states, "To get the special fare you must purchase your tickets between January 1 and February 15 and fly round trip between March 1 and April 1. You must depart on a Monday, Tuesday, or Wednesday, and return on a Tuesday, Wednesday, or Thursday, and stay over at least one Saturday."

a) Determine which of the following individuals will qualify for the special fare.

b) If the person does not qualify for the special fare, explain why.

   Wing Park plans to purchase his ticket on January 15, depart on Monday, March 3, and return on Tuesday, March 18.

   Gina Vela plans to purchase her ticket on February 1, depart on Wednesday, March 12, and return on Thursday, April 3.

   Kara Shavo plans to purchase her ticket on February 14, depart on Tuesday, March 4, and return on Monday, March 19.

   Christos Supernaw plans to purchase his ticket on January 4, depart on Monday, March 10, and return on Thursday, March 13.

   Alex Chang plans to purchase his ticket on January 1, depart on Monday, March 3, and return on Monday, March 10.

▲ See Exercise 69

PROBLEM SOLVING/GROUP ACTIVITIES

In Exercises 70 and 71, construct a truth table for the symbolic statement.

70. $\sim[(\sim(p \vee q)) \vee (q \wedge r)]$

71. $[(q \wedge \sim r) \wedge (\sim p \vee \sim q)] \vee (p \vee \sim r)$

72. On page 120, we indicated that a compound statement consisting of n simple statements had 2^n distinct true–false cases.

a) How many distinct true–false cases does a truth table containing simple statements p, q, r, and s have?

b) List all possible true–false cases for a truth table containing the simple statements p, q, r, and s.

c) Use the list in part (b) to construct a truth table for $(q \wedge p) \vee (\sim r \wedge s)$.

d) Construct a truth table for $(\sim r \wedge \sim s) \wedge (\sim p \vee q)$.

73. Must $(p \wedge \sim q) \vee r$ and $(q \wedge \sim r) \vee p$ have the same number of trues in their answer columns? Explain.

INTERNET/RESEARCH ACTIVITIES

74. Do research and write a report on each of the following.

a) The relationship between *negation* in logic and *complement* in set theory.

b) The relationship between *conjunction* in logic and *intersection* in set theory.

c) The relationship between *disjunction* in logic and *union* in set theory.

3.3   TRUTH TABLES FOR THE CONDITIONAL AND BICONDITIONAL

Suppose I said to you, "If you get an A, then I will buy you a car." As we discussed in Section 3.1, this statement is called a *conditional* statement. In this section, we will discuss under what conditions a conditional statement is true and under what conditions a conditional statement is false.

Conditional

In Section 3.1, we mentioned that the statement preceding the conditional symbol is called the *antecedent* and that the statement following the conditional symbol is called the *consequent*. For example, consider $(p \vee q) \rightarrow [\sim(q \wedge r)]$. In this statement, $(p \vee q)$ is the antecedent and $[\sim(q \wedge r)]$ is the consequent.

To develop a truth table for the conditional statement, consider the statement "If you get an A, then I will buy you a car." Assume this statement is true except when I have actually broken my promise to you.

Let

p: You get an A.

q: I buy you a car.

Translated into symbolic form, the statement becomes $p \rightarrow q$. Let's examine the four cases shown in Table 3.12.

CASE 1: (T, T) You get an A, and I buy a car for you. I have met my commitment, and the statement is true.

CASE 2: (T, F) You get an A, and I do not buy a car for you. I have broken my promise, and the statement is false.

What happens if you don't get an A? If you don't get an A, I no longer have a commitment to you, and therefore I cannot break my promise.

CASE 3: (F, T) You do not get an A, and I buy you a car. I have not broken my promise, and therefore the statement is true.

CASE 4: (F, F) You do not get an A, and I don't buy you a car. I have not broken my promise, and therefore the statement is true.

The conditional statement is false when the antecedent is true and the consequent is false. In every other case the conditional statement is true.

> The **conditional statement** $p \rightarrow q$ is true in every case except when p is a true statement and q is a false statement.

▲ Under what conditions is the statement *If you get an A, then I will buy you a car* true?

Table 3.12 Conditional

p	q	$p \rightarrow q$
T	T	T
T	F	F
F	T	T
F	F	T

┌ **EXAMPLE ❶** *A Truth Table with a Conditional*
│ Construct a truth table for the statement $\sim p \rightarrow \sim q$.

Table 3.13

p	q	$\sim p$	$\rightarrow$	$\sim q$
T	T	F	T	F
T	F	F	T	T
F	T	T	F	F
F	F	T	T	T
		1	3	2

SOLUTION Because this statement is a conditional, the answer will lie under the $\rightarrow$. Fill out the truth table by placing the appropriate truth values under $\sim p$, column 1, and under $\sim q$, column 2 (see Table 3.13). Then, using the information given in the truth table for the conditional and the truth values in columns 1 and 2, determine the solution, column 3. In row 1, the antecedent, $\sim p$, is false and the consequent, $\sim q$, is also false. Row 1 is F $\rightarrow$ F, which according to row 4 of Table 3.12, is T. Likewise, row 2 of Table 3.13 is F $\rightarrow$ T, which is T. Row 3 is T $\rightarrow$ F, which is F. Row 4 is T $\rightarrow$ T, which is T. ●

EXAMPLE ❷ *A Conditional Truth Table with Three Simple Statements*

Construct a truth table for the statement $p \rightarrow (\sim q \wedge r)$.

SOLUTION Because this statement is a conditional, the answer will lie under the $\rightarrow$. Work within the parentheses first. Place the truth values under $\sim q$, column 1, and r, column 2 (Table 3.14). Then take the conjunction of columns 1 and 2 to obtain column 3. Next, place the truth values under p in column 4. To determine the answer, column 5, use columns 3 and 4 and the information of the conditional statement given in Table 3.12. Column 4 represents the truth values of the antecedent, and column 3 represents the truth values of the consequent. Remember that the conditional is false only when the antecedent is true and the consequent is false, as in cases (rows) 1, 2, and 4 of column 5. ●

Table 3.14

p	q	r	p	$\rightarrow$	$(\sim q$	$\wedge$	$r)$
T	T	T	T	F	F	F	T
T	T	F	T	F	F	F	F
T	F	T	T	T	T	T	T
T	F	F	T	F	T	F	F
F	T	T	F	T	F	F	T
F	T	F	F	T	F	F	F
F	F	T	F	T	T	T	T
F	F	F	F	T	T	F	F
			4	5	1	3	2

EXAMPLE ❸ *Examining an Advertisement*

An advertisement for Perky Morning coffee makes the following claim: "If you drink Perky Morning coffee, then you will not be sluggish and you will have a great day." Translate the statement into symbolic form and construct a truth table.

SOLUTION Let

p: You drink Perky Morning coffee.
q: You will be sluggish.
r: You will have a great day.

In symbolic form, the claim is

$$p \rightarrow (\sim q \wedge r)$$

This symbolic statement is identical to the statement in Table 3.14, and the truth tables are the same. Column 3 represents the truth values of $(\sim q \wedge r)$, which corresponds to the statement "You will not be sluggish and you will have a great day." Note that column 3 is true in cases (rows) 3 and 7. In case 3, since p is true, you drank Perky Morning coffee. In case 7, however, since p is false, you did not drink Perky Morning coffee. From this information we can conclude that it is possible for you to not be sluggish and for you to have a great day without drinking Perky Morning coffee. ●

A truth table alone cannot tell us whether a statement is true or false. It can, however, be used to examine the various possibilities.

Biconditional

The *biconditional statement*, $p \leftrightarrow q$, means that $p \rightarrow q$ and $q \rightarrow p$, or, symbolically, $(p \rightarrow q) \wedge (q \rightarrow p)$. To determine the truth table for $p \leftrightarrow q$, we will construct the truth table for $(p \rightarrow q) \wedge (q \rightarrow p)$ (Table 3.15). Table 3.16 shows the truth values for the biconditional statement.

Table 3.15

p	q	$(p$	$\rightarrow$	$q)$	$\wedge$	$(q$	$\rightarrow$	$p)$
T	T	T	T	T	T	T	T	T
T	F	T	F	F	F	F	T	T
F	T	F	T	T	F	T	F	F
F	F	F	T	F	T	F	T	F
		1	3	2	7	4	6	5

Table 3.16 Biconditional

p	q	$p \leftrightarrow q$
T	T	T
T	F	F
F	T	F
F	F	T

From Table 3.16 we see that the biconditional statement is true when the antecedent and the consequent have the same truth value and false when the antecedent and consequent have different truth values.

> The **biconditional statement**, $p \leftrightarrow q$, is true only when p and q have the same truth value, that is, when both are true or both are false.

EXAMPLE ④ *A Truth Table Using a Biconditional*

Construct a truth table for the statement $p \leftrightarrow (q \rightarrow \sim r)$.

SOLUTION Since there are three letters, there must be eight cases. The parentheses indicate that the answer must be under the biconditional (Table 3.17). Use columns 3 and 4 to obtain the answer in column 5. When columns 3 and 4 have the same truth values, place a T in column 5. When columns 3 and 4 have different truth values, place an F in column 5.

Table 3.17

p	q	r	p	$\leftrightarrow$	$(q$	$\rightarrow$	$\sim r)$
T	T	T	T	F	T	F	F
T	T	F	T	T	T	T	T
T	F	T	T	T	F	T	F
T	F	F	T	T	F	T	T
F	T	T	F	T	T	F	F
F	T	F	F	F	T	T	T
F	F	T	F	F	F	T	F
F	F	F	F	F	F	T	T
			4	5	1	3	2

●

In Section 3.2, we showed that finding the truth value of a compound statement for a specific case does not require constructing an entire truth table. Examples 5 and 6 illustrate this technique for the conditional and the biconditional.

EXAMPLE ❺ *Determine the Truth Value of a Compound Statement*

Determine the truth value of the statement $(\sim p \leftrightarrow q) \rightarrow (\sim q \leftrightarrow r)$ when p is false, q is true, and r is false.

SOLUTION Substitute the truth value for each simple statement:

$$(\sim p \leftrightarrow q) \rightarrow (\sim q \leftrightarrow r)$$
$$(T \leftrightarrow T) \rightarrow (F \leftrightarrow F)$$
$$T \qquad \rightarrow \qquad T$$
$$T$$

For this specific case, the statement is true.

EXAMPLE ❻ *Determine the Truth Value of a Compound Statement*

Determine the truth value for each simple statement. Then use the truth values to determine the truth value of the compound statement.

a) If 15 is an even number, then 29 is an even number.

b) Northwestern University is in Illinois and Marquette University is in Alaska, if and only if Purdue University is in Alabama.

SOLUTION

a) Let

p: 15 is an even number.

q: 29 is an even number.

Then the statement "If 15 is an even number, then 29 is an even number" can be written $p \rightarrow q$. Since 15 is not an even number, p is a false statement. Also, since 29 is not an even number, q is a false statement. We substitute F for p and F for q and evaluate the statement:

$$p \rightarrow q$$
$$F \rightarrow F$$
$$T$$

Therefore, "If 15 is an even number, then 29 is an even number" is a true statement.

b) Let

p: Northwestern University is in Illinois.

q: Marquette University is in Alaska.

r: Purdue University is in Alabama.

The original compound statement can be written $(p \wedge q) \leftrightarrow r$. By checking the Internet or other references we can find that Northwestern University is in Illinois, Marquette University is in Wisconsin, and Purdue University is in Indiana. Therefore, p is a true statement, but q and r are false statements. We will substitute T for p, F for q, and F for r and evaluate the compound statement:

$$(p \wedge q) \leftrightarrow r$$
$$(T \wedge F) \leftrightarrow F$$
$$F \quad \leftrightarrow F$$
$$T$$

Therefore, the original compound statement is true.   ●

EXAMPLE ❼ *Using Real Data in Compound Statements*

The graph in Fig. 3.2 represents the student population by age group in 2005 for Everett Community College (EvCC). Use this graph to determine the truth value of the following compound statements.

a) If 18- to 21-year-old students account for 25% of the EvCC student population, then 22- to 25-year-old students account for 12% of the EvCC student population and 31- to 35-year-old students account for 16% of the EvCC student population.

b) Students under age 18 years old account for 10% of the EvCC student population, if and only if 26- to 30-year-old students account for 25% of the EvCC student population or 36- to 40-year-old students account for 9% of the EvCC student population.

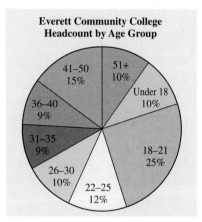

Source: Everett Community College Web Site

Figure 3.2

SOLUTION

a) Let

p: 18- to 21-year-old students account for 25% of the EvCC student population.

q: 22- to 25-year-old students account for 12% of the EvCC student population.

r: 31- to 35-year-old students account for 16% of the EvCC student population.

Then the original statement can be written $p \rightarrow (q \wedge r)$. We can see from Fig. 3.2 that both p and q are true statements and that r is a false statement. Substitute T for p, T for q, and F for r and evaluate the statement:

$$
\begin{array}{c}
p \rightarrow (q \wedge r) \\
T \rightarrow (T \wedge F) \\
T \rightarrow \quad F \\
F
\end{array}
$$

Therefore, "If 18- to 21-year-old students account for 25% of the EvCC student population, then 22- to 25-year-old students account for 12% of the EvCC student population and 31- to 35-year-old students account for 16% of the EvCC student population" is a false statement.

b) Let

 p: Students under age 18 years old account for 10% of the EvCC student population.

 q: 26- to 30-year-old students account for 25% of the EvCC student population.

 r: 36- to 40-year-old students account for 9% of the EvCC student population.

Then the original statement can be written $p \leftrightarrow (q \vee r)$. We can see from Fig. 3.2 that p is true, q is false, and r is true. Substitute T for p, F for q, and T for r and evaluate the statement:

$$
\begin{array}{c}
p \leftrightarrow (q \vee r) \\
T \leftrightarrow (F \vee T) \\
T \leftrightarrow \quad T \\
T
\end{array}
$$

Therefore, the original statement, "Students under age 18 years old account for 10% of the EvCC student population, if and only if 26- to 30-year-old students account for 25% of the EvCC student population or 36- to 40-year-old students account for 9% of the EvCC student population" is a true statement. ●

Self-Contradictions, Tautologies, and Implications

Two special situations can occur in the truth table of a compound statement: The statement may always be false, or the statement may always be true. We give such statements special names.

> A **self-contradiction** is a compound statement that is always false.

When every truth value in the answer column of the truth table is false, then the statement is a self-contradiction.

Table 3.18

p	q	$(p \leftrightarrow q)$	$\wedge$	$(p$	$\leftrightarrow$	$\sim q)$
T	T	T	F	T	F	F
T	F	F	F	T	T	T
F	T	F	F	F	T	F
F	F	T	F	F	F	T
		1	5	2	4	3

EXAMPLE **8** *All Falses, a Self-Contradiction*

Construct a truth table for the statement $(p \leftrightarrow q) \wedge (p \leftrightarrow \sim q)$.

SOLUTION See Table 3.18. In this example, the truth values are false in each case of column 5. This statement is an example of a self-contradiction or a *logically false statement.* ●

A **tautology** is a compound statement that is always true.

When every truth value in the answer column of the truth table is true, the statement is a tautology.

EXAMPLE ❾ *All Trues, a Tautology*

Construct a truth table for the statement $(p \wedge q) \rightarrow (p \vee r)$.

SOLUTION The answer is given in column 3 of Table 3.19. The truth values are true in every case. Thus, the statement is an example of a tautology or a *logically true statement*.

Table 3.19

p	q	r	$(p \wedge q)$	$\rightarrow$	$(p \vee r)$
T	T	T	T	T	T
T	T	F	T	T	T
T	F	T	F	T	T
T	F	F	F	T	T
F	T	T	F	T	T
F	T	F	F	T	F
F	F	T	F	T	T
F	F	F	F	T	F
			1	3	2

The conditional statement $(p \wedge q) \rightarrow (p \vee r)$ is a tautology. Conditional statements that are tautologies are called *implications*. In Example 9, we can say that $p \wedge q$ implies $p \vee r$.

An **implication** is a conditional statement that is a tautology.

In any implication the antecedent of the conditional statement implies the consequent. In other words, if the antecedent is true, then the consequent must also be true. That is, the consequent will be true whenever the antecedent is true.

▲ "Heads I win, tails you lose." Do you think that this statement is a tautology, self-contradiction, or neither? See Problem-Solving Exercise 81.

Table 3.20

p	q	$[(p \wedge q)$	$\wedge$	$p]$	$\rightarrow$	q
T	T	T	T	T	T	T
T	F	F	F	T	T	F
F	T	F	F	F	T	T
F	F	F	F	F	T	F
		1	3	2	5	4

EXAMPLE ❿ *An Implication?*

Determine whether the conditional statement $[(p \wedge q) \wedge p] \rightarrow q$ is an implication.

SOLUTION If the conditional statement is a tautology, the conditional statement is an implication. Because the conditional statement is a tautology (see Table 3.20), the conditional statement is an implication. The antecedent $[(p \wedge q) \wedge p]$ implies the consequent q. Note that the antecedent is true only in case 1 and that the consequent is also true in case 1.

SECTION 3.3 EXERCISES

CONCEPT/WRITING EXERCISES

1. a) Construct the truth table for the conditional statement $p \rightarrow q$.

   b) Explain when the conditional statement is true and when it is false.

   c) Construct the truth table for the biconditional statement $p \leftrightarrow q$.

   d) Explain when the biconditional statement is true and when it is false.

2. a) What is the statement preceding the conditional symbol called?

   b) What is the statement following the conditional symbol called?

3. a) Explain the procedure to determine the truth value of a compound statement when specific truth values are provided for the simple statements.

   b) Follow the procedure in part (a) and determine the truth value of the symbolic statement

   $$[(p \leftrightarrow q) \vee (\sim r \rightarrow q)] \rightarrow \sim r$$

   when p is true, q is true, and r is false.

4. What is a tautology?

5. What is a self-contradiction?

6. What is an implication?

PRACTICE THE SKILLS

In Exercises 7–16, construct a truth table for the statement.

7. $\sim p \rightarrow q$

8. $\sim p \rightarrow \sim q$

9. $\sim(p \rightarrow \sim q)$

10. $\sim(\sim p \leftrightarrow q)$

11. $\sim q \leftrightarrow p$

12. $(p \leftrightarrow q) \rightarrow p$

13. $p \leftrightarrow (q \vee p)$

14. $(\sim q \wedge p) \rightarrow \sim q$

15. $q \rightarrow (p \rightarrow \sim q)$

16. $(p \vee q) \leftrightarrow (p \wedge q)$

In Exercises 17–26, construct a truth table for the statement.

17. $\sim p \rightarrow (q \wedge r)$

18. $q \vee (p \rightarrow \sim r)$

19. $p \leftrightarrow (\sim q \rightarrow r)$

20. $(\sim p \rightarrow \sim r) \rightarrow q$

21. $(q \vee \sim r) \leftrightarrow \sim p$

22. $(p \wedge r) \rightarrow (q \vee r)$

23. $(\sim r \vee \sim q) \rightarrow p$

24. $[r \wedge (q \vee \sim p)] \leftrightarrow \sim p$

25. $(p \rightarrow q) \leftrightarrow (\sim q \rightarrow \sim r)$

26. $(\sim p \leftrightarrow \sim q) \rightarrow (\sim q \leftrightarrow r)$

In Exercises 27–32, write the statement in symbolic form. Then construct a truth table for the symbolic statement.

27. If I take niacin, then I will stay healthy and I will have lower cholesterol.

28. The goalie will make the save if and only if the stopper is in position, or the forward cannot handle the ball.

29. The election was fair if and only if the polling station stayed open until 8 P.M., or we will request a recount.

30. If the dam holds then we can go fishing, if and only if the pole is not broken.

31. If Mary Andrews does not send me an e-mail then we can call her, or we can write to Mom.

32. It is false that if Eileen Jones went to lunch, then she cannot take a message and we will have to go home.

In Exercises 33–38, determine whether the statement is a tautology, self-contradiction, or neither.

33. $p \rightarrow \sim p$

34. $(p \wedge \sim q) \leftrightarrow \sim p$

35. $p \wedge (q \wedge \sim p)$

36. $(p \land \sim q) \rightarrow q$

37. $(\sim q \rightarrow p) \lor \sim q$

38. $[(p \rightarrow q) \lor r] \leftrightarrow [(p \land q) \rightarrow r]$

In Exercises 39–44, determine whether the statement is an implication.

39. $\sim p \rightarrow (p \lor q)$

40. $(p \land q) \rightarrow (\sim p \lor q)$

41. $(q \land p) \rightarrow (p \land q)$

42. $(p \lor q) \rightarrow (p \lor \sim r)$

43. $[(p \rightarrow q) \land (q \rightarrow p)] \rightarrow (p \leftrightarrow q)$

44. $[(p \lor q) \land r] \rightarrow (p \lor q)$

In Exercises 45–56, if p is true, q is false, and r is true, find the truth value of the statement.

45. $\sim p \rightarrow (q \rightarrow r)$

46. $(p \lor q) \rightarrow \sim r$

47. $q \leftrightarrow (\sim p \lor r)$

48. $r \rightarrow (\sim p \leftrightarrow \sim q)$

49. $(\sim p \land \sim q) \lor \sim r$

50. $\sim[p \rightarrow (q \land r)]$

51. $(p \land r) \leftrightarrow (p \lor \sim q)$

52. $(\sim p \lor q) \rightarrow \sim r$

53. $(\sim p \leftrightarrow r) \lor (\sim q \leftrightarrow r)$

54. $(r \rightarrow \sim p) \land (q \rightarrow \sim r)$

55. $\sim[(p \lor q) \leftrightarrow (p \rightarrow \sim r)]$

56. $[(\sim r \rightarrow \sim q) \lor (p \land \sim r)] \rightarrow q$

PROBLEM SOLVING

In Exercises 57–64, determine the truth value for each simple statement. Then, using the truth values, determine the truth value of the compound statement.

57. If $2 + 7 = 9$, then $15 - 3 = 12$.

58. If $\frac{3}{4} < 1$ and $\frac{7}{8} > 1$, then $\frac{15}{16} < 1$.

59. A cat has whiskers or a fish can swim, and a chicken lays eggs.

60. Tallahassee is in Florida and Atlanta is in Georgia, or Chicago is in Mississippi.

61. Apple makes computers, if and only if Nike makes sports shoes or Rolex makes watches.

62. Spike Lee is a movie director, or if Halle Berry is a school-teacher then George Clooney is a circus clown.

63. Valentine's Day is in February or President's Day is in March, and Thanksgiving is in November.

64. Honda makes automobiles or Honda makes motorcycles, if and only if Toyota makes cereal.

In Exercises 65–68, use the information provided about the moons for the planets Jupiter and Saturn to determine the truth values of the simple statements. Then determine the truth value of the compound statement.

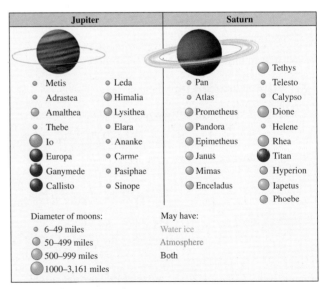

Source: Data from *Time* Magazine

65. *Jupiter's Moons* Io has a diameter of 1000–3161 miles or Thebe may have water, and Io may have atmosphere.

66. *Moons of Saturn* Titan may have water and Titan may have atmosphere, if and only if Janus may have water.

67. *Moon Comparisons* Phoebe has a larger diameter than Rhea if and only if Callisto may have water ice, and Calypso has a diameter of 6–49 miles.

68. *Moon Comparisons* If Jupiter has 16 moons or Saturn does not have 18 moons, then Saturn has 7 moons that may have water ice.

In Exercises 69 and 70, use the graphs to determine the truth values of each simple statement. Then determine the truth value of the compound statement.

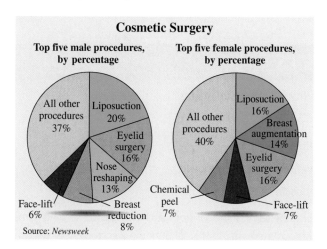

Cosmetic Surgery

Top five male procedures, by percentage

Top five female procedures, by percentage

Source: *Newsweek*

69. *Most Common Cosmetic Surgery* The most common cosmetic surgery procedure for females is liposuction or the most common procedure for males is eyelid surgery, and 20% of male cosmetic surgery is for nose reshaping.

70. *Face-lifts and Eyelid Surgeries* 7% of female cosmetic surgeries are for face-lifts and 10% of male cosmetic surgeries are for face-lifts, if and only if males have a higher percent of eyelid surgeries than females.

In Exercises 71–76, suppose both of the following statements are true.

p: Muhundan spoke at the teachers' conference.

q: Muhundan received the outstanding teacher award.

Find the truth values of each compound statement.

71. If Muhundan received the outstanding teacher award, then Muhundan spoke at the teachers' conference.

72. If Muhundan spoke at the teachers' conference, then Muhundan did not receive the outstanding teacher award.

73. If Muhundan did not receive the outstanding teacher award, then Muhundan spoke at the teachers' conference.

74. Muhundan did not receive the outstanding teacher award if and only if Muhundan spoke at the teachers' conference.

75. Muhundan received the outstanding teacher award if and only if Muhundan spoke at the teachers' conference.

76. If Muhundan did not receive the outstanding teacher award, then Muhundan did not speak at the teachers' conference.

77. *A New Computer* Your parents make the following statement to your sister, "If you get straight A's this semester, then we will buy you a new computer." At the end of the semester your parents buy your sister a new computer. Can you conclude that your sister got straight A's? Explain.

78. *Job Interview* Consider the statement "If your interview goes well, then you will be offered the job." If you are interviewed and then offered the job, can you conclude that your interview went well? Explain.

CHALLENGE PROBLEMS/GROUP ACTIVITIES

In Exercises 79 and 80, construct truth tables for the symbolic statement.

79. $[p \lor (q \rightarrow \sim r)] \leftrightarrow (p \land \sim q)$

80. $[(r \rightarrow \sim q) \rightarrow \sim p] \lor (q \leftrightarrow \sim r)$

81. Is the statement "Heads I win, tails you lose" a tautology, a self-contradiction, or neither? Explain your answer.

82. Construct a truth table for

a) $(p \lor q) \rightarrow (r \land s)$.

b) $(q \rightarrow \sim p) \lor (r \leftrightarrow s)$.

RECREATIONAL MATHEMATICS

83. *Cat Puzzle* Solve the following puzzle. The Joneses have four cats. The parents are Tiger and Boots, and the kittens are Sam and Sue. Each cat insists on eating out of its own bowl. To complicate matters, each cat will eat only its own brand of cat food. The colors of the bowls are red, yellow, green, and blue. The different types of cat food are Whiskas, Friskies, Nine Lives, and Meow Mix. Tiger will eat Meow Mix if and only if it is in a yellow bowl. If Boots is to eat her food, then it must be in a yellow bowl. Mrs. Jones knows that the label on the can containing Sam's food is the same color as his bowl. Boots eats Whiskas. Meow Mix and Nine Lives are packaged in a brown paper bag. The color of Sue's bowl is green if and only if she eats Meow Mix. The label on the Friskies can is red. Match each cat with its food and the bowl of the correct color.

84. *The Youngest Triplet* The Barr triplets have an annoying habit: Whenever a question is asked of the three of them, two tell the truth and the third lies. When I asked them which of them was born last, they replied as follows.

Mary: Katie was born last.
Katie: I am the youngest.
Annie: Mary is the youngest.

Which of the Barr triplets was born last?

INTERNET/RESEARCH ACTIVITY

85. Select an advertisement from the Internet, a newspaper, or a magazine that makes or implies a conditional statement. Analyze the advertisement to determine whether the consequent necessarily follows from the antecedent. Explain your answer. (See Example 3.)

3.4   EQUIVALENT STATEMENTS

▲ Which statement is equivalent to *If it is sunny, then we go to the beach?*

Suppose your friend makes the following statements:

If it is sunny, then we go to the beach.

If we go to the beach, then it is sunny.

If it is not sunny, then we do not go to the beach.

If we do not go to the beach, then it is not sunny.

Are these statements all saying the same thing, or does each one say something completely different from each other statement? In this section, we will study how we can answer this question by using logic symbols and truth tables. We also will learn to identify variations of conditional statements.

Equivalent statements are an important concept in the study of logic.

> Two statements are **equivalent,** symbolized ⇔ ,* if both statements have exactly the same truth values in the answer columns of the truth tables.

Sometimes the words *logically equivalent* are used in place of the word *equivalent.*

To determine whether two statements are equivalent, construct a truth table for each statement and compare the answer columns of the truth tables. If the answer columns are identical, the statements are equivalent. If the answer columns are not identical, the statements are not equivalent.

---

*The symbol ≡ is also used to indicate equivalent statements.

EXAMPLE ❶ *Equivalent Statements*

Determine whether the following two statements are equivalent.

$$p \wedge (q \vee r)$$
$$(p \wedge q) \vee (p \wedge r)$$

SOLUTION Construct a truth table for each statement (see Table 3.21).

Table 3.21

p	q	r	p	$\wedge$	$(q \vee r)$	$(p \wedge q)$	$\vee$	$(p \wedge r)$
T	T	T	T	T	T	T	T	T
T	T	F	T	T	T	T	T	F
T	F	T	T	T	T	F	T	T
T	F	F	T	F	F	F	F	F
F	T	T	F	F	T	F	F	F
F	T	F	F	F	T	F	F	F
F	F	T	F	F	T	F	F	F
F	F	F	F	F	F	F	F	F
			1	3	2	1	3	2

Because the truth tables have the same answer (column 3 for both tables), the statements are equivalent. Therefore, we can write

$$p \wedge (q \vee r) \Leftrightarrow (p \wedge q) \vee (p \wedge r)$$ •

EXAMPLE ❷ *Are the Following Equivalent Statements?*

Determine whether the following statements are equivalent.

a) If you work hard and obey all of the rules, then you will succeed in life.

b) If you do not work hard or do not obey all of the rules, then you will not succeed in life.

SOLUTION First write each statement in symbolic form, then construct a truth table for each statement. If the answer columns of both truth tables are identical, then the statements are equivalent. If the answer columns are not identical, then the statements are not equivalent.

Let

p: You work hard.

q: You obey all of the rules.

r: You will succeed in life.

In symbolic form, the statements are

a) $(p \wedge q) \rightarrow r$. b) $(\sim p \vee \sim q) \rightarrow \sim r$.

The truth tables for these statements are given in Tables 3.22 and 3.23, respectively. The answers in the columns labeled 5 are not identical, so the statements are not equivalent.

Table 3.22

p	q	r	(p	$\wedge$	q)	$\rightarrow$	r
T	T	T	T	T	T	T	T
T	T	F	T	T	T	F	F
T	F	T	T	F	F	T	T
T	F	F	T	F	F	T	F
F	T	T	F	F	T	T	T
F	T	F	F	F	T	T	F
F	F	T	F	F	F	T	T
F	F	F	F	F	F	T	F
			1	3	2	5	4

Table 3.23

p	q	r	($\sim p$	$\vee$	$\sim q$)	$\rightarrow$	$\sim r$
T	T	T	F	F	F	T	F
T	T	F	F	F	F	T	T
T	F	T	F	T	T	F	F
T	F	F	F	T	T	T	T
F	T	T	T	T	F	F	F
F	T	F	T	T	F	T	T
F	F	T	T	T	T	F	F
F	F	F	T	T	T	T	T
			1	3	2	5	4

●

EXAMPLE ❸ *Which Statements Are Logically Equivalent?*

Determine which statement is logically equivalent to "It is not true that the tire is both out of balance and flat."

a) If the tire is not flat, then the tire is not out of balance.
b) The tire is not out of balance or the tire is not flat.
c) The tire is not flat and the tire is not out of balance.
d) If the tire is not out of balance, then the tire is not flat.

SOLUTION　To determine whether any of the choices are equivalent to the given statement, first write the given statements and the choices in symbolic form. Then construct truth tables and compare the answer columns of the truth tables.

Let

p:　The tire is out of balance.
q:　The tire is flat.

The given statement may be written "It is not true that the tire is out of balance and the tire is flat." The statement is expressed in symbolic form as $\sim(p \wedge q)$. Using p and q as indicated, choices (a) through (d) may be expressed symbolically as

a) $\sim q \rightarrow \sim p$.　　b) $\sim p \vee \sim q$.　　c) $\sim q \wedge \sim p$.　　d) $\sim p \rightarrow \sim q$.

Now construct a truth table for the given statement (Table 3.24) and for each statement (a) through (d), given in Table 3.25(a) through (d). By examining the truth tables, we see that the given statement, $\sim(p \wedge q)$, is logically equivalent to choice (b), $\sim p \vee \sim q$. Therefore, the correct answer is "The tire is not out of balance or the tire is not flat." This statement is logically equivalent to the statement "It is not true that the tire is both out of balance and flat."

Table 3.24

p	q	~	(p	∧	q)
T	T	F	T	T	T
T	F	T	T	F	F
F	T	T	F	F	T
F	F	T	F	F	F
		4	1	3	2

Table 3.25

		(a)			(b)			(c)			(d)		
p	q	~q	→	~p	~p	∨	~q	~q	∧	~p	~p	→	~q
T	T	F	T	F	F	F	F	F	F	F	F	T	F
T	F	T	F	F	F	T	T	T	F	F	F	T	T
F	T	F	T	T	T	T	F	F	F	T	T	F	F
F	F	T	T	T	T	T	T	T	T	T	T	T	T

De Morgan's Laws

Example 3 showed that a statement of the form $\sim(p \wedge q)$ is equivalent to a statement of the form $\sim p \vee \sim q$. Thus, we may write $\sim(p \wedge q) \Leftrightarrow \sim p \vee \sim q$. This equivalent statement is one of two special laws called De Morgan's laws. The laws, named after Augustus De Morgan, an English mathematician, were first introduced in Section 2.4, where they applied to sets.

DE MORGAN'S LAWS

1. $\sim(p \wedge q) \Leftrightarrow \sim p \vee \sim q$
2. $\sim(p \vee q) \Leftrightarrow \sim p \wedge \sim q$

You can demonstrate that De Morgan's second law is true by constructing and comparing truth tables for $\sim(p \vee q)$ and $\sim p \wedge \sim q$. Do so now.

When using De Morgan's laws, if it becomes necessary to negate an already negated statement, use the fact that $\sim(\sim p)$ is equivalent to p. For example, the negation of the statement "Today is not Monday" is "Today is Monday."

EXAMPLE ❹ *Use De Morgan's Laws*

Select the statement that is logically equivalent to "I do not have investments, but I do not have debts."

a) I do not have investments or I do not have debts.

b) It is false that I have investments and I have debts.

c) It is false that I have investments or I have debts.

d) I have investments or I have debts.

SOLUTION To determine which statement is equivalent, write each statement in symbolic form.

Let

p: I have investments.

q: I have debts.

The statement, "I do not have investments, but I do not have debts" written symbolically is $\sim p \wedge \sim q$. Recall that the word *but* means the same thing as *and*. Now, write parts (a) through (d) symbolically.

a) $\sim p \vee \sim q$ b) $\sim(p \wedge q)$ c) $\sim(p \vee q)$ d) $p \vee q$

De Morgan's law shows that $\sim p \wedge \sim q$ is equivalent to $\sim(p \vee q)$. Therefore, the answer is (c): "It is false that I have investments or I have debts." ●

EXAMPLE ❺ Using De Morgan's Laws to Write an Equivalent Statement

Write a statement that is logically equivalent to "It is not true that tomatoes are poisonous or eating peppers cures the common cold."

SOLUTION Let

p: Tomatoes are poisonous.
q: Eating peppers cures the common cold.

The given statement is of the form $\sim(p \vee q)$. Using the second of De Morgan's laws, we see that an equivalent statement in symbols is $\sim p \wedge \sim q$. Therefore, an equivalent statement in words is "Tomatoes are not poisonous and eating peppers does not cure the common cold." ●

Consider $\sim(p \wedge q) \Leftrightarrow \sim p \vee \sim q$, one of De Morgan's laws. To go from $\sim(p \wedge q)$ to $\sim p \vee \sim q$, we negate both the p and the q within parentheses; change the conjunction, $\wedge$, to a disjunction, $\vee$; and remove the negation symbol preceding the left parentheses and the parentheses themselves. We can use a similar procedure to obtain equivalent statements. For example,

$$\sim(\sim p \wedge q) \Leftrightarrow p \vee \sim q$$
$$\sim(p \wedge \sim q) \Leftrightarrow \sim p \vee q$$

We can use a similar procedure to obtain equivalent statements when a disjunction is within parentheses. Note that

$$\sim(\sim p \vee q) \Leftrightarrow p \wedge \sim q$$
$$\sim(p \vee \sim q) \Leftrightarrow \sim p \wedge q$$

EXAMPLE ❻ Using De Morgan's Laws to Write an Equivalent Statement

Use De Morgan's laws to write a statement logically equivalent to "Benjamin Franklin was not a U.S. president, but he signed the Declaration of Independence."

SOLUTION Let

p: Benjamin Franklin was a U.S. president.
q: Benjamin Franklin signed the Declaration of Independence.

The statement written symbolically is $\sim p \wedge q$. Earlier we showed that

$$\sim p \wedge q \Leftrightarrow \sim(p \vee \sim q)$$

Therefore, the statement "It is false that Benjamin Franklin was a U.S. president or Benjamin Franklin did not sign the Declaration of Independence" is logically equivalent to the given statement. ●

There are strong similarities between the topics of sets and logic. We can see them by examining De Morgan's laws for sets and logic.

De Morgan's laws: set theory	De Morgan's laws: logic
$(A \cap B)' = A' \cup B'$	$\sim(p \wedge q) \Leftrightarrow \sim p \vee \sim q$
$(A \cup B)' = A' \cap B'$	$\sim(p \vee q) \Leftrightarrow \sim p \wedge \sim q$

The complement in set theory, $'$, is similar to the negation, $\sim$, in logic. The intersection, $\cap$, is similar to the conjunction, $\wedge$; and the union, $\cup$, is similar to the disjunction, $\vee$. If we were to interchange the set symbols with the logic symbols, De Morgan's laws would remain, but in a different form.

　　Both $'$ and $\sim$ can be interpreted as *not*.

　　Both $\cap$ and $\wedge$ can be interpreted as *and*.

　　Both $\cup$ and $\vee$ can be interpreted as *or*.

For example, the set statement $A' \cup B$ can be written as a statement in logic as $\sim a \vee b$.

Statements containing connectives other than *and* and *or* may have equivalent statements. To illustrate this point, construct truth tables for $p \rightarrow q$ and for $\sim p \vee q$. The truth tables will have the same answer columns and therefore the statements are equivalent. That is,

$$p \rightarrow q \Leftrightarrow \sim p \vee q$$

　　With these equivalent statements, we can write a conditional statement as a disjunction or a disjunction as a conditional statement. For example, the statement "If the game is polo, then you ride a horse" can be equivalently stated as "The game is not polo or you ride a horse."

　　To change a conditional statement to a disjunction, negate the antecedent, change the conditional symbol to a disjunction symbol, and keep the consequent the same. To change a disjunction statement to a conditional statement, negate the first statement, change the disjunction symbol to a conditional symbol, and keep the second statement the same.

EXAMPLE ❼ *Rewriting a Disjunction as a Conditional Statement*

Write a conditional statement that is logically equivalent to "The Oregon Ducks will win or the Oregon State Beavers will lose." Assume that the negation of winning is losing.

SOLUTION Let

> p: The Oregon Ducks will win.
> q: The Oregon State Beavers will win.

The original statement may be written symbolically as $p \vee \sim q$. To write an equivalent conditional statement, negate the first statement, p, change the disjunction symbol to a conditional symbol, and keep the second statement the same. Symbolically, the equivalent statement is $\sim p \rightarrow \sim q$. The equivalent statement in words is "If the Oregon Ducks lose, then the Oregon State Beavers will lose."   ●

Negation of the Conditional Statement

Now we will discuss how to negate a conditional statement. To negate a statement we use the fact that $p \rightarrow q \Leftrightarrow \sim p \vee q$ and De Morgan's laws. Examples 8 and 9 show the process.

EXAMPLE ❽ *The Negation of a Conditional Statement*

Determine a statement equivalent to $\sim (p \rightarrow q)$.

SOLUTION Begin with $p \rightarrow q \Leftrightarrow \sim p \vee q$, negate both statements, and use De Morgan's laws.

$$p \rightarrow q \Leftrightarrow \sim p \vee q$$
$$\sim (p \rightarrow q) \Leftrightarrow \sim (\sim p \vee q) \qquad \text{Negate both statements}$$
$$\Leftrightarrow p \wedge \sim q \qquad \text{De Morgan's laws}$$

Therefore, $\sim (p \rightarrow q)$ is equivalent to $p \wedge \sim q$.   ●

EXAMPLE ❾ *Write an Equivalent Statement*

Write a statement that is equivalent to "It is false that if it is snowing then we cannot go to the basketball game."

SOLUTION Let

> p: It is snowing.
> q: We can go to the basketball game.

The given statement can then be represented symbolically as $\sim (p \rightarrow \sim q)$. Using the procedure illustrated in Example 8 we can determine that $\sim (p \rightarrow \sim q)$ is equivalent to $p \wedge q$. You should verify that the two statements are equivalent now. Therefore, an equivalent statement is "It is snowing and we can go to the basketball game."   ●

Variations of the Conditional Statement

We know that $p \to q$ is equivalent to $\sim p \vee q$. Are any other statements equivalent to $p \to q$? Yes, there are many. Now let's look at the variations of the conditional statement to determine whether any are equivalent to the conditional statement. *The variations of the conditional statement are made by switching and/or negating the antecedent and the consequent of a conditional statement.* The variations of the conditional statement are the *converse* of the conditional, the *inverse* of the conditional, and the *contrapositive* of the conditional.

Listed here are the variations of the conditional with their symbolic form and the words we say to read each one.

VARIATIONS OF THE CONDITIONAL STATEMENT

Name	Symbolic form	Read
Conditional	$p \to q$	"If p, then q"
Converse of the conditional	$q \to p$	"If q, then p"
Inverse of the conditional	$\sim p \to \sim q$	"If not p, then not q"
Contrapositive of the conditional	$\sim q \to \sim p$	"If not q, then not p"

To write the converse of the conditional statement, switch the order of the antecedent and the consequent. To write the inverse, negate both the antecedent and the consequent. To write the contrapositive, switch the order of the antecedent and the consequent and then negate both of them.

Are any of the variations of the conditional statement equivalent? To determine the answer, we can construct a truth table for each variation, as shown in Table 3.26. It reveals that the conditional statement is equivalent to the contrapositive statement and that the converse statement is equivalent to the inverse statement.

Table 3.26

p	q	Conditional $p \to q$	Contrapositive $\sim q \to \sim p$	Converse $q \to p$	Inverse $\sim p \to \sim q$
T	T	T	T	T	T
T	F	F	F	T	T
F	T	T	T	F	F
F	F	T	T	T	T

EXAMPLE ⑩ *The Converse, Inverse and Contrapositive*

For the conditional statement "If the song contains sitar music, then the song was written by George Harrison," write the

a) converse.　　b) inverse.　　c) contrapositive.

▲ A sitar

SOLUTION

a) Let

      p: The song contains sitar music.

      q: The song was written by George Harrison.

The conditional statement is of the form $p \rightarrow q$, so the converse must be of the form $q \rightarrow p$. Therefore, the converse is "If the song was written by George Harrison, then the song contains sitar music."

b) The inverse is of the form $\sim p \rightarrow \sim q$. Therefore, the inverse is "If the song does not contain sitar music, then the song was not written by George Harrison."

c) The contrapositive is of the form $\sim q \rightarrow \sim p$. Therefore, the contrapositive is "If the song was not written by George Harrison, then the song does not contain sitar music."   ●

EXAMPLE ⑪ *Determine the Truth Values*

Let

      p: The number is divisible by 9.

      q: The number is divisible by 3.

Write the following statements and determine which are true.

a) The conditional statement, $p \rightarrow q$

b) The converse of $p \rightarrow q$

c) The inverse of $p \rightarrow q$

d) The contrapositive of $p \rightarrow q$

SOLUTION

a) *Conditional statement:* $(p \rightarrow q)$
   If the number is divisible by 9, then the number is divisible by 3. This statement is true. A number divisible by 9 must also be divisible by 3, since 3 is a divisor of 9.

b) *Converse of the conditional:* $(q \rightarrow p)$
   If the number is divisible by 3, then the number is divisible by 9. This statement is false. For instance, 6 is divisible by 3, but 6 is not divisible by 9.

c) *Inverse of the conditional:* $(\sim p \rightarrow \sim q)$
   If the number is not divisible by 9, then the number is not divisible by 3. This statement is false. For instance, 6 is not divisible by 9, but 6 is divisible by 3.

d) *Contrapositive of the conditional:* $(\sim q \rightarrow \sim p)$
   If the number is not divisible by 3, then the number is not divisible by 9. The statement is true, since any number that is divisible by 9 must be divisible by 3.   ●

EXAMPLE ⑫ *Use the Contrapositive*

Use the contrapositive to write a statement logically equivalent to "If you don't eat your meat, then you can't have pudding."

SOLUTION Let

p: You do eat your meat.

q: You can have pudding.

The given statement written symbolically is

$$\sim p \rightarrow \sim q$$

The contrapositive of the statement is

$$q \rightarrow p$$

Therefore, an equivalent statement is "If you can have pudding, then you do eat your meat." •

The contrapositive of the conditional is very important in mathematics. Consider the statement "If a^2 is not a whole number, then a is not a whole number." Is this statement true? You may find this question difficult to answer. Writing the statement's contrapositive may enable you to answer the question. The contrapositive is "If a is a whole number, then a^2 is a whole number." Since the contrapositive is a true statement, the original statement must also be true.

EXAMPLE ⓭ *Which Are Equivalent?*

Determine which, if any, of the following statements are equivalent. You may use De Morgan's laws, the fact that $p \rightarrow q \Leftrightarrow \sim p \vee q$, information from the variations of the conditional, or truth tables.

a) If you leave by 9 A.M., then you will get to your destination on time.
b) You do not leave by 9 A.M. or you will get to your destination on time.
c) It is false that you will get to your destination on time or you did not leave by 9 A.M.
d) If you do not get to your destination on time, then you did not leave by 9 A.M.

SOLUTION Let

p: You leave by 9 A.M.
q: You will get to your destination on time.

In symbolic form, the four statements are

a) $p \rightarrow q$.   b) $\sim p \vee q$.   c) $\sim(q \vee \sim p)$.   d) $\sim q \rightarrow \sim p$.

Which of these statements are equivalent? Earlier in this section, you learned that $p \rightarrow q$ is equivalent to $\sim p \vee q$. Therefore, statements (a) and (b) are equivalent. Statement (d) is the contrapositive of statement (a). Therefore, statement (d) is also equivalent to statement (a) and statement (b). All these statements have the same truth table (Table 3.27 on page 146).

Table 3.27

		(a)	(b)	(d)
p	q	$p \rightarrow q$	$\sim p \vee q$	$\sim q \rightarrow \sim p$
T	T	T	T	T
T	F	F	F	F
F	T	T	T	T
F	F	T	T	T

Now let's look at statement (c). To determine whether $\sim(q \vee \sim p)$ is equivalent to the other statements, we will construct its truth table (Table 3.28) and compare the answer column with the answer columns in Table 3.27.

Table 3.28

				(c)	
p	q	$\sim$	$(q$	$\vee$	$\sim p)$
T	T	F	T	T	F
T	F	T	F	F	F
F	T	F	T	T	T
F	F	F	F	T	T
		4	1	3	2

None of the three answer columns of the truth table in Table 3.27 is the same as the answer column of the truth table in Table 3.28. Therefore $\sim(q \vee \sim p)$ is not equivalent to any of the other statements. Therefore, statements (a), (b), and (d) are equivalent to each other.   ●

SECTION 3.4   EXERCISES

CONCEPT/WRITING EXERCISES

1. a) What are equivalent statements?

   b) Explain how you can determine whether two statements are equivalent.

2. Suppose two statements are connected with the biconditional and the truth table is constructed. If the answer column of the truth table has all trues, what must be true about these two statements? Explain.

3. Write De Morgan's laws for logic.

4. For a statement of the form $p \rightarrow q$, symbolically indicate the form of the

   a) converse.

   b) inverse.

   c) contrapositive.

5. Which of the following are equivalent statements?

   a) The converse

   b) The contrapositive

   c) The inverse

   d) The conditional

6. Write a disjunction that is logically equivalent to $p \rightarrow q$.

7. If $p \rightarrow q$ is equivalent to $\sim p \vee q$, write a conjunction that is equivalent to $\sim(p \rightarrow q)$.

8. Write a conjunction involving two conditional statements that is logically equivalent to $p \leftrightarrow q$. See page 128.

PRACTICE THE SKILLS

In Exercises 9–18, use De Morgan's laws to determine whether the two statements are equivalent.

9. $\sim(p \wedge q), \sim p \wedge q$

10. $\sim(p \vee \sim q), \sim p \wedge q$

11. $\sim(p \wedge q), \sim(q \vee \sim p)$

12. $\sim p \vee \sim q, \sim(p \wedge q)$

13. $\sim(p \vee q), \sim p \wedge \sim q$

14. $\sim(p \wedge q), \sim p \wedge \sim q$

15. $(\sim p \vee \sim q) \rightarrow r, \sim(p \wedge q) \rightarrow r$

16. $q \rightarrow \sim(p \wedge \sim r), q \rightarrow \sim p \vee r$

17. $\sim(p \rightarrow \sim q), p \wedge q$

18. $\sim(\sim p \rightarrow q), \sim p \wedge \sim q$

In Exercises 19–30, use a truth table to determine whether the two statements are equivalent.

19. $p \rightarrow q, \sim p \vee q$

20. $\sim(p \rightarrow q), p \wedge \sim q$

21. $\sim q \rightarrow \sim p, p \rightarrow q$

22. $q \rightarrow p, \sim p \rightarrow \sim q$

23. $(p \vee q) \vee r, p \vee (q \vee r)$

24. $p \vee (q \wedge r), \sim p \rightarrow (q \wedge r)$

25. $p \wedge (q \vee r), (p \wedge q) \vee r$

26. $\sim(q \rightarrow p) \vee r, (p \vee q) \wedge \sim r$

27. $(p \rightarrow q) \wedge (q \rightarrow r), (p \rightarrow q) \rightarrow r$

28. $\sim q \rightarrow (p \wedge r), \sim(p \vee r) \rightarrow q$

29. $(p \rightarrow q) \wedge (q \rightarrow p), (p \leftrightarrow q)$

30. $[\sim(p \rightarrow q)] \wedge [\sim(q \rightarrow p)], \sim(p \leftrightarrow q)$

PROBLEM SOLVING

In Exercises 31–38, use De Morgan's laws to write an equivalent statement for the sentence.

31. It is false that the Rocky Mountains are in the East and the Appalachian Mountains are in the West.

32. It is false that Johanna Chan is the secretary or Davidson Pierre is the treasurer.

33. The watch was neither a Swatch watch nor was the watch a Swiss Army watch.

34. The pot roast is hot, but it is not well done.

35. The hotel does not have a weight room or the conference center does not have an auditorium.

36. Robert Farinelli is an authorized WedgCor dealer or he is not going to work for Prism Construction Company.

37. If Ashley Tabai takes the new job, then she will not move or she will buy a new house in town.

38. If Phil Murphy buys us dinner, then we will not go to the top of the CN Tower but we will be able to walk to the Red Bistro Restaurant.

In Exercises 39–46, use the fact that $p \rightarrow q$ is equivalent to $\sim p \vee q$ to write an equivalent form of the given statement.

39. If Ena Salter selects a new textbook, then she will have to write a new syllabus.

40. Maxwell's favorite TV show is *Jimmy Neutron* or his favorite TV show is *Danny Phantom*.

▲ Jimmy Neutron

41. Bob the Tomato visited the nursing home or he did not visit the Cub Scout meeting.

42. If Joanne Ernst goes to the Lightning game, then she will not go to the Devil Rays game.

43. If the plumbers meet in Kansas City then the Rainmakers will provide the entertainment.

44. Mary Beth Headlee organized the conference or John Waters does not work at Sinclair Community College.

45. Chase is not hiding or the pitcher is broken.

46. If Weezer is not on the radio, then Tim Ollendick is working.

In Exercises 47–54, use the fact that $\sim(p \to q)$ is equivalent to $p \land \sim q$ to write the statement in an equivalent form.

47. It is false that if we go to Cincinnati, then we will go to the zoo.

48. It is false that if General Electric makes the telephone, then the telephone is made in the United States.

49. I am cold and the heater is not working.

50. The Badgers beat the Nittany Lions and the Bucks beat the 76ers.

51. It is not true that if Borders has a sale then we will buy $100 worth of books.

52. Thompson is sick today but Allen didn't go to school.

53. John Deere will hire new workers and the city of Dubuque will retrain the workers.

54. My cell phone is not made by Motorola and my carrier is Alltel.

In Exercises 55–60, write the converse, inverse, and contrapositive of the statement. (For Exercise 60, use De Morgan's laws.)

55. If we work every night, then we can finish the quilt in 1 week.

56. If your cell phone is beeping, then you need to charge the battery.

57. If I go to Mexico, then I buy silver jewelry.

58. If Bob Dylan records a new CD, then he will go on tour.

59. If that annoying paper clip shows up on my computer screen, then I will scream.

60. If the sun is shining, then we will go down to the marina and we will take out the sailboat.

In Exercises 61–66, write the contrapositive of the statement. Use the contrapositive to determine whether the conditional statement is true or false.

61. If a natural number is not divisible by 5, then the natural number is not divisible by 10.

62. If the opposite sides of the quadrilateral are not parallel, then the quadrilateral is not a parallelogram.

63. If a natural number is divisible by 3, then the natural number is divisible by 6.

64. If $1/n$ is not a natural number, then n is not a natural number.

65. If two lines do not intersect in at least one point, then the two lines are parallel.

66. If $\dfrac{m \cdot a}{m \cdot b} \neq \dfrac{a}{b}$, then m is not a counting number.

In Exercises 67–82, determine which, if any, of the three statements are equivalent (see Example 13).

67. a) Bill Rush is not the editor or Bill Rush is the vice president.

b) If Bill Rush is the vice president, then Bill Rush is not the editor.

c) If Bill Rush is the editor, then Bill Rush is not the vice president.

68. a) If Fido is our dog's name, then Rex is not our dog's name.

b) It is false that Fido is our dog's name and Rex is not our dog's name.

c) Fido is not our dog's name or Rex is our dog's name.

69. a) The office is not cool and the copier is jammed.

b) If the office is not cool, then the copier is not jammed.

c) It is false that the office is cool or the copier is not jammed.

70. a) The test is not written or the review sheet is not ready.

b) If the test is written, then the review sheet is ready.

c) It is false that the review sheet is ready and the test is not written.

71. a) Today is not Sunday or the library is open.

b) If today is Sunday, then the library is not open.

c) If the library is open, then today is not Sunday.

72. a) If you are fishing at 1 P.M., then you are driving a car at 1 P.M.

b) You are not fishing at 1 P.M. or you are driving a car at 1 P.M.

c) It is false that you are fishing at 1 P.M. and you are not driving a car at 1 P.M.

▲ See Exercise 72

73. a) The grass grows and the trees are blooming.

b) If the trees are blooming, then the grass does not grow.

c) The trees are not blooming or the grass does not grow.

74. a) Johnny Patrick is chosen as department chair if and only if he is the only candidate.

b) If Johnny Patrick is chosen as department chair then he is the only candidate, and if Johnny Patrick is the only candidate then he is chosen as department chair.

c) Johnny Patrick is not chosen as department chair and he is not the only candidate.

75. a) It is false that if you do not drink milk then your cholesterol count will be lower.

b) Your cholesterol count will be lower if and only if you drink milk.

c) It is false that if you drink milk then your cholesterol count will not be lower.

76. a) Bruce Springsteen will not go on tour if and only if Clarence Clemmons does not play the saxophone in his band.

b) It is false that Bruce Springsteen will go on tour if and only if Clarence Clemmons does not play the saxophone in his band.

c) If Bruce Springsteen goes on tour, then Clarence Clemmons plays saxophone in his band.

77. a) If the pay is good and today is Monday, then I will take the job.

b) If I do not take the job, then it is false that the pay is good or today is Monday.

c) The pay is good and today is Monday, or I will take the job.

78. a) If you are 18 years old and a citizen of the United States, then you can vote in the presidential election.

b) You can vote in the presidential election, if and only if you are a citizen of the United States and you are 18 years old.

c) You cannot vote in the presidential election, or you are 18 years old and you are not a citizen of the United States.

79. a) The package was sent by Federal Express, or the package was not sent by United Parcel Service but the package arrived on time.

b) The package arrived on time, if and only if it was sent by Federal Express or it was not sent by United Parcel Service.

c) If the package was not sent by Federal Express, then the package was not sent by United Parcel Service but the package arrived on time.

80. a) If we put the dog outside or we feed the dog, then the dog will not bark.

b) If the dog barks, then we did not put the dog outside and we did not feed the dog.

c) If the dog barks, then it is false that we put the dog outside or we feed the dog.

81. a) The car needs oil, and the car needs gas or the car is new.

b) The car needs oil, and it is false that the car does not need gas and the car is not new.

c) If the car needs oil, then the car needs gas or the car is not new.

82. a) The mortgage rate went down, if and only if Tim purchased the house and the down payment was 10%.

b) The down payment was 10%, and if Tim purchased the house then the mortgage rate went down.

c) If Tim purchased the house, then the mortgage rate went down and the down payment was not 10%.

83. If p and q represent two simple statements, and if $p \rightarrow q$ is a false statement, what must be the truth value of the converse, $q \rightarrow p$? Explain.

84. If p and q represent two simple statements, and if $p \rightarrow q$ is a false statement, what must be the truth value of the inverse, $\sim p \rightarrow \sim q$? Explain.

85. If p and q represent two simple statements, and if $p \rightarrow q$ is a false statement, what must be the truth value of the contrapositive, $\sim q \rightarrow \sim p$? Explain.

86. If p and q represent two simple statements, and if $p \rightarrow q$ is a true statement, what must be the truth value of the contrapositive, $\sim q \rightarrow \sim p$? Explain.

CHALLENGE PROBLEMS/GROUP ACTIVITIES

87. We learned that $p \rightarrow q \Leftrightarrow \sim p \vee q$. Determine a conjunction that is equivalent to $p \rightarrow q$. (*Hint:* There are many answers.)

88. Determine whether $\sim[\sim(p \vee \sim q)] \Leftrightarrow p \vee \sim q$. Explain the method(s) you used to determine your answer.

89. In an appliance or device that uses fuzzy logic, a change in one condition causes a change in a second condition. For example, in a camera, if the brightness increases, the lens aperture automatically decreases to get the proper exposure on the film. Name at least 10 appliances or devices that make use of fuzzy logic and explain how fuzzy logic is used in each appliance or device. See the Did You Know? on page 146.

90. In symbolic logic, a statement is either true or false (consider true to have a value of 1 and false a value of 0). In fuzzy logic, nothing is true or false, but everything is a matter of degree. For example, consider the statement "The sun is shining." In fuzzy logic, this statement may have a value between 0 and 1 and may be constantly changing. For example, if the sun is partially blocked by clouds, the value of this statement may be 0.25. In fuzzy logic, the values of connective statements are found as follows for statements p and q.

> Not p has a truth value of $1 - p$.
> $p \wedge q$ has a truth value equal to the lesser of p and q.
> $p \vee q$ has a truth value equal to the greater of p and q.
> $p \rightarrow q$ has a truth value equal to the lesser of 1 and $1 - p + q$.
> $p \leftrightarrow q$ has a truth value equal to $1 - |p - q|$, that is, 1 minus the absolute value* of p minus q.

Suppose the statement "p: The sun is shining" has a truth value of 0.25 and the statement "q: Mary is getting a tan" has a truth value of 0.20. Find the truth value of

a) $\sim p$.   **b)** $\sim q$.

c) $p \wedge q$.   **d)** $p \vee q$.

e) $p \rightarrow q$.   **f)** $p \leftrightarrow q$.

INTERNET/RESEARCH ACTIVITIES

91. Do research and write a report on fuzzy logic.

92. Read one of Lewis Carroll's books and write a report on how he used logic in the book. Give at least five specific examples.

93. Do research and write a report on the life and achievements of Augustus De Morgan. Indicate in your report his contributions to sets and logic.

3.5 SYMBOLIC ARGUMENTS

▲ John Mellencamp and his band sing the national anthem.

Consider the following statements.

> If John Mellencamp sings the national anthem, then U2 will play at halftime.
>
> John Mellencamp sings the national anthem.

If you accept these two statements as true, then what logical conclusion could you draw? Do you agree that you can logically conclude that U2 will play at halftime? In this section, we will use our knowledge of logic to study the structure of such statements to draw logical conclusions.

Previously in this chapter, we used symbolic logic to determine the truth value of a compound statement. We now extend those basic ideas to determine whether we can draw logical conclusions from a set of given statements. Consider once again the two statements.

> If John Mellencamp sings the national anthem, then U2 will play at halftime.
>
> John Mellencamp sings the national anthem.

—————

*Absolute values are discussed in Section 13.8.

The Dog Did Nothing

I n the case of the disappearance of the racehorse Silver Blaze, Sherlock Holmes demonstrated that sometimes the absence of a clue is itself a clue. The local police inspector asked him, "Is there any point to which you would wish to draw my attention?" Holmes replied, "To the curious incident of the dog in the nighttime." The inspector, confused, asked: "The dog did nothing in the nighttime." "That was the curious incident," remarked Sherlock Holmes. From the lack of the dog's bark, Holmes concluded that the horse had been "stolen" by a stablehand. How did Holmes reach his conclusion?

These statements in the following form constitute what we will call a *symbolic argument*.

Premise 1: If John Mellencamp sings the national anthem, then U2 will play at halftime.

Premise 2: John Mellencamp sings the national anthem.

Conclusion: U2 will play at halftime.

A *symbolic argument* consists of a set of *premises* and a *conclusion*. It is called a symbolic argument because we generally write it in symbolic form to determine its validity.

> An **argument is valid** when its conclusion necessarily follows from a given set of premises.
>
> An **argument is invalid** or a **fallacy** when the conclusion does not necessarily follow from the given set of premises.

An argument that is not valid is invalid. The argument just presented is an example of a valid argument, as the conclusion necessarily follows from the premises. Now we will discuss a procedure to determine whether an argument is valid or invalid. We begin by writing the argument in symbolic form. To write the argument in symbolic form, we let p and q be

p: John Mellencamp sings the national anthem.

q: U2 will play at halftime.

Symbolically, the argument is written

Premise 1: $p \to q$

Premise 2: p

Conclusion: $\therefore q$ (The three-dot triangle is read "therefore.")

Write the argument in the following form.

If [*premise 1* **and** *premise 2*] **then** *conclusion*

$[(p \to q) \quad \wedge \quad p] \quad \to \quad q$

Then construct a truth table for the statement $[(p \to q) \wedge p] \to q$ (Table 3.29). *If the truth table answer column is true in every case, then the statement is a tautology, and the argument is valid. If the truth table is not a tautology, then the argument is invalid.* Since the statement is a tautology (see column 5), the conclusion necessarily follows from the premises and the argument is valid.

Table 3.29

p	q	$[(p \to q)$	$\wedge$	$p]$	$\to$	q
T	T	T	T	T	T	T
T	F	F	F	T	T	F
F	T	T	F	F	T	T
F	F	T	F	F	T	F
		1	3	2	5	4

Once we have demonstrated that an argument in a particular form is valid, all arguments with exactly the same form will also be valid. In fact, many of these forms have been assigned names. The argument form just discussed,

$$p \rightarrow q$$
$$\underline{p}$$
$$\therefore q$$

is called the *law of detachment*, or *modus ponens*.

▲ Alaska

EXAMPLE ❶ *Determining Validity without a Truth Table*

Determine whether the following argument is valid or invalid.

If Canada is north of the United States, then Alaska is in Mexico.
Canada is north of the United States.
∴ Alaska is in Mexico.

SOLUTION Translate the argument into symbolic form.

Let

c: Canada is north of the United States.
a: Alaska is in Mexico.

In symbolic form, the argument is

$$c \rightarrow a$$
$$\underline{c}$$
$$\therefore a$$

This argument is also the law of detachment. Therefore, it is a valid argument. ●

Note that the argument in Example 1 is valid even though the conclusion, "Alaska is in Mexico," is a false statement. It is also possible to have an invalid argument in which the conclusion is a true statement. When an argument is valid, the conclusion necessarily follows from the premises. It is not necessary for the premises or the conclusion to be true statements in an argument.

PROCEDURE TO DETERMINE WHETHER AN ARGUMENT IS VALID

1. Write the argument in symbolic form.
2. Compare the form of the argument with forms that are known to be valid or invalid. If there are no known forms to compare it with, or you do not remember the forms, go to step 3.
3. If the argument contains two premises, write a conditional statement of the form

    $$[(\text{premise } 1) \wedge (\text{premise } 2)] \rightarrow \text{conclusion}$$

4. Construct a truth table for the statement in step 3.
5. If the answer column of the truth table has all trues, the statement is a tautology, and the argument is valid. If the answer column does not have all trues, the argument is invalid.

Examples 1 through 4 contain two premises. When an argument contains more than two premises, step 3 of the procedure will change slightly, as will be explained shortly.

EXAMPLE ❷ *Determining Validity with a Truth Table*

Determine whether the following argument is valid or invalid.

> If you score 90% on the final exam, then you will get an A in the course.
> You will not get an A in the course.
> ∴ You do not score 90% on the final exam.

SOLUTION We first write the argument in symbolic form.

Let

p: You score 90% on the final exam.
q: You will get an A in the course.

In symbolic form, the argument is

$$p \rightarrow q$$
$$\sim q$$
$$\therefore \sim p$$

As we have not tested an argument in this form, we will construct a truth table to determine whether the argument is valid or invalid. We write the argument in the form $[(p \rightarrow q) \wedge \sim q] \rightarrow \sim p$, and construct a truth table (Table 3.30). Since the answer, column 5, has all T's, the argument is valid.

Table 3.30

p	q	$[(p \rightarrow q)$	$\wedge$	$\sim q]$	$\rightarrow$	$\sim p$
T	T	T	F	F	T	F
T	F	F	F	T	T	F
F	T	T	F	F	T	T
F	F	T	T	T	T	T
		1	3	2	5	4

The argument form in Example 2 is an example of the *law of contraposition*, or *modus tollens*.

EXAMPLE ❸ *Another Symbolic Argument*

Determine whether the following argument is valid or invalid.

> The grass is green or the grass is full of weeds.
> The grass is not green.
> ∴ The grass is full of weeds.

SOLUTION Let

> p: The grass is green.
>
> q: The grass is full of weeds.

In symbolic form, the argument is

$$p \lor q$$
$$\underline{\sim p}$$
$$\therefore q$$

As this form is not one of those we are familiar with, we will construct a truth table. We write the argument in the form $[(p \lor q) \land \sim p] \to q$. Next we construct a truth table, as shown in Table 3.31. The answer to the truth table, column 5, is true in *every case*. Therefore, the statement is a tautology, and the argument is valid.

Table 3.31

p	q	$[(p \lor q)$	$\land$	$\sim p]$	$\to$	q
T	T	T	F	F	T	T
T	F	T	F	F	T	F
F	T	T	T	T	T	T
F	F	F	F	T	T	F
		1	3	2	5	4

The argument form in Example 3 is an example of *disjunctive syllogism*. Other standard forms of arguments are given in the following chart.

STANDARD FORMS OF ARGUMENTS				
Valid Arguments	*Law of Detachment*	*Law of Contraposition*	*Law of Syllogism*	*Disjunctive Syllogism*
	$p \to q$	$p \to q$	$p \to q$	$p \lor q$
	$\underline{p}$	$\underline{\sim q}$	$\underline{q \to r}$	$\underline{\sim p}$
	$\therefore q$	$\therefore \sim p$	$\therefore p \to r$	$\therefore q$
Invalid Arguments	*Fallacy of the Converse*	*Fallacy of the Inverse*		
	$p \to q$	$p \to q$		
	$\underline{q}$	$\underline{\sim p}$		
	$\therefore p$	$\therefore \sim q$		

As we saw in Example 1, it is not always necessary to construct a truth table to determine whether or not an argument is valid. The next two examples will show how we can identify an argument as one of the standard arguments given in the chart above.

EXAMPLE ➍ *Identifying the Law of Syllogism in an Argument*

Determine whether the following argument is valid or invalid.

> If my laptop battery is dead, then I use my home computer.
> If I use my home computer, then my kids will play outside.
> ∴ If my laptop battery is dead, then my kids will play outside.

SOLUTION Let

> p: My laptop battery is dead.
> q: I use my home computer.
> r: My kids will play outside.

In symbolic form, the argument is

$$\begin{array}{l} p \to q \\ \underline{q \to r} \\ \therefore\ p \to r \end{array}$$

The argument is in the form of the law of syllogism. Therefore, the argument is valid, and there is no need to construct a truth table.   ●

EXAMPLE ➎ *Identifying Common Fallacies in Arguments*

Determine whether the following arguments are valid or invalid.

a)

> If it is snowing, then we put salt on the driveway.
> We put salt on the driveway.
> ∴ It is snowing.

b)

> If it is snowing, then we put salt on the driveway.
> It is not snowing.
> ∴ We do not put salt on the driveway.

SOLUTION

a) Let

> p: It is snowing.
> q: We put salt on the driveway.

In symbolic form, the argument is

$$\begin{array}{l} p \to q \\ \underline{q } \\ \therefore\ p \end{array}$$

This argument is in the form of the fallacy of the converse. Therefore, the argument is a fallacy, or invalid.

b) Using the same symbols defined in the solution to part (a), in symbolic form, the argument is

$$p \rightarrow q$$
$$\underline{\sim p}$$
$$\therefore \sim q$$

This argument is in the form of the fallacy of the inverse. Therefore, the argument is a fallacy, or invalid. ●

TIMELY TIP If you are not sure whether an argument with two premises is one of the standard forms or if you do not remember the standard forms, you can always determine whether a given argument is valid or invalid by using a truth table. To do so, follow the boxed procedure on page 153.

In Example 5(b), if you did not recognize that this argument was of the same form as the fallacy of the inverse you could construct the truth table for the conditional statement

$$[(p \rightarrow q) \wedge \sim p] \rightarrow \sim q$$

The true–false values under the conditional column, →, would be T, T, F, T. Since the statement is not a tautology, the argument is invalid.

Now we consider an argument that has more than two premises. When an argument contains more than two premises, the statement we test, using a truth table, is formed by taking the conjunction of all the premises as the antecedent of a conditional statement and the conclusion as the consequent of the conditional statement. One example is an argument of the form

$$p_1$$
$$p_2$$
$$\underline{p_3}$$
$$\therefore c$$

We evaluate the truth table for $[p_1 \wedge p_2 \wedge p_3] \rightarrow c$. When we evaluate $[p_1 \wedge p_2 \wedge p_3]$, it makes no difference whether we evaluate $[(p_1 \wedge p_2) \wedge p_3]$ or $[p_1 \wedge (p_2 \wedge p_3)]$ because both give the same answer. In Example 6, we evaluate $[p_1 \wedge p_2 \wedge p_3]$ from left to right; that is, $[(p_1 \wedge p_2) \wedge p_3]$.

┌**EXAMPLE ❻** *An Argument with Three Premises*

Use a truth table to determine whether the following argument is valid or invalid.

If my cell phone company is Verizon, then I can call you free of charge.
I can call you free of charge or I can send you an e-mail.
I can send you an e-mail or my cell phone company is Verizon.
∴ My cell phone company is Verizon.

SOLUTION This argument contains three simple statements.

Let

p: My cell phone company is Verizon.
q: I can call you free of charge.
r: I can send you an e-mail.

In symbolic form, the argument is

$$p \rightarrow q$$
$$q \vee r$$
$$\underline{r \vee p}$$
$$\therefore p$$

Write the argument in the form

$$[(p \rightarrow q) \wedge (q \vee r) \wedge (r \vee p)] \rightarrow p.$$

Now construct the truth table (Table 3.32). The answer, column 7, is not true in every case. Thus, the argument is a fallacy, or invalid.

Table 3.32

p	q	r	$[(p \rightarrow q)$	$\wedge$	$(q \vee r)$	$\wedge$	$(r \vee p)]$	$\rightarrow$	p
T	T	T	T	T	T	T	T	T	T
T	T	F	T	T	T	T	T	T	T
T	F	T	F	F	T	F	T	T	T
T	F	F	F	F	F	F	T	T	T
F	T	T	T	T	T	T	T	F	F
F	T	F	T	T	T	F	F	T	F
F	F	T	T	T	T	T	T	F	F
F	F	F	T	F	F	F	F	T	F
			1	3	2	5	4	7	6

●

Let's now investigate how we can arrive at a valid conclusion from a given set of premises.

EXAMPLE ❼ *Determine a Logical Conclusion*

Determine a logical conclusion that follows from the given statements. "If the price of gas is below $3.00 per gallon, then we will drive to the Offspring concert. We will not drive to the Offspring concert. Therefore . . . "

SOLUTION If you recognize a specific form of an argument, you can use your knowledge of that form to draw a logical conclusion.

Let

p: The price of gas is below \$3.00 per gallon.

q: We will drive to the Offspring concert.

The argument is of the following form.

$$\frac{\begin{array}{c} p \rightarrow q \\ \sim q \end{array}}{\therefore \ ?}$$

If the question mark is replaced with a $\sim p$, this argument is of the form of the law of contraposition. Thus, a logical conclusion is "Therefore, the price of gas is not below \$3.00 per gallon." ●

SECTION 3.5   EXERCISES

CONCEPT/WRITING EXERCISES

1. a) What does it mean when an argument is a valid argument?

    b) What does it mean when an argument is a fallacy?

2. Explain how to determine whether an argument with premises p_1 and p_2 and conclusion c is a valid argument or an invalid argument.

3. Is it possible for an argument to be invalid if its conclusion is true? Explain your answer.

4. Is it possible for an argument to be valid if its conclusion is false? Explain your answer.

5. Is it possible for an argument to be invalid if the premises are all true? Explain your answer.

6. Is it possible for an argument to be valid if the premises are all false? Explain your answer.

In Exercises 7–10, (a) indicate the form of the valid argument and (b) write an original argument in words for each form.

7. Law of detachment   8. Law of syllogism

9. Law of contraposition   10. Disjunctive syllogism

In Exercises 11 and 12, (a) indicate the form of the fallacy, and (b) write an original argument in words for each form.

11. Fallacy of the converse   12. Fallacy of the inverse

PRACTICE THE SKILLS

In Exercises 13–32, determine whether the argument is valid or invalid. You may compare the argument to a standard form, given on page 155, or use a truth table.

13. $\dfrac{\begin{array}{c} a \rightarrow b \\ \sim a \end{array}}{\therefore \ \sim b}$
       14. $\dfrac{\begin{array}{c} c \vee d \\ \sim c \end{array}}{\therefore \ d}$
       15. $\dfrac{\begin{array}{c} e \rightarrow f \\ e \end{array}}{\therefore \ f}$

16. $\dfrac{\begin{array}{c} g \rightarrow h \\ h \rightarrow i \end{array}}{\therefore \ g \rightarrow i}$
       17. $\dfrac{\begin{array}{c} j \vee k \\ \sim k \end{array}}{\therefore \ j}$
       18. $\dfrac{\begin{array}{c} l \rightarrow m \\ \sim m \end{array}}{\therefore \ \sim l}$

19. $\dfrac{\begin{array}{c} n \rightarrow o \\ o \end{array}}{\therefore \ n}$
       20. $\dfrac{\begin{array}{c} r \rightarrow s \\ r \end{array}}{\therefore \ s}$
       21. $\dfrac{\begin{array}{c} t \rightarrow u \\ \sim u \end{array}}{\therefore \ \sim t}$

22. $\dfrac{\begin{array}{c} v \rightarrow w \\ w \end{array}}{\therefore \ v}$
       23. $\dfrac{\begin{array}{c} x \rightarrow y \\ y \rightarrow z \end{array}}{\therefore \ x \rightarrow z}$
       24. $\dfrac{\begin{array}{c} x \rightarrow y \\ \sim x \end{array}}{\therefore \ \sim y}$

25. $\dfrac{\begin{array}{c} p \leftrightarrow q \\ q \wedge r \end{array}}{\therefore \ p \vee r}$
       26. $\dfrac{\begin{array}{c} p \leftrightarrow q \\ q \rightarrow r \end{array}}{\therefore \ \sim r \rightarrow \sim p}$
       27. $\dfrac{\begin{array}{c} r \leftrightarrow p \\ \sim p \wedge q \end{array}}{\therefore \ p \wedge r}$

28. $\dfrac{\begin{array}{c} p \vee q \\ r \wedge p \end{array}}{\therefore \ q}$
       29. $\dfrac{\begin{array}{c} p \rightarrow q \\ q \vee r \\ r \vee p \end{array}}{\therefore \ p}$
       30. $\dfrac{\begin{array}{c} p \rightarrow q \\ q \rightarrow r \\ r \rightarrow p \end{array}}{\therefore \ q \rightarrow p}$

31. $p \rightarrow q$
$r \rightarrow \sim p$
$p \vee r$
$\therefore q \vee \sim p$

32. $p \leftrightarrow q$
$p \vee r$
$q \rightarrow r$
$\therefore q \vee r$

PROBLEM SOLVING

In Exercises 33–50, (a) translate the argument into symbolic form and (b) determine if the argument is valid or invalid. You may compare the argument to a standard form or use a truth table.

33. If Will Smith wins an Academy Award, then he will retire from acting.
Will Smith did not win an Academy Award.

∴ Will Smith will not retire from acting.

34. If the car is a Road Runner, then the car is fast.
The car is fast.

∴ The car is a Road Runner.

35. If the baby is a boy, then we will name him Alexander Martin.
The baby is a boy.

∴ We will name him Alexander Martin.

36. If I can get my child to preschool by 8:45 A.M., then I can take the 9:00 A.M. class.
If I can take the 9:00 A.M. class, then I can be done by 2:00 P.M.

∴ If I can get my child to preschool by 8:45 A.M., then I can be done by 2:00 P.M.

37. If the guitar is a Les Paul model, then the guitar is made by Gibson.
The guitar is not made by Gibson.

∴ The guitar is not a Les Paul model.

▲ Les Paul

38. We will go for a bike ride or we will go shopping.
We will not go shopping.

∴ We will go for a bike ride.

39. If we planted the garden by the first Friday in April, then we will have potatoes by the Fourth of July.
We will have potatoes by the Fourth of July.

∴ We planted the garden by the first Friday in April.

40. If you pass general chemistry, then you can take organic chemistry.
You pass general chemistry.

∴ You can take organic chemistry.

41. Sarah Hughes will win an Olympic gold medal in figure skating or Joey Cheek will win an Olympic gold medal in speed skating.
Sarah Hughes will not win an Olympic gold medal in figure skating.

∴ Joey Cheek will win an Olympic gold medal in speed skating.

42. If Nicholas Thompson teaches this course, then I will get a passing grade.
I did not get a passing grade.

∴ Nicholas Thompson did not teach the course.

43. If it is cold, then graduation will be held indoors.
If graduation is held indoors, then the fireworks will be postponed.

∴ If it is cold, then the fireworks will be postponed.

44. If the canteen is full, then we can go for a walk.
We can go for a walk and we will not get thirsty.

∴ If we go for a walk, then the canteen is not full.

45. Marie works at the post office and Jim works for Target.
If Jim works for Target, then Tommy gets an internship.

∴ If Tommy gets an internship, then Marie works at the post office.

46. Vitamin C helps your immune system or niacin helps reduce cholesterol.
If niacin helps reduce cholesterol, then vitamin E enhances your skin.
∴ Vitamin C helps your immune system and vitamin E enhances your skin.

47. It is snowing and I am going skiing.
If I am going skiing, then I will wear a coat.
∴ If it is snowing, then I will wear a coat.

48. The garden has vegetables or the garden has flowers.
If the garden does not have flowers, then the garden has vegetables.
∴ The garden has flowers or the garden has vegetables.

49. If the house has electric heat, then the Flynns will buy the house.
If the price is not less than $100,000, then the Flynns will not buy the house.
∴ If the house has electric heat, then the price is less than $100,000.

50. If there is an atmosphere, then there is gravity.
If an object has weight, then there is gravity.
∴ If there is an atmosphere, then an object has weight.

In Exercises 51–60, translate the argument into symbolic form. Then determine whether the argument is valid or invalid.

51. If the prescription was called in to Walgreen's, then you can pick it up by 4:00 P.M. You cannot pick it up by 4:00 P.M. Therefore, the prescription was not called in to Walgreen's.

52. The printer has a clogged nozzle or the printer does not have toner. The printer has toner. Therefore, the printer has a clogged nozzle.

53. Max is playing Game Boy with the sound off or Max is wearing headphones. Max is not playing Game Boy with the sound off. Therefore, Max is wearing headphones.

54. If the cat is in the room, then the mice are hiding. The mice are not hiding. Therefore, the cat is not in the room.

55. The test was easy and I received a good grade. The test was not easy or I did not receive a good grade. Therefore, the test was not easy.

56. If Bonnie passes the bar exam, then she will practice law. Bonnie will not practice law. Therefore, Bonnie did not pass the bar exam.

57. The baby is crying but the baby is not hungry. If the baby is hungry then the baby is crying. Therefore, the baby is hungry.

58. If the car is new, then the car has air conditioning. The car is not new and the car has air conditioning. Therefore, the car is not new.

59. If the football team wins the game, then Dave played quarterback. If Dave played quarterback, then the team is not in second place. Therefore, if the football team wins the game, then the team is in second place.

60. The engineering courses are difficult and the chemistry labs are long. If the chemistry labs are long, then the art tests are easy. Therefore, the engineering courses are difficult and the art tests are not easy.

In Exercises 61–67, using the standard forms of arguments and other information you have learned, supply what you believe is a logical conclusion to the argument. Verify that the argument is valid for the conclusion you supplied.

61. If you eat an entire bag of M & M's, then your face will break out.
You eat an entire bag of M & M's.
Therefore, . . .

62. If the temperature hits 100°, then we will go swimming.
We did not go swimming.
Therefore, . . .

63. I am stressed out or I have the flu.
I do not have the flu.
Therefore, . . .

64. If I can get Nick to his piano lesson by 3:30 P.M., then I can do my shopping.
If I can do my shopping, then we do not need to order pizza again.
Therefore, . . .

65. If you close the deal, then you will get a commission.
You did not get a commission.
Therefore, . . .

66. If Katherine is at her ballet lesson, then Allyson will take a nap.
Katherine is at her ballet lesson.
Therefore, . . .

67. If you do not pay off your credit card bill, then you will have to pay interest.
If you have to pay interest, then the bank makes money.
Therefore, . . .

CHALLENGE PROBLEMS/GROUP ACTIVITIES

68. Determine whether the argument is valid or invalid.

If Lynn wins the contest or strikes oil, then she will be rich.
If Lynn is rich, then she will stop working.
∴ If Lynn does not stop working, she did not win the contest.

69. Is it possible for an argument to be invalid if the conjunction of the premises is false in every case of the truth table? Explain your answer.

RECREATIONAL MATHEMATICS

70. René Descartes was a seventeenth-century French mathematician and philosopher. One of his most memorable statements is, "I think, therefore, I am." This statement is the basis for the following joke.

Descartes walks into an inn. The innkeeper asks Descartes if he would like something to drink. Descartes replies, "I think not," and promptly vanishes into thin air!

This joke can be summarized in the following argument: If I think, then I am. I think not. Therefore, I am not.

a) Represent this argument symbolically.

b) Is it a valid argument?

c) Explain your answer using either a standard form of argument or using a truth table.

INTERNET/RESEARCH ACTIVITIES

71. Show how logic is used in advertising. Discuss several advertisements and show how logic is used to persuade the reader.

72. Find examples of valid (or invalid) arguments in printed matter such as newspaper or magazine articles. Explain why the arguments are valid (or invalid).

3.6 EULER DIAGRAMS AND SYLLOGISTIC ARGUMENTS

▲ Adam Brody and Meg Ryan in the movie *In the Land of Women*.

While trying to choose a movie to watch on Valentine's Day, your spouse says to you, "All Meg Ryan movies are romantic comedies. Meg Ryan is in the movie *In the Land of Women*. Therefore, *In the Land of Women* is a romantic comedy." This argument uses a type of argument that we will discuss and analyze in this section.

In Section 3.5, we showed how to determine the validity of *symbolic arguments* using truth tables and comparing the arguments to standard forms. This section presents another form of argument called a *syllogistic argument*, better known by the shorter name *syllogism*. The validity of a syllogistic argument is determined by using Euler (pronounced "oiler") diagrams, as is explained shortly.

Syllogistic logic, a deductive process of arriving at a conclusion, was developed by Aristotle in about 350 B.C. Aristotle considered the relationships among the four types of statements that follow.

All _____ are _____.

No _____ are _____.

Some _____ are _____.

Some _____ are not _____.

Examples of these statements are: *All doctors are tall. No doctors are tall. Some doctors are tall. Some doctors are not tall.* Since Aristotle's time, other types of statements have been added to the study of syllogistic logic, two of which are

_____ is a _____.

_____ is not a _____.

Examples of these statements are: *Maria is a doctor. Maria is not a doctor.*

The difference between a symbolic argument and a syllogistic argument can be seen in the following chart. Symbolic arguments use the connectives *and, or, not, if–then,* and *if and only if.* Syllogistic arguments use the quantifiers *all, some,* and *none,* which were discussed in Section 3.1.

SYMBOLIC ARGUMENTS VERSUS SYLLOGISTIC ARGUMENTS		
	Words or phrases used	*Method of determining validity*
Symbolic argument	and, or, not, if–then, if and only if	Truth tables or by comparison with standard forms of arguments
Syllogistic argument	all are, some are, none are, some are not	Euler diagrams

As with symbolic logic, the premises and the conclusion together form an argument. An example of a syllogistic argument is

All German shepherds are dogs.

All dogs bark.

∴ All German shepherds bark.

This is an example of a valid argument. Recall from Section 3.5 that an argument is *valid* when its conclusion necessarily follows from a given set of premises. Recall that an argument in which the conclusion does not necessarily follow from the given premises is said to be an *invalid argument* or a *fallacy.*

Before we give another example of a syllogism, let's review the Venn diagrams discussed in Section 2.3 in relationship with Aristotle's four statements.

All *A*s are *B*s

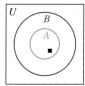

If an element is in set *A*, then it is in set *B*.

No *A*s are *B*s

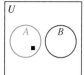

If an element is in set *A*, then it is not in set *B*.

Some *A*s are *B*s

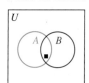

There is at least one element that is in both set *A* and set *B*.

Some *A*s are not *B*s

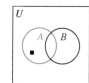

There is at least one element that is in set *A* that is not in set *B*.

One method used to determine whether an argument is valid or is a fallacy is by means of an *Euler diagram*, named after Leonhard Euler, who used circles to represent sets in syllogistic arguments. The technique of using Euler diagrams is illustrated in Example 1.

EXAMPLE ❶ *Using an Euler Diagram*

Determine whether the following syllogism is valid or invalid.

> All keys are made of brass.
> All things made of brass are valuable.
> ∴ All keys are valuable.

Figure 3.3

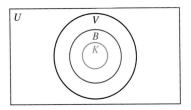

Figure 3.4

SOLUTION To determine whether this syllogism is valid or not valid, we will construct an Euler diagram. We begin with the first premise, "All keys are made of brass." As shown in Fig. 3.3, the inner blue circle labeled *K* represents the set of all keys and the outer red circle labeled *B* represents the set of all brass objects. The first premise requires that the inner blue circle must be entirely contained within the outer red circle. Next, we will represent the second premise, "All things made of brass are valuable." As shown in Fig. 3.4, the outermost black circle labeled *V* represents the set of all valuable objects. The second premise dictates that the red circle, representing the set of brass objects, must be entirely contained within the black circle, representing the set of valuable objects. Now, examine the completed Euler diagram in Fig. 3.4. Note that the premises force the set of keys to be within the set of valuable objects. Therefore, the argument is valid since the conclusion, "All keys are valuable," necessarily follows from the set of premises. ●

The syllogism in Example 1 is valid even though the conclusion, "All keys are valuable," is not a true statement. Similarly, a syllogism can be invalid, or a fallacy, even if the conclusion is a true statement.

When we determine the validity of an argument, we are determining whether the conclusion necessarily follows from the premises. When we say that an argument is valid, we are saying that if all the premises are true statements, then the conclusion must also be a true statement.

The form of the argument determines its validity, not the particular statements. For example, consider the syllogism

> All Earth people have two heads.
> All people with two heads can fly.
> ∴ All Earth people can fly.

The form of this argument is the same as that of the previous valid argument in Example 1. Therefore, this argument is also valid.

EXAMPLE ❷ *Analyzing a Syllogism*

Determine whether the following syllogism is valid or invalid.

> All judges are lawyers.
> William is a judge.
> ∴ William is a lawyer.

Figure 3.5

Figure 3.6

SOLUTION The statement "All judges are lawyers" is illustrated in Fig. 3.5. Note that the *J* circle must be completely inside the *L* circle. The second premise, "William is a judge," tells us that William must be placed in the inner circle labeled *J* (Fig. 3.6). The Euler diagram illustrates that by placing William inside the *J* circle William must also be inside the *L* circle. Therefore, the conclusion "William is a lawyer" necessarily follows from the premises and the argument is valid. ●

In both Example 1 and Example 2, we had no choice as to where the second premise was to be placed in the Euler diagram. In Example 1, the set of brass objects had to be placed inside the set of valuable objects. In Example 2, William had to be placed inside the set of judges. Often when determining the truth value of a syllogism, a premise can be placed in more than one area in the diagram. *We always try to draw the Euler diagram so that the conclusion does not necessarily follow from the premises. If that can be done, then the conclusion does not necessarily follow from the premises and the argument is invalid.* If we cannot show that the argument is invalid, only then do we accept the argument as valid. We illustrate this process in Example 3.

EXAMPLE ❸ *Ballerinas and Athletes*

Determine whether the following syllogism is valid or is invalid.

> All ballerinas are athletic.
> Keyshawn is athletic.
> ∴ Keyshawn is a ballerina.

SOLUTION The premise "All ballerinas are athletic" is illustrated in Fig. 3.7(a). The next premise, "Keyshawn is athletic," tells us that Keyshawn must be placed in the set of athletic people. Two diagrams in which both premises are satisfied are shown in Fig. 3.7(b) and (c). By examining Fig. 3.7(b), however, we see that Keyshawn is not a ballerina. Therefore, the conclusion "Keyshawn is a ballerina" does not necessarily follow from the set of premises. Thus, the argument is invalid, or a fallacy.

　　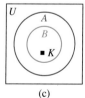

(a)　　　　　(b)　　　　　(c)

Figure 3.7

●

EXAMPLE ❹ *Parrots and Chickens*

Determine whether the following syllogism is valid or invalid.

> No parrots eat chicken.
> Fletch does not eat chicken.
> ∴ Fletch is a parrot.

Figure 3.8

Figure 3.9

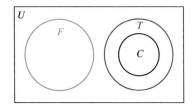

Figure 3.11

SOLUTION The first premise tells us that parrots and things that eat chicken are disjoint sets—that is, sets that do not intersect. The diagram in Fig. 3.8 satisfies the two given premises and also shows that Fletch is not a parrot. Therefore, the argument is invalid, or is a fallacy.   ●

TIMELY TIP Note that in Example 4 if we placed Fletch in circle *P*, the argument would appear to be valid. Remember that *whenever testing the validity of an argument, always try to show that the argument is invalid*. If there is any way of showing that the conclusion does not necessarily follow from the premises, then the argument is invalid.

EXAMPLE ❺ *A Syllogism Involving the Word Some*

Determine whether the following syllogism is valid or invalid.

> All *A*s are *B*s.
> Some *B*s are *C*s.
> ∴ Some *A*s are *C*s.

SOLUTION The premise "All *A*s are *B*s" is illustrated in Fig. 3.9. The premise "Some *B*s are *C*s" means that there is at least one *B* that is a *C*. We can illustrate this set of premises in four ways, as illustrated in Fig. 3.10.

(a)   (b)   (c)   (d)

Figure 3.10

In all four illustrations, we see that (1) all *A*s are *B*s and (2) some *B*s are *C*s. The conclusion is "Some *A*s are *C*s." Since at least one of the illustrations, Fig. 3.10(a), shows that the conclusion does not necessarily follow from the given premises, the argument is invalid.   ●

EXAMPLE ❻ *Frogs and Cats*

Determine whether the following syllogism is valid or invalid.

> No frogs eat tuna.
> All cats eat tuna.
> ∴ No frogs are cats.

SOLUTION The first premise tells us that frogs and the things that eat tuna are disjoint sets as shown in Fig. 3.11. The second premise tells us that the set of cats is a subset of the things that eat tuna. Therefore, the circle representing the set of cats must go within the circle representing the set of the things that eat tuna.

Note that the set of frogs and the set of cats cannot be made to intersect without violating a premise. Thus, the conclusion "No frogs are cats" necessarily follows from the premises and the argument is valid. Note that we did not say that this conclusion is true, only that the argument is valid. ●

SECTION 3.6 EXERCISES

CONCEPT/WRITING EXERCISES

1. **a)** If an Euler diagram can only be drawn in a way in which the conclusion necessarily follows from the premises, what can be said about the syllogistic argument under consideration?

   b) If an Euler diagram can be drawn in a way in which the conclusion does not necessarily follow from the premises, what can be said about the syllogistic argument under consideration?

2. Explain the difference between a symbolic argument and a syllogistic argument.

3. Represent the given statement with two circles as you would when constructing an Euler diagram.

   a) All *A*s are *B*s.

   b) Some *C*s are *D*s.

   c) No *E*s are *F*s.

4. **a)** What does it mean when we determine that an argument is a valid argument?

   b) What does it mean when we determine that an argument is invalid?

5. Can an argument be valid if the conclusion is a false statement? Explain your answer.

6. Can an argument be invalid if the conclusion is a true statement? Explain.

PRACTICE THE SKILLS/PROBLEM SOLVING

In Exercises 7–30, use an Euler diagram to determine whether the syllogism is valid or invalid.

7. All gingerbread men can sing.
   Gingy is a gingerbread man.
   ∴ Gingy can sing.

▲ See Exercise 7

8. All cats have whiskers.
   All things with whiskers are dogs.
   ∴ All cats are dogs.

9. No candy bars are health foods.
   All Snickers are candy bars.
   ∴ No Snickers are health foods.

10. All dolphins are mammals.
    All mammals are vertebrates.
    ∴ All dolphins are vertebrates.

11. All theme parks have walkways.
    Metropolitan Community College has walkways.
    ∴ Metropolitan Community College is a theme park.

12. All golfers have rain gear.
    John Pearse has rain gear.
    ∴ John Pearse is a golfer.

13. No horses buck.
    Palominos are horses.
    ∴ Palominos do not buck.

14. No jockeys weigh more than 200 pounds.
    Deb Otto is not a jockey.
    ∴ Deb Otto weighs more than 200 pounds.

15. Some mushrooms are poisonous.
A morel is a mushroom.

∴ A morel is poisonous.

16. Some policemen are polite.
Jarod Harshbarger is a policeman.

∴ Jarod Harshbarger is not polite.

17. Some farmers are politicians.
Some politicians are senators.

∴ Some farmers are senators.

18. Some professional golfers give golf lessons.
All people who belong to the PGA are professional golfers.

∴ All people who belong to the PGA give golf lessons.

19. No lawn weeds are flowers.
Sedge is not a flower.

∴ Sedge is a lawn weed.

20. Some caterpillars are furry.
All furry things are mammals.

∴ Some caterpillars are mammals.

21. Some flowers love sunlight.
All things that love sunlight love water.

∴ Some flowers love water.

22. Some CD players are MP3 players.
All iPods are MP3 players.

∴ Some CD players are iPods.

23. No scarecrows are tin men.
No tin men are lions.

∴ No scarecrows are lions.

24. All pilots can fly.
All astronauts can fly.

∴ Some pilots are astronauts.

25. Some dogs wear glasses.
Fido wears glasses.

∴ Fido is a dog.

26. All rainy days are cloudy.
Today it is cloudy.

∴ Today is a rainy day.

27. All sweet things taste good.
All things that taste good are fattening.
All things that are fattening put on pounds.

∴ All sweet things put on pounds.

28. All books have red covers.
All books that have red covers contain 200 pages.
Some books that contain 200 pages are novels.

∴ All books that contain 200 pages are novels.

29. All country singers play the guitar.
All country singers play the drums.
Some people who play the guitar are rock singers.

∴ Some country singers are rock singers.

30. Some hot dogs are made of turkey.
All things made of turkey are edible.
Some things that are made of beef are edible.

∴ Some hot dogs are made of beef.

CHALLENGE PROBLEM/GROUP ACTIVITY

31. *Sets and Logic* Statements in logic can be translated into set statements: for example, $p \wedge q$ is similar to $P \cap Q$; $p \vee q$ is similar to $P \cup Q$; and $p \rightarrow q$ is equivalent to $\sim p \vee q$, which is similar to $P' \cup Q$. Euler diagrams can also be used to show that arguments similar to those discussed in Section 3.5 are valid or invalid. Use Euler diagrams to show that the following symbolic argument is invalid.

$$p \rightarrow q$$
$$\underline{p \vee q}$$
$$\therefore \sim p$$

3.7 SWITCHING CIRCUITS

▲ What must be true for a lamp to produce light?

Suppose you are sitting at your desk and want to turn on a lamp so you can read a book. If a wall switch controls the power to the outlet that the lamp is plugged into, and the lamp has an on/off switch, what conditions must be true for the lamp's bulb to light? The wiring in our homes can be explained and described using logic. In this section, we will learn how to analyze electrical circuits using logic.

Using Symbolic Statements to Represent Circuits

A common application of logic is switching circuits. To understand the basic concepts of switching circuits, let us examine a few simple circuits that are common in most homes. The typical lamp has a cord, which is plugged into a wall outlet. Somewhere between the bulb in the lamp and the outlet is a switch to turn the lamp on and off. A switch is often referred to as being on or off. When the switch is in the *on* position, the current flows through the switch and the bulb lights up. When the switch is in the *on* position, we can say that the switch is *closed* and that current will flow through the switch. When the switch is in the *off* position, the current does not flow through the switch and the bulb does not light. When the switch is in the *off* position, we can say that the switch is *open*, and the current does not flow through the switch. The basic configuration of a switch is shown in Fig. 3.12.

Electric circuits can be expressed as logical statements. We represent switches as letters, using T to represent a closed switch (or current flow) and F to represent an open switch (or no current flow). This relationship is indicated in Table 3.33.

Occasionally, we have a wall switch connected to a wall outlet and a lamp plugged into the wall outlet (Fig. 3.13). We then say that the wall switch and the switch on the lamp are in *series,* meaning that for the bulb in the lamp to light, both switches must be on at the same time (Fig. 3.14 on page 170). On the other hand, the bulb will not light if either switch is off or if both switches are off. In either of these conditions, the electricity will not flow through the circuit. In a *series circuit,* the current can take only one path. If any switch in the path is open, the current cannot flow.

To illustrate this situation symbolically, let *p* represent the wall switch and *q* the lamp switch. The letter T will be used to represent both a closed switch and the bulb lighting. The letter F will represent an open switch and the bulb not lighting. Thus, we have four possible cases.

Figure 3.12

Table 3.33

Switch	Lightbulb
T	on (switch closed)
F	off (switch open)

Figure 3.13

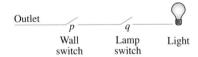

Figure 3.14

Table 3.34

p	q	Light	p ∧ q
T	T	on (T)	T
T	F	off (F)	F
F	T	off (F)	F
F	F	off (F)	F

Figure 3.15

Table 3.35

p	q	Light	p ∨ q
T	T	on (T)	T
T	F	on (T)	T
F	T	on (T)	T
F	F	off (F)	F

Figure 3.16

CASE 1: Both switches are closed; that is, *p* is T and *q* is T. The light is on, T.

CASE 2: Switch *p* is closed and switch *q* is open; that is, *p* is T and *q* is F. The light is off, F.

CASE 3: Switch *p* is open and switch *q* is closed; that is, *p* is F and *q* is T. The light is off, F.

CASE 4: Both switches are open; that is, *p* is F and *q* is F. The light is off, F.

Table 3.34 summarizes the results.

The on–off results are the same as the truth table for the conjunction *p* ∧ *q* if we think of "on" as true and "off" as false.

> Switches in series will always be represented with a conjunction, ∧.

Another type of electric circuit used in the home is the *parallel circuit*, in which there are two or more paths that the current can take. If the current can pass through either path or both (see Fig. 3.15), the light will go on. The letter T will be used to represent both a closed switch and the bulb lighting. The letter F will represent an open switch and the bulb not lighting. Thus, we have four possible cases.

CASE 1: Both switches are closed; that is, *p* is T and *q* is T. The light is on, T.

CASE 2: Switch *p* is closed and switch *q* is open; that is, *p* is T and *q* is F. The light is on, T.

CASE 3: Switch *p* is open and switch *q* is closed; that is, *p* is F and *q* is T. The light is on, T.

CASE 4: Both switches are open; that is, *p* is F and *q* is F. The light is off, F.

Table 3.35 summarizes the results. The on–off results are the same as the *p* ∨ *q* truth table if we think of "on" as true and "off" as false.

> Switches in parallel will always be represented with a disjunction, ∨.

Sometimes it is necessary to have two or more switches in the same circuit that will both be open at the same time and both be closed at the same time. In such circuits, we will use the same letter to represent both switches. For example, in the circuit shown in Fig. 3.16, there are two switches labeled *p*. Therefore, both of these switches must be open at the same time and both must be closed at the same time. One of the *p* switches cannot be open at the same time the other *p* switch is closed.

We can now combine some of these basic concepts to analyze more circuits.

┌EXAMPLE ❶ *Representing a Switching Circuit with Symbolic Statements*

a) Write a symbolic statement that represents the circuit shown in Fig. 3.16.

b) Construct a truth table to determine when the light will be on.

Figure 3.17

Table 3.36

p	q	p	∧	(p ∨ q)
T	T	T	T	T
T	F	T	T	T
F	T	F	F	T
F	F	F	F	F
		1	3	2

SOLUTION

a) In Fig. 3.16, there is a blue switch p on the left, and to the right there is a branch containing a red switch p and a switch q. The current must flow through the blue switch p into the branch on its right. Therefore, the blue switch p is in series with the branch containing the red switch p and switch q. After the current flows through the blue switch p it reaches the branch on its right. At this point, the current has the option of flowing into the red switch p, switch q, or both of these switches. Therefore, the branch containing the red switch p and switch q is a parallel branch. We say these two switches are in parallel. The entire circuit in symbolic form is $p \wedge (p \vee q)$. Note that the parentheses are very important. Without parentheses, the symbolic statement could be interpreted as $(p \wedge p) \vee q$. The diagram for $(p \wedge p) \vee q$ is illustrated in Fig. 3.17. We shall see later in this section that the circuits in Fig. 3.16 and Fig. 3.17 are not the same.

b) The truth table for the statement (Table 3.36) indicates that the light will be on only in the cases in which p is true or when switch p is closed.

EXAMPLE ❷ *Representing a Switching Circuit with Symbolic Statements*

a) Write a symbolic statement that represents the circuit in Fig. 3.18.
b) Construct a truth table to determine when the light will be on.

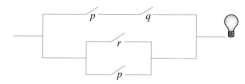

Figure 3.18

Table 3.37

p	q	r	(p ∧ q)	∨	(r ∨ p)
T	T	T	T	T	T
T	T	F	T	T	T
T	F	T	F	T	T
T	F	F	F	T	T
F	T	T	F	T	T
F	T	F	F	F	F
F	F	T	F	T	T
F	F	F	F	F	F
			1	3	2

SOLUTION

a) The upper branch of the circuit contains two switches, p and q, in series. We can represent that branch with the statement $p \wedge q$. The lower branch of the circuit has two switches, r and p, in parallel. We can represent that branch with the statement $r \vee p$. The upper branch is in parallel with the lower branch. Putting the two branches together, we get the statement $(p \wedge q) \vee (r \vee p)$.

b) The truth table for the statement (Table 3.37) shows that cases 6 and 8 are false. Thus, the light will be off in these cases and on in all the other cases. In examining the truth values, we see that the statement is false only in the cases in which both p and r are false or when both switches are open. By examining the diagram or truth table, we can see that the current will flow if switch p is closed or switch r is closed.

Drawing Switching Circuits That Represent Symbolic Statements

We will next study how to draw a switching circuit that represents a given symbolic statement. Suppose we are given the statement $(p \wedge q) \vee r$ and are asked to construct a circuit corresponding to it. Remember that $\wedge$ indicates a series branch

Figure 3.19

Figure 3.20

Figure 3.21

Figure 3.22

and ∨ indicates a parallel branch. Working first within parentheses, we see that switches *p* and *q* are in series. This series branch is in parallel with switch *r*, as indicated in Fig. 3.19.

Occasionally, it is necessary to have two switches in the same circuit such that when one switch is open, the other switch is closed; and when one switch is closed, the other switch is open. Therefore, the two switches will never both be open together and never be closed together. This situation can be represented by using *p* for one of the switches and $\overline{p}$ for the other switch. The switch labeled $\overline{p}$ corresponds to ~*p* in a logic statement. For example, in a series circuit, *p* ∧ ~*p* would be represented by Fig. 3.20. In this case, the light would never go on; the switches would counteract each other. When switch *p* is closed, switch $\overline{p}$ is open; and when *p* is open, $\overline{p}$ is closed.

EXAMPLE ❸ *Representing a Symbolic Statement as a Switching Circuit*

Draw a switching circuit that represents [(*p* ∧ ~*q*) ∨ (*r* ∨ *q*)] ∧ *s*.

SOLUTION In the statement, *p* and ~*q* have ∧ between them, so switches *p* and $\overline{q}$ are in series, as represented in Fig. 3.21. Also in the statement, *r* and *q* have ∨ between them, so switches *r* and *q* are in parallel, as represented in Fig. 3.22.

Because (*p* ∧ ~*q*) and (*r* ∨ *q*) are connected with ∨, the two branches are in parallel with each other. The parallel branches are represented in Fig. 3.23. Finally, *s* in the statement is connected to the rest of the statement with ∧. Therefore, switch *s* is in series with the entire rest of the circuit, as illustrated in Fig. 3.24.

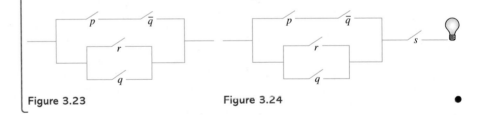

Figure 3.23   **Figure 3.24**   ●

Equivalent Circuits

Sometimes two circuits that look very different will actually have the exact same conditions under which the light will be on. If we were to analyze the truth tables for the corresponding symbolic statements for such circuits, we would find that they have identical answer columns. In other words, the corresponding symbolic statements are equivalent.

Equivalent circuits are two circuits that have equivalent corresponding symbolic statements.

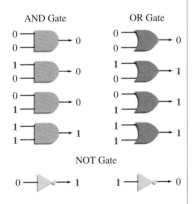
To determine whether two circuits are equivalent, we will analyze the answer columns of the truth tables of their corresponding symbolic statements. If the answer columns from their corresponding symbolic statements are identical, then the circuits are equivalent.

EXAMPLE ❹ *Are the Circuits Equivalent?*

Determine whether the two circuits are equivalent.

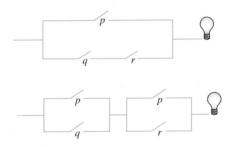

SOLUTION The symbolic statement that represents the first circuit is $p \lor (q \land r)$. The symbolic statement that represents the second circuit is $(p \lor q) \land (p \lor r)$. The truth tables for these statements are shown in Table 3.38.

Table 3.38

p	q	r	p	$\lor$	$(q \land r)$	$(p \lor q)$	$\land$	$(p \lor r)$
T	T	T	T	T	T	T	T	T
T	T	F	T	T	F	T	T	T
T	F	T	T	T	F	T	T	T
T	F	F	T	T	F	T	T	T
F	T	T	F	T	T	T	T	T
F	T	F	F	F	F	T	F	F
F	F	T	F	F	F	F	F	T
F	F	F	F	F	F	F	F	F

Note that the answer columns for the two statements are identical. Therefore, $p \lor (q \land r)$ is equivalent to $(p \lor q) \land (p \lor r)$ and the two circuits are equivalent.

SECTION 3.7   EXERCISES

CONCEPT/WRITING EXERCISES

1. a) In your own words, describe a series circuit.

   b) Which type of logical connective can be used to represent a series circuit?

2. a) In your own words, describe a parallel circuit.

   b) Which type of logical connective can be used to represent a parallel circuit?

3. Explain why the lightbulb will never go on in the following circuit.

4. Explain why the lightbulb will always be on in the following circuit.

PRACTICE THE SKILLS/PROBLEM SOLVING

In Exercises 5–12, (a) write a symbolic statement that represents the circuit and (b) construct a truth table to determine when the lightbulb will be on. That is, determine which switches must be open and which switches must be closed for the lightbulb to be on.

5.

6.

7.

8.

9.

10.

11.

12.

In Exercises 13–20, draw a switching circuit that represents the symbolic statement.

13. $p \vee q$ 14. $p \vee (q \wedge r)$

15. $(p \vee q) \wedge r$ 16. $\sim p \wedge (q \vee r)$

17. $(p \vee q) \wedge (r \vee s)$ 18. $(p \wedge q) \vee (r \wedge s)$

19. $[(p \vee q) \vee (r \wedge q)] \wedge (\sim p)$

20. $[(p \vee q) \wedge r] \vee (\sim p \wedge q)$

In Exercises 21–26, represent each circuit with a symbolic statement. Then use a truth table to determine if the circuits are equivalent.

21.

22.

23.

24.

25.

26.

CHALLENGE PROBLEMS/GROUP ACTIVITIES

27. Design a circuit that can be represented by

  a) $p \rightarrow q$

  b) $\sim(p \rightarrow q)$

  (*Hint:* See Section 3.4.)

28. Design two circuits that each involve switches labeled p, q, r, and s that appear to be different but are actually equivalent circuits.

INTERNET/RESEARCH ACTIVITIES

29. Digital computers use gates that work like switches to perform calculations. (See Did You Know? on page 173.) Information is fed into the gates and information leaves the gates, according to the type of gate. The three basic gates used in computers are the NOT gate, the AND gate, and the OR gate. Do research on the three types of gates.

  a) Explain how each gate works.

  b) Explain the relationship between each gate and the corresponding logic connectives *not, and,* and *or.*

  c) Illustrate how two or more gates can be combined to form a more complex gate.

CHAPTER ❸ SUMMARY

Important Facts

Quantifiers

Form of Statement	Form of Negation
All are.	Some are not.
None are.	Some are.
Some are.	None are.
Some are not.	All are.

Summary of connectives

Formal Name	Symbol	Read	Symbolic Form
Negation	$\sim$	not	$\sim p$
Conjunction	$\wedge$	and	$p \wedge q$
Disjunction	$\vee$	or	$p \vee q$
Conditional	$\rightarrow$	if–then	$p \rightarrow q$
Biconditional	$\leftrightarrow$	if and only if	$p \leftrightarrow q$

Basic truth tables

	Negation			Conjunction	Disjunction	Conditional	Biconditional
p	$\sim p$	p	q	$p \wedge q$	$p \vee q$	$p \rightarrow q$	$p \leftrightarrow q$
T	F	T	T	T	T	T	T
F	T	T	F	F	T	F	F
		F	T	F	T	T	F
		F	F	F	F	T	T

De Morgan's laws

$$\sim(p \wedge q) \Leftrightarrow \sim p \vee \sim q$$
$$\sim(p \vee q) \Leftrightarrow \sim p \wedge \sim q$$

Other equivalent forms

$$p \rightarrow q \Leftrightarrow \sim p \vee q$$
$$\sim(p \rightarrow q) \Leftrightarrow p \wedge \sim q$$
$$p \leftrightarrow q \Leftrightarrow [(p \rightarrow q) \wedge (q \rightarrow p)]$$

Variations of the conditional statement

Name	Symbolic Form	Read
Conditional	$p \rightarrow q$	If p, then q.
Converse of the conditional	$q \rightarrow p$	If q, then p.
Inverse of the conditional	$\sim p \rightarrow \sim q$	If not p, then not q.
Contrapositive of the conditional	$\sim q \rightarrow \sim p$	If not q, then not p.

Standard forms of arguments
Valid arguments

Law of Detachment	Law of Contra-position	Law of Syllogism	Disjunctive Syllogism
$p \rightarrow q$	$p \rightarrow q$	$p \rightarrow q$	$p \vee q$
p	$\sim q$	$q \rightarrow r$	$\sim p$
$\therefore q$	$\therefore \sim p$	$\therefore p \rightarrow r$	$\therefore q$

Invalid arguments

Fallacy of the Converse	Fallacy of the Inverse
$p \rightarrow q$	$p \rightarrow q$
q	$\sim p$
$\therefore p$	$\therefore \sim q$

Symbolic argument vs. syllogistic argument

	Words or Phrases Used	Method of Determining Validity
Symbolic argument	and, or, not, if–then, if and only if	Truth tables or by comparison with standard forms of arguments
Syllogistic argument	all are, some are, none are, some are not	Euler diagrams

Switching Circuits as Symbolic Statements

Switches in series will always be represented with a conjunction, $\wedge$.

Switches in parallel will always be represented with a disjunction, $\vee$.

CHAPTER ③ REVIEW EXERCISES

3.1

In Exercises 1–4, write the negation of the statement.

1. No gift cards are exchangeable.

2. All bears are mammals.

3. Some women are presidents.

▲ Ellen Johnson-Sirleaf, President of Liberia

4. Some pine trees are not green.

In Exercises 5–10, write each compound statement in words.

> p: The coffee is Maxwell House.
> q: The coffee is hot.
> r: The coffee is strong.

5. $p \vee q$

6. $\sim q \wedge r$

7. $q \rightarrow (r \wedge \sim p)$

8. $p \leftrightarrow \sim r$

9. $\sim p \leftrightarrow (r \wedge \sim q)$

10. $(p \vee \sim q) \wedge \sim r$

3.2

In Exercises 11–16, use the statements for p, q, and r as in Exercises 5–10 to write the statement in symbolic form.

11. The coffee is strong and the coffee is hot.

12. If the coffee is Maxwell House, then the coffee is strong.

13. If the coffee is strong then the coffee is hot, or the coffee is not Maxwell House.

14. The coffee is hot if and only if the coffee is Maxwell House, and the coffee is not strong.

15. The coffee is strong and the coffee is hot, or the coffee is not Maxwell House.

16. It is false that the coffee is strong and the coffee is hot.

In Exercises 17–22, construct a truth table for the statement.

17. $(p \vee q) \wedge \sim p$

18. $q \leftrightarrow (p \vee \sim q)$

19. $(p \vee q) \leftrightarrow (p \vee r)$

20. $p \wedge (\sim q \vee r)$

21. $p \rightarrow (q \wedge \sim r)$

22. $(p \wedge q) \rightarrow \sim r$

3.2, 3.3

In Exercises 23–26, determine the truth value of the statement.

23. If the IRS collects taxes or the Postal Service delivers mail, then the FBI investigates crop damage.

24. $17 + 4 = 21$, and $\dfrac{15}{5} = -7$ or $3 - 9 = -6$.

25. If Oregon borders the Pacific Ocean or California borders the Atlantic Ocean, then Minnesota is south of Texas.

26. $15 - 7 = 22$ or $4 + 9 = 13$, and $9 - 8 = 1$.

3.3

In Exercises 27–30, determine the truth value of the statement when p is T, q is F, and r is F.

27. $(p \rightarrow \sim r) \vee (p \wedge q)$

28. $(p \vee q) \leftrightarrow (\sim r \wedge p)$

29. $\sim r \leftrightarrow [(p \vee q) \leftrightarrow \sim p]$

30. $\sim[(q \wedge r) \rightarrow (\sim p \vee r)]$

3.4

In Exercises 31–34, determine whether the pairs of statements are equivalent. You may use De Morgan's laws, the fact that $(p \rightarrow q) \Leftrightarrow (\sim p \vee q)$, the fact that $\sim(p \rightarrow q) \Leftrightarrow (p \wedge \sim q)$, truth tables, or equivalent forms of the conditional statement.

31. $\sim p \vee \sim q, \qquad \sim p \leftrightarrow q$

32. $\sim p \rightarrow \sim q, \qquad p \vee \sim q$

33. $\sim p \vee (q \wedge r), \quad (\sim p \vee q) \wedge (\sim p \vee r)$

34. $(\sim q \rightarrow p) \wedge p, \quad \sim(\sim p \leftrightarrow q) \vee p$

In Exercises 35–39, use De Morgan's laws, the fact that $(p \rightarrow q) \Leftrightarrow (\sim p \vee q)$, or the fact that $\sim(p \rightarrow q) \Leftrightarrow (p \wedge \sim q)$, to write an equivalent statement for the given statement.

35. Bobby Darin sang *Mack the Knife* and Elvis did not write *Memphis*.

▲ Bobby Darin and Sandra Dee

36. Lynn Swann played for the Steelers or Jack Tatum played for the Raiders.

37. It is not true that Altec Lansing only produces speakers or Harman Kardon only produces stereo receivers.

38. Travis Tritt did not win an Academy Award and Randy Jackson does not do commercials for Milk Bone Dog Biscuits.

39. If the temperature is not above 32°, then we will go ice fishing at O'Leary's Lake.

In Exercises 40–44, write the (a) converse, (b) inverse, and (c) contrapositive for the given statement.

40. If you hear a new voice today, then you soften your opinion.

41. If we take the table to *Antiques Roadshow,* then we will learn the table's value.

42. If Maureen Gerald is not in attendance, then she is helping at the school.

43. If the desk is made by Winner's Only and the desk is in the Rose catalog, then we will not buy a desk at Miller's Furniture.

44. If you get straight A's on your report card, then I will let you attend the prom.

In Exercises 45–48, determine which, if any, of the three statements are equivalent.

45. a) If you read 10 books in the summer, then you reach your goal.

b) You do not read 10 books in the summer or you reach your goal.

c) It is false that you read 10 books in the summer and you do not reach your goal.

46. a) The screwdriver is on the workbench if and only if the screwdriver is not on the counter.

b) If the screwdriver is not on the counter, then the screwdriver is not on the workbench.

c) It is false that the screwdriver is on the counter and the screwdriver is not on the workbench.

47. a) If $2 + 3 = 6$, then $3 + 1 = 5$.

b) $2 + 3 = 6$ if and only if $3 + 1 \neq 5$.

c) If $3 + 1 \neq 5$, then $2 + 3 \neq 6$.

48. a) If the sale is on Tuesday and I have money, then I will go to the sale.

b) If I go to the sale, then the sale is on Tuesday and I have money.

c) I go to the sale, or the sale is on Tuesday and I have money.

3.5

In Exercises 49–52, determine whether the argument is valid or invalid. You may compare the argument to a standard form or use a truth table.

49. $p \rightarrow q$
$\underline{\sim p}$
$\therefore q$

50. $p \wedge q$
$\underline{q \rightarrow r}$
$\therefore p \rightarrow r$

51. If Jose Macias is the manager, then Kevin Geis is the coach.
If Kevin Geis is the coach, then Tim Weisman is the umpire.
$\therefore$ If Jose Macias is the manager, then Tim Weisman is the umpire.

52. If we eat at Joe's, then we will get heartburn. We get heartburn or we take Tums. Therefore, we do not get heartburn.

3.6

In Exercises 53–56, use an Euler diagram to determine whether the argument is valid or invalid.

53. All plumbers wear overalls.
Some electricians wear overalls.
$\therefore$ Some electricians are plumbers.

54. Some submarines are yellow.
All dandelions are yellow.
$\therefore$ Some dandelions are submarines.

55. No Beatles are Rolling Stones.
Some Rolling Stones are Small Faces.
$\therefore$ No Beatles are Small Faces.

56. All bears are furry.
Teddy is furry.
$\therefore$ Teddy is a bear.

3.7

57. a) Write the corresponding symbolic statement of the circuit shown.

b) Construct a truth table to determine when the lightbulb will be on.

58. Construct a diagram of a circuit that corresponds to the symbolic statement $(p \vee q) \vee (p \wedge q)$.

59. Determine whether the circuits shown are equivalent.

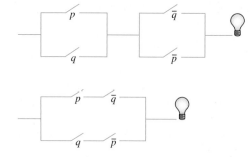

CHAPTER ③ TEST

In Exercises 1–3, write the statement in symbolic form.

 p: Ann is the secretary.

 q: Dick is the vice president.

 r: Elaine is the president.

1. Ann is the secretary but Elaine is the president, or Dick is not the vice president.

2. If Elaine is the president then Dick is the vice president, or Ann is not the secretary.

3. It is false that Elaine is the president if and only if Dick is not the vice president.

In Exercises 4 and 5, use p, q, and r as above to write each symbolic statement in words.

4. $(\sim p \wedge r) \leftrightarrow \sim q$ 5. $(p \vee \sim q) \rightarrow r$

In Exercises 6 and 7, construct a truth table for the given statement.

6. $[\sim(p \rightarrow r)] \wedge q$ 7. $(q \leftrightarrow \sim r) \vee p$

In Exercises 8 and 9, find the truth value of the statement.

8. $2 + 6 = 8$ or $7 - 12 = 5$.

9. Harrison Ford is an actor or Gerald Ford was a president, if and only if the Mississippi is a river.

In Exercises 10 and 11, given that p is true, q is false, and r is true, determine the truth value of the statement.

10. $(r \vee q) \leftrightarrow (p \wedge \sim q)$

11. $[\sim(r \rightarrow \sim p)] \wedge (q \rightarrow p)$

12. Determine whether the pair of statements are equivalent.

$$\sim p \vee q, \qquad \sim(p \wedge \sim q)$$

In Exercises 13 and 14, determine which, if any, of the three statements are equivalent.

13. a) If the bird is red, then it is a cardinal.

 b) The bird is not red or it is a cardinal.

 c) If the bird is not red, then it is not a cardinal.

14. a) It is not true that the test is today or the concert is tonight.

 b) The test is not today and the concert is not tonight.

 c) If the test is not today, then the concert is not tonight.

15. *Translate the following argument into symbolic form. Determine whether the argument is valid or invalid by comparing the argument to a recognized form or by using a truth table.*

 If the soccer team wins the game, then Sue played fullback. If Sue played fullback, then the team is in second place. Therefore, if the soccer team wins the game, then the team is in second place.

16. *Use an Euler diagram to determine whether the syllogism is valid or is a fallacy.*

 All living things contain carbon.
 Roger contains carbon.

 ∴ Therefore, Roger is a living thing.

In Exercises 17 and 18, write the negation of the statement.

17. All highways are roads.

18. Nick played football and Max played baseball.

19. Write the converse, inverse, and contrapositive of the conditional statement, "If the garbage truck comes, then today is Saturday."

20. Construct a diagram of a circuit that corresponds to $(p \wedge q) \vee (\sim p \vee \sim q)$

G R O U P   P R O J E C T S

COMPUTER GATES

1. Gates in computers work on the same principles as switching circuits. The three basic types of gate are the NOT gate, the AND gate, and the OR gate. Each is illustrated along with a table that indicates current flow entering and exiting the gate. If current flows into a NOT gate, then no current exits, and vice versa. Current exits an AND gate only when both inputs have a current flow. Current exits an OR gate if current flows through either, or both, inputs. In the table, a 1 represents a current flow and a 0 indicates no current flow. For example, in the AND gate, if there is a current flow in input A (I_a has a value of 1) and no current flow in input B (I_b has a value of 0), there is no current flow in the output (O has a value of 0); see row 2 of the AND Gate table.

NOT gate

Input ──▷○── Output

NOT gate

I	O
1	0
0	1

AND gate

Input A ──╮
Input B ──╯ ──── Output

AND gate

I_a	I_b	O
1	1	1
1	0	0
0	1	0
0	0	0

OR gate

Input A ──╮
Input B ──╯ ──── Output

OR gate

I_a	I_b	O
1	1	1
1	0	1
0	1	1
0	0	0

a) If 1 is considered true and 0 is considered false, explain how these tables are similar to the *not, and,* and *or* truth tables.

    For the inputs indicated in the following figures determine whether the output is 1 or 0.

b)
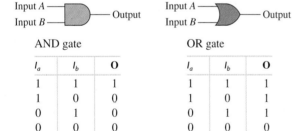
I_a 1, I_b 0 ── Output

c)
I_a 1, I_b 0 ── Output

d)
I_a 1, I_b 1 ── Output

e) What values for I_a and I_b will give an output of 1 in the figure in part (d)? Explain how you determined your answer.

f) Construct a truth table using 1's and 0's for the following gate. Your truth table should have columns I_a, I_b, and O and should indicate the four possible cases for the inputs and each corresponding output.

I_a ──▷○──╮
I_b ──────╯ ──── Output

LOGIC GAME

2. a) Shown is a photograph of a logic game at the Ontario Science Centre. There are 12 balls on top of the game board, numbered from left to right, with ball 1 on the extreme left and ball 12 on the extreme right. On the platform in front of the players are 12 buttons. Button 1 corresponds to ball 1, button 2 corresponds to ball 2, and so on. When 6 buttons are pushed, the 6 respective balls are released. When 1 or 2 balls reach an *and* gate or an *or* gate, a single ball may or may not pass through the gate. The object of the game is to select a proper combination of 6 buttons that will allow 1 ball to reach the bottom. Using your knowledge of *and* and *or*, select a combination of 6 buttons that will result in a win. (There is more than one answer.) Explain how you determined your answer.

b) Construct a game similar to this one where 15 balls are at the top and 8 balls must be selected to allow 1 ball to reach the bottom.

c) Indicate all solutions to the game you constructed in part (c).

Systems of Numeration

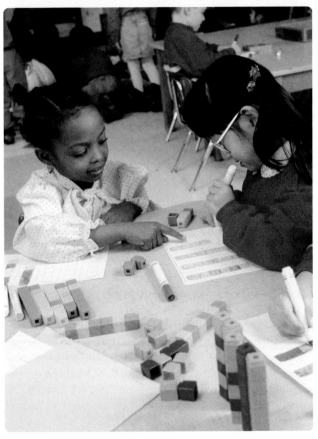

▲ There are many ways that we can communicate numerically.

WHAT YOU WILL LEARN

- Additive, multiplicative, and ciphered systems of numeration
- Place-value systems of numeration
- Egyptian, Hindu–Arabic, Roman, Chinese, Ionic Greek, Babylonian, and Mayan numerals
- Converting base 10 numerals to numerals in other bases
- Converting numerals in other bases to base 10 numerals
- Performing addition, subtraction, multiplication, and division in other bases
- Other computational methods such as duplation and mediation, the lattice method, and Napier's rods

WHY IT IS IMPORTANT

The number system most of the world uses today, called the Hindu–Arabic system, is only one way to communicate numerically. In this chapter, we will study other systems of numeration. We will also study how to perform basic arithmetic in other bases and with methods other than those most of us were taught when we were children.

4.1 ADDITIVE, MULTIPLICATIVE, AND CIPHERED SYSTEMS OF NUMERATION

▲ Hines Ward in front of the 2006 Super Bowl XL Logo

We are all familiar with the numerals we grew up with: 1, 2, 3, and so forth. Now think of another set of numerals. You probably thought of Roman numerals: I, II, III, and so forth. Some common modern-day uses of Roman numerals include clock and watch numbers, the copyright year on movie and television credits, outlines, and page numbers at the beginning of books. In fact, if you look at the pages in this textbook that precede page 1, you will see Roman numerals in use. Perhaps an even more famous use of Roman numerals is their use in the logo for the Super Bowl each year. In this section, we will study several sets of numerals besides our familiar 1, 2, 3, ...

Just as the first attempts to write were made long after the development of speech, the first representation of numbers by symbols came long after people had learned to count. A tally system using physical objects, such as scratch marks in the soil or on a stone, notches on a stick, pebbles, or knots on a vine, was probably the earliest method of recording numbers.

In primitive societies, such a tally system adequately served the limited need for recording livestock, agriculture, or whatever was counted. As civilization developed, however, more efficient and accurate methods of calculating and keeping records were needed. Because tally systems are impractical and inefficient, societies developed symbols to replace them. The Egyptians, for example, used the symbol ∩, and the Babylonians used the symbol ≺ to represent the number we symbolize by 10.

A *number* is a quantity, and it answers the question "How many?" A *numeral* is a symbol such as ∩, ≺, or 10 used to represent the number. We think a number but write a numeral. The distinction between number and numeral will be made here only if it is helpful to the discussion.

In language, relatively few letters of the alphabet are used to construct a large number of words. Similarly, in arithmetic, a small variety of numerals can be used to represent all numbers. In general, when representing a number, we use as few numerals as possible. One of the greatest accomplishments of humankind has been the development of systems of numeration, whereby all numbers are "created" from a few symbols. Without such systems, mathematics would not have developed to its present level.

> A **system of numeration** consists of a set of numerals and a scheme or rule for combining the numerals to represent numbers.

Four types of numeration systems used by different cultures are the topic of this chapter. They are additive (or repetitive), multiplicative, ciphered, and place-value systems. You do not need to memorize all the symbols, but you should understand the principles behind each system. By the end of this chapter, we hope that you better understand the system we use, the *Hindu–Arabic system*, and its relationship to other types of systems.

Additive Systems

An additive system is one in which the number represented by a particular set of numerals is simply the sum of the values of the numerals. The additive system of numeration is one of the oldest and most primitive types of numeration systems. One of the first additive systems, the Egyptian hieroglyphic system, dates back to about 3000 B.C. The Egyptians used symbols for the powers of 10: 10^0 or 1, 10^1 or 10, 10^2 or $10 \cdot 10$, 10^3 or $10 \cdot 10 \cdot 10$, and so on. Table 4.1 lists the Egyptian hieroglyphic numerals with the equivalent Hindu–Arabic numerals.

To write the number 600 in Egyptian hieroglyphics, we write the numeral for 100 six times: 999999.

Table 4.1 Egyptian Hieroglyphics

Hindu–Arabic Numerals	Egyptian Numerals	Description
1	I	Staff (vertical stroke)
10	∩	Heel bone (arch)
100	9	Scroll (coiled rope)
1000	⌇	Lotus flower
10,000	⌐	Pointing finger
100,000	⌒	Tadpole (or whale)
1,000,000	⚡	Astonished person

EXAMPLE ❶ *From Egyptian to Hindu–Arabic Numerals*

Write the following numeral as a Hindu–Arabic numeral.

ﻻﻻﻻ999∩IIII

SOLUTION

$$10{,}000 + 10{,}000 + 10{,}000 + 100 + 100 + 100 + 10 + 1 + 1 + 1 + 1$$
$$= 30{,}314$$ ●

EXAMPLE ❷ *From Hindu–Arabic to Egyptian Numerals*

Write 43,628 as an Egyptian numeral.

SOLUTION

$$43{,}628 = 40{,}000 + 3000 + 600 + 20 + 8$$

ﻻﻻﻻﻻ⌇⌇⌇999999∩∩IIIIIIII ●

In the Egyptian hieroglyphic system, the order of the symbols is not important. For example, ⌒ II / 99 ∩ and II99⌒∩ both represent 100,212.

Users of additive systems easily accomplished addition and subtraction by combining or removing symbols. Multiplication and division were more difficult; they were performed by a process called *duplation and mediation* (see Section 4.5). The Egyptians had no symbol for zero, but they did have an understanding of fractions. The symbol ◡ was used to take the reciprocal of a number; thus, ⌢ meant $\frac{1}{3}$ and ⌢

was $\frac{1}{11}$. Writing large numbers in the Egyptian hieroglyphic system takes longer than in other systems because so many symbols have to be listed. For example, 45 symbols are needed to represent 99,999.

The Roman numeration system, a second example of an additive system, was developed later than the Egyptian system. Roman numerals (Table 4.2) were used in most European countries until the eighteenth century. They are still commonly seen on buildings, on clocks, and in books. Roman numerals are selected letters of the Roman alphabet.

Table 4.2 Roman Numerals

Roman numerals	I	V	X	L	C	D	M
Hindu–Arabic numerals	1	5	10	50	100	500	1000

The Roman system has two advantages over the Egyptian system. The first is that it uses the subtraction principle as well as the addition principle. Starting from the left, we add each numeral unless its value is smaller than the value of the numeral to its right. In that case, we subtract its value from the value of the numeral to its right. Only the numbers 1, 10, 100, 1000, ... can be subtracted, and they can only be subtracted from the next two higher numbers. For example, C (100) can be subtracted only from D (500) or M (1000). The symbol DC represents 500 + 100, or 600, and CD represents 500 − 100, or 400. Similarly, MC represents 1000 + 100, or 1100, and CM represents 1000 − 100, or 900.

EXAMPLE 3 *From Roman Numerals to Hindu–Arabic Numerals*

Write MMCCCLXII as a Hindu–Arabic numeral.

SOLUTION Since each numeral is larger than the one on its right, no subtraction is necessary.

$$MMCCCLXII = 1000 + 1000 + 100 + 100 + 100 + 50 + 10 + 1 + 1$$
$$= 2362$$

EXAMPLE 4 *From Roman Numerals to Hindu–Arabic Numerals*

Write CMLXIV as a Hindu–Arabic numeral.

SOLUTION As you read the numerals from left to right, you will note that C (100) has a smaller value than M (1000) and that C is to the left of M. Therefore, CM represents 1000 − 100, or 900. Also note that I (1) has a smaller value than V (5) and that I is to the left of V. Therefore, IV represents 5 − 1, or 4. The rest of the numerals can be added from left to right.

$$CMLXIV = (1000 − 100) + 50 + 10 + (5 − 1) = 964$$

EXAMPLE 5 *Writing a Roman Numeral*

Write 289 as a Roman numeral.

SOLUTION

$$289 = 200 + 80 + 9 = 100 + 100 + 50 + 10 + 10 + 10 + 9$$

(Nine is treated as 10 − 1.)

$$289 = CCLXXXIX$$

In the Roman numeration system, a symbol is not repeated more than three consecutive times. For example, the number 646 would be written DCXLVI instead of DCXXXXVI.

The second advantage of the Roman numeration system over the Egyptian numeration system is that it makes use of the multiplication principle for numerals greater than 1000. A bar above a symbol or group of symbols indicates that the symbol or symbols are to be multiplied by 1000. Thus, $\overline{V} = 5 \times 1000 = 5000$, $\overline{X} = 10 \times 1000 = 10,000$, and $\overline{CD} = 400 \times 1000 = 400,000$. Other examples are $\overline{VI} = 6 \times 1000 = 6000$, $\overline{XIX} = 19 \times 1000 = 19,000$, and $\overline{XCIV} = 94 \times 1000 = 94,000$. This practice greatly reduces the number of symbols needed to write large numbers.

EXAMPLE ❻ *Writing a Large Roman Numeral*

Write 12,345 as a Roman numeral.

SOLUTION

$$12,345 = (12 \times 1000) + 300 + 40 + 5$$
$$= [(10 + 1 + 1) \times 1000] + (100 + 100 + 100) + (50 - 10) + 5$$
$$= \overline{XII}CCCXLV$$

Multiplicative Systems

Multiplicative numeration systems are more similar to our Hindu–Arabic system than are additive systems. In a multiplicative system, 642 might be written (6) (100) (4) (10) (2) or

$$6$$
$$100$$
$$4$$
$$10$$
$$2$$

Note that no addition signs are needed in the representation. From this illustration, try to formulate a rule explaining how multiplicative systems work.

The principle example of a multiplicative system is the traditional Chinese system. The numerals used in this system are given in Table 4.3.

Table 4.3 Traditional Chinese Numerals

Traditional Chinese numerals	零	一	二	三	四	五	六	七	八	九	十	百	千
Hindu–Arabic numerals	0	1	2	3	4	5	6	7	8	9	10	100	1000

Chinese numerals are always written vertically. The numeral on top will be from 1 to 9 inclusive. This numeral is to be multiplied by the power of 10 below it. Thus, 20 is written

$$\left.\begin{matrix} 二 \\ 十 \end{matrix}\right\} 2 \times 10 = 20$$

and 400 is written

$$\left.\begin{matrix} 四 \\ 百 \end{matrix}\right\} 4 \times 100 = 400$$

EXAMPLE ❼ *A Traditional Chinese Numeral*

Write 538 as a Chinese numeral.

SOLUTION

$$538 = \begin{cases} 500 = & \begin{cases} 5 & 五 \\ 100 & 百 \end{cases} \\ 30 = & \begin{cases} 3 & 三 \\ 10 & 十 \end{cases} \\ 8 = & \quad 8 \quad 八 \end{cases}$$

Note that in Example 7 the units digit, the 8, is not multiplied by a power of the base.

When writing Chinese numerals, there are some special cases that need to be considered. When writing a numeral between 11 and 19 it is not necessary to include the 1 before the 10. Thus, 18 would be written 十八 rather than ─十八. Another special case involves the use of zero.

When more than one consecutive zero occurs (except at the end of a numeral) you need to write a zero, but only once for two or more consecutive zeros. Zeros are not included at the end of numerals. The top two illustrations that follow show how zeros are used within a numeral, and the bottom two show that zeros are not used at the end of a numeral.

$$406 = \begin{array}{l} \left.\begin{array}{l} 四 \\ 百 \end{array}\right\} 4 \times 100 = 400 \\ \left.\begin{array}{l} 零 \end{array}\right\} 0 \times 10 = \quad 0 \\ \left.\begin{array}{l} 六 \end{array}\right\} 6 \qquad = \quad 6 \end{array}$$

$$4006 = \begin{array}{l} \left.\begin{array}{l} 四 \\ 千 \end{array}\right\} 4 \times 1000 = 4000 \\ \left.\begin{array}{l} 零 \end{array}\right\} \begin{array}{l} 0 \times 100, \\ 0 \times 10 \end{array} = \quad 0 \\ \left.\begin{array}{l} 六 \end{array}\right\} 6 \qquad = \quad 6 \end{array}$$

$$460 = \begin{array}{l} \left.\begin{array}{l} 四 \\ 百 \end{array}\right\} 4 \times 100 = 400 \\ \left.\begin{array}{l} 六 \\ 十 \end{array}\right\} 6 \times 10 = \quad 60 \end{array}$$

$$4600 = \begin{array}{l} \left.\begin{array}{l} 四 \\ 千 \end{array}\right\} 4 \times 1000 = 4000 \\ \left.\begin{array}{l} 六 \\ 百 \end{array}\right\} 6 \times 100 = \quad 600 \end{array}$$

EXAMPLE ❽ *Traditional Chinese Numerals*

Write the following as traditional Chinese numerals.
a) 7080   b) 7008

SOLUTION   In part (a), there is one zero between the 7 and the 8. In part (b), there are two zeros between the 7 and the 8. As just mentioned, the symbol for zero is used only once in each of these numerals.

a) $$7080 = \begin{array}{l} \left.\begin{array}{l} 七 \\ 千 \end{array}\right\} 7 \times 1000 \\ \left.\begin{array}{l} 零 \end{array}\right\} 0 \times 100 \\ \left.\begin{array}{l} 八 \\ 十 \end{array}\right\} 8 \times 10 \end{array}$$

b) $$7008 = \begin{array}{l} \left.\begin{array}{l} 七 \\ 千 \end{array}\right\} 7 \times 1000 \\ \left.\begin{array}{l} 零 \end{array}\right\} \begin{array}{l} 0 \times 100, \\ 0 \times 10 \end{array} \\ \left.\begin{array}{l} 八 \end{array}\right\} 8 \end{array}$$

TIMELY TIP Notice the difference between our Hindu–Arabic numeration system, which is a positional numeration system, and the Chinese system, which is a multiplicative numeration system. Below we show how to write 5678 as a Chinese numeral if the Chinese system was a positional value system similar to ours.

Multiplicative		Positional Value	
五	5	五	5
千	1000	六	6
六	6	七	7
百	100	八	8
七	7		
十	10		
八	8		

Note that the multiples of base 10 are removed when writing positional value numerals. We will discuss positional value systems in more detail shortly.

Ciphered Systems

A ciphered numeration system is one in which there are numerals for numbers up to and including the base and for multiples of the base. The numbers represented by a particular set of numerals is the sum of the values of the numerals.

Ciphered numeration systems require the memorization of many different symbols but have the advantage that numerals can be written in a compact form. The ciphered numeration system that we discuss is the Ionic Greek system (Table 4.4). The Ionic Greek system was developed in about 3000 B.C., and it used letters of the Greek alphabet for numerals. Other ciphered systems include the Hebrew, Coptic, Hindu, Brahmin, Syrian, Egyptian Hieratic, and early Arabic systems.

Table 4.4 Ionic Greek Numerals

1	α	alpha	60	ξ	xi
2	β	beta	70	o	omicron
3	γ	gamma	80	π	pi
4	δ	delta	90	φ[*]	koppa
5	ε	epsilon	100	ρ	rho
6	f[*]	digamma	200	σ	sigma
7	ζ	zeta	300	τ	tau
8	η	eta	400	υ	upsilon
9	θ	theta	500	ϕ	phi
10	ι	iota	600	χ	chi
20	κ	kappa	700	ψ	psi
30	λ	lambda	800	ω	omega
40	μ	mu	900	∂[*]	sampi
50	ν	nu			

[*]Ancient Greek letters not used in the classic or modern Greek language.

The classic Greek alphabet contains only 24 letters, however 27 symbols were needed. Thus, the Greeks used three obsolete Greek letters—f, Ϙ, and ℈—that are not part of the classic Greek alphabet. To distinguish words from numerals, the Greeks would place a mark similar to our apostrophe to the right and above each letter that was used as a numeral. In this text, we will not use this mark because we will not be using Greek words and numerals together.

We can write 45 as $40 + 5$. When 45 is written as a Greek numeral, the plus sign is omitted:

$$45 = \mu\varepsilon$$

Similarly, 768 written as a Greek numeral is $\psi\xi\eta$.

When the Greek letter iota, ι, is placed to the left and above a numeral, it represents the numeral multiplied by 1000. For example,

$$'\theta = 9 \times 1000 = 9000$$
$$'\zeta = 7 \times 1000 = 7000$$

EXAMPLE ❾ *The Ionic Greek System: A Ciphered System*

Write $\phi\, \nu\, \gamma$ as a Hindu–Arabic numeral.

SOLUTION $\phi = 500$, $\nu = 50$, and $\gamma = 3$. The sum is 553. ●

EXAMPLE ❿ *Writing an Ionic Greek Numeral*

Write 1654 as an Ionic Greek numeral.

SOLUTION

$$
\begin{aligned}
1654 &= 1000 + 600 + 50 + 4 \\
&= (1 \times 1000) + 600 + 50 + 4 \\
&= {}'\alpha \qquad\quad \chi \qquad \nu \qquad \delta \\
&= {}'\alpha\, \chi\, \nu\, \delta
\end{aligned}
$$

●

SECTION 4.1   EXERCISES

CONCEPT/WRITING EXERCISES

1. What is the difference between a number and a numeral?

2. List four numerals given in this section that may be used to represent the number ten.

3. What is a system of numeration?

4. List four numerals given in this section that may be used to represent the number one hundred.

5. What is the name of the system of numeration that we presently use?

6. Explain how numbers are represented in an additive numeration system.

7. Explain how numbers are represented in a multiplicative numeration system.

8. Explain how numbers are represented in a ciphered numeration system.

PRACTICE THE SKILLS

In Exercises 9–14, write the Egyptian numeral as a Hindu–Arabic numeral.

9. �999∩∩∩||||| 10. ꟿ99∩||||

11. ꟿꟿ99999∩∩||| 12. ꟿꟿꟿꟿꟿꟿꟿ99∩

13. ꙮꙮ999999∩∩||||| 14. ꙮꙮꙮ999999∩∩∩|

In Exercises 15–20, write the numeral as an Egyptian numeral.

15. 53   16. 224

17. 2045   18. 1812

19. 173,845   20. 3,235,614

In Exercises 21–32, write the Roman numeral as a Hindu–Arabic numeral.

21. VIII   22. XXIX

23. XLIII   24. CDLXXIV

25. MCCXXXVI   26. MCMLXIV

27. MMCMXLVI   28. MDCCXLVI

29. $\overline{\text{X}}$MMDCLXVI   30. $\overline{\text{L}}$MCMXLIV

31. $\overline{\text{IX}}$CDLXIV   32. $\overline{\text{V}}$MCCCXXXIII

In Exercises 33–44, write the numeral as a Roman numeral.

33. 27   34. 89

35. 341   36. 477

37. 2005   38. 4285

39. 4793   40. 6274

41. 9999   42. 14,315

43. 20,644   44. 99,999

In Exercises 45–52, write the Chinese numeral as a Hindu–Arabic numeral.

45. 七十四   46. 六十二   47. 四千零八十一   48. 三千零二十九

49. 八千五百五十   50. 三千四百八十七

78. Make up your ... dicate the sy... Using yo... a) yo...

In Exercises 53–60, ... Chinese numeral.

53. 53   54. 178   55. 378

56. 2001   57. 4260   58. 6905

59. 7056   60. 3009

In Exercises 61–66, write the numeral as a Hindu–Arabic numeral.

61. $\kappa\ \digamma$ 62. $\rho\ \lambda\ \eta$ 63. $\sigma\ o\ \theta$

64. $\vartheta\ \varepsilon$ 65. $'\beta\ \omega\ \pi\ \gamma$ 66. $'\zeta\ \upsilon\ \xi\ \zeta$

In Exercises 67–72, write the numeral as an Ionic Greek numeral.

67. 59   68. 178   69. 726

70. 2008   71. 5005   72. 9999

In Exercises 73–76, write the numeral as numerals in the indicated systems of numeration.

73. ꟿ∩∩| in Hindu–Arabic, Roman, traditional Chinese, and Greek

74. MCMXXXVI in Hindu–Arabic, Egyptian, Greek, and traditional Chinese

75. 五百二十七 in Hindu–Arabic, Egyptian, Roman, and Greek

76. $\upsilon\kappa\beta$ in Hindu–Arabic, Egyptian, Roman, and traditional Chinese

CHALLENGE PROBLEMS/GROUP ACTIVITIES

77. Write the Roman numeral for 999,999.

own additive system of numeration and in-
mbols and rules used to represent numbers.
r system of numeration, write

ur age.

b) the year you were born.

c) the current year.

In Exercises 79–81, compare the advantages and disadvantages of a ciphered system of numeration with those of the named system.

79. An additive system

80. A multiplicative system

81. The Hindu–Arabic system

RECREATIONAL MATHEMATICS

82. Without using any type of writing instrument, what can you do to make the following incorrect statement a correct statement?

$$XI + I = X$$

83. Words and numerals that read the same both backward and forward are called *palindromes*. Some examples are the words CIVIC and RACECAR, and the numerals 121 and 32523. Using Roman numerals, list the last year that was a palindrome.

84. Which year in the past 2000 years required the most Roman numerals to write? Write out the year in Roman numerals.

INTERNET/RESEARCH ACTIVITIES

85. ***Egypt, China, and Greece*** In this section, we discussed the numeration systems of Egypt, China, and Greece. Choose one of these countries and write a paper that does the following.

a) State the current numerals used in your chosen country.

b) Explain how that country's current system of numeration works. If more than one system is used, discuss the system used most commonly.

▲ Great Pyramid of Khufu in Giza, Egypt

<table>
<tr><td>4.2</td><td>

PLACE-VALUE OR POSITIONAL-VALUE NUMERATION SYSTEMS

</td></tr>
</table>

▲ Place value is used in many games on *The Price is Right*.

Have you ever watched the television game show *The Price is Right*? Chances are that you have seen one of the many games that require contestants to place digits in the correct order to guess the price of a car, appliance, or other fabulous prize. Such games require the contestant to have an understanding of the concept of *place value* that is used in our modern number system. In this section, we will study two other number systems that also rely on place value.

Today the most common type of numeration system is the place-value system. The Hindu–Arabic numeration system, used in the United States and many other countries, is an example of a place-value system. In a *place-value system*, which is also

Eighteenth-century mathematician Pierre Simon, Marquis de Laplace, speaking of the positional principle, said: "The idea is so simple that this very simplicity is the reason for our not being sufficiently aware of how much attention it deserves."

called a *positional value system*, the value of the symbol depends on its position in the representation of the number. For example, the 2 in 20 represents 2 tens, and the 2 in 200 represents 2 hundreds. A true positional-value system requires a *base* and a set of symbols, including a symbol for zero and one for each counting number less than the base. Although any number can be written in any base, the most common positional system is the base 10 system which is called the *decimal number system*.

The Hindus in India are credited with the invention of zero and the other symbols used in our system. The Arabs, who traded regularly with the Hindus, also adopted the system, thus the name Hindu–Arabic. Not until the middle of the fifteenth century, however, did the Hindu–Arabic numerals take the form we know today.

The Hindu–Arabic numerals and the positional system of numeration revolutionized mathematics by making addition, subtraction, multiplication, and division much easier to learn and very practical to use. Merchants and traders no longer had to depend on the counting board or abacus. The first group of mathematicians, who computed with the Hindu–Arabic system rather than with pebbles or beads on a wire, were known as the "algorists."

In the Hindu–Arabic system, the symbols 0, 1, 2, 3, 4, 5, 6, 7, 8, and 9 are called *digits*. The base 10 system was developed from counting on fingers, and the word *digit* comes from the Latin word for fingers.

The positional values in the Hindu–Arabic system are

$$\ldots, (10)^5, (10)^4, (10)^3, (10)^2, 10, 1$$

To evaluate a numeral in the Hindu–Arabic system, we multiply the first digit on the right by 1. We multiply the second digit from the right by the base, 10. We multiply the third digit from the right by the base squared, 10^2 or 100. We multiply the fourth digit from the right by the base cubed, 10^3 or 1000, and so on. In general, we multiply the digit n places from the right by 10^{n-1}. Therefore, we multiply the digit eight places from the right by 10^7. Using the place-value rule, we can write a number in *expanded form*. In expanded form, 1234 is written

$$1234 = (1 \times 10^3) + (2 \times 10^2) + (3 \times 10) + (4 \times 1)$$

or

$$(1 \times 1000) + (2 \times 100) + (3 \times 10) + 4$$

DID YOU KNOW?

Babylonian Numerals

The form Babylonian numerals took is directly related to their writing materials. Babylonians used a reed (later a stylus) to make their marks in wet clay. The end could be used to make a thin wedge, ❡, which represents a unit, or a wider wedge, ◀, which represents 10 units. The clay dried quickly, so the writings tended to be short but extremely durable.

Babylonian Numerals

The oldest known numeration system that resembled a place-value system was developed by the Babylonians in about 2500 B.C. Their system resembled a place-value system with a base of 60, a sexagesimal system. It was not a true place-value system because it lacked a symbol for zero. The lack of a symbol for zero led to a great deal of ambiguity and confusion. Table 4.5 gives the Babylonian numerals.

The positional values in the Babylonian system are

$$\ldots, (60)^3, (60)^2, 60, 1$$

In a Babylonian numeral, a gap is left between the characters to distinguish between the various place values. From right to left, the sum of the first group of numerals is multiplied by 1. The sum of the second group is multiplied by 60. The sum of the third group is multiplied by $(60)^2$, and so on.

Table 4.5 Babylonian Numerals

Babylonian Numerals	❡	◀
Hindu–Arabic numerals	1	10

EXAMPLE ❶ *The Babylonian System: A Positional Value System*

Write ⟨⟨⟨ ⟨⟨❚❚ as a Hindu–Arabic numeral.

SOLUTION

$$\underbrace{⟨⟨⟨}_{60\text{'s}} \qquad\qquad \underbrace{⟨⟨❚❚}_{\text{units}}$$

$$\underbrace{10 + 10 + 10}_{60\text{'s}} \qquad \underbrace{10 + 10 + 1 + 1}_{\text{units}}$$

$$(30 \times 60) + (22 \times 1)$$
$$1800 + 22 = 1822$$

The Babylonians used the symbol ❟ to indicate subtraction. The numeral ⟨❟❚❚ represents $10 - 2$, or 8. The numeral ⟨⟨⟨⟨ ❟❚❚❚ represents $40 - 3$, or 37 in base 10 or decimal notation.

EXAMPLE ❷ *From Babylonian to Hindu–Arabic Numerals*

Write ❚❚ ⟨❚ ⟨⟨❟❚❚ as a Hindu–Arabic numeral.

SOLUTION The place value of these three groups of numerals from left to right is

$$(60)^2, \quad 60, \quad 1$$

or

$$3600, \quad 60, \quad 1$$

The numeral in the group on the right has a value of $20 - 2$, or 18. The numeral in the center group has a value of $10 + 1$, or 11. The numeral on the left represents $1 + 1$, or 2. Multiplying each group by its positional value gives

$$(2 \times 60^2) + (11 \times 60) + (18 \times 1)$$
$$= (2 \times 3600) + (11 \times 60) + (18 \times 1)$$
$$= 7200 + 660 + 18$$
$$= 7878$$

To explain the procedure used to convert from a Hindu–Arabic numeral to a Babylonian numeral, we will consider a length of time. How can we change 9820 seconds into hours, minutes, and seconds? Since there are 3600 seconds in an hour (60 seconds to a minute and 60 minutes to an hour), we can find the number of hours in 9820 seconds by dividing 9820 by 60^2, or 3600.

$$\begin{array}{r} 2 \quad \leftarrow \text{Hours} \\ 3600\overline{)9820} \\ \underline{7200} \\ 2620 \quad \leftarrow \text{Remaining seconds} \end{array}$$

Now we can determine the number of minutes by dividing the remaining seconds by 60, the number of seconds in a minute.

$$\begin{array}{r} 43 \quad \leftarrow \text{Minutes} \\ 60\overline{)2620} \\ \underline{2400} \\ 220 \\ \underline{180} \\ 40 \quad \leftarrow \text{Remaining seconds} \end{array}$$

Since the remaining number of seconds, 40, is less than the number of seconds in a minute, our task is complete.

$$9820 \text{ sec} = 2 \text{ hr}, 43 \text{ min, and } 40 \text{ sec}$$

The same procedure is used to convert a decimal (base 10) numeral to a Babylonian numeral or any numeral in a different base.

EXAMPLE ❸ *From Hindu–Arabic to Babylonian Numerals*

Write 2519 as a Babylonian numeral.

SOLUTION The Babylonian numeration system has positional values of

$$\ldots, 60^3, 60^2, 60, 1$$

which can be expressed as

$$\ldots, 216000, 3600, 60, 1$$

The largest positional value less than or equal to 2519 is 60. To determine how many groups of 60 are in 2519, divide 2519 by 60.

$$\begin{array}{r} 41 \quad \leftarrow \text{Groups of 60} \\ 60\overline{)2519} \\ \underline{240} \\ 119 \\ \underline{60} \\ 59 \quad \leftarrow \text{Units remaining} \end{array}$$

Thus, $2519 \div 60 = 41$ with remainder 59. There are 41 groups of 60 and 59 units remaining. Because the remainder, 59, is less than the base, 60, no further division is necessary. The remainder represents the number of units when the numeral is written in expanded form. Therefore, $2519 = (41 \times 60) + (59 \times 1)$. When written as a Babylonian numeral, 2519 is

<<<<∣ <<<<<<∣∣

EXAMPLE ❹ *Using Division to Determine a Babylonian Numeral*

Write 6270 as a Babylonian numeral.

SOLUTION Divide 6270 by the largest positional value less than or equal to 6270. That value is 3600.

$$6270 \div 3600 = 1 \text{ with remainder } 2670$$

There is one group of 3600 in 6270. Next divide the remainder 2670 by 60 to determine the number of groups of 60 in 2670.

$$2670 \div 60 = 44 \text{ with remainder } 30$$

There are 44 groups of 60 and 30 units remaining.

$$6270 = (1 \times 60^2) + (44 \times 60) + (30 \times 1)$$

Thus, 6270 written as a Babylonian numeral is

$$\text{᛭ ᚜᚜᚜᚜᛭᛭᛭᛭ ᚜᚜᚜}$$

Mayan Numerals

Another place-value system is the Mayan numeration system. The Mayans, who lived on the Yucatan Peninsula in present day Mexico, developed a sophisticated numeration system based on their religious and agricultural calendar. The numerals in this system are written vertically rather than horizontally, with the units position on the bottom. In the Mayan system, the numeral in the bottom row is to be multiplied by 1. The numeral in the second row from the bottom is to be multiplied by 20. The numeral in the third row is to be multiplied by 18×20, or 360. You probably expected the numeral in the third row to be multiplied by 20^2 rather than 18×20. It is believed that the Mayans used 18×20 so that their numeration system would conform to their calendar of 360 days. The positional values above 18×20 are 18×20^2, 18×20^3, and so on.

Positional Values in the Mayan System

$\ldots 18 \times (20)^3,$	$18 \times (20)^2,$	$18 \times 20,$	$20,$	1
or$\ldots 144{,}000,$	$7200,$	$360,$	$20,$	1

The digits $0, 1, 2, 3, \ldots, 19$ of the Mayan systems are formed by a simple grouping of dots and lines, as shown in Table 4.6.

Table 4.6 Mayan Numerals

0	1	2	3	4	5	6	7	8	9
⬭	•	••	•••	••••	—	•̲	••̲	•••̲	••••̲

10	11	12	13	14	15	16	17	18	19
═	•̳	••̳	•••̳	••••̳	≡	•̿	••̿	•••̿	••••̿

EXAMPLE ❺ *The Mayan System: A Positional Value System*

Write
•̲
•••
•̲
 as a Hindu–Arabic numeral.

SOLUTION In the Mayan numeration system, the first three positional values are

$$18 \times 20$$
$$20$$
$$1$$

•̲	$= 11 \times (18 \times 20)$	$= 3960$
•••	$= 3 \times 20$	$= 60$
•̲	$= 16 \times 1$	$= \underline{\ \ 16}$
		4036

EXAMPLE 6 *From Mayan to Hindu–Arabic Numerals*

Write •• / ••• / ÷ / •••• as a Hindu–Arabic numeral.

SOLUTION

$$
\begin{aligned}
\bullet\bullet &= 2 \times [18 \times (20)^2] = 14{,}400 \\
\bullet\bullet\bullet &= 8 \times (18 \times 20) = 2880 \\
\doteq &= 11 \times 20 = 220 \\
\bullet\bullet\bullet\bullet &= 4 \times 1 = \underline{4} \\
& \hspace{4.5cm} 17{,}504
\end{aligned}
$$

EXAMPLE 7 *From Hindu–Arabic to Mayan Numerals*

Write 4025 as a Mayan numeral.

SOLUTION To convert from a Hindu–Arabic to a Mayan numeral, we use a procedure similar to the one used to convert to a Babylonian numeral. The Mayan positional values are . . . , 7200, 360, 20, 1. The greatest positional value less than or equal to 4025 is 360. Divide 4025 by 360.

$$4025 \div 360 = 11 \text{ with remainder } 65$$

There are 11 groups of 360 in 4025. Next, divide the remainder, 65, by 20.

$$65 \div 20 = 3 \text{ with remainder } 5$$

There are 3 groups of 20 with five units remaining.

$$4025 = (11 \times 360) + (3 \times 20) + (5 \times 1)$$

4025 written as a Mayan numeral is

$$
\left.
\begin{aligned}
11 \times 360 \\
3 \times 20 \\
5 \times 1
\end{aligned}
\right\} =
\begin{aligned}
\doteq \\
\bullet\bullet\bullet \\
\text{—}
\end{aligned}
$$

TIMELY TIP Notice that changing a numeral from the Babylonian or Mayan numeration system *to the Hindu–Arabic* (or decimal or base 10) system involves *multiplication*. Changing a numeral *from the Hindu–Arabic system* to the Babylonian or Mayan numeration system involves *division*.

SECTION 4.2   EXERCISES

CONCEPT/WRITING EXERCISES

1. What is another name for a place-value system?

2. What is the most common type of numeration system used in the world today?

3. Consider the numerals 40 and 400 in the Hindu–Arabic numeration system. What does the 4 represent in each numeral?

4. What is the most common base used for a positional value system? Explain why you believe the base you named is the most common base.

5. a) What is the base in the Hindu–Arabic numeration system?

   b) What are the digits in the Hindu–Arabic numeration system?

6. In a true positional-value system, what symbols are required?

7. Explain how to represent a number in expanded form in a positional-value numeration system.

8. Why was the Babylonian system not a true place-value system?

9. a) The Babylonian system did not have a symbol for zero. Why did this lead to some confusion?

   b) Write 133 and 7980 as Babylonian numerals.

10. Consider the Babylonian numeral ◁▔.
    Give two Hindu–Arabic numerals this numeral may represent. Explain your answer.

11. List the first five positional values, starting with the units position, for the Mayan numeration system.

12. Describe two ways that the Mayan place-value system differs from the Hindu–Arabic place-value system.

PRACTICE THE SKILLS

In Exercises 13–24, write the Hindu–Arabic numeral in expanded form.

13. 23	14. 84	15. 359
16. 562	17. 897	18. 3769
19. 4387	20. 23,468	21. 16,402

22. 125,678   23. 346,861   24. 3,765,934

In Exercises 25–30, write the Babylonian numeral as a Hindu–Arabic numeral.

25. ◁▔▔▔▔   26. ◁◁▔▔

27. ◁▔▔▔ ▔▔▔▔   28. ◁▔ ◁◁▔▔▔▔

29. ▔ ◁◁▔ ◁▔▔▔   30. ◁ ◁◁▔▔▔▔ ▔▔

In Exercises 31–36, write the numeral as a Babylonian numeral.

31. 35	32. 129	33. 471
34. 512	35. 3685	36. 12,435

In Exercises 37–42, write the Mayan numeral as a Hindu–Arabic numeral.

37. ••
    ⬓⬓⬓

38. ⬓
    ••

39. •••
    ◯

    ••

40. ••
    ••••

    ••

41. •

    ••

    ◯

42. ••
    ⬓
    ⬓

In Exercises 43–48, write the numeral as a Mayan numeral.

43. 17	44. 257	45. 297
46. 406	47. 2163	48. 17,708

In Exercises 49 and 50, write the numeral in the indicated systems of numeration.

49. ⬓   in Hindu–Arabic and Babylonian
    ••

    ••••

50. ◁◁◁▔▔▔ in Hindu–Arabic and Mayan

In Exercises 51 and 52, suppose a place-value numeration system has base ◯, with digits represented by the symbols △, ◇, ☐, and ⌀. Write each expression in expanded form.

51. △ ☐ ◇   52. ⌀ △ ◇ ☐

CHALLENGE PROBLEMS/GROUP ACTIVITIES

53. **a)** Is there a largest numeral in the Mayan numeration system? Explain.

   b) Write the Mayan numeral for 999,999.

54. **a)** Is there a largest numeral in the Babylonian numeration system? Explain.

   b) Write the Babylonian numeral for 999,999.

In Exercises 55–58, first convert each numeral to a Hindu–Arabic numeral and then perform the indicated operation. Finally, convert the answer back to a numeral in the original numeration system.

55. 𝕋𝕋 ≪≪ 𝕋𝕋𝕋 + ≪≪𝕋𝕋𝕋   56. 𝕋𝕋𝕋 ≪≪≪𝕋𝕋𝕋 − ≪≪≪𝕋𝕋

57. (Mayan numeral) + (Mayan numeral)   58. (Mayan numeral) − (Mayan numeral)

59. *Comparisons of Systems* Compare the advantages and disadvantages of a place-value system with those of

   a) additive numeration systems.

   b) multiplicative numeration systems.

   c) ciphered numeration systems.

60. *Your Own System* Create your own place-value system. Write 2009 in your system.

INTERNET/RESEARCH ACTIVITIES

61. Investigate and write a report on the development of the Hindu–Arabic system of numeration. Start with the earliest records of this system in India.

62. The Arabic numeration system currently in use is a base 10 positional-value system, which uses different symbols than the Hindu–Arabic numeration system. Write the symbols used in the Arabic system of numeration and their equivalent symbols in the Hindu–Arabic numeration system. Write 54, 607, and 2000 in Arabic numerals.

4.3   OTHER BASES

▲ Electronic devices have computer chips that use numeral bases other than base 10.

Have you done any of the following activities today: read your e-mail, browsed the Internet, watched television, listened to a music CD or MP3, driven a car, or gone grocery shopping? Chances are that while doing any of these or many other daily activities you used a device with a computer chip that uses numerals other than our familiar base 10 (decimal) numerals. In this section, we will study base 2 (binary), base 8 (octal), and base 16 (hexadecimal) numerals, all of which are used in most modern computers as well as in many other modern electronic devices.

The positional values in the Hindu–Arabic numeration system are

$$\ldots, (10)^4, (10)^3, (10)^2, 10, 1$$

The positional values in the Babylonian numeration system are

$$\ldots, (60)^4, (60)^3, (60)^2, 60, 1$$

Thus, 10 and 60 are called the *bases* of the Hindu–Arabic and Babylonian systems, respectively

Any counting number greater than 1 may be used as a base for a positional-value numeration system. If a positional-value system has a base b, then its positional values will be

$$\ldots, b^4, b^3, b^2, b, 1$$

The positional values in a base 8 system are

$$\ldots, 8^4, 8^3, 8^2, 8, 1$$

and the positional values in a base 2 system are

$$\ldots, 2^4, 2^3, 2^2, 2, 1$$

As we indicated in Section 4.2, the Mayan numeration system is based on the number 20. It is not, however, a true base 20 positional-value system. Why not?

The reason for the almost universal acceptance of base 10 numeration systems is that most human beings have 10 fingers. Even so, there are still some positional-value numeration systems that use bases other than 10. Some societies are still using a base 2 numeration system. They include some groups of people in Australia, New Guinea, Africa, and South America. Bases 3 and 4 are also used in some areas of South America. Base 5 systems were used by some primitive tribes in Bolivia, but the tribes are now extinct. The pure base 6 system occurs only sparsely in Northwest Africa. Base 6 also occurs in other systems in combination with base 12, the *duodecimal system*.

Our society still contains remnants of other base systems from other cultures. For example, there are 12 inches in a foot and 12 months in a year. Base 12 is also evident in the dozen, the 24-hour day, and the gross (12×12). Remains of base 60 are found in measurements of time (60 seconds to 1 minute, 60 minutes to 1 hour) and angles (60 seconds to 1 minute, 60 minutes to 1 degree).

Computers and many other electronic devices make use of three numeration systems: the *binary* (base 2), the *octal* (base 8), and the *hexadecimal* (base 16) numeration systems. One familiar example of the binary numeration system is the bar codes found on most items purchased in stores today. Computers can use the binary numeration system because the binary number system consists of only the digits 0 and 1. These digits can be represented with electronic switches that are either off (0) or on (1). All data that we enter into a computer can be converted into a series of single binary digits. Each binary digit is known as a *bit*. The octal numeration system is used when eight bits of data are grouped together to form a *byte*. In the American Standard Code for Information Interchange (ASCII) code, the byte 01000001 represents the character A, and 01100001 represents the character a. Other characters along with their decimal, binary, octal, and hexadecimal representation can be found at the web site www.asciitable.com. The hexadecimal numeration system is used to create computer languages. Two such examples of computer languages that rely on the hexadecimal system are HTML and CSS, both of which are used heavily in creating Internet web pages. Computers can easily convert between binary (base 2), octal (base 8), and hexadecimal (base 16) numbers.

▲ Computers make use of the binary, octal, and hexadecimal number systems.

Bases Less Than 10

A place-value system with base b must have b distinct symbols, one for zero and one for each numeral less than the base. A base 6 system must have symbols for 0, 1, 2, 3, 4, and 5. All numerals in base 6 are constructed from these 6 symbols. A base 8 system must have symbols for 0, 1, 2, 3, 4, 5, 6, and 7. All numerals in base 8 are constructed from these 8 symbols, and so on.

A numeral in a base other than base 10 will be indicated by a subscript to the right of the numeral. Thus, 123_5 represents a base 5 numeral. The numeral 123_6 represents a base 6 numeral. The value of 123_5 is not the same as the value of 123_{10}, and the value of 123_6 is not the same as the value of 123_{10}. A base 10 numeral may be written without a subscript. For example, 123 means 123_{10} and 456 means 456_{10}. For clarity in certain problems, we will use the subscript 10 to indicate a numeral in base 10.

The symbols that represent the base itself, in any base b, are 10_b. For example, in base 5, the symbols 10_5 represent the number five. Note that $10_5 = 1 \times 5 + 0 \times 1 = 5 + 0 = 5_{10}$, or the number five in base 10. The symbols 10_5 mean one group of 5 and no units. In base 6, the symbols 10_6 represent the number six. The symbols 10_6 represent one group of 6 and no units, and so on.

To change a numeral in a base other than 10 to a base 10 numeral, we follow the same procedure we used in Section 4.2 to change the Babylonian and Mayan numerals to base 10 numerals. Multiply each digit by its respective positional value. Then find the sum of the products.

EXAMPLE ❶ *Converting from Base 6 to Base 10*

Convert 453_6 to base 10.

SOLUTION In base 6, the positional values are $\ldots, 6^3, 6^2, 6, 1$. In expanded form,

$$
\begin{aligned}
453_6 &= (4 \times 6^2) + (5 \times 6) + (3 \times 1) \\
&= (4 \times 36) + (5 \times 6) + (3 \times 1) \\
&= \quad 144 \quad + \quad 30 \quad + \quad 3 \\
&= 177
\end{aligned}
$$

In Example 1, the units digit in 453_6 is 3. Notice that 3_6 has the same value as 3_{10} since both are equal to 3 units. That is, $3_6 = 3_{10}$. If n is a digit less than the base b, and the base b is less than or equal to 10, then $n_b = n_{10}$.

EXAMPLE ❷ *Converting from Base 8 to Base 10*

Convert 3615_8 to base 10.

SOLUTION

$$
\begin{aligned}
3615_8 &= (3 \times 8^3) \ + (6 \times 8^2) + (1 \times 8) + (5 \times 1) \\
&= (3 \times 512) + (6 \times 64) + (1 \times 8) \ + (5 \times 1) \\
&= \quad 1536 \quad + \quad 384 \quad + \quad 8 \quad + \quad 5 \\
&= 1933
\end{aligned}
$$

EXAMPLE ❸ *Converting from Base 2 to Base 10*

Convert 101101_2 to base 10.

SOLUTION

$$
\begin{aligned}
101101_2 &= (1 \times 2^5) + (0 \times 2^4) + (1 \times 2^3) + (1 \times 2^2) + (0 \times 2) + (1 \times 1) \\
&= \quad 32 \quad + \quad 0 \quad + \quad 8 \quad + \quad 4 \quad + \quad 0 \quad + \quad 1 \\
&= 45
\end{aligned}
$$

To change a base 10 numeral to a different base, we will use a procedure similar to the one we used to convert base 10 numerals to Babylonian and Mayan numerals, as was explained in Section 4.2. Divide the base 10 numeral by the highest power of the new base that is less than or equal to the given base 10 numeral. Record this quotient. Then divide the remainder by the next smaller power of the new base and record this quotient. Repeat this procedure until the remainder is less than the new base. The answer is the set of quotients listed from left to right, with the remainder on the far right. This procedure is illustrated in Examples 4 and 5.

EXAMPLE ❹ *Converting from Base 10 to Base 8*

Convert 486 to base 8.

SOLUTION We are converting a numeral in base 10 to a numeral in base 8. The positional values in the base 8 system are . . . , $8^3, 8^2, 8, 1$, or . . . , 512, 64, 8, 1. The highest power of 8 that is less than or equal to 486 is 8^2, or 64. Divide 486 by 64.

First digit in answer
$$486 \div 64 = 7 \text{ with remainder } 38$$

Therefore, there are 7 groups of 8^2 in 486. Next divide the remainder, 38, by 8.

Second digit in answer
$$38 \div 8 = 4 \text{ with remainder } 6$$
Third digit in answer

There are 4 groups of 8 in 38 and 6 units remaining. Since the remainder, 6, is less than the base, 8, no further division is required.

$$= (7 \times 64) + (4 \times 8) + (6 \times 1)$$
$$= (7 \times 8^2) + (4 \times 8) + (6 \times 1)$$
$$= 746_8$$

Notice that we placed the subscript 8 to the right of 746 to show that it is a base 8 numeral.   ●

EXAMPLE ❺ *Converting from Base 10 to Base 3*

Convert 273 to base 3.

SOLUTION The place values in the base 3 system are . . . , $3^6, 3^5, 3^4, 3^3, 3^2, 3, 1$, or . . . , 729, 243, 81, 27, 9, 3, 1. The highest power of the base that is less than or equal to 273 is 3^5, or 243. Successive divisions by the powers of the base give the following result.

$$273 \div 243 = 1 \text{ with remainder } 30$$
$$30 \div 81 = 0 \text{ with remainder } 30$$
$$30 \div 27 = 1 \text{ with remainder } 3$$
$$3 \div 9 = 0 \text{ with remainder } 3$$
$$3 \div 3 = 1 \text{ with remainder } 0$$

The remainder, 0, is less than the base, 3, so no further division is necessary. To obtain the answer, list the quotients from top to bottom followed by the remainder in the last division.

We can represent 273 as one group of 243, no groups of 81, one group of 27, no groups of 9, one group of 3, and no units.

$$273 = (1 \times 243) + (0 \times 81) + (1 \times 27) + (0 \times 9) + (1 \times 3) + (0 \times 1)$$
$$= (1 \times 3^5) + (0 \times 3^4) + (1 \times 3^3) + (0 \times 3^2) + (1 \times 3) + (0 \times 1)$$
$$= 101010_3 \qquad \bullet$$

Bases Greater Than 10

Recall that a place-value system with base b must have symbols for the digits from 0 up to one less than the base. For example, base 6 uses the digits 0, 1, 2, 3, 4, and 5; and base 8 uses the digits 0, 1, 2, 3, 4, 5, 6, and 7. What happens if the base is larger than ten? We will need single digit symbols to represent the numbers ten, eleven, twelve, . . . up to one less than the base. *In this textbook, whenever a base larger than ten is used we will use the capital letter A to represent ten, the capital letter B to represent eleven, the capital letter C to represent twelve, and so on.* For example, for base 12, known as the *duodecimal system*, we use the symbols 0, 1, 2, 3, 4, 5, 6, 7, 8, 9, A, and B, where A represents ten and B represents eleven. For base 16, known as the hexadecimal system, we use the symbols 0, 1, 2, 3, 4, 5, 6, 7, 8, 9, A, B, C, D, E, and F.

EXAMPLE 6 *Converting to and from Base 12*

a) Convert $39BA_{12}$ to base 10.   b) Convert 6893 to base 12.

SOLUTION

a) In base 12, the positional values are . . . 12^3, 12^2, 12, 1 or . . . 1728, 144, 12, 1. Since B has the value of eleven and A has the value of ten, we perform the following calculation.

$$39BA_{12} = (3 \times 12^3) + (9 \times 12^2) + (11 \times 12) + (10 \times 1)$$
$$= (3 \times 1728) + (9 \times 144) + 132 + 10$$
$$= 5184 + 1296 + 132 + 10 = 6622$$

b) The highest power of base 12 that is less than or equal to 6893 is 12^3, or 1728. To convert 6893 to base 12, we will use a process similar to the one we used in Examples 4 and 5. Remember that in base 12 we represent ten with the numeral A and eleven with the numeral B.

$$6893 \div 1728 = 3 \text{ with remainder } 1709$$
$$1709 \div 144 = B \text{ with remainder } 125 \quad \text{Note that B has a value of eleven}$$
$$125 \div 12 = A \text{ with remainder } 5 \quad \text{Note that A has a value of ten}$$

The remainder, 5, is less than the base, 12, so no further division is necessary. Thus, the answer is $3BA5_{12}$. To check this answer, perform the calculation $(3 \times 12^3) + (11 \times 12^2) + (10 \times 12) + (5 \times 1)$ to verify that you obtain 6893 in base 10. $\qquad \bullet$

EXAMPLE 7 *Converting to and from Base 16*

a) Convert $7DE_{16}$ to base 10.   b) Convert 6713 to base 16.

SOLUTION

a) In a base 16 system, the positional values are ..., 16^3, 16^2, 16, 1 or ... 4096, 256, 16, 1. Since D has a value of thirteen and E has a value of fourteen, we perform the following calculation.

$$7DE_{16} = (7 \times 16^2) + (D \times 16) + (E \times 1)$$
$$= (7 \times 256) + (13 \times 16) + (14 \times 1)$$
$$= 1792 + 208 + 14$$
$$= 2014$$

b) The highest power of base 16 less than or equal to 6713 is 16^3, or 4096. If we obtain a quotient greater than nine but less than sixteen, we will use the corresponding letter A through F.

$$6713 \div 4096 = 1 \text{ with remainder } 2617$$
$$2617 \div 256 = A \text{ with remainder } 57 \quad \text{Note that A has a value of ten}$$
$$57 \div 16 = 3 \text{ with remainder } 9$$

Thus, $6713 = 1A39_{16}$.

TIMELY TIP It is important to remember the following items presented in this section.

- If a numeral is shown without a base, we assume the numeral is a base 10 numeral.
- When converting a base 10 numeral to a different base, your answer should never contain a digit greater than or equal to that different base.
- When changing a numeral given in a base other than 10 to a numeral in base 10, use multiplication. When changing a base 10 numeral to a numeral in a different base, use division.

SECTION 4.3   EXERCISES

CONCEPT/WRITING EXERCISES

1. a) In your own words, explain how to convert a numeral in a base other than 10 to a base 10 numeral.

   b) In your own words, explain how to convert a base 10 numeral to a numeral in a base other than 10.

2. List the digits that are used when writing numerals in each of the following bases.

   a) 2   b) 8

   c) 12   d) 16

3. For each of the following, explain why the numeral is written incorrectly.

   a) 234_4 b) 1030001_2

   c) $4A8LB4A_{12}$ d) $BCFGA_{16}$

4. a) What is 10_2 equal to in base 10?

   b) What is 10_8 equal to in base 10?

   c) What is 10_{16} equal to in base 10?

   d) What is 10_{32} equal to in base 10?

PRACTICE THE SKILLS

In Exercises 5–22, convert the given numeral to a numeral in base 10.

5. 1_2 **6.** 10_2 **7.** 23_5

8. 41_5 **9.** 270_8 **10.** 306_8

11. 309_{12} **12.** 1005_{12} **13.** 573_{16}

14. 423_{16} **15.** 110101_2 **16.** 1110111_2

17. 7654_8 **18.** 5032_8 **19.** $A91_{12}$

20. $B52_{12}$ **21.** $C679_{16}$ **22.** $D20E_{16}$

In Exercises 23–40, convert each of the following to a numeral in the base indicated.

23. 6 to base 2 **24.** 25 to base 2

25. 347 to base 3 **26.** 513 to base 3

27. 53 to base 4 **28.** 591 to base 4

29. 102 to base 5 **30.** 549 to base 7

31. 1098 to base 8 **32.** 2921 to base 8

33. 1432 to base 9 **34.** 2170 to base 9

35. 9004 to base 12 **36.** 13,312 to base 12

37. 493 to base 16 **38.** 3243 to base 16

39. 9455 to base 16 **40.** 39,885 to base 16

In Exercises 41–50, convert 2009 to a numeral in the base indicated.

41. 3 **42.** 4 **43.** 5 **44.** 6

45. 7 **46.** 9 **47.** 11 **48.** 13

49. 15 **50.** 17

PROBLEM SOLVING

In Exercises 51–54, assume the numerals given are in a base 5 numeration system. The numerals in this system and their equivalent Hindu–Arabic numerals are

$\bigcirc$ = 0 $\ominus$ = 1 $\ovoid$ = 2 $\ominus$ = 3 $\oslash$ = 4

Write the Hindu–Arabic numerals equivalent to each of the following.

51. $\oslash\ominus_5$ **52.** $\ovoid\ominus_5$ **53.** $\ominus\bigcirc\ominus_5$ **54.** $\ovoid\oslash\ominus_5$

In Exercises 55–58, write the Hindu–Arabic numerals in the numeration system discussed in Exercises 51–54.

55. 19 **56.** 23 **57.** 74 **58.** 85

In Exercises 59–62, suppose colors as indicated below represent numerals in a base 4 numeration system.

$\bullet$ = 0 $\bullet$ = 1 $\bullet$ = 2 $\bullet$ = 3

Write the Hindu–Arabic numerals equivalent to each of the following.

59. $\bullet\bullet_4$ **60.** $\bullet\bullet_4$ **61.** $\bullet\bullet\bullet_4$ **62.** $\bullet\bullet\bullet_4$

In Exercises 63–66, write the Hindu–Arabic numerals in the base 4 numeration system discussed in Exercises 59–62. You will need to use the colors indicated above to write the answer.

63. 10 **64.** 15 **65.** 60 **66.** 56

67. *Another Conversion Method* There is an alternative method for changing a base 10 numeral to a different base. This method will be used to convert 328 to base 5. Dividing 328 by 5 gives a quotient of 65 and a remainder of 3. Write the quotient below the dividend and the remainder on the right, as shown.

$$5\overline{)328} \quad \text{remainder}$$
$$65 \qquad 3$$

Continue this process of division by 5.

$$
\begin{array}{r|l}
5\overline{)328} & \text{remainder} \\
5\overline{)65} & 3 \\
5\overline{)13} & 0 \\
5\overline{)2} & 3 \\
0 & 2
\end{array}
$$

(In the last division, since the dividend, 2, is smaller than the divisor, 5, the quotient is 0 and the remainder is 2.)

Note that the division continues until the quotient is zero. The answer is read from the bottom number to the top number in the remainder column. Thus, $328 = 2303_5$.

a) Explain why this procedure results in the proper answer.

b) Convert 683 to base 5 by this method.

c) Convert 763 to base 8 by this method.

CHALLENGE PROBLEMS/GROUP ACTIVITIES

68. a) Use the numerals 0, 1, and 2 to write the first 20 numbers in the base 3 numeration system.

b) What is the next numeral after 222_3?

69. a) *Your Own Numeration System* Make up your own base 20 positional-value numeration system. Indicate the 20 numerals you will use to represent the 20 numbers less than the base.

b) Write 523 and 5293 in your base 20 numeration system.

70. *Computer Code* The ASCII code used by most computers uses the last seven positions of an eight-bit byte to represent all the characters on a standard keyboard. How many different orderings of 0's and 1's (or how many different characters) can be made by using the last seven positions of an eight-bit byte?

RECREATIONAL MATHEMATICS

71. Find b if $111_b = 43$.

72. Find d if $ddd_5 = 124$.

73. Suppose a base 4 place-value system has its digits represented by colors as follows:

a) Determine the value of ● ● ● ● ● $_4$ in base 10.

b) Write 177 in the base 4 system using only the four colors given in the exercise.

INTERNET/RESEARCH ACTIVITIES

74. Investigate and write a report on the use of the duodecimal (base 12) system of numeration. You may wish to contact the Dozenal Society (see the Did You Know? on page 201) for more information.

75. We mentioned at the beginning of this section that some societies still use a base 2 and base 3 numeration system. These societies are in Australia, New Guinea, Africa, and South America. Write a report on these societies, covering the symbols they use and how they combine these symbols to represent numbers in their numeration system.

4.4 COMPUTATION IN OTHER BASES

You may recall learning how to add, subtract, multiply, and divide when you were in grade school. You may recall "carrying" when adding and "borrowing" when subtracting. For example, when performing 75 + 18, you would add 5 and 8 to get 13. You would write down the 3 and "carry the 1." When performing 43 − 19 you would "borrow from the 4" so you could perform 13 − 9 to get 4. In this section, we will use similar procedures when working with numerals that have bases other than base 10. By working in other bases, we will gain a better understanding of our own base 10 place-value system.

▲ Students do arithmetic by "carrying" and "borrowing."

Addition

When computers perform calculations, they do so in base 2, the binary system. In this section, we will explain how to perform calculations in base 2 and other bases.

In a base 2 system, the only digits are 0 and 1, and the place values are

$$\ldots, 2^4, 2^3, 2^2, 2, 1$$
or $\ldots, 16, 8, 4, 2, 1$

Suppose we want to add $1_2 + 1_2$. The subscript 2 indicates that we are adding in base 2. Remember that the answer to $1_2 + 1_2$ must be written using only the digits 0 and 1. The sum of $1_2 + 1_2$ is 10_2, which represents 1 group of two and 0 units in base 2. Recall that 10_2 means $1(2) + 0(1)$.

If we wanted to find the sum of $10_2 + 1_2$, we would add the digits in the right-hand, or units, column. Since $0_2 + 1_2 = 1_2$, the sum of $10_2 + 1_2 = 11_2$.

We are going to work additional examples and exercises in base 2, so rather than performing individual calculations in every problem, we can construct and use an addition table, Table 4.7, for base 2 (just as we used an addition table in base 10 when we first learned to add in base 10).

Table 4.7 Base 2 Addition Table

+	0	1
0	0	1
1	1	10

EXAMPLE ❶ *Adding in Base 2*

Add 1101_2
     $+\ 111_2$

SOLUTION Begin by adding the digits in the right-hand, or units, column. From previous discussion, and as can be seen in Table 4.7, $1_2 + 1_2 = 10_2$. Place the 0 under the units column and carry the 1 to the 2's column, the second column from the right.

$$2^3\ 2^2\ 2\ 1 \quad \text{Place value of columns}$$
$$\downarrow\ \downarrow\ \downarrow\ \downarrow$$
$$1\ 1\ {}^1 0\ 1$$
$$+\ \ \ 1\ 1\ 1$$
$$\overline{\qquad\quad 0_2}$$

Now add the three digits in the 2's column, $1_2 + 0_2 + 1_2$. Treat it as $(1_2 + 0_2) + 1_2$. Therefore, add $1_2 + 0_2$ to get 1_2, then add $1_2 + 1_2$ to get 10_2. Place the 0 under the 2's column and carry the 1 to the 2^2 column (the third column from the right).

$$1\ {}^1 1\ {}^1 0\ 1$$
$$+\ \ \ 1\ \ 1\ 1$$
$$\overline{\qquad\ 0\ \ 0_2}$$

Now add the three 1's in the 2^2 column to get $(1_2 + 1_2) + 1_2 = 10_2 + 1_2 = 11_2$. Place the 1 under the 2^2 column and carry the 1 to the 2^3 column (the fourth column from the right).

$$ {}^1 1\ {}^1 1\ {}^1 0\ 1$$
$$+\ \ \ 1\ \ 1\ 1$$
$$\overline{\quad 1\ \ 0\ \ 0_2}$$

Now add the two 1's in the 2^3 column, $1_2 + 1_2 = 10_2$. Place the 10 as follows.

$$ {}^1 1\ {}^1 1\ {}^1 0\ 1$$
$$+\ \ \ \ \ 1\ \ 1\ 1$$
$$\overline{1\ 0\ 1\ \ 0\ 0_2}$$

Therefore, the sum is 10100_2.

Table 4.8 Base 5 Addition Table

+	0	1	2	3	4
0	0	1	2	3	4
1	1	2	3	4	10
2	2	3	4	10	11
3	3	4	10	11	12
4	4	10	11	⑫	13

DID YOU KNOW?

Speaking to Machines

Pit   Land

On digital video discs (DVDs) and compact discs (CDs), video and sound are digitally encoded on the underside of the disc in a binary system of pits and "lands" (nonpits). To play the disc, a laser beam tracks along the spiral and is reflected when it hits a land (signal sent = 1), but it is not reflected by the pits (no signal = 0). The binary sequence is then converted into video images and music.

For the past 600 years, we have used the Hindu–Arabic system of numeration without change. Our base 10 numeration system seems so obvious to us, perhaps because of our 10 fingers and 10 toes, but it would be rash to think that numbers in other bases are not useful. In fact, one of the most significant numeration systems is the binary system, or base 2. This system, with its elemental simplicity, is what is used by computers to process information and "talk" to one another. When a computer receives a command or data, every character in the command or data must first be converted into a binary numeral for the computer to understand and use it.

Let's now look at addition in a base 5 system. In base 5, the only digits are 0, 1, 2, 3, and 4, and the positional values are

$$\ldots, \quad 5^4, \quad 5^3, 5^2, 5, 1$$
$$\text{or} \quad \ldots, 625, 125, 25, 5, 1$$

What is the sum of $4_5 + 3_5$? We can consider this to mean $(1 + 1 + 1 + 1) + (1 + 1 + 1)$. We can regroup the seven 1's into one group of five and two units as $(1 + 1 + 1 + 1 + 1) + (1 + 1)$. Thus, the sum of $4_5 + 3_5 = 12_5$ (circled in Table 4.8). Recall that 12_5 means $1(5) + 2(1)$. We can use this same procedure in obtaining the remaining values in the base 5 addition table.

EXAMPLE ❷ Use the Base 5 Addition Table

Add 44_5
     $+ 43_5$

SOLUTION First determine from Table 4.8 that $4_5 + 3_5$ is 12_5. Record the 2 and carry the 1 to the 5's column.

$$
\begin{array}{r}
{}^{1}4\ 4_5 \\
+\ 4\ 3_5 \\
\hline
2_5
\end{array}
$$

Add the numbers in the second column, $(1_5 + 4_5) + 4_5 = 10_5 + 4_5 = 14_5$. Record the 14.

$$
\begin{array}{r}
{}^{1}4\ 4_5 \\
+\ \ \ 4\ 3_5 \\
\hline
1\ 4\ 2_5
\end{array}
$$

The sum is 142_5. ●

EXAMPLE ❸ Add in Base 5

Add 1234_5
     $+ 2042_5$

SOLUTION

$$
\begin{array}{r}
1\ \ {}^{1}2\ {}^{1}3\ 4_5 \\
+\ 2\ \ 0\ \ 4\ 2_5 \\
\hline
3\ \ 3\ \ 3\ 1_5
\end{array}
$$
●

You can develop an addition table for any base and use it to add in that base. As you get more comfortable with addition in other bases, however, you may prefer to add in other bases by using mental arithmetic. To do so, convert the sum from the given base to base 10 and then convert the base 10 numeral back into the given base. You must clearly understand how to convert from base 10 to the given base, as discussed in Section 4.3. For example, to add $7_9 + 8_9$, add $7 + 8$ in base 10 to get 15_{10} and then mentally convert 15_{10} to 16_9 using the procedure given earlier. Remember,

16_9 when converted to base 10 becomes $1(9) + 6(1)$, or 15. Addition using this procedure is illustrated in Examples 4 and 5.

EXAMPLE ❹ *Adding in Base 10; Converting to Base 3*

Add 1022_3
       $+ 2121_3$

SOLUTION To solve this problem, make the necessary conversions by using mental arithmetic. $2 + 1 = 3_{10} = 10_3$. Record the 0 and carry the 1.

$$\begin{array}{r} 1\;0\;^12\;2_3 \\ +\,2\;1\;\;2\;1_3 \\ \hline 0_3 \end{array}$$

$1 + 2 + 2 = 5_{10} = 12_3$. Record the 2 and carry the 1.

$$\begin{array}{r} 1\;\;^10\;^12\;2_3 \\ +\,2\;\;1\;\;2\;1_3 \\ \hline 2\;\;0_3 \end{array}$$

$1 + 0 + 1 = 2_{10} = 2_3$. Record the 2.

$$\begin{array}{r} 1\;\;^10\;^12\;2_3 \\ +\,2\;\;1\;\;2\;1_3 \\ \hline 2\;\;2\;\;0_3 \end{array}$$

$1 + 2 = 3_{10} = 10_3$. Record the 10.

$$\begin{array}{r} 1022_3 \\ +\;2121_3 \\ \hline 10220_3 \end{array}$$

The sum is 10220_3. ●

EXAMPLE ❺ *Adding in Base 10; Converting to Base 4*

Add 322_4
         223_4
         230_4
       $+ 132_4$

SOLUTION Adding the digits in the right-hand column gives $2 + 3 + 0 + 2 = 7_{10} = 13_4$. Record the 3 and carry the 1. Adding the 1 with the digits in the next column gives $1 + 2 + 2 + 3 + 3 = 11_{10} = 23_4$. Record the 3 and carry the 2. Adding the 2 with the digits in the next column gives $2 + 3 + 2 + 2 + 1 = 10_{10} = 22_4$. Record both digits. The sum is 2233_4.

$$\begin{array}{r} ^23\;^12\;2_4 \\ 2\;\;2\;3_4 \\ 2\;\;3\;0_4 \\ +\;\;1\;\;3\;2_4 \\ \hline 2\;2\;\;3\;3_4 \end{array}$$

●

Subtraction

Subtraction can also be performed in bases other than base 10. Always remember that when you "borrow," you borrow the amount of the base given in the subtraction problem. For example, if subtracting in base 5, when you borrow, you borrow 5. If subtracting in base 12, when you borrow, you borrow 12.

EXAMPLE ❻ *Subtracting in Base 5*

Subtract 3032_5
               -1004_5

SOLUTION We will perform the subtraction in base 10 and convert the results to base 5. Since 4 is greater than 2, we must borrow one group of 5 from the preceding column. This action gives a sum of $5 + 2$, or 7 in base 10. Now we subtract 4 from 7; the difference is 3. We complete the problem in the usual manner. The 3 in the second column becomes a 2, $2 - 0 = 2$. In the third column, $0 - 0 = 0$. Finally, in the fourth column, $3 - 1 = 2$.

$$\begin{array}{r} 3032_5 \\ -1004_5 \\ \hline 2023_5 \end{array}$$

In the next example, we will subtract base 12 numerals. Recall from Section 4.3 that we use the capital letter A to represent the number ten and the capital letter B to represent the number eleven in base 12.

EXAMPLE ❼ *Subtracting in Base 12*

Subtract $8\ 2\ B_{12}$
               $-3\ A\ 6_{12}$

SOLUTION Recall that B represents eleven. Therefore, in the units column we have $11 - 6 = 5$. Next, recall that A represents ten. Therefore, in the next column we must subtract 10 from 2. Since 10 is greater than 2, borrowing is necessary. We must borrow one group of 12 from the preceding column. We then have a sum of $12 + 2$, or 14. We can now subtract 10 from 14 and the difference is 4. The 8 in the third column becomes 7, and $7 - 3 = 4$.

$$\begin{array}{r} 8\ 2\ B_{12} \\ -3\ A\ 6_{12} \\ \hline 4\ 4\ 5_{12} \end{array}$$

Multiplication

Multiplication can also be performed in bases other than base 10. Doing so is helped by forming a multiplication table for the base desired. Suppose we want to determine the product of $4_5 \times 3_5$. In base 10, 4×3 means there are four groups of three units. Similarly, in a base 5 system, $4_5 \times 3_5$ means there are four groups of three units, or

$$(1 + 1 + 1) + (1 + 1 + 1) + (1 + 1 + 1) + (1 + 1 + 1)$$

Table 4.9 Base 5 Multiplication
Table

×	0	1	2	3	4
0	0	0	0	0	0
1	0	1	2	3	4
2	0	2	4	11	13
3	0	3	11	14	22
4	0	4	13	(22)	31

Regrouping the 12 units above into groups of five gives

$$(1 + 1 + 1 + 1 + 1) + (1 + 1 + 1 + 1 + 1) + (1 + 1)$$

or two groups of five, and two units. Thus, $4_5 \times 3_5 = 22_5$.

We can construct other values in the base 5 multiplication table in the same way. You may, however, find it easier to multiply the values in the base 10 system and then change the product to base 5 by using the procedure discussed in Section 4.3. Multiplying 4×3 in base 10 gives 12, and converting 12 from base 10 to base 5 gives 22_5.

The product of $4_5 \times 3_5$ is circled in Table 4.9, the base 5 multiplication table. The other values in the table may be found by either method discussed.

┌─**EXAMPLE** ❽ *Using the Base 5 Multiplication Table*

Multiply 13_5
            $\times\ 3_5$

SOLUTION Use the base 5 multiplication table to find the products. When the product consists of two digits, record the right digit and carry the left digit. Multiplying gives $3_5 \times 3_5 = 14_5$. Record the 4 and carry the 1.

$$\begin{array}{r} {}^113_5 \\ \times\ \ 3_5 \\ \hline 4 \end{array}$$

$(3_5 \times 1_5) + 1_5 = 4_5$. Record the 4.

$$\begin{array}{r} {}^113_5 \\ \times\ \ 3_5 \\ \hline 44_5 \end{array}$$

The product is 44_5.   ●

Constructing a multiplication table is often tedious, especially when the base is large. To multiply in a given base without the use of a table, multiply in base 10 and convert the products to the appropriate base before recording them. This procedure is illustrated in Example 9.

┌─**EXAMPLE** ❾ *Multiplying in Base 7*

Multiply 43_7
            $\times 25_7$

SOLUTION $5 \times 3 = 15_{10} = 2(7) + 1(1) = 21_7$. Record the 1 and carry the 2.

$$\begin{array}{r} {}^243_7 \\ \times\ 25_7 \\ \hline 1 \end{array}$$

$(5 \times 4) + 2 = 20 + 2 = 22_{10} = 3(7) + 1(1) = 31_7$. Record the 31.

$$\begin{array}{r} {}^243_7 \\ \times\ 25_7 \\ \hline 311 \end{array}$$

$2 \times 3 = 6_{10} = 6_7$. Record the 6.

$$\begin{array}{r} {}^{2}43_7 \\ \times\ 25_7 \\ \hline 311 \\ 6 \end{array}$$

$2 \times 4 = 8_{10} = 1(7) + 1(1) = 11_7$. Record the 11. Now add in base 7 to determine the answer. Remember, in base 7, there are no digits greater than 6.

$$\begin{array}{r} {}^{2}43_7 \\ \times\ 25_7 \\ \hline 311 \\ 116 \\ \hline 1501_7 \end{array}$$

Division

Division is performed in much the same manner as long division in base 10. A detailed example of a division in base 5 is illustrated in Example 10. The same procedure is used for division in any other base.

EXAMPLE ⑩ Dividing in Base 5

Divide $2_5\overline{)143_5}$.

SOLUTION Using the multiplication table for base 5, Table 4.9 on page 209, we list the multiples of the divisor, 2.

$$2_5 \times 1_5 = 2_5$$
$$2_5 \times 2_5 = 4_5$$
$$2_5 \times 3_5 = 11_5$$
$$2_5 \times 4_5 = 13_5$$

Since $2_5 \times 4_5 = 13_5$, which is the largest product less than 14_5, 2_5 divides into 14_5 four times.

$$\begin{array}{r} 4 \\ 2_5\overline{)143_5} \\ 13 \\ \hline 1 \end{array}$$

Subtract 13_5 from 14_5. The difference is 1_5. Record the 1. Now bring down the 3 as when dividing in base 10.

$$\begin{array}{r} 4 \\ 2_5\overline{)143_5} \\ 13 \\ \hline 13 \end{array}$$

We see that $2_5 \times 4_5 = 13_5$. Use this information to complete the problem.

$$
\begin{array}{r}
44_5 \\
2_5 \overline{)143_5} \\
13 \\
\hline
13 \\
13 \\
\hline
0
\end{array}
$$

Therefore, $143_5 \div 2_5 = 44_5$ with remainder 0_5.   •

A division problem can be checked by multiplication. If the division was performed correctly, (quotient $\times$ divisor) + remainder = dividend. We can check Example 10 as follows.

$$(44_5 \times 2_5) + 0_5 = 143_5$$

$$
\begin{array}{r}
44_5 \\
\times\ \ 2_5 \\
\hline
143_5 \quad \text{Check}
\end{array}
$$

EXAMPLE ⑪ *Dividing in Base 8*

Divide $6_8 \overline{)4071_8}$.

SOLUTION The multiples of 6 in base 8 are

$$6_8 \times 1_8 = 6_8 \qquad 6_8 \times 2_8 = 14_8 \qquad 6_8 \times 3_8 = 22_8 \qquad 6_8 \times 4_8 = 30_8$$
$$6_8 \times 5_8 = 36_8 \qquad 6_8 \times 6_8 = 44_8 \qquad 6_8 \times 7_8 = 52_8$$

$$
\begin{array}{r}
536_8 \\
6_8 \overline{)4071_8} \\
36 \\
\hline
27 \\
22 \\
\hline
51 \\
44 \\
\hline
5
\end{array}
$$

Thus, the quotient is 536_8, with remainder 5_8.

Be careful when subtracting! In the above division problem, we had to borrow when subtracting 6 from 0 and when subtracting 4 from 1. Remember that we borrow 10_8, which is the same as 8 in base 10.

CHECK: Does $(536_8 \times 6_8) + 5_8 = 4071_8$?

$$
\begin{array}{r}
536_8 \\
\times\ \ \ 6_8 \\
\hline
4064_8 + 5_8 = 4071_8 \quad \text{True}
\end{array}
$$
                                                                        •

SECTION 4.4 EXERCISES

CONCEPT/WRITING EXERCISES

1. a) What are the first five positional values, from right to left, in base b?

   b) What are the first five positional values, from right to left, in base 2?

2. In the addition

$$367_8$$
$$+24_8$$

   what are the positional values of the first column on the right, the second column from the right, and the third column from the right? Explain how you determined your answer.

3. Suppose you add two base 3 numerals and you obtain an answer of 2032_3. Can your answer be correct? Explain.

4. Suppose you add two base 5 numerals and you obtain an answer of 463_5. Can your answer be correct? Explain.

5. In your own words, explain how to subtract two numerals in a given base. Include in your explanation what you do when, in one column, you must subtract a larger numeral from a smaller numeral.

6. In your own words, explain how to add two numerals in a given base. In your explanation, answer the question, "What happens when the sum of the numerals in a column is greater than the base?"

PRACTICE THE SKILLS

In Exercises 7–18, add in the indicated base.

7. 21_3
   $+ 20_3$

8. 23_4
   $+ 13_4$

9. 1234_5
   $+ 341_5$

10. 1101_2
    $+ 111_2$

11. 799_{12}
    $+ 218_{12}$

12. 222_3
    $+ 22_3$

13. 1112_3
    $+ 1011_3$

14. 470_{12}
    $+ 347_{12}$

15. 14631_7
    $+ 6040_7$

16. 1341_8
    $+ 341_8$

17. 1110_2
    $+ 110_2$

18. $43A_{16}$
    $+ 496_{16}$

In Exercises 19–30, subtract in the indicated base.

19. 201_3
    -120_3

20. 512_6
    -421_6

21. 2138_9
    -1207_9

22. $AB32_{12}$
    $- 207_{12}$

23. 1101_2
    $- 111_2$

24. 1221_3
    $- 202_3$

25. 1001_2
    $- 110_2$

26. 2173_8
    -1654_8

27. 4223_7
    $- 304_7$

28. 4232_5
    -2341_5

29. 2100_3
    -1012_3

30. $4E7_{16}$
    -189_{16}

In Exercises 31–42, multiply in the indicated base.

31. 22_3
    $\times 2_3$

32. 124_5
    $\times 3_5$

33. 647_8
    $\times 5_8$

34. 12_4
    $\times 30_4$

35. 512_6
    $\times 23_6$

36. 124_{12}
    $\times 6_{12}$

37. $B12_{12}$
    $\times 83_{12}$

38. $6A3_{12}$
    $\times 24_{12}$

39. 111_2
    $\times 101_2$

40. 584_9
    $\times 24_9$

41. 316_7
    $\times 16_7$

42. $7A9_{12}$
    $\times 12_{12}$

In Exercises 43–54, divide in the indicated base.

43. $1_2 \overline{)101_2}$

44. $2_3 \overline{)120_3}$

45. $2_6 \overline{)451_6}$

46. $7_9 \overline{)123_9}$

47. $2_4 \overline{)312_4}$

48. $6_{12} \overline{)431_{12}}$

49. $2_4 \overline{)213_4}$

50. $5_6 \overline{)214_6}$

51. $3_5 \overline{)224_5}$

52. $4_6 \overline{)210_6}$

53. $6_7 \overline{)404_7}$

54. $3_7 \overline{)2101_7}$

PROBLEM SOLVING

In Exercises 55–58, the numerals in a base 5 numeration system are as illustrated with their equivalent Hindu–Arabic numerals.

$\bigcirc = 0$ $\ominus = 1$ $\bigcirc\!\!| = 2$ $\ominus\!\!| = 3$ $\bigcirc\!\!) = 4$

Add the following base 5 numerals.

55.

56.

57.

58.

In Exercises 59–66, assume the numerals given are in a base 4 numeration system. In this system, suppose colors are used as numerals, as indicated below.

● = 0, ● = 1, ● = 2, ● = 3

Add the following base 4 numerals. Your answers will contain a variety of the colors indicated.

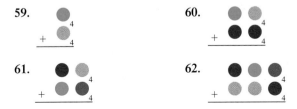

59.   60.

61.   62.

Subtract the following in base 4. Your answer will contain a variety of the colors indicated.

63.   64.

65.   66.

For Exercises 67 and 68, study the pattern in the boxes. The numeral in the bottom row of each box represents the value of each dot in the box directly above it. For example, the following box represents $(3 \times 7^2) + (2 \times 7) + (4 \times 1)$, or the numeral 324_7. This numeral in base 10 is 165.

• • •	• •	• • • •
7^2	7	1

67. Determine the base 5 numeral represented by the dots in the top row of the boxes. Then convert the base 5 numeral to a numeral in base 10.

• •	• • • •		• •
5^3	5^2	5	1

68. Fill in the correct amount of dots in the columns above the base values if the numeral represented by the dots is to equal 327 in base 10.

9^2	9	1

CHALLENGE PROBLEMS/GROUP ACTIVITIES

In Exercises 69–72, perform the indicated operation.

69. FAB_{16}
     $\times \quad 4_{16}$

70. $D_{16}\overline{)FACE_{16}}$

71. $20_4\overline{)223_4}$

72. $14_5\overline{)242_5}$

73. Consider the multiplication

$$462_8$$
$$\times\ 35_8$$

a) Multiply the numerals in base 8.

b) Convert 462_8 and 35_8 to base 10.

c) Multiply the base 10 numerals determined in part (b).

d) Convert the answer obtained in base 8 in part (a) to base 10.

e) Are the answers obtained in parts (c) and (d) the same? Why or why not?

RECREATIONAL MATHEMATICS

74. Determine b, by trial and error, if $1304_b = 204$.

75. In a base 4 system, each of the four numerals is represented by one of the following colors:

Determine the value of each color if the following addition is true in base 4.

INTERNET/RESEARCH ACTIVITIES

76. Write a report on how computers use the binary (base 2), octal (base 8), and hexadecimal (base 16) numeration systems.

77. One method used by computers to perform subtraction is the "end around carry method." Do research and write a report explaining, with specific examples, how a computer performs subtraction by using the end around carry method.

4.5   EARLY COMPUTATIONAL METHODS

▲ Multiplication can be used to calculate the amount of your paycheck.

Suppose that last week you worked at your job for 27 hours at a rate of $8 per hour. How much money did you make? You may find the answer by multiplying 27×8 "by hand."

$$\begin{array}{r} {}^5 27 \\ \times\ \ 8 \\ \hline 216 \end{array}$$

Although most of us would use this method, it is not the only method for multiplying two numbers together. Early civilizations used various other methods. In this section, we will study three other methods of multiplication.

Duplation and Mediation

The first method of multiplication we will study is *duplation and mediation*. This method, also known as *Russian peasant multiplication,* is still used in parts of Russia today. Duplation refers to doubling one number, and mediation refers to halving one number. The duplation and mediation method is similar to a method used by the ancient Egyptians as described on the Rhind Papyrus. Example 1 illustrates this method of multiplication.

EXAMPLE ❶ *Using Duplation and Mediation*

Multiply 21×32 using duplation and mediation.

SOLUTION Write 21 and 32 with a dash between the two numbers. Divide the number on the left, 21, by 2, drop the remainder, and place the quotient, 10, under the 21. Double the number on the right, 32, obtaining 64, and place it under the 32. You will then have the following number pairs.

$$21 — 32$$
$$10 — 64$$

Continue this process, dividing the number in the left-hand column by 2, disregarding the remainder, and doubling the number in the right-hand column, as shown below. When a 1 appears in the left-hand column, stop.

$$21 — 32$$
$$10 — 64$$
$$5 — 128$$
$$2 — 256$$
$$1 — 512$$

Next, cross out all the even numbers in the left-hand column and the corresponding numbers in the right-hand column.

$$21 — 32$$
$$\cancel{10} \text{—} \cancel{64}$$
$$5 — 128$$
$$\cancel{2} \text{—} \cancel{256}$$
$$1 — 512$$

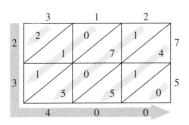

Figure 4.3

Now add the remaining numbers in the right-hand column to obtain $32 + 128 + 512 = 672$. The number obtained, 672, is the product of 21 and 32. Thus, $21 \times 32 = 672$. ●

Lattice Multiplication

Another method of multiplication is *lattice multiplication*. This method's name comes from the use of a grid, or lattice, when multiplying two numbers. It is also known as the *gelosia* method. The lattice method was brought to Europe by Fibonacci, in the 1200s who likely learned it from the Egyptians, Arabs, or Hindus (see the *Profile in Mathematics* on page 298).

┌ **EXAMPLE ②** *Using Lattice Multiplication*

Multiply 312×75 using lattice multiplication.

SOLUTION To multiply 312×75 using lattice multiplication, first construct a rectangle consisting of three columns (one for each digit of 312) and two rows (one for each digit of 75).

Place the digits 3, 1, 2 above the boxes and the digits 7, 5 on the right of the boxes, as shown in Fig. 4.1. Then place a diagonal in each box.

Complete each box by multiplying the number on top of the box by the number to the right of the box (Fig. 4.2). Place the units digit of the product below the diagonal and the tens digit of the product above the diagonal.

Figure 4.1

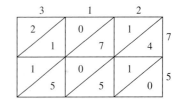

Figure 4.2

Add the numbers along the diagonals, as shown with the blue shaded arrows in Fig. 4.3, starting with the bottom right diagonal. If the sum in a diagonal is 10 or greater, record the units digit below the rectangle and carry the tens digit to the next diagonal to the left.

For example, when adding 4, 1, and 5 (along the second blue diagonal from the right), the sum is 10. Record the 0 below the rectangle and carry the 1 to the next blue diagonal. The sum of $1 + 1 + 7 + 0 + 5$ is 14. Record the 4 and carry the 1 to the next blue diagonal. The sum of the numbers in the next blue diagonal is $1 + 0 + 1 + 1$ or 3.

The answer is read down the left-hand column and along the bottom, as shown by the purple arrow in Fig. 4.3. Therefore, $312 \times 75 = 23,400$. ●

Napier's Rods

The third method used to multiply numbers was developed from the lattice method by John Napier in the early 1600s and is known as *Napier's rods* or *Napier's bones*. This method is similar to that used in modern computers. Napier developed a system of separate rods (which were often made out of bones) numbered 0 through 9 and an additional rod for an index, numbered vertically 1 through 9 (Fig. 4.4). Each rod is divided into 10 blocks. Each block below the first block contains a multiple of the number in the first block, with a diagonal separating the digits. The units digits are placed to the right of the diagonals and the tens digits to the left. Example 3 explains how Napier's rods are used to multiply numbers.

INDEX	0	1	2	3	4	5	6	7	8	9
1	0/0	0/1	0/2	0/3	0/4	0/5	0/6	0/7	0/8	0/9
2	0/0	0/2	0/4	0/6	0/8	1/0	1/2	1/4	1/6	1/8
3	0/0	0/3	0/6	0/9	1/2	1/5	1/8	2/1	2/4	2/7
4	0/0	0/4	0/8	1/2	1/6	2/0	2/4	2/8	3/2	3/6
5	0/0	0/5	1/0	1/5	2/0	2/5	3/0	3/5	4/0	4/5
6	0/0	0/6	1/2	1/8	2/4	3/0	3/6	4/2	4/8	5/4
7	0/0	0/7	1/4	2/1	2/8	3/5	4/2	4/9	5/6	6/3
8	0/0	0/8	1/6	2/4	3/2	4/0	4/8	5/6	6/4	7/2
9	0/0	0/9	1/8	2/7	3/6	4/5	5/4	6/3	7/2	8/1

Figure 4.4

EXAMPLE ❸ *Using Napier's Rods*

Multiply 8 × 365, using Napier's rods.

SOLUTION To multiply 8 × 365, line up the rods 3, 6, and 5 to the right of the index 8, as shown in Fig. 4.5. Below the 3, 6, and 5 place the blocks that contain the products of 8 × 3, 8 × 6, and 8 × 5, respectively. To obtain the answer, add along the diagonals as in the lattice method.

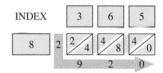

Thus, 8 × 365 = 2920.   ●

Example 4 illustrates the procedure to follow to multiply numbers containing more than one digit, using Napier's rods.

EXAMPLE ❹ *Using Napier's Rods to Multiply Two- and Three-Digit Numbers*

Multiply 48 × 365, using Napier's rods.

SOLUTION 48 × 365 = (40 + 8) × 365

Write (40 + 8) × 365 = (40 × 365) + (8 × 365). To find 40 × 365, determine 4 × 365 and multiply the product by 10. To evaluate 4 × 365, set up

INDEX	3	6	5
1	0/3	0/6	0/5
2	0/6	1/2	1/0
3	0/9	1/8	1/5
4	1/2	2/4	2/0
5	1/5	3/0	2/5
6	1/8	3/6	3/0
7	2/1	4/2	3/5
8	2/4	4/8	4/0
9	2/7	5/4	4/5

Figure 4.5

Napier's rods for 3, 6, and 5 with index 4, and then evaluate along the diagonals, as indicated.

Therefore, $4 \times 365 = 1460$. Then $40 \times 365 = 1460 \times 10 = 14,600$.

$$48 \times 365 = (40 \times 365) + (8 \times 365)$$
$$= 14,600 + 2920$$
$$= 17,520$$

8 × 365 = 2920
from Example 3

SECTION 4.5   EXERCISES

CONCEPT/WRITING EXERCISES

1. What are the three early computational methods discussed in this section?

2. a) Explain in your own words how multiplication by duplation and mediation is performed.

   b) Using the procedure given in part (a), multiply 73×21.

3. a) Explain in your own words how lattice multiplication is performed.

   b) Using the procedure given in part (a), multiply 143×26.

4. a) Explain in your own words how multiplication using Napier's rods is performed.

   b) Using the procedure given in part (a), multiply 34×7.

PRACTICE THE SKILLS

In Exercises 5–12, multiply using duplation and mediation.

5. 9×171
6. 39×57
7. 27×53
8. 138×41
9. 35×236
10. 96×53
11. 93×93
12. 49×124

In Exercises 13–20, use lattice multiplication to determine the product.

13. 5×191
14. 6×227
15. 8×469
16. 9×509
17. 75×12
18. 47×259
19. 314×652
20. 634×832

In Exercises 21–28, multiply using Napier's rods.

21. 3×49
22. 4×54
23. 5×63
24. 6×171
25. 5×125
26. 75×125
27. 9×6742
28. 7×3456

PROBLEM SOLVING

In Exercises 29 and 30, we show lattice multiplications.
(a) Determine the numbers being multiplied. Explain how you determined your answer. (b) Determine the product.

29.

30.
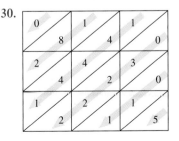

In Exercises 31 and 32, we solve a multiplication problem using Napier's rods. (a) Determine the numbers being multiplied. Each empty box contains a single digit. Explain how you determined your answer. (b) Determine the product.

31.

32.

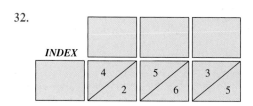

CHALLENGE PROBLEMS/GROUP ACTIVITIES

In Exercises 33 and 34, use the method of duplation and mediation to perform the multiplication. See Section 4.1 for Egyptian and Roman numerals. Write the answer in the numeration system in which the exercise is given.

33. $(\cap \text{III}) \cdot (\cap \cap \text{II})$ **34.** $(\text{XXVI}) \cdot (\text{LXVII})$

In Exercises 35 and 36, (a) use lattice multiplication to perform the multiplication. (Hint: Be sure not to list any number greater than or equal to the base within the box.) Write the answer in the base in which the exercise is given. (b) Multiply the numbers as explained in Section 4.4. If you do not obtain the results obtained in part (a), explain why.

35. $21_3 \times 21_3$ **36.** $24_5 \times 234_5$

RECREATIONAL MATHEMATICS

37. Obtain a clean U.S. $1 bill. On the back side of the bill is a circle containing a pyramid. On the base of the pyramid is a Roman numeral. a) Determine the value of that Roman numeral in our Hindu–Arabic system.

While you have that bill out, let's consider something else. For many people the number 13 is considered an unlucky number. In fact, many hotels do not have a thirteenth floor because many guests refuse to stay on the thirteenth floor of a hotel. Yet, if you look at the back of a $1 bill, you will find

13 steps on the pyramid.
13 letters in the Latin words *Annuit Coeptis.*
13 letters in "E Pluribus Unum."
13 stars above the eagle.
13 plum feathers on each span of the eagle's wing.
13 bars on the eagle's shield.
13 leaves on the olive branch.
13 fruits.
13 arrows.

In addition, the U.S. flag has 13 stripes, there were 13 original colonies, and don't forget the important Thirteenth Amendment that abolished slavery. b) So why, in your opinion, do we as a society fear the number 13? By the way, the official name for the fear of the number 13 is *triskaidekaphobia.*

INTERNET/RESEARCH ACTIVITIES

38. In addition to Napier's rods, John Napier is credited with making other important contributions to mathematics. Write a report on John Napier and his contributions to mathematics.

39. Write a paper explaining why the duplation and mediation method works.

CHAPTER ❹ SUMMARY

IMPORTANT FACTS

TYPES OF NUMERATION SYSTEMS

Additive (Egyptian hieroglyphics, Roman)

Multiplicative (traditional Chinese)

Ciphered (Ionic Greek)

Place-value (Babylonian, Mayan, Hindu–Arabic)

EARLY COMPUTATIONAL METHODS

Duplation and mediation

Lattice multiplication

Napier's rods

POSITIONAL VALUES

Hindu–Arabic (base 10) . . . , $(10)^4$, $(10)^3$, $(10)^2$, 10, 1
Base b . . . , b^4, b^3, b^2, b, 1

CHAPTER ④ REVIEW EXERCISES

4.1, 4.2

In Exercises 1–6, assume an additive numeration system in which $a = 1, b = 10, c = 100,$ and $d = 1000$. Find the value of the numeral.

1. ddcbba   **2.** dccbaaaa   **3.** bcccad

4. cbdadaaa   **5.** ddcccbaaaa   **6.** ccbaddac

In Exercises 7–12, assume the same additive numeration system as in Exercises 1–6. Write the numeral in terms of a, b, c, and d.

7. 31   **8.** 125   **9.** 293

10. 2008   **11.** 6851   **12.** 2314

In Exercises 13–18, assume a multiplicative numeration system in which $a = 1, b = 2, c = 3, d = 4, e = 5,$ $f = 6, g = 7, h = 8, i = 9, x = 10, y = 100,$ and $z = 1000$. Find the value of the numeral.

13. cxb   **14.** hxe

15. gydxi   **16.** dzfxh

17. ezfydxh   **18.** fziye

In Exercises 19–24, assume the same multiplicative numeration system as in Exercises 13–18. Write the Hindu–Arabic numeral in that system.

19. 37   **20.** 157   **21.** 862

22. 3094   **23.** 6004   **24.** 2001

In Exercises 25–36, use the following ciphered numeration system.

Decimal	1	2	3	4	5	6	7	8	9
Units	a	b	c	d	e	f	g	h	i
Tens	j	k	l	m	n	o	p	q	r
Hundreds	s	t	u	v	w	x	y	z	A
Thousands	B	C	D	E	F	G	H	I	J
Ten thousands	K	L	M	N	O	P	Q	R	S

Convert the numeral to a Hindu–Arabic numeral.

25. rc   **26.** tc   **27.** woh

28. NGzqc   **29.** PEvqa   **30.** Pwki

Write each of the following in the ciphered numeration system.

31. 31   **32.** 274   **33.** 493

34. 1997   **35.** 53,467   **36.** 75,496

In Exercises 37–42, convert 1776 to a numeral in the indicated numeration system.

37. Egyptian   **38.** Roman   **39.** Chinese

40. Ionic Greek   **41.** Babylonian   **42.** Mayan

In Exercises 43–48, convert the numeral to a Hindu–Arabic numeral.

43. ⌐⌐⌐𝄞𝄞∩∩∩|||||

44.

45. χπε   **46.** MCMXCI

47. ⟨⟨𝌆 ⟨⟨𝌆|||| **48.** •• ••• ═══

4.3

In Exercises 49–54, convert the numeral to a Hindu–Arabic numeral.

49. 47_8 **50.** 101_2 **51.** 130_4

52. 3425_7 **53.** $A94_{12}$ **54.** 20220_3

In Exercises 55–60, convert 463 to a numeral in the base indicated.

55. base 2   **56.** base 3   **57.** base 4

58. base 8   **59.** base 12   **60.** base 16

4.4

In Exercises 61–66, add in the base indicated.

61. 52_7
$+ 55_7$

62. 10110_2
$+ 11001_2$

63. $9B_{12}$
$+ 87_{12}$

64. $2B9_{16}$
$+ 456_{16}$

65. 3024_5
$+ 4023_5$

66. 3407_8
$+ 7014_8$

76. 476_8
$\times 23_8$

77. 126_{12}
$\times 47_{12}$

78. $1A3_{16}$
$\times 12_{16}$

In Exercises 67–72, subtract in the base indicated.

67. 4032_7
$- 321_7$

68. 1001_2
$- 101_2$

69. $A7B_{12}$
$- 95_{12}$

In Exercises 79–84, divide in the base indicated.

79. $2_3\overline{)120_3}$

80. $2_4\overline{)320_4}$

81. $3_5\overline{)130_5}$

82. $4_6\overline{)3020_6}$

83. $3_6\overline{)2034_6}$

84. $6_8\overline{)5072_8}$

70. 4321_5
$- 442_5$

71. 1713_8
$- 1243_8$

72. $F64_{16}$
$- 2A3_{16}$

4.5

85. Multiply 142×24, using the duplation and mediation method.

In Exercises 73–78, multiply in the base indicated.

86. Multiply 142×24, using lattice multiplication.

73. 1011_2
$\times 101_2$

74. 221_3
$\times 22_3$

75. 34_5
$\times 21_5$

87. Multiply 142×24, using Napier's rods.

CHAPTER 4 TEST

1. Explain the difference between a numeral and a number.

In Exercises 2–7, convert the numeral to a Hindu–Arabic numeral.

2. MMCDLXXXIV

3. ⟨⟨⟨Ⅰ ⟨ⅠⅠⅠⅠⅠ

4. 八
千
零
九
十

5. ••
••
•••
••••

6. ◁𓏠𓏪𓍯9∩∩∩ΙΙ

7. 'βψμε

In Exercises 8–12, convert the base 10 numbered to a numeral in the numeration system indicated.

8. 2124 to Egyptian

9. 2476 to Ionic Greek

10. 1434 to Mayan

11. 1596 to Babylonian

12. 3706 to Roman

In Exercises 13–16, describe briefly each of the systems of numeration. Explain how each type of numeration system is used to represent numbers.

13. Additive system

14. Multiplicative system

15. Ciphered system

16. Place-value system

In Exercises 17–20, convert the numeral to a Hindu–Arabic numeral

17. 23_4

18. 403_5

19. 101101_2

20. $3A7_{12}$

In Exercises 21–24, convert each of the following to a numeral in the base indicated.

21. 36 to base 2

22. 93 to base 8

23. 2356 to base 12

24. 2938 to base 16

In Exercises 25–28, perform the indicated operations.

25. 1101_2
$+ 1011_2$

26. 324_6
$- 142_6$

27. 45_6
$\times 23_6$

28. $3_5\overline{)1210_5}$

29. Multiply 35×28, using duplation and mediation.

30. Multiply 43×196, using lattice multiplication.

GROUP PROJECTS

U.S. POSTAL SERVICE BAR CODES

Wherever we look nowadays, we see bar codes. We find them on items we buy at grocery stores and department stores and on many pieces of mail we receive. There are various types of bar codes, but each can be considered a type of numeration system. Although bar codes may vary in design, most are made up of a series of long and short bars. (New bar codes now being developed use a variety of shapes.) In this group project, we explain how postal bar codes are used.

The U.S. Postal Service introduced a bar coding system for zip codes in 1976. The system became known as Postnet (*post*al *n*umeric *e*ncoding *t*echnique), and it has been refined over the years. Our basic zip code consists of five digits. The post office would like us to use the basic zip code followed by a hyphen and four additional digits. The post office refers to this nine-digit zip code as "zip + 4."

The Postnet bar code uses a series of long and short bars. A bar code may contain either 52 or 62 bars. The code designates the location to which the letter is being sent. The following bar code, with 52 bars, is for an address in Pittsburgh, Pennsylvania.

|ıllılılııılılılılllıılıılılılılllıııllıılıııl

15250-7406 (Pittsburgh, PA)

In bar codes, each short bar represents 0 and each long bar represents 1. Each code starts and ends with a long bar that is *not* used in determining the zip + 4. If the code contains 52 bars, the code represents the zip + 4 and an extra digit referred to as a check digit. If the code contains 62 bars, it contains the zip + 4, the last two digits of the address number, and a check digit. If the code contains 52 bars, the sum of the zip + 4 and the check digit must equal a number that is divisible by 10. If the code contains 62 bars, the sum of the zip + 4, the last two digits of the address number, and the check digit must equal a number that is divisible by 10. The check digit is added to make each sum divisible by 10.

In a postal bar code, each of the digits 0 through 9 is represented by a series of five digits containing zeros and ones:

> 11000 (0) 00011 (1) 00101 (2) 00110 (3) 01001 (4)
> 01010 (5) 01100 (6) 10001 (7) 10010 (8) 10100 (9)

Consider the postal code from Pittsburgh given earlier. If you disregard the bar on the left, the next five bars are ıllll. Since each small bar represents a 0 and each large bar represents a 1, these five bars can be represented as 00011. From the chart, we see that this represents the number 1. The first five bars (after the bar on the far left has been excluded) tell the region of the country in which the address is located on the map shown below. Notice that Pennsylvania is located, along with New York and Delaware, in the region marked 1 on the map.

National zip code areas

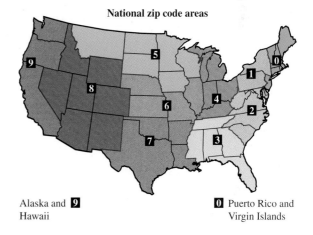

Alaska and **9**
Hawaii

0 Puerto Rico and
Virgin Islands

The second five lines in the bar code, ılılı, represent 01010 and have a value of 5. The other digits in the zip + 4 are determined in a similar manner. This code has 52 bars. The 45 bars, after the first bar, give the zip + 4. If you add the digits in the zip + 4, you get $1 + 5 + 2 + 5 + 0 + 7 + 4 + 0 + 6 = 30$. Since 30 is divisible by 10, the five bars to the left of the bar on the very right should be 0. Note that llııı is represented as 11000 and has a value of 0. If, for example, the sum of the nine digits in the zip + 4 were 36, then the last five digits would need to represent the number 4, to make the sum of the digits divisible by 10. The five bars to the left of the last bar on the right are always used as a check.

    Now let's work some problems.

a) For the Postnet code

determine the zip + 4 and the check digit. Then check by adding the zip + 4 and the check digit. Is the sum divisible by 10?

b) For each of the following Postnet codes, determine the zip + 4, the last two numbers of the address number (if applicable), and the check digit.

  i)

  ii)

c) Construct the Postnet code of long and short bars for each of the numbers. The numbers represent the zip + 4 and the last two digits of the address number. Do not forget the check digit.

  i) 32226-8600-34   **ii)** 20794-1063-50

d) Construct the 52-bar Postnet code for your college's zip + 4. Don't forget to include the check digit.

CHAPTER 5

Number Theory and the Real Number System

▲ Number theory plays an important role in cell phone technology.

WHAT YOU WILL LEARN

- An introduction to number theory
- Prime numbers
- Integers, rational numbers, irrational numbers, and real numbers
- Properties of real numbers
- Rules of exponents and scientific notation
- Arithmetic and geometric sequences
- The Fibonacci sequence

WHY IT IS IMPORTANT

Every time we use a computer, make a telephone call, or watch television we use a device that relies on numbers to operate. In addition to playing many roles in our everyday lives, numbers are also used to describe the natural world, communicate information, and model problems facing scientists and researchers. In this chapter, we will see how number theory, the study of numbers and their properties, makes all these roles possible.

5.1   NUMBER THEORY

▲ There are many ways to display 48 pieces of chocolate.

Suppose you decide to start your own chocolate candy company. You would like to sell boxes containing 48 pieces of candy. You must decide how to arrange the pieces within the box. You could have six rows with eight pieces in each row, or four rows with twelve pieces in each row, or three layers with four rows each with four pieces in each row. Each of these possibilities involves writing the number 48 as the product of two or more smaller numbers. In this section, we will discuss many other similar problems related to the writing of numbers as the product of smaller numbers.

This chapter introduces *number theory*, the study of numbers and their properties. The numbers we use to count are called the *counting numbers* or *natural numbers*. Because we begin counting with the number 1, the set of natural numbers begins with 1. The set of natural numbers is frequently denoted by N:

$$N = \{1, 2, 3, 4, 5, \dots\}$$

Any natural number can be expressed as a product of two or more natural numbers. For example, $8 = 2 \times 4$, $16 = 4 \times 4$, and $19 = 1 \times 19$. The natural numbers that are multiplied together are called factors of the product. For example,

$$2 \times 4 = 8$$
$$\uparrow \quad \uparrow$$
Factors

A natural number may have many factors. For example, what pairs of numbers have a product of 18?

$$1 \cdot 18 = 18$$
$$2 \cdot 9 = 18$$
$$3 \cdot 6 = 18$$

The numbers 1, 2, 3, 6, 9, and 18 are all factors of 18. Each of these numbers divides 18 without a remainder.

If a and b are natural numbers, we say that a is a *divisor* of b or a *divides* b, symbolized $a \mid b$, if the quotient of b divided by a has a remainder of 0. If a divides b, then b is *divisible* by a. For example, 4 divides 12, symbolized $4 \mid 12$, since the quotient of 12 divided by 4 has a remainder of 0. Note that 12 is divisible by 4. The notation $7 \nmid 12$ means that 7 does not divide 12. Note that every factor of a natural number is also a divisor of the natural number. *Caution:* Do not confuse the symbols $a \mid b$ and a / b; $a \mid b$ means "a divides b" and a / b means "a divided by b" ($a \div b$). The symbols a / b and $a \div b$ indicate that the operation of division is to be performed, and b may or may not be a divisor of a.

Prime and Composite Numbers

Every natural number greater than 1 can be classified as either a prime number or a composite number.

A **prime number** is a natural number greater than 1 that has exactly two factors (or divisors), itself and 1.

The number 5 is a prime number because it is divisible only by the factors 1 and 5. The first eight prime numbers are 2, 3, 5, 7, 11, 13, 17, and 19. The number 2 is the only even prime number. All other even numbers have at least three divisors: 1, 2, and the number itself.

A **composite number** is a natural number that is divisible by a number other than itself and 1.

Any natural number greater than 1 that is not prime is composite. The first eight composite numbers are 4, 6, 8, 9, 10, 12, 14, and 15.

The number 1 is neither prime nor composite; it is called a *unit*. The number 38 has at least three divisors, 1, 2, and 38, and hence is a composite number. In contrast, the number 23 is a prime number since its only divisors are 1 and 23.

More than 2000 years ago, the ancient Greeks developed a technique for determining which numbers are prime numbers and which are not. This technique is known as the *sieve of Eratosthenes*, for the Greek mathematician Eratosthenes of Cyrene who first used it.

To find the prime numbers less than or equal to any natural number, say, 50, using this method, list the first 50 counting numbers (Fig. 5.1). Cross out 1 since it is not a prime number. Circle 2, the first prime number. Then cross out all the multiples of 2: 4, 6, 8, . . . , 50. Circle the next prime number, 3. Cross out all multiples of 3 that are not already crossed out. Continue this process of crossing out multiples of prime numbers until you reach the prime number p, such that $p \cdot p$, or p^2, is greater than the last number listed, in this case 50. Therefore, we next circle 5 and cross out its multiples. Then circle 7 and cross out its multiples. The next prime number is 11, and $11 \cdot 11$, or 121, is greater than 50, so you are done. At this point, circle all the remaining numbers to obtain the prime numbers less than or equal to 50. The prime numbers less than or equal to 50 are 2, 3, 5, 7, 11, 13, 17, 19, 23, 29, 31, 37, 41, 43, and 47.

1̸	②	③	4̸	⑤	6̸	⑦	8̸	9̸	1̸0̸
⑪	1̸2̸	⑬	1̸4̸	1̸5̸	1̸6̸	⑰	1̸8̸	⑲	2̸0̸
2̸1̸	2̸2̸	㉓	2̸4̸	2̸5̸	2̸6̸	2̸7̸	2̸8̸	㉙	3̸0̸
㉛	3̸2̸	3̸3̸	3̸4̸	3̸5̸	3̸6̸	㊲	3̸8̸	3̸9̸	4̸0̸
㊶	4̸2̸	㊸	4̸4̸	4̸5̸	4̸6̸	㊼	4̸8̸	4̸9̸	5̸0̸

Figure 5.1

Now we turn our attention to composite numbers and their factors. The rules of divisibility given in the chart on page 226 are helpful in finding divisors (or factors) of composite numbers.

Rules of Divisibility

Divisible by	Test	Example
2	The number is even.	924 is divisible by 2 since 924 is even.
3	The sum of the digits of the number is divisible by 3.	924 is divisible by 3 since the sum of the digits, $9 + 2 + 4 = 15$, and 15 is divisible by 3.
4	The number formed by the last two digits of the number is divisible by 4.	924 is divisible by 4 since the number formed by the last two digits, 24, is divisible by 4.
5	The number ends in 0 or 5.	265 is divisible by 5 since the number ends in 5.
6	The number is divisible by both 2 and 3.	924 is divisible by 6 since it is divisible by both 2 and 3.
8	The number formed by the last three digits of the number is divisible by 8.	5824 is divisible by 8 since the number formed by the last three digits, 824, is divisible by 8.
9	The sum of the digits of the number is divisible by 9.	837 is divisible by 9 since the sum of the digits, 18, is divisible by 9.
10	The number ends in 0.	290 is divisible by 10 since the number ends in 0.

The test for divisibility by 6 is a particular case of the general statement that the product of two prime divisors of a number is a divisor of the number. Thus, for example, if both 3 and 7 divide a number, then 21 will also divide the number.

Note that the chart does not list rules for the divisibility for the number 7. The rule for 7 is given in Exercise 92 at the end of this section. You may find after using the rule that the easiest way to check divisibility by 7 is simply to perform the division.

⌐EXAMPLE ❶ *Using the Divisibility Rules*

Determine whether 384,540 is divisible by

a) 2   b) 3   c) 4   d) 5   e) 6   f) 8   g) 9   h) 10

SOLUTION

a) Since 384,540 is even, it is divisible by 2.

b) The sum of the digits of 384,540 is $3 + 8 + 4 + 5 + 4 + 0 = 24$. Since 24 is divisible by 3, the number 384,540 is divisible by 3.

c) The number formed by the last two digits is 40. Since 40 is divisible by 4, the number 384,540 is divisible by 4.

d) Since 384,540 has 0 as the last digit, the number 384,540 is divisible by 5.

e) Since 384,540 is divisible by both 2 and 3, the number 384,540 is divisible by 6.

f) The number formed by the last three digits is 540. Since 540 is not divisible by 8, the number 384,540 is not divisible by 8.

g) The sum of the digits of 384,540 is 24. Since 24 is not divisible by 9, the number 384,540 is not divisible by 9.

h) Since 384,540 has 0 as the last digit, the number 384,540 is divisible by 10. ●

Every composite number can be expressed as a product of prime numbers. The process of breaking a given composite number down into a product of prime numbers is called *prime factorization*. The prime factorization of 18 is $3 \times 3 \times 2$. No other natural number listed as a product of primes will have the same prime factorization as 18. The *fundamental theorem of arithmetic* states this concept formally. (A *theorem* is a statement or proposition that can be proven true.)

THE FUNDAMENTAL THEOREM OF ARITHMETIC
Every composite number can be expressed as a *unique* product of prime numbers.

In writing the prime factorization of a number, the order of the factors is immaterial. For example, we may write the prime factors of 18 as $3 \times 3 \times 2$ or $2 \times 3 \times 3$ or $3 \times 2 \times 3$.

A number of techniques can be used to find the prime factorization of a number. Two methods are illustrated.

Method 1: Branching

To find the prime factorization of a number, select any two numbers whose product is the number to be factored. If the factors are not prime numbers, continue factoring each composite number until all factors are prime.

EXAMPLE ❷ *Prime Factorization by Branching*

Write 2100 as a product of primes.

SOLUTION Select any two numbers whose product is 2100. Among the many choices, two possibilities are $21 \cdot 100$ and $30 \cdot 70$. Let us consider $21 \cdot 100$. Since neither 21 nor 100 are prime numbers, find any two numbers whose product is 21 and any two numbers whose product is 100. Continue branching as shown in Fig. 5.2 until the numbers in the last row are all prime numbers. To determine the answer, write the product of all the prime factors. The branching diagram is sometimes called a *factor tree*.

Figure 5.2

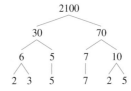

Figure 5.3

We see that the numbers in the last row of factors in Fig. 5.2 are all prime numbers. Thus, the prime factorization of 2100 is $3 \cdot 7 \cdot 2 \cdot 5 \cdot 2 \cdot 5 = 2 \cdot 2 \cdot 3 \cdot 5 \cdot 5 \cdot 7 = 2^2 \cdot 3 \cdot 5^2 \cdot 7$. Note from Fig. 5.3 that had we chosen 30 and 70 as the first pair of factors, the prime factorization would still be $2^2 \cdot 3 \cdot 5^2 \cdot 7$. ●

Method 2: Division

To obtain the prime factorization of a number by this method, divide the given number by the smallest prime number by which it is divisible. Place the quotient under the given number. Then divide the quotient by the smallest prime number by which it is divisible and again record the quotient. Repeat this process until the quotient is a prime number. The prime factorization is the product of all the prime divisors and the prime (or last) quotient. This procedure is illustrated in Example 3.

┌EXAMPLE ❸ *Prime Factorization by Division*

Write 2100 as a product of prime numbers.

SOLUTION Because 2100 is an even number, the smallest prime number that divides it is 2. Divide 2100 by 2. Place the quotient, 1050, below the 2100. Repeat this process of dividing each quotient by the smallest prime number that divides it.

2	2100
2	1050
3	525
5	175
5	35
	7

The final quotient, 7, is a prime number, so we stop. The prime factorization of 2100 is

$$2 \cdot 2 \cdot 3 \cdot 5 \cdot 5 \cdot 7 = 2^2 \cdot 3 \cdot 5^2 \cdot 7$$ ●

Note that despite the different methods used in Examples 2 and 3, the answer is the same.

Greatest Common Divisor

The discussion in Section 5.3 of how to reduce fractions makes use of the greatest common divisor (GCD). We will now discuss how to determine the GCD of a set of numbers. One technique of finding the GCD is to use prime factorization.

The **greatest common divisor (GCD)** of a set of natural numbers is the largest natural number that divides (without remainder) every number in that set.

What is the GCD of 12 and 18? One way to determine it is to list the divisors (or factors) of 12 and 18:

Divisors of 12 $\{\mathbf{1}, \mathbf{2}, \mathbf{3}, 4, \mathbf{6}, 12\}$

Divisors of 18 $\{\mathbf{1}, \mathbf{2}, \mathbf{3}, \mathbf{6}, 9, 18\}$

The common divisors are 1, 2, 3, and 6. Therefore, the greatest common divisor is 6.

If the numbers are large, this method of finding the GCD is not practical. The GCD can be found more efficiently by using prime factorization.

> ## TO DETERMINE THE GREATEST COMMON DIVISOR OF TWO OR MORE NUMBERS
>
> 1. Determine the prime factorization of each number.
> 2. List each prime factor with the smallest exponent that appears in each of the prime factorizations.
> 3. Determine the product of the factors found in step 2.

Example 4 illustrates this procedure.

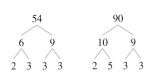

Figure 5.4

EXAMPLE ❹ *Using Prime Factorization to Determine the GCD*

Determine the GCD of 54 and 90.

> **SOLUTION** The branching method of finding the prime factors of 54 and 90 is illustrated in Fig. 5.4.
>
> a) The prime factorization of 54 is $2 \cdot 3^3$, and the prime factorization of 90 is $2 \cdot 3^2 \cdot 5$.
>
> b) The prime factors with the smallest exponents that appear in each of the factorizations of 54 and 90 are 2 and 3^2.
>
> c) The product of the factors found in step 2 is $2 \cdot 3^2 = 2 \cdot 9 = 18$. The GCD of 54 and 90 is 18. Eighteen is the largest natural number that divides both 54 and 90. ●

EXAMPLE ❺ *Determining the GCD*

Determine the GCD of 315 and 450.

> **SOLUTION**
>
> a) The prime factorization of 315 is $3^2 \cdot 5 \cdot 7$, and the prime factorization of 450 is $2 \cdot 3^2 \cdot 5^2$. You should verify these answers using either the branching method or the division method.
>
> b) The prime factors with the smallest exponents that appear in each of the factorizations of 315 and 450 are 3^2 and 5.
>
> c) The product of the factors found in step 2 is $3^2 \cdot 5 = 9 \cdot 5 = 45$. The greatest common divisor of 315 and 450 is 45. It is the largest natural number that will evenly divide both 315 and 450. ●

Least Common Multiple

To perform addition and subtraction of fractions (Section 5.3), we use the least common multiple (LCM). We will now discuss how to determine the LCM of a set of numbers.

> The **least common multiple (LCM)** of a set of natural numbers is the smallest natural number that is divisible (without remainder) by each element of the set.

What is the least common multiple of 12 and 18? One way to determine the LCM is to list the multiples of each number:

Multiples of 12 $\{12, 24, \mathbf{36}, 48, 60, \mathbf{72}, 84, 96, \mathbf{108}, 120, 132, \mathbf{144}, \dots\}$

Multiples of 18 $\{18, \mathbf{36}, 54, \mathbf{72}, 90, \mathbf{108}, 126, \mathbf{144}, 162, \dots\}$

Some common multiples of 12 and 18 are 36, 72, 108, and 144. The least common multiple, 36, is the smallest number that is divisible by both 12 and 18. Usually, the most efficient method of finding the LCM is to use prime factorization.

TO DETERMINE THE LEAST COMMON MULTIPLE OF TWO OR MORE NUMBERS

1. Determine the prime factorization of each number.
2. List each prime factor with the greatest exponent that appears in any of the prime factorizations.
3. Determine the product of the factors found in step 2.

Example 6 illustrates this procedure.

EXAMPLE ⑥   *Using Prime Factorization to Determine the LCM*

Determine the LCM of 54 and 90.

SOLUTION

a) Determine the prime factors of each number. In Example 4, we determined that

$$54 = 2 \cdot 3^3 \quad \text{and} \quad 90 = 2 \cdot 3^2 \cdot 5$$

b) List each prime factor with the greatest exponent that appears in either of the prime factorizations: $2, 3^3, 5$.

c) Determine the product of the factors found in step 2:

$$2 \cdot 3^3 \cdot 5 = 2 \cdot 27 \cdot 5 = 270$$

Thus, 270 is the LCM of 54 and 90. It is the smallest natural number that is divisible by both 54 and 90. ●

EXAMPLE ⑦   *Determining the LCM*

Determine the LCM of 315 and 450.

SOLUTION

a) Determine the prime factorization of each number. In Example 5, we determined that

$$315 = 3^2 \cdot 5 \cdot 7 \quad \text{and} \quad 450 = 2 \cdot 3^2 \cdot 5^2.$$

b) List each prime factor with the greatest exponent that appears in either of the prime factorizations: $2, 3^2, 5^2, 7$.

c) Determine the product of the factors from step (b)

$$2 \cdot 3^2 \cdot 5^2 \cdot 7 = 2 \cdot 9 \cdot 25 \cdot 7 = 3150$$

Thus, 3150 is the least common multiple of 315 and 450. It is the smallest natural number that is evenly divisible by both 315 and 450. ●

The Search For Larger Prime Numbers

More than 2000 years ago, the Greek mathematician Euclid proved that there is no largest prime number. Mathematicians, however, continue to strive to find larger and larger prime numbers.

Marin Mersenne (1588–1648), a seventeenth-century monk, found that numbers of the form $2^n - 1$ are often prime numbers when n is a prime number. For example,

$$2^2 - 1 = 4 - 1 = 3 \qquad 2^3 - 1 = 8 - 1 = 7$$
$$2^5 - 1 = 32 - 1 = 31 \qquad 2^7 - 1 = 128 - 1 = 127$$

Numbers of the form $2^n - 1$ that are prime are referred to as *Mersenne primes*. The first 10 Mersenne primes occur when $n = 2, 3, 5, 7, 13, 17, 19, 31, 61, 89$. The first time the expression $2^n - 1$ does not generate a prime number, for prime number n, is when n is 11. The number $2^{11} - 1$ is a composite number (see Exercise 90).

Scientists frequently use Mersenne primes in their search for larger and larger primes. The largest prime number found to date was discovered on September 4, 2006, by Curtis Cooper and Stephen Boone, professors at Central Missouri State University. Cooper and Boone had worked in conjunction with the Great Internet Mersenne Prime Search (GIMPS; see Mathematics Today at left) to use the university's computers during down time to conduct the search. The number is the Mersenne prime $2^{32,582,657} - 1$. This record prime number is the 44th known Mersenne prime, and when written out it is 9,808,358 digits long, which is more than 12 miles long if written using a standard 12-point font.

More About Prime Numbers

Another mathematician who studied prime numbers was Pierre de Fermat (1601–1665). A lawyer by profession, Fermat became interested in mathematics as a hobby. He became one of the finest mathematicians of the seventeenth century. Fermat conjectured that each number of the form $2^{2^n} + 1$, now referred to as a *Fermat number*, was prime for each natural number n. Recall that a *conjecture* is a supposition that has not been proved nor disproved. In 1732, Leonhard Euler proved that for $n = 5$, $2^{32} + 1$ was a composite number, thus disproving Fermat's conjecture.

Since Euler's time, mathematicians have only been able to evaluate the sixth, seventh, eighth, ninth, tenth, and eleventh Fermat numbers to determine whether they are prime or composite. Each of these numbers has been shown to be composite. The eleventh Fermat number was factored by Richard Brent and François Morain in 1988. The sheer magnitude of the numbers involved makes it difficult to test these numbers, even with supercomputers.

In 1742, Christian Goldbach conjectured in a letter to Euler that every even number greater than or equal to 4 can be represented as the sum of two (not necessarily distinct) prime numbers (for example, $4 = 2 + 2$, $6 = 3 + 3$, $8 = 3 + 5$, $10 = 5 + 5$, $12 = 5 + 7$). This conjecture became known as *Goldbach's conjecture*, and it remains unproven to this day. The *twin prime conjecture* is another famous long-standing conjecture. *Twin primes* are primes of the form p and $p + 2$ (for example, 3 and 5, 5 and 7, 11 and 13). This conjecture states that there are an infinite number of pairs of twin primes. At the time of this writing, the largest twin primes are of the form $16,869,987,339,975 \cdot 2^{171,960} \pm 1$, which was found by a group of Hungarian mathematicians in September 2005.

SECTION 5.1   EXERCISES

CONCEPT/WRITING EXERCISES

1. What is number theory?

2. What does "*a* and *b* are factors of *c*" mean?

3. **a)** What does "*a* divides *b*" mean?

   b) What does "*a* is divisible by *b*" mean?

4. What is a prime number?

5. What is a composite number?

6. What does the fundamental theorem of arithmetic state?

7. **a)** What is the least common multiple (LCM) of a set of natural numbers?

   b) In your own words, explain how to find the LCM of a set of natural numbers by using prime factorization.

   c) Find the LCM of 16 and 40 by using the procedure given in part (b).

8. **a)** What is the greatest common divisor (GCD) of a set of natural numbers?

   b) In your own words, explain how to find the GCD of a set of natural numbers by using prime factorization.

   c) Find the GCD of 16 and 40 by using the procedure given in part (b).

9. What are Mersenne primes?

10. What is a conjecture?

11. What is Goldbach's conjecture?

12. What are twin primes?

PRACTICE THE SKILLS

13. Use the sieve of Eratosthenes to find the prime numbers up to 100.

14. Use the sieve of Eratosthenes to find the prime numbers up to 150.

In Exercises 15–26, determine whether the statement is true or false. Modify each false statement to make it a true statement.

15. 7 is a factor of 28.

16. $7 \mid 24$.

17. 13 is a multiple of 26.

18. 6 is a divisor of 18.

19. 8 is divisible by 56.

20. 15 is a factor of 45.

21. If a number is not divisible by 5, then it is not divisible by 10.

22. If a number is not divisible by 10, then it is not divisible by 5.

23. If a number is divisible by 3, then every digit of the number is divisible by 3.

24. If every digit of a number is divisible by 3, then the number itself is divisible by 3.

25. If a number is divisible by 2 and 3, then the number is divisible by 6.

26. If a number is divisible by 3 and 5, then the number is divisible by 15.

In Exercises 27–32, determine whether the number is divisible by each of the following numbers: 2, 3, 4, 5, 6, 8, 9, and 10.

27. 11,115

28. 33,813

29. 474,138

30. 170,820

31. 1,882,320

32. 3,941,221

33. Determine a number that is divisible by 2, 3, 4, 5, and 6.

34. Determine a number that is divisible by 3, 4, 5, 9, and 10.

In Exercises 35–46, find the prime factorization of the number.

35. 48

36. 58

37. 168

38. 315

39. 332

40. 399

41. 513

42. 663

43. 1336

44. 1313

45. 2001

46. 3190

In Exercises 47–56, find (a) the greatest common divisor (GCD) and (b) the least common multiple (LCM).

47. 6 and 21

48. 15 and 24

49. 20 and 35

50. 32 and 224

51. 40 and 900

52. 120 and 240

53. 96 and 212

54. 240 and 285

55. 24, 48, and 128

56. 18, 78, and 198

PROBLEM SOLVING

57. Find the next two sets of twin primes that follow the set 11, 13.

58. The primes 2 and 3 are consecutive natural numbers. Is there another pair of consecutive natural numbers both of which are prime? Explain.

59. For each pair of numbers, determine whether the numbers are relatively prime. Write yes or no as your answer.

   a) 10, 21

   b) 22, 26

   c) 27, 28

   d) 85, 119

60. Find the first five Mersenne prime numbers.

61. Find the first three Fermat numbers and determine whether they are prime or composite.

62. Show that Goldbach's conjecture is true for the even numbers 4 through 20.

63. *Setting Up Chairs* Jerrett Dumouchel is setting up chairs for his school band concert. He needs to put 120 chairs on the gymnasium floor in rows of equal size, and there must be at least 2 chairs in each row. List the number of rows and the number of chairs in each row that are possible.

64. *U.S. Senate Committees* The U.S. Senate consists of 100 members. Senate committees are to be formed so that each of the committees contains the same number of senators and each senator is a member of exactly one committee. The committees are to have more than 2 members but fewer than 50 members. There are various ways that these committees can be formed.

   a) What size committees are possible?

   b) How many committees are there for each size?

65. *Taking Medicine* Sue Chen-Harper takes the medicine bisphosphonate once every 30 days and the medicine pegaspargase once every 14 days. If Sue took both medicines on June 1, how many days would it be before she has to take both medicines on the same day again?

66. *Barbie and Ken* Mary Lois King collects Barbie dolls and Ken dolls. She has 390 Barbie dolls and 468 Ken dolls. Mary Lois wishes to display the dolls in groups so that the same number of dolls are in each group and that each doll belongs to one group. If each group is to consist only of Barbie dolls or only of Ken dolls, what is the largest number of dolls Mary Lois can have in each group?

67. *Toy Car Collection* Martha Goshaw collects Matchbox® and HotWheels® toy cars. She has 70 red cars and 175 blue cars. She wants to line up her cars in groups so that each group has the same number of cars and each group contains only red cars or only blue cars. What is the largest number of cars she can have in a group?

68. *Stacking Trading Cards* Desmond Freeman collects trading cards. He has 432 baseball cards and 360 football cards. He wants to make stacks of cards on a table so that each stack contains the same number of cards and each card belongs to one stack. If the baseball and football cards must not be mixed in the stacks, what is the largest number of cards that he can have in a stack?

69. *Tree Rows* Elizabeth Dwyer is the manager at Queen Palm Nursery and is in charge of displaying potted trees in rows.

Elizabeth has 150 citrus trees and 180 palm trees. She wants to make rows of trees so that each row has the same number of trees and each tree is in a row. If the citrus trees and the palm trees must not be mixed in the rows, what is the largest number of trees that she can have in a row?

70. *Car Maintenance* For many sport utility vehicles, it is recommended that the oil be changed every 3500 miles and that the tires be rotated every 6000 miles. If Carmella Gonzalez just had the oil changed and tires rotated on her SUV during the same visit to her mechanic, how many miles will she drive before she has the oil changed and tires rotated again during the same visit?

71. *Work Schedules* Sara Pappas and Harry Kinnan both work the 3:00 P.M. to 11:00 P.M. shift. Sara has every fifth night off and Harry has every sixth night off. If they both have tonight off, how many days will pass before they have the same night off again?

72. *Prime Numbers* Consider the first eight prime numbers greater than 3. The numbers are 5, 7, 11, 13, 17, 19, 23, and 29.

   a) Determine which of these prime numbers differs by 1 from a multiple of the number 6.

   b) Use inductive reasoning and the results obtained in part (a) to make a conjecture regarding prime numbers.

   c) Select a few more prime numbers and determine whether your conjecture appears to be correct.

73. State a procedure that defines a divisibility test for 15.

74. State a procedure that defines a divisibility test for 22.

Euclidean Algorithm *Another method that can be used to find the greatest common divisor is known as the Euclidean algorithm. We illustrate this procedure by finding the GCD of 60 and 220.*

   First divide 220 by 60 as shown below. Disregard the quotient 3 and then divide 60 by the remainder 40. Continue this process of dividing the divisors by the remainders until you obtain a remainder of 0. The divisor in the last division, in which the remainder is 0, is the GCD.

$$\begin{array}{r} 3 \\ 60\overline{)220} \\ 180 \\ \hline 40 \end{array} \qquad \begin{array}{r} 1 \\ 40\overline{)60} \\ 40 \\ \hline 20 \end{array} \qquad \begin{array}{r} 2 \\ 20\overline{)40} \\ 40 \\ \hline 0 \end{array}$$

Since 40/20 had a remainder of 0, the GCD is 20.

   In Exercises 75–80, use the Euclidean algorithm to find the GCD.

75. 15, 40

76. 12, 28

77. 35, 105

78. 78, 104

79. 150, 180

80. 210, 560

Perfect Numbers A number whose proper factors (factors other than the number itself) add up to the number is called a perfect number. For example, 6 is a perfect number because its proper factors are 1, 2, and 3, and 1 + 2 + 3 = 6. Determine which, if any, of the following numbers are perfect.

81. 12

82. 28

83. 48

84. 496

CHALLENGE PROBLEMS/GROUP ACTIVITIES

85. *Number of Factors* The following procedure can be used to determine the *number of factors* (or *divisors*) of a composite number. Write the number in prime factorization form. Examine the exponents on the prime numbers in the prime factorization. Add 1 to each exponent and then find the product of these numbers. This product gives the number of positive divisors of the composite number.

   a) Use this procedure to determine the number of divisors of 60.

   b) To check your answer, list all the divisors of 60. You should obtain the same number of divisors found in part (a).

86. Recall that if a number is divisible by both 2 and 3, then the number is divisible by 6. If a number is divisible by both 2 and 4, is the number necessarily divisible by 8? Explain your answer.

87. The product of any three consecutive natural numbers is divisible by 6. Explain why.

88. A number in which each digit except 0 appears exactly three times is divisible by 3. For example, 888,444,555 and 714,714,714 are both divisible by 3. Explain why this outcome must be true.

89. Use the fact that if $a|b$ and $a|c$, then $a|(b + c)$ to determine whether 54,036 is divisible by 18. (*Hint:* Write 54,036 as 54,000 + 36.)

90. Show that the $2^n - 1$ is a (Mersenne) prime for $n = 2, 3, 5$, and 7 but composite for $n = 11$.

91. Goldbach also conjectured in his letter to Euler that *every* integer greater than 5 is the sum of three prime numbers. For example, $6 = 2 + 2 + 2$ and $7 = 2 + 2 + 3$. Show that this conjecture is true for integers 8 through 20.

92. *Divisibility by Seven* The following describes a procedure to determine whether a number is divisible by 7. We will demonstrate the procedure with the number 203.

   i) Remove the units digit from the number, double the units digit, and subtract it from the remaining number. For 203, we will remove the 3, double it to get 6, and then subtract 20 − 6 to get 14.

   ii) If this new number is divisible by 7, then so is the original number. In our case, since 14 is divisible by 7, then 203 is also divisible by 7.

   iii) If you are not sure if the new number is divisible by 7, you can repeat the process described in the first step.

Use the procedure described above to determine whether the following numbers are divisible by 7.

   a) 329   **b)** 553   **c)** 583   **d)** 4823

RECREATIONAL MATHEMATICS

93. *Country, Animal, Fruit* Select a number from 1 to 10. Multiply your selected number by 9. Add the digits of the product together (if the product has more than one digit).

Now subtract 5. Determine which letter of the alphabet corresponds to the number you ended with (for example, 1 = A, 2 = B, 3 = C, and so on). Think of a country whose name begins with that letter. Remember the last letter of the name of that country. Think of an animal whose name begins with that letter. Remember the last letter in the name of that animal. Think of a fruit whose name begins with that letter. a) What country, animal, and fruit did you select? Turn to the answer section to see if your response matches the responses of over 90% of the people who attempt this activity. b) Can you explain why most people select the given answer?

RESEARCH ACTIVITIES

94. Conduct an Internet search on the GIMPS project. (See Mathematics Today on page 231.) Write a report describing the history and development of the project. Include a current update of the project's findings.

95. Do research and explain what *deficient numbers* and *abundant numbers* are. Give an example of each type of number. References include history of mathematics books, encyclopedias, and the Internet.

5.2   THE INTEGERS

Answer the question, What is the temperature, to the nearest degree Fahrenheit, outside right now? Depending on where you are and what time of year it is, you may have answered with a positive number, a negative number, or zero. These numbers are examples of the group of numbers called the *integers* that we will study in this section. We will also study the operations of addition, subtraction, multiplication, and division using the integers.

▲ Fahrenheit temperatures can be positive, negative, or zero.

In Section 5.1, we introduced the natural or counting numbers:

$$N = \{1, 2, 3, 4, \dots\}$$

Another important set of numbers, the *whole numbers*, help to answer the question, How many?

$$\text{Whole numbers} = \{0, 1, 2, 3, 4, \dots\}$$

Note that the set of whole numbers contains the number 0 but that the set of counting numbers does not. If a farmer were asked how many chickens were in a coop, the answer would be a whole number. If the farmer had no chickens, he or she would answer zero. Although we use the number 0 daily and take it for granted, the number 0 as we know it was not used and accepted until the sixteenth century.

If the temperature is 12°F and drops 20°, the resulting temperature is −8°F. This type of problem shows the need for negative numbers. The set of *integers* consists of the negative integers, 0, and the positive integers.

$$\text{Integers} = \{\ldots, -4, -3, -2, -1, 0, 1, 2, 3, \ldots\}$$

$$\underbrace{}_{\text{Negative integers}} \qquad \underbrace{}_{\text{Positive integers}}$$

The term *positive integers* is yet another name for the natural numbers or counting numbers.

An understanding of addition, subtraction, multiplication, and division of the integers is essential in understanding algebra (Chapter 6). To aid in our explanation of addition and subtraction of integers, we introduce the real number line (Fig. 5.5). To construct the real number line, arbitrarily select a point for zero to serve as the starting point. Place the positive integers to the right of 0, equally spaced from one another. Place the negative integers to the left of 0, using the same spacing. The real number line contains the integers and all the other real numbers that are not integers. Some examples of real numbers that are not integers are indicated in Fig. 5.5, namely $-\frac{5}{2}$, $\frac{1}{2}$, $\sqrt{2}$, and π. We discuss real numbers that are not integers in the next two sections.

Figure 5.5

The arrows at the ends of the real number line indicate that the line continues indefinitely in both directions. Note that for any natural number, n, on the number line, the *opposite of* that number, $-n$, is also on the number line. This real number line was drawn horizontally, but it could just as well have been drawn vertically. In fact, in the next chapter, we show that the axes of a graph are the union of two number lines, one horizontal and the other vertical.

The number line can be used to determine the greater (or lesser) of two integers. Two *inequality symbols* that we will use in this chapter are $>$ and $<$. The symbol $>$ is read "is greater than," and the symbol $<$ is read "is less than." Expressions that contain an inequality symbol are called *inequalities*. On the number line, the numbers increase from left to right. The number 3 is greater than 2, written $3 > 2$. Observe that 3 is to the right of 2. Similarly, we can see that $0 > -1$ by observing that 0 is to the right of -1 on the number line.

Instead of stating that 3 is greater than 2, we could state that 2 is less than 3, written $2 < 3$. Note that 2 is to the left of 3 on the number line. We can also see that $-1 < 0$ by observing that -1 is to the left of 0. The inequality symbol always points to the smaller of the two numbers when the inequality is true.

EXAMPLE ❶ *Writing an Inequality*

Insert either $>$ or $<$ in the shaded area between the paired numbers to make the statement correct.

a) $-3 \;\blacksquare\; 1$ b) $-3 \;\blacksquare\; -5$ c) $-6 \;\blacksquare\; -4$ d) $0 \;\blacksquare\; -7$

SOLUTION

a) $-3 < 1$ since -3 is to the left of 1 on the number line.

b) $-3 > -5$ since -3 is to the right of -5 on the number line.

c) $-6 < -4$ since -6 is to the left of -4 on the number line.

d) $0 > -7$ since 0 is to the right of -7 on the number line. ●

Addition of Integers

Addition of integers can be represented geometrically using a number line. To do so, begin at 0 on the number line. Represent the first addend (the first number to be added) by an arrow starting at 0. Draw the arrow to the right if the addend is positive. If the addend is negative, draw the arrow to the left. From the tip of the first arrow, draw a second arrow to represent the second addend. Draw the second arrow to the right or left, as just explained. The sum of the two integers is found at the tip of the second arrow.

┌ **EXAMPLE ❷** *Adding Integers*

Evaluate the following using the number line.

a) $3 + (-5)$ b) $-1 + (-4)$ c) $-6 + 4$ d) $3 + (-3)$

SOLUTION

a)

Thus, $3 + (-5) = -2$.

b)

Thus, $-1 + (-4) = -5$.

c)

Thus, $-6 + 4 = -2$.

d)

Thus, $3 + (-3) = 0$.   ●

In Example 2(d), the number -3 is said to be the additive inverse of 3 and 3 is the additive inverse of -3 because their sum is 0. In general, the *additive inverse* of the number n is $-n$ since $n + (-n) = 0$. Inverses are discussed more formally in Chapter 10.

Subtraction of Integers

Any subtraction problem can be rewritten as an addition problem. To do so, we use the following definition of subtraction.

SUBTRACTION

$$a - b = a + (-b)$$

The rule for subtraction indicates that to subtract b from a, *add* the additive inverse of b to a. For example,

$$3 - 5 = 3 + (-5)$$

Subtraction   Addition   Additive inverse of 5

Now we can determine the value of $3 + (-5)$.

Thus, $3 - 5 = 3 + (-5) = -2$.

EXAMPLE ❸ Subtracting Integers

Evaluate $-4 - (-3)$ using the number line.

SOLUTION We are subtracting -3 from -4. The additive inverse of -3 is 3; therefore, we add 3 to -4. We now add $-4 + 3$ on the number line to obtain the answer -1.

Thus, $-4 - (-3) = -4 + 3 = -1$.   ●

In Example 3, we found that $-4 - (-3) = -4 + 3$. In general, $a - (-b) = a + b$. As you get more proficient in working with integers, you should be able to answer questions involving them without drawing a number line.

EXAMPLE ❹ Subtracting: Adding the Inverse

Evaluate.
a) $-7 - 3$ b) $-7 - (-3)$ c) $7 - (-3)$ d) $7 - 3$

SOLUTION
a) $-7 - 3 = -7 + (-3) = -10$ b) $-7 - (-3) = -7 + 3 = -4$
c) $7 - (-3) = 7 + 3 = 10$ d) $7 - 3 = 7 + (-3) = 4$ ●

EXAMPLE ❺ Elevation Difference

The highest point on Earth is Mount Everest, in the Himalayas, at a height of 29,035 ft above sea level. The lowest point on Earth is the Mariana Trench, in the Pacific Ocean, at a depth of 36,198 ft below sea level ($-36,198$ ft). Find the vertical height difference between Mount Everest and the Mariana Trench.

SOLUTION We obtain the vertical difference by subtracting the lower elevation from the higher elevation.

$$29{,}035 - (-36{,}198) = 29{,}035 + 36{,}198 = 65{,}233$$

The vertical difference is 65,233 ft.   ●

Multiplication of Integers

The multiplication property of zero is important in our discussion of multiplication of integers. It indicates that the product of 0 and any number is 0.

> **MULTIPLICATION PROPERTY OF ZERO**
> $$a \cdot 0 = 0 \cdot a = 0$$

We will develop the rules for multiplication of integers using number patterns. The four possible cases are

1. positive integer $\times$ positive integer,
2. positive integer $\times$ negative integer,
3. negative integer $\times$ positive integer, and
4. negative integer $\times$ negative integer.

CASE 1: *POSITIVE INTEGER $\times$ POSITIVE INTEGER* The product of two positive integers can be defined as repeated addition of a positive integer. Thus, $3 \cdot 2$ means $2 + 2 + 2$. This sum will always be positive. Thus, a positive integer times a positive integer is a positive integer.

CASE 2: *POSITIVE INTEGER $\times$ NEGATIVE INTEGER* Consider the following patterns:

$$3(3) = 9$$
$$3(2) = 6$$
$$3(1) = 3$$

Note that each time the second factor is reduced by 1, the product is reduced by 3. Continuing the process gives

$$3(0) = 0$$

What comes next?

$$3(-1) = -3$$
$$3(-2) = -6$$

The pattern indicates that a positive integer times a negative integer is a negative integer.

We can confirm this result by using the number line. The expression $3(-2)$ means $(-2) + (-2) + (-2)$. Adding $(-2) + (-2) + (-2)$ on the number line, we obtain a sum of -6.

CASE 3: *NEGATIVE INTEGER × POSITIVE INTEGER* A procedure similar to that used in case 2 will indicate that a negative integer times a positive integer is a negative integer.

CASE 4: *NEGATIVE INTEGER × NEGATIVE INTEGER* We have illustrated that a positive integer times a negative integer is a negative integer. We make use of this fact in the following pattern:

$$4(-4) = -16$$
$$3(-4) = -12$$
$$2(-4) = -8$$
$$1(-4) = -4$$

In this pattern, each time the first term is decreased by 1, the product is increased by 4. Continuing this process gives

$$0(-4) = 0$$
$$(-1)(-4) = 4$$
$$(-2)(-4) = 8$$

This pattern illustrates that a negative integer times a negative integer is a positive integer.

The examples were restricted to integers. The rules for multiplication, however, can be used for any numbers. We summarize them as follows.

RULES FOR MULTIPLICATION

1. The product of two numbers with *like signs* (positive × positive or negative × negative) is a *positive number.*
2. The product of two numbers with *unlike signs* (positive × negative or negative × positive) is a *negative number.*

┌ **EXAMPLE ❻** *Multiplying Integers*

Evaluate.

a) $4 \cdot 8$ b) $4 \cdot (-8)$ c) $(-4) \cdot 8$ d) $(-4) \cdot (-8)$

SOLUTION

a) $4 \cdot 8 = 32$ b) $4 \cdot (-8) = -32$

c) $(-4) \cdot 8 = -32$ d) $(-4) \cdot (-8) = 32$ ●

Division of Integers

You may already realize that a relationship exists between multiplication and division.

$$6 \div 2 = 3 \quad \text{means that} \quad 3 \cdot 2 = 6$$

$$\frac{20}{10} = 2 \quad \text{means that} \quad 2 \cdot 10 = 20$$

These examples demonstrate that division is the reverse process of multiplication.

DIVISION

For any a, b, and c where $b \neq 0$, $\dfrac{a}{b} = c$ means that $c \cdot b = a$.

We discuss the four possible cases for division of integers, which are similar to those for multiplication.

CASE 1: *POSITIVE INTEGER ÷ POSITIVE INTEGER* A positive integer divided by a positive integer is positive.

$$\frac{6}{2} = 3 \qquad \text{since} \qquad 3(2) = 6$$

CASE 2: *POSITIVE INTEGER ÷ NEGATIVE INTEGER* A positive integer divided by a negative integer is negative.

$$\frac{6}{-2} = -3 \qquad \text{since} \qquad (-3)(-2) = 6$$

CASE 3: *NEGATIVE INTEGER ÷ POSITIVE INTEGER* A negative integer divided by a positive integer is negative.

$$\frac{-6}{2} = -3 \qquad \text{since} \qquad (-3)(2) = -6$$

CASE 4: *NEGATIVE INTEGER ÷ NEGATIVE INTEGER* A negative integer divided by a negative integer is positive.

$$\frac{-6}{-2} = 3 \qquad \text{since} \qquad 3(-2) = -6$$

The examples were restricted to integers. The rules for division, however, can be used for any numbers. You should realize that division of integers does not always result in an integer. The rules for division are summarized as follows.

RULES FOR DIVISION

1. The quotient of two numbers with *like signs* (positive ÷ positive or negative ÷ negative) is a *positive number*.
2. The quotient of two numbers with *unlike signs* (positive ÷ negative or negative ÷ positive) is a *negative number*.

EXAMPLE 7 *Dividing Integers*

Evaluate.

a) $\dfrac{42}{7}$ b) $\dfrac{-42}{7}$ c) $\dfrac{42}{-7}$ d) $\dfrac{-42}{-7}$

SOLUTION

a) $\dfrac{42}{7} = 6$ b) $\dfrac{-42}{7} = -6$ c) $\dfrac{42}{-7} = -6$ d) $\dfrac{-42}{-7} = 6$ ●

In the definition of division, we stated that the denominator could not be 0. Why not? Suppose we are trying to find the quotient $\frac{5}{0}$. Let's say that this quotient is equal to some number x. Then we would have $\frac{5}{0} = x$. If true, it would mean that $5 = x \cdot 0$. The right side of the equation is $x \cdot 0$, which is equal to 0 for any real value of x. That leads us to conclude that $5 = 0$, which is false. Thus, there is no number that can replace x that makes the equation $\frac{5}{0} = x$ true. Therefore, in mathematics, division by 0 is not allowed and we say that a quotient of any number divided by zero is *undefined*.

SECTION 5.2 EXERCISES

CONCEPT/WRITING EXERCISES

1. Explain how to add numbers using a number line.

2. What is the additive inverse of a number n?

3. Explain how to rewrite a subtraction problem as an addition problem.

4. Explain the rules for division of real numbers.

5. Explain the rules for multiplication of real numbers.

6. Explain why the quotient of a number divided by 0 is undefined.

PRACTICE THE SKILLS

In Exercises 7–16, evaluate the expression.

7. $-4 + 7$

8. $3 + (-7)$

9. $-9 + 5$

10. $-2 + (-2)$

11. $[6 + (-11)] + 0$

12. $(2 + 5) + (-4)$

13. $[(-3) + (-4)] + 9$

14. $[8 + (-3)] + (-2)$

15. $[(-23) + (-9)] + 11$

16. $[5 + (-13)] + 18$

In Exercises 17–26, evaluate the expression.

17. $1 - 7$

18. $-4 - 8$

19. $-5 - 4$

20. $6 - (-3)$

21. $-5 - (-3)$

22. $-4 - 4$

23. $14 - 20$

24. $8 - (-3)$

25. $[5 + (-3)] - 4$

26. $6 - (8 + 6)$

In Exercises 27–36, evaluate the expression.

27. $-5 \cdot 6$

28. $9(-2)$

29. $(-8)(-8)$

30. $-4(11)$

31. $[(-8)(-2)] \cdot 6$

32. $4(-5)(-6)$

33. $(5 \cdot 6)(-2)$

34. $(-9)(-1)(-2)$

35. $[(-3)(-6)] \cdot [(-5)(8)]$

36. $[(-8 \cdot 4) \cdot 5](-2)$

In Exercises 37–46, evaluate the expression.

37. $-28 \div (-4)$

38. $-72 \div 9$

39. $11 \div (-11)$

40. $-100 \div 20$

41. $56/-8$

42. $-75/15$

43. $-210/14$

44. $186/-6$

45. $144 \div (-3)$

46. $(-900) \div (-4)$

In Exercises 47–56, determine whether the statement is true or false. Modify each false statement to make it a true statement.

47. Every whole number is an integer.

48. Every integer is a whole number.

49. The difference of any two negative integers is a negative integer.

50. The sum of any two negative integers is a negative integer.

51. The product of any two positive integers is a positive integer.

52. The difference of a positive integer and a negative integer is always a negative integer.

53. The quotient of a negative integer and a positive integer is always a negative number.

54. The quotient of any two negative integers is a negative number.

55. The sum of a positive integer and a negative integer is always a positive integer.

56. The product of a positive integer and a negative integer is always a positive integer.

In Exercises 57–66, evaluate the expression.

57. $(7 + 11) \div 3$

58. $(-8) \div [64 \div (-8)]$

59. $[(-7)(-6)] - 22$

60. $[8(-2)] - 15$

61. $(4 - 8)(3)$

62. $[18 \div (-2)](-3)$

63. $[2 + (-17)] \div 3$

64. $(5 - 9) \div (-4)$

65. $[(-22)(-3)] \div (2 - 13)$

66. $[15(-4)] \div (-6)$

In Exercises 67–70, write the numbers in increasing order from left to right.

67. $0, -3, 3, -6, 6, -9$

68. $100, -10, 1, 0, -1, 10$

69. $-5, -2, -3, -1, -4, -6$

70. $106, 33, -47, -108, 72, -76$

PROBLEM SOLVING

71. *Dow Jones Industrial Average* On March 28, 2006, the Dow Jones Industrial Average (DJIA) opened at 11,250 points. During that day it lost 95 points. On March 29, 2006, it gained 61 points. On March 30, 2006, it lost 65 points. On March 31, 2006, it lost 42 points. Determine the closing DJIA on March 31, 2006.

72. *Pit Score* While playing the game of Pit, John Pearse began with a score of zero points. He then gained 100 points, lost 40 points, gained 90 points, lost 20 points, and gained 80 points on his next five rounds. What is John's score after five rounds?

73. *Elevation Difference* Mount Whitney, in the Sierra Nevada mountains of California, is the highest point in the contiguous United States. It is 14,495 ft above sea level. Death Valley, in California and Nevada, is the lowest point in the United States, 282 ft below sea level. Find the vertical height difference between Mount Whitney and Death Valley.

74. *Submarine Depth* The *USS Toledo* submarine is moving at an underwater depth of 1000 feet. It then rises 600 feet, dives 700 feet, and finally dives 400 feet. What is its final depth?

75. *Football Yardage* In the first four plays of a football game, the New Orleans Saints lost 6 yards, gained 4 yards, lost 1 yard, and gained 12 yards. What is the total number of yards gained in the first four plays? Did the Saints make a first down? (Ten yards are needed for a first down.)

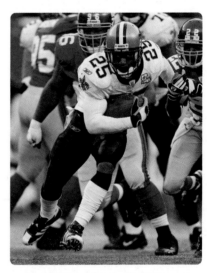

▲ Reggie Bush of the New Orleans Saints

76. *Extreme Temperatures* The hottest temperature ever recorded in the United States was 134°F, which occurred at Greenland Ranch, California, in Death Valley on July 10, 1913. The coldest temperature ever recorded in the United States was −79.8°F, which occurred at Prospect Creek Camp, Alaska, in the Endicott Mountains on January 23, 1971. Determine the difference between these two temperatures.

77. *Time Zone Calculations* Part of a World Standard Time Zones chart used by airlines and the United States Navy is shown. The scale along the bottom is just like a number line with the integers $-12, -11, \ldots, 11, 12$ on it.

   a) Find the difference in time between Amsterdam (zone $+1$) and Los Angeles (zone -8).

   b) Find the difference in time between Boston (zone -5) and Puerto Vallarta (zone -7).

78. Explain why $\dfrac{a}{b} = \dfrac{-a}{-b}$.

CHALLENGE PROBLEMS/GROUP ACTIVITIES

79. Find the quotient:

$$\frac{-1 + 2 - 3 + 4 - 5 + \cdots - 99 + 100}{1 - 2 + 3 - 4 + 5 - \cdots + 99 - 100}$$

80. *Pentagonal Numbers* Triangular numbers and square numbers were introduced in the Section 1.1 Exercises. There are also **pentagonal numbers**, which were also studied by the Greeks. Four pentagonal numbers are 1, 5, 12, and 22.

   a) Determine the next three pentagonal numbers.

   b) Describe a procedure to determine the next five pentagonal numbers without drawing the figures.

   c) Is 72 a pentagonal number? Explain how you determined your answer.

81. Place the appropriate plus or minus signs between each digit so that the total will equal 1.

$$0 \quad 1 \quad 2 \quad 3 \quad 4 \quad 5 \quad 6 \quad 7 \quad 8 \quad 9 = 1$$

RECREATIONAL MATHEMATICS

82. *Four 4's* The game of *Four 4's* is a challenging way to learn about some of the operations on integers. In this game, you must use exactly four 4's, and no other digits, along with one or more of the operations of addition, subtraction, multiplication, and division[†] to write the expressions. The object of the game is to write an expression that is equal to 0. Then write a second expression that is equal to 1, write a third that is equal to 2, and so on up to 9. You may use as many grouping symbols (that is, parentheses and brackets) as you wish, but you must use exactly four 4's. For example, one way to obtain 1 is as follows: $(4 + 4) \div (4 + 4) = 8 \div 8 = 1$. One way to obtain 2 is as follows: $(4 \div 4) + (4 \div 4) = 1 + 1 = 2$.

   a) Use the rules as defined above to obtain each whole number 0 through 9.

   b) Use the rules as defined above to obtain the following whole numbers: 12, 15, 16, 17, 20.

   c) We will now change our rules to allow the number 44 to count as two of the four 4's. Use the number 44 and two other fours to obtain the whole number 10.

RESEARCH ACTIVITY

83. Do research and write a report on the history of the number 0 in the Hindu–Arabic numeration system.

[†]This game will be expanded in future exercise sets to include other operations such as exponents and square roots.

5.3   THE RATIONAL NUMBERS

Among the ingredients called for in a recipe for carrot cake are $1\frac{1}{3}$ cups flour, $\frac{1}{2}$ teaspoon salt, $1\frac{1}{3}$ teaspoons baking powder, $1\frac{1}{3}$ teaspoons baking soda, $1\frac{1}{3}$ teaspoons cinnamon, $\frac{1}{2}$ teaspoon cloves, and $\frac{1}{2}$ teaspoon ginger. How much of each of these ingredients would you need to use if you wish to cut the recipe in half? In this section, we will review the use of fractions such as those in the carrot cake recipe. We will also learn about the operations of addition, subtraction, multiplication, and division using fractions.

▲ Rational numbers are frequently used in recipes.

We introduced the number line in Section 5.1 and discussed the integers in Section 5.2. The numbers that fall between the integers on the number line are either rational or irrational numbers. In this section, we discuss the rational numbers, and in Section 5.4, we discuss the irrational numbers.

Any number that can be expressed as a quotient of two integers (denominator not 0) is a rational number.

> The set of **rational numbers**, denoted by Q, is the set of all numbers of the form p/q, where p and q are integers and $q \neq 0$.

The following numbers are examples of rational numbers:

$$\frac{1}{3}, \quad \frac{3}{4}, \quad -\frac{7}{8}, \quad 1\frac{2}{3}, \quad 2, \quad 0, \quad \frac{15}{7}$$

The integers 2 and 0 are rational numbers because each can be expressed as the quotient of two integers: $2 = \frac{2}{1}$ and $0 = \frac{0}{1}$. In fact, every integer n is a rational number because it can be written in the form of $\frac{n}{1}$.

Numbers such as $\frac{1}{3}$ and $-\frac{7}{8}$ are also called *fractions*. The number above the fraction line is called the *numerator*, and the number below the fraction line is called the *denominator*.

Reducing Fractions

Sometimes the numerator and denominator in a fraction have a common divisor (or common factor). For example, both the numerator and denominator of the fraction $\frac{6}{10}$ have the common divisor 2. When a numerator and denominator have a common divisor, we can *reduce the fraction to its lowest terms*.

A fraction is said to be in its lowest terms (or reduced) when the numerator and denominator are relatively prime (that is, have no common divisors other than 1). To reduce a fraction to its lowest terms, divide both the numerator and the denominator by the greatest common divisor. Recall that a procedure for finding the greatest common divisor was discussed in Section 5.1.

The fraction $\frac{6}{10}$ is reduced to its lowest terms as follows.

$$\frac{6}{10} = \frac{6 \div 2}{10 \div 2} = \frac{3}{5}$$

EXAMPLE ❶ *Reducing a Fraction to Lowest Terms*

Reduce $\dfrac{54}{90}$ to its lowest terms.

SOLUTION On page 229 in Example 4 of Section 5.1, we determined that the GCD of 54 and 90 is 18. Divide the numerator and the denominator by GCD, 18.

$$\frac{54}{90} = \frac{54 \div 18}{90 \div 18} = \frac{3}{5}$$

Since there are no common divisors of 3 and 5 other than 1, the fraction $\frac{3}{5}$ is in its lowest terms. ●

Mixed Numbers and Improper Fractions

Consider the number $2\frac{3}{4}$. It is an example of a *mixed number*. It is called a mixed number because it consists of an integer, 2, and a fraction, $\frac{3}{4}$. The mixed number $2\frac{3}{4}$ means $2 + \frac{3}{4}$. The mixed number $-4\frac{1}{4}$ means $-(4 + \frac{1}{4})$. Rational numbers greater than 1 or less than -1 that are not integers may be represented as mixed numbers, or as *improper fractions*. An improper fraction is a fraction whose numerator is greater than its denominator. An example of an improper fraction is $\frac{8}{5}$. Figure 5.6 shows both mixed numbers and improper fractions indicated on a number line. In this section, we show how to convert mixed numbers to improper fractions and vice versa.

Figure 5.6

We begin by limiting our discussion to positive mixed numbers and positive improper fractions.

CONVERTING A POSITIVE MIXED NUMBER TO AN IMPROPER FRACTION
1. Multiply the denominator of the fraction in the mixed number by the integer preceding it.
2. Add the product obtained in step 1 to the numerator of the fraction in the mixed number. This sum is the numerator of the improper fraction we are seeking. The denominator of the improper fraction we are seeking is the same as the denominator of the fraction in the mixed number.

EXAMPLE ❷ *Converting Mixed Numbers to Improper Fractions*

Convert the following mixed numbers to improper fractions.

a) $1\frac{2}{3}$ b) $4\frac{5}{8}$

SOLUTION

a) $1\dfrac{2}{3} = \dfrac{3 \cdot 1 + 2}{3} = \dfrac{3 + 2}{3} = \dfrac{5}{3}$

b) $4\dfrac{5}{8} = \dfrac{8 \cdot 4 + 5}{8} = \dfrac{32 + 5}{8} = \dfrac{37}{8}$

Notice that both $\frac{5}{3}$ and $\frac{37}{8}$ have numerators larger than their denominators; therefore, both numbers are improper fractions. ●

Now let's discuss converting an improper fraction to a mixed number.

CONVERTING A POSITIVE IMPROPER FRACTION TO A MIXED NUMBER

1. Divide the numerator by the denominator. Identify the quotient and the remainder.

2. The quotient obtained in step 1 is the integer part of the mixed number. The remainder is the numerator of the fraction in the mixed number. The denominator in the fraction of the mixed number will be the same as the denominator in the original fraction.

EXAMPLE ❸ *From Improper Fraction to Mixed Number*

Convert the following improper fractions to mixed numbers.

a) $\dfrac{8}{5}$ b) $\dfrac{225}{8}$

SOLUTION

a) Divide the numerator, 8, by the denominator, 5.

$$
\begin{array}{r}
1 \quad \leftarrow \text{Quotient} \\
\text{Divisor} \rightarrow \quad 5\overline{)8} \quad \leftarrow \text{Dividend} \\
\underline{5} \\
3 \quad \leftarrow \text{Remainder}
\end{array}
$$

Therefore,

$$
\overset{\text{Quotient}}{\underset{\downarrow}{}}
$$

$$
\dfrac{8}{5} = 1\dfrac{3}{5} \quad \begin{array}{l} \leftarrow \text{Remainder} \\ \leftarrow \text{Divisor} \end{array}
$$

The mixed number is $1\frac{3}{5}$.

b) Divide the numerator, 225, by the denominator, 8.

$$
\begin{array}{r}
28 \quad \leftarrow \text{Quotient} \\
\text{Divisor} \rightarrow \quad 8\overline{)225} \quad \leftarrow \text{Dividend} \\
\underline{16} \\
65 \\
\underline{64} \\
1 \quad \leftarrow \text{Remainder}
\end{array}
$$

Therefore,

Quotient
↓

$$\frac{225}{8} = 28\frac{1}{8} \quad \begin{matrix} \leftarrow \text{Remainder} \\ \leftarrow \text{Divisor} \end{matrix}$$

The mixed number is $28\frac{1}{8}$. ●

Up to this point, we have only worked with positive mixed numbers and positive improper fractions. When converting a negative mixed number to an improper fraction, or a negative improper fraction to a mixed number, it is best to ignore the negative sign temporarily. Perform the calculation as described earlier and then reattach the negative sign.

EXAMPLE ❹ *Negative Mixed Numbers and Improper Fractions*

a) Convert $-1\frac{2}{3}$ to an improper fraction.

b) Convert $-\frac{8}{5}$ to a mixed number.

SOLUTION

a) First, ignore the negative sign and examine $1\frac{2}{3}$. We learned in Example 2(a) that $1\frac{2}{3} = \frac{5}{3}$. To convert $-1\frac{2}{3}$ to an improper fraction, we reattach the negative sign. Thus, $-1\frac{2}{3} = -\frac{5}{3}$.

b) We learned in Example 3(a) that $\frac{8}{5} = 1\frac{3}{5}$. Therefore, $-\frac{8}{5} = -1\frac{3}{5}$. ●

Terminating or Repeating Decimal Numbers

Note the following important property of the rational numbers.

Every *rational number* when expressed as a decimal number will be either a terminating or a repeating decimal number.

Examples of terminating decimal numbers are 0.5, 0.75, and 4.65. Examples of repeating decimal numbers are 0.333..., 0.2323..., and 8.13456456.... One way to indicate that a number or group of numbers repeat is to place a bar above the number or group of numbers that repeat. Thus, 0.333... may be written $0.\overline{3}$, 0.2323... may be written $0.\overline{23}$, and 8.13456456... may be written $8.13\overline{456}$.

EXAMPLE ❺ *Terminating Decimal Numbers*

Show that the following rational numbers can be expressed as terminating decimal numbers.

a) $\frac{3}{4}$ b) $-\frac{7}{20}$ c) $\frac{13}{8}$

SOLUTION To express the rational number in decimal form, divide the numerator by the denominator. If you use a calculator or long division, you will obtain the following results.

a) $\dfrac{3}{4} = 0.75$ b) $-\dfrac{7}{20} = -0.35$ c) $\dfrac{13}{8} = 1.625$ ●

EXAMPLE ⑥ *Repeating Decimal Numbers*

Show that the following rational numbers can be expressed as repeating decimal numbers.

a) $\dfrac{2}{3}$ b) $\dfrac{14}{99}$ c) $1\dfrac{5}{36}$

SOLUTION If you use a calculator or long division, you will see that each fraction results in a repeating decimal number.

a) $2 \div 3 = 0.6666\ldots$ or $0.\overline{6}$

b) $14 \div 99 = 0.141414\ldots$ or $0.\overline{14}$

c) $1\dfrac{5}{36} = \dfrac{41}{36} = 1.138888\ldots$ or $1.13\overline{8}$ ●

Note that in each part of Example 6, the quotient when expressed as a decimal number has no final digit and continues indefinitely. Each number is a repeating decimal number.

When a fraction is converted to a decimal number, the maximum number of digits that can repeat is $n - 1$, where n is the denominator of the fraction. For example, when $\frac{2}{7}$ is converted to a decimal number, the maximum number of digits that can repeat is $7 - 1$, or 6.

Converting Decimal Numbers to Fractions

We can convert a terminating or repeating decimal number into a quotient of integers. The explanation of the procedure will refer to the positional values to the right of the decimal point, as illustrated here:

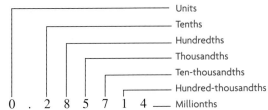

Example 7 demonstrates how to convert from a decimal number to a fraction.

EXAMPLE ⑦ *Converting Decimal Numbers to Fractions*

Convert the following terminating decimal numbers to a quotient of integers. If necessary, reduce the quotient to lowest terms.

a) 0.7   b) 0.35   c) 0.016   d) 3.41

SOLUTION When converting a terminating decimal number to a quotient of integers, we observe the last digit to the right of the decimal point. The position of this digit will indicate the denominator of the quotient of integers.

a) $0.7 = \frac{7}{10}$ because the 7 is in the tenths position.

b) $0.35 = \frac{35}{100}$ because the right-most digit, 5, is in the hundredths position. We finish by reducing the fraction to lowest terms:

$$\frac{35}{100} = \frac{35 \div 5}{100 \div 5} = \frac{7}{20}$$

c) $0.016 = \frac{16}{1000}$ because the right-most digit, 6, is in the thousandths position. We finish by reducing the fraction to lowest terms:

$$\frac{16}{1000} = \frac{16 \div 8}{1000 \div 8} = \frac{2}{125}$$

d) We begin by writing 3.41 as the mixed number, $3\frac{41}{100}$ because the right-most digit, 1, is in the hundredths position. To write the mixed number as a ratio of two integers, we follow the procedure shown in Example 2, on page 247:

$$3\frac{41}{100} = \frac{100 \cdot 3 + 41}{100} = \frac{341}{100}$$

The fraction $\frac{341}{100}$ cannot be reduced further.   ●

Converting a repeating decimal number to a quotient of integers is more difficult than converting a terminating decimal number to a quotient of integers. To do so, we must "create" another repeating decimal number with the same repeating digits so that when one repeating decimal number is subtracted from the other repeating decimal number, the difference will be a whole number. To create a number with the same repeating digits, multiply the original repeating decimal number by 10 if one digit repeats, by 100 if two digits repeat, by 1000 if three digits repeat, and so on. Examples 8 through 10 demonstrate this procedure.

EXAMPLE ❽ Converting a Repeating Decimal Number to a Fraction

Convert $0.\overline{3}$ to a quotient of integers.

SOLUTION $0.\overline{3} = 0.3\overline{3} = 0.33\overline{3}$, and so on.

Let the original repeating decimal number be n; thus, $n = 0.\overline{3}$. Because one digit repeats, we multiply both sides of the equation by 10, which gives $10n = 3.\overline{3}$. Then we subtract.

$$\begin{aligned} 10n &= 3.\overline{3} \\ -\quad n &= 0.\overline{3} \\ \hline 9n &= 3.0 \end{aligned}$$

Note that $10n - n = 9n$ and $3.\overline{3} - 0.\overline{3} = 3.0$.

Next, we solve for n by dividing both sides of the equation by 9.

$$\frac{9n}{9} = \frac{3.0}{9}$$

$$n = \frac{3}{9} = \frac{1}{3}$$

Therefore, $0.\overline{3} = \frac{1}{3}$. Evaluate $1 \div 3$ on a calculator now and see what value you get.   ●

EXAMPLE ❾ *Converting a Repeating Decimal Number to a Fraction*

Convert $0.\overline{35}$ to a quotient of integers.

SOLUTION Let $n = 0.\overline{35}$. Since two digits repeat, multiply both sides of the equation by 100. Thus, $100n = 35.\overline{35}$. Now we subtract n from $100n$.

$$
\begin{array}{r}
100n = 35.\overline{35} \\
-\quad n = \ 0.\overline{35} \\
\hline
99n = 35
\end{array}
$$

Finally, we divide both sides of the equation by 99.

$$\frac{99n}{99} = \frac{35}{99}$$

$$n = \frac{35}{99}$$

Therefore, $0.\overline{35} = \frac{35}{99}$. Evaluate $35 \div 99$ on a calculator now and see what value you get.   ●

EXAMPLE ❿ *Converting a Repeating Decimal Number to a Fraction*

Convert $12.14\overline{2}$ to a quotient of integers.

SOLUTION This example is different from the two preceding examples in that the repeating digit, 2, is not directly to the right of the decimal point. When this situation arises, move the decimal point to the right until the repeating terms are directly to its right. For each place the decimal point is moved, the number is multiplied by 10. In this example, the decimal point must be moved two places to the right. Thus, the number must be multiplied by 100.

$$n = 12.14\overline{2}$$
$$100n = 100 \times 12.14\overline{2} = 1214.\overline{2}$$

Now proceed as in the previous two examples. Since one digit repeats, multiply both sides by 10.

$$100n = 1214.\overline{2}$$
$$10 \times 100n = 10 \times 1214.\overline{2}$$
$$1000n = 12142.\overline{2}$$

Now subtract $100n$ from $1000n$ so that the repeating part will drop out.

$$
\begin{array}{r}
1000n = 12142.\overline{2} \\
-\ 100n = \ 1214.\overline{2} \\
\hline
900n = 10928
\end{array}
$$

$$n = \frac{10,928}{900} = \frac{2732}{225}$$

Therefore, $12.14\overline{2} = \frac{2732}{225}$. Evaluate $2732 \div 225$ on a calculator now and see what value you get.   ●

Multiplication and Division of Fractions

The product of two fractions is found by multiplying the numerators together and multiplying the denominators together.

MULTIPLICATION OF FRACTIONS

$$\frac{a}{b} \cdot \frac{c}{d} = \frac{a \cdot c}{b \cdot d} = \frac{ac}{bd}, \quad b \neq 0, \quad d \neq 0$$

EXAMPLE ⑪ *Multiplying Fractions*

Evaluate.

a) $\dfrac{3}{5} \cdot \dfrac{7}{8}$ b) $\left(\dfrac{-2}{3}\right)\left(\dfrac{-4}{9}\right)$ c) $\left(1\dfrac{7}{8}\right)\left(2\dfrac{1}{4}\right)$

SOLUTION

a) $\dfrac{3}{5} \cdot \dfrac{7}{8} = \dfrac{3 \cdot 7}{5 \cdot 8} = \dfrac{21}{40}$

b) $\left(\dfrac{-2}{3}\right)\left(\dfrac{-4}{9}\right) = \dfrac{(-2)(-4)}{(3)(9)} = \dfrac{8}{27}$

c) $\left(1\dfrac{7}{8}\right)\left(2\dfrac{1}{4}\right) = \dfrac{15}{8} \cdot \dfrac{9}{4} = \dfrac{135}{32} = 4\dfrac{7}{32}$

The *reciprocal* of any number is 1 divided by that number. The product of a number and its reciprocal must equal 1. Examples of some numbers and their reciprocals follow.

Number		Reciprocal		Product
3	·	$\dfrac{1}{3}$	=	1
$\dfrac{3}{5}$	·	$\dfrac{5}{3}$	=	1
-6	·	$-\dfrac{1}{6}$	=	1

To find the quotient of two fractions, multiply the first fraction by the reciprocal of the second fraction.

DIVISION OF FRACTIONS

$$\frac{a}{b} \div \frac{c}{d} = \frac{a}{b} \cdot \frac{d}{c} = \frac{ad}{bc}, \quad b \neq 0, \quad d \neq 0, \quad c \neq 0$$

EXAMPLE 12 *Dividing Fractions*

Evaluate.

a) $\dfrac{5}{7} \div \dfrac{3}{4}$ b) $\left(-\dfrac{3}{5}\right) \div \dfrac{7}{8}$

SOLUTION

a) $\dfrac{5}{7} \div \dfrac{3}{4} = \dfrac{5}{7} \cdot \dfrac{4}{3} = \dfrac{5 \cdot 4}{7 \cdot 3} = \dfrac{20}{21}$

b) $\left(-\dfrac{3}{5}\right) \div \dfrac{7}{8} = \dfrac{-3}{5} \cdot \dfrac{8}{7} = \dfrac{-3 \cdot 8}{5 \cdot 7} = \dfrac{-24}{35} = -\dfrac{24}{35}$

Addition and Subtraction of Fractions

Before we can add or subtract fractions, the fractions must have a common denominator. A common denominator is another name for a common multiple of the denominators. The *lowest common denominator (LCD)* is the least common multiple of the denominators.

To add or subtract two fractions with a common denominator, we add or subtract their numerators and retain the common denominator.

> **ADDITION AND SUBTRACTION OF FRACTIONS**
> $$\dfrac{a}{c} + \dfrac{b}{c} = \dfrac{a + b}{c}, \quad c \neq 0; \quad \dfrac{a}{c} - \dfrac{b}{c} = \dfrac{a - b}{c}, \quad c \neq 0$$

EXAMPLE 13 *Adding and Subtracting Fractions*

Evaluate.

a) $\dfrac{5}{16} + \dfrac{7}{16}$ b) $\dfrac{13}{21} - \dfrac{8}{21}$

SOLUTION

a) $\dfrac{5}{16} + \dfrac{7}{16} = \dfrac{5 + 7}{16} = \dfrac{12}{16} = \dfrac{3}{4}$ b) $\dfrac{13}{21} - \dfrac{8}{21} = \dfrac{13 - 8}{21} = \dfrac{5}{21}$

Note that in Example 13, the denominators of the fractions being added or subtracted were the same; that is, they have a common denominator. *When adding or subtracting two fractions with unlike denominators, first rewrite each fraction with a common denominator. Then add or subtract the fractions.*

Writing fractions with a common denominator is accomplished with the *fundamental law of rational numbers.*

> **FUNDAMENTAL LAW OF RATIONAL NUMBERS**
> If a, b, and c are integers, with $b \neq 0$ and $c \neq 0$, then
> $$\dfrac{a}{b} = \dfrac{a}{b} \cdot \dfrac{c}{c} = \dfrac{a \cdot c}{b \cdot c}$$

The terms $\dfrac{a}{b}$ and $\dfrac{a \cdot c}{b \cdot c}$ are called *equivalent fractions*. For example, since $\dfrac{5}{12} = \dfrac{5 \cdot 5}{12 \cdot 5} = \dfrac{25}{60}$, the fractions $\dfrac{5}{12}$ and $\dfrac{25}{60}$ are equivalent fractions. We will see the importance of equivalent fractions in the next two examples.

EXAMPLE ⓮ Subtracting Fractions with Unlike Denominators

Evaluate $\dfrac{5}{12} - \dfrac{3}{10}$.

SOLUTION Using prime factorization (Section 5.1), we find that the LCM of 12 and 10 is 60. We will therefore express each fraction as an equivalent fraction with a denominator of 60. Sixty divided by 12 is 5. Therefore, the denominator, 12, must be multiplied by 5 to get 60. If the denominator is multiplied by 5, the numerator must also be multiplied by 5 so that the value of the fraction remains unchanged. Multiplying both numerator and denominator by 5 is the same as multiplying by 1.

We follow the same procedure for the other fraction, $\frac{3}{10}$. Sixty divided by 10 is 6. Therefore, we multiply both the denominator, 10, and the numerator, 3, by 6 to obtain an equivalent fraction with a denominator of 60.

$$\frac{5}{12} - \frac{3}{10} = \left(\frac{5}{12} \cdot \frac{5}{5}\right) - \left(\frac{3}{10} \cdot \frac{6}{6}\right)$$

$$= \frac{25}{60} - \frac{18}{60}$$

$$= \frac{7}{60}$$

EXAMPLE ⓯ Adding Fractions with Unlike Denominators

Evaluate $\dfrac{1}{54} + \dfrac{1}{90}$.

SOLUTION On page 230, in Example 6 of Section 5.1, we determined that the LCM of 54 and 90 is 270. Rewrite each fraction as an equivalent fraction using the LCM as the common denominator.

$$\frac{1}{54} + \frac{1}{90} = \left(\frac{1}{54} \cdot \frac{5}{5}\right) + \left(\frac{1}{90} \cdot \frac{3}{3}\right)$$

$$= \frac{5}{270} + \frac{3}{270}$$

$$= \frac{8}{270}$$

Now we reduce $\frac{8}{270}$ by dividing both 8 and 270 by 2, their greatest common factor.

$$\frac{8}{270} = \frac{8 \div 2}{270 \div 2} = \frac{4}{135}$$

EXAMPLE ⓰ *Rice Preparation*

Following are the instructions given on a box of Minute Rice. Determine the amount of (a) rice and water, (b) salt, and (c) butter or margarine needed to make 3 servings of rice.

DIRECTIONS

1. Bring water, salt, and butter (or margarine) to a boil.
2. Stir in rice. Cover; remove from heat. Let stand 5 minutes. Fluff with fork.

To Make	Rice & Water (Equal Measures)	Salt	Butter or Margarine (If Desired)
2 servings	$\frac{2}{3}$ cup	$\frac{1}{4}$ tsp	1 tsp
4 servings	$1\frac{1}{3}$ cups	$\frac{1}{2}$ tsp	2 tsp

SOLUTION Since 3 is halfway between 2 and 4, we can find the amount of each ingredient by finding the average of the amount for 2 and 4 servings. To do so, we add the amounts for 2 servings and 4 servings and divide the sum by 2.

a) Rice and water: $\dfrac{\frac{2}{3} + 1\frac{1}{3}}{2} = \dfrac{\frac{2}{3} + \frac{4}{3}}{2} = \dfrac{\frac{6}{3}}{2} = \dfrac{2}{2} = 1$ cup

b) Salt: $\dfrac{\frac{1}{4} + \frac{1}{2}}{2} = \dfrac{\frac{1}{4} + \frac{2}{4}}{2} = \dfrac{\frac{3}{4}}{2} = \dfrac{3}{4} \cdot \dfrac{1}{2} = \dfrac{3}{8}$ tsp

c) Butter or margarine: $\dfrac{1 + 2}{2} = \dfrac{3}{2}$, or $1\frac{1}{2}$ tsp

The solution to Example 16 can be found in other ways. Suggest two other procedures for solving the same problem.

SECTION 5.3 EXERCISES

CONCEPT/WRITING EXERCISES

1. Describe the set of rational numbers.

2. a) Explain how to write a terminating decimal number as a fraction.

   b) Write 0.213 as a fraction.

3. a) Explain how to reduce a fraction to lowest terms.

   b) Reduce $\frac{28}{35}$ to lowest terms using the procedure in part (a).

4. Explain how to convert an improper fraction to a mixed number.

5. Explain how to convert a mixed number to an improper fraction.

6. a) Explain how to multiply two fractions.

   b) Multiply $\frac{14}{15} \cdot \frac{25}{28}$ using the procedure in part (a).

7. a) Explain how to determine the reciprocal of a number.

   b) Determine the reciprocal of -2 by using the procedure in part(a).

8. a) Explain how to divide two fractions.

   b) Divide $\frac{14}{33} \div \frac{10}{21}$ by using the procedure in part (a).

9. **a)** Explain how to add or subtract two fractions having a common denominator.

   b) Add $\frac{11}{24} + \frac{3}{24}$ using the procedure in part (a).

   c) Subtract $\frac{37}{48} - \frac{13}{48}$ using the procedure in part (a).

10. **a)** Explain how to add or subtract two fractions having unlike denominators.

   b) Add $\frac{2}{3} + \frac{5}{8}$ using the procedure in part (a).

   c) Subtract $\frac{5}{6} - \frac{2}{15}$ using the procedure in part (a).

11. In your own words, state the fundamental law of rational numbers.

12. Are $\frac{7}{11}$ and $\frac{21}{33}$ equivalent fractions? Explain your answer.

PRACTICE THE SKILLS

In Exercises 13–22, reduce each fraction to lowest terms.

13. $\dfrac{5}{10}$ 14. $\dfrac{12}{16}$ 15. $\dfrac{24}{54}$

16. $\dfrac{36}{56}$ 17. $\dfrac{95}{125}$ 18. $\dfrac{13}{221}$

19. $\dfrac{112}{176}$ 20. $\dfrac{120}{135}$ 21. $\dfrac{45}{495}$

22. $\dfrac{124}{148}$

In Exercises 23–28, convert each mixed number to an improper fraction.

23. $3\dfrac{5}{8}$ 24. $2\dfrac{1}{4}$ 25. $-1\dfrac{15}{16}$

26. $-7\dfrac{1}{5}$ 27. $-4\dfrac{15}{16}$ 28. $11\dfrac{9}{16}$

In Exercises 29–32, write the number of inches indicated by the arrows as an improper fraction.

29.

30.

31.

32.

In Exercises 33–38, convert each improper fraction to a mixed number.

33. $\dfrac{13}{5}$ 34. $\dfrac{23}{4}$ 35. $-\dfrac{73}{6}$

36. $-\dfrac{157}{12}$ 37. $-\dfrac{878}{15}$ 38. $\dfrac{1028}{21}$

In Exercises 39–48, express each rational number as terminating or repeating decimal number.

39. $\dfrac{7}{10}$ 40. $\dfrac{15}{16}$ 41. $\dfrac{2}{9}$

42. $\dfrac{1}{6}$ 43. $\dfrac{3}{8}$ 44. $\dfrac{23}{7}$

45. $\dfrac{13}{6}$ 46. $\dfrac{115}{15}$ 47. $\dfrac{85}{15}$

48. $\dfrac{100}{9}$

In Exercises 49–58, express each terminating decimal number as a quotient of two integers. If possible, reduce the quotient to lowest terms.

49. 0.75   50. 0.29   51. 0.045

52. 0.375   53. 0.2   54. 0.251

55. 0.0131   56. 0.2345   57. 0.0001

58. 0.0053

In Exercises 59–68, express each repeating decimal number as a quotient of two integers. If possible, reduce the quotient to lowest terms.

59. $0.\overline{1}$ 60. $0.\overline{5}$ 61. $1.\overline{9}$

62. $0.\overline{23}$ 63. $1.\overline{36}$ 64. $0.1\overline{35}$

65. $2.0\overline{5}$ 66. $2.4\overline{9}$ 67. $3.4\overline{78}$

68. $4.1\overline{19}$

In Exercises 69–78, perform the indicated operation and reduce your answer to lowest terms.

69. $\dfrac{1}{2} \cdot \dfrac{4}{5}$ 70. $\dfrac{2}{9} \div \dfrac{4}{7}$

71. $\left(\dfrac{-3}{8}\right)\left(\dfrac{-16}{15}\right)$　　72. $\left(-\dfrac{3}{5}\right) \div \dfrac{10}{21}$

73. $\dfrac{7}{8} \div \dfrac{8}{7}$　　74. $\dfrac{3}{7} \div \dfrac{3}{7}$

75. $\left(\dfrac{3}{5} \cdot \dfrac{4}{7}\right) \div \dfrac{1}{3}$　　76. $\left(\dfrac{4}{7} \div \dfrac{4}{5}\right) \cdot \dfrac{1}{7}$

77. $\left[\left(-\dfrac{2}{3}\right)\left(\dfrac{5}{8}\right)\right] \div \left(-\dfrac{7}{16}\right)$

78. $\left(\dfrac{7}{15} \cdot \dfrac{5}{8}\right) \div \left(\dfrac{7}{9} \cdot \dfrac{5}{2}\right)$

In Exercises 79–88, perform the indicated operation and reduce your answer to lowest terms.

79. $\dfrac{1}{4} + \dfrac{2}{3}$　　80. $\dfrac{7}{8} - \dfrac{1}{6}$

81. $\dfrac{2}{11} + \dfrac{5}{22}$　　82. $\dfrac{5}{12} + \dfrac{7}{36}$

83. $\dfrac{5}{9} - \dfrac{7}{54}$　　84. $\dfrac{17}{25} - \dfrac{43}{100}$

85. $\dfrac{1}{12} + \dfrac{1}{48} + \dfrac{1}{72}$　　86. $\dfrac{3}{5} + \dfrac{7}{15} + \dfrac{9}{75}$

87. $\dfrac{1}{30} - \dfrac{3}{40} - \dfrac{7}{50}$　　88. $\dfrac{4}{25} - \dfrac{9}{100} - \dfrac{7}{40}$

PROBLEM SOLVING

Alternative methods for adding and subtracting two fractions are shown. These methods may not result in a solution in its lowest terms.

$$\frac{a}{b} + \frac{c}{d} = \frac{ad + bc}{bd} \quad \text{and} \quad \frac{a}{b} - \frac{c}{d} = \frac{ad - bc}{bd}$$

In Exercises 89–94, use the appropriate formula to evaluate the expression.

89. $\dfrac{2}{3} + \dfrac{1}{8}$　　90. $\dfrac{1}{5} + \dfrac{4}{9}$

91. $\dfrac{5}{6} - \dfrac{7}{8}$　　92. $\dfrac{7}{3} - \dfrac{5}{12}$

93. $\dfrac{3}{8} + \dfrac{5}{12}$　　94. $\left(\dfrac{2}{3} + \dfrac{1}{4}\right) - \dfrac{3}{5}$

In Exercises 95–100, evaluate each expression.

95. $\left(\dfrac{2}{3} \cdot \dfrac{9}{10}\right) + \dfrac{2}{5}$　　96. $\left(\dfrac{7}{6} \div \dfrac{4}{3}\right) - \dfrac{11}{12}$

97. $\left(\dfrac{3}{4} + \dfrac{1}{6}\right) \div \left(2 - \dfrac{7}{6}\right)$

98. $\left(\dfrac{1}{3} \cdot \dfrac{3}{7}\right) + \left(\dfrac{3}{5} \cdot \dfrac{10}{11}\right)$

99. $\left(3 - \dfrac{4}{9}\right) \div \left(4 + \dfrac{2}{3}\right)$

100. $\left(\dfrac{2}{5} \div \dfrac{4}{9}\right)\left(\dfrac{3}{5} \cdot 6\right)$

In Exercises 101–112, write an expression that will solve the problem and then evaluate the expression.

101. **Height Increase** When David Conway finished his first year of high school, his height was $69\frac{7}{8}$ inches. When he returned to school after the summer, David's height was $71\frac{5}{8}$ inches. How much did David's height increase over the summer?

102. **Thistles** Diane Helbing has four different varieties of thistles invading her pasture. She estimates that of these thistles, $\frac{1}{2}$ are Canada thistles, $\frac{1}{4}$ are bull thistles, $\frac{1}{6}$ are plumeless thistles, and the rest are musk thistles. What fraction of the thistles are musk thistles?

103. **Stairway Height** A stairway consists of 14 stairs, each $8\frac{5}{8}$ inches high. What is the vertical height of the stairway?

104. **Alphabet Soup** Margaret Cannata's recipe for alphabet soup calls for (among other items) $\frac{1}{4}$ cup snipped parsley, $\frac{1}{8}$ teaspoon pepper, and $\frac{1}{2}$ cup sliced carrots. Margaret is expecting company and needs to multiply the amounts of the ingredients by $1\frac{1}{2}$ times. Determine the amount of (a) snipped parsley, (b) pepper, and (c) sliced carrots she needs for the soup.

105. *Sprinkler System* To repair his sprinkler system, Tony Gambino needs a total of $20\frac{5}{16}$ inches of PVC pipe. He has on hand pieces that measure $2\frac{1}{4}$ inches, $3\frac{7}{8}$ inches, and $4\frac{1}{4}$ inches in length. If he can combine these pieces and use them in the repair, how long of a piece of PVC pipe will Tony need to purchase to repair his sprinkler system?

106. *Crop Storage* Todd Schroeder has a silo on his farm in which he can store silage made from his various crops. He currently has a silo that is $\frac{1}{4}$ full of corn silage, $\frac{2}{5}$ full of hay silage, and $\frac{1}{3}$ full of oats silage. What fraction of Todd's silo is currently in use?

107. *Department Budget* Jaime Bailey is chair of the humanities department at Santa Fe Community College. Jaime has a budget in which $\frac{1}{2}$ of the money is for photocopying, $\frac{2}{5}$ of the money is for computer-related expenses, and the rest of the money is for student tutors in the foreign languages lab. What fraction of Jaime's budget is for student tutors?

108. *Art Supplies* Denise Viale teaches kindergarten and is buying supplies for her class to make papier-mâché piggy banks. Each piggy bank to be made requires $1\frac{1}{4}$ cups of flour. If Denise has 15 students who are going to make piggy banks, how much flour does Denise need to purchase?

109. *Height of a Computer Stand* The instructions for assembling a computer stand include a diagram illustrating its dimensions. Find the total height of the stand.

110. *Width of a Picture* The width of a picture is $24\frac{7}{8}$ in., as shown in the diagram. Find x, the distance from the edge of the frame to the center.

111. *Traveling Interstate 5* While on vacation, Janet Mazerella traveled the full length of Interstate Highway 5 from the U.S.-Canadian border to the U.S.-Mexican border. She recorded the following travel times between cities: Blaine, Washington, to Seattle: 1 hour 49 minutes; Seattle to Portland: 2 hours 48 minutes; Portland to Sacramento: 9 hours 6 minutes; Sacramento to Los Angeles: 6 hours 3 minutes; Los Angeles to San Diego: 2 hours 9 minutes; San Diego to San Ysidro, California: 22 minutes.

a) Write each of these times as a mixed number with minutes represented as a fraction with a denominator of 60. Do not reduce the fractional part of the mixed number.

b) What is Janet's total driving time from Blaine, Washington, to San Ysidro, California? Give your answer as a mixed number and in terms of hours and minutes.

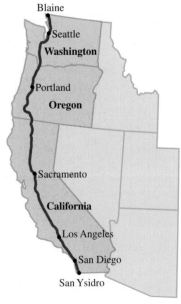

▲ Interstate Highway 5

112. *Traveling Across Texas* While on vacation, Omar Rodriguez traveled the entire length of Interstate Highway 10 in Texas. Omar recorded the following travel times between cities on his trip. Anthony to El Paso: 25 minutes; El Paso to San Antonio: 7 hours 54 minutes; San Antonio to Houston: 3 hours 1 minute; Houston to Beaumont: 1 hour 23 minutes; Beaumont to Orange: 28 minutes.

  a) Write each of these times as a mixed number with minutes represented as a fraction with a denominator of 60. Do not reduce the fractional part of the mixed number.

  b) What is Omar's total driving time from Anthony to Orange? Give your answer as a mixed number and in terms of hours and minutes.

▲ Interstate Highway 10

▲ The Alamo, San Antonio, Texas

CHALLENGE PROBLEMS/GROUP ACTIVITIES

113. *Cutting Lumber* If a piece of wood $8\frac{3}{4}$ ft long is to be cut into four equal pieces, find the length of each piece. (Allow $\frac{1}{8}$ in. for each saw cut.)

114. *Dimensions of a Room* A rectangular room measures 8 ft 3 in. by 10 ft 8 in. by 9 ft 2 in. high.

  a) Determine the perimeter of the room in feet. Write your answer as a mixed number.

  b) Calculate the area of the floor of the room in square feet.

  c) Calculate the volume of the room in cubic feet. Use volume = length × width × height. Write your answer as a mixed number.

115. *Hanging a Picture* The back of a framed picture that is to be hung is shown. A nail is to be hammered into the wall, and the picture will be hung by the wire on the nail.

  a) If the center of the wire is to rest on the nail and a side of the picture is to be 20 in. from the window, how far from the window should the nail be placed?

  b) If the top of the frame is to be $26\frac{1}{4}$ in. from the ceiling, how far from the ceiling should the nail be placed? (Assume the wire will not stretch.)

  c) Repeat part (b) if the wire will stretch $\frac{1}{4}$ in. when the picture is hung.

116. *Increasing a Book Size* The dimensions of the cover of a book have been increased from $8\frac{1}{2}$ in. by $9\frac{1}{4}$ in. to $8\frac{1}{2}$ in. by $10\frac{1}{4}$ in. By how many square inches has the surface area increased? Use area = length × width.

Dense Set of Numbers A set of numbers is said to be a *dense set* if between any two distinct members of the set there exists a third distinct member of the set. The set of integers is not dense since between any two consecutive integers there is not another integer. For example, between 1 and 2 there are no other integers. The set of rational numbers is dense because between any two distinct rational numbers there exists a third distinct rational number. For example, we can find a rational number between 0.243 and 0.244. The number 0.243 can be written as 0.2430, and 0.244 can be written as 0.2440. There are many numbers between these two.

Some of them are 0.2431, 0.2435, and 0.243912. In Exercises 117–122, find a rational number between the two numbers in each pair.

117. 0.10 and 0.11   **118.** 5.03 and 5.003

119. -2.176 and -2.175 **120.** 1.3457 and 1.34571

121. 4.872 and 4.873   **122.** -3.7896 and -3.7895

Halfway Between Two Numbers To find a rational number halfway between any two rational numbers given in fraction form, add the two numbers together and divide their sum by 2. In Exercises 123–128, find a rational number halfway between the two fractions in each pair.

123. $\frac{1}{3}$ and $\frac{2}{3}$ **124.** $\frac{2}{7}$ and $\frac{3}{7}$

125. $\frac{1}{100}$ and $\frac{1}{10}$ **126.** $\frac{7}{13}$ and $\frac{8}{13}$

127. $\frac{1}{10}$ and $\frac{1}{100}$ **128.** $\frac{1}{2}$ and $\frac{2}{3}$

129. *Cooking Oatmeal* Following are the instructions given on a box of oatmeal. Determine the amount of water (or milk) and oats needed to make $1\frac{1}{2}$ servings by:

a) Adding the amount of each ingredient needed for 1 serving to the amount needed for 2 servings and dividing by 2.

b) Adding the amount of each ingredient needed for 1 serving to half the amount needed for 1 serving.

DIRECTIONS

1. Boil water or milk and salt (if desired).
2. Stir in oats.
3. Stirring occasionally, cook over medium heat for 5 minutes.

Servings	1	2
Water (or milk)	1 cup	$1\frac{3}{4}$ cup
Oats	$\frac{1}{2}$ cup	1 cup
Salt (optional)	dash	$\frac{1}{8}$ tsp

130. Consider the rational number $0.\overline{9}$.

a) Use the method from Example 8 on page 250 to convert $0.\overline{9}$ to a quotient of integers.

b) Find a number halfway between $0.\overline{9}$ and 1 by adding the two numbers and dividing by 2.

c) Find $\frac{1}{3} + \frac{2}{3}$. Express $\frac{1}{3}$ and $\frac{2}{3}$ as repeating decimals. Now find the same sum using the repeating decimal representation of $\frac{1}{3}$ and $\frac{2}{3}$.

d) What conclusion can you draw from parts (a), (b), and (c)?

RECREATIONAL MATHEMATICS

131. *Paper Folding* Fold a sheet of paper in half. Now unfold the paper. You will see that this one crease divided the paper into two equal regions. Each of these regions represents $\frac{1}{2}$ of the area of the entire sheet of paper. Next, fold a sheet of paper in half and then fold it in half again. Now unfold this piece of paper. You will see that these two creases divided the paper into four equal regions. Each of these regions can be considered $\frac{1}{4}$ of the area of the sheet of paper. Continue this process and answer the following questions.

a) If you fold the paper in half three times, each region will be what fraction of the area of the sheet of paper?

b) If you fold the paper in half four times, each region will be what fraction of the area of the sheet of paper?

c) How many creases will you need to form regions that are $\frac{1}{32}$ of the area of the sheet of paper?

d) How many creases will you need to form regions that are $\frac{1}{64}$ of the area of the sheet of paper?

INTERNET/RESEARCH ACTIVITY

132. The ancient Greeks are often considered the first true mathematicians. Write a report summarizing the ancient Greeks' contributions to rational numbers. Include in your report what they learned and believed about the rational numbers. References include encyclopedias, history of mathematics books, and Internet web sites.

5.4 THE IRRATIONAL NUMBERS AND THE REAL NUMBER SYSTEM

▲ The distance from the northwest corner of a football field to the southeast corner is an irrational number.

A football field is 160 ft wide and 360 ft long. A football player catches the football from a kickoff in the northwest corner of one end zone and runs across the field to the southeast corner of the other end zone, as shown by the dashed line in the diagram.

How far did the football player run? The answer to this question is an example of an *irrational* number. In this section, we will introduce irrational numbers along with the operations of addition, subtraction, multiplication, and division of irrational numbers. The answer to the above question is $40\sqrt{97}$ feet, or about 394 feet.

Pythagoras (ca. 585–500 B.C.), a Greek mathematician, is credited with providing a written proof that in any *right triangle* (a triangle with a 90° angle; see Fig. 5.7), the square of the length of one side (a^2) added to the square of the length of the other side (b^2) equals the square of the length of the hypotenuse (c^2). The formula $a^2 + b^2 = c^2$ is now known as the *Pythagorean theorem*.* Pythagoras found that the solution of the formula, where $a = 1$ and $b = 1$, is not a rational number.

$$a^2 + b^2 = c^2$$
$$1^2 + 1^2 = c^2$$
$$1 + 1 = c^2$$
$$2 = c^2$$

Hypotenuse (longest side of right triangle)

b c

a

$a^2 + b^2 = c^2$

Figure 5.7

There is no rational number that when squared will equal 2. This fact prompted a need for a new set of numbers, the irrational numbers.

In Section 5.2, we introduced the real number line. The points on the real number line that are not rational numbers are referred to as irrational numbers. Recall that every rational number is either a terminating or a repeating decimal number. Therefore, irrational numbers, when represented as decimal numbers, will be nonterminating, nonrepeating decimal numbers.

> An **irrational number** is a real number whose decimal representation is a nonterminating, nonrepeating decimal number.

A nonrepeating decimal number such as 5.12639537. . . can be used to indicate an irrational number. Notice that no number or set of numbers repeat on a continuous basis, and the three dots at the end of the number indicate that the number continues

---

*The Pythagorean theorem is discussed in more detail in Section 9.3.

indefinitely. Nonrepeating number patterns can be used to indicate irrational numbers. For example, 6.1011011101111 … and 0.525225222 … are both irrational numbers.

The expression $\sqrt{2}$ is read "the square root of 2" or "radical 2." The symbol $\sqrt{}$ is called the *radical sign*, and the number or expression inside the radical sign is called the *radicand*. In $\sqrt{2}$, 2 is the radicand.

The square roots of some numbers are rational, whereas the square roots of other numbers are irrational. The *principal* (or *positive*) *square root* of a number n, written $\sqrt{n}$, is the positive number that when multiplied by itself, gives n. Whenever we mention the term "square root" in this text, we mean the principal square root. For example,

$$\sqrt{9} = 3 \quad \text{since} \quad 3 \cdot 3 = 9$$
$$\sqrt{36} = 6 \quad \text{since} \quad 6 \cdot 6 = 36$$

Both $\sqrt{9}$ and $\sqrt{36}$ are examples of numbers that are rational numbers because their square roots, 3 and 6 respectively, are terminating decimal numbers.

Returning to the problem faced by Pythagoras: If $c^2 = 2$, then c has a value of $\sqrt{2}$, but what is $\sqrt{2}$ equal to? The $\sqrt{2}$ is an irrational number, and it cannot be expressed as a terminating or repeating decimal number. It can only be approximated by a decimal number: $\sqrt{2}$ is approximately 1.4142135 (to seven decimal places). Later in this section, we will discuss using a calculator to approximate irrational numbers.

Other irrational numbers include $\sqrt{3}$, $\sqrt{5}$, and $\sqrt{37}$. Another important irrational number used to represent the ratio of a circle's circumference to its diameter is pi, symbolized π. Pi is approximately 3.1415926.

We have discussed procedures for performing the arithmetic operations of addition, subtraction, multiplication, and division with rational numbers. We can perform the same operations with the irrational numbers. Before we can proceed, however, we must understand the numbers called perfect squares. Any number that is the square of a natural number is said to be a *perfect square*. Some perfect squares are shown in the following chart.

Natural numbers	1,	2,	3,	4,	5,	6, …
Squares of the natural numbers	1^2,	2^2,	3^2,	4^2,	5^2,	6^2, …
or perfect squares	1,	4,	9,	16,	25,	36, …

The numbers 1, 4, 9, 16, 25, and 36 are some of the perfect square numbers. Can you determine the next two perfect square numbers? How many perfect square numbers are there? The square root of a perfect square number will be a natural number. For example, $\sqrt{1} = 1$, $\sqrt{4} = 2$, $\sqrt{9} = 3$, $\sqrt{16} = 4$, $\sqrt{25} = 5$, and so on.

The number that multiplies a radical is called the radical's *coefficient*. For example, in $3\sqrt{5}$, the 3 is the coefficient of the radical.

Some irrational numbers can be simplified by determining whether there are any perfect square factors in the radicand. If there are, the following rule can be used to simplify the radical.

PRODUCT RULE FOR RADICALS

$$\sqrt{a \cdot b} = \sqrt{a} \cdot \sqrt{b}, \qquad a \geq 0, \qquad b \geq 0$$

To simplify a radical, write the radical as a product of two radicals. One of the radicals should contain the greatest perfect square that is a factor of the radicand in the original expression. Then simplify the radical containing the perfect square factor.

A Piece of the Pi

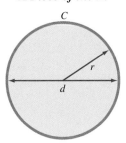

$$C = \pi d$$
$$\text{or } \pi = \frac{C}{d}$$

American philosopher and psychologist William James wrote in 1909: "The thousandth decimal of π sleeps there though no one may ever try to compute it." It sleeps no more. We now know π, the ratio of a circle's circumference to its diameter, to over 1.2 trillion digits.

In December 2002, Yasumasa Kanada and others at the University of Tokyo announced that they had calculated π to 1,241,100,000,000 decimal places, beating their previous record set in 1999. Their computation of π consumed more than 600 hours of time on a Hitachi SR8000 supercomputer.

This record, like the record for the largest prime number, will most likely be broken in the near future (it might already be broken as you read this). Mathematicians and computer scientists continue to improve both their computers and their methods used to find numbers like the most accurate approximation for π or the largest prime number.

For example,

$$\sqrt{18} = \sqrt{9 \cdot 2} = \sqrt{9} \cdot \sqrt{2} = 3 \cdot \sqrt{2} = 3\sqrt{2}$$

and

$$\sqrt{75} = \sqrt{25 \cdot 3} = \sqrt{25} \cdot \sqrt{3} = 5 \cdot \sqrt{3} = 5\sqrt{3}$$

EXAMPLE ❶ *Simplifying Radicals*

Simplify.
a) $\sqrt{18}$ b) $\sqrt{80}$

SOLUTION

a) Since 9 is a perfect square factor of 18, we write

$$\sqrt{18} = \sqrt{9 \cdot 2} = \sqrt{9} \cdot \sqrt{2} = 3 \cdot \sqrt{2} = 3\sqrt{2}$$

Since 2 has no perfect square factors, $\sqrt{2}$ cannot be simplified.

b) Since 16 is a perfect square factor of 80, we write

$$\sqrt{80} = \sqrt{16 \cdot 5} = \sqrt{16} \cdot \sqrt{5} = 4 \cdot \sqrt{5} = 4\sqrt{5} \qquad \bullet$$

In Example 1(b), you can obtain the correct answer if you start out factoring differently:

$$\sqrt{80} = \sqrt{4 \cdot 20} = \sqrt{4} \cdot \sqrt{20} = 2 \cdot \sqrt{20} = 2\sqrt{20}$$

Note that 20 has 4 as a perfect square factor.

$$2\sqrt{20} = 2\sqrt{4 \cdot 5} = 2 \cdot \sqrt{4} \cdot \sqrt{5} = 2 \cdot 2 \cdot \sqrt{5} = 4\sqrt{5}$$

The second method will eventually give the same answer, but it requires more work. It is best to try to factor out the *largest* perfect square factor from the radicand.

Addition and Subtraction of Irrational Numbers

To add or subtract two or more square roots with the same radicand, add or subtract their coefficients while keeping the common radicand. The answer is the sum or difference of the coefficients multiplied by the common radical.

EXAMPLE ❷ *Adding and Subtracting Radicals with the Same Radicand*

Simplify.
a) $4\sqrt{3} + 5\sqrt{3}$ b) $2\sqrt{5} - 7\sqrt{5} + \sqrt{5}$

SOLUTION
a) $4\sqrt{3} + 5\sqrt{3} = (4 + 5)\sqrt{3} = 9\sqrt{3}$
b) $2\sqrt{5} - 7\sqrt{5} + \sqrt{5} = (2 - 7 + 1)\sqrt{5} = -4\sqrt{5}$

Note that $\sqrt{5} = 1\sqrt{5}$. $\qquad \bullet$

EXAMPLE ❸ *Subtracting Radicals with Different Radicands*

Simplify $5\sqrt{3} - \sqrt{12}$.

SOLUTION These radicals cannot be subtracted in their present form because they contain different radicands. When that occurs, determine whether one or more of the radicals can be simplified so that they have the same radicand.

$$\begin{aligned}
5\sqrt{3} - \sqrt{12} &= 5\sqrt{3} - \sqrt{4 \cdot 3} \\
&= 5\sqrt{3} - \sqrt{4} \cdot \sqrt{3} \\
&= 5\sqrt{3} - 2\sqrt{3} \\
&= (5 - 2)\sqrt{3} = 3\sqrt{3}
\end{aligned}$$

Multiplication of Irrational Numbers

When multiplying irrational numbers, we again make use of the product rule for radicals. After the radicands are multiplied, simplify the remaining radical when possible.

EXAMPLE ❹ *Multiplying Radicals*

Simplify.
a) $\sqrt{3} \cdot \sqrt{27}$ b) $\sqrt{3} \cdot \sqrt{7}$ c) $\sqrt{6} \cdot \sqrt{10}$

SOLUTION
a) $\sqrt{3} \cdot \sqrt{27} = \sqrt{3 \cdot 27} = \sqrt{81} = 9$
b) $\sqrt{3} \cdot \sqrt{7} = \sqrt{3 \cdot 7} = \sqrt{21}$
c) $\sqrt{6} \cdot \sqrt{10} = \sqrt{6 \cdot 10} = \sqrt{60} = \sqrt{4 \cdot 15} = \sqrt{4} \cdot \sqrt{15} = 2\sqrt{15}$

Division of Irrational Numbers

To divide irrational numbers, use the following rule. After performing the division, simplify when possible.

> **QUOTIENT RULE FOR RADICALS**
> $$\frac{\sqrt{a}}{\sqrt{b}} = \sqrt{\frac{a}{b}}, \qquad a \geq 0, \qquad b > 0$$

EXAMPLE ❺ *Dividing Radicals*

Divide.
a) $\dfrac{\sqrt{8}}{\sqrt{2}}$ b) $\dfrac{\sqrt{96}}{\sqrt{2}}$

SOLUTION

a) $\dfrac{\sqrt{8}}{\sqrt{2}} = \sqrt{\dfrac{8}{2}} = \sqrt{4} = 2$

b) $\dfrac{\sqrt{96}}{\sqrt{2}} = \sqrt{\dfrac{96}{2}} = \sqrt{48} = \sqrt{16 \cdot 3} = \sqrt{16} \cdot \sqrt{3} = 4\sqrt{3}$

Rationalizing the Denominator

A denominator of a fraction is *rationalized* when it contains no radical expressions. To rationalize a denominator that contains only a square root, multiply both the numerator and denominator of the fraction by a number that will result in the radicand in the denominator becoming a perfect square. (This action is the equivalent of multiplying the fraction by 1 because the value of the fraction does not change.) Then simplify the fractions when possible.

EXAMPLE ❻ *Rationalizing the Denominator*

Rationalize the denominator of the following.

a) $\dfrac{5}{\sqrt{2}}$ b) $\dfrac{5}{\sqrt{12}}$ c) $\dfrac{\sqrt{5}}{\sqrt{10}}$

SOLUTION

a) Multiply the numerator and denominator by a number that will make the radicand in the denominator a perfect square.

$$\frac{5}{\sqrt{2}} = \frac{5}{\sqrt{2}} \cdot \frac{\sqrt{2}}{\sqrt{2}} = \frac{5\sqrt{2}}{\sqrt{4}} = \frac{5\sqrt{2}}{2}$$

Note that the 2's in the answer cannot be divided out because one 2 is a radicand and the other is not.

b) $\dfrac{5}{\sqrt{12}} = \dfrac{5}{\sqrt{12}} \cdot \dfrac{\sqrt{3}}{\sqrt{3}} = \dfrac{5\sqrt{3}}{\sqrt{36}} = \dfrac{5\sqrt{3}}{6}$

You could have obtained the same answer to this problem by multiplying both the numerator and denominator by $\sqrt{12}$ and then simplifying. Try to do so now.

c) Write $\dfrac{\sqrt{5}}{\sqrt{10}}$ as $\sqrt{\dfrac{5}{10}}$ and reduce the fraction to obtain $\sqrt{\dfrac{1}{2}}$. By the quotient rule for radicals, $\sqrt{\dfrac{1}{2}} = \dfrac{\sqrt{1}}{\sqrt{2}}$ or $\dfrac{1}{\sqrt{2}}$. Now rationalize the denominator of $\dfrac{1}{\sqrt{2}}$.

$$\frac{1}{\sqrt{2}} = \frac{1}{\sqrt{2}} \cdot \frac{\sqrt{2}}{\sqrt{2}} = \frac{\sqrt{2}}{2}$$

●

Approximating Square Roots on a Scientific Calculator

Consider the irrational number the square root of two. We use the symbol $\sqrt{2}$ to represent the *exact value* of this number. Although exact values are important, approximations are also important, especially when working with application problems. We can use a scientific calculator to obtain approximations for square roots. Scientific calculators generally have one of the following square root keys:*

$$\boxed{\sqrt{}} \quad \text{or} \quad \boxed{\sqrt{x}}$$

---

*If your calculator has the $\sqrt{}$ symbol printed *above* the key instead of on the face of the key, you can access the square root function by first pressing the "2nd" or the "inverse" key.

For simplicity, we will refer to the square root key with the $\boxed{\sqrt{}}$ symbol. To approximate $\sqrt{2}$, perform the following keystrokes:

$$\boxed{2}\quad\boxed{\sqrt{}}$$

or, depending on your model of calculator, you may have to do the following:

$$\boxed{\sqrt{}}\quad\boxed{2}\quad\boxed{\text{ENTER}}$$

The display on your calculator may read 1.414213562. Your calculator may display more or fewer digits. It is important to realize that 1.414213562 is a rational number *approximation* for the irrational number $\sqrt{2}$. The symbol $\approx$ means *is approximately equal to,* and we write

$$\sqrt{2} \approx 1.414213562$$

Exact value (irrational number)   Approximation (rational number)

EXAMPLE ❼ *Approximating Square Roots*

Use a scientific calculator to approximate the following square roots. Round each answer to two decimal places.

a) $\sqrt{3}$ b) $\sqrt{19}$ c) $\sqrt{79}$ d) $\sqrt{371}$

SOLUTION

a) $\sqrt{3} \approx 1.73$ b) $\sqrt{19} \approx 4.36$

c) $\sqrt{79} \approx 8.89$ d) $\sqrt{371} \approx 19.26$ •

SECTION 5.4   EXERCISES

CONCEPT/WRITING EXERCISES

1. Explain the difference between a rational number and an irrational number.

2. What is the principal square root of a number?

3. What is a perfect square?

4. a) State the product rule for radicals

   b) State the quotient rule for radicals.

5. a) Explain how to add or subtract square roots that have the same radicand.

   b) Using the procedure in part (a), add $3\sqrt{5} + 8\sqrt{5} - 6\sqrt{5}$.

6. What does it mean to rationalize the denominator?

7. a) Explain how to rationalize a denominator that contains a square root.

   b) Using the procedure in part (a), rationalize the denominator of $\dfrac{2}{\sqrt{3}}$.

8. a) Explain how to approximate square roots on your calculator.

   b) Using the procedure in part (a), approximate $\sqrt{11}$. Round your answer to the nearest hundredth.

PRACTICE THE SKILLS

In Exercises 9–18, determine whether the number is rational or irrational.

9. $\sqrt{25}$

10. $\sqrt{50}$

11. $\dfrac{3}{5}$

12. $0.212112111\ldots$

13. $3.575775777\ldots$

14. π

15. $\dfrac{22}{7}$

16. 3.14159

17. 3.14159...

18. $\dfrac{\sqrt{5}}{\sqrt{5}}$

In Exercises 19–28, evaluate the expression.

19. $\sqrt{16}$

20. $\sqrt{144}$

21. $\sqrt{100}$

22. $-\sqrt{121}$

23. $-\sqrt{169}$

24. $\sqrt{25}$

25. $-\sqrt{81}$

26. $-\sqrt{36}$

27. $-\sqrt{100}$

28. $\sqrt{289}$

In Exercises 29–38, classify the number as a member of one or more of the following sets: the rational numbers, the integers, the natural numbers, the irrational numbers.

29. 2

30. -3

31. $\sqrt{25}$

32. $\dfrac{3}{8}$

33. 0.040040004

34. 2.718

35. $-\dfrac{7}{8}$

36. 0.123123123

37. $0.\overline{123}$

38. 0.123112311123...

In Exercises 39–48, simplify the radical.

39. $\sqrt{12}$

40. $\sqrt{20}$

41. $\sqrt{48}$

42. $\sqrt{50}$

43. $\sqrt{63}$

44. $\sqrt{75}$

45. $\sqrt{84}$

46. $\sqrt{90}$

47. $\sqrt{162}$

48. $\sqrt{200}$

In Exercises 49–58, perform the indicated operation.

49. $3\sqrt{5} + 2\sqrt{5}$

50. $4\sqrt{6} - 7\sqrt{6}$

51. $5\sqrt{18} - 7\sqrt{8}$

52. $2\sqrt{18} - 3\sqrt{50}$

53. $4\sqrt{12} - 7\sqrt{27}$

54. $2\sqrt{7} + 5\sqrt{28}$

55. $5\sqrt{3} + 7\sqrt{12} - 3\sqrt{75}$

56. $13\sqrt{2} + 2\sqrt{18} - 5\sqrt{32}$

57. $\sqrt{8} - 3\sqrt{50} + 9\sqrt{32}$

58. $\sqrt{63} + 13\sqrt{98} - 5\sqrt{112}$

In Exercises 59–68, perform the indicated operation. Simplify the answer when possible.

59. $\sqrt{3}\sqrt{27}$

60. $\sqrt{5}\sqrt{15}$

61. $\sqrt{6}\sqrt{10}$

62. $\sqrt{2}\sqrt{10}$

63. $\sqrt{10}\sqrt{20}$

64. $\sqrt{11}\sqrt{33}$

65. $\dfrac{\sqrt{20}}{\sqrt{5}}$

66. $\dfrac{\sqrt{125}}{\sqrt{5}}$

67. $\dfrac{\sqrt{72}}{\sqrt{8}}$

68. $\dfrac{\sqrt{145}}{\sqrt{5}}$

In Exercises 69–78, rationalize the denominator.

69. $\dfrac{1}{\sqrt{5}}$

70. $\dfrac{3}{\sqrt{3}}$

71. $\dfrac{\sqrt{3}}{\sqrt{7}}$

72. $\dfrac{\sqrt{2}}{\sqrt{7}}$

73. $\dfrac{\sqrt{20}}{\sqrt{3}}$

74. $\dfrac{\sqrt{50}}{\sqrt{14}}$

75. $\dfrac{\sqrt{5}}{\sqrt{3}}$

76. $\dfrac{\sqrt{15}}{\sqrt{3}}$

77. $\dfrac{\sqrt{10}}{\sqrt{6}}$

78. $\dfrac{\sqrt{2}}{\sqrt{27}}$

PROBLEM SOLVING

Approximating Radicals *The following diagram is a sketch of a 16 in. ruler marked using $\frac{1}{2}$ inches.*

In Exercises 79–84, without using a calculator, indicate between which two adjacent ruler marks each of the following irrational numbers will fall. Explain how you obtained your answer. Support your answer by obtaining an approximation with a calculator.

79. $\sqrt{5}$ in.

80. $\sqrt{17}$ in.

81. $\sqrt{107}$ in.

82. $\sqrt{135}$ in.

83. $\sqrt{170}$ in.

84. $\sqrt{200}$ in.

In Exercises 85–90, determine whether the statement is true or false. Rewrite each false statement to make it a true statement. A false statement can be modified in more than one way to be made a true statement.

85. $\sqrt{c}$ is a rational number for any composite number c.

86. $\sqrt{p}$ is a rational number for any prime number p.

87. The sum of any two rational numbers is always a rational number.

88. The product of any two rational numbers is always a rational number.

89. The product of an irrational and a rational number is always an irrational number.

90. The product of any two irrational numbers is always an irrational number.

In Exercises 91–94, give an example to show that the stated case can occur.

91. The sum of two irrational numbers may be an irrational number.

92. The sum of two irrational numbers may be a rational number.

93. The product of two irrational numbers may be an irrational number.

94. The product of two irrational numbers may be a rational number.

95. Without doing any calculations, determine whether $\sqrt{2} = 1.414$. Explain your answer.

96. Without doing any calculations, determine whether $\sqrt{17} = 4.123$. Explain your answer.

97. The number π is an irrational number. Often the values 3.14 or $\frac{22}{7}$ are used for π. Does π equal either 3.14 or $\frac{22}{7}$? Explain your answer.

98. Give an example to show that $\sqrt{a + b} \neq \sqrt{a} + \sqrt{b}$.

99. Give an example to show that $\sqrt{a \cdot b} = \sqrt{a} \cdot \sqrt{b}$.

100. *A Swinging Pendulum* The time T required for a pendulum to swing back and forth may be found by the formula

$$T = 2\pi\sqrt{\frac{l}{g}}$$

where l is the length of the pendulum and g is the acceleration of gravity. Find the time in seconds for the pendulum to swing back and forth if $l = 35$ cm and $g = 980$ cm/sec^2. Round your answer to the nearest tenth of a second.

101. *Estimating Speed of a Vehicle* The speed that a vehicle was traveling, s, in miles per hour, when the brakes were first applied, can be estimated using the formula $s = \sqrt{\dfrac{d}{0.04}}$ where d is the length of the vehicle's skid marks, in feet.

   a) Determine the speed of a car that made skid marks 4 ft long.

   b) Determine the speed of a car that made skid marks 16 ft long.

   c) Determine the speed of a car that made skid marks 64 ft long.

   d) Determine the speed of a car that made skid marks 256 ft long.

102. *Dropping an Object* The formula $t = \dfrac{\sqrt{d}}{4}$ can be used to estimate the time, t, in seconds it takes for an object dropped to travel d feet.

   a) Determine the time it takes for an object to drop 100 ft.

   b) Determine the time it takes for an object to drop 400 ft.

   c) Determine the time it takes for an object to drop 900 ft.

   d) Determine the time it takes for an object to drop 1600 ft.

CHALLENGE PROBLEMS/GROUP ACTIVITIES

103. **a)** Is $\sqrt{0.04}$ rational or irrational? Explain.

    b) Is $\sqrt{0.\overline{7}}$ rational or irrational? Explain.

104. One way to find a rational number between two distinct rational numbers is to add the two distinct rational numbers and divide by 2. Do you think that this method will work for finding an irrational number between two distinct irrational numbers? Explain.

RECREATIONAL MATHEMATICS

105. *More Four 4's* In Exercise 82 on page 244, we introduced some of the basic rules of the game Four 4's. We now expand our operations to include square roots. For example, one way to obtain the whole number 8 is $\sqrt{4} + \sqrt{4} + \sqrt{4} + \sqrt{4} = 2 + 2 + 2 + 2 = 8$. Using the rules given on page 244, and using at least

one square root of 4, $\sqrt{4}$, play Four 4's to obtain the following whole numbers:

    a) 11

    b) 13

    c) 14

    d) 18

INTERNET/RESEARCH ACTIVITIES

In Exercises 106 and 107, references include history of mathematics books, encyclopedias, and Internet web sites.

106. Write a report on the history of the development of the irrational numbers.

107. Write a report on the history of pi. In your report, indicate when the symbol π was first used and list the first 10 digits of π.

5.5   REAL NUMBERS AND THEIR PROPERTIES

▲ Real number properties are often learned as we learn arithmetic facts.

Evaluate $10 + 5$ and then evaluate $5 + 10$. Evaluate 10×5 and then evaluate 5×10. Evaluate $10 - 5$ and then evaluate $5 - 10$. Finally, evaluate $10 \div 5$ and then evaluate $5 \div 10$. You will notice that $10 + 5 = 5 + 10$ and that $10 \times 5 = 5 \times 10$. On the other hand, $10 - 5 \neq 5 - 10$ and $10 \div 5 \neq 5 \div 10$. These simple examples demonstrate a certain property that holds for addition and multiplication, but not for subtraction and division. In this section, we will study the set of *real* numbers and their properties.

Now that we have discussed both the rational and irrational numbers, we can discuss the real numbers and the properties of the real number system. The union of the rational numbers and the irrational numbers is the *set of real numbers*, symbolized by $\mathbb{R}$.

Figure 5.8 on page 270 illustrates the relationship among various sets of numbers. It shows that the natural numbers are a subset of the whole numbers, the integers, the rational numbers, and the real numbers. For example, since the number 3 is a natural or counting number, it is also a whole number, an integer, a rational number, and a real number. Since the rational number $\frac{1}{4}$ is outside the set of integers, it is not an integer, a whole number, or a natural number. The number $\frac{1}{4}$ is a real number, however, as is the irrational number $\sqrt{2}$. Note that the real numbers are the union of the rational numbers and the irrational numbers.

Some years ago, a group of French mathematicians who worked under the collective pseudonym "Monsieur Nicholas Bourbaki" embarked on the development of an encyclopedic description of all mathematics. They devoted 200 pages simply to introduce the innocent-looking concept, the number 1.

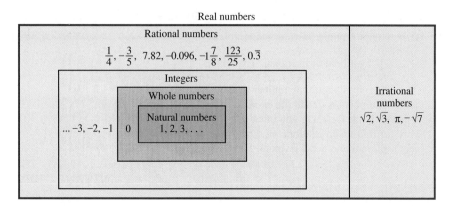

Figure 5.8

The relationship between the various sets of numbers in the real number system can also be illustrated with a tree diagram, as in Fig. 5.9.

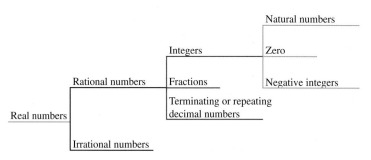

Figure 5.9

Figure 5.9 shows that, for example, the natural numbers are a subset of the integers, the rational numbers, and the real numbers. We can also see, for example, the natural numbers, zero, and the negative integers together form the integers.

Properties of the Real Number System

We are now prepared to consider the properties of the real number system. The first property we will discuss is *closure.*

> If an operation is performed on any two elements of a set and the result is an element of the set, we say that the set is **closed** under that given operation.

Is the sum of any two natural numbers a natural number? The answer is yes. Thus, we say that the natural numbers are closed under the operation of addition.

Are the natural numbers closed under the operation of subtraction? If we subtract one natural number from another natural number, must the difference always be a natural number? The answer is no. For example, $3 - 5 = -2$, and -2 is not a natural number. Therefore, the natural numbers are not closed under the operation of subtraction.

EXAMPLE ❶ *Closure of Sets*

Determine whether the set of whole numbers are closed under the operations of (a) multiplication and (b) division.

SOLUTION

a) If we multiply any two whole numbers, will the product always be a whole number? You may want to try multiplying several whole numbers. You will see that yes, the product of any two whole numbers is a whole number. Thus, the set of whole numbers are closed under the operation of multiplication.

b) If we divide any two whole numbers, will the quotient always be a whole number? The answer is no. For example $1 \div 2 = \frac{1}{2}$, and $\frac{1}{2}$ is not a whole number. Therefore, the set of whole numbers are not closed under the operation of division.   ●

Next we will discuss three important properties: the commutative property, the associative property, and the distributive property. A knowledge of these properties is essential for the understanding of algebra. We begin with the commutative property.

COMMUTATIVE PROPERTY

Addition	Multiplication
$a + b = b + a$	$a \cdot b = b \cdot a$

for any real numbers a and b.

The commutative property states that the *order* in which two numbers are added or multiplied is not important. For example, $4 + 5 = 5 + 4 = 9$ and $3 \cdot 6 = 6 \cdot 3 = 18$. Note that the commutative property does not hold for the operations of subtraction or division. For example,

$$4 - 7 \neq 7 - 4 \quad \text{and} \quad 9 \div 3 \neq 3 \div 9$$

Now we introduce the associative property.

ASSOCIATIVE PROPERTY

Addition	Multiplication
$(a + b) + c = a + (b + c)$	$(a \cdot b) \cdot c = a \cdot (b \cdot c)$

for any real numbers a, b, and c.

The associative property states that when adding or multiplying three real numbers, we may place parentheses around any two adjacent numbers. The result is the same regardless of the placement of parentheses. For example,

$$(3 + 4) + 5 = 3 + (4 + 5) \qquad (3 \cdot 4) \cdot 5 = 3 \cdot (4 \cdot 5)$$
$$7 + 5 = 3 + 9 \qquad\qquad 12 \cdot 5 = 3 \cdot 20$$
$$12 = 12 \qquad\qquad\qquad 60 = 60$$

The associative property does not hold for the operations of subtraction and division. For example,

$$(10 - 6) - 2 \neq 10 - (6 - 2) \qquad \text{and} \qquad (27 \div 9) \div 3 \neq 27 \div (9 \div 3)$$

Note the difference between the commutative property and the associative property. The commutative property involves a change in *order,* whereas the associative property involves a change in *grouping* (or the *association* of numbers that are grouped together).

Another property of the real numbers is the distributive property of multiplication over addition.

DISTRIBUTIVE PROPERTY OF MULTIPLICATION OVER ADDITION

$$a \cdot (b + c) = a \cdot b + a \cdot c$$

for any real numbers a, b, and c.

For example, if $a = 3$, $b = 4$, and $c = 5$, then

$$3 \cdot (4 + 5) = (3 \cdot 4) + (3 \cdot 5)$$
$$3 \cdot 9 = 12 + 15$$
$$27 = 27$$

This result indicates that, when using the distributive property, you may either add first and then multiply or multiply first and then add. Note that the distributive property involves two operations, addition and multiplication. Although positive integers were used in the example, any real numbers could have been used.

We frequently use the commutative, associative, and distributive properties without realizing that we are doing so. To add $13 + 4 + 6$, we may add the $4 + 6$ first to get 10. To this sum we then add 13 to get 23. Here we have done the equivalent of placing parentheses around the $4 + 6$. We can do so because of the associative property of addition.

To multiply 102×11 in our heads, we might multiply $100 \times 11 = 1100$ and $2 \times 11 = 22$ and add these two products to get 1122. We are permitted to do so because of the distributive property.

$$102 \times 11 = (100 + 2) \times 11 = (100 \times 11) + (2 \times 11)$$
$$= 1100 + 22 = 1122$$

EXAMPLE ❷ *Identifying Properties of Real Numbers*

Name the property illustrated.
a) $2 + 5 = 5 + 2$
b) $(x + 3) + 5 = x + (3 + 5)$
c) $4 \cdot (3 \cdot y) = (4 \cdot 3) \cdot y$
d) $9(w + 3) = 9 \cdot w + 9 \cdot 3$
e) $5 + (z + 3) = 5 + (3 + z)$
f) $(2p) \cdot 5 = 5 \cdot (2p)$

SOLUTION
a) Commutative property of addition
b) Associative property of addition

c) Associative property of multiplication

d) Distributive property of multiplication over addition

e) The only change between the left and right sides of the equal sign is the order of the z and 3 within the parentheses. The order is changed from $z + 3$ to $3 + z$ using the commutative property of addition.

f) The order of 5 and $(2p)$ is changed by using the commutative property of multiplication.

EXAMPLE ❸ *Simplifying by Using the Distributive Property*

Use the distributive property to simplify

a) $2(3 + \sqrt{5})$ b) $\sqrt{2}(7 + \sqrt{3})$

SOLUTION

a) $2(3 + \sqrt{5}) = (2 \cdot 3) + (2 \cdot \sqrt{5})$
$$= 6 + 2\sqrt{5}$$

b) $\sqrt{2}(7 + \sqrt{3}) = (\sqrt{2} \cdot 7) + (\sqrt{2} \cdot \sqrt{3})$
$$= 7\sqrt{2} + \sqrt{6}$$

Note that $\sqrt{2} \cdot 7$ is written $7\sqrt{2}$.

EXAMPLE ❹ *Distributive Property*

Use the distributive property to multiply $5(z + 3)$. Then simplify the result.

SOLUTION

$$5(z + 3) = 5 \cdot z + 5 \cdot 3$$
$$= 5z + 15$$

We summarize the properties mentioned in this section as follows, where a, b, and c are any real numbers.

Commutative property of addition	$a + b = b + a$
Commutative property of multiplication	$a \cdot b = b \cdot a$
Associative property of addition	$(a + b) + c = a + (b + c)$
Associative property of multiplication	$(a \cdot b) \cdot c = a \cdot (b \cdot c)$
Distributive property of multiplication over addition	$a \cdot (b + c) = a \cdot b + a \cdot c$

SECTION 5.5 EXERCISES

CONCEPT/WRITING EXERCISES

1. What are the real numbers?

2. What symbol is used to represent the set of real numbers?

3. What does it mean if a set is closed under a given operation?

4. Give the commutative property of addition, explain what it means, and give an example illustrating it.

5. Give the commutative property of multiplication, explain what it means, and give an example illustrating it.

6. Give the associative property of multiplication, explain what it means, and give an example illustrating it.

7. Give the associative property of addition, explain what it means, and give an example illustrating it.

8. Give the distributive property of multiplication over addition, explain what it means, and give an example illustrating it.

PRACTICE THE SKILLS

In Exercises 9–12, determine whether the natural numbers are closed under the given operation.

9. Subtraction   10. Addition

11. Multiplication   12. Division

In Exercises 13–16, determine whether the integers are closed under the given operation.

13. Addition   14. Subtraction

15. Multiplication   16. Division

In Exercises 17–20, determine whether the rational numbers are closed under the given operation.

17. Subtraction   18. Addition

19. Division   20. Multiplication

In Exercises 21–24, determine whether the irrational numbers are closed under the given operation.

21. Addition   22. Subtraction

23. Division   24. Multiplication

In Exercises 25–28, determine whether the real numbers are closed under the given operation.

25. Subtraction   26. Addition

27. Multiplication   28. Division

29. Does $(5 + x) + 6 = (x + 5) + 6$ illustrate the commutative property or the associative property? Explain your answer.

30. Does $(x + 5) + 6 = x + (5 + 6)$ illustrate the commutative property or the associative property? Explain your answer.

31. Give an example to show that the commutative property of addition may be true for the negative integers.

32. Give an example to show that the commutative property of multiplication may be true for the negative integers.

33. Does the commutative property hold for the integers under the operation of subtraction? Give an example to support your answer.

34. Does the commutative property hold for the rational numbers under the operation of division? Give an example to support your answer.

35. Give an example to show that the associative property of addition may be true for the negative integers.

36. Give an example to show that the associative property of multiplication may be true for the negative integers.

37. Does the associative property hold for the integers under the operation of division? Give an example to support your answer.

38. Does the associative property hold for the integers under the operation of subtraction? Give an example to support your answer.

39. Does the associative property hold for the real numbers under the operation of division? Give an example to support your answer.

40. Does $a + (b \cdot c) = (a + b) \cdot (a + c)$? Give an example to support your answer.

In Exercises 41–56, state the name of the property illustrated.

41. $3(y + 5) = 3 \cdot y + 3 \cdot 5$

42. $6 + 7 = 7 + 6$

43. $(7 \cdot 8) \cdot 9 = 7 \cdot (8 \cdot 9)$

44. $c + d = d + c$

45. $(24 + 7) + 3 = 24 + (7 + 3)$

46. $4 \cdot (11 \cdot x) = (4 \cdot 11) \cdot x$

47. $\sqrt{3} \cdot 7 = 7 \cdot \sqrt{3}$

48. $\dfrac{3}{8} + \left(\dfrac{1}{8} + \dfrac{3}{2} \right) = \left(\dfrac{3}{8} + \dfrac{1}{8} \right) + \dfrac{3}{2}$

49. $-1(x + 4) = (-1) \cdot x + (-1) \cdot 4$

50. $(p + q) + r = p + (q + r)$

51. $\sqrt{5} \cdot 2 = 2 \cdot \sqrt{5}$

52. $(r + s) + t = t + (r + s)$

53. $(r + s) \cdot t = (r \cdot t) + (s \cdot t)$

54. $g \cdot (h + i) = (h + i) \cdot g$

55. $(f \cdot g) + (j \cdot h) = (g \cdot f) + (j \cdot h)$

56. $(k + l) \cdot (m + n) = (l + k) \cdot (m + n)$

In Exercises 57–68, use the distributive property to multiply. Then, if possible, simplify the resulting expression.

57. $4(z + 1)$

58. $-1(a + b)$

59. $-\dfrac{3}{4}(x - 12)$

60. $-x(-y + z)$

61. $6\left(\dfrac{x}{2} + \dfrac{2}{3}\right)$

62. $24\left(\dfrac{x}{3} - \dfrac{1}{8}\right)$

63. $32\left(\dfrac{1}{16}x - \dfrac{1}{32}\right)$

64. $15\left(\dfrac{2}{3}x - \dfrac{4}{5}\right)$

65. $\sqrt{2}(\sqrt{8} - \sqrt{2})$

66. $-3(2 - \sqrt{3})$

67. $5(\sqrt{2} + \sqrt{3})$

68. $\sqrt{5}(\sqrt{15} - \sqrt{20})$

In Exercises 69–74, name the property used to go from step to step. You only need to name the properties indicated by an a), b), c), or d). For example, in Exercise 69, you need to supply an answer for part a), and an answer to go from the second step of part a) to part b).

69. a) $3(a + 2) + 5 = (3 \cdot a + 3 \cdot 2) + 5$
$= (3a + 6) + 5$

   b) $\qquad\qquad = 3a + (6 + 5)$
$= 3a + 11$

70. a) $2(b + 4) + 1 = (2 \cdot b + 2 \cdot 4) + 1$
$= (2b + 8) + 1$

   b) $\qquad\qquad = 2b + (8 + 1)$
$= 2b + 9$

71. a) $7(k + 1) + 2k = (7 \cdot k + 7 \cdot 1) + 2k$
$= (7k + 7) + 2k$

   b) $\qquad\qquad = 7k + (7 + 2k)$

   c) $\qquad\qquad = 7k + (2k + 7)$

   d) $\qquad\qquad = (7k + 2k) + 7$
$= 9k + 7$

72. a) $11(h + 6) + 5h = (11 \cdot h + 11 \cdot 6) + 5h$
$= (11h + 66) + 5h$

   b) $\qquad\qquad = 11h + (66 + 5h)$

   c) $\qquad\qquad = 11h + (5h + 66)$

   d) $\qquad\qquad = (11h + 5h) + 66$
$= 16h + 66$

73. a) $9 + 2(t + 3) + 7t = 9 + (2 \cdot t + 2 \cdot 3) + 7t$
$= 9 + (2t + 6) + 7t$

   b) $\qquad\qquad = 9 + (6 + 2t) + 7t$

   c) $\qquad\qquad = (9 + 6) + 2t + 7t$
$= 15 + 9t$

   d) $\qquad\qquad = 9t + 15$

74. a) $7 + 5(s + 4) + 3s = 7 + (5 \cdot s + 5 \cdot 4) + 3s$
$= 7 + (5s + 20) + 3s$

   b) $\qquad\qquad = 7 + (20 + 5s) + 3s$

   c) $\qquad\qquad = (7 + 20) + 5s + 3s$
$= 27 + 8s$

   d) $\qquad\qquad = 8s + 27$

In Exercises 75–80, determine whether the activity can be used to illustrate the commutative property. For the property to hold, the end result must be identical, regardless of the order in which the actions are performed.

75. Feeding your dog and giving your dog water

76. Putting your wallet in your back pocket and putting your keys in your front pocket

77. Washing clothes and drying clothes

78. Turning on a computer and typing a term paper on the computer

79. Filling your car with gasoline and washing the windshield

80. Turning on a lamp and reading a book

rcises 81–88, determine whether the activity can be ... to illustrate the associative property. For the property ... hold, doing the first two actions followed by the third would produce the same end result as doing the second and third actions followed by the first.

81. Putting batteries in your calculator, pressing the $\boxed{\text{ON}}$ key, and pressing the $\boxed{7}$ key

82. Filling your car with gas, washing the windshield, and checking the tire pressure

83. Sending a holiday card to your grandmother, sending one to your parents, and sending one to your teacher

84. Mowing the lawn, trimming the bushes, and removing dead limbs from trees

85. Brushing your teeth, washing your face, and combing your hair

86. Cracking an egg, pouring out the egg, and cooking the egg

87. Taking a bath, brushing your teeth, and taking your vitamins

88. While making meatloaf, mixing in the milk, mixing in the spices, and mixing in the bread crumbs

CHALLENGE PROBLEMS/GROUP ACTIVITIES

89. Describe two other activities that can be used to illustrate the commutative property (see Exercises 75–80).

90. Describe three other activities that can be used to illustrate the associative property (see Exercises 81–88).

91. Does $0 \div a = a \div 0$ (assume $a \neq 0$)? Explain.

RECREATIONAL MATHEMATICS

92. a) Consider the three words *man eating tiger*. Does (*man eating*) *tiger* mean the same as *man* (*eating tiger*)?

   b) Does (*horse riding*) *monkey* mean the same as *horse* (*riding monkey*)?

   c) Can you find three other nonassociative word triples?

INTERNET/RESEARCH ACTIVITY

93. A set of numbers that was not discussed in this chapter is the set of *complex numbers*. Write a report on complex numbers. Include their relationship to the real numbers.

5.6   RULES OF EXPONENTS AND SCIENTIFIC NOTATION

▲ The diameter of this galaxy may be given in scientific notation.

What is the current world population? What is the diameter of an atom? What is the diameter of a galaxy? What is the wavelength of an x-ray? What is the current national federal debt? All these questions have answers that may involve either very large numbers or very small numbers. One way to accurately represent such numbers is with *scientific notation*. In this section, we will study how to use scientific notation when answering questions such as those posed above.

An understanding of exponents is important in solving problems in algebra. In the expression 5^2, the 2 is referred to as the *exponent* and the 5 is referred to as the *base*. We read 5^2 as 5 to the second power, or 5 squared, which means

$$5^2 = \underbrace{5 \cdot 5}_{\text{2 factors of 5}}$$

The number 5 to the third power, or 5 cubed, written 5^3, means

$$5^3 = \underbrace{5 \cdot 5 \cdot 5}_{3 \text{ factors of } 5}$$

In general, the number b to the nth power, written b^n, means

$$b^n = \underbrace{b \cdot b \cdot b \cdot \cdots \cdot b}_{n \text{ factors of } b}$$

EXAMPLE ❶ Evaluating the Power of a Number

Evaluate.

a) 3^2 b) $(-7)^2$ c) 4^3 d) 1^{100} e) 100^1

SOLUTION

a) $3^2 = 3 \cdot 3 = 9$

b) $(-7)^2 = (-7) \cdot (-7) = 49$

c) $4^3 = 4 \cdot 4 \cdot 4 = 64$

d) $1^{100} = 1$

   The number 1 multiplied by itself any number of times equals 1.

e) $100^1 = 100$

   Any number with an exponent of 1 equals the number itself.   ●

EXAMPLE ❷ The Importance of Parentheses

Evaluate.

a) $(-2)^4$ b) -2^4 c) $(-2)^5$ d) -2^5

SOLUTION

a) $(-2)^4 = (-2)(-2)(-2)(-2) = 4(-2)(-2) = -8(-2) = 16$

b) -2^4 means take the opposite of 2^4 or $-1 \cdot 2^4$.

   $-1 \cdot 2^4 = -1 \cdot 2 \cdot 2 \cdot 2 \cdot 2 = -1 \cdot 16 = -16$

c) $(-2)^5 = (-2)(-2)(-2)(-2)(-2) = 4(-2)(-2)(-2) = -8(-2)(-2)$

   $= 16(-2) = -32$

d) $-2^5 = -1 \cdot 2^5 = -1 \cdot 2 \cdot 2 \cdot 2 \cdot 2 \cdot 2 = -1 \cdot 32 = -32$ ●

From Example 2, we can see that $(-x)^n \neq -x^n$, where n is an even natural number.

Rules of Exponents

Now that we know how to evaluate powers of numbers we can discuss the rules of exponents. Consider

$$2^2 \cdot 2^3 = \underbrace{2 \cdot 2}_{2 \text{ factors}} \cdot \underbrace{2 \cdot 2 \cdot 2}_{3 \text{ factors}} = 2^5$$

This example illustrates the product rule for exponents.

PRODUCT RULE FOR EXPONENTS

$$a^m \cdot a^n = a^{m+n}$$

Therefore, by using the product rule, $2^2 \cdot 2^3 = 2^{2+3} = 2^5$.

EXAMPLE ❸ *Using the Product Rule for Exponents*

Use the product rule to simplify.

a) $2^3 \cdot 2^7$ b) $5^2 \cdot 5^6$

SOLUTION

a) $2^3 \cdot 2^7 = 2^{3+7} = 2^{10}$ b) $5^2 \cdot 5^6 = 5^{2+6} = 5^8$ •

Consider

$$\frac{2^5}{2^2} = \frac{2 \cdot 2 \cdot 2 \cdot \cancel{2} \cdot \cancel{2}}{\cancel{2} \cdot \cancel{2}} = 2 \cdot 2 \cdot 2 = 2^3$$

This example illustrates the quotient rule for exponents.

QUOTIENT RULE FOR EXPONENTS

$$\frac{a^m}{a^n} = a^{m-n}, \qquad a \neq 0$$

Therefore, $\dfrac{2^5}{2^2} = 2^{5-2} = 2^3$.

EXAMPLE ❹ *Using the Quotient Rule for Exponents*

Use the quotient rule to simplify.

a) $\dfrac{5^8}{5^5}$ b) $\dfrac{8^{12}}{8^5}$

SOLUTION

a) $\dfrac{5^8}{5^5} = 5^{8-5} = 5^3$ b) $\dfrac{8^{12}}{8^5} = 8^{12-5} = 8^7$ •

Consider $2^3 \div 2^3$. The quotient rule gives

$$\frac{2^3}{2^3} = 2^{3-3} = 2^0$$

But $\dfrac{2^3}{2^3} = \dfrac{8}{8} = 1$. Therefore, 2^0 must equal 1. This example illustrates the zero exponent rule.

ZERO EXPONENT RULE

$$a^0 = 1, \qquad a \neq 0$$

Note that 0^0 is not defined by the zero exponent rule.

EXAMPLE ❺   *The Zero Power*

Use the zero exponent rule to simplify.

a) 2^0 b) $(-2)^0$ c) -2^0 d) $(5x)^0$ e) $5x^0$

SOLUTION

a) $2^0 = 1$ b) $(-2)^0 = 1$ c) $-2^0 = -1 \cdot 2^0 = -1 \cdot 1 = -1$

d) $(5x)^0 = 1$ e) $5x^0 = 5 \cdot x^0 = 5 \cdot 1 = 5$

Consider $2^3 \div 2^5$. The quotient rule yields

$$\frac{2^3}{2^5} = 2^{3-5} = 2^{-2}$$

But $\dfrac{2^3}{2^5} = \dfrac{2 \cdot 2 \cdot 2}{2 \cdot 2 \cdot 2 \cdot 2 \cdot 2} = \dfrac{1}{2^2}$. Since $\dfrac{2^3}{2^5}$ equals both 2^{-2} and $\dfrac{1}{2^2}$, then 2^{-2} must equal $\dfrac{1}{2^2}$. This example illustrates the negative exponent rule.

NEGATIVE EXPONENT RULE

$$a^{-m} = \frac{1}{a^m}, \qquad a \neq 0$$

EXAMPLE ❻   *Using the Negative Exponent Rule*

Use the negative exponent rule to simplify.

a) 2^{-3} b) 5^{-1}

SOLUTION

a) $2^{-3} = \dfrac{1}{2^3} = \dfrac{1}{8}$ b) $5^{-1} = \dfrac{1}{5^1} = \dfrac{1}{5}$

Consider $(2^3)^2$.

$$(2^3)^2 = (2^3)(2^3) = 2^{3+3} = 2^6$$

This example illustrates the power rule for exponents.

POWER RULE FOR EXPONENTS

$$(a^m)^n = a^{m \cdot n}$$

Thus, $(2^3)^2 = 2^{3 \cdot 2} = 2^6$.

EXAMPLE ❼   *Evaluating a Power Raised to Another Power*

Use the power rule to simplify.

a) $(5^4)^3$ b) $(7^2)^5$

SOLUTION

a) $(5^4)^3 = 5^{4 \cdot 3} = 5^{12}$ b) $(7^2)^5 = 7^{2 \cdot 5} = 7^{10}$

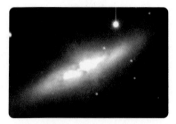

The diameter of a galaxy may be 1×10^5 light-years

The diameter of an atom may be 1×10^{-10} meter

O ur everyday activities don't require us to deal with quantities much above those in the thousands: $6.95 for lunch, a 100 lap race, a $15,000 car loan, and so on. Yet as modern technology has developed, so has our ability to study all aspects of the universe we live in, from the very large to the very small. Modern technology can be used with the rules of exponents and scientific notation to study everything from the diameter of a galaxy to the diameter of an atom.

SUMMARY OF THE RULES OF EXPONENTS

$a^m \cdot a^n = a^{m+n}$	Product rule for exponents
$\dfrac{a^m}{a^n} = a^{m-n}, \quad a \neq 0$	Quotient rule for exponents
$a^0 = 1, \quad a \neq 0$	Zero exponent rule
$a^{-m} = \dfrac{1}{a^m}, \quad a \neq 0$	Negative exponent rule
$(a^m)^n = a^{m \cdot n}$	Power rule for exponents

Scientific Notation

Often scientific problems deal with very large and very small numbers. For example, the distance from Earth to the sun is about 93,000,000 miles. The wavelength of a yellow color of light is about 0.0000006 meter. Because working with many zeros is difficult, scientists developed a notation that expresses such numbers with exponents. For example, consider the distance from Earth to the sun, 93,000,000 miles.

$$93{,}000{,}000 = 9.3 \times 10{,}000{,}000$$
$$= 9.3 \times 10^7$$

The wavelength of a yellow color of light is about 0.0000006 meter.

$$0.0000006 = 6.0 \times 0.0000001$$
$$= 6.0 \times 10^{-7}$$

The numbers 9.3×10^7 and 6.0×10^{-7} are written in a form called *scientific notation*. Each number written in scientific notation is written as a number greater than or equal to 1 and less than 10 multiplied by some power of 10.

Some examples of numbers in scientific notation are

$$3.7 \times 10^3, \quad 2.05 \times 10^{-3}, \quad 5.6 \times 10^8, \quad \text{and} \quad 1.00 \times 10^{-5}$$

The following is a procedure for writing a number in scientific notation.

TO WRITE A NUMBER IN SCIENTIFIC NOTATION

1. Move the decimal point in the original number to the right or left until you obtain a number greater than or equal to 1 and less than 10.

2. Count the number of places you have moved the decimal point to obtain the number in step 1. If the decimal point was moved to the left, the count is to be considered positive. If the decimal point was moved to the right, the count is to be considered negative.

3. Multiply the number obtained in step 1 by 10 raised to the count found in step 2. (Note that the count determined in step 2 is the exponent on the base 10.)

EXAMPLE ⑧ *Converting from Decimal Notation to Scientific Notation*

Write each number in scientific notation.

a) In 2006, the population of the United States was about 299,000,000.
b) In 2006, the population of China was about 1,306,000,000.
c) In 2006, the population of the world was about 6,522,000,000
d) The diameter of a hydrogen atom nucleus is about 0.0000000000011 millimeter.
e) The wavelength of an x-ray is about 0.000000000492 meter.

SOLUTION

a) $299,000,000 = 2.99 \times 10^8$
b) $1,306,000,000 = 1.306 \times 10^9$
c) $6,522,000,000 = 6.522 \times 10^9$
d) $0.0000000000011 = 1.1 \times 10^{-12}$
e) $0.000000000492 = 4.92 \times 10^{-10}$ ●

To convert from a number given in scientific notation to decimal notation we reverse the procedure.

TO CHANGE A NUMBER IN SCIENTIFIC NOTATION TO DECIMAL NOTATION

1. Observe the exponent on the 10.
2. a) If the exponent is positive, move the decimal point in the number to the right the same number of places as the exponent. Adding zeros to the number might be necessary.

    b) If the exponent is negative, move the decimal point in the number to the left the same number of places as the exponent. Adding zeros might be necessary.

EXAMPLE ⑨ *Converting from Scientific Notation to Decimal Notation*

Write each number in decimal notation.

a) The average distance from Mars to the sun is about 1.4×10^8 miles.
b) The half-life of uranium-235 is about 4.5×10^9 years.
c) The average grain size in siltstone is 1.35×10^{-3} inch.
d) A *millimicron* is a unit of measure used for very small distances. One millimicron is about 3.94×10^{-8} inch.

SOLUTION

a) $1.4 \times 10^8 = 140,000,000$
b) $4.5 \times 10^9 = 4,500,000,000$
c) $1.35 \times 10^{-3} = 0.00135$
d) $3.94 \times 10^{-8} = 0.0000000394$ ●

In scientific journals and books, we occasionally see numbers like 10^{15} and 10^{-6}. We interpret these numbers as 1×10^{15} and 1×10^{-6}, respectively, when converting the numbers to decimal form.

EXAMPLE ⑩ *Multiplying Numbers in Scientific Notation*

Multiply $(2.1 \times 10^5)(9 \times 10^{-3})$. Write the answer in scientific notation and in decimal notation.

SOLUTION

$$
\begin{aligned}
(2.1 \times 10^5)(9 \times 10^{-3}) &= (2.1 \times 9)(10^5 \times 10^{-3}) \\
&= 18.9 \times 10^2 \\
&= 1.89 \times 10^3 \qquad \text{Scientific notation} \\
&= 1.890 \qquad\qquad \text{Decimal notation}
\end{aligned}
$$

EXAMPLE ⑪ *Dividing Numbers Using Scientific Notation*

Divide $\dfrac{0.000000000048}{24{,}000{,}000{,}000}$. Write the answer in scientific notation.

SOLUTION First write each number in scientific notation.

$$
\begin{aligned}
\frac{0.000000000048}{24{,}000{,}000{,}000} = \frac{4.8 \times 10^{-11}}{2.4 \times 10^{10}} &= \left(\frac{4.8}{2.4}\right)\left(\frac{10^{-11}}{10^{10}}\right) \\
&= 2.0 \times 10^{-11-10} \\
&= 2.0 \times 10^{-21}
\end{aligned}
$$

Scientific Notation on the Scientific Calculator

One advantage of using scientific notation when working with very large and very small numbers is the ease with which you can perform operations. Performing these operations is even easier with the use of a scientific calculator. Most scientific calculators have a scientific notation key labeled "Exp," "EXP," or "EE." We will refer to the scientific notation key as EXP . The following keystrokes reading downwards can be used to enter the number 4.3×10^6.

Keystroke(s)	Calculator display
4.3	4.3
EXP	4.3^{00}
6	4.3^{06}

Your calculator may have some slight variations to the display shown here. The display 4.3^{06} means 4.3×10^6. We now will use our calculators to perform some computations using scientific notation.

EXAMPLE ⑫ *Use Scientific Notation on a Calculator to Find a Product*

Multiply $(4.3 \times 10^6)(2 \times 10^{-4})$ using a scientific calculator. Write the answer in decimal notation.

SOLUTION Our sequence of keystrokes is as follows:

Keystroke(s)	Display
4.3	4.3
EXP	4.3^{00}
6	4.3^{06}
$\times$	4300000

4.3×10^6 is now entered in the calculator. Most calculators will convert to decimal notation (if the display can show the number)

2	2.
EXP	$2.^{00}$
4	$2.^{04}$
+/−	$2.^{-04}$
=	860.

4 — $2.^{04}$ — Enter the positive form of the exponent
+/− — $2.^{-04}$ — Make the exponent negative*
= — 860. — Press = to obtain the answer of 860[†]

EXAMPLE ⓭ Use Scientific Notation on a Calculator to Find a Quotient

Divide $\dfrac{0.000000000048}{24,000,000,000}$ using a scientific calculator. Write the answer in scientific notation.

SOLUTION We first rewrite the numerator and denominator using scientific notation. See Example 11.

$$\frac{0.000000000048}{24,000,000,000} = \frac{4.8 \times 10^{-11}}{2.4 \times 10^{10}}$$

Next we use a scientific calculator to perform the computation. The keystrokes are as follows:

4.8 [EXP] 11 [+/−] [÷] 2.4 [EXP] 10 [=]

The display on the calculator is $2.^{-21}$, which means 2.0×10^{-21}.

EXAMPLE ⓮ U.S. Debt per Person

On June 15, 2006, the U.S. Department of the Treasury estimated the U.S. federal debt to be about $8.38 trillion. On this same day, the U.S. Bureau of the Census estimated the U.S. population to be about 299 million people. Determine the average debt, per person, by dividing the U.S. federal debt by the U.S. population.

---

*Some calculators will require you to enter the negative sign before entering the exponent.

[†]Some calculators will display the answer in scientific notation. In this case, the display will show $8.6^{\,02}$, which means 8.6×10^2 and equals 860.

SOLUTION We will first write the numbers involved using decimal notation. Then we will convert them to scientific notation.

$$8.38 \text{ trillion} = 8,380,000,000,000 = 8.38 \times 10^{12}$$
$$299 \text{ million} = 299,000,000 = 2.99 \times 10^{8}$$

Now we will divide 8.38×10^{12} by 2.99×10^{8} using a scientific calculator. The keystrokes are

8.38 | EXP | 12 | ÷ | 2.99 | EXP | 8 | = |

The display then shows 28026.7558. This number indicates that on June 15, 2006, the U.S. government owed about $28,026.76 per man, woman, and child living in the United States.   ●

SECTION 5.6 EXERCISES

CONCEPT/WRITING EXERCISES

1. In the expression 2^3, what is the name given to the 2, and what is the name given to the 3?

2. Explain the meaning of b^n.

3. **a)** Explain the product rule for exponents.

   b) Use the product rule to simplify $2^3 \cdot 2^4$.

4. **a)** Explain the quotient rule for exponents.

   b) Use the quotient rule to simplify $\dfrac{5^6}{5^4}$.

5. **a)** Explain the zero exponent rule.

   b) Use the zero exponent rule to simplify 7^0.

6. **a)** Explain the negative exponent rule.

   b) Use the negative exponent rule to simplify 2^{-3}.

7. **a)** Explain the power rule for exponents.

   b) Use the power rule to simplify $(3^2)^4$.

8. Explain how you can simplify the expression 1^{500}.

9. Explain how you can simplify the following expressions and then simplify the expression.

   a) -1^{500} **b)** $(-1)^{500}$

   c) -1^{501} **d)** $(-1)^{501}$

10. **a)** In your own words, explain how to change a number in decimal notation to scientific notation.

    b) Using the procedure in part (a), change 0.000739 to scientific notation.

    c) Using the procedure in part (a), change 1,940,000,000 to scientific notation.

11. **a)** In your own words, explain how to change a number in scientific notation to decimal notation.

    b) Using the procedure in part (a), change 2.91×10^{-5} to decimal notation.

    c) Using the procedure in part (a), change 7.02×10^{6} to decimal notation.

12. A number is given in scientific notation. What does it indicate about the number when the exponent on the 10 is (a) positive, (b) zero, and (c) negative?

PRACTICE THE SKILLS

In Exercises 13–40, evaluate the expression.

13. **a)** 3^2 **b)** 2^3 14. **a)** 2^5 **b)** 5^2

15. **a)** $(-5)^2$ **b)** -5^2 16. **a)** -3^2 **b)** $(-3)^2$

17. **a)** -2^4 **b)** $(-2)^4$ 18. **a)** $(-3)^4$ **b)** -3^4

19. **a)** -4^3 **b)** $(-4)^3$ 20. **a)** $(-2)^7$ **b)** -2^7

21. a) $\left(\dfrac{1}{8}\right)^2$ **b)** $\left(-\dfrac{3}{4}\right)^2$

22. a) $-\left(\dfrac{2}{3}\right)^2$ **b)** $\left(\dfrac{1}{6}\right)^3$

23. a) 1000^1 **b)** 1^{1000} **24. a)** 2008^1 **b)** 1^{2008}

25. a) $3^2 \cdot 3^3$ **b)** $(-3)^2 \cdot (-3)^3$

26. a) $2^3 \cdot 2$ **b)** $(-2)^3 \cdot (-2)$

27. a) $\dfrac{5^7}{5^5}$ **b)** $\dfrac{(-5)^7}{(-5)^5}$ **28. a)** $\dfrac{4^5}{4^2}$ **b)** $\dfrac{(-4)^5}{(-4)^2}$

29. a) 6^0 **b)** -6^0 **30. a)** $(-6)^0$ **b)** $-(-6)^0$

31. a) $(6x)^0$ **b)** $6x^0$ **32. a)** $-6x^0$ **b)** $(-6x)^0$

33. a) 3^{-3} **b)** 7^{-2} **34. a)** 5^{-1} **b)** 2^{-4}

35. a) -9^{-2} **b)** $(-9)^{-2}$

36. a) -5^{-2} **b)** $(-5)^{-2}$

37. a) $(2^2)^3$ **b)** $(2^3)^2$

38. a) $\left[\left(\dfrac{1}{2}\right)^2\right]^3$ **b)** $\left[\left(-\dfrac{1}{2}\right)^2\right]^3$

39. a) $4^3 \cdot 4^{-2}$ **b)** $2^{-2} \cdot 2^{-2}$

40. a) $(-5)^{-2}(-5)^{-2}$ **b)** $(-1)^{-5}(-1)^{-5}$

In Exercises 41–56, express the number in scientific notation.

41. 175,000 **42.** 496,000,000 **43.** 0.00023

44. 0.000034 **45.** 0.56 **46.** 0.00467

47. 19,000 **48.** 1,260,000,000 **49.** 0.000186

50. 0.0003 **51.** 0.00000423 **52.** 54,000

53. 711 **54.** 0.02 **55.** 0.153

56. 416,000

In Exercises 57–72, express the number in decimal notation.

57. 1.7×10^2 **58.** 2.03×10^4 **59.** 1.097×10^{-4}

60. 4.003×10^7 **61.** 8.62×10^{-5} **62.** 2.19×10^{-4}

63. 3.12×10^{-1} **64.** 4.6×10^1 **65.** 9×10^6

66. 7.3×10^4 **67.** 2.31×10^2 **68.** 1.04×10^{-2}

69. 3.5×10^4 **70.** 2.17×10^{-6}

71. 1×10^4 **72.** 1×10^{-3}

In Exercises 73–82, (a) perform the indicated operation without the use of a calculator and express each answer in decimal notation. (b) Confirm your answer from part (a) by using a scientific calculator to perform the operations. If the calculator displays the answer in scientific notation, convert the answer to decimal notation.

73. $(3 \times 10^2)(2.5 \times 10^3)$

74. $(2 \times 10^8)(3.9 \times 10^{-5})$

75. $(5.1 \times 10^1)(3 \times 10^{-4})$

76. $(1.6 \times 10^{-2})(4 \times 10^{-3})$

77. $\dfrac{7.5 \times 10^6}{3 \times 10^4}$ **78.** $\dfrac{5.2 \times 10^7}{4 \times 10^9}$

79. $\dfrac{8.4 \times 10^{-6}}{4 \times 10^{-3}}$ **80.** $\dfrac{25 \times 10^3}{5 \times 10^{-2}}$

81. $\dfrac{4 \times 10^5}{2 \times 10^4}$ **82.** $\dfrac{16 \times 10^3}{8 \times 10^{-3}}$

In Exercises 83–92, (a) perform the indicated operation without the use of a calculator and express each answer in scientific notation. (b) Confirm your answer from part (a) by using a scientific calculator to perform the operations. If the calculator displays the answer in decimal notation, convert the answer to scientific notation.

83. $(5,000,000)(20,000)$ **84.** $(0.0015)(4,000,000)$

85. $(0.003)(0.00015)$ **86.** $(230,000)(3000)$

87. $\dfrac{5,600,000}{80,000}$ **88.** $\dfrac{28,000}{0.004}$

89. $\dfrac{0.00004}{200}$ **90.** $\dfrac{0.0012}{0.000006}$

91. $\dfrac{150,000}{0.0005}$ **92.** $\dfrac{24,000}{8,000,000}$

PROBLEM SOLVING

In Exercises 93–96, list the numbers from smallest to largest.

93. 1.03×10^4; 1.7; 3.6×10^{-3}; 9.8×10^2

94. 8.5×10^{-5}; 8.2×10^3; 1.3×10^{-1}; 6.2×10^4

),000; 4.1×10^3; 0.00079; 8.3×10^{-5}

). 267,000,000; 3.14×10^7; 1,962,000; 4.79×10^6

In Exercises 97–118, you may use a scientific calculator to perform the necessary operations.

97. **U.S. Population** On June 16, 2006, the U.S. Bureau of the Census estimated that the population of the United States was about 2.990×10^8 people and the population of the world was about 6.522×10^9 people. Determine the ratio of the U.S. population to the world's population by dividing the U.S. population by the world's population. Write your answer as a decimal number rounded to the nearest thousandth.

98. **China's Population** On June 16, 2006, the U.S. Bureau of the Census estimated that the population of China was about 1.307×10^9 people and the population of the world was about 6.522×10^9 people. Determine the ratio of China's population to the world's population by dividing China's population by the world's population. Write your answer as a decimal number rounded to the nearest thousandth.

99. **India's Population** On June 16, 2006, the U.S. Bureau of the Census estimated that the population of India was about 1.095×10^9 people and the population of the world was about 6.522×10^9 people. Determine the ratio of India's population to the world's population by dividing India's population by the world's population. Write your answer as a decimal number rounded to the nearest thousandth.

100. **U.S. Debt per Person: 2000 versus 2006** In Example 14, the U.S. government debt was discussed. It was found that the debt in 2006 was about $28,026.76 per person. In 2000, the U.S. government debt was about $5.67 trillion and the population of the United States was about 273 million.

   a) Determine the amount of debt per person in the United States in 2000.

   b) How much more per person did the U.S. government owe in 2006 than in 2000?

101. **Computer Speed** As of December 26, 2006, the fastest computer in the world was IBM's BlueGene/L System located at Lawrence Livermore National Laboratory. This computer was capable of performing about 280.6 trillion calculations per second. At this rate, how long would it take to perform a task requiring 7.6×10^{65} calculations? Write your answer in scientific notation.

▲ IBM's BlueGene/L System Computer, see Exercise 101

102. **Gross Domestic Product** The gross domestic product (GDP) of a country is the total national output of goods and services produced within that country. In 2005, the GDP of the United States was about $12.4 trillion and the U.S. population was about 299 million people. Determine the U.S. GDP per person by dividing the GDP by the population. Round your answer to the nearest whole dollar.

103. **China's GDP** In 2005, the GDP (see Exercise 102) of China was about $8.86 trillion and the population of China was about 1.307 billion people. Determine China's GDP per person by dividing the GDP by the population. Round your answer to the nearest whole dollar.

104. **Traveling to Jupiter** The distance from Earth to the planet Jupiter is approximately 4.5×10^8 mi. If a spacecraft traveled at a speed of 25,000 mph, how many hours would the spacecraft need to travel from Earth to Jupiter? Use distance = rate $\times$ time.

105. *Traveling to the Moon* The distance from Earth to the moon is approximately 239,000 mi. If a spacecraft travels at a speed of 20,000 mph, how many hours would the spacecraft need to travel from Earth to the moon? Use distance = rate × time.

106. *Bucket Full of Molecules* A drop of water contains about 40 billion molecules. If a bucket has half a million drops of water in it, how many molecules of water are in the bucket? Write your answer in scientific notation.

107. *Blood Cells in a Cubic Millimeter* If a cubic millimeter of blood contains 5,800,000 red blood cells, how many red blood cells are contained in 50 cubic millimeters of blood? Write your answer in scientific notation.

108. *Radioactive Isotopes* The half-life of a radioactive isotope is the time required for half the quantity of the isotope to decompose. The half-life of uranium 238 is 4.5×10^9 years, and the half-life of uranium 234 is 2.5×10^5 years. How many times greater is the half-life of uranium 238 than uranium 234?

109. *1950 Niagara Treaty* The 1950 Niagara Treaty between the United States and Canada requires that during the tourist season a minimum of 100,000 cubic feet of water per second (ft^3/sec) flow over Niagara Falls (another 130,000–160,000 ft^3/sec are diverted for power generation). Find the minimum amount of water that will flow over the falls in a 24-hour period during the tourist season. Write your answer in scientific notation.

110. *Disposable Diaper Quantity* Laid end to end, the 18 billion disposable diapers thrown away in the United States each year would reach the moon and back seven times. If the distance from Earth to the moon is 2.38×10^5 miles, what is the length of all these diapers placed end to end? Write your answer in decimal notation.

111. *Mutual Fund Manager* Lauri Mackey is the fund manager for the Mackey Mutual Fund. This mutual fund has total assets of $1.2 billion. Lauri wants to maintain the investments in this fund according to the following pie chart.

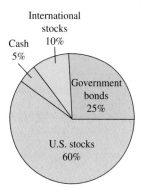

a) How much of the $1.2 billion should be invested in U.S. stocks?

b) How much should be invested in government bonds?

c) How much should be invested in international stocks?

d) How much should remain in cash?

112. *Another Mutual Fund Manager* Susan Dratch is the fund manager for the Dratch Mutual Fund. This mutual fund has total assets of $3.4 billion. Susan wants to maintain the investments in this fund according to the following pie chart.

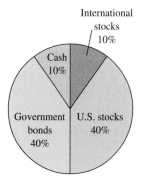

a) How much of the $3.4 billion should be invested in U.S. stocks?

b) How much should be invested in government bonds?

c) How much should be invested in international stocks?

d) How much should remain in cash?

Metric System Comparison In the metric system,
1 meter $= 10^3$ millimeters. How many times greater is
a meter than a millimeter? Explain how you determined
your answer.

114. In the metric system, 1 gram $= 10^3$ milligrams and
1 gram $= 10^{-3}$ kilogram. What is the relationship between milligrams and kilograms? Explain how you determined your answer.

115. *Earth to Sun Comparison* The mass of the sun is approximately 2×10^{30} kilograms, and the mass of Earth
is approximately 6×10^{24} kilograms. How many times
greater is the mass of the sun than the mass of Earth?
Write your answer in decimal notation.

CHALLENGE PROBLEMS/GROUP ACTIVITIES

116. *Comparing a Million to a Billion* Many people have no
idea of the difference in size between a million
(1,000,000), a billion (1,000,000,000), and a trillion
(1,000,000,000,000).

a) Write a million, a billion, and a trillion in scientific
notation.

b) Determine how long it would take to spend one
million dollars if you spent $1000 a day.

c) Repeat part (b) for one billion dollars.

d) Repeat part (b) for one trillion dollars.

e) How many times greater is one billion dollars than
one million dollars?

117. *Speed of Light*

a) Light travels at a speed of 1.86×10^5 mi/sec. A
light-year is the distance that light travels in 1 year.
Determine the number of miles in a light year.

b) Earth is approximately 93,000,000 mi from the sun.
How long does it take light from the sun to reach Earth?

118. *Bacteria in a Culture* The exponential function
$E(t) = 2^{10} \cdot 2^t$ approximates the number of bacteria in a
certain culture after t hours.

a) The initial number of bacteria is determined when
$t = 0$. What is the initial number of bacteria?

b) How many bacteria are there after $\frac{1}{2}$ hour?

INTERNET/RESEARCH ACTIVITIES

119. John Allen Paulos of Temple University has written many
entertaining books about mathematics for nonmathemati-
cians. Included among these are *Mathematics and Humor*
(1980); *I Think, Therefore I Laugh* (1985); *Innumeracy—
Mathematical Illiteracy and Its Consequences* (1989);
Beyond Numeracy—Ruminations of a Numbers Man
(1991); *A Mathematician Reads the Newspaper* (1995);
Once Upon a Number (1998); *A Mathematician Plays the
Stock Market* (2003); and *Music and Gender* (2003). Read
one of Paulos's books and write a 500-word report on it.

120. Obtain data from the U.S. Department of the Treasury
and from the U.S. Bureau of the Census Internet web
sites to calculate the current U.S. government debt per
person. Write a report in which you compare your figure
with those obtained in Exercise 100 and Example 14. Include in your report definitions of the following terms:
revenues, expenditures, deficit, and surplus.

121. Find an article in a newspaper or magazine that contains
scientific notation. Write a paragraph explaining how scientific notation was used. Attach a copy of the article to
your report.

5.7 ARITHMETIC AND GEOMETRIC SEQUENCES

▲ The modern study of diseases involves the use of mathematics to predict the size of bacteria populations.

Under optimal laboratory conditions, the number of *Escherichia coli* (*E. coli*) bacteria will double every 17 minutes. If an experiment conducted under optimal conditions begins with 100 bacteria, how many bacteria will be present after 24 hours? Questions such as these are important to the modern study of diseases. In this section, we will study such questions by using lists of numbers known as *sequences*.

Now that you can recognize the various sets of real numbers and know how to add, subtract, multiply, and divide real numbers, we can discuss sequences. A *sequence* is a list of numbers that are related to each other by a rule. The numbers that form the sequence are called its *terms*. If your salary increases or decreases by a fixed amount over a period of time, the listing of the amounts, over time, would form an arithmetic sequence. When interest in a savings account is compounded at regular intervals, the listing of the amounts in the account over time will be a geometric sequence.

Arithmetic Sequences

A sequence in which each term after the first term differs from the preceding term by a constant amount is called an *arithmetic sequence*. The amount by which each pair of successive terms differs is called the *common difference*, *d*. The common difference can be found by subtracting any term from the term that directly follows it.

Examples of arithmetic sequences	Common differences
$1, 5, 9, 13, 17, \ldots$	$d = 5 - 1 = 4$
$-7, -5, -3, -1, 1, \ldots$	$d = -5 - (-7) = -5 + 7 = 2$
$\dfrac{5}{2}, \dfrac{3}{2}, \dfrac{1}{2}, -\dfrac{1}{2}, \ldots$	$d = \dfrac{3}{2} - \dfrac{5}{2} = -\dfrac{2}{2} = -1$

┌ **EXAMPLE ❶** *The First Five Terms of an Arithmetic Sequence*

Write the first five terms of the arithmetic sequence with first term 4 and common difference of 6.

SOLUTION The first term is 4. The second term is 4 + 6, or 10. The third term is 10 + 6, or 16. The fourth term is 16 + 6, or 22. The fifth term is 22 + 6, or 28. Thus, the first five terms of the arithmetic sequence are 4, 10, 16, 22, 28.   ●

┌ **EXAMPLE ❷** *An Arithmetic Sequence with a Negative Difference*

Write the first five terms of the arithmetic sequence with first term 10 and common difference of −3.

SOLUTION The sequence is

$$10, 7, 4, 1, -2$$ ●

When discussing a sequence, we often represent the first term as a_1 (read "*a* sub 1"), the second term as a_2, the fifteenth term as a_{15}, and so on. We use the notation a_n

to represent the general or *n*th term of a sequence. Thus, a sequence may be symbolized as

$$a_1, a_2, a_3, a_4, \ldots, a_n, \ldots$$

For example, in the sequence 2, 5, 8, 11, 14, . . . , we have

$$a_1 = 2, a_2 = 5, a_3 = 8, a_4 = 11, a_5 = 14, \ldots.$$

When we know the first term of an arithmetic sequence and the common difference, we can use the following formula to find the value of any specific term.

GENERAL OR *N*TH TERM OF AN ARITHMETIC SEQUENCE

$$a_n = a_1 + (n - 1)d$$

EXAMPLE ❸ *Determining the Twelfth Term of an Arithmetic Sequence*

Determine the twelfth term of the arithmetic sequence whose first term is -5 and whose common difference is 3.

SOLUTION To determine the twelfth term, or a_{12}, replace *n* in the formula with 12, *a* with -5, and *d* with 3.

$$a_n = a_1 + (n - 1)d$$
$$a_{12} = -5 + (12 - 1)3$$
$$= -5 + (11)3$$
$$= -5 + 33$$
$$= 28$$

The twelfth term is 28. As a check, we have listed the first 12 terms of the sequence: $-5, -2, 1, 4, 7, 10, 13, 16, 19, 22, 25, 28$. ●

EXAMPLE ❹ *Determining an Expression for the nth Term*

Write an expression for the general or *n*th term, a_n, for the sequence 1, 6, 11, 16, ...

SOLUTION In this sequence, the first term, a_1, is 1 and the common difference, *d*, is 5. We substitute these values into $a_n = a_1 + (n - 1)d$ to obtain an expression for the *n*th term, a_n.

$$a_n = a_1 + (n - 1)d$$
$$= 1 + (n - 1)5$$
$$= 1 + 5n - 5$$
$$= 5n - 4$$

The expression is $a_n = 5n - 4$. Note that when $n = 1$, the first term is $5(1) - 4 = 1$. When $n = 2$, the second term is $5(2) - 4 = 6$, and so on. ●

We can use the following formula to find the sum of the first n terms in an arithmetic sequence.

SUM OF THE FIRST *N* TERMS IN AN ARITHMETIC SEQUENCE

$$s_n = \frac{n(a_1 + a_n)}{2}$$

In this formula, s_n represents the sum of the first n terms, a_1 is the first term, a_n is the nth term, and n is the number of terms in the sequence from a_1 to a_n.

┌─ **EXAMPLE ❺** *Determining the Sum of an Arithmetic Sequence*

Determine the sum of the first 25 even natural numbers.

SOLUTION The sequence we are discussing is

$$2, 4, 6, 8, 10, \ldots, 50$$

In this sequence the first term is 1, the 25th term is 50, and there are 25 terms; therefore, $a_1 = 1$, $a_{25} = 50$, and $n = 25$. Thus, the sum of the first 25 terms is

$$s_n = \frac{n(a_1 + a_n)}{2}$$

$$s_{25} = \frac{25(2 + 50)}{2}$$

$$= \frac{25(52)}{2}$$

$$= \frac{1300}{2}$$

$$= 650$$

The sum of the first 25 even natural numbers, $2 + 4 + 6 + 8 + \cdots + 50$, is 650. ●

Geometric Sequences

The next type of sequence we will discuss is the geometric sequence. A *geometric sequence* is one in which the ratio of any term to the term that directly precedes it is a constant. This constant is called the *common ratio*. The common ratio, r, can be found by taking any term except the first and dividing that term by the preceding term.

Examples of geometric sequences	Common ratios
$2, 4, 8, 16, 32, \ldots$	$r = 4 \div 2 = 2$
$-3, 6, -12, 24, -48, \ldots$	$r = 6 \div (-3) = -2$
$\dfrac{2}{3}, \dfrac{2}{9}, \dfrac{2}{27}, \dfrac{2}{81}, \ldots$	$r = \dfrac{2}{9} \div \dfrac{2}{3} = \left(\dfrac{2}{9}\right)\left(\dfrac{3}{2}\right) = \dfrac{1}{3}$

To construct a geometric sequence when the first term, a_1, and common ratio are known, multiply the first term by the common ratio to get the second term. Then multiply the second term by the common ratio to get the third term, and so on.

EXAMPLE 6 *The First Five Terms of a Geometric Sequence*

Write the first five terms of the geometric sequence whose first term, a_1, is 3 and whose common ratio, r, is 4.

SOLUTION The first term is 3. The second term, found by multiplying the first term by 4, is $3 \cdot 4$, or 12. The third term is $12 \cdot 4$, or 48. The fourth term is $48 \cdot 4$, or 192. The fifth term is $192 \cdot 4$, or 768. Thus, the first five terms of the sequence are 3, 12, 48, 192, 768. ●

When we know the first term of a geometric sequence and the common ratio, we can use the following formula to find the value of the general or nth term, a_n.

GENERAL OR NTH TERM OF A GEOMETRIC SEQUENCE

$$a_n = a_1 r^{n-1}$$

EXAMPLE 7 *Determining the Twelfth Term of a Geometric Sequence*

Determine the twelfth term of the geometric sequence whose first term is -4 and whose common ratio is 2.

SOLUTION In this sequence, $a_1 = -4$, $r = 2$, and $n = 12$. Substituting the values into the formula, we obtain

$$
\begin{aligned}
a_n &= a_1 r^{n-1} \\
a_{12} &= -4 \cdot 2^{12-1} \\
&= -4 \cdot 2^{11} \\
&= -4 \cdot 2048 \\
&= -8192
\end{aligned}
$$

As a check, we have listed the first twelve terms of the sequence:

$$-4, -8, -16, -32, -64, -128, \quad 256, -512, -1024, -2048, -4096, -8192 \ ●$$

EXAMPLE 8 *Determining an Expression for the nth Term*

Write an expression for the general or nth term, a_n, of the sequence 2, 6, 18, 54, ...

SOLUTION In this sequence, the first term is 2, so $a_1 = 2$. Note that $\frac{6}{2} = 3$, $\frac{18}{6} = 3$, $\frac{54}{18} = 3$, and so on; therefore, $r = 3$. We substitute $a_1 = 2$ and $r = 3$ into $a_n = a_1 r^{n-1}$ to obtain an expression for the nth term, a_n.

$$
\begin{aligned}
a_n &= a_1 r^{n-1} \\
&= 2(3)^{n-1}
\end{aligned}
$$

The expression is $a_n = 2(3)^{n-1}$. Note than when $n = 1$, $a_1 = 2(3)^0 = 2(1) = 2$. When $n = 2$, $a_2 = 2(3)^1 = 6$, and so on. ●

We can use the following formula to find the sum of the first n terms of a geometric sequence.

SUM OF THE FIRST *N* TERMS OF A GEOMETRIC SEQUENCE

$$s_n = \frac{a_1(1 - r^n)}{1 - r}, \qquad r \neq 1$$

EXAMPLE ⑨ *Adding the First n Terms of a Geometric Sequence*

Determine the sum of the first five terms in the geometric sequence whose first term is 4 and whose common ratio is 2.

SOLUTION In this sequence, $a_1 = 4$, $r = 2$, and $n = 5$. Substituting these values into the formula, we get

$$s_n = \frac{a_1(1 - r^n)}{1 - r}$$

$$s_5 = \frac{4[1 - (2)^5]}{1 - 2}$$

$$= \frac{4(1 - 32)}{-1}$$

$$= \frac{4(-31)}{-1} = \frac{-124}{-1} = 124$$

The sum of the first five terms of the sequence is 124. The first five terms of the sequence are 4, 8, 16, 32, 64. If you add these five numbers, you will obtain the sum 124. ●

EXAMPLE ⑩ *Pounds and Pounds of Silver*

As a reward for saving his kingdom from a band of thieves, a king offered a knight one of two options. The knight's first option was to be paid 100,000 pounds of silver all at once. The second option was to be paid over the course of a month. On the first day, he would receive one pound of silver. On the second day, he would receive two pounds of silver. On the third day, he would receive four pounds of silver, and so on, each day receiving double the amount given on the previous day. Assuming the month has 30 days, which option would provide the knight with more silver?

SOLUTION The first option pays the knight 100,000 pounds of silver. The second option pays according to the geometric sequence 1, 2, 4, 8, 16, ... In this sequence,

$a_1 = 1$, $r = 2$, and $n = 30$. The sum of this sequence can be found by substituting these values into the formula to obtain

$$s_n = \frac{a_1(1 - r^n)}{1 - r}$$

$$s_{30} = \frac{1(1 - 2^{30})}{1 - 2}$$

$$= \frac{1 - 1,073,741,824}{-1}$$

$$= \frac{-1,073,741,823}{-1}$$

$$= 1,073,741,823$$

Thus, the knight would get 1,073,741,823 pounds of silver with the second option. The second option would result in an additional 1,073,641,823 pounds of silver than with the first option.   ●

SECTION 5.7   EXERCISES

CONCEPT/WRITING EXERCISES

1. State the definition of *sequence* and give an example.

2. What are the numbers that make up a sequence called?

3. a) State the definition of *arithmetic sequence* and give an example.

   b) State the definition of *geometric sequence* and give an example.

4. a) In the arithmetic sequence 1, 6, 11, 16, 21, . . . , state the common difference, d.

   b) In the geometric sequence $-1, 3, -9, 27, -81, \ldots$, state the common ratio, r.

5. For an arithmetic sequence, state the meaning of each of the following symbols.

   a) a_n b) a_1 c) d d) s_n

6. For a geometric sequence, state the meaning of each of the following symbols.

   a) a_n b) a_1 c) r d) s_n

PRACTICE THE SKILLS

In Exercises 7–14, write the first five terms of the arithmetic sequence with the first term, a_1, and common difference, d.

7. $a_1 = 5$, $d = 1$

8. $a_1 = 4$, $d = 2$

9. $a_1 = 12$, $d = -2$

10. $a_1 = -11$, $d = 5$

11. $a_1 = 5$, $d = -2$

12. $a_1 = -3$, $d = -4$

13. $a_1 = \frac{3}{4}$, $d = \frac{1}{4}$

14. $a_1 = \frac{5}{2}$, $d = -\frac{3}{2}$

In Exercises 15 22, determine the indicated term for the arithmetic sequence with the first term, a_1, and common difference, d.

15. Determine a_5 when $a_1 = 4$, $d = 1$.

16. Determine a_8 when $a_1 = -5$, $d = 4$.

17. Determine a_{10} when $a_1 = -5$, $d = 2$.

18. Determine a_{12} when $a_1 = 7$, $d = -3$.

19. Determine a_{20} when $a_1 = \frac{4}{5}$, $d = -1$.

20. Determine a_{14} when $a_1 = \frac{1}{3}$, $d = \frac{2}{3}$.

21. Determine a_{22} when $a_1 = -23$, $d = \frac{5}{7}$.

22. Determine a_{15} when $a_1 = \frac{4}{3}$, $d = \frac{1}{3}$.

In Exercises 23–30, write an expression for the general or nth term, a_n, for the arithmetic sequence.

23. $1, 2, 3, 4, \ldots$

24. $1, 3, 5, 7, \ldots$

25. $2, 4, 6, 8, \ldots$

26. $-3, 0, 3, 6, \ldots$

27. $\frac{1}{4}, 1, \frac{7}{4}, \frac{5}{2}, \ldots$

28. $-\frac{1}{4}, 0, \frac{1}{4}, \frac{1}{2}, \ldots$

29. $-3, -\frac{3}{2}, 0, \frac{3}{2}, \ldots$

30. $-5, -2, 1, 4, \ldots$

In Exercises 31–38, determine the sum of the terms of the arithmetic sequence. The number of terms, n, is given.

31. $1, 2, 3, 4, \ldots, 50$; $n = 50$

32. $2, 4, 6, 8, \ldots, 100$; $n = 50$

33. $1, 3, 5, 7, \ldots, 99$; $n = 50$

34. $-4, -7, -10, -13, \ldots, -28$; $n = 9$

35. $11, 6, 1, -4, \ldots, -24$; $n = 8$

36. $-\frac{1}{16}, 0, \frac{1}{16}, \frac{1}{8}, \ldots, \frac{17}{16}$; $n = 19$

37. $\frac{1}{5}, \frac{2}{5}, \frac{3}{5}, \frac{4}{5}, \ldots, \frac{24}{5}$; $n = 24$

38. $-\frac{3}{4}, -\frac{1}{2}, -\frac{1}{4}, 0, \ldots, \frac{17}{4}$; $n = 21$

In Exercises 39–46, write the first five terms of the geometric sequence with the first term, a_1, and common ratio, r.

39. $a_1 = 1$, $r = 5$

40. $a_1 = 1$, $r = 2$

41. $a_1 = 2$, $r = -2$

42. $a_1 = 8$, $r = \frac{1}{2}$

43. $a_1 = -3$, $r = -1$

44. $a_1 = -6$, $r = -2$

45. $a_1 = 81$, $r = -\frac{1}{3}$

46. $a_1 = \frac{16}{15}$, $r = \frac{1}{2}$

In Exercises 47–54, determine the indicated term for the geometric sequence with the first term, a_1, and common ratio, r.

47. Determine a_6 when $a_1 = 5$, $r = 2$.

48. Determine a_6 when $a_1 = 1$, $r = 3$.

49. Determine a_3 when $a_1 = 3$, $r = \frac{1}{2}$.

50. Determine a_7 when $a_1 = -3$, $r = -3$.

51. Determine a_7 when $a_1 = -5$, $r = 3$.

52. Determine a_8 when $a_1 = \frac{1}{2}$, $r = -2$.

53. Determine a_{10} when $a_1 = -2$, $r = 3$.

54. Determine a_{18} when $a_1 = -5$, $r = -2$.

In Exercises 55–62, write an expression for the general or nth term, a_n, for the geometric sequence.

55. $1, 2, 4, 8, \ldots$

56. $1, 3, 9, 27, \ldots$

57. $-1, 1, -1, 1, \ldots$

58. $-16, -8, -4, -2, \ldots$

59. $2, 1, \frac{1}{2}, \frac{1}{4}, \ldots$

60. $-3, 6, -12, 24, \ldots$

61. $9, 3, 1, \frac{1}{3}, \frac{1}{9}, \ldots$

62. $-4, -\frac{8}{3}, -\frac{16}{9}, -\frac{32}{27}, \ldots$

In Exercises 63–70, determine the sum of the first n terms of the geometric sequence for the values of a_1 and r.

63. $n = 5$, $a_1 = 6$, $r = 2$

64. $n = 4$, $a_1 = 4$, $r = 3$

65. $n = 6$, $a_1 = -3$, $r = 4$

66. $n = 9$, $a_1 = -3$, $r = 5$

67. $n = 11$, $a_1 = -7$, $r = 3$

68. $n = 15$, $a_1 = -1$, $r = 2$

69. $n = 15$, $a_1 = -1$, $r = -2$

70. $n = 10$, $a_1 = 512$, $r = \frac{1}{2}$

PROBLEM SOLVING

71. Determine the sum of the first 100 natural numbers.

72. Determine the sum of the first 100 odd natural numbers.

73. Determine the sum of the first 100 even natural numbers.

74. Determine the sum of the first 50 multiples of 3.

75. *Pendulum Movement* Each swing of a pendulum (from far left to far right) is 3 in. shorter than the preceding swing. The first swing is 8 ft.

   a) Find the length of the twelfth swing.

   b) Determine the total distance traveled by the pendulum during the first 12 swings.

76. *Clock Strikes* A clock strikes once at 1 o'clock, twice at 2 o'clock, three times at 3 o'clock, and so on. How many times does it strike over a 12 hr period?

77. *Annual Pay Raises* Rita Fernandez is given a starting salary of $35,000 and promised a $1400 raise per year after each of the next 8 years.

   a) Determine her salary during her eighth year of work.

   b) Determine the total salary she received over the 8 years.

78. *A Bouncing Ball* Each time a ball bounces, the height attained by the ball is 6 in. less than the previous height attained. If on the first bounce the ball reaches a height of 6 ft, find the height attained on the eleventh bounce.

79. *Squirrels and Pinecones* A tree squirrel cut down 1 pinecone on the first day of October, 2 pinecones on the second day, 3 pinecones on the third day, and so on. How many pinecones did this squirrel cut down during the month of October, which contains 31 days?

80. *Enrollment Increase* The current enrollment at Loras College is 8000 students. If the enrollment increases by 8% per year, determine the enrollment 10 years later.

81. *Decomposing Substance* A certain substance decomposes and loses 20% of its weight each hour. If there are originally 200 g of the substance, how much remains after 6 hr?

82. *Samurai Sword Construction* While making a traditional Japanese samurai sword, the master sword maker prepares the blade by heating a bar of iron until it is white hot. He then folds it over and pounds it smooth. Therefore, after each folding, the number of layers of steel is doubled. Assuming the sword maker starts with a bar of one layer and folds it 15 times, how many layers of steel will the finished sword contain?

83. *Salary Increase* If your salary were to increase at a rate of 6% per year, find your salary during your 15th year if your original salary is $31,000.

84. *A Bouncing Ball* When dropped, a ball rebounds to four-fifths of its original height. How high will the ball rebound after the fourth bounce if it is dropped from a height of 30 ft?

85. *Value of a Stock* Ten years ago, Nancy Hart purchased $2000 worth of shares in RCF, Inc. Since then, the price of the stock has roughly tripled every 2 years. Approximately how much are Nancy's shares worth today?

86. *A Baseball Game* During a baseball game, the visiting team scored 1 run in the first inning, 2 runs in the second inning, 3 runs in the third inning, 4 runs in the fourth inning, and so on. The home team scored 1 run in the first inning, 2 runs in the second inning, 4 runs in the third inning, 8 runs in the fourth inning, and so on. What is the score of the game after eight innings?

Inning	1	2	3	4	5	6	7	8	9
Visitors	1	2	3	4					
Home	1	2	4	8					

CHALLENGE PROBLEMS/GROUP ACTIVITIES

87. A geometric sequence has $a_1 = 82$ and $r = \frac{1}{2}$; find s_6.

88. *Sums of Interior Angles* The sums of the interior angles of a triangle, a quadrilateral, a pentagon, and a hexagon are 180°, 360°, 540°, and 720°, respectively. Use this pattern to find a formula for the general term, a_n, where a_n represents the sum of the interior angles of an n-sided polygon.

89. *Divisibility by 6* Determine how many numbers between 7 and 1610 are divisible by 6.

90. Determine r and a_1 for the geometric sequence with $a_2 = 24$ and $a_5 = 648$.

91. *Total Distance Traveled by a Bouncing Ball* A ball is dropped from a height of 30 ft. On each bounce, it attains a height four-fifths of its original height (or of the previous bounce). Find the total vertical distance traveled by the ball after it has completed its fifth bounce (has therefore hit the ground six times).

RECREATIONAL MATHEMATICS

92. *A Wagering Strategy* The following is a strategy used by some people involved in games of chance. A player begins by betting a standard bet, say \$1. If the player wins, the player again bets \$1 in the next round. If the player loses, the player bets \$2 in the next round. Next, if the player wins, the player again bets \$1; if the player loses, the player now bets \$4 in the next round. The process continues as long as the player keeps playing, betting \$1 after a win or doubling the previous bet after a loss.

a) Assume a player is using a \$1 standard bet and loses five times in a row. How much money should the player bet in the sixth round? How much money has the player lost at the end of the fifth round?

b) Assume a player is using a \$10 standard bet and loses five times in a row. How much money should the player bet in the sixth round? How much money has the player lost at the end of the fifth round?

c) Assume a player is using a \$1 standard bet and loses 10 times in a row. How much money should the player bet in the 11th round? How much money has the player lost at the end of the 10th round?

d) Assume a player is using a \$10 standard bet and loses 10 times in a row. How much money should the player bet in the 11th round? How much money has the player lost at the end of the 10th round?

e) Why is this strategy dangerous?

INTERNET/RESEARCH ACTIVITY

93. A topic generally associated with sequences is *series*.

a) Research *series* and explain what a series is and how it differs from a sequence. Also write a formal definition of series. Give examples of different kinds of series.

b) Write the arithmetic series associated with the arithmetic sequence 1, 4, 7, 10, 13, ...

c) Write the geometric series associated with the geometric sequence 3, 6, 12, 24, 48, ...

d) What is an infinite geometric series?

e) Determine the sum of the terms of the infinite geometric series $1 + \dfrac{1}{2} + \dfrac{1}{4} + \dfrac{1}{8} + \dfrac{1}{16} + \cdots$.

5.8   FIBONACCI SEQUENCE

▲ The number of ancestors of a male bee is among the numbers found in the Fibonacci sequence.

Consider the following numbers found in nature: the number of petals on a daisy, the number of sunflower seeds in a sunflower, the number of scales on a pineapple, the number of ancestors of a male bee. In this section, we will find that all these numbers are part of a special sequence of numbers called the Fibonacci sequence.

Our discussion of sequences would not be complete without mentioning a sequence known as the *Fibonacci sequence*. The sequence is named after Leonardo of Pisa, also known as Fibonacci. He was one of the most distinguished mathematicians of the Middle Ages. This sequence is first mentioned in his book *Liber Abacci* (Book of the Abacus), which contained many interesting problems, such as: "A certain man put a pair of rabbits in a place surrounded on all sides by a wall. How many pairs of rabbits can be produced from that pair in a year if it is assumed that every month each pair begets a new pair which from the second month becomes productive?"

The solution to this problem (Fig. 5.10) led to the development of the sequence that bears its author's name: the Fibonacci sequence. The sequence is shown in Table 5.1. The numbers in the columns titled *Pairs of Adults* form the Fibonacci sequence.

PROFILE IN MATHEMATICS

Fibonacci

L eonardo of Pisa (1170–1250) is considered one of the most distinguished mathematicians of the Middle Ages. He was born in Italy and was sent by his father to study mathematics with an Arab master. When he began writing, he referred to himself as Fibonacci, or "son of Bonacci," the name by which he is known today. In addition to the famous sequence bearing his name, Fibonacci is also credited with introducing the Hindu–Arabic number system into Europe. His 1202 book, *Liber Abacci* (Book of the Abacus), explained the use of this number system and emphasized the importance of the number zero.

FIBONACCI SEQUENCE
1, 1, 2, 3, 5, 8, 13, 21, . . .

Table 5.1

Month	Pairs of Adults	Pairs of Babies	Total Pairs
1	1	0	1
2	1	1	2
3	2	1	3
4	3	2	5
5	5	3	8
6	8	5	13
7	13	8	21
8	21	13	34
9	34	21	55
10	55	34	89
11	89	55	144
12	144	89	233

Figure 5.10

▲ The head of a sunflower

Table 5.2

Numbers	Ratio
1, 1	$\frac{1}{1} = 1$
1, 2	$\frac{2}{1} = 2$
2, 3	$\frac{3}{2} = 1.5$
3, 5	$\frac{5}{3} = 1.666\ldots$
5, 8	$\frac{8}{5} = 1.6$
8, 13	$\frac{13}{8} = 1.625$
13, 21	$\frac{21}{13} \approx 1.615$
21, 34	$\frac{34}{21} \approx 1.619$
34, 55	$\frac{55}{34} \approx 1.618$
55, 89	$\frac{89}{55} \approx 1.618$

In the Fibonacci sequence, the first and second terms are 1. The sum of these two terms is the third term. The sum of the second and third terms is the fourth term, and so on.

In the middle of the nineteenth century, mathematicians made a serious study of this sequence and found strong similarities between it and many natural phenomena. Fibonacci numbers appear in the seed arrangement of many species of plants and in the petal counts of various flowers. For example, when the flowering head of the sunflower matures to seed, the seeds' spiral arrangement becomes clearly visible. A typical count of these spirals may give 89 steeply curving to the right, 55 curving more shallowly to the left, and 34 again shallowly to the right. The largest known specimen to be examined had spiral counts of 144 right, 89 left, and 55 right. These numbers, like the other three mentioned, are consecutive terms of the Fibonacci sequence.

On the heads of many flowers, petals surrounding the central disk generally yield a Fibonacci number. For example, some daisies contain 21 petals, and others contain 34, 55, or 89 petals. (People who use a daisy to play the "love me, love me not" game will likely pluck 21, 34, 55, or 89 petals before arriving at an answer.)

Fibonacci numbers are also observed in the structure of pinecones and pineapples. The tablike or scalelike structures called bracts that make up the main body of the pinecone form a set of spirals that start from the cone's attachment to the branch. Two sets of oppositely directed spirals can be observed, one steep and the other more gradual. A count on the steep spiral will reveal a Fibonacci number, and a count on the gradual one will be the adjacent smaller Fibonacci number, or if not, the next smaller Fibonacci number. One investigation of 4290 pinecones from 10 species of pine trees found in California revealed that only 74 cones, or 1.7%, deviated from this Fibonacci pattern.

Like pinecone bracts, pineapple scales are patterned into spirals, and because they are roughly hexagonal in shape, three distinct sets of spirals can be counted. Generally, the number of pineapple scales in each spiral are Fibonacci numbers.

Fibonacci Numbers and Divine Proportions

In 1753, while studying the Fibonacci sequence, Robert Simson, a mathematician at the University of Glasgow, noticed that when he took the ratio of any term to the term that immediately preceded it, the value he obtained remained in the vicinity of one specific number. To illustrate this, we indicate in Table 5.2 the ratio of various pairs of sequential Fibonacci numbers.

The ratio of the 50th term to the 49th term is 1.6180. Simson proved that the ratio of the $(n + 1)$ term to the nth term as n gets larger and larger is the irrational number $(\sqrt{5} + 1)/2$, which begins $1.61803\ldots$ This number was already well known to mathematicians at that time as the *golden number*.

Many years earlier, Bavarian astronomer and mathematician Johannes Kepler wrote that for him the golden number symbolized the Creator's intention "to create like from like." The golden number $(\sqrt{5} + 1)/2$ is frequently referred to as "phi," symbolized by the Greek letter Φ.

Figure 5.11

The ancient Greeks, in about the sixth century B.C., sought unifying principles of beauty and perfection, which they believed could be described by using mathematics. In their study of beauty, the Greeks used the term *golden ratio*. To understand the golden ratio, let's consider the line segment AB in Fig. 5.11. When this line segment is

▲ The Great Pyramid of Gizeh

Figure 5.12

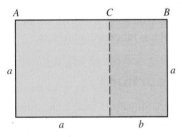

Figure 5.13

divided at a point C such that the ratio of the whole, AB, to the larger part, AC, is equal to the ratio of the larger part, AC, to the smaller part, CB, each ratio AB/AC and AC/CB is referred to as a *golden ratio*. The proportion these ratios form, $AB/AC = AC/CB$, is called the *golden proportion*. Furthermore, each ratio in the proportion will have a value equal to the golden number, $(\sqrt{5} + 1)/2$.

$$\frac{AB}{AC} = \frac{AC}{CB} = \frac{\sqrt{5} + 1}{2} \approx 1.618$$

The Great Pyramid of Gizeh in Egypt, built about 2600 B.C., is the earliest known example of use of the golden ratio in architecture. The ratio of any of its sides of the square base (775.75 ft) to its altitude (481.4 ft) is about 1.611. Other evidence of the use of the golden ratio appears in other Egyptian buildings and tombs.

In medieval times, people referred to the golden proportion as the *divine proportion*, reflecting their belief in its relationship to the will of God.

Twentieth-century architect Le Corbusier developed a scale of proportions for the human body that he called the Modulor (Fig. 5.12). Note that the navel separates the entire body into golden proportions, as do the neck and knee.

From the golden proportion, the *golden rectangle* can be formed, as shown in Fig. 5.13.

$$\frac{\text{Length}}{\text{Width}} = \frac{a + b}{a} = \frac{a}{b} = \frac{\sqrt{5} + 1}{2}$$

Note that when a square is cut off one end of a golden rectangle, as in Fig. 5.13, the remaining rectangle has the same properties as the original golden rectangle (creating "like from like" as Johannes Kepler had written) and is therefore itself a golden rectangle. Interestingly, the curve derived from a succession of diminishing golden rectangles, as shown in Fig. 5.14, is the same as the spiral curve of the chambered nautilus. The same curve appears on the horns of rams and some other animals. It is the same curve that is observed in the plant structures mentioned earlier: sunflowers, other flower heads, pinecones, and pineapples. You will recall that Fibonacci numbers were observed in each of these plant structures. The curve shown in Fig. 5.14 closely approximates what mathematicians call a *logarithmic spiral*.

Figure 5.14

▲ The Parthenon

Ancient Greek civilization used the golden rectangle in art and architecture. The main measurements of many buildings of antiquity, including the Parthenon in Athens, are governed by golden ratios and rectangles. Greek statues, vases, urns, and so on also exhibit characteristics of the golden ratio. It is for Phidas, considered the greatest of Greek sculptors, that the golden ratio was named "phi." The proportions can be found abundantly in his work.

The proportions of the golden rectangle can be found in the work of many artists, from the old masters to the moderns. For example, the golden rectangle can be seen in the painting *Invitation to the Sideshow (La Parade de Cirque),* 1887, by George Seurat, a French neoimpressionist artist.

▲ *Invitation to the Sideshow (La Parade de Cirque),* 1887, by George Seurat

▲ *Fibonacci's Garden* by Caryl Bryer Fallert

In addition to using the golden rectangle in art, several artists have used Fibonacci numbers in art. One contemporary example is the 1995 work by Caryl Bryer Fallert called *Fibonacci's Garden*. This artwork is a quilt constructed from two separate fabrics that are put together in a pattern based on the Fibonacci sequence.

Fibonacci numbers are also found in another form of art, namely music. Perhaps the most obvious link between Fibonacci numbers and music can be found on the piano keyboard. An octave (Fig. 5.15) on a keyboard has 13 keys: 8 white keys and 5 black keys (the 5 black keys are in one group of 2 and one group of 3).

Figure 5.15

In Western music, the most complete scale, the chromatic scale, consists of 13 notes (from C to the next higher C). Its predecessor, the diatonic scale, contains 8

Fibonacci and the Male Bee's Ancestors

The most frequent example given to introduce the Fibonacci sequence involves rabbits producing offspring, two at a time. Although this example makes for a nice introduction to Fibonacci sequences, it is not at all realistic. A much better example comes from the breeding practices of bees. Female or worker bees are produced when the queen bee mates with a male bee. Male bees are produced from the queen's unfertilized eggs. In essence, then, female bees have two parents, whereas male bees only have one parent. The family tree of a male bee would look like this:

From this tree, we can see that the **1** male bee (circled) has **1** parent, **2** grandparents, **3** great-grandparents, **5** great-great-grandparents, **8** great-great-great-grandparents, and so on. We see the Fibonacci sequence as we move back through the male bees' generations.

notes (an octave). The diatonic scale was preceded by a 5-note pentatonic scale (*penta* is Greek for "five"). Each number is a Fibonacci number.

The visual arts deal with what is pleasing to the eye, whereas musical composition deals with what is pleasing to the ear. Whereas art achieves some of its goals by using division of planes and area, music achieves some of its goals by a similar division of time, using notes of various duration and spacing. The musical intervals considered by many to be the most pleasing to the ear are the major sixth and minor sixth. A major sixth, for example, consists of the note C, vibrating at about 264[†] vibrations per second, and note A, vibrating at about 440 vibrations per second. The ratio of 440 to 264 reduces to 5 to 3, or $\frac{5}{3}$, a ratio of two consecutive Fibonacci numbers. An example of a minor sixth is E (about 330 vibrations per second) and C (about 528 vibrations per second). The ratio 528 to 330 reduces to 8 to 5, or $\frac{8}{5}$, the next ratio of two consecutive Fibonacci numbers. The vibrations of any sixth interval reduce to a similar ratio.

Patterns that can be expressed mathematically in terms of Fibonacci relationships have been found in Gregorian chants and works of many composers, including Bach, Beethoven, and Bartók. A number of twentieth-century musical works, including Ernst Krenek's *Fibonacci Mobile*, were deliberately structured by using Fibonacci proportions.

A number of studies have tried to explain why the Fibonacci sequence and related items are linked to so many real-life situations. It appears that the Fibonacci numbers are a part of natural harmony that is pleasing to both the eye and the ear. In the nineteenth century, German physicist and psychologist Gustav Fechner tried to determine which dimensions were most pleasing to the eye. Fechner, along with psychologist Wilhelm Wundt, found that most people do unconsciously favor golden dimensions when purchasing greeting cards, mirrors, and other rectangular objects. This discovery has been widely used by commercial manufacturers in their packaging and labeling designs, by retailers in their store displays, and in other areas of business and advertising.

[†]Frequencies of notes vary in different parts of the world and change over time.

SECTION 5.8   EXERCISES

CONCEPT/WRITING EXERCISES

1. Explain how to construct the Fibonacci sequence.

2. **a)** Write out the first ten terms of the Fibonacci sequence.

   b) Divide the ninth term by the eighth term, rounding to the nearest thousandth.

   c) Divide the tenth term by the ninth term, rounding to the nearest thousandth.

   d) Try a few more divisions and then make a conjecture about the result.

3. a) What is the value of the golden number?

 b) What is the golden ratio?

 c) What is the golden proportion?

 d) What is the golden rectangle?

4. Describe three examples of where the golden ratio can be found

 a) in nature.

 b) in manufactured items.

5. In your own words, explain the relationship between the golden number, golden ratio, golden proportion, and golden rectangle.

6. Describe three examples of where Fibonacci numbers can be found

 a) in nature.

 b) in manufactured items.

PRACTICE THE SKILLS/PROBLEM SOLVING

7. a) To what decimal value is $(\sqrt{5} + 1)/2$ aproximately equal?

 b) To what decimal value is $(\sqrt{5} - 1)/2$ approximately equal?

 c) By how much do the results in parts (a) and (b) differ?

8. The eleventh Fibonacci number is 89. Examine the first six digits in the decimal expression of its reciprocal, $\frac{1}{89}$. What do you find?

9. Find the ratio of the second to the first term of the Fibonacci sequence. Then find the ratio of the third to the second term of the sequence and determine whether this ratio was an increase or decrease from the first ratio. Continue this process for 10 ratios and then make a conjecture regarding the increasing or decreasing values in consecutive ratios.

10. A musical composition is described as follows. Explain why this piece is based on the golden ratio.

Entire Composition

34 measures	55 measures	21 measures	34 measures
Theme	Fast, Loud	Slow	Repeat of theme

11. The sum of any 10 consecutive Fibonacci numbers is always divisible by 11. Select any 10 consecutive Fibonacci numbers and show that for your selection this statement is true.

12. The greatest common factor of any two consecutive Fibonacci numbers is 1. Show that this statement is true for the first 15 Fibonacci numbers.

13. For any four consecutive Fibonacci numbers, the difference of the squares of the middle two numbers equals the product of the smallest and largest numbers. Select four consecutive Fibonacci numbers and show that this pattern holds for the numbers you selected.

14. Twice any Fibonacci number minus the next Fibonacci number equals the second number preceding the original number. Select a number in the Fibonacci sequence and show that this pattern holds for the number selected.

15. Determine the ratio of the length to width of various photographs and compare these ratios to Φ.

16. Determine the ratio of the length to the width of a 6 inch by 4 inch standard index card, and compare the ratio to Φ.

17. Determine the ratio of the length to width of several picture frames and compare these ratios to Φ.

18. Determine the ratio of the length to the width of your television screen and compare this ratio to Φ.

19. Determine the ratio of the length to the width of a computer screen and compare this ratio to Φ.

20. Determine the ratio of the length to width of a desktop in your classroom and compare this ratio to Φ.

21. Determine the ratio of the length to the width of this textbook and compare this ratio to Φ.

22. Find three physical objects whose dimensions are very close to a golden rectangle.

 a) List the articles and record the dimensions.

 b) Compute the ratios of their lengths to their widths.

 c) Find the difference between the golden ratio and the ratio you obtain in part (b)—to the nearest tenth—for each object.

In Exercises 23–30, determine whether the sequence is a Fibonacci-type sequence (each term is the sum of the two preceding terms). If it is, determine the next two terms of the sequence.

23. 1, 2, 3, 5, 6, 7, . . .   24. 1, 3, 4, 7, 11, 18, . . .

, 1, 0, 1, 1, 2, . . .   **26.** 1, 4, 9, 16, 25, 36, . . .

5, 10, 15, 25, 40, 65, . . .   **28.** $\frac{1}{4}, \frac{1}{4}, \frac{1}{2}, \frac{3}{4}, 1\frac{1}{4}, 2, \ldots$

29. $-5, 3, -2, 1, -1, 0, \ldots$ **30.** $-4, 5, 1, 6, 7, 13, \ldots$

31. a) Select any two nonzero digits and add them to obtain a third digit. Continue adding the two previous terms to get a Fibonacci-type sequence.

b) Form ratios of successive terms to show how they will eventually approach the golden number.

32. Repeat Exercise 31 for two different nonzero numbers.

33. a) Select any three consecutive terms of a Fibonacci sequence. Subtract the product of the terms on each side of the middle term from the square of the middle term. What is the difference?

b) Repeat part (a) with three different consecutive terms of the sequence.

c) Make a conjecture about what will happen when you repeat this process for any three consecutive terms of a Fibonacci sequence.

34. *Pascal's Triangle* One of the most famous number patterns involves *Pascal's triangle*. The Fibonacci sequence can be found by using Pascal's triangle. Can you explain how that can be done? A hint is shown.

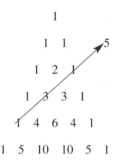

35. *Lucas Sequence* a) A sequence related to the Fibonacci sequence is the *Lucas sequence*. The Lucas sequence is formed in a manner similar to the Fibonacci sequence. The first two numbers of the Lucas sequence are 1 and 3. Write the first eight terms of the Lucas sequence.

b) Complete the next two lines of the following chart.

$$1 + 2 = 3$$
$$1 + 3 = 4$$
$$2 + 5 = 7$$
$$3 + 8 = 11$$
$$5 + 13 = 18$$

c) What do you observe about the first column in the chart in part (b)?

CHALLENGE PROBLEMS/GROUP ACTIVITIES

36. *Fibonacci-Type Sequence* The following sequence represents a Fibonacci-type sequence (each term is the sum of the two preceding terms). Here x represents any natural number from 1 to 10:

$$-10, x, -10 + x, -10 + 2x, -20 + 3x, -30 + 5x, \ldots$$

For example, if $x = 2$, the first 10 terms of the sequence would be $-10, 2, -8, -6, -14, -20, -34, -54, -88, -142$.

Write out the first 10 terms of this Fibonacci-type sequence for x equal to the following.

a) 4   **b)** 5   **c)** 6   **d)** 7   **e)** 8

f) For values of x of 4, 5, and 6, you should have found that each term after the seventh term in the sequence was a negative number. For values of x of 7 and 8, you should have found that each term after the seventh term in the sequence was a positive number. Do you believe that for any value of x greater than or equal to 7, each term after the seventh term of the sequence will always be a positive number? Explain.

37. Draw a line of length 5 in. Determine and mark the point on the line that will create the golden ratio. Explain how you determined your answer.

38. The divine proportion is $(a + b)/a = a/b$ (see Fig. 5.13), which can be written $1 + (b/a) = a/b$. Now let $x = a/b$, which gives $1 + (1/x) = x$. Multiply both sides of this equation by x to get a quadratic equation and then use the quadratic formula (Section 6.8) to show that one answer is $x = (1 + \sqrt{5})/2$ (the golden ratio).

39. *Pythagorean Triples* A Pythagorean triple is a set of three whole numbers, $\{a, b, c\}$, such that $a^2 + b^2 = c^2$.

For example, since $6^2 + 8^2 = (10)^2$, $\{6, 8, 10\}$ is a Pythagorean triple. The following steps show how to find Pythagorean triples using any four consecutive Fibonacci numbers. Here we will demonstrate the process with the Fibonacci numbers 3, 5, 8, and 13.

1. Determine the product of 2 and the two inner Fibonacci numbers. We have $2(5)(8) = 80$, which is the first number in the Pythagorean triple. So $a = 80$.
2. Determine the product of the two outer numbers. We have $3(13) = 39$, which is the second number in the Pythagorean triple. So $b = 39$.
3. Determine the sum of the squares of the inner two numbers. We have $5^2 + 8^2 = 25 + 64 = 89$, which is the third number in the Pythagorean triple. So $c = 89$.

This process has produced the Pythagorean triple, $\{80, 39, 89\}$. To verify,

$$(80)^2 + (39)^2 = (89)^2$$
$$6400 + 1521 = 7921$$
$$7921 = 7921$$

Use this process to produce four other Pythagorean triples.

40. *Reflections* When two panes of glass are placed face to face, four interior reflective surfaces exist labeled 1, 2, 3, and 4. If light is not reflected, it has just one path through the glass (see the figure below). If it has one reflection, it can be reflected in two ways. If it has two reflections, it can be reflected in three ways. Use this information to answer parts (a) through (c).

| 0 reflections | 1 reflection | 2 reflections |
| 1 path | 2 paths | 3 paths |

a) If a ray is reflected three times, there are five paths it can follow. Show the paths.

b) If a ray is reflected four times, there are eight paths it can follow. Show the paths.

c) How many paths can a ray follow if it is reflected five times? Explain how you determined your answer.

INTERNET/RESEARCH ACTIVITIES

41. Write a report on the history and mathematical contributions of Fibonacci.

42. The digits 1 through 9 have evolved considerably since they appeared in Fibonacci's book *Liber Abacci*. Write a report tracing the history of the evolution of the digits 1 through 9 since Fibonacci's time.

43. Write a report indicating where the golden ratio and golden rectangle have been used in art and architecture. You may wish to include information on art and architecture related to the golden ratio and Fibonacci sequences.

CHAPTER 5 SUMMARY

IMPORTANT FACTS

FUNDAMENTAL THEOREM OF ARITHMETIC

Every composite number can be expressed as a unique product of prime numbers.

SETS OF NUMBERS

Natural or counting numbers: $\{1, 2, 3, 4, \dots\}$

Whole numbers: $\{0, 1, 2, 3, 4, \dots\}$

Integers: $\{\dots, -3, -2, -1, 0, 1, 2, 3, \dots\}$

Rational numbers: Numbers of the form p/q, where p and q are integers, $q \neq 0$. Every rational number when expressed as a decimal number will be either a terminating or repeating decimal number.

Irrational number: A real number whose representation is a nonterminating, nonrepeating decimal number (not a rational number).

DEFINITION OF SUBTRACTION

$$a - b = a + (-b)$$

FUNDAMENTAL LAW OF RATIONAL NUMBERS

$$\frac{a}{b} = \frac{a}{b} \cdot \frac{c}{c} = \frac{ac}{bc}, \qquad b \neq 0, \qquad c \neq 0$$

RULES OF RADICALS

Product rule for radicals:

$$\sqrt{a \cdot b} = \sqrt{a} \cdot \sqrt{b}, \qquad a \geq 0, \qquad b \geq 0$$

Quotient rule for radicals:

$$\frac{\sqrt{a}}{\sqrt{b}} = \sqrt{\frac{a}{b}}, \qquad a \geq 0, \qquad b > 0$$

PROPERTIES OF REAL NUMBERS

Commutative property of addition: $a + b = b + a$

Commutative property of multiplication: $a \cdot b = b \cdot a$

Associative property of addition:

$$(a + b) + c = a + (b + c)$$

Associative property of multiplication:

$$(a \cdot b) \cdot c = a \cdot (b \cdot c)$$

Distributive property: $a \cdot (b + c) = ab + ac$

RULES OF EXPONENTS

Product rule for exponents: $a^m \cdot a^n = a^{m+n}$

Quotient rule for exponents: $\dfrac{a^m}{a^n} = a^{m-n}, \qquad a \neq 0$

Zero exponent rule: $a^0 = 1,$ $a \neq 0$

Negative exponent rule: $a^{-m} = \dfrac{1}{a^m},$ $a \neq 0$

Power rule: $(a^m)^n = a^{m \cdot n}$

ARITHMETIC SEQUENCE

General or nth term: $a_n = a_1 + (n-1)d$

Sum of the first n terms: $s_n = \dfrac{n(a_1 + a_n)}{2}$

GEOMETRIC SEQUENCE

General or nth term: $a_n = a_1 r^{n-1}$

Sum of the first n terms: $s_n = \dfrac{a_1(1 - r^n)}{1 - r},$ $r \neq 1$

FIBONACCI SEQUENCE

$$1, 1, 2, 3, 5, 8, 13, 21, \ldots$$

GOLDEN NUMBER

$$\dfrac{\sqrt{5} + 1}{2} \approx 1.618$$

GOLDEN PROPORTION

$$\dfrac{a + b}{a} = \dfrac{a}{b}$$

CHAPTER ⑤ REVIEW EXERCISES

5.1

In Exercises 1 and 2, determine whether the number is divisible by each of the following numbers: 2, 3, 4, 5, 6, 8, 9, and 10.

1. 158,340

2. 400,644

In Exercises 3–7, determine the prime factorization of the number.

3. 540   **4.** 693   **5.** 840

6. 882   **7.** 1452

In Exercises 8–13, determine the GCD and LCM of the numbers.

8. 30, 105   **9.** 63, 108   **10.** 45, 250

11. 90, 300   **12.** 60, 40, 96   **13.** 36, 108, 144

14. *Train Stops* From 1912 to 1971, the Milwaukee Road Railroad Company had a train stop every 15 days in Dubuque, Iowa. During this same period, the same train also stopped in Des Moines, Iowa, every 9 days. If on April 18, 1964, the train made a stop in Dubuque and a stop in Des Moines, how many days was it until the train again stopped in both cities on the same day?

▲ See Exercise 14

5.2

In Exercises 15–22, use a number line to evaluate the expression.

15. $4 + (-7)$ **16.** $-2 + 5$

17. $-2 + (-4)$ **18.** $4 - 8$

19. $-5 - 4$ **20.** $-3 - (-6)$

21. $(-3 + 7) - 4$ **22.** $-1 + (9 - 4)$

In Exercises 23–30, evaluate the expression.

23. $-5 \cdot 3$ **24.** $(-2)(-12)$

25. $(14)(-4)$

26. $\dfrac{-35}{-7}$

27. $\dfrac{12}{-6}$

28. $[8 \div (-4)](-3)$

29. $-20 \div (-4 - 1)$

30. $[(-30) \div (10)] \div (-1)$

5.3

In Exercises 31–39, express the fraction as a terminating or repeating decimal.

31. $\dfrac{3}{10}$ **32.** $\dfrac{11}{25}$ **33.** $\dfrac{15}{40}$

34. $\dfrac{13}{4}$ **35.** $\dfrac{6}{7}$ **36.** $\dfrac{7}{12}$

37. $\dfrac{3}{8}$ **38.** $\dfrac{11}{16}$ **39.** $\dfrac{5}{7}$

In Exercises 40–46, express the decimal number as a quotient of two integers.

40. 0.225 **41.** 1.4 **42.** $0.\overline{6}$ **43.** $0.\overline{51}$

44. 0.083 **45.** 0.0073 **46.** $2.3\overline{4}$

In Exercises 47–50, express each mixed number as an improper fraction.

47. $1\frac{3}{4}$ **48.** $4\frac{1}{6}$ **49.** $-3\frac{1}{4}$ **50.** $-35\frac{3}{8}$

In Exercises 51–54, express each improper fraction as a mixed number.

51. $\dfrac{11}{5}$ **52.** $\dfrac{75}{8}$ **53.** $-\dfrac{12}{7}$ **54.** $-\dfrac{136}{5}$

In Exercises 55–63, perform the indicated operation and reduce your answer to lowest terms.

55. $\dfrac{1}{3} + \dfrac{3}{4}$ **56.** $\dfrac{11}{12} - \dfrac{2}{3}$

57. $\dfrac{1}{6} + \dfrac{5}{4}$ **58.** $\dfrac{7}{16} \cdot \dfrac{12}{21}$

59. $\dfrac{5}{9} \div \dfrac{6}{7}$ **60.** $\left(\dfrac{4}{5} + \dfrac{5}{7}\right) \div \dfrac{4}{5}$

61. $\left(\dfrac{2}{3} \cdot \dfrac{1}{7}\right) \div \dfrac{4}{7}$ **62.** $\left(\dfrac{1}{5} + \dfrac{2}{3}\right)\left(\dfrac{3}{8}\right)$

63. $\left(\dfrac{1}{5} \cdot \dfrac{2}{3}\right) + \left(\dfrac{1}{5} \div \dfrac{1}{2}\right)$

64. *Cajun Turkey* A recipe for Roasted Cajun Turkey calls for $\frac{1}{8}$ teaspoon of cayenne pepper per pound of turkey. If Jennifer Sabo is preparing a turkey that weighs $17\frac{3}{4}$ pounds, how much cayenne pepper does she need?

5.4

In Exercises 65–80, simplify the expression. Rationalize the denominator when necessary.

65. $\sqrt{45}$ **66.** $\sqrt{200}$ **67.** $\sqrt{5} + 7\sqrt{5}$

68. $\sqrt{2} - 4\sqrt{2}$ **69.** $\sqrt{8} + 6\sqrt{2}$ **70.** $\sqrt{3} - 7\sqrt{27}$

71. $\sqrt{28} + \sqrt{63}$ **72.** $\sqrt{3} \cdot \sqrt{6}$ **73.** $\sqrt{8} \cdot \sqrt{6}$

74. $\dfrac{\sqrt{300}}{\sqrt{3}}$ **75.** $\dfrac{\sqrt{56}}{\sqrt{2}}$ **76.** $\dfrac{4}{\sqrt{3}}$

77. $\dfrac{\sqrt{7}}{\sqrt{5}}$ **78.** $3(2 + \sqrt{7})$

79. $\sqrt{3}(4 + \sqrt{6})$ **80.** $\sqrt{3}(\sqrt{6} + \sqrt{15})$

5.5

In Exercises 81–90, state the name of the property illustrated.

81. $5 + 7 = 7 + 5$

82. $5 \cdot m = m \cdot 5$

83. $(1 + 2) + 3 = 1 + (2 + 3)$

84. $9(x + 1) = 9 \cdot x + 9 \cdot 1$

85. $4 + (5 + 6) = (4 + 5) + 6$

86. $(3 + 5) + (4 + 3) = (4 + 3) + (3 + 5)$

87. $(3 \cdot a) \cdot b = 3 \cdot (a \cdot b)$

88. $a \cdot (2 + 3) = (2 + 3) \cdot a$

89. $2(x + 3) = (2 \cdot x) + (2 \cdot 3)$

90. $x \cdot 2 + 6 = 2 \cdot x + 6$

In Exercises 91–96, determine whether the set of numbers is closed under the given operation.

91. Natural numbers, subtraction

92. Whole numbers, multiplication

93. Integers, division

94. Real numbers, subtraction

95. Irrational numbers, multiplication

96. Rational numbers, division

5.6

In Exercises 97–104, evaluate each expression.

97. 5^2 **98.** 5^{-2} **99.** $\dfrac{9^5}{9^3}$ **100.** $5^2 \cdot 5$

101. 7^0 **102.** 4^{-3} **103.** $(2^3)^2$ **104.** $(3^2)^2$

In Exercises 105–108, write each number in scientific notation.

105. $8,200,000,000$ **106.** 0.0000158

107. 0.02309 **108.** $4,950,000$

In Exercises 109–112, express each number in decimal notation.

109. 2.8×10^5 **110.** 1.39×10^{-4}

111. 1.75×10^{-4} **112.** 1×10^7

In Exercises 113–116, (a) perform the indicated operation and write your answer in scientific notation. (b) Confirm the result found in part (a) by performing the calculation on a scientific calculator.

113. $(3 \times 10^4)(2 \times 10^{-9})$

114. $(5 \times 10^6)(7.5 \times 10^5)$

115. $\dfrac{8.4 \times 10^3}{4 \times 10^2}$

116. $\dfrac{1.5 \times 10^{-3}}{5 \times 10^{-4}}$

In Exercises 117–121, (a) perform the indicated calculation by first converting each number to scientific notation. Write your answer in decimal notation. (b) Confirm the result found in part (a) by performing the calculation on a scientific calculator.

117. $(550,000)(2,000,000)$ **118.** $(35,000)(0.00002)$

119. $\dfrac{8,400,000}{70,000}$ **120.** $\dfrac{0.000002}{0.0000004}$

121. *Space Distances* The distance from Earth to the sun is about 1.49×10^{11} meters. The distance from Earth to the moon is about 3.84×10^8 meters. The distance from Earth to the sun is how many times larger than the distance from Earth to the moon? Use a scientific calculator and round your answer to the nearest whole number.

122. *Outstanding Debt* As a result of a recent water and sewer system improvement, the city of Galena, Illinois, has an outstanding debt of $20,000,000. If the population of Galena is 3600 people, how much would each person have to contribute to pay off the outstanding debt?

5.7

In Exercises 123–128, determine whether the sequence is arithmetic or geometric. Then determine the next two terms of the sequence.

123. $1, 5, 9, 13, \ldots$ **124.** $\frac{1}{2}, 1, 2, 4, \ldots$

125. $-3, -6, -9, -12, \ldots$ **126.** $\frac{1}{2}, \frac{1}{4}, \frac{1}{8}, \frac{1}{16}, \ldots$

127. $1, 4, 7, 10, 13, \ldots$ **128.** $\frac{1}{2}, -\frac{1}{2}, \frac{1}{2}, -\frac{1}{2}, \ldots$

In Exercises 129–134, determine the indicated term of the sequence with the given first term, a_1, and common difference, d, or common ratio, r.

129. Determine a_6 when $a_1 = 2, d = 5$.

130. Determine a_9 when $a_1 = -6, d = 2$.

131. Determine a_{10} when $a_1 = -20, d = 5$.

132. Determine a_5 when $a_1 = 3, r = 2$.

133. Determine a_5 when $a_1 = 4, r = \frac{1}{2}$.

134. Determine a_4 when $a_1 = -6, r = 2$.

In Exercises 135–138, determine the sum of the arithmetic sequence. The number of terms, n, is given.

135. $3, 6, 9, 12, \ldots, 150; n = 50$

136. $-4, -3\frac{3}{4}, -3\frac{1}{2}, -3\frac{1}{4}, \ldots, -2\frac{1}{4}; n = 8$

137. $100, 94, 88, 82, \ldots, 58; n = 8$

138. $0.5, 0.75, 1.00, 1.25, \ldots, 5.25; n = 20$

In Exercises 139–142, determine the sum of the first n terms of the geometric sequence for the values of a_1 and r.

139. $n = 5, a_1 = 2, r = 4$

140. $n = 4, a_1 = 3, r = 2$

141. $n = 5, a_1 = 3, r = -2$

142. $n = 6, a_1 = 1, r = -2$

In Exercises 143–148, first determine whether the sequence is arithmetic or geometric; then write an expression for the general or nth term, a_n.

143. $3, 6, 9, 12, \ldots$

144. $1, 4, 7, 10, \ldots$

145. $4, \frac{5}{2}, 1, -\frac{1}{2}, \ldots$

146. $3, 6, 12, 24, \ldots$

147. $2, -2, 2, -2, \ldots$

148. $5, \frac{5}{3}, \frac{5}{9}, \frac{5}{27}, \ldots$

5.8

In Exercises 149–152, determine whether the sequence is a Fibonacci-type sequence. If so, determine the next two terms.

149. $0, 1, 1, 2, 2, 3, 3, 4, 4, \ldots$

150. $-1, 0, -1, -1, -2, -3, -5, \ldots$

151. $1, 4, 3, -1, -4, -5, \ldots$

152. $-10, 10, 0, 10, 20, \ldots$

CHAPTER ⑤ TEST

1. Which of the numbers 2, 3, 4, 5, 6, 8, 9, and 10 divide 48,395?

2. Determine the prime factorization of 414.

3. Evaluate $[(-3) + 7] - (-4)$.

4. Evaluate $-7 - 13$.

5. Evaluate $[(-70)(-5)] \div (8 - 10)$.

6. Convert $4\frac{5}{8}$ to an improper fraction.

7. Convert $\frac{176}{9}$ to a mixed number.

8. Write $\frac{5}{8}$ as a terminating or repeating decimal.

9. Express 6.45 as a quotient of two integers.

10. Evaluate $\left(\frac{5}{16} \div 3\right) + \left(\frac{4}{5} \cdot \frac{1}{2}\right)$.

11. Perform the operation and reduce the answer to lowest terms: $\frac{17}{24} - \frac{7}{12}$.

12. Simplify $\sqrt{75} + \sqrt{48}$.

13. Rationalize the denominator $\dfrac{\sqrt{2}}{\sqrt{7}}$.

14. Determine whether the integers are closed under the operation of multiplication. Explain your answer.

Name the property illustrated.

15. $(x + 3) + 5 = x + (3 + 5)$

16. $4(x + 3) = 4x + 12$

Evaluate.

17. $\dfrac{4^5}{4^2}$ **18.** $4^3 \cdot 4^2$ **19.** 3^{-4}

20. Perform the operation by first converting the numerator and denominator to scientific notation. Write the answer in scientific notation.

$$\frac{7,200,000}{0.000009}$$

21. Write an expression for the general or *n*th term, a_n, of the sequence $-2, -6, -10, -14, \ldots$.

22. Determine the sum of the terms of the arithmetic sequence. The number of terms, *n,* is given.

$$-2, -5, -8, -11, \ldots, -32; n = 11$$

23. Determine a_6 when $a_1 = 2$ and $r = 3$.

24. Determine the sum of the first eight terms of the sequence when $a_1 = -2$ and $r = 2$.

25. Write an expression for the general or *n*th term, a_n, of the sequence $3, 6, 12, 24, \ldots$.

26. Write the first 10 terms of the Fibonacci sequence.

GROUP PROJECTS

1. *Making Rice* The amount of ingredients needed to make 3 and 5 servings of rice are:

To Make	Rice and Water	Salt	Butter
3 servings	1 cup	$\frac{3}{8}$ tsp	$1\frac{1}{2}$ tsp
5 servings	$1\frac{2}{3}$ cup	$\frac{5}{8}$ tsp	$2\frac{1}{2}$ tsp

Find the amount of each ingredient needed to make (a) 2 servings, (b) 1 serving, and (c) 29 servings. Explain how you determined your answers.

2. *Finding Areas*

a) Determine the area of the trapezoid shown by finding the area of the three parts indicated and finding the sum of the three areas. The necessary geometric formulas are given in Chapter 9.

b) Determine the area of the trapezoid by using the formula for the area of a trapezoid given in Chapter 9.

c) Compare your answers from parts (a) and (b). Are they the same? If not, explain why they are different.

3. *Medical Insurance* On a medical insurance policy (such as Blue Cross/Blue Shield), the policyholder may need to make copayments for prescription drugs, office visits, and procedures until the total of all copayments reaches a specified amount. Suppose that the copayment on the Gattelaro's medical policy for prescription drugs is 50% of the cost; the copayment for office visits is $10; and the copayment for all medical tests, x-rays, and other procedures is 20% of the cost. After the family's copayment totals $500 in a calendar year, all medical and prescription bills are paid in full by the insurance company. The Gattelaros had the following medical expenses from January 1 through April 30.

Date	Reason	Cost before Copayment
January 10	Office visit	$40
	Prescription	$44
February 27	Office visit	$40
	Medical tests	$188
April 19	Office visit	$40
	X-rays	$348
	Prescription	$76

a) How much had the Gattelaros paid in copayments from January 1 through April 30?

b) How much had the medical insurance company paid?

c) What is the remaining copayment that must be paid by the Gattelaros before the $500 copayment limit is reached?

4. *A Branching Plant* A plant grows for two months and then adds a new branch. Each new branch grows for two months and then adds another branch. After the second month, each branch adds a new branch every month. Assume the growth begins in January.

a) How many branches will there be in February?

b) How many branches will there be in May?

c) How many branches will there be after 12 months?

d) How is this problem similar to the problem involving the rabbits that appeared in Fibonacci's book *Liber Abacci* (see page 298)?

CHAPTER 6

Algebra, Graphs, and Functions

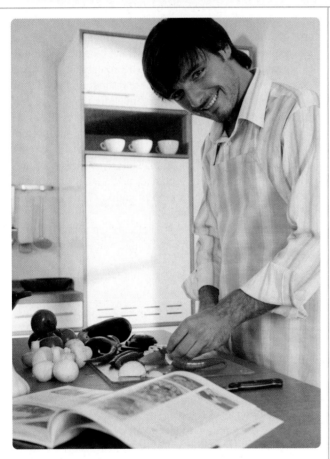

▲ Algebra is a tool for solving everyday problems such as changing a recipe to increase or decrease the number of servings.

WHAT YOU WILL LEARN

- Order of operations
- Solving linear and quadratic equations, and linear inequalities in one variable
- Evaluating a formula
- Solving application problems involving linear, quadratic, and exponential equations
- Solving application problems dealing with variation
- Graphing equations and functions, including linear, quadratic, and exponential equations

WHY IT IS IMPORTANT

Algebra is one of the most practical tools for solving everyday problems. You probably use algebra in your daily life without realizing it. For example, you use a coordinate system when you consult a map to find directions. You solve simple equations when you change a recipe to increase or decrease the number of servings. To evaluate how much interest you will earn in a savings account or to figure out how long it will take you to travel a given distance, you use common formulas that are algebraic equations. The symbolic language of algebra makes it an excellent tool for solving problems because it allows us to write lengthy expressions in compact form. English philosopher Alfred North Whitehead explained the power of algebra when he stated, "By relieving the brain of unnecessary work, a good notation sets the mind free to concentrate on more advanced problems."

6.1   ORDER OF OPERATIONS

▲ You can use algebra to determine the sales tax rate when purchasing items such as tires.

Suppose you see an advertisement for tires that states: "Purchase Michelin All-Season Tires for $135 each. This price includes mounting." You purchase four tires and have them mounted on your car. Your total price, including sales tax, comes to $583.50. What is the sales tax rate? You may be able to solve this problem and other problems in everyday life by using arithmetic or trial and error. With a knowledge of algebra, however, you can often find the solution with much less effort. In this section, we will discuss how to use algebra to answer questions like the one above.

Algebra is a generalized form of arithmetic. The word *algebra* is derived from the Arabic word *al-jabr* (meaning "reunion of broken parts"), which was the title of a book written by the mathematician Muhammed ibn-Musa al Khwarizmi in about A.D. 825.

Algebra uses letters of the alphabet called *variables* to represent numbers. Often the letters x and y are used to represent variables. However, any letter may be used as a variable. A symbol that represents a specific quantity is called a *constant*.

Multiplication of numbers and variables may be represented in several different ways in algebra. Because the "times" sign might be confused with the variable x, a dot between two numbers or variables indicates multiplication. Thus, $3 \cdot 4$ means 3 times 4, and $x \cdot y$ means x times y. Placing two letters or a number and a letter next to one another, with or without parentheses, also indicates multiplication. Thus, $3x$ means 3 times x, xy means x times y, and $(x)(y)$ means x times y.

An *algebraic expression* (or simply an *expression*) is a collection of variables, numbers, parentheses, and operation symbols. Some examples of algebraic expressions are

$$x, \qquad x + 2, \qquad 3(2x + 3), \qquad \frac{3x + 1}{2x - 3}, \qquad \text{and} \qquad x^2 + 7x + 3$$

Two algebraic expressions joined by an equal sign form an *equation*. Some examples of equations are

$$x + 2 = 4, \qquad 3x + 4 = 1, \qquad \text{and} \qquad x + 3 = 2x$$

The *solution to an equation* is the number or numbers that replace the variable to make the equation a true statement. For example, the solution to the equation $x + 3 = 4$ is 1. When we find the solution to an equation, we *solve the equation*.

We can determine if any number is a solution to an equation by *checking the solution*. To check the solution, we substitute the number for the variable in the equation. If the resulting statement is a true statement, that number is a solution to the equation. If the resulting statement is a false statement, the number is not a solution to the equation. To check the number 1 in the equation $x + 3 = 4$, we do the following.

$$
\begin{aligned}
x + 3 &= 4 \\
1 + 3 &= 4 \qquad \text{Substitute 1 for } x \\
4 &= 4 \qquad \text{True}
\end{aligned}
$$

The same number is obtained on both sides of the equal sign, so 1 is the solution. For the equation $x + 3 = 4$, the only solution is 1. Any other value of x would result in the check being a false statement.

To *evaluate an expression* means to find the value of the expression for a given value of the variable. To evaluate expressions and solve equations, you must have an understanding of exponents. Exponents (Section 5.6) are used to abbreviate repeated multiplication. For example, the expression 5^2 means $5 \cdot 5$. The 2 in the expression 5^2 is the *exponent*, and the 5 is the *base*. We read 5^2 as "5 to the second power" or "5 squared," and 5^2 means $5 \cdot 5$ or 25.

In general, the number b to the nth power, written b^n, means

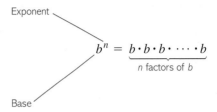

An exponent refers only to its base. In the expression -5^2, the base is 5. In the expression $(-5)^2$, the base is -5.

$$-5^2 = -(5)^2 = -1(5)^2 = -1(5)(5) = -25$$
$$(-5)^2 = (-5)(-5) = 25$$

Note that $-5^2 \neq (-5)^2$ since $-25 \neq 25$.

Order of Operations

To evaluate an expression or to check the solution to an equation, we need to know the *order of operations* to follow. For example, suppose we want to evaluate the expression $2 + 3x$ when $x = 4$. Substituting 4 for x, we obtain $2 + 3 \cdot 4$. What is the value of $2 + 3 \cdot 4$? Does it equal 20, or does it equal 14? Some standard rules, called the order of operations, have been developed to ensure that there is only one correct answer. In mathematics, unless parentheses indicate otherwise, always perform multiplication before addition. Thus, the correct answer is 14.

$$2 + 3 \cdot 4 = 2 + (3 \cdot 4) = 2 + 12 = 14$$

The order of operations for evaluating an expression is as follows.

ORDER OF OPERATIONS

1. First, perform all operations within parentheses or other grouping symbols (according to the following order).
2. Next, perform all exponential operations (that is, raising to powers or finding roots).
3. Next, perform all multiplications and divisions from left to right.
4. Finally, perform all additions and subtractions from left to right.

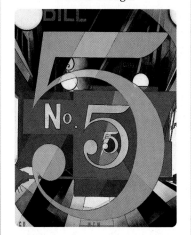
TIMELY TIP Some students use the phrase "*P*lease *E*xcuse *M*y *D*ear *A*unt *S*ally" or the word "PEMDAS" (*P*arentheses, *E*xponents, *M*ultiplication, *D*ivision, *A*ddition, *S*ubtraction) to remind them of the order of operations. Remember: Multiplication and division are of the same order, and addition and subtraction are of the same order.

EXAMPLE ❶ *Evaluating an Expression*

Evaluate the expression $-x^2 + 4x + 16$ for $x = 3$.

SOLUTION Substitute 3 for each x and use the order of operations to evaluate the expression.

$$-x^2 + 4x + 16$$
$$= -(3)^2 + 4(3) + 16$$
$$= -9 + 12 + 16$$
$$= 3 + 16$$
$$= 19$$

EXAMPLE ❷ *Finding the Height*

A ball is thrown upward off a bridge 40 feet above ground. Its height, in feet, above ground, t seconds after it is thrown, can be determined by the expression $-16t^2 + 30t + 40$. Find the height of the ball, above ground, 2 seconds after it is thrown.

SOLUTION Substitute 2 for each t.

$$-16t^2 + 30t + 40$$
$$= -16(2)^2 + 30(2) + 40$$
$$= -16(4) + 30(2) + 40$$
$$= -64 + 60 + 40$$
$$= -4 + 40$$
$$= 36$$

The ball is 36 feet above ground 2 seconds after it is thrown.

EXAMPLE ❸ *Substituting for Two Variables*

Evaluate $-3x^2 + 2xy - 2y^2$ when $x = 2$ and $y = 3$.

SOLUTION Substitute 2 for each x and 3 for each y; then evaluate using the order of operations.

$$-3x^2 + 2xy - 2y^2$$
$$= -3(2)^2 + 2(2)(3) - 2(3)^2$$
$$= -3(4) + 2(2)(3) - 2(9)$$
$$= -12 + 12 - 18$$
$$= 0 - 18$$
$$= -18$$

EXAMPLE **4** *Is 2 a Solution?*

Determine whether 2 is a solution to the equation $3x^2 + 7x - 11 = 15$.

SOLUTION To determine whether 2 is a solution to the equation, substitute 2 for each x in the equation. Then evaluate the left-hand side of the equation using the order of operations. If the left-hand side of the equation equals 15, then both sides of the equation have the same value and 2 is a solution. When checking the solution, we use $\stackrel{?}{=}$, which means we are not sure if the statement is true.

$$3(2)^2 + 7(2) - 11 \stackrel{?}{=} 15$$
$$3(4) + 14 - 11 \stackrel{?}{=} 15$$
$$12 + 14 - 11 \stackrel{?}{=} 15$$
$$26 - 11 \stackrel{?}{=} 15$$
$$15 = 15 \quad \text{True}$$

Because 2 makes the equation a true statement, 2 is a solution to the equation.   •

SECTION 6.1 EXERCISES

CONCEPT/WRITING EXERCISES

1. What is a *variable*?

2. What is a *constant*?

3. What does it mean when we state that *a number is a solution to an equation*?

4. What is an algebraic expression? Illustrate an algebraic expression with an example.

5. a) For the term 4^5, identify the base and the exponent.

   b) In your own words, explain how to evaluate 4^5.

6. In your own words, explain the order of operations.

7. a) Evaluate $8 + 16 \div 4$ using the order of operations.

   b) Evaluate $9 + 6 \cdot 3$ using the order of operations.

8. a) Explain why $-x^2$ will always be a negative number for any nonzero real number selected for x.

   b) Explain why $(-x)^2$ will always be a positive number for any nonzero real number selected for x.

PRACTICE THE SKILLS

In Exercises 9–28, evaluate the expression for the given value(s) of the variable(s).

9. x^2, $x = -5$

10. x^2, $x = 6$

11. $-x^2$, $x = -2$

12. $-x^2$, $x = -3$

13. $-2x^3$, $x = -7$

14. $-x^3$, $x = -4$

15. $x - 7$, $x = 4$

16. $8x - 3$, $x = \frac{5}{2}$

17. $-4x + 4$, $x = -2$

18. $x^2 - 3x + 8$, $x = 5$

19. $-x^2 + 3x - 10$, $x = -2$

20. $7x^2 + 5x - 11$, $x = -1$

21. $\frac{1}{2}x^2 - 5x + 2$, $x = \frac{2}{3}$

22. $\frac{2}{3}x^2 + x - 1$, $x = \frac{1}{2}$

23. $8x^3 - 4x^2 + 7$, $x = \frac{1}{2}$

24. $-x^2 + 5xy$, $x = 2$, $y = -3$

25. $2x^2 + xy + 3y^2$, $x = -2$, $y = 1$

26. $3x^2 + \frac{2}{5}xy - \frac{1}{5}y^2$, $x = 2$, $y = 5$

27. $4x^2 - 10xy + 3y^2$, $x = 2$, $y = -1$

28. $(x - 2y)^2$, $x = 4$, $y = -2$

In Exercises 29–38, determine whether the value(s) is (are) a solution to the equation.

29. $8x + 3 = 23$, $x = 3$

30. $6x - 7 = -31$, $x = -4$

31. $x - 3y = 0$, $x = 6$, $y = 3$

32. $4x + 2y = -2$, $x = -2$, $y = 3$

33. $x^2 - 3x + 6 = 5$, $x = 2$

34. $-2x^2 + x + 5 = 20$, $x = 3$

35. $2x^2 + x = 28$, $x = -4$

36. $y = x^2 + 4x - 5$, $x = -1$, $y = -8$

37. $y = -x^2 + 3x - 1$, $x = 3$, $y = -1$

38. $y = x^3 - 3x^2 + 1$, $x = 2$, $y = -3$

PROBLEM SOLVING

39. *Sales Tax* If the sales tax on an item is 7%, the sales tax, in dollars, on an item costing d dollars can be found by using the expression $0.07d$. Determine the sales tax on a refrigerator costing $899.

40. *8 Trillion Calculations* If a computer can do a calculation in 0.000002 sec, the time required to do n calculations can be determined by the expression $0.000002n$. Determine the number of seconds needed for the computer to do 8 trillion (8,000,000,000,000) calculations.

41. *Electronic Filing* The graph shows the number of taxpayers, in millions, filing their taxes electronically for the years 2000 through 2005. The number of taxpayers, in millions, filing their taxes electronically per year can be approximated by the expression $6.2x + 34.8$, where x

is the number of years since 2000. For example, the year 2001 would correspond to $x = 1$. Assuming this trend continues, use the expression to approximate the number of taxpayers who will file their taxes electronically in 2010.

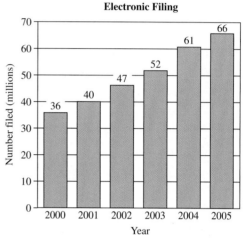

Electronic Filing

Source: Internal Revenue Service

42. *Cost of a Tour* The cost, in dollars, for Crescent City Tours to provide a tour for x people can be determined by the expression $220 + 2.75x$. Determine the cost for Crescent City Tours to provide a tour for 75 people.

43. *Drying Time* The time, in minutes, needed for clothes hanging on a line outdoors to dry, at a specific temperature and wind speed, depends on the humidity, h. The time can be approximated by the expression $2h^2 + 80h + 40$, where h is the percent humidity expressed as a decimal number. Find the length of time required for clothing to dry if there is 60% humidity.

44. *Stopping Distance* A typical car's stopping distance, in feet, on wet pavement can be approximated by the expression $0.08s^2 + 0.24s - 7.10$, where s is the speed of the car, in miles per hour (mph), before braking and $60 \le s \le 80$ miles per hour. Use the expression to approximate the car's stopping distance on wet pavement if the speed of the car before braking is 72 mph.

45. *Grass Growth* The rate of growth of grass, in inches per week, depends on a number of factors, including rainfall and temperature. For a certain area, this rate can be approximated by the expression $0.2R^2 + 0.003RT + 0.0001T^2$, where R is the weekly rainfall, in inches, and T is the average weekly temperature, in degrees Fahrenheit. Find the amount of growth of grass for a week in which the rainfall is 2 in. and the average temperature is 70°F.

CHALLENGE PROBLEMS/GROUP ACTIVITIES

46. Explain why $(-1)^n = 1$ for any even number n.

47. Does $(x + y)^2 = x^2 + y^2$? Complete the table and state your conclusion.

x	y	$(x + y)^2$	$x^2 + y^2$
2	3		
-2	-3		
-2	3		
2	-3		

48. Suppose n represents any natural number. Explain why 1^n equals 1.

INTERNET/RESEARCH ACTIVITY

49. When were exponents first used? Write a paper explaining how exponents were first used and when mathematicians began writing them in the present form.

6.2 LINEAR EQUATIONS IN ONE VARIABLE

▲ A linear equation can be used to approximate the number of adults in the United States age 65 or older.

**Number of Adults
Age 65 or Older in the U. S.
(projected for years after 2000)**

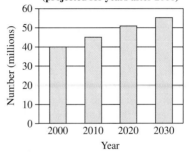

Source: U.S. Census Bureau

Figure 6.1

The graph in Figure 6.1 shows that the number of adults in the United States age 65 or older is projected to increase steadily from 2000 through 2030. The number of adults in the United States age 65 or older can be approximated by the equation $A = 0.51x + 40.1$, where A represents the number of adults in the United States age 65 or older, in millions, and x represents the number of years after 2000. Suppose we wish to determine the year when there will be 60 million adults in the United States age 65 or older. How would we do so? We could substitute 60 for A in the equation to obtain the equation $60 = 0.51x + 40.1$. We would then need to solve this equation for x. In this section, we explain how to solve equations such as $60 = 0.51x + 40.1$.

In Section 6.1, we stated that two algebraic expressions joined by an equal sign form an equation. The solution to some equations, such as $x + 3 = 4$, can be easily found by trial and error. Solving more difficult equations such as $60 = 0.51x + 40.1$ or $2x - 3 = 4(x + 3)$, however, requires us to understand some basic properties of equations.

The parts that are added or subtracted in an algebraic expression are called *terms*. The expression $0.51x + 40.1$ contains two terms: $0.51x$ and $+ 40.1$. The expression $4x - 3y - 5$ contains three terms: $4x$, $-3y$, and -5. The $+$ and $-$ signs that break the expression into terms are part of the terms. When listing the terms of an expression, however, it is not necessary to include the $+$ sign at the beginning of the term.

The numerical part of a term is called its *numerical coefficient* or, simply, its *coefficient*. In the term $4x$, the 4 is the numerical coefficient. In the term $-4y$, the -4 is the numerical coefficient.

Like terms are terms that have the same variables with the same exponents on the variables. *Unlike terms* have different variables or different exponents on the variables.

Like Terms	Unlike Terms
$2x$, $7x$ (same variable, x)	$2x$, 9 (only first term has a variable)
$-8y$, $3y$ (same variable, y)	$5x$, $6y$ (different variables)
-4, 10 (both constants)	x, 8 (only first term has a variable)
$-5x^2$, $6x^2$ (same variable with same exponent)	$2x^3$, $3x^2$ (different exponents)

To *simplify an expression* means to combine like terms by using the commutative, associative, and distributive properties discussed in Chapter 5. For convenience, we list these properties below.

PROPERTIES OF THE REAL NUMBERS

$a(b + c) = ab + ac$	Distributive property
$a + b = b + a$	Commutative property of addition
$ab = ba$	Commutative property of multiplication
$(a + b) + c = a + (b + c)$	Associative property of addition
$(ab)c = a(bc)$	Associative property of multiplication

EXAMPLE ❶ *Combining Like Terms*

Combine like terms in each expression.

a) $7x + 3x$ b) $6y - 2y$

c) $x + 12 - 3x + 7$ d) $-2x + 4 - 6y - 11 - 5y + 3x$

SOLUTION

a) We use the distributive property (in reverse) to combine like terms.

$$7x + 3x = (7 + 3)x \qquad \text{Distributive property}$$
$$= 10x$$

b) $6y - 2y = (6 - 2)y = 4y$

c) $x + 12 - 3x + 7 = x - 3x + 12 + 7$ Rearrange terms, place like terms together.
$$= -2x + 19 \qquad \text{Combine like terms.}$$

d) $-2x + 4 - 6y - 11 - 5y + 3x$
$$= -2x + 3x - 6y - 5y + 4 - 11 \qquad \text{Rearrange terms, place like terms together.}$$
$$= x - 11y - 7 \qquad \text{Combine like terms.} \qquad ●$$

We are able to rearrange the terms of an expression, as was done in Example 1(c) and (d) by the commutative and associative properties that were discussed in Section 5.5.

The order of the terms in an expression is not crucial. However, when listing the terms of an expression we generally list the terms in alphabetical order with the constant, the term without a variable, last.

Solving Equations

Recall that to solve an equation means to find the value or values for the variable that make(s) the equation true. In this section, we discuss solving *linear (or first-degree) equations*. A linear equation in one variable is one in which the exponent on the variable is 1. Examples of linear equations are $5x - 1 = 3$ and $2x + 4 = 6x - 5$.

Equivalent equations are equations that have the same solution. The equations $2x - 5 = 1$, $2x = 6$, and $x = 3$ are all equivalent equations since they all have the same solution, 3. When we solve an equation, we write the given equation as a series of simpler equivalent equations until we obtain an equation of the form $x = c$, where c is some real number.

To solve any equation, we have to *isolate the variable*. That means getting the variable by itself on one side of the equal sign. The four properties of equality that we are about to discuss are used to isolate the variable. The first is the addition property.

ADDITION PROPERTY OF EQUALITY
If $a = b$, then $a + c = b + c$ for all real numbers a, b, and c.

The addition property of equality indicates that the same number can be added to both sides of an equation without changing the solution.

EXAMPLE ❷ *Using the Addition Property of Equality*

Find the solution to the equation $x - 9 = 15$.

SOLUTION To isolate the variable, add 9 to both sides of the equation.

$$x - 9 = 15$$
$$x - 9 + 9 = 15 + 9$$
$$x + 0 = 24$$
$$x = 24$$

CHECK:
$$x - 9 = 15$$
$$24 - 9 \overset{?}{=} 15 \quad \text{Substitute 24 for } x.$$
$$15 = 15 \quad \text{True}$$

In Example 2, we showed the step $x + 0 = 24$. This step is usually done mentally, and the step is usually not listed.

SUBTRACTION PROPERTY OF EQUALITY
If $a = b$, then $a - c = b - c$ for all real numbers a, b, and c.

The subtraction property of equality indicates that the same number can be subtracted from both sides of an equation without changing the solution.

EXAMPLE ❸ *Using the Subtraction Property of Equality*

Find the solution to the equation $x + 11 = 19$.

SOLUTION To isolate the variable, subtract 11 from both sides of the equation.

$$x + 11 = 19$$
$$x + 11 - 11 = 19 - 11$$
$$x = 8$$

Note in Example 3 that we did not subtract 19 from both sides of the equation, because doing so would not result in getting x on one side of the equal sign by itself. Now we discuss the multiplication property.

MULTIPLICATION PROPERTY OF EQUALITY
If $a = b$, then $a \cdot c = b \cdot c$ for all real numbers a, b, and c, where $c \neq 0$.

The multiplication property of equality indicates that both sides of the equation can be multiplied by the same nonzero number without changing the solution.

EXAMPLE ❹ *Using the Multiplication Property of Equality*

Find the solution to $\dfrac{x}{6} = 3$.

SOLUTION To solve this equation, multiply both sides of the equation by 6.

$$\frac{x}{6} = 3$$
$$6\left(\frac{x}{6}\right) = 6(3)$$
$$\frac{\overset{1}{6}x}{\underset{1}{6}} = 18$$
$$1x = 18$$
$$x = 18$$

In Example 4, we showed the steps $\dfrac{6x}{6} = 18$ and $1x = 18$. Usually, we will not illustrate these steps.

DIVISION PROPERTY OF EQUALITY
If $a = b$, then $\dfrac{a}{c} = \dfrac{b}{c}$ for all real numbers a, b, and c, $c \neq 0$.

The division property of equality indicates that both sides of an equation can be divided by the same nonzero number without changing the solution. Note that the divisor, c, cannot be 0 because division by 0 is not permitted.

EXAMPLE ❺ *Using the Division Property of Equality*

Solve the equation $4x = 16$.

SOLUTION To solve this equation, divide both sides of the equation by 4.

$$4x = 16$$

$$\frac{4x}{4} = \frac{16}{4}$$

$$x = 4$$

An *algorithm* is a general procedure for accomplishing a task. The following general procedure is an algorithm for solving linear (or first-degree) equations. Sometimes the solution to an equation may be found more easily by using a variation of this general procedure. Remember that the primary objective in solving any equation is to isolate the variable.

A GENERAL PROCEDURE FOR SOLVING LINEAR EQUATIONS

1. If the equation contains fractions, multiply both sides of the equation by the lowest common denominator (or least common multiple). This step will eliminate all fractions from the equation.

2. Use the distributive property to remove parentheses when necessary.

3. Combine like terms on the same side of the equal sign when possible.

4. Use the addition or subtraction property to collect all terms with a variable on one side of the equal sign and all constants on the other side of the equal sign. It may be necessary to use the addition or subtraction property more than once. This process will eventually result in an equation of the form $ax = b$, where a and b are real numbers.

5. Solve for the variable using the division or multiplication property. The result will be an answer in the form $x = c$, where c is a real number.

EXAMPLE ❻ *Using the General Procedure*

Solve the equation $2x - 9 = 19$.

SOLUTION Our goal is to isolate the variable; therefore, we start by getting the term $2x$ by itself on one side of the equation.

$$2x - 9 = 19$$

$$2x - 9 + 9 = 19 + 9 \quad \text{Add 9 to both sides of the equation (addition property) (step 4).}$$

$$2x = 28$$

$$\frac{2x}{2} = \frac{28}{2} \quad \text{Divide both sides of the equation by 2 (division property) (step 5).}$$

$$x = 14$$

A check will show that 14 is the solution to $2x - 9 = 19$.

EXAMPLE ❼ *Solving a Linear Equation*

Solve the equation $4 = 5 + 2(t + 1)$ for t.

SOLUTION Our goal is to isolate the variable t. To do so, follow the general procedure for solving equations.

$$4 = 5 + 2(t + 1)$$

$$4 = 5 + 2t + 2 \qquad \text{Distributive property (step 2)}$$

$$4 = 2t + 7 \qquad \text{Combine like terms (step 3).}$$

$$4 - 7 = 2t + 7 - 7 \qquad \text{Subtraction property (step 4)}$$

$$-3 = 2t$$

$$\frac{-3}{2} = \frac{2t}{2} \qquad \text{Division property (step 5)}$$

$$-\frac{3}{2} = t$$

EXAMPLE ❽ *Solving an Equation Containing Fractions*

Solve the equation $\dfrac{2x}{3} + \dfrac{1}{3} = \dfrac{3}{4}$.

SOLUTION When an equation contains fractions, we generally begin by multiplying each term of the equation by the lowest common denominator, LCD (see Chapter 5). In this example, the LCD is 12 because 12 is the smallest number that is divisible by both 3 and 4.

$$12\left(\frac{2x}{3} + \frac{1}{3}\right) = 12\left(\frac{3}{4}\right) \qquad \text{Multiply both sides of the equation by the LCD (step 1).}$$

$$12\left(\frac{2x}{3}\right) + 12\left(\frac{1}{3}\right) = 12\left(\frac{3}{4}\right) \qquad \text{Distributive property (step 2)}$$

$$\overset{4}{\cancel{12}}\left(\frac{2x}{\underset{1}{\cancel{3}}}\right) + \overset{4}{\cancel{12}}\left(\frac{1}{\underset{1}{\cancel{3}}}\right) = \overset{3}{\cancel{12}}\left(\frac{3}{\underset{1}{\cancel{4}}}\right) \qquad \text{Divide out common factors.}$$

$$8x + 4 = 9$$

$$8x + 4 - 4 = 9 - 4 \qquad \text{Subtraction property (step 4)}$$

$$8x = 5$$

$$\frac{8x}{8} = \frac{5}{8} \qquad \text{Division property (step 5)}$$

$$x = \frac{5}{8}$$

A check will show that $\frac{5}{8}$ is the solution to the equation. You could have worked the problem without first multiplying both sides of the equation by the LCD. Try it!

EXAMPLE ❾ *Variables on Both Sides of the Equation*

Solve the equation $6x + 8 = 10x + 12$.

SOLUTION Note that the equation has an x on both sides of the equal sign. In equations of this type, you might wonder what to do first. It actually does not matter as long as you do not forget the goal of isolating the variable x. Let's collect the terms containing a variable on the left-hand side of the equation.

$$6x + 8 = 10x + 12$$
$$6x + 8 - 8 = 10x + 12 - 8 \qquad \text{Subtraction property (step 4)}$$
$$6x = 10x + 4$$
$$6x - 10x = 10x - 10x + 4 \qquad \text{Subtraction property (step 4)}$$
$$-4x = 4$$
$$\frac{-4x}{-4} = \frac{4}{-4} \qquad \text{Division property (step 5)}$$
$$x = -1$$

In the solution to Example 9, the terms containing the variable were collected on the left-hand side of the equal sign. Now work Example 9, collecting the terms with the variable on the right-hand side of the equal sign. If you do so correctly, you will get the same result.

EXAMPLE ⑩ *Solving an Equation Containing Decimals*

Solve the equation $4x - 0.48 = 0.8x + 4$ and check your solution.

SOLUTION This equation may be solved with the decimals, or you may multiply each term by 100 and eliminate the decimals. We will solve the equation with the decimals.

$$4x - 0.48 = 0.8x + 4$$
$$4x - 0.48 + 0.48 = 0.8x + 4 + 0.48 \qquad \text{Addition property}$$
$$4x = 0.8x + 4.48$$
$$4x - 0.8x = 0.8x - 0.8x + 4.48 \qquad \text{Subtraction property}$$
$$3.2x = 4.48$$
$$\frac{3.2x}{3.2} = \frac{4.48}{3.2} \qquad \text{Division property}$$
$$x = 1.4$$

CHECK:
$$4x - 0.48 = 0.8x + 4$$
$$4(1.4) - 0.48 \stackrel{?}{=} 0.8(1.4) + 4 \qquad \text{Substitute 1.4 for each } x \text{ in the equation.}$$
$$5.6 - 0.48 \stackrel{?}{=} 1.12 + 4$$
$$5.12 = 5.12 \qquad \text{True}$$

In Chapter 5, we explained that $a - b$ can be expressed as $a + (-b)$. We use this principle in Example 11.

EXAMPLE ⑪ *Using the Definition of Subtraction*

Solve $12 = 2(3r - 4) - 4r$ for r.

SOLUTION Our goal is to isolate the variable r. To do so, follow the general procedure for solving equations.

$$12 = 2(3r - 4) - 4r$$
$$12 = 2[3r + (-4)] - 4r \quad \text{Definition of subtraction}$$
$$12 = 2(3r) + 2(-4) - 4r \quad \text{Distributive property}$$
$$12 = 6r - 8 - 4r$$
$$12 = 2r - 8 \quad \text{Combine like terms.}$$
$$12 + 8 = 2r - 8 + 8 \quad \text{Addition property}$$
$$20 = 2r$$
$$\frac{20}{2} = \frac{2r}{2} \quad \text{Division property}$$
$$10 = r$$

●

TIMELY TIP Remember that the goal in solving an equation is to get the variable alone on one side of the equal sign by using the general procedure for solving equations.

So far, every equation has had exactly one solution. Some equations, however, have no solution, and others have more than one solution. Example 12 illustrates an equation that has no solution, and Example 13 illustrates an equation that has an infinite number of solutions.

EXAMPLE ⓬ An Equation with No Solution

Solve $3(x - 4) + x + 2 = 6x - 2(x + 3)$.

SOLUTION

$$3(x - 4) + x + 2 = 6x - 2(x + 3)$$
$$3x - 12 + x + 2 = 6x - 2x - 6 \quad \text{Distributive property}$$
$$4x - 10 = 4x - 6 \quad \text{Combine like terms.}$$
$$4x - 4x - 10 = 4x - 4x - 6 \quad \text{Subtraction property}$$
$$-10 = -6 \quad \text{False}$$

During the process of solving an equation, if you obtain a false statement like $-10 = -6$, or $-4 = 0$, the equation has *no solution*. An equation that has no solution is called a *contradiction*. The equation $3(x - 4) + x + 2 = 6x - 2(x + 3)$ is a contradiction and thus has no solution. ●

EXAMPLE ⓭ An Equation with Infinitely Many Solutions

Solve $2(x + 4) - 3(x - 5) = -x + 23$.

SOLUTION

$$2(x + 4) - 3(x - 5) = -x + 23$$
$$2x + 8 - 3x + 15 = -x + 23 \quad \text{Distributive property}$$
$$-x + 23 = -x + 23 \quad \text{Combine like terms.}$$

Note that at this point both sides of the equation are the same. Every real number will satisfy this equation. This equation has an infinite number of solutions. An equation of this type is called an *identity*. When solving an equation, you can tell that the equation is an identity if you notice that the same expression appears on both sides of the equal sign. The solution to any linear equation with one variable that is an identity is *all real numbers*. If you continue to solve an equation that is an identity, you will end up with $0 = 0$, as follows.

$$-x + 23 = -x + 23$$
$$-x + x + 23 = -x + x + 23 \quad \text{Addition property}$$
$$23 = 23 \quad \text{Combine like terms.}$$
$$23 - 23 = 23 - 23 \quad \text{Subtraction property}$$
$$0 = 0 \quad \text{True for any value of } x$$

Proportions

A *ratio* is a quotient of two quantities. An example is the ratio of 2 to 5, which can be written $2 : 5$ or $\frac{2}{5}$ or 2/5. Ratios are used in proportions.

A **proportion** is a statement of equality between two ratios.

An example of a proportion is $\dfrac{a}{b} = \dfrac{c}{d}$, where $b \neq 0$ and $d \neq 0$. Consider the proportion

$$\frac{x + 2}{5} = \frac{x + 5}{8}$$

We can solve this proportion by first multiplying both sides of the equation by the LCD, 40.

$$\frac{x + 2}{5} = \frac{x + 5}{8}$$
$$\overset{8}{\cancel{40}}\left(\frac{x + 2}{\cancel{5}}\right) = \overset{5}{\cancel{40}}\left(\frac{x + 5}{\cancel{8}}\right) \quad \text{Multiplication property}$$
$$8(x + 2) = 5(x + 5)$$
$$8x + 16 = 5x + 25$$
$$3x + 16 = 25$$
$$3x = 9$$
$$x = 3$$

A check will show that 3 is the solution.

Proportions can often be solved more easily by using cross multiplication.

CROSS MULTIPLICATION

If $\dfrac{a}{b} = \dfrac{c}{d}$, then $ad = bc$, where $b \neq 0, d \neq 0$.

Let's use cross multiplication to solve the proportion $\dfrac{x+2}{5} = \dfrac{x+5}{8}$.

$$\frac{x+2}{5} = \frac{x+5}{8}$$

$$8(x+2) = 5(x+5) \quad \text{Cross multiplication}$$

$$8x + 16 = 5x + 25$$

$$3x + 16 = 25$$

$$3x = 9$$

$$x = 3$$

Many practical application problems can be solved using proportions.

TO SOLVE APPLICATION PROBLEMS USING PROPORTIONS

1. Represent the unknown quantity by a variable.

2. Set up the proportion by listing the given ratio on the left-hand side of the equal sign and the unknown and other given quantity on the right-hand side of the equal sign. When setting up the right-hand side of the proportion, the same respective quantities should occupy the same respective positions on the left and right. For example, an acceptable proportion might be

$$\frac{\text{miles}}{\text{hour}} = \frac{\text{miles}}{\text{hour}}$$

3. Once the proportion is properly written, drop the units and use cross multiplication to solve the equation.

4. Answer the question or questions asked using the appropriate units.

EXAMPLE ⑭ *Water Usage*

The cost for water in Orange County is $1.64 per 750 gallons (gal) of water used. What is the water bill if 30,000 gallons are used?

SOLUTION This problem may be solved by setting up a proportion. One proportion that can be used is

$$\frac{\text{cost of 750 gal}}{750 \text{ gal}} = \frac{\text{cost of 30,000 gal}}{30,000 \text{ gal}}$$

The unknown quantity is the cost for 30,000 gallons of water, so we will call this quantity x. The proportion then becomes

Given ratio $\left\{ \dfrac{1.64}{750} = \dfrac{x}{30,000} \right.$

Now we solve for x by using cross multiplication.

$$(1.64)(30,000) = 750x$$

$$49,200 = 750x$$

$$\frac{49,200}{750} = \frac{750x}{750}$$

$$\$65.60 = x$$

The cost of 30,000 gallons of water is $65.60.

EXAMPLE ⑮ *Determining the Amount of Insulin*

Insulin comes in 10 cubic centimeter (cc) vials labeled in the number of units of insulin per cubic centimeter of fluid. A vial of insulin marked U40 has 40 units of insulin per cubic centimeter of fluid. If a patient needs 30 units of insulin, how much fluid should be drawn into the syringe from the U40 vial?

SOLUTION The unknown quantity, x, is the number of cubic centimeters of fluid to be drawn into the syringe. Following is one proportion that can be used to find that quantity.

$$\text{Given ratio} \left\{ \frac{40 \text{ units}}{1 \text{ cc}} = \frac{30 \text{ units}}{x \text{ cc}} \right.$$

$$40x = 30(1)$$
$$40x = 30$$
$$x = \frac{30}{40} = 0.75$$

The nurse or doctor putting the insulin in the syringe should draw 0.75 cc of the fluid.

SECTION 6.2 EXERCISES

CONCEPT/WRITING EXERCISES

1. Define and give an example of a *term*.

2. Define and give an example of *like terms*.

3. Define and give an example of a *numerical coefficient*.

4. Define and give an example of a *linear equation*.

5. Explain how to simplify an expression. Give an example.

6. State the addition property of equality. Give an example.

7. State the multiplication property of equality. Give an example.

8. State the subtraction property of equality. Give an example.

9. State the division property of equality. Give an example.

10. Define *algorithm*.

11. Define and give an example of a *ratio*.

12. Define and give an example of a *proportion*.

13. Are $3x$ and $\frac{1}{2}x$ like terms? Explain.

14. Are $4x$ and $4y$ like terms? Explain.

PRACTICE THE SKILLS

In Exercises 15–38, combine like terms.

15. $4x + 6x$

16. $-7x - 5x$

17. $5x - 3x + 12$

18. $9x - 6x + 21$

19. $7x + 3y - 4x + 8y$

20. $x - 4x + 3$

21. $-3x + 2 - 5x$

22. $-3x + 4x - 2 + 5$

23. $2 - 3x - 2x + 1$

24. $-0.2x + 1.7x - 4$

25. $6.2x - 8.3 + 7.1x$

26. $\frac{2}{3}x + \frac{1}{6}x - 5$

27. $\frac{1}{5}x - \frac{1}{3}x \quad 4$

28. $7s + 4t + 9 - 2t - 2s - 15$

29. $6x - 4y - 5y + 4x + 3$

30. $2p - 4q - 3p + 4q - 15$

31. $2(s + 3) + 6(s - 4) + 1$

32. $6(r - 3) - 2(r + 5) + 10$

33. $0.2(x + 4) + 1.2(x - 3)$

34. $\frac{3}{5}(x + 1) - \frac{3}{10}x$

35. $\frac{1}{4}x + \frac{4}{5} - \frac{2}{3}x$

36. $n - \frac{3}{4} + \frac{5}{9}n - \frac{1}{6}$

37. $0.5(2.6x - 4) + 2.3(1.4x - 5)$

38. $\frac{2}{3}(3x + 9) - \frac{1}{4}(2x + 5)$

In Exercises 39–64, solve the equation.

39. $y - 7 = 10$ **40.** $3y + 6 = 24$

41. $15 = 9 - 6x$ **42.** $18 = 8 + 2x$

43. $\dfrac{3}{x} = \dfrac{7}{8}$ **44.** $\dfrac{x - 1}{5} = \dfrac{x + 5}{15}$

45. $\frac{1}{2}x + \frac{1}{3} = \frac{2}{3}$ **46.** $\frac{1}{2}y + \frac{1}{3} = \frac{1}{4}$

47. $0.9x - 1.2 = 2.4$ **48.** $5x + 0.050 = -0.732$

49. $6t - 8 = 4t - 2$ **50.** $\dfrac{x}{4} + 2x = \dfrac{1}{3}$

51. $\dfrac{x - 3}{2} = \dfrac{x + 4}{3}$ **52.** $\dfrac{x - 5}{4} = \dfrac{x - 9}{3}$

53. $6t - 7 = 8t + 9$

54. $12x - 1.2 = 3x + 1.5$

55. $2(x + 3) - 4 = 2(x - 4)$

56. $3(x + 2) + 2(x - 1) = 5x - 7$

57. $4(x - 4) + 12 = 4(x - 1)$

58. $6(t + 2) - 14 = 6t - 2$

59. $\frac{1}{3}(x + 3) = \frac{2}{5}(x + 2)$

60. $\frac{2}{3}(x - 4) = \frac{1}{4}(x + 1)$

61. $3x + 2 - 6x = -x - 15 + 8 - 5x$

62. $6x + 8 - 22x = 28 + 14x - 10 + 12x$

63. $4(t - 3) + 8 = 4(2t - 6)$

64. $8.3y - 3.1(y - 4) = 29.04$

PROBLEM SOLVING

In Exercises 65 and 66, use the water rate for Newton, New Jersey, which is $7.75 per 1000 gallons of water.

65. *Water Bill* What is the water bill if a resident of Newton uses 27,000 gal?

66. *Limiting the Cost* How many gallons of water can a customer use if the water bill is not to exceed $200?

67. *Concrete Sealer* A gallon of concrete sealer covers 360 ft^2. How much sealer is needed to cover a basement with a surface area of 1440 ft^2?

68. *Fajitas* A recipe for six servings of beef fajitas requires 16 oz of beef sirloin.

   a) If the recipe were to be made for nine servings, how many ounces of beef sirloin would be needed?

   b) How many servings of beef fajitas can be made with 32 oz of beef sirloin?

69. *Watching Television* Nielson Media Research determines the number of people who watch a television show. One rating point means that about 1,102,000 households watched the show. The top-rated television show for the week of October 16, 2006, was *Grey's Anatomy* with a rating of 14.4. About how many households watched *Grey's Anatomy* that week?

70. *Topsoil* A 40-pound bag of topsoil will cover a surface area of 12 ft^2.

   a) How many pounds of topsoil are needed to cover a surface area of 480 ft^2?

   b) How many bags of topsoil must be purchased to cover a surface area of 480 ft^2?

71. *Speed Limit* When Jacob Abbott crossed over from Niagara Falls, New York, to Niagara Falls, Canada, he saw a sign that said 50 miles per hour (mph) is equal to 80 kilometers per hour (kph).

▲ Niagara Falls, Canada

a) How many kilometers per hour are equal to 1 mph?

b) On a stretch of the Queen Elizabeth Way, the speed limit is 90 kph. What is the speed limit in miles per hour?

72. *The Proper Dosage* A doctor asks a nurse to give a patient 250 milligrams (mg) of the drug Simethicone. The drug is available only in a solution whose concentration is 40 mg Simethicone per 0.6 millimeter (mm) of solution. How many millimeters of solution should the nurse give the patient?

Amount of Insulin In Exercises 73 and 74, how much insulin (in cc) would be given for the following doses? (Refer to Example 15 on page 327.)

73. 15 units of insulin from a vial marked U40

74. 35 units of insulin from a vial marked U40

75. a) In your own words, summarize the procedure to use to solve an equation.

b) Solve the equation $2(x + 3) = 4x + 3 - 5x$ with the procedure you outlined in part (a).

76. a) What is an identity?

b) When solving an equation, how will you know if the equation is an identity?

77. a) What is a contradiction?

b) When solving an equation, how will you know if the equation is a contradiction?

CHALLENGE PROBLEMS/GROUP ACTIVITIES

78. *Depth of a Submarine* The pressure, P, in pounds per square inch (psi) exerted on an object x ft below the sea is given by the formula $P = 14.70 + 0.43x$. The 14.70 represents the weight in pounds of the column of air (from sea level to the top of the atmosphere) standing over a 1 in. by 1 in. square of seawater. The $0.43x$ represents the weight in pounds of a column of water 1 in. by 1 in. by x ft (see Fig. 6.2).

This column of air weighs 14.7 lb

This column of water weighs 0.43x lb

x ft

1 in. by 1 in. square

Figure 6.2

a) A submarine is built to withstand a pressure of 148 psi. How deep can that submarine go?

b) If the pressure gauge in the submarine registers a pressure of 128.65 psi, how deep is the submarine?

79. a) *Gender Ratios* If the ratio of males to females in a class is 2 : 3, what is the ratio of males to all the students in the class? Explain your answer.

b) If the ratio of males to females in a class is $m : n$, what is the ratio of males to all the students in the class?

80. Determine whether the given equations are equivalent.

a) $\dfrac{x}{3} = \dfrac{x + 2}{6}, \quad 6x = 3x + 6$

b) $2x - 9 = 19, \quad \dfrac{2x - 9}{4} = \dfrac{19}{4}$

INTERNET/RESEARCH ACTIVITIES

81. Ratio and proportion are used in many different ways in everyday life. Submit two articles from newspapers, magazines, or the Internet in which ratios and/or proportions are used. Write a brief summary of each article explaining how ratio and/or proportion were used.

82. Write a report explaining how the ancient Egyptians used equations. Include in your discussion the forms of the equations used.

6.3   FORMULAS

▲ A formula is used to determine the amount of carpet needed in a room.

Suppose you need to purchase new carpet for your family room, which is in the shape of a rectangle. To determine the amount of carpet you need, you would need to determine the area of the room by multiplying the length of the room by the width of the room. In this section, we will discuss many applications that can be solved by using a special kind of equation called a *formula*.

A *formula* is an equation that typically has a real-life application. To *evaluate a formula*, substitute the given values for their respective variables and then evaluate using the order of operations given in Section 6.1. Many of the formulas given in this section are discussed in greater detail in other parts of the book.

EXAMPLE ❶ *Simple Interest*

The simple interest formula,* interest = principal × rate × time, or $i = prt$, is used to find the interest you must pay on a simple interest loan when you borrow principal, p, at simple interest rate, r, in decimal form, for time, t. Chris Campbell borrows $4000 at a simple interest rate of 7% for 3 years.

a) How much will Chris Campbell pay in interest at the end of 3 years?
b) What is the total amount he will repay the bank at the end of 3 years?

SOLUTION
a) Substitute the values of p, r, and t into the formula; then evaluate.

$$i = prt$$
$$= 4000(0.07)(3)$$
$$= 840$$

Thus, Chris must pay $840 interest.

b) The total he must pay at the end of 3 years is the principal, $4000, plus the $840 interest, for a total of $4840.   ●

EXAMPLE ❷ *Volume of an Ice-Cream Box*

The formula for the volume of a rectangular box[†] is volume = length × width × height, or $V = lwh$. Use the formula $V = lwh$ to find the width of a rectangular box of ice cream if $l = 7$ in., $h = 3.5$ in., and $V = 122.5$ in.[3].

SOLUTION We substitute the appropriate values into the volume formula and solve for the desired quantity, w.

$$V = lwh$$
$$122.5 = (7)w(3.5)$$
$$122.5 = 24.5w$$

---
*The simple interest formula is discussed in Section 11.2.
[†]The volume formula is discussed in Section 9.4.

$$\frac{122.5}{24.5} = \frac{24.5w}{24.5}$$

$$5 = w$$

Therefore, the width of the ice-cream box is 5 in.   ●

In Example 1, we used the formula $i = prt$. In Example 2, we used the formula $V = lwh$. In these examples, we used a mathematical equation to represent real phenomena. When we represent real phenomena, such as finding simple interest, mathematically we say we have created a *mathematical model* or simply a *model* to represent the situation. A model may be a single formula, or equation, or a system of many equations. By using models we gain insight into real-life situations, such as how much interest you will accumulate in your savings account. We will use mathematical models throughout this chapter and elsewhere in the text. In some exercises in this and the next chapter, when you are asked to determine an equation to represent a real-life situation, we will sometimes write the word *modeling* in the instructions.

Many formulas contain Greek letters, such as μ (mu), σ (sigma), Σ (capital sigma), δ (delta), ϵ (epsilon), π (pi), θ (theta), and λ (lambda). Example 3 makes use of Greek letters.

EXAMPLE ❸ *A Statistics Formula*

A formula used in the study of statistics to find the standard score (or z-score) is

$$z = \frac{\bar{x} - \mu}{\frac{\sigma}{\sqrt{n}}}$$

Find the value of z when $\bar{x}$ (read "x bar") $= 120$, $\mu = 100$, $\sigma = 16$, and $n = 4$.

SOLUTION

$$z = \frac{\bar{x} - \mu}{\frac{\sigma}{\sqrt{n}}} = \frac{120 - 100}{\frac{16}{\sqrt{4}}} = \frac{20}{\frac{16}{2}} = \frac{20}{8} = 2.5$$ ●

Some formulas contain *subscripts*. Subscripts are numbers (or letters) placed below and to the right of variables. They are used to help clarify a formula. For example, if two different amounts are used in a problem, they may be symbolized as A and A_0, or A_1 and A_2. Subscripts are read using the word *sub*; for example, A_0 is read "A sub zero" and A_1 is read "A sub one."

Exponential Equations

Many real-life problems, including population growth, growth of bacteria, and decay of radioactive substances, increase or decrease at a very rapid rate. For example, in Fig. 6.3 on page 332, which shows the amount of money borrowed by college students through private loans, in billions of dollars, from 1999–2000 to 2004–2005, the graph is increasing rapidly. This is an example in which the

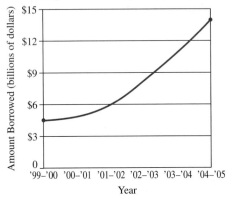

Amount of Money Borrowed by College Students Through Private Loans, in Billions of Dollars

Source: College Board

Figure 6.3

graph is increasing *exponentially*. The equation of a graph that increases or decreases exponentially is called an *exponential equation*. An exponential equation is of the form $y = a^x, a > 0, a \neq 1$. We often use exponential equations to model real-life problems. In Section 6.10, we will discuss exponential equations in more detail.

In an exponential equation, letters other than x and y may be used to represent the variables. The following are examples of exponential equations: $y = 2^x$, $A = (\frac{1}{2})^x$, and $P = 2.3^t$. Note in exponential equations that the variable is the exponent of some positive constant that is not equal to 1. In many real-life applications, the variable t will be used to represent time. Problems involving exponential equations can be evaluated much more easily if you use a calculator containing a $\boxed{y^x}$, $\boxed{x^y}$, or $\boxed{\wedge}$ key.

The following equation, referred to as the *exponential growth* or *decay formula*, is used to solve many real-life problems.

$$P = P_0 a^{kt}, \quad a > 0, \quad a \neq 1$$

In the formula, P_0 represents the original amount present, P represents the amount present after t years, and a and k are constants.

When $k > 0$, P increases as t increases and we have exponential growth. When $k < 0$, P decreases as t increases and we have exponential decay.

EXAMPLE ❹ *Using an Exponential Decay Formula*

Carbon dating is used by scientists to find the age of fossils, bones, and other items. The formula used in carbon dating is

$$P = P_0 2^{-t/5600}$$

where P_0 represents the original amount of carbon 14 (C_{14}) present and P represents the amount of C_{14} present after t years. If 10 mg of C_{14} is present in an animal bone recently excavated, how many milligrams will be present in 3000 years?

SOLUTION Substituting the values in the formula gives

$$P = P_0 2^{-t/5600}$$
$$P = 10(2)^{-3000/5600}$$
$$P \approx 10(2)^{-0.54} \qquad \text{(Recall that } \approx \text{ means "is approximately equal to.")}$$
$$P \approx 10(0.69)$$
$$P \approx 6.9 \text{ mg}$$

Thus, in 3000 years, approximately 6.9 mg of the original 10 mg of C_{14} will remain. ●

In Example 4, we used a calculator to evaluate $(2)^{-3000/5600}$. The steps used to find this quantity on a calculator with a $\boxed{y^x}$ key are

$$2 \boxed{y^x} \boxed{(} 3000 \boxed{+/-} \boxed{\div} 5600 \boxed{)} \boxed{=} .6898170602$$

After the $\boxed{=}$ key is pressed, the calculator displays the answer 0.6898170602. To evaluate $10(2)^{-3000/5600}$ on a scientific calculator, we can press the following keys.

$$10 \boxed{\times} 2 \boxed{y^x} \boxed{(} 3000 \boxed{+/-} \boxed{\div} 5600 \boxed{)} \boxed{=} 6.898170602$$

Notice that the answer obtained using the calculator steps shown above is a little more accurate than the answer we gave when we rounded the values before the final answer in Example 4.

When the a in the formula $P = P_0 a^{kt}$ is replaced with the very special letter e, we get the *natural exponential growth or decay formula*

$$P = P_0 e^{kt}$$

The letter e represents an irrational number whose value is approximately 2.7183. The number e plays an important role in mathematics and is used in finding the solution to many application problems.

To evaluate $e^{(0.04)5}$ on a calculator, as will be needed in Example 5, press*

$$\boxed{(} .04 \boxed{\times} 5 \boxed{)} \boxed{\text{inv}} \boxed{\ln} \boxed{=} 1.221402758$$

⬈ Inverse key   ⬉ Natural logarithm key

After the $\boxed{\ln}$ key is pressed, the calculator displays the answer 1.221402758.

To evaluate $10{,}000e^{(0.04)5}$ on a calculator, press

$$10000 \boxed{\times} \boxed{(} .04 \boxed{\times} 5 \boxed{)} \boxed{\text{inv}} \boxed{\ln} \boxed{-} 12214.02758$$

In this calculation, after the $\boxed{=}$ key is pressed, the calculator displays the answer 12214.02758.

---

*Keys to press may vary on some calculators.

EXAMPLE ❺ *Using an Exponential Growth Formula*

Banks often credit compound interest continuously. When that is done, the principal amount in the account, P, at any time t can be calculated by the natural exponential formula $P = P_0 e^{kt}$, where P_0 is the initial principal invested, k is the interest rate in decimal form, and t is the time.

Suppose \$10,000 is invested in a savings account at a 4% interest rate compounded continuously. What will be the balance (or principal) in the account in 5 years?

SOLUTION

$$P = P_0 e^{kt}$$
$$= 10{,}000 e^{(0.04)5}$$
$$= 10{,}000 e^{(0.20)}$$
$$\approx 10{,}000(1.221402758)$$
$$\approx 12{,}214$$

Thus, after 5 years, the account's value will have grown from \$10,000 to about \$12,214, an increase of about \$2214.   ●

Graphing calculators are a tool that can be used to graph equations. Figure 6.4 shows the graph of $P = 10{,}000 e^{0.04t}$ as it appears on the screen (or window) of a Texas Instrument TI-84 Plus graphing calculator. To obtain this screen, the domain (or the x-values) and range (or the y-values) of the window need to be set to selected values. We will speak a little more about graphing calculators shortly. In Section 6.10, we will explain how to graph exponential equations by plotting points.

$P = 10{,}000 e^{0.04t}$

Figure 6.4

EXAMPLE ❻ *Population of Arizona*

The population of Arizona is growing exponentially at the rate of about 3.5% per year. In 2005, the population of Arizona was about 5,939,292 people. Arizona's expected population, t years after 2005, is given by the formula $P = 5{,}939{,}292 e^{0.035t}$. Assuming that the population of Arizona continues to grow at the same rate, determine the expected population of Arizona in 2012.

SOLUTION Since 2012 is 7 years after 2005, $t = 7$ years.

$$P = 5{,}939{,}292 e^{0.035t}$$
$$= 5{,}939{,}292 e^{0.035(7)} \qquad \text{Substitute 7 for } t.$$
$$= 5{,}939{,}292 e^{0.245}$$
$$\approx 5{,}939{,}292(1.277621313)$$
$$\approx 7{,}588{,}166.045$$

Thus, in 2012, the population of Arizona is expected to be about 7,588,166 people.   ●

TIMELY TIP When doing calculations on the calculator, do not round any value before obtaining the final answer. By not rounding, you will obtain a more accurate answer. For example, if we work Example 6 on a calculator and rounded $e^{0.245}$ to 1.28, we would determine that the population of Arizona in 2012 is expected to be about 7,602,294, which is a less accurate answer.

Solving for a Variable in a Formula or Equation

Often in mathematics and science courses, you are given a formula or an equation expressed in terms of one variable and asked to express it in terms of a different variable. For example, you may be given the formula $P = i^2r$ and asked to solve the formula for r. To do so, treat each of the variables, except the one you are solving for, as if it were a constant. Then solve for the variable desired, using the properties previously discussed. Examples 7 through 9 show how to do this task.

When graphing equations in Section 6.7, you will sometimes have to solve the equation for the variable y as is done in Example 7.

EXAMPLE 7 Solving for a Variable in an Equation

Solve the equation $2x + 5y - 10 = 0$ for y.

SOLUTION We need to isolate the term containing the variable y. Begin by moving the constant, -10, and the term $2x$ to the right-hand side of the equation.

$$2x + 5y - 10 = 0$$

$$2x + 5y - 10 + 10 = 0 + 10 \qquad \text{Addition property}$$

$$2x + 5y = 10$$

$$-2x + 2x + 5y = -2x + 10 \qquad \text{Subtraction property}$$

$$5y = -2x + 10$$

$$\frac{5y}{5} = \frac{-2x + 10}{5} \qquad \text{Division property}$$

$$y = \frac{-2x + 10}{5}$$

$$y = -\frac{2x}{5} + \frac{10}{5}$$

$$y = -\frac{2}{5}x + 2$$

Note that once you have found $y = \frac{-2x + 10}{5}$, you have solved the equation for y. The solution can also be expressed in the form $y = -\frac{2}{5}x + 2$. This form of the equation is convenient for graphing equations, as will be explained in Section 6.7. Example 7 can also be solved by moving the y term to the right-hand side of the equal sign. Do so now and note that you obtain the same answer.

EXAMPLE 8 Solving for a Variable in a Formula

An important formula used in statistics is

$$z = \frac{x - \mu}{\sigma}$$

Solve this formula for x.

SOLUTION To isolate the term x, use the general procedure for solving linear equations given in Section 6.2. Treat each letter, except x, as if it were a constant.

$$z = \frac{x - \mu}{\sigma}$$

$$z \cdot \sigma = \frac{x - \mu}{\sigma} \cdot \sigma \qquad \text{Multiply both sides of the equation by } \sigma$$

$$z\sigma = x - \mu$$

$$z\sigma + \mu = x - \mu + \mu \qquad \text{Add } \mu \text{ to both sides of the equation.}$$

$$z\sigma + \mu = x$$

$$\text{or} \quad x = z\sigma + \mu$$

EXAMPLE ❾ *The Tax-Free Yield Formula*

A formula that may be important to you now or sometime in the future is the tax-free yield formula, $T_f = T_a(1 - F)$. This formula can be used to convert a taxable yield, T_a, into its equivalent tax-free yield, T_f, where F is the federal income tax bracket of the individual. A taxable yield is an interest rate for which income tax is paid on the interest made. A tax-free yield is an interest rate for which income tax does not have to be paid on the interest made.

a) For someone in a 25% tax bracket, find the equivalent tax-free yield of a 4% taxable investment.

b) Solve this formula for T_a. That is, write a formula for taxable yield in terms of tax-free yield.

SOLUTION

a) $T_f = T_a(1 - F)$

$\qquad = 0.04(1 - 0.25) = 0.04(0.75) = 0.03, \quad \text{or} \quad 3\%$

Thus, a taxable investment of 4% is equivalent to a tax-free investment of 3% for a person in a 25% income tax bracket.

b) $\qquad T_f = T_a(1 - F)$

$$\frac{T_f}{1 - F} = \frac{T_a(1 - F)}{1 - F} \qquad \text{Divide both sides of the equation by } 1 - F.$$

$$\frac{T_f}{1 - F} = T_a, \quad \text{or} \quad T_a = \frac{T_f}{1 - F}$$

SECTION 6.3 EXERCISES

CONCEPT/WRITING EXERCISES

1. What is a formula?

2. Explain how to evaluate a formula.

3. What are subscripts?

4. What is the simple interest formula?

5. What is an exponential equation?

6. a) In an exponential equation of the form $y = a^x$, what are the restrictions on a?

   b) In an exponential equation of the form $y = P_0 a^{kt}$, what does P_0 represent?

PRACTICE THE SKILLS

In Exercises 7–40, use the formula to find the value of the indicated variable for the values given. Use a calculator when one is needed. When necessary, round answers to the nearest hundredth.

7. $A = lw$; determine A when $l = 4$ and $w = 14$ (geometry).

8. $P = a + b + c$; determine P when $a = 25$, $b = 53$, and $c = 32$ (geometry).

9. $P = 2l + 2w$; determine P when $l = 12$ and $w = 16$ (geometry).

10. $F = ma$; determine m when $F = 40$ and $a = 5$ (physics).

11. $K = \frac{1}{2} mv^2$; determine m when $v = 30$ and $K = 4500$ (physics).

12. $p = i^2 r$; determine r when $p = 62{,}500$ and $i = 5$ (electronics).

13. $S = \pi r(r + h)$; determine S when $r = 8$, $\pi = 3.14$, and $h = 2$ (geometry).

14. $B = \frac{703w}{h^2}$; determine B when $w = 130$ and $h = 67$ (body mass index).

15. $z = \frac{x - \mu}{\sigma}$; determine μ when $z = 2.5$, $x = 42.1$, and $\sigma = 2$ (statistics).

16. $S = 2hw + 2lw + 2lh$; determine l when $S = 122$, $h = 4$, and $w = 3$ (geometry).

17. $T = \frac{PV}{k}$; determine P when $T = 80$, $V = 20$, and $k = 0.5$ (physics).

18. $m = \frac{a + b + c}{3}$; determine a when $m = 70$, $b = 60$, and $c = 90$ (statistics).

19. $A = P(1 + rt)$; determine P when $A = 3600$, $r = 0.04$, and $t = 5$ (economics).

20. $m = \frac{a + b}{2}$; determine a when $m = 70$ and $b = 77$ (statistics).

21. $V = \pi r^2 h$; determine h when $V = 942$, $\pi = 3.14$, and $r = 5$ (geometry).

22. $F = \frac{9}{5}C + 32$; determine F when C = 7 (temperature conversion).

23. $C = \frac{5}{9}(F - 32)$; determine C when F = 77 (temperature conversion).

24. $K = \frac{F - 32}{1.8} + 273.1$; determine K when F = 100 (chemistry).

25. $m = \frac{y_2 - y_1}{x_2 - x_1}$; determine m when $y_2 = 8$, $y_1 = -4$, $x_2 = -3$, and $x_1 = -5$ (mathematics).

26. $z = \frac{\bar{x} - \mu}{\frac{\sigma}{\sqrt{n}}}$; determine z when $\bar{x} = 66$, $\mu = 60$, $\sigma = 15$, and $n = 25$ (statistics).

27. $S = R - rR$; determine R when $S = 186$ and $r = 0.07$ (for determining sale price when an item is discounted).

28. $S = C + rC$; determine C when $S = 115$ and $r = 0.15$ (for determining selling price when an item is marked up).

29. $E = a_1 p_1 + a_2 p_2 + a_3 p_3$; determine E when $a_1 = 5$, $p_1 = 0.2$, $a_2 = 7$, $p_2 = 0.6$, $a_3 = 10$, and $p_3 = 0.2$ (probability).

30. $x = \frac{-b + \sqrt{b^2 - 4ac}}{2a}$; determine x when $a = 2$, $b = -5$, and $c = -12$ (mathematics).

31. $s = -16t^2 + v_0 t + s_0$;   determine s when $t = 4$, $v_0 = 30$, and $s_0 = 150$ (physics).

32. $R = O + (V - D)r$;   determine O when $R = 670$, $V = 100$, $D = 10$, and $r = 4$ (economics).

33. $P = \dfrac{f}{1 + i}$;   determine f when $i = 0.08$ and $P = 3000$ (investment banking).

34. $Q_w = m_w c_w (T_f - T_w)$;   determine Q_w when $m_w = 0.5$, $c_w = 4186$, $T_f = 22$, and $T_w = 19$ (physics).

35. $R_T = \dfrac{R_1 R_2}{R_1 + R_2}$;   determine R_T when $R_1 = 100$ and $R_2 = 200$ (electronics).

36. $P = \dfrac{nRT}{V}$;   determine V when $P = 12$, $n = 10$, $R = 60$, and $T = 8$ (chemistry).

37. $a_n = a_1 + (n - 1)d$;   determine a_n when $a_1 = 15$, $n = 4$, and $d = 8$ (mathematics).

38. $A = P\left(1 + \dfrac{r}{n}\right)^{nt}$;   determine A when $P = 100$, $r = 6\%$, $n = 1$, and $t = 3$ (banking).

39. $P = P_0 e^{kt}$;   determine P when $P_0 = 5000$, $k = 0.06$, and $t = 7$ (economics).

40. $S = S_0 e^{-0.028t}$;   determine S when $S_0 = 1000$ and $t = 35$ (chemistry).

In Exercises 41–50, solve the equation for y.

41. $4x - 9y = 14$

42. $11x - 15y = 23$

43. $8x + 7y = 21$

44. $-9x + 4y = 11$

45. $2x - 3y + 6 = 0$

46. $3x + 4y = 0$

47. $-2x + 3y + z = 15$

48. $5x + 3y - 2z = 22$

49. $9x + 4z - 7 + 8y$

50. $2x - 3y + 5z = 0$

In Exercises 51–70, solve for the variable indicated.

51. $d = rt$ for r

52. $V = lwh$ for w

53. $p = a + b + c$ for a

54. $p = a + b + s_1 + s_2$ for s_1

55. $V = \frac{1}{3} Bh$ for B

56. $V = \pi r^2 h$ for h

57. $C = 2\pi r$ for r

58. $y = mx + b$ for m

59. $y = mx + b$ for b

60. $V = \dfrac{1}{3} \pi r^2 h$ for h

61. $P = 2l + 2w$ for w

62. $A = \dfrac{d_1 d_2}{2}$ for d_2

63. $A = \dfrac{a + b + c}{3}$ for c

64. $\dfrac{w_1}{w_2} = \dfrac{f_2 - f}{f - f_1}$ for w_1

65. $P = \dfrac{KT}{V}$ for T

66. $\dfrac{P_1 V_1}{T_1} = \dfrac{P_2 V_2}{T_2}$ for V_2

67. $F = \frac{9}{5} C + 32$ for C

68. $C = \frac{5}{9}(F - 32)$ for F

69. $S = 2\pi rh + 2\pi r^2$ for h

70. $A = \dfrac{1}{2} h(b_1 + b_2)$ for b_2

PROBLEM SOLVING

In Exercises 71–78, when appropriate, round answers to the nearest hundredth.

71. **Savings Account** Christine Northrup borrowed $4500 from a bank at a simple interest rate of 2.5% for one year.

   a) Determine how much interest Christine paid at the end of 1 year.

   b) Determine the total amount Christine will repay the bank at the end of 1 year.

72. **Interest on a Loan** Jeff Hubbard borrowed $800 from his brother for 2 years. At the end of 2 years, he repaid the $800 plus $128 in interest. What simple interest rate did he pay?

73. **Volume of an Ice-Cream Cone** An ice-cream cone is filled with ice cream to the top of the cone. Determine the volume, in cubic inches, of ice cream in the cone if the cone's radius is 1 in. and the height is 4 in. The formula for the volume of a cone is $V = \dfrac{1}{3} \pi r^2 h$. Use the pi key, $\boxed{\pi}$, on your calculator, or 3.14 for π if your calculator does not have a pi key.

74. *Body Mass Index* A person's body mass index (BMI) is found by the formula $B = \dfrac{703w}{h^2}$, where w is the person's weight, in pounds, and h is the person's height, in inches. Lance Bass is 6 ft tall and weighs 200 lb.

   a) Determine his BMI.

   b) If Lance would like to have a BMI of 26, how much weight would he need to gain or lose?

75. *Bacteria* The number of a certain type of bacteria, y, present in a culture is determined by the equation $y = 2000(3)^x$, where x is the number of days the culture has been growing. Find the number of bacteria present after 5 days.

76. *Adjusting for Inflation* If P is the price of an item today, the price of the same item n years from today, P_n, is $P_n = P(1 + r)^n$, where r is the constant rate of inflation. Determine the price of a movie ticket 10 years from today if the price today is \$8.00 and the annual rate of inflation is constant at 3%.

77. *Value of New York City* Assume the value of the island of Manhattan has grown at an exponential rate of 8% per year since 1626 when Peter Minuit of the Dutch West India Company purchased the island for \$24. The value of the island, V, at any time, t, in years after 1626, can be found by the formula $V = 24e^{0.08t}$. What is the value of the island in 2008, 382 years after Minuit purchased it? Use a scientific calculator and give your answer in scientific notation as provided by your calculator.

78. *Radioactive Decay* Radium-226 is a radioactive isotope that decays exponentially at a rate of 0.0428% per year. The amount of radium-226, R, remaining after t years can be found by the formula $R = R_0 e^{-0.000428t}$, where R_0 is the original amount present. If there are originally 10 grams of radium-226, determine the amount of radium-226 remaining after 1000 years.

CHALLENGE PROBLEM/GROUP ACTIVITY

79. Determine the volume of the block shown in Fig. 6.5, excluding the hole. The formula for the volume of a rectangular solid is $V = lwh$. The formula for the volume of a cylinder is $V = \pi r^2 h$.

12"
8"
4"
12"

Figure 6.5

6.4 APPLICATIONS OF LINEAR EQUATIONS IN ONE VARIABLE

▲ A mathematical equation can be used to determine a sales representative's weekly salary.

Suppose your weekly salary as a sales representative consists of a weekly base salary of \$400 and a 6% commission on the total dollar amount of all sales you make. How many dollars in sales would you need to make in a week to earn a total of \$1000 a week? To answer this question you must first express the written problem as a mathematical equation. Then you can use the general procedure for solving linear equations discussed in Section 6.2 to obtain the answer. In this section, we will explain how to write mathematical equations to represent word problems.

One reason to study algebra is that it can be used to solve everyday problems. In this section, we will do two things: (1) show how to translate a written problem into a mathematical equation and (2) show how linear equations can be used in solving everyday problems. We begin by illustrating how English phrases can be written as mathematical expressions. When writing a mathematical expression, we may use any letter to represent the variable. In the following illustrations, we use the letter x.

Phrase	Mathematical expression
Six more than a number	$x + 6$
A number increased by 3	$x + 3$
Four less than a number	$x - 4$
A number decreased by 9	$x - 9$
Twice a number	$2x$
Four times a number	$4x$
3 decreased by a number	$3 - x$
The difference between a number and 5	$x - 5$

Sometimes the phrase that must be converted to a mathematical expression involves more than one operation.

Phrase	Mathematical expression
Four less than 3 times a number	$3x - 4$
Ten more than twice a number	$2x + 10$
The sum of 5 times a number and 3	$5x + 3$
Eight times a number, decreased by 7	$8x - 7$

The word *is* often represents the equal sign.

Phrase	Mathematical equation
Six more than a number is 10.	$x + 6 = 10$
Five less than a number is 20.	$x - 5 = 20$
Twice a number, decreased by 6 is 12.	$2x - 6 = 12$
A number decreased by 13 is 6 times the number.	$x - 13 = 6x$

The following is a general procedure for solving word problems.

TO SOLVE A WORD PROBLEM

1. Read the problem carefully at least twice to be sure that you understand it.
2. If possible, draw a sketch to help visualize the problem.
3. Determine which quantity you are being asked to find. Choose a letter to represent this unknown quantity. Write down exactly what this letter represents.
4. Write the word problem as an equation.
5. Solve the equation for the unknown quantity.
6. Answer the question or questions asked.
7. Check the solution.

This general procedure for solving word problems is illustrated in Examples 1 through 4. In these examples, the equations we obtain are mathematical models of the given situations.

EXAMPLE ❶ *How Much Can You Purchase?*

Roberto Raynor spent $339.95 on textbooks at the bookstore. In addition to textbooks, he wanted to purchase as many notebooks as possible, but he only has a total of $350 to spend. If a notebook, including tax, costs $0.99, how many notebooks can he purchase?

SOLUTION In this problem, the unknown quantity is the number of notebooks Roberto can purchase. Let's select n to represent the number of notebooks he can purchase. Then we construct an equation using the given information that will allow us to solve for n.

Let

$$n = \text{number of notebooks Roberto can purchase}$$

Then

$$\$0.99n = \text{cost for } n \text{ notebooks at } \$0.99 \text{ per notebook}$$

Cost of textbooks + cost of notebooks = total amount to spend
$$\$339.95 \quad + \quad \$0.99n \quad = \quad \$350$$

Now solve the equation.

$$339.95 + 0.99n = 350$$
$$339.95 - 339.95 + 0.99n = 350 - 339.95$$
$$0.99n = 10.05$$
$$\frac{0.99n}{0.99} = \frac{10.05}{0.99}$$
$$n \approx 10.15$$

Therefore, Roberto can purchase 10 notebooks. When we solve the equation, we obtain $n = 10.15$ (to the nearest hundredth). Since he cannot purchase a part of a notebook, only 10 notebooks can be purchased.

CHECK: The check is made with the information given in the original problem.

Total amount to spend = cost of textbooks + cost of notebooks
$$= 339.95 + 0.99(10)$$
$$= 339.95 + 9.90$$
$$= 349.85$$

This result would leave 15 cents change from the $350 he has to spend, which is not enough to purchase another notebook. Therefore, this answer checks. ●

EXAMPLE ❷ *Dividing Land*

Mark IV Construction Company purchased 100 acres of land to be split into three parcels of land on which to build houses. One parcel of land will be three times as large as each of the other two. How many acres of land will each parcel contain?

SOLUTION Two parcels will contain the same number of acres, and the third parcel will contain three times that amount.

Let

$$x = \text{number of acres in the first parcel}$$
$$x = \text{number of acres in the second parcel}$$
$$3x = \text{number of acres in the third parcel}$$

Then

$$x + x + 3x = \text{total amount of land}$$
$$x + x + 3x = 100$$
$$5x = 100$$
$$x = 20$$

Thus, two parcels will contain 20 acres of land. The third parcel will contain 3(20), or 60, acres of land. A check in the original problem will verify that this answer is correct.   ●

EXAMPLE ❸ *Dimensions of an Exercise Area*

Dr. Christine Seidel, a veterinarian, wants to fence in a large rectangular region in the yard behind her office for exercising dogs that are boarded overnight. She has 130 ft of fencing to use for the perimeter of the region. What should the dimensions of the region be if she wants the length to be 15 ft greater than the width?

SOLUTION The formula for finding the perimeter of a rectangle is $P = 2l + 2w$, where P is the perimeter, l is the length, and w is the width. A diagram, such as the one shown below, is often helpful in solving problems of this type.

Let w equal the width of the region. The length is 15 ft more than the width, so $l = w + 15$. The total distance around the region P, is 130 ft.
Substitute the known quantities in the formula.

$$P = 2l + 2w$$
$$130 = 2(w + 15) + 2w$$
$$130 = 2w + 30 + 2w$$
$$130 = 4w + 30$$
$$100 = 4w$$
$$25 = w$$

The width of the region is 25 ft and the length of the region is $25 + 15 = 40$ ft.   ●

In shopping and other daily activities, we are occasionally asked to solve problems using percents. The word *percent* means "per hundred." Thus, for example, 7%

means 7 per hundred, or $\frac{7}{100}$. When $\frac{7}{100}$ is converted to a decimal number, we obtain 0.07. Thus, 7% = 0.07.

Let's look at one example involving percent. (See Section 11.1 for a more detailed discussion of percent.)

EXAMPLE ④ *Art Show*

Kim Stappenbeck is selling her original paintings at an art show. Determine the cost of a painting before tax if the total cost of a painting, including an 8% sales tax, is to be $145.80.

SOLUTION We are asked to find the cost of a painting before sales tax.

Let

$$x = \text{cost of a painting before sales tax.}$$

Then

$$0.08x = 8\% \text{ of the cost of the painting (the sales tax)}$$

$$\text{Cost of a painting before tax} + \text{tax on a painting} = 145.80$$
$$x + 0.08x = 145.80$$
$$1.08x = 145.80$$
$$\frac{1.08x}{1.08} = \frac{145.80}{1.08}$$
$$x = \frac{145.80}{1.08}$$
$$x = 135$$

Thus, the cost of a painting before tax is $135.   ●

SECTION 6.4 EXERCISES

CONCEPT/WRITING EXERCISES

1. What is the difference between a mathematical expression and an equation?

2. Give an example of a mathematical expression and an example of a mathematical equation.

In Exercises 3–6, write the mathematical expression as a phrase. (There are several acceptable answers.)

3. $x + 4$

4. $x - 7$

5. $2x - 3$

6. $3x + 5$

PRACTICE THE SKILLS

In Exercises 7–18, write the phrase as a mathematical expression.

7. 8 more than x

8. 5 less than y

9. 3 increased by 2 times z

10. 5 times x decreased by 8

11. The sum of 6 times w and 9

12. 4 more than twice a number

13. 6 more than 4 times x

14. 8 increased by 5 times x

15. 18 decreased by s, divided by 4

16. The sum of 8 and t, divided by 2

17. Three times the sum of a number and 7

18. The quotient of 8 and y, decreased by 3 times x

In Exercises 19–30, write an equation and solve.

19. 5 more than a number is 11.

20. A number decreased by 4 is 9.

21. The difference between a number and 4 is 20.

22. Seven multiplied by a number is 56.

23. Ten less than four times a number is 42.

24. Sixteen increased by 8 times a number is 88.

25. Twelve more than 4 times a number is 32.

26. Eight less than 3 times a number is 6 times the number increased by 10.

27. A number increased by 6 is 3 less than twice the number.

28. A number divided by 3 is 4 less than the number.

29. A number increased by 10 is 2 times the sum of the number and 3.

30. Twice a number decreased by 3 is 4 more than the number.

PROBLEM SOLVING

In Exercises 31–50, set up an equation that can be used to solve the problem. Solve the equation and determine the desired value(s).

31. MODELING - *Reimbursed Expenses* When sales representatives for Pfizer Pharmaceuticals drive to out-of-town meetings that require an overnight stay, they receive $150 for lodging plus $0.42 per mile driven. How many miles did Joe Kotaska drive if Pfizer reimbursed him $207.54 for an overnight trip?

32. MODELING - *Sales Commission* Tito Sanchez receives a weekly salary of $400 at Anderson's Appliances. He also receives a 6% commission on the total dollar amount of all sales he makes. What must his total sales be in a week if he is to make a total of $790?

33. MODELING - *Jet Skiing* At Action Water Sports of Ocean City, Maryland, the cost of renting a Jet Ski is $42 per half hour, which includes a 5% sales tax. Determine the cost of a half hour Jet Ski rental before tax.

34. MODELING - *Pet Supplies* PetSmart has a sale offering 10% off of all pet supplies. If Amanda Miller spent $15.72 on pet supplies before tax, what was the price of the pet supplies she purchased before the discount?

35. MODELING - *Copying* Ronnie McNeil pays 8¢ to make a copy of a page at a copy shop. She is considering purchasing a photocopy machine that is on sale for $250, including tax. How many copies would Ronnie have to make in the copy shop for her cost to equal the purchase price of the photocopy machine she is considering buying?

36. MODELING - *Number of CDs* Samantha Silverstone and Josie Appleton together receive 12 free compact discs by joining a compact disc club. How many CDs will each receive if Josie is to have three times as many as Samantha?

37. MODELING - *Scholarship Donation* Each year, Andrea Choi donates a total of $1000 for scholarships at Mercer County Community College. This year, she wants the amount she donates for scholarships for liberal arts to be three times the amount she donates for scholarships for business. Determine the amount she will donate for each type of scholarship.

38. MODELING - *Homeowners Association* Cross Creek Townhouses Homeowners Association needs to charge each homeowner a supplemental assessment to help pay for some unexpected repairs to the townhouses. The association has $2000 in its reserve fund that it will use to help pay for the repairs. How much must the association charge each of the 50 homeowners if the total cost for the repairs is $13,350?

39. MODELING - *Dimensions of a Deck* Jim Yuhas is building a rectangular deck and wants the length to be 3 ft greater than the width. What will be the dimensions of the deck if the perimeter is to be 54 ft?

▲ See Exercise 39 on
page 344

40. MODELING - *Floor Area* The total floor space in three barns is 45,000 ft². The two smaller ones have the same area, and the largest one has three times the area of the smaller ones.

a) Determine the floor space for each barn.

b) Can merchandise that takes up 8500 ft² of floor space fit into either of the smaller barns?

41. MODELING - *Wild Mustangs* According to the Bureau of Land Management, the wild mustang population of Nevada in 2004 was 2515 more than six times the wild mustang population in Utah. The total number of wild mustangs in Nevada and Utah was 21,730. Determine the number of wild mustangs in each state in 2004.

42. MODELING - *Counting Calories* According to *The Step Diet Book*, the number of steps it takes for a 150 lb person to burn off the calories from eating a cheeseburger and drinking a 12 oz soda is 11,040. The number of steps it takes to burn off the calories from eating a cheeseburger is 690 more than twice the number of steps it takes to burn off the calories from drinking a 12 oz soda. Determine the number of steps needed for a 150 lb person to burn off the calories from eating a cheeseburger and from drinking a 12 oz soda.

43. MODELING - *Tornados* The typical number of tornados in the United States in May is 11 more than 14 times the typical number of tornados in December. The sum of the typical number of tornados in December and in May is 341. Determine the typical number of tornados in the month of December and also in the month of May.

44. MODELING - *Car Purchase* The Gilberts purchased a car. If the total cost, including a 5% sales tax, was $14,512, find the cost of the car before tax.

45. MODELING - *Enclosing Two Pens* Chuck Salvador has 140 ft of fencing in which he wants to build two connecting, adjacent square pens (see the figure). What will be the dimensions if the length of the entire enclosed region is to be twice the width?

46. MODELING - *Dimensions of a Bookcase* A bookcase with three shelves, as shown in the diagram, is to be built by a woodworking student. If the height of the bookcase is to be 2 ft longer than the length of a shelf and the total amount of wood to be used is 32 ft, find the dimensions of the bookcase.

47. MODELING - *Laundry Cost* The cost of doing the family laundry for a month at a local laundromat is $70. A new washer and dryer cost a total of $760. How many months would it take for the cost of doing the laundry at the laundromat to equal the cost of a new washer and dryer?

48. MODELING - *Renting Videos* Hollywood Video offers customers two options to rent DVDs. The Unlimited Option costs $28 a month for unlimited DVD rentals. The Single Option costs $3.50 per DVD rental. How many DVDs must be rented in a month for the cost of the Unlimited Option to equal the cost of the Single Option?

49. MODELING - *Airfare* Rachel James has been told that with her half-off airfare coupon, her airfare from New York to San Diego will be $257.00. The $257.00 includes a 7% tax *on the regular fare*. On the way to the airport, Rachel realizes that she has lost her coupon. What will her regular fare be before tax?

▲ Hotel del Coronado, San Diego, California, see Exercise 49 on page 345

50. MODELING - *Truck Rentals* The cost of renting a small truck at the U-Haul rental agency is $35 per day plus 20¢ a mile. The cost of renting the same truck at the Ryder rental agency is $25 per day plus 32¢ a mile. How far would you have to drive in one day for the cost of renting from U-Haul to equal the cost of renting from Ryder?

CHALLENGE PROBLEMS/GROUP ACTIVITIES

51. *Income Tax* Some states allow a husband and wife to file individual tax returns (on a single form) even though they have filed a joint federal tax return. It is usually to the taxpayers' advantage to do so when both husband and wife work. The smallest amount of tax owed (or the largest refund) will occur when the husband's and wife's taxable incomes are the same.

Mr. McAdams's 2007 taxable income was $34,200, and Mrs. McAdams's taxable income for that year was $36,400. The McAdams's total tax deduction for the year was $3640. This deduction can be divided between Mr. and Mrs. McAdams any way they wish. How should the $3640 be divided between them to result in each individual having the same taxable income and therefore the greatest tax refund?

52. Write each equation as a sentence. There are many correct answers.

a) $x + 3 = 13$ **b)** $3x + 5 = 8$

c) $3x - 8 = 7$

53. Show that the sum of any three consecutive integers is 3 less than 3 times the largest.

54. *Auto Insurance* A driver education course at the East Lake School of Driving costs $45 but saves those under 25 years of age 10% of their annual insurance premiums until they are 25. Dan has just turned 18, and his insurance costs $100.00 per month.

a) When will the amount saved from insurance equal the price of the course?

b) Including the cost of the course, when Dan turns 25, how much will he have saved?

RECREATIONAL MATHEMATICS

55. The relationship between Fahrenheit temperature (F) and Celsius temperature (C) is shown by the formula $F = \dfrac{9}{5}C + 32$. At what temperature will a Fahrenheit thermometer read the same as a Celsius thermometer?

6.5 VARIATION

▲ Variation can be used to estimate the property tax on a home.

Your real estate agent tells you the property tax on a $170,000 home in a certain town is $2550. Suppose you are about to purchase a home in this town for $190,000. We can use variation to estimate the property tax on the home you are about to purchase. In this section, we will discuss four different types of variation and show how variation can be used to solve real-life problems.

Direct Variation

Many scientific formulas are expressed as variations. A *variation* is an equation that relates one variable to one or more other variables through the operations of multiplication or division (or both operations). There are essentially four types of variation problems: direct, inverse, joint, and combined variation.

In *direct variation*, the values of the two related variables increase together or decrease together; that is, as one increases so does the other, and as one decreases so does the other.

Consider a car traveling at 40 miles an hour. The car travels 40 miles in 1 hour, 80 miles in 2 hours, and 120 miles in 3 hours. Note that, as the time increases, the distance traveled increases, and, as the time decreases, the distance traveled decreases.

The formula used to calculate distance traveled is

$$\text{Distance} = \text{rate} \cdot \text{time}$$

Since the rate is a constant 40 miles per hour, the formula can be written

$$d = 40t$$

We say that distance *varies directly* as time or that distance is *directly proportional* to time.

The preceding equation is an example of direct variation.

DIRECT VARIATION
If a variable y varies directly with a variable x, then

$$y = kx$$

where k is the **constant of proportionality** (or the variation constant).

Examples 1 through 4 illustrate direct variation.

EXAMPLE ❶ *Direct Variation in Physics*

The length that a spring will stretch, S, varies directly with the force (or weight) F, attached to the spring. Write the equation for the length that a spring will stretch, S, if the constant of proportionality is 0.07.

SOLUTION

$$S = kF \qquad \text{\small{S varies directly as F.}}$$
$$S = 0.07F \qquad \text{\small{Constant of proportionality is 0.07.}}$$

EXAMPLE ❷ *Direct Variation in Medicine*

The recommended dosage, d, of the antibiotic drug vancomycin is directly proportional to a person's weight, w.

a) Write this variation as an equation.

b) Find the recommended dosage, in milligrams, for Doug Kulzer, who weighs 192 lb. Assume the constant of proportionality for the dosage is 18.

SOLUTION
a) $d = kw$
b) $d = 18(192) = 3456$

The recommended dosage for Doug Kulzer is 3456 mg.

In certain variation problems, the constant of proportionality, k, may not be known. In such cases, we can often find it by substituting the given values in the variation formula and solving for k.

┌─ **EXAMPLE ❸** *Finding the Constant of Proportionality*

Suppose w varies directly as the square of y. If w is 60 when y is 20, find the constant of proportionality.

SOLUTION Since w varies directly as the *square of y*, we begin with the formula $w = ky^2$. Since the constant of proportionality is not given, we must find k using the given information. Substitute 60 for w and 20 for y.

$$w = ky^2$$
$$60 = k(20)^2$$
$$60 = 400k$$
$$\frac{60}{400} = \frac{400k}{400}$$
$$0.15 = k$$

Thus, the constant of proportionality is 0.15.   ●

┌─ **EXAMPLE ❹** *Using the Constant of Proportionality*

The area, a, of a picture projected on a movie screen varies directly as the square of the distance, d, from the projector to the screen. If a projector at a distance of 25 feet projects a picture with an area of 100 square feet, what is the area of the projected picture when the projector is at a distance of 40 feet?

SOLUTION We begin with the formula $a = kd^2$. Since the constant of proportionality is not given, we must determine k, using the given information.

$$a = kd^2$$
$$100 = k(25)^2$$
$$100 = k(625)$$
$$\frac{100}{625} = k$$
$$0.16 = k$$

We now use $k = 0.16$ to determine a when $d = 40$.

$$a = kd^2$$
$$a = 0.16d^2$$
$$a = 0.16(40)^2$$
$$a = 0.16(1600)$$
$$a = 256 \text{ ft}^2$$

Thus, the area of a projected picture is 256 ft^2 when the projector is at a distance of 40 ft.   ●

Inverse Variation

A second type of variation is *inverse variation*. When two quantities vary inversely, as one quantity increases the other quantity decreases, and vice versa.

To explain inverse variation, we use the formula, distance = rate · time. If we solve for time, we get time = distance/rate. Assume the distance is fixed at 100 miles; then

$$\text{Time} = \frac{100}{\text{rate}}$$

At 100 miles per hour, it would take 1 hour to cover this distance. At a rate of 50 miles an hour, it would take 2 hours. At a rate of 25 miles an hour, it would take 4 hours. Note that as the rate (or speed) decreases, the time increases and vice versa.

The preceding equation can be written

$$t = \frac{100}{r}$$

This equation is an example of an inverse variation. The time and rate are inversely proportional. The constant of proportionality in this case is 100.

INVERSE VARIATION
If a variable y varies inversely with a variable x, then

$$y = \frac{k}{x}$$

where k is the constant of proportionality.

Two quantities *vary inversely*, or are *inversely proportional*, when as one quantity increases the other quantity decreases and vice versa. Examples 5 and 6 illustrate inverse variation.

EXAMPLE ❺ *Inverse Variation in Astronomy*

The velocity, v, of a meteor approaching Earth varies inversely as the square root of its distance from the center of Earth. Assuming the velocity is 2 miles per second at a distance, d, of 6400 miles from the center of Earth, determine the equation that expresses the relationship between the velocity of a meteor and its distance from the center of Earth.

SOLUTION Since the velocity of the meteor varies inversely as the *square root* of its distance from the center of Earth, the general form of the equation is

$$v = \frac{k}{\sqrt{d}}$$

To determine k, we substitute the given values for v and d.

$$2 = \frac{k}{\sqrt{6400}}$$

$$2 = \frac{k}{80}$$

$$(2)(80) = k$$

$$160 = k$$

Thus, the formula is $v = \frac{160}{\sqrt{d}}$.

EXAMPLE ❻ *Using the Constant of Proportionality*

Suppose y varies inversely as x. If $y = 8$ when $x = 15$, find y when $x = 18$.

SOLUTION First write the inverse variation, then solve for k.

$$y = \frac{k}{x}$$

$$8 = \frac{k}{15}$$

$$120 = k$$

Now substitute 120 for k in $y = \dfrac{k}{x}$ and find y when $x = 18$.

$$y = \frac{120}{x} = \frac{120}{18} = 6.7 \quad \text{(to the nearest tenth)}$$

Joint Variation

One quantity may vary directly as a product of two or more other quantities. This type of variation is called *joint variation*.

> **JOINT VARIATION**
> The general form of a joint variation, where y varies directly as x and z, is
>
> $$y = kxz$$
>
> where k is the constant of proportionality.

EXAMPLE ❼ *Joint Variation in Geometry*

The area, A, of a triangle varies jointly as its base, b, and height, h. If the area of a triangle is 48 in.^{2} when its base is 12 in. and its height is 8 in., find the area of a triangle whose base is 15 in. and whose height is 20 in.

SOLUTION First write the joint variation, then substitute the known values and solve for k.

$$A = kbh$$

$$48 = k(12)(8)$$

$$48 = k(96)$$

$$\frac{48}{96} = k$$

$$\tfrac{1}{2} = k$$

Now solve for the area of the given triangle.

$$A = kbh$$

$$= \tfrac{1}{2}(15)(20)$$

$$= 150 \text{ in.}^2$$

SUMMARY OF VARIATIONS

Direct	Inverse	Joint
$y = kx$	$y = \dfrac{k}{x}$	$y = kxz$

Combined Variation

Often in real-life situations, one variable varies as a combination of variables. The following examples illustrate the use of *combined variations*.

EXAMPLE ❽ *Combined Variation in Engineering*

The load, L, that a horizontal beam can safely support varies jointly as the width, w, and the square of the depth, d, and inversely as the length, l. Express L in terms of w, d, l, and the constant of proportionality, k.

SOLUTION

$$L = \frac{kwd^2}{l}$$

EXAMPLE ❾ *Hot Dog Price, Combined Variation*

The owners of Henrietta Hots find their weekly sales of hot dogs, S, vary directly with their advertising budget, A, and inversely with their hot dog price, P. When their advertising budget is $600 and the price of a hot dog is $1.50, they sell 5600 hot dogs a week.

a) Write a variation expressing S in terms of A and P. Include the value of the constant of proportionality.

b) Find the expected sales if the advertising budget is $800 and the hot dog price is $1.75.

SOLUTION

a) Since S varies directly as A and inversely as P, we begin with the equation

$$S = \frac{kA}{P}$$

We now find k using the known values.

$$5600 = \frac{k(600)}{1.50}$$
$$5600 = 400k$$
$$14 = k$$

Therefore, the equation for the weekly sales of hot dogs is $S = \dfrac{14A}{P}$.

b)
$$S = \frac{14A}{P}$$
$$= \frac{14(800)}{1.75}$$
$$= 6400$$

Henrietta Hots can expect to sell 6400 hot dogs a week if the advertising budget is $800 and the hot dog price is $1.75.   •

EXAMPLE ⑩ *Combined Variation*

A varies jointly as *B* and *C* and inversely as the square of *D*. If $A = 1$ when $B = 9, C = 4$, and $D = 6$, find *A* when $B = 8, C = 12$, and $D = 5$.

SOLUTION   We begin with the equation

$$A = \frac{kBC}{D^2}$$

We must first find the constant of proportionality, *k*, by substituting the known values for *A*, *B*, *C*, and *D* and solving for *k*.

$$1 = \frac{k(9)(4)}{6^2}$$

$$1 = \frac{36k}{36}$$

$$1 = k$$

Thus, the constant of proportionality equals 1. Now we find *A* for the corresponding values of *B*, *C*, and *D*.

$$A = \frac{kBC}{D^2}$$

$$A = \frac{(1)(8)(12)}{5^2} = \frac{96}{25} = 3.84$$ •

SECTION 6.5   EXERCISES

CONCEPT/WRITING EXERCISES

In Exercises 1–4, use complete sentences to answer the question.

1. Describe direct variation.

2. Describe inverse variation.

3. Describe joint variation.

4. Describe combined variation.

In Exercises 5–20, use your intuition to determine whether the variation between the indicated quantities is direct or inverse.

5. The distance between two cities on a map and the actual distance between the two cities

6. The time required to fill a pool with a hose and the volume of water coming from the hose

7. The time required to boil water on a burner and the temperature of the burner

8. A person's salary and the amount of money withheld from his or her salary for federal income taxes

9. The interest earned on an investment and the interest rate

10. The volume of a balloon and its radius

11. A person's speed and the time needed for the person to complete the race

12. The time required to cool a room and the temperature of the room

13. The number of workers hired to install a fence and the time required to install the fence

14. The number of calories in a slice of pizza and the size of the slice

15. The time required to defrost a frozen hamburger in a room and the temperature of the room

16. On Earth, the weight and mass of an object

17. The number of people in line at a bank and the amount of time required for the last person to reach the teller

18. The number of books that can be placed upright on a shelf 3 ft long and the width of the books

19. The amount of fertilizer needed to fertilize a lawn and the area of the lawn

20. The percent of light that filters through water and the depth of the water

In Exercises 21 and 22, use Exercises 5–20 as a guide.

21. Name two items that have not been mentioned in this section that have a direct variation.

22. Name two items that have not been mentioned in this section that have an inverse variation.

PRACTICE THE SKILLS

In Exercises 23–40, (a) write the variation and (b) determine the quantity indicated.

23. y varies directly as x. Determine y when $x = 15$ and $k = 8$.

24. x varies inversely as y. Determine x when $y = 7$ and $k = 14$.

25. m varies inversely as the square of n. Determine m when $n = 8$ and $k = 16$.

26. r varies directly as the square of s. Determine r when $s = 2$ and $k = 13$.

27. A varies directly as B and inversely as C. Determine A when $B = 5$, $C = 10$, and $k = 5$.

28. M varies directly as J and inversely as C. Determine M when $J = 10$, $C = 20$, and $k = 8$.

29. F varies jointly as D and E. Determine F when $D = 3$, $E = 10$, and $k = 7$.

30. A varies jointly as R_1 and R_2 and inversely as the square of L. Determine A when $R_1 = 120$, $R_2 = 8$, $L = 5$, and $k = \frac{3}{2}$.

31. t varies directly as the square of d and inversely as f. If $t = 192$ when $d = 8$ and $f = 4$, determine t when $d = 10$ and $f = 6$.

32. y varies directly as the square root of t and inversely as s. If $y = 12$ when $t = 36$ and $s = 2$, determine y when $t = 81$ and $s = 4$.

33. Z varies jointly as W and Y. If $Z = 12$ when $W = 9$ and $Y = 4$, determine Z when $W = 50$ and $Y = 6$.

34. y varies directly as the square of R. If $y = 4$ when $R = 4$, determine y when $R = 8$.

35. H varies directly as L. If $H = 15$ when $L = 50$, determine H when $L = 10$.

36. C varies inversely as J. If $C = 7$ when $J = 0.7$, determine C when $J = 12$.

37. A varies directly as the square of B. If $A = 245$ when $B = 7$, determine A when $B = 12$.

38. F varies jointly as M_1 and M_2 and inversely as the square of d. If $F = 20$ when $M_1 = 5$, $M_2 = 10$, and $d = 0.2$, determine F when $M_1 = 10$, $M_2 = 20$, and $d = 0.4$.

39. F varies jointly as q_1 and q_2 and inversely as the square of d. If $F = 80$ when $q_1 = 4$, $q_2 = 16$, and $d = 0.4$, determine F when $q_1 = 12$, $q_2 = 20$, and $d = 0.2$.

40. S varies jointly as I and the square of T. If $S = 4$ when $I = 10$ and $T = 4$, determine S when $I = 4$ and $T = 6$.

PROBLEM SOLVING

In Exercises 41–49, (a) write the variation and (b) determine the quantity indicated.

41. *Property Tax* The property tax, t, on a home is directly proportional to the assessed value, v, of the home. If the property tax on a home with an assessed value of $140,000 is $2100, what is the property tax on a home with an assessed value of $180,000?

42. *Resistance* The resistance, R, of a wire varies directly as its length, L. If the resistance of a 30 ft length of wire is 0.24 ohm, determine the resistance of a 40 ft length of wire.

43. *Speaker Loudness* The loudness of a stereo speaker, l, measured in decibels (dB), is inversely proportional to the square of the distance, d, of the listener from the speaker. If the loudness is 20 dB when the listener is 6 ft from the speaker, what is the loudness when the listener is 3 ft from the speaker?

44. *Melting an Ice Cube* The time, t, for an ice cube to melt is inversely proportional to the temperature, T, of the water in which the ice cube is placed. If it takes an ice cube 2 minutes to melt in 75°F water, how long will it take an ice cube of the same size to melt in 80°F water?

▲ See Exercise 44

45. *Video Rentals* The number of weekly videotape rentals, R, at Busterblock Video varies directly with the advertising budget, A, and inversely with the daily rental price, P. When the video store's advertising budget is $600 and the rental price is $3 per day, it rents 4800 tapes per week. How many tapes would it rent per week if the store increased its advertising budget to $700 and raised its rental price to $3.50?

46. *Stopping Distance of a Car* The stopping distance, d, of a car after the brakes are applied varies directly as the square of the speed, s, of the car. If a car traveling at a speed of 40 mph can stop in 80 ft, what is the stopping distance of a car traveling at 65 mph?

47. *Guitar Strings* The number of vibrations per second, v, of a guitar string varies directly as the square root of the tension, t, and inversely as the length of the string, l. If the number of vibrations per second is 5 when the tension is 225 kg and the length of the string is 0.60 m, determine the number of vibrations per second when the tension is 196 kg and the length of the string is 0.70 m.

48. *Electrical Resistance* The electrical resistance of a wire, R, varies directly as its length, L, and inversely as its cross-sectional area, A. If the resistance of a wire is 0.2 ohm when the length is 200 ft and its cross-sectional area is 0.05 in.^{2}, what is the resistance of a wire whose length is 5000 ft with a cross-sectional area of 0.01 in.^{2}?

49. *Phone Calls* The number of phone calls between two cities during a given time period, N, varies directly as the populations p_1 and p_2 of the two cities and inversely to the distance, d, between them. If 100,000 calls are made between two cities 300 mi apart and the populations of the cities are 60,000 and 200,000, how many calls are made between two cities with populations of 125,000 and 175,000 that are 450 mi apart?

50. a) If y varies directly as x and the constant of proportionality is 2, does x vary directly or inversely as y? Explain.

b) Give the new constant of proportionality for x as a variation of y.

51. a) If y varies inversely as x and the constant of proportionality is 0.3, does x vary directly or inversely as y? Explain.

b) Give the new constant of proportionality for x as a variation of y.

CHALLENGE PROBLEMS/GROUP ACTIVITIES

52. *Photography* An article in the magazine *Outdoor and Travel Photography* states, "If a surface is illuminated by a point-source of light, the intensity of illumination produced is inversely proportional to the square of the distance separating them. In practical terms, this means that foreground objects will be grossly overexposed if your background subject is properly exposed with a flash. Thus direct flash will not offer pleasing results if there are any intervening objects between the foreground and the subject."

    If the subject you are photographing is 4 ft from the flash and the illumination on this subject is $\frac{1}{16}$ of the light

of the flash, what is the intensity of illumination on an intervening object that is 3 ft from the flash?

53. *Water Cost* In a specific region of the country, the amount of a customer's water bill, W, is directly proportional to the average daily temperature for the month, T, the lawn area, A, and the square root of F, where F is the family size, and inversely proportional to the number of inches of rain, R.

    In one month, the average daily temperature is 78°F and the number of inches of rain is 5.6. If the average family of four who has a thousand square feet of lawn pays $72.00 for water for that month, estimate the water bill in the same month for the average family of six who has 1500 ft^2 of lawn.

6.6   LINEAR INEQUALITIES

▲ An inequality can be used to determine how many textbooks must be sold so that a bookstore's revenue is greater than its cost.

Suppose your campus bookstore needs to determine how many textbooks for a particular course must be sold so that the bookstore's textbook revenue is greater than its cost. To determine this number, an inequality would be set up. In this section, we will discuss how to set up and how to solve an inequality.

The symbols of inequality are as follows.

SYMBOLS OF INEQUALITY
$a < b$ means that a is less than b.
$a \leq b$ means that a is less than or equal to b.
$a > b$ means that a is greater than b.
$a \geq b$ means that a is greater than or equal to b.

An *inequality* consists of two (or more) expressions joined by an inequality sign.

Examples of inequalities
$$3 < 5, \qquad x < 2, \qquad 3x - 2 \geq 5$$

    A statement of inequality can be used to indicate a set of real numbers. For example, $x < 2$ represents the set of all real numbers less than 2. Listing all these numbers is impossible, but some are $-2, -1.234, -1, -\frac{1}{2}, 0, \frac{97}{163}, 1$.

To indicate all real numbers less than 2, we can use the number line. The number line was discussed in Chapter 5.

Solving Inequalities

To indicate the solution set of $x < 2$ on the number line, we draw an open circle at 2 and a line to the left of 2 with an arrow at its end. This technique indicates that all points to the left of 2 are part of the solution set. The open circle indicates that the solution set does not include the number 2.

To indicate the solution set of $x \leq 2$ on the number line, we draw a closed (or darkened) circle at 2 and a line to the left of 2 with an arrow at its end. The closed circle indicates that the 2 is part of the solution.

┌─**EXAMPLE ❶** *Graphing a Less Than or Equal to Inequality*

Graph the solution set of $x \leq -1$, where x is a real number, on the number line.

SOLUTION The numbers less than or equal to -1 are all the points on the number line to the left of -1 and -1 itself. The closed circle at -1 shows that -1 is included in the solution set.

The inequality statements $x < 2$ and $2 > x$ have the same meaning. Note that the inequality symbol points to the x in both cases. Thus, one inequality may be written in place of the other. Likewise, $x > 2$ and $2 < x$ have the same meaning. Note that the inequality symbol points to the 2 in both cases. We make use of this fact in Example 2.

┌─**EXAMPLE ❷** *Graphing a Less Than Inequality*

Graph the solution set of $3 < x$, where x is a real number, on the number line.

SOLUTION We can restate $3 < x$ as $x > 3$. Both statements have identical solutions. Any number that is greater than 3 satisfies the inequality $x > 3$. The graph includes all the points to the right of 3 on the number line. To indicate that 3 is not part of the solution set, we place an open circle at 3.

We can find the solution to an inequality by adding, subtracting, multiplying, or dividing both sides of the inequality by the same number or expression. We use the procedure discussed in Section 6.2 to isolate the variable, with one important exception: *When both sides of an inequality are multiplied or divided by a negative number, the direction of the inequality symbol is reversed.*

EXAMPLE ❸ *Multiplying by a Negative Number*

Solve the inequality $-x > 3$ and graph the solution set on the number line.

SOLUTION To solve this inequality, we must eliminate the negative sign in front of the x. To do so, we multiply both sides of the inequality by -1 and change the direction of the inequality symbol.

$$-x > 3$$
$$-1(-x) < -1(3)$$

Multiply both sides of the inequality by -1 and change the direction of the inequality symbol.

$$x < -3$$

The solution set is graphed on the number line as follows.

EXAMPLE ❹ *Dividing by a Negative Number*

Solve the inequality $-5x < 20$ and graph the solution set on the number line.

SOLUTION Solving the inequality requires making the coefficient of the x term 1. To do so, divide both sides of the inequality by -5 and change the direction of the inequality symbol.

$$-5x < 20$$
$$\frac{-5x}{-5} > \frac{20}{-5}$$

Divide both sides of the inequality by -5 and change the direction of the inequality symbol.

$$x > -4$$

The solution set is graphed on the number line as follows.

EXAMPLE ❺ *Solving an Inequality*

Solve the inequality $3x - 5 > 13$ and graph the solution set on the number line.

SOLUTION To find the solution set, isolate x on one side of the inequality symbol.

$$3x - 5 > 13$$
$$3x - 5 + 5 > 13 + 5 \quad \text{Add 5 to both sides of the inequality.}$$
$$3x > 18$$
$$\frac{3x}{3} > \frac{18}{3} \quad \text{Divide both sides of the inequality by 3.}$$
$$x > 6$$

Thus, the solution set to $3x - 5 > 13$ is all real numbers greater than 6.

Note that in Example 5, the *direction of the inequality symbol* did not change when both sides of the inequality were divided by the positive number 3.

EXAMPLE ❻ A Solution of Only Integers

Solve the inequality $x + 5 < 9$, where x is an integer, and graph the solution set on the number line.

SOLUTION To find the solution set, isolate x on one side of the inequality symbol.

$$x + 5 < 9$$
$$x + 5 - 5 < 9 - 5 \quad \text{Subtract 5 from both sides of the inequality.}$$
$$x < 4$$

Since x is an integer and is less than 4, the solution set is the set of integers less than 4, or $\{\ldots, -2, -1, 0, 1, 2, 3\}$. To graph the solution set, we make solid dots at the corresponding points on the number line. The ellipsis (the three dots) to the left of -3 indicate that all the integers to the left of -3 are included.

TIMELY TIP Some linear inequalities have no solution. When solving an inequality, if you obtain a statement that is always false, such as $3 > 5$, the answer is *no solution*. There is no real number that makes the statement true.

Some linear inequalities have all real numbers as their solution. When solving an inequality, if you obtain a statement that is always true, such as $2 > 1$, the solution is *all real numbers*.

Solving Compound Inequalities

An inequality of the form $a < x < b$ is called a *compound inequality*. Consider the compound inequality $-3 < x \leq 2$, which means that $-3 < x$ *and* $x \leq 2$.

EXAMPLE ❼ A Compound Inequality

Graph the solution set of the inequality $-3 < x \leq 2$
a) where x is an integer.   b) where x is a real number.

SOLUTION

a) The solution set is all the integers between -3 and 2, including the 2 but not including the -3, or $\{-2, -1, 0, 1, 2\}$.

b) The solution set consists of all the real numbers between -3 and 2, including the 2 but not including the -3.

EXAMPLE ❽ Solving a Compound Inequality

Solve the compound inequality for x and graph the solution set.

$$-4 < \frac{x + 3}{2} \leq 5$$

"In mathematics the art of posing problems is easier than that of solving them."

Georg Cantor

SOLUTION To solve this compound inequality, we must isolate the x in the middle term. To do so, we use the same principles used to solve inequalities.

$$-4 < \frac{x + 3}{2} \leq 5$$

$$2(-4) < 2\left(\frac{x + 3}{2}\right) \leq 2(5) \qquad \text{Multiply each part of the inequality by 2.}$$

$$-8 < x + 3 \leq 10$$

$$-8 - 3 < x + 3 - 3 \leq 10 - 3 \qquad \text{Subtract 3 from each part of the inequality.}$$

$$-11 < x \leq 7$$

The solution set is graphed on the number line as follows.

$-11 < x \leq 7$

EXAMPLE ❾ *Average Grade*

A student must have an average (the mean) on five tests that is greater than or equal to 80% but less than 90% to receive a final grade of B. Devon's grades on the first four tests were 98%, 76%, 86%, and 92%. What range of grades on the fifth test would give him a B in the course?

SOLUTION The unknown quantity is the range of grades on the fifth test. First construct an inequality that can be used to find the range of grades on the fifth test. The average (mean) is found by adding the grades and dividing the sum by the number of tests.

Let x = the fifth grade. Then

$$\text{Average} = \frac{98 + 76 + 86 + 92 + x}{5}$$

For Devon to obtain a B in this course, his average must be greater than or equal to 80 but less than 90.

$$80 \leq \frac{98 + 76 + 86 + 92 + x}{5} < 90$$

$$80 \leq \frac{352 + x}{5} < 90$$

$$5(80) \leq 5\left(\frac{352 + x}{5}\right) < 5(90) \qquad \begin{array}{l}\text{Multiply the three terms}\\\text{of the inequality by 5.}\end{array}$$

$$400 \leq 352 + x < 450$$

$$400 - 352 \leq 352 - 352 + x < 450 - 352 \qquad \text{Subtract 352 from all three terms.}$$

$$48 \leq x < 98$$

Thus, a grade of 48% up to but not including a grade of 98% on the fifth test will result in a grade of B in this course.

TIMELY TIP Remember to change the direction of the inequality symbol when multiplying or dividing both sides of an inequality by a negative number.

SECTION 6.6 EXERCISES

CONCEPT/WRITING EXERCISES

1. Give the four inequality symbols we use in this section and indicate how each is read.

2. a) What is an inequality?

   b) Give an example of three inequalities.

3. When solving an inequality, under what conditions do you need to change the direction of the inequality symbol?

4. Does $x < 2$ have the same meaning as $2 > x$? Explain.

5. Does $x > -3$ have the same meaning as $-3 < x$? Explain.

6. When graphing the solution set to an inequality on the number line, when should you use an open circle and when should you use a closed circle?

7. a) What is a compound inequality?

   b) Give an example of a compound inequality.

8. a) Give an example of an inequality whose solution on a number line would contain a closed circle.

   b) Give an example of an inequality whose solution on a number line would contain an open circle.

PRACTICE THE SKILLS

In Exercises 9–26, graph the solution set of the inequality, where x is a real number, on the number line.

9. $x \geq 4$

10. $x < 7$

11. $x - 8 \geq 3$

12. $x + 5 > -2$

13. $-3x \leq 18$

14. $-4x < 12$

15. $\dfrac{x}{5} < 4$

16. $\dfrac{x}{6} > 2$

17. $\dfrac{-x}{3} \geq 3$

18. $\dfrac{x}{2} \geq -4$

19. $2x + 6 \geq 14$

20. $3x + 12 < 5x + 14$

21. $4(2x - 1) < 2(4x - 3)$

22. $-2(x - 3) \leq 4x + 6$

23. $-1 \leq x \leq 3$

24. $-4 < x \leq 1$

25. $3 < x - 7 \leq 6$

26. $\dfrac{1}{2} < \dfrac{x + 4}{2} \leq 4$

In Exercises 27–46, graph the solution set of the inequality, where x is an integer, on the number line.

27. $x > 1$

28. $-2 < x$

29. $-3x \leq 27$

30. $3x \geq 27$

31. $x - 2 < 4$

32. $-5x \leq 15$

33. $\dfrac{x}{3} \leq -2$

34. $\dfrac{x}{4} \geq -3$

35. $-\dfrac{x}{4} \geq 2$

36. $\dfrac{3x}{5} \leq 3$

37. $-15 < -4x - 3$

38. $-5x - 1 < 17 - 2x$

39. $3(x + 4) \geq 4x + 13$

40. $-2(x - 1) < 3(x - 4) + 5$

41. $5(x + 4) - 6 \leq 2x + 8$

42. $-3 \leq x < 5$

43. $1 > -x > -5$

44. $-2 < 2x + 3 < 6$

45. $0.3 \leq \dfrac{x + 2}{10} \leq 0.5$

46. $-\dfrac{1}{3} < \dfrac{x - 2}{12} \leq \dfrac{1}{4}$

PROBLEM SOLVING

47. *Federal Drug Costs* The following bar graph shows the government's projected cost, in billions of dollars, for the Medicare prescription drug benefit for the years 2005–2015.

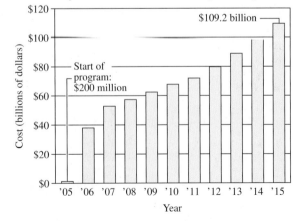

Projected Cost of Medicare Prescription Drug Benefit

Source: CMS

In which years is the projected cost

a) > \$60 billion? **b)** ≤ \$60 billion?

c) > \$20 billion? **d)** ≤ \$40 billion?

48. *Work Stoppages* The following bar graph shows the number of work stoppages involving at least 1000 workers, including strikes and lockouts, in the United States for the years 1999–2004.

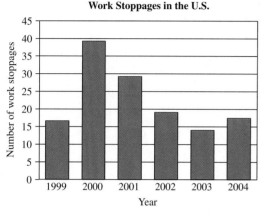

Work Stoppages in the U.S.

Source: Bureau of Labor Statistics, U.S. Department of Labor

In which years was the number of work stoppages in the United States

a) <25? **b)** >30? **c)** <40? **d)** >20?

49. *Video Rental* Movie Mania offers two rental plans. One has an annual fee, and the other has no annual fee. The annual membership fee and the daily charge per video for each plan are shown in the table. Determine the maximum number of videos that can be rented for the no fee plan to cost less than the annual fee plan.

Rental Plan	Yearly Fee	Daily Charge per Video
Annual fee	\$30	\$1.49
No fee	None	\$2.99

50. *Salary Plans* Bobby Exler recently accepted a sales position in Portland, Oregon. He can select between the two salary plans shown in the table. Determine the dollar amount of weekly sales that would result in Bobby earning more with Plan B than with Plan A.

Salary Plan	Weekly Salary	Commission on Sales
Plan A	\$500	6%
Plan B	\$400	8%

51. *College Employment* To be eligible for financial assistance for college, Samantha Lavin can earn no more than \$3000 during her summer employment. She will earn

\$725 cutting lawns. She is also earning \$6.50 per hour at an evening job as a cashier at a grocery store. What is the maximum number of hours Samantha can work at the grocery store without jeopardizing her financial assistance?

52. *Moving Boxes* A janitor must move a large shipment of books from the first floor to the fifth floor. Each box of books weighs 60 lb, and the janitor weighs 180 lb. The sign on the elevator reads, "Maximum weight 1200 lb."

a) Write a statement of inequality to determine the maximum number of boxes of books the janitor can place on the elevator at one time. (The janitor must ride in the elevator with the books.)

b) Determine the maximum number of boxes that can be moved in one trip.

53. *Parking Costs* A parking garage in Richmond, Virginia, charges \$1.75 for the first hour and \$0.50 for each additional hour or part of an hour. What is the maximum length of time Tom Shalhoub can park in the garage if he wishes to pay no more than \$4.25?

54. *Making a Profit* For a business to realize a profit, its revenue, R, must be greater than its costs, C; that is, a profit will only result if $R > C$ (the company breaks even when $R = C$). A book publishing company has a weekly cost equation of $C = 2x + 2000$ and a weekly revenue equation of $R = 12x$, where x is the number of books produced and sold in a week. How many books must be sold weekly for the company to make a profit?

55. *Finding Velocity* The velocity, v, in feet per second, t sec after a tennis ball is projected directly upward is given by the formula $v = 84 - 32t$. How many seconds after being projected upward will the velocity be between 36 ft/sec and 68 ft/sec?

56. *Speed Limit* The minimum speed for vehicles on a highway is 40 mph, and the maximum speed is 55 mph. If Philip Rowe has been driving nonstop along the highway for 4 hr, what range in miles could he have legally traveled?

57. *A Grade of B* In Example 9 on page 359, what range of grades on the fifth test would result in Devon receiving a grade of B if his grades on the first four tests were 78%, 64%, 88%, and 76%?

58. *Tent Rental* The Cuhayoga Community College Planning Committee wants to rent tents for the spring job fair. Rent-a-Tent charges $325 for setup and delivery of its tents. This fee is charged regardless of the number of tents delivered and set up. In addition, Rent-a-Tent charges $125 for each tent rented. If the minimum amount the planning committee wishes to spend is $950 and the maximum amount they wish to spend is $1200, determine the minimum and maximum number of tents the committee can rent.

CHALLENGE PROBLEMS/GROUP ACTIVITIES

59. *Painting a House* Donovan Davis is painting the exterior of his house. The instructions on the paint can indicate that 1 gal covers from 250 to 400 ft². The total surface of the house to be painted is 2750 ft². Determine the number of gallons of paint he could use and express the answer as an inequality.

60. *Final Exam* Teresa's five test grades for the semester are 86%, 78%, 68%, 92%, and 72%. Her final exam counts one-third of her final grade. What range of grades on her final exam would result in Teresa receiving a final grade of B in the course? (See Example 9.)

61. A student multiplied both sides of the inequality $-\frac{1}{3}x \le 4$ by -3 and forgot to reverse the direction of the inequality symbol. What is the relation between the student's incorrect solution set and the correct solution set? Is there any number in both the correct solution set and the student's incorrect solution set? If so, what is it?

INTERNET/RESEARCH ACTIVITY

62. Find a newspaper or a magazine article that contains the mathematical concept of inequality.

a) From the information in the article write a statement of inequality.

b) Summarize the article and explain how you arrived at the inequality statement in part (a).

6.7   GRAPHING LINEAR EQUATIONS

▲ The cost of mailing a package may depend on the weight of the package.

The profit, *p*, of a company may depend on the amount of sales, *s*. The cost, *c*, of mailing a package may depend on the weight, *w*, of the package. Real-life problems, such as these two examples, often involve two or more variables. In this section, we will discuss how to graph equations with two variables.

To be able to work with equations with two variables requires understanding the *Cartesian* (or *rectangular*) *coordinate system*, named after the French mathematician René Descartes. The rectangular coordinate system consists of two perpendicular number lines (Fig. 6.6 on page 363). The horizontal line is the *x-axis*, and the vertical line is the *y-axis*. The point of intersection of the *x*-axis and *y*-axis is called the *origin*. The numbers on the axes to the right and above the origin are positive. The numbers on the axes on the left and below the origin are negative. The axes divide the plane into four parts: the first, second, third, and fourth *quadrants*.

We indicate the location of a point in the rectangular coordinate system by means of an *ordered pair* of the form (x, y). The x-coordinate is always placed first, and the y-coordinate is always placed second in the ordered pair. Consider the point illustrated in Fig. 6.7. Since the x-coordinate of the point is 5 and the y-coordinate is 3, the ordered pair that represents this point is $(5, 3)$.

Figure 6.6

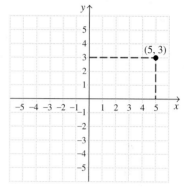

Figure 6.7

The origin is represented by the ordered pair $(0, 0)$. Every point on the plane can be represented by one and only one ordered pair (x, y), and every ordered pair (x, y) represents one and only one point on the plane.

EXAMPLE ❶ Plotting Points

Plot the points $A(-2, 4)$, $B(-3, -1)$, $C(4, 0)$, $D(2, 1)$, and $E(0, -3)$.

SOLUTION Point A has an x-coordinate of -2 and a y-coordinate of 4. Project a vertical line up from -2 on the x-axis and a horizontal line to the left from 4 on the y-axis. The two lines intersect at the point denoted A (Fig. 6.8). The other points are plotted in a similar manner.

Figure 6.8

EXAMPLE ❷ *A Parallelogram*

The points, A, B, and C are three vertices of a parallelogram with two sides parallel to the x-axis. Plot the three points below and determine the coordinates of the fourth vertex, D.

$$A(1, 2) \qquad B(2, 4) \qquad C(7, 4)$$

SOLUTION A parallelogram is a figure that has opposite sides that are of equal length and are parallel. (Parallel lines are two lines in the same plane that do not intersect.) The horizontal distance between points B and C is 5 units (see Fig. 6.9). Therefore, the horizontal distance between points A and D must also be 5 units. This problem has two possible solutions, as illustrated in Fig. 6.9. In each figure, we have indicated the given points in red.

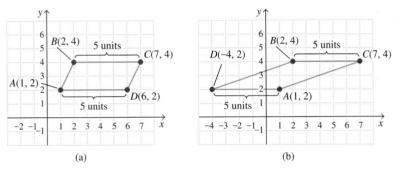

(a)   (b)

Figure 6.9

The solutions are the points $(6, 2)$ and $(-4, 2)$.

Graphing Linear Equations by Plotting Points

Consider the following equation in two variables: $y = x + 1$. Every ordered pair that makes the equation a true statement is a solution to, or satisfies, the equation. We can mentally find some ordered pairs that satisfy the equation $y = x + 1$ by picking some values of x and solving the equation for y. For example, suppose we let $x = 1$; then $y = 1 + 1 = 2$. The ordered pair $(1, 2)$ is a solution to the equation $y = x + 1$. We can make a chart of other ordered pairs that are solutions to the equation.

x	y	Ordered Pair
1	2	(1, 2)
2	3	(2, 3)
3	4	(3, 4)
4.5	5.5	(4.5, 5.5)
−3	−2	(−3, −2)

How many other ordered pairs satisfy the equation? Infinitely many ordered pairs satisfy the equation. Since we cannot list all the solutions, we show them by means of a graph. A *graph* is an illustration of all the points whose coordinates satisfy an equation.

The points $(1, 2)$, $(2, 3)$, $(3, 4)$, $(4.5, 5.5)$, and $(-3, -2)$ are plotted in Fig. 6.10. With a straightedge we can draw one line that contains all these points. This line, when extended indefinitely in either direction, passes through all the points in the plane that satisfy the equation $y = x + 1$. The arrows on the ends of the line indicate that the line extends indefinitely.

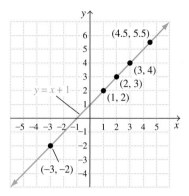

Figure 6.10

All equations of the form $ax + by = c$, $a \neq 0$ and $b \neq 0$, will be straight lines when graphed. Thus, such equations are called *linear equations in two variables*. The exponents on the variables x and y must be 1 for the equation to be linear. Since only two points are needed to draw a line, only two points are needed to graph a linear equation. It is always a good idea to plot a third point as a checkpoint. If no error has been made, all three points will be in a line, or *collinear*. One method that can be used to obtain points is to solve the equation for y, substitute values for x, and find the corresponding values of y.

EXAMPLE ❸ *Graphing an Equation by Plotting Points*

Graph $y = 2x + 2$.

SOLUTION Since the equation is already solved for y, select values for x and find the corresponding values for y. The table indicates values arbitrarily selected for x and the corresponding values for y. The ordered pairs are $(0, 2)$, $(1, 4)$, and $(-1, 0)$. The graph is shown in Fig. 6.11.

x	y
0	2
1	4
−1	0

Figure 6.11

TO GRAPH LINEAR EQUATIONS BY PLOTTING POINTS

1. Solve the equation for y.
2. Select at least three values for x and find their corresponding values of y.
3. Plot the points.
4. The points should be in a straight line. Draw a line through the set of points and place arrow tips at both ends of the line.

In step 4 of the procedure, if the points are not in a straight line, recheck your calculations and find your error.

Graphing by Using Intercepts

Example 3 contained two special points on the graph, $(-1, 0)$ and $(0, 2)$. At these points, the line crosses the x-axis and the y-axis, respectively. The ordered pairs $(-1, 0)$ and $(0, 2)$ represent the *x-intercept* and the *y-intercept*, respectively. Another method that can be used to graph linear equations is to find the x- and y-intercepts of the graph.

FINDING THE *X*- AND *Y*-INTERCEPTS
To find the x-intercept, set $y = 0$ and solve the equation for x.
To find the y-intercept, set $x = 0$ and solve the equation for y.

A linear equation may be graphed by finding the x- and y-intercepts, plotting the intercepts, and drawing a straight line through the intercepts. When graphing by this method, you should always plot a checkpoint before drawing your graph. To obtain a checkpoint, select a nonzero value for x and find the corresponding value of y. The checkpoint should be collinear with the x- and y-intercepts.

EXAMPLE ❹ *Graphing by Using Intercepts*

Graph $2x - 4y = 8$ by using the x- and y-intercepts.

SOLUTION To find the x-intercept, set $y = 0$ and solve for x.

$$2x - 4y = 8$$
$$2x - 4(0) = 8$$
$$2x = 8$$
$$x = 4$$

The x-intercept is $(4, 0)$. To find the y-intercept, set $x = 0$ and solve for y.

$$2x - 4y = 8$$
$$2(0) - 4y = 8$$
$$-4y = 8$$
$$y = -2$$

The y-intercept is $(0, -2)$. As a checkpoint, try $x = 2$ and find the corresponding value for y.

$$2x - 4y = 8$$
$$2(2) - 4y = 8$$
$$4 - 4y = 8$$
$$-4y = 4$$
$$y = -1$$

The checkpoint is the ordered pair $(2, -1)$.

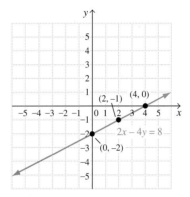

Figure 6.13

Since all three points in Fig. 6.13 are collinear, draw a line through the three points to obtain the graph. ●

Figure 6.14

Slope

Another useful concept when you are working with straight lines is slope, which is a measure of the "steepness" of a line. The *slope of a line* is a ratio of the vertical change to the horizontal change for any two points on the line. Consider Fig. 6.14. Point A has coordinates (x_1, y_1), and point B has coordinates (x_2, y_2). The vertical change between points A and B is $y_2 - y_1$. The horizontal change between points A and B is $x_2 - x_1$. Thus, the slope, which is often symbolized with the letter m, can be found as follows.

$$\textbf{Slope} = \frac{\text{vertical change}}{\text{horizontal change}}$$

$$m = \frac{y_2 - y_1}{x_2 - x_1}$$

The Greek capital letter delta, Δ, is often used to represent the words "the change in." Therefore, slope may be defined as

$$m = \frac{\Delta y}{\Delta x}$$

A line may have a positive slope, a negative slope, or a zero slope, or the slope may be undefined, as indicated in Fig. 6.15. A line with a positive slope rises from left to right, as shown in Fig. 6.15(a). A line with a negative slope falls from left to right, as shown in Fig. 6.15(b). A horizontal line, which neither rises nor falls, has a slope of zero, as shown in Fig. 6.15(c). Since a vertical line does not have any horizontal change (the x value remains constant) and since we cannot divide by 0, the slope of a vertical line is undefined, as shown in Fig. 6.15(d).

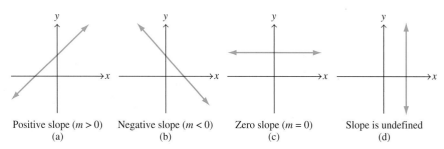

Positive slope ($m > 0$)   Negative slope ($m < 0$)   Zero slope ($m = 0$)   Slope is undefined
    (a)   (b)   (c)   (d)

Figure 6.15

EXAMPLE ❺ *Finding the Slope of a Line*

Determine the slope of the line that passes through the points $(-1, -3)$ and $(1, 5)$.

SOLUTION Let's begin by drawing a sketch, illustrating the points and the line. See Fig. 6.16(a).

We will let (x_1, y_1) be $(-1, -3)$ and (x_2, y_2) be $(1, 5)$. Then

$$\text{Slope} = \frac{y_2 - y_1}{x_2 - x_1} = \frac{5 - (-3)}{1 - (-1)} = \frac{5 + 3}{1 + 1} = \frac{8}{2} = \frac{4}{1} = 4$$

The slope of 4 means that there is a vertical change of 4 units for each horizontal change of 1 unit; see Fig. 6.16(b). The slope is positive, and the line rises from left to right. Note that we would have obtained the same result if we let (x_1, y_1) be $(1, 5)$ and (x_2, y_2) be $(-1, -3)$. Try this method now and see.

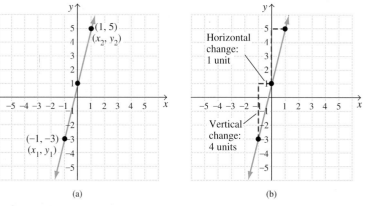

        (a)   (b)

Figure 6.16

Graphing Linear Equations by Using the Slope and y-Intercept

A linear equation given in the form $y = mx + b$ is said to be in slope–intercept form.

Figure 6.17

SLOPE–INTERCEPT FORM OF THE EQUATION OF A LINE

$$y = mx + b$$

where m is the slope of the line and $(0, b)$ is the y-intercept of the line.

In the equation $y = mx + b$, b represents the value of y where the graph of the equation $y = mx + b$ crosses the y-axis.

Consider the graph of the equation $y = 3x + 4$, which appears in Fig. 6.17. By examining the graph, we can see that the y-intercept is $(0, 4)$. We can also see that the graph has a positive slope because it rises from left to right. Because the vertical change is 3 units for every 1 unit of horizontal change, the slope must be $\frac{3}{1}$ or 3.

We could graph this equation by marking the y-intercept at $(0, 4)$ and then moving *up* 3 units and to the *right* 1 unit to get another point. If the slope were -3, which means $\frac{-3}{1}$, we could start at the y-intercept and move *down* 3 units and to the *right* 1 unit. Thus, if we know the slope and y-intercept of a line, we can graph the line.

TO GRAPH LINEAR EQUATIONS BY USING THE SLOPE AND Y-INTERCEPT

1. Solve the equation for y to place the equation in slope–intercept form.
2. Determine the slope and y-intercept from the equation.
3. Plot the y-intercept.
4. Obtain a second point using the slope.
5. Draw a straight line through the points.

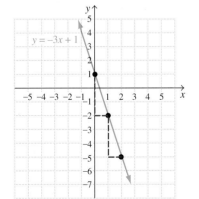

Figure 6.18

EXAMPLE ❻ *Graphing an Equation by Using the Slope and y-Intercept*

Graph $y = -3x + 1$ by using the slope and y-intercept.

SOLUTION The slope is -3 or $\dfrac{-3}{1}$ and the y-intercept is $(0, 1)$ (see Fig. 6.18). Plot $(0, 1)$ on the y-axis. Then plot the next point by moving *down* 3 units and to the *right* 1 unit. A third point has been plotted in the same way. The graph of $y = -3x + 1$ is the line drawn through these three points. ●

EXAMPLE ❼ *Writing an Equation in Slope–Intercept Form*

a) Write $4x - 3y = 9$ in slope–intercept form.
b) Graph the equation.

Figure 6.19

Figure 6.20

Figure 6.21

Figure 6.22

SOLUTION

a) To write $4x - 3y = 9$ in slope–intercept from, we solve the given equation for y.

$$4x - 3y = 9$$
$$4x - 4x - 3y = -4x + 9$$
$$-3y = -4x + 9$$
$$\frac{-3y}{-3} = \frac{-4x + 9}{-3}$$
$$y = \frac{-4x}{-3} + \frac{9}{-3} \quad \text{or} \quad y = \frac{4}{3}x - 3$$

Thus, in slope–intercept form, the equation is $y = \frac{4}{3}x - 3$.

b) The y-intercept is $(0, -3)$, and the slope is $\frac{4}{3}$. Plot a point at $(0, -3)$ on the y-axis, then move *up* 4 units and to the *right* 3 units to obtain the second point (see Fig. 6.19). Draw a line through the two points.   •

EXAMPLE 8 *Determining the Equation of a Line from Its Graph*

Determine the equation of the line in Fig. 6.20.

SOLUTION If we determine the slope and the y-intercept of the line, we can write the equation using slope–intercept form, $y = mx + b$. We see from the graph that the y-intercept is $(0, 2)$; thus, $b = 2$. The slope of the line is negative because the graph falls from left to right. The change in y is 2 units for every 3-unit change in x. Thus, m, the slope of the line, is $-\frac{2}{3}$.

$$y = mx + b$$
$$y = -\frac{2}{3}x + 2$$

The equation of the line is $y = -\frac{2}{3}x + 2$.   •

EXAMPLE 9 *Horizontal and Vertical Lines*

In the Cartesian coordinate system, graph (a) $y = 2$ and (b) $x = -3$.

SOLUTION

a) For any value of x, the value of y is 2. Therefore, the graph will be a horizontal line through $y = 2$ (Fig. 6.21).

b) For any value of y, the value of x is -3. Therefore, the graph will be a vertical line through $x = -3$ (Fig. 6.22).

Note that the graph of $y = 2$ has a slope of 0. The slope of the graph of $x = -3$ is undefined.   •

In graphing the equations in this section, we labeled the horizontal axis the x-axis and the vertical axis the y-axis. For each equation, we can determine values for y by substituting values for x. Since the value of y depends on the value of x, we refer to y as the *dependent variable* and x as the *independent variable*. We label the *vertical axis* with the *dependent variable* and the *horizontal axis* with the *independent variable*. For the equation $C = 3n + 5$, the C is the dependent variable and n is the independent variable. Thus, to graph this equation, we label the vertical axis C and the horizontal axis n.

In many graphs, the values to be plotted on one axis are much greater than the values to be plotted on the other axis. When that occurs, we can use different scales on the horizontal and the vertical axes, as illustrated in Examples 10 and 11. The next two examples illustrate applications of graphing.

EXAMPLE ⑩ *Using a Graph to Determine Area*

The Professional Patio Company installs brick patios. The area of brick, a, in square feet, the company can install in t hours can be approximated by the formula $a = 5t$.

a) Graph $a = 5t$, for $t \leq 6$.

b) Use the graph to estimate the area of brick the company can install in 4 hours.

SOLUTION

a) Since $a = 5t$ is a linear equation, its graph will be a straight line. Select three values for t, find the corresponding values for a, and then draw the graph (Fig. 6.23).

$$a = 5t$$

Let $t = 0$, $a = 5(0) = 0$

Let $t = 2$, $a = 5(2) = 10$

Let $t = 6$, $a = 5(6) = 30$

t	a
0	0
2	10
6	30

b) By drawing a vertical line from $t = 4$ on the time axis up to the graph and then drawing a horizontal line across to the area axis, we can determine that the area installed in 4 hours is 20 ft². ●

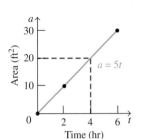

Figure 6.23

EXAMPLE ⑪ *Using a Graph to Determine Profits*

Javier Mungro owns a business that sells mountain bicycles. He believes the profit (or loss) from the bicycles sold can be estimated by the formula $P = 60S - 90,000$, where S is the number of bicycles sold.

a) Graph $P = 60S - 90,000$, for $S \leq 5000$ bicycles.

b) From the graph, estimate the number of bicycles that must be sold for the business to break even (Fig. 6.24).

c) From the graph, estimate the number of bicycles sold if the profit from selling bicycles is $150,000.

SOLUTION

a) Select values for S and find the corresponding values of P.

S	P
0	$-90,000$
2000	30,000
5000	210,000

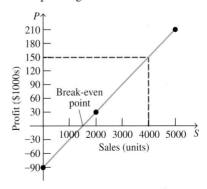

Figure 6.24

b) The break-even point is the number of bicycles that must be sold for the company to have neither a profit nor a loss. The break-even point is where the graph intercepts the horizontal, or *S*-axis, for that is where the profit, *P*, is 0. To break even, 1500 bicycles must be sold.

c) We can obtain the answer by drawing a horizontal line from 150 on the profit axis. Since the horizontal line cuts the graph at 4 on the *S* axis, 4000 bicycles were sold.

SECTION 6.7 EXERCISES

CONCEPT/WRITING EXERCISES

1. What is a graph?

2. Explain how to find the *x*-intercept of a linear equation.

3. Explain how to find the *y*-intercept of a linear equation.

4. What is the slope of a line?

5. **a)** Explain in your own words how to find the slope of a line between two points.

   b) Based on your explanation in part (a), find the slope of the line through the points $(6, 2)$ and $(-3, 5)$.

6. Describe the three methods used to graph a linear equation given in this section.

7. **a)** In which quadrant is the point $(1, 4)$ located?

   b) In which quadrant is the point $(-2, -5)$ located?

8. What is the minimum number of points needed to graph a linear equation?

PRACTICE THE SKILLS

In Exercises 9–16, plot all the points on the same axes.

9. $(2, 4)$	**10.** $(-4, 2)$	**11.** $(-5, -1)$
12. $(4, 0)$	**13.** $(0, 2)$	**14.** $(0, 0)$
15. $(0, -5)$	**16.** $(2\frac{1}{2}, 5\frac{1}{2})$	

In Exercises 17–24, plot all the points on the same axes.

17. $(1, 4)$	**18.** $(-3, 0)$	**19.** $(-2, -5)$
20. $(0, -1)$	**21.** $(0, 2)$	**22.** $(-4, 5)$
23. $(4, -1)$	**24.** $(4.5, 3.5)$	

In Exercises 25–34 (indicated on Fig. 6.25), write the coordinates of the corresponding point.

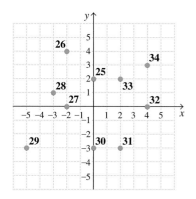

Figure 6.25

In Exercises 35–42, determine which ordered pairs satisfy the given equation.

35. $x + 2y = 9$ $(5, 2)$ $(1, 4)$ $(0, 5)$

36. $4x - y = 4$ $(0, -4)$ $(1, 0)$ $(2, -3)$

37. $2x - 3y = 10$ $(8, 2)$ $(-1, 4)$ $(0, -\frac{10}{3})$

38. $3y = 4x + 2$ $(2, 1)$ $(1, 2)$ $(0, \frac{2}{3})$

39. $7y = 3x - 5$ $(1, -1)$, $(-3, -2)$, $(2, 5)$

40. $3x - 5y = -12$ $(1, 3)$, $(-1, 2)$, $(-4, 0)$

41. $\frac{x}{4} + \frac{2y}{3} = 2$ $(8, 0)$, $(1, \frac{1}{2})$, $(0, 3)$

42. $\frac{x}{2} + 3y = 4$ $(0, \frac{4}{3})$, $(8, 0)$, $(10, -2)$

In Exercises 43–46, graph the equation and state the slope of the line if the slope exists (see Example 9).

43. $x = -1$ **44.** $x = 3$ **45.** $y = 2$ **46.** $y = -4$

In Exercises 47–56, graph the equation by plotting points, as in Example 3.

47. $y = x - 1$ **48.** $y = x + 4$

49. $y = 3x - 1$ **50.** $y = -x + 4$

51. $y + 3x = 6$ **52.** $y - 4x = 8$

53. $y = \frac{1}{2}x + 4$ **54.** $3y = 2x - 3$

55. $3y = x + 6$ **56.** $y = -\frac{2}{3}x$

In Exercises 57–66, graph the equation, using the x- and y-intercepts, as in Example 4.

57. $x + y = 4$ **58.** $x - y = 5$

59. $2x + y = 6$ **60.** $3x - y = -3$

61. $2x = -4y - 8$ **62.** $y = 4x + 4$

63. $y = 3x + 5$ **64.** $3x + 6y = 9$

65. $3x - y = 5$ **66.** $6x = 3y - 9$

In Exercises 67–76, determine the slope of the line through the given points. If the slope is undefined, so state.

67. $(6, 2)$ and $(8, 6)$ **68.** $(5, 3)$ and $(3, 5)$

69. $(-5, 6)$ and $(7, -9)$ **70.** $(-5, 6)$ and $(8, 8)$

71. $(5, 2)$ and $(-3, 2)$ **72.** $(-3, -5)$ and $(-1, -2)$

73. $(8, -3)$ and $(8, 3)$ **74.** $(2, 6)$ and $(2, -3)$

75. $(-3, -1)$ and $(8, -9)$ **76.** $(2, 1)$ and $(-5, -9)$

In Exercises 77–86, graph the equation using the slope and y-intercept, as in Examples 6 and 7.

77. $y = x - 3$ **78.** $y = -x - 4$

79. $y = 2x + 3$ **80.** $y = 3x - 2$

81. $y = -\frac{1}{3}x + 2$ **82.** $y = -x - 2$

83. $7y = 4x - 7$ **84.** $3x + 2y = 6$

85. $3x - 2y + 6 = 0$ **86.** $3x + 4y - 8 = 0$

In Exercises 87–90, determine the equation of the graph.

87.

88.

89.

90.

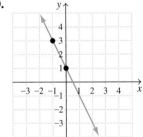

PROBLEM SOLVING

In Exercises 91 and 92, points A, B, and C are three vertices of a rectangle (the points where two sides meet). Plot the three points. (a) Find the coordinates of the fourth point, D, to complete the rectangle. (b) Find the area of the rectangle; use A = lw.

91. $A(-1, 3)$, $B(3, 3)$, $C(-1, -2)$

92. $A(-4, 2)$, $B(7, 2)$, $C(7, 8)$

In Exercises 93 and 94, points A, B, and C are three vertices of a parallelogram with sides parallel to the x-axis. Plot the three points. Find the coordinates of the fourth point, D, to complete the parallelogram. Note: There are two possible answers for point D.

93. $A(3, 2)$, $B(5, 5)$, $C(9, 5)$

94. $A(-2, 2)$, $B(3, 2)$, $C(6, -1)$

In Exercises 95–98, for what value of b will the line joining the points P and Q be parallel to the indicated axis?

95. $P(2, -3)$, $Q(6, b)$;   x-axis

96. $P(4, 7)$, $Q(b, -2)$;   y-axis

97. $P(3b - 1, 5)$, $Q(8, 4)$;   y-axis

98. $P(-6, 2b + 3)$, $Q(7, -1)$;   x-axis

99. *Hanging Wallpaper* Tanisha Vizquez owns a wallpaper hanging business. Her charge, C, for hanging wallpaper is $40 plus $0.30 per square foot of wallpaper she hangs, or $C = 40 + 0.30s$, where s is the number of square feet of wallpaper she hangs.

a) Graph $C = 40 + 0.30s$, for $s \leq 500$.

b) From the graph, estimate her charge if she hangs 300 square feet of wallpaper.

c) If her charge is $70, use the equation for C to determine how many square feet of wallpaper she hung.

100. *Selling Chocolates* Ryan Stewart sells chocolate on the Internet. His monthly profit, p, in dollars, can be estimated by $p = 15n - 300$, where n is the number of dozens of chocolates he sells in a month.

a) Graph $p = 15n - 300$, for $n \leq 60$.

b) From the graph, estimate his profit if he sells 40 dozen chocolates in a month.

c) How many dozens of chocolates must he sell in a month to break even?

101. *Cable Cost* The Schwimmer's cost for cable television is given by $C = 75 + 25x$, where x is the number of months of service.

a) Graph $C = 75 + 25x$, for $x \leq 50$.

b) From the graph, estimate the Schwimmer's cost for 30 months of cable television.

c) If the Schwimmer's total cost for cable television is $975, estimate the number of months of service.

102. *Earning Simple Interest* When $1000 is invested in a savings account paying simple interest for a year, the interest, i, in dollars, earned can be found by the formula $i = 1000r$, where r is the rate in decimal form.

a) Graph $i = 1000r$, for r up to and including a rate of 15%.

b) If the rate is 4%, what is the simple interest?

c) If the rate is 6%, what is the simple interest?

In Exercises 103 and 104, a set of points is plotted. Also shown is a straight line through the set of points that is called the line of best fit *(or a regression line, as will be discussed in Chapter 13, Statistics.)*

103. *Determining the Number of Defects* The graph on the top of page 375 shows the daily number of workers absent from the assembly line at J. B. Davis Corporation and the number of defects coming off the assembly line for 8 days. (The two points indicated on the line do not represent any of the 8 days.) The line of best fit, the blue line on the graph, can be used to approximate the number of defects coming off the assembly line per day for a given number of workers absent.

a) Determine the slope of the line of best fit using the two points indicated.

b) Using the slope determined in part (a) and the y-intercept, (0, 9), determine the equation of the line of best fit.

c) Using the equation you determined in part (b), determine the approximate number of defects for a day if 3 workers are absent.

d) Using the equation you determined in part (b), approximate the number of workers absent for a day if there are 17 defects that day.

**Daily Number of Defects at
J. B. Davis Corporation**

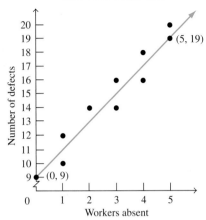

104. *Determining a Test Grade* The graph shows the hours studied and the test grades on a biology test for six students. (The two points indicated on the line do not represent any of the six students.) The line of best fit, the red line on the graph, can be used to approximate the test grade the average student receives for the number of hours he or she studies.

Test Grades on Biology Exam

a) Determine the slope of the line of best fit using the two points indicated.

b) Using the slope determined in part (a) and the *y*-intercept, (0, 53), determine the equation of the line of best fit.

c) Using the equation you determined in part (b), determine the approximate test grade for a student who studied for 3 hours.

d) Using the equation you determined in part (b), determine the amount of time a student would need to study to receive a grade of 80 on the biology test.

105. *Drug Sales* The green graph shows U.S. sales of pain relief drugs, in billions of dollars, for the years 2000–2004. The red dashed line can be used to approximate the U.S. sales of these pain relief drugs. If we let 0 represent 2000, 1 represent 2001, and so on, then 2004 would be represented by 4. Using the ordered pairs (0, 2.60) and (4, 2.41),

a) determine the slope of the red dashed line.

b) determine the equation of the red dashed line using (0, 2.60) as the *y*-intercept.

c) Using the equation you determined in part (b), estimate the U.S. sales of these pain relief drugs in 2002.

d) Using the equation you determined in part (b), determine the year in which U.S. sales of these pain relief drugs was $2.4 billion.

U.S. Sales of Pain Relief Drugs

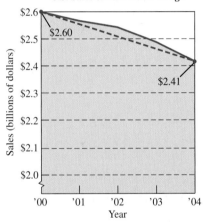

Source: AC Nielsen

106. *Airline Passengers* The blue graph on page 376 shows the actual number of passengers, in millions, who departed from Greater Rochester International Airport (GRIA) for the years 2001–2005. The red dashed line can be used to approximate the number of passengers, in millions, who departed from GRIA. If we let 0 represent 2001 and 1 represent 2002, and so on, then 2005 would be represented by 4. Using the ordered pairs (0, 1.15) and (4, 1.45),

a) determine the slope of the red dashed line.

b) determine the equation of the red dashed line using (0, 1.15) as the *y*-intercept of the graph.

c) Using the equation you determined in part (b), estimate the number of passengers who departed from GRIA in 2003.

d) Assuming this trend continues, use the equation you determined in part (b) to determine the year in which there will be 1.55 million passengers departing from GRIA.

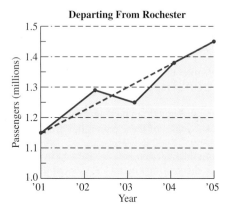

Departing From Rochester

Source: Monroe County

CHALLENGE PROBLEMS/GROUP ACTIVITIES

107. a) Two lines are parallel when they do not intersect no matter how far they are extended. Explain how you can determine, without graphing the equations, whether two equations will be parallel lines when graphed.

  b) Determine whether the graphs of the equations $2x - 3y = 6$ and $4x = 6y + 6$ are parallel lines.

108. In which quadrants will the set of points that satisfy the equation $x + y = 1$ lie? Explain.

INTERNET/RESEARCH ACTIVITY

109. René Descartes is known for his contributions to algebra. Write a paper on his life and his contributions to algebra.

6.8   LINEAR INEQUALITIES IN TWO VARIABLES

▲ A linear inequality in two variables may be used to represent the number of grills that can be shipped to each of two stores.

Suppose a manufacturer of gas grills needs to determine how many grills to ship to each of two stores. Also suppose the manufacturer can ship at most 300 grills in total. A linear inequality in two variables can be used to represent the number of grills that can be shipped to each store. In this section, we will discuss how to solve linear inequalities in two variables.

In Section 6.6, we introduced linear inequalities in one variable. Now we will introduce linear inequalities in two variables. Some examples of linear inequalities in two variables are $2x + 3y \leq 7$, $x + 7y \geq 5$, and $x - 3y < 6$.

The solution set of a linear inequality in one variable may be indicated on a number line. The solution set of a linear inequality in two variables is indicated on a coordinate plane.

An inequality that is strictly less than ($<$) or greater than ($>$) will have as its solution set a *half-plane*. A half-plane is the set of all the points on one side of a line. An inequality that is less than or equal to ($\leq$) or greater than or equal to ($\geq$) will have as its solution set the set of points that consists of a half-plane and a line. To indicate that the line is part of the solution set, we draw a solid line. To indicate that the line is not part of the solution set, we draw a dashed line.

TO GRAPH LINEAR INEQUALITIES IN TWO VARIABLES

1. Mentally substitute the equal sign for the inequality sign and plot points as if you were graphing the equation.

2. If the inequality is $<$ or $>$, draw a dashed line through the points. If the inequality is $\leq$ or $\geq$, draw a solid line through the points.

3. Select a test point not on the line and substitute the x- and y-coordinates into the inequality. If the substitution results in a true statement, shade in the area on the same side of the line as the test point. If the test point results in a false statement, shade in the area on the opposite side of the line as the test point.

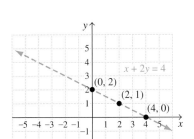

Figure 6.26

EXAMPLE ❶ *Graphing an Inequality*

Draw the graph of $x + 2y < 4$.

SOLUTION To obtain the solution set, start by graphing $x + 2y = 4$. Since the original inequality is strictly "less than," draw a dashed line (Fig 6.26). The dashed line indicates that the points on the line are not part of the solution set.

The line $x + 2y = 4$ divides the plane into three parts, the line itself and two *half-planes*. The line is the boundary between the two half-planes. The points in one half-plane will satisfy the inequality $x + 2y < 4$. The points in the other half-plane will satisfy the inequality $x + 2y > 4$.

To determine the solution set to the inequality $x + 2y < 4$, pick any point on the plane that is not on the line. The simplest point to work with is the origin, $(0, 0)$. Substitute $x = 0$ and $y = 0$ into $x + 2y < 4$.

$$x + 2y < 4$$
$$\text{Is } 0 + 2(0) < 4?$$
$$0 + 0 < 4$$
$$0 < 4 \quad \text{True}$$

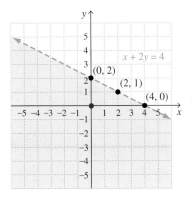

Figure 6.27

Since 0 is less than 4, the point $(0, 0)$ is part of the solution set. All the points on the same side of the graph of $x + 2y = 4$ as the point $(0, 0)$ are members of the solution set. We indicate that by shading the half-plane that contains $(0, 0)$. The graph is shown in Fig. 6.27. ●

EXAMPLE ❷ *Graphing an Inequality*

Draw the graph of $3x - 2y \leq -6$.

SOLUTION First draw the graph of the equation $3x - 2y = -6$. Use a solid line because the points on the boundary line are included in the solution set (see Fig. 6.28). Now pick a point that is not on the line. Take $(0, 0)$ as the test point.

$$3x - 2y \leq -6$$
$$\text{Is } 3(0) - 2(0) \leq -6?$$
$$0 \leq -6 \quad \text{False}$$

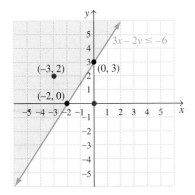

Figure 6.28

Since 0 is not less than or equal to $-6 (0 \nleq -6)$, the solution set is the line and the half-plane that does not contain the point $(0, 0)$.

If you had arbitrarily selected the test point $(-3, 2)$ from the other half-plane, you would have found that the inequality would be true: $3(-3) - 2(2) \leq -6$, or $-13 \leq -6$. Because -13 is less than or equal to -6, the point $(-3, 2)$ would be in the half-plane containing the solution set.

EXAMPLE ❸ *Graphing an Inequality*

Draw the graph of $y < x$.

SOLUTION The inequality is strictly "less than," so the boundary line is not part of the solution set. In graphing the equation $y = x$, draw a dashed line (Fig. 6.29). Since $(0, 0)$ is *on* the line, it cannot serve as a test point. Let's pick the point $(1, -1)$.

$$y < x$$
$$-1 < 1 \quad \text{True}$$

Since $-1 < 1$ is true, the solution set is the half-plane containing the point $(1, -1)$.

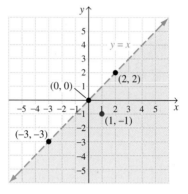

Figure 6.29

SECTION 6.8   EXERCISES

CONCEPT/WRITING EXERCISES

1. Outline the procedure used to graph inequalities in two variables.

2. Explain why we use a solid line when graphing an inequality containing $\leq$ or $\geq$ and we use a dashed line when graphing an inequality containing $<$ or $>$.

3. What is a half-plane?

4. **a)** How do the graphs of $x < 1$ and $x > 1$ differ?

   b) How do the graphs of $y > x + 3$ and $y \geq x + 3$ differ?

5. For each inequality, determine whether $(1, 3)$ is a solution to the inequality.

   a) $2x + 3y < 11$ **b)** $2x + 3y \leq 11$

   c) $2x + 3y \geq 11$ **d)** $2x + 3y > 11$

6. For each inequality, determine whether $(-2, 4)$ is a solution to the inequality.

   a) $4x - 2y < -10$ **b)** $4x - 2y \leq -10$

   c) $4x - 2y > -10$ **d)** $4x - 2y \geq -10$

PRACTICE THE SKILLS

In Exercises 7–28, draw the graph of the inequality.

7. $x \leq 3$

8. $y \geq -3$

9. $y > x - 4$

10. $y < x + 1$

11. $y \geq 2x - 6$

12. $y < -2x + 2$

13. $3x - 4y > 12$

14. $x + 2y > 4$

15. $3x - 4y \leq 9$

16. $4y - 3x \geq 9$

17. $3x + 2y < 6$

18. $-x + 2y < 2$

19. $x + y > 0$

20. $x + 2y \leq 0$

21. $5x + 2y \geq 10$

22. $y \geq -2x + 1$

23. $3x - 2y < 12$

24. $y \leq 3x - 4$

25. $\frac{2}{5}x - \frac{1}{2}y \leq 1$

26. $\frac{1}{2}x + \frac{2}{3}y \geq 1$

27. $0.1x + 0.3y \leq 0.4$

28. $0.2x + 0.5y \leq 0.3$

PROBLEM SOLVING

29. *Gas Grills* A manufacturer of gas grills must produce and ship x gas grills to one store and y gas grills to a second store. The maximum number of gas grills the manufacturer can produce and ship is 300.

   a) Write an inequality in two variables that represents this problem.

   b) Graph the inequality.

30. *Buying CDs and DVDs* Reggie Wayne is allowed to spend a maximum of $100 for purchasing x DVDs and y CDs. Each DVD costs $20, and each CD costs $10.

a) State this problem as an inequality in two variables.

b) Graph the inequality.

CHALLENGE PROBLEMS/GROUP ACTIVITIES

31. *Men's Shirts* The Tommy Hilfiger Company must ship x men's shirts to one outlet and y men's shirts to a second outlet. The maximum number of shirts the manufacturer can produce and ship is 250. We can represent this situation with the inequality $x + y \leq 250$.

a) Can x or y be negative? Explain.

b) Graph the inequality.

c) Write one or two paragraphs interpreting the information that the graph provides.

32. Which of the following inequalities have the same graph? Explain how you determined your answer.

a) $2x - y < 8$ **b)** $-2x + y > -8$

c) $2x - 4y < 16$ **d)** $y > 2x - 8$

6.9 SOLVING QUADRATIC EQUATIONS BY USING FACTORING AND BY USING THE QUADRATIC FORMULA

▲ A quadratic equation can be used to determine the width of a border around a rectangular swimming pool.

You are adding a brick border of uniform width around your rectangular swimming pool. If you know the surface area of the pool and the area you would like the brick border to cover, you can determine the width of the border by solving a quadratic equation. In this section, we will learn two methods that can be used to solve quadratic equations.

We begin this section by discussing binomials.

A *binomial* is an expression that contains two terms in which each exponent that appears on a variable is a whole number.

Examples of binomials

$x + 3$ $x - 5$

$3x + 5$ $4x - 2$

To multiply two binomials, we can use the *FOIL method*. The name of the method, FOIL, is an acronym to help its users remember it as a method that obtains the products of the *F*irst, *O*uter, *I*nner, and *L*ast terms of the binomials.

$$(a + b)(c + d) = \overset{\text{First}}{a \cdot c} + \overset{\text{Outer}}{a \cdot d} + \overset{\text{Inner}}{b \cdot c} + \overset{\text{Last}}{b \cdot d}$$

After multiplying the first, outer, inner, and last terms, combine all like terms.

EXAMPLE ❶ *Multiplying Binomials*

Multiply $(x + 4)(x + 7)$.

SOLUTION The FOIL method of multiplication yields

$$
\begin{aligned}
(x + 4)(x + 7) &= \overset{F}{x \cdot x} + \overset{O}{x \cdot 7} + \overset{I}{4 \cdot x} + \overset{L}{4 \cdot 7} \\
&= x^2 + 7x + 4x + 28 \\
&= x^2 + 11x + 28
\end{aligned}
$$

Note that the like terms $7x$ and $4x$ were combined to get $11x$.   ●

EXAMPLE ❷ *Multiplying Binomials*

Multiply $(2x - 1)(x + 4)$.

SOLUTION

$$
\begin{aligned}
(2x - 1)(x + 4) &= \overset{F}{2x \cdot x} + \overset{O}{2x \cdot 4} + \overset{I}{(-1) \cdot x} + \overset{L}{(-1) \cdot 4} \\
&= 2x^2 + 8x - x - 4 \\
&= 2x^2 + 7x - 4
\end{aligned}
$$

●

Factoring Trinomials of the Form $x^2 + bx + c$

The expression $x^2 + 8x + 15$ is an example of a trinomial. A *trinomial* is an expression containing three terms in which each exponent that appears on a variable is a whole number.

Example 1 showed that

$$(x + 4)(x + 7) = x^2 + 11x + 28$$

Since the product of $x + 4$ and $x + 7$ is $x^2 + 11x + 28$, we say that $x + 4$ and $x + 7$ are *factors* of $x^2 + 11x + 28$. To *factor* an expression means to write the expression as a product of its factors. For example, to factor $x^2 + 11x + 28$ we write

$$x^2 + 11x + 28 = (x + 4)(x + 7)$$

Let's look at the factors more closely.

$$4 + 7 = 11$$
$$4 \cdot 7 = 28$$
$$x^2 + 11x + 28 = (x + 4)(x + 7)$$

Note that the sum of the two numbers in the factors is $4 + 7$, or 11. The 11 is the coefficient of the x term. Also note that the product of the numbers in the two factors is $4 \cdot 7$, or 28. The 28 is the constant in the trinomial. In general, when factoring an expression of the form $x^2 + bx + c$, we need to find two numbers whose product is c and whose sum is b. When we determine the two numbers, the factors will be of the form

$$(x +)\ (x +)$$

One number   Other number

EXAMPLE ❸ *Factoring a Trinomial*

Factor $x^2 + 5x + 6$.

SOLUTION We need to find two numbers whose product is 6 and whose sum is 5. Since the product is $+6$, the two numbers must both be positive or both be negative. Because the coefficient of the x term is positive, only the positive factors of 6 need to be considered. Can you explain why? We begin by listing the positive numbers whose product is 6.

Factors of 6	Sum of Factors
1(6)	$1 + 6 = 7$
(2(3)	$2 + 3 = 5$)

Since $2 \cdot 3 = 6$ and $2 + 3 = 5$, the numbers we are seeking are 2 and 3. Thus, we write

$$x^2 + 5x + 6 = (x + 2)(x + 3)$$

Note that $(x + 3)(x + 2)$ is also an acceptable answer.   ●

TO FACTOR TRINOMIAL EXPRESSIONS OF THE FORM
$x^2 + bx + c$

1. Find two numbers whose product is c and whose sum is b.

2. Write factors in the form

$$(x +)\ (x +)$$

One number   Other number
from step 1   from step 1

3. Check your answer by multiplying the factors using the FOIL method.

If, for example, the numbers found in step 1 of the above procedure were 6 and -4, the factors would be written $(x + 6)(x - 4)$.

EXAMPLE ❹ Factoring a Trinomial

Factor $x^2 - 6x - 16$.

SOLUTION We must find two numbers whose product is -16 and whose sum is -6. Begin by listing the factors of -16.

Factors of -16	Sum of Factors
$-16(1)$	$-16 + 1 = -15$
$-8(2)$	$-8 + 2 = -6$
$-4(4)$	$-4 + 4 = 0$
$-2(8)$	$-2 + 8 = 6$
$-1(16)$	$-1 + 16 = 15$

The table lists all the factors of -16. The only factors listed whose product is -16 and whose sum is -6 are -8 and 2. We listed all factors in this example so that you could see, for example, that $-8(2)$ is a different set of factors than $-2(8)$. Once you find the factors you are looking for, there is no need to go any further. The trinomial can be written in factored form as

$$x^2 - 6x - 16 = (x - 8)(x + 2)$$

Factoring Trinomials of the Form $ax^2 + bx + c, a \neq 1$

Now we discuss how to factor an expression of the form $ax^2 + bx + c$, where a, the coefficient of the squared term, is not equal to 1.

Consider the multiplication problem $(2x + 1)(x + 3)$.

$$(2x + 1)(x + 3) = 2x \cdot x + 2x \cdot 3 + 1 \cdot x + 1 \cdot 3$$
$$= 2x^2 + 6x + x + 3$$
$$= 2x^2 + 7x + 3$$

Since $(2x + 1)(x + 3) = 2x^2 + 7x + 3$, the factors of $2x^2 + 7x + 3$ are $2x + 1$ and $x + 3$.

Let's study the coefficients more closely.

$$2x^2 + 7x + 3 \qquad (2x + 1)(1x + 3)$$

$$\mathbf{F} = 2 \cdot 1 = 2 \qquad \mathbf{O + I} = (2 \cdot 3) + (1 \cdot 1) = 7 \qquad \mathbf{L} = 1 \cdot 3 = 3$$

Note that the product of the coefficient of the first terms in the multiplication of the binomials equals 2, the coefficient of the squared term. The sum of the products of the coefficients of the outer and inner terms equals 7, the coefficient of the x-term. The product of the last terms equals 3, the constant.

A procedure to factor expressions of the form $ax^2 + bx + c$, $a \neq 1$, follows.

TO FACTOR TRINOMIAL EXPRESSIONS OF THE FORM
$$ax^2 + bx + c, a \neq 1$$

1. Write all pairs of factors of the coefficient of the squared term, a.

2. Write all pairs of factors of the constant, c.

3. Try various combinations of these factors until the sum of the products of the outer and inner terms is bx.

4. Check your answer by multiplying the factors using the FOIL method.

EXAMPLE ❺ *Factoring a Trinomial, a ≠ 1*

Factor $3x^2 + 17x + 10$.

SOLUTION The only positive factors of 3 are 3 and 1. Therefore, we write

$$3x^2 + 17x + 10 = (3x \quad)(x \quad)$$

The number 10 has both positive and negative factors. However, since both the constant, 10, and the sum of the products of the outer and inner terms, 17, are positive, the two factors must be positive. Why? The positive factors of 10 are 1(10) and 2(5). The following is a list of the possible factors.

Possible Factors	Sum of Products of Outer and Inner Terms
$(3x + 1)(x + 10)$	$31x$
$(3x + 10)(x + 1)$	$13x$
$(3x + 2)(x + 5)$	$17x$ ← Correct middle term
$(3x + 5)(x + 2)$	$11x$

Thus, $3x^2 + 17x + 10 = (3x + 2)(x + 5)$.   ●

Note that factoring problems of this type may be checked by using the FOIL method of multiplication. We will check the results to Example 5:

$$(3x + 2)(x + 5) = 3x \cdot x + 3x \cdot 5 + 2 \cdot x + 2 \cdot 5$$
$$= 3x^2 + 15x + 2x + 10$$
$$= 3x^2 + 17x + 10$$

Since we obtained the expression we started with, our factoring is correct.

EXAMPLE 6   *Factoring a Trinomial, a ≠ 1*

Factor $6x^2 - 11x - 10$.

SOLUTION The factors of 6 will be either $6 \cdot 1$ or $2 \cdot 3$. Therefore, the factors may be of the form $(6x \quad)(x \quad)$ or $(2x \quad)(3x \quad)$. When there is more than one set of factors for the first term, we generally try the medium-sized factors first. If that does not work, we try the other factors. Thus, we write

$$6x^2 - 11x - 10 = (2x \quad)(3x \quad)$$

The factors of -10 are $(-1)(10)$, $(1)(-10)$, $(-2)(5)$, and $(2)(-5)$. There will be eight different pairs of possible factors of the trinomial $6x^2 - 11x - 10$. Can you list them?

The correct factoring is $6x^2 - 11x - 10 = (2x - 5)(3x + 2)$.   ●

Note that in Example 6 we first tried factors of the form $(2x \quad)(3x \quad)$. If we had not found the correct factors using them, we would have tried $(6x \quad)(x \quad)$.

Solving Quadratic Equations by Factoring

In Section 6.2, we solved linear, or first-degree, equations. In those equations, the exponent on all variables was 1. Now we deal with the *quadratic equation*. The standard form of a quadratic equation in one variable is shown in the box.

STANDARD FORM OF A QUADRATIC EQUATION

$$ax^2 + bx + c = 0, \qquad a \neq 0$$

Note that in the standard form of a quadratic equation, the greatest exponent on x is 2 and the right side of the equation is equal to zero. *To solve a quadratic equation* means to find the value or values that make the equation true. In this section, we will solve quadratic equations by factoring and by the quadratic formula.

To solve a quadratic equation by factoring, set one side of the equation equal to 0 and then use the *zero-factor* property.

ZERO-FACTOR PROPERTY

If $a \cdot b = 0$, then $a = 0$ or $b = 0$.

The zero-factor property indicates that if the product of two factors is 0, then one (or both) of the factors must have a value of 0.

EXAMPLE 7   *Using the Zero-Factor Property*

Solve the equation $(x + 2)(x - 5) = 0$.

SOLUTION When we use the zero-factor property, either $(x + 2)$ or $(x - 5)$ must equal 0 for the product to equal 0. Thus, we set each individual factor equal to 0 and solve each resulting equation for x.

$$(x + 2)(x - 5) = 0$$

$$x + 2 = 0 \quad \text{or} \quad x - 5 = 0$$

$$x = -2 \qquad\qquad x = 5$$

Thus, the solutions are -2 and 5.

CHECK: $\qquad\qquad x = -2 \qquad\qquad\qquad\qquad\qquad x = 5$

$$(x + 2)(x - 5) = 0 \qquad\qquad (x + 2)(x - 5) = 0$$

$$(-2 + 2)(-2 - 5) = 0 \qquad\qquad (5 + 2)(5 - 5) = 0$$

$$0(-7) = 0 \qquad\qquad\qquad 7(0) = 0$$

$$0 = 0 \quad \text{True} \qquad\qquad\qquad 0 = 0 \quad \text{True} \qquad\qquad ●$$

TO SOLVE A QUADRATIC EQUATION BY FACTORING

1. Use the addition or subtraction property to make one side of the equation equal to 0.

2. Factor the side of the equation not equal to 0.

3. Use the zero-factor property to solve the equation.

Examples 8 and 9 illustrate this procedure.

EXAMPLE ❽ *Solving a Quadratic Equation by Factoring*

Solve the equation $x^2 - 8x = -15$.

SOLUTION First add 15 to both sides of the equation to make the right-hand side of the equation equal to 0.

$$x^2 - 8x = -15$$

$$x^2 - 8x + 15 = -15 + 15$$

$$x^2 - 8x + 15 = 0$$

Factor the left-hand side of the equation. The object is to find two numbers whose product is 15 and whose sum is -8. Since the product of the numbers is positive and the sum of the numbers is negative, the two numbers must both be negative. The numbers are -3 and -5. Note that $(-3)(-5) = 15$ and $-3 + (-5) = -8$.

$$x^2 - 8x + 15 = 0$$

$$(x - 3)(x - 5) = 0$$

Now use the zero-factor property to find the solution.

$$x - 3 = 0 \quad \text{or} \quad x - 5 = 0$$

$$x = 3 \qquad\qquad x = 5$$

The solutions are 3 and 5.   ●

EXAMPLE 9   *Solving a Quadratic Equation by Factoring*

Solve the equation $3x^2 - 13x + 4 = 0$.

SOLUTION $3x^2 - 13x + 4$ factors into $(3x - 1)(x - 4)$. Thus, we write

$$3x^2 - 13x + 4 = 0$$
$$(3x - 1)(x - 4) = 0$$
$$3x - 1 = 0 \quad \text{or} \quad x - 4 = 0$$
$$3x = 1 \quad\quad\quad\quad x = 4$$
$$x = \frac{1}{3}$$

The solutions are $\frac{1}{3}$ and 4.

TIMELY TIP   As stated on page 383, every factoring problem can be checked by multiplying the factors. If you have factored correctly, the product of the factors will be identical to the original expression that was factored. If we wished to check the factoring of Example 9, we would multiply $(3x - 1)(x - 4)$. Since the product of the factors is $3x^2 - 13x + 4$, which is the expression we started with, our factoring is correct.

Solving Quadratic Equations by Using the Quadratic Formula

Not all quadratic equations can be solved by factoring. When a quadratic equation cannot be easily solved by factoring, we can solve the equation with the *quadratic formula*. The quadratic formula can be used to solve any quadratic equation.

QUADRATIC FORMULA

For a quadratic equation in standard form, $ax^2 + bx + c = 0$, $a \neq 0$, the quadratic formula is

$$x = \frac{-b \pm \sqrt{b^2 - 4ac}}{2a}$$

In the quadratic formula, the plus or minus symbol, $\pm$, is used. If, for example, $x = 2 \pm 3$, then $x = 2 + 3 = 5$ or $x = 2 - 3 = -1$.

It is possible for a quadratic equation to have no real solution. In solving an equation, if the radicand (the expression inside the square root) is a negative number, the quadratic equation has *no real solution*.

To use the quadratic formula, first write the quadratic equation in standard form. Then determine the values for a (the coefficient of the squared term), b (the coefficient of the x term), and c (the constant). Finally, substitute the values of a, b, and c into the quadratic formula and evaluate the expression.

EXAMPLE ❿ *Solving a Quadratic Equation by Using the Quadratic Formula*

Solve the equation $x^2 - 5x - 14 = 0$ by using the quadratic formula.

SOLUTION In this equation, $a = 1$, $b = -5$, and $c = -14$.

$$x = \frac{-b \pm \sqrt{b^2 - 4ac}}{2a} = \frac{-(-5) \pm \sqrt{(-5)^2 - 4(1)(-14)}}{2(1)}$$

$$= \frac{5 \pm \sqrt{25 + 56}}{2}$$

$$= \frac{5 \pm \sqrt{81}}{2}$$

$$= \frac{5 \pm 9}{2}$$

$$\frac{5 + 9}{2} = \frac{14}{2} = 7 \quad \text{or} \quad \frac{5 - 9}{2} = \frac{-4}{2} = -2$$

The solutions are 7 and -2.

Note that Example 10 can also be solved by factoring. We suggest you do so now.

EXAMPLE ⓫ *Irrational Solutions to a Quadratic Equation*

Solve $4x^2 - 8x = -1$ by using the quadratic formula.

SOLUTION Begin by writing the equation in standard form by adding 1 to both sides of the equation, which gives the following.

$$4x^2 - 8x + 1 = 0$$

$$a = 4, \qquad b = -8, \qquad c = 1$$

$$x = \frac{-b \pm \sqrt{b^2 - 4ac}}{2a} = \frac{-(-8) \pm \sqrt{(-8)^2 - 4(4)(1)}}{2(4)}$$

$$= \frac{8 \pm \sqrt{64 - 16}}{8}$$

$$= \frac{8 \pm \sqrt{48}}{8}$$

Since $\sqrt{48} = \sqrt{16}\sqrt{3} = 4\sqrt{3}$ (see Section 5.4), we write

$$\frac{8 \pm \sqrt{48}}{8} = \frac{8 \pm 4\sqrt{3}}{8} = \frac{\overset{1}{\cancel{4}}(2 \pm \sqrt{3})}{\underset{2}{\cancel{8}}} = \frac{2 \pm \sqrt{3}}{2}$$

The solutions are $\dfrac{2 + \sqrt{3}}{2}$ and $\dfrac{2 - \sqrt{3}}{2}$.

Note that the solutions to Example 11 are irrational numbers.

Figure 6.30

EXAMPLE ⓬ *Brick Border*

Diane Cecero and her husband recently installed an inground rectangular swimming pool measuring 40 ft by 30 ft. They want to add a brick border of uniform width around all sides of the pool. How wide can they make the brick border if they purchased enough brick to cover an area of 296 ft²?

SOLUTION Let's make a diagram of the pool and the brick border (Fig. 6.30). Let x = the uniform width of the brick border. Then the total length of the larger rectangular area, the pool plus the border, is $2x + 40$. The total width of the larger rectangular area is $2x + 30$.

The area of the brick border can be found by subtracting the area of the pool from the area of the pool plus the brick border.

$$\text{Area of pool} = l \cdot w = (40)(30) = 1200 \text{ ft}^2$$
$$\text{Area of pool plus brick border} = l \cdot w = (2x + 40)(2x + 30)$$
$$= 4x^2 + 140x + 1200$$
$$\text{Area of the brick border} = \text{area of pool plus brick border} - \text{area of pool}$$
$$= (4x^2 + 140x + 1200) - 1200$$
$$= 4x^2 + 140x$$

The area of the brick border must be 296 ft². Therefore,

$$296 = 4x^2 + 140x$$

or

$$4x^2 + 140x - 296 = 0$$
$$4(x^2 + 35x - 74) = 0 \quad \text{Factor out 4 from each term.}$$
$$\frac{4}{4}(x^2 + 35x - 74) = \frac{0}{4} \quad \text{Divide both sides of the equation by 4.}$$
$$x^2 + 35x - 74 = 0$$
$$(x + 37)(x - 2) = 0 \quad \text{Factor trinomial.}$$
$$x + 37 = 0 \quad \text{or} \quad x - 2 = 0$$
$$x = -37 \qquad\qquad x = 2$$

Since lengths are positive, the only possible answer is $x = 2$. Thus, they can make a brick border 2 ft wide all around the pool.   ●

SECTION ⑥.⑨   EXERCISES

CONCEPT/WRITING EXERCISES

1. What is a *binomial*? Give three examples of binomials.

2. What is a *trinomial*? Give three examples of trinomials.

3. In your own words, explain the FOIL method used to multiply two binomials.

4. In your own words, state the zero-factor property.

5. Give the standard form of a quadratic equation.

6. Have you memorized the quadratic formula? If not, you need to do so. Without looking at the book, write the quadratic formula.

PRACTICE THE SKILLS

In Exercises 7–22, factor the trinomial. If the trinomial cannot be factored, so state.

7. $x^2 + 8x + 15$ **8.** $x^2 + 9x + 14$

9. $x^2 - x - 2$ **10.** $x^2 + 3x - 4$

11. $x^2 + 2x - 24$ **12.** $x^2 - 6x + 8$

13. $x^2 - 2x - 3$ **14.** $x^2 - 5x - 6$

15. $x^2 - 10x + 21$ **16.** $x^2 - x - 20$

17. $x^2 - 49$ **18.** $x^2 - 64$

19. $x^2 + 3x - 28$ **20.** $x^2 + 4x - 32$

21. $x^2 + 2x - 63$ **22.** $x^2 - 2x - 48$

In Exercises 23–34, factor the trinomial. If the trinomial cannot be factored, so state.

23. $2x^2 - x - 6$ **24.** $3x^2 - 7x - 6$

25. $5x^2 + 16x + 3$ **26.** $2x^2 + 3x - 20$

27. $5x^2 + 12x + 4$ **28.** $2x^2 - 9x + 10$

29. $4x^2 + 11x + 6$ **30.** $4x^2 + 20x + 21$

31. $4x^2 - 11x + 6$ **32.** $6x^2 - 11x + 4$

33. $8x^2 - 10x - 3$ **34.** $6x^2 - 19x + 3$

In Exercises 35–38, solve each equation, using the zero-factor property.

35. $(x - 3)(x + 6) = 0$ **36.** $(2x + 1)(x - 5) = 0$

37. $(3x + 4)(2x - 1) = 0$ **38.** $(x - 6)(5x - 4) = 0$

In Exercises 39–58, solve each equation by factoring.

39. $x^2 + 7x + 10 = 0$ **40.** $x^2 + x - 12 = 0$

41. $x^2 - 7x + 6 = 0$ **42.** $x^2 - 10x + 21 = 0$

43. $x^2 - 15 = 2x$ **44.** $x^2 - 7x = -6$

45. $x^2 = 4x - 3$ **46.** $x^2 - 13x + 40 = 0$

47. $x^2 - 81 = 0$ **48.** $x^2 - 64 = 0$

49. $x^2 + 5x - 36 = 0$ **50.** $x^2 + 12x + 20 = 0$

51. $3x^2 + 10x = 8$ **52.** $3x^2 - 5x = 2$

53. $5x^2 + 11x = -2$ **54.** $2x^2 = -5x + 3$

55. $3x^2 - 4x = -1$ **56.** $5x^2 + 16x + 12 = 0$

57. $6x^2 - 11x + 3 = 0$ **58.** $4x^2 + 11x - 3 = 0$

In Exercises 59–78, solve the equation, using the quadratic formula. If the equation has no real solution, so state.

59. $x^2 + 2x - 15 = 0$ **60.** $x^2 + 12x + 27 = 0$

61. $x^2 - 3x - 18 = 0$ **62.** $x^2 - 6x - 16 = 0$

63. $x^2 - 8x = 9$ **64.** $x^2 = -8x + 15$

65. $x^2 - 2x + 3 = 0$ **66.** $2x^2 - x - 3 = 0$

67. $x^2 - 4x + 2 = 0$ **68.** $2x^2 - 5x - 2 = 0$

69. $3x^2 - 8x + 1 = 0$ **70.** $2x^2 + 4x + 1 = 0$

71. $3x^2 - 6x - 5 = 0$ **72.** $4x^2 - 5x - 3 = 0$

73. $2x^2 + 7x + 5 = 0$ **74.** $3x^2 = 9x - 5$

75. $3x^2 - 10x + 7 = 0$ **76.** $4x^2 + 7x - 1 = 0$

77. $4x^2 + 6x = -5$ **78.** $5x^2 - 4x = -2$

CHALLENGE PROBLEMS/GROUP ACTIVITIES

79. *Flower Garden* Karen and Kurt Ohliger's backyard has a width of 20 meters and a length of 30 meters. Karen and Kurt want to put a rectangular flower garden in the middle of the backyard leaving a strip of grass of uniform width around all sides of the flower garden. If they want to have 336 square meters of grass, what will be the width and length of the garden?

80. *Lawn Maintenance Business* The weekly profit, p, of a lawn maintenance business, in dollars, can be approximated by the equation

$$p = x^2 + 40x - 110$$

where x is the number of lawns maintained in a week. How many lawns must be maintained to have a weekly profit of $1090?

81. a) Explain why solving $(x - 4)(x - 7) = 6$ by setting each factor equal to 6 is not correct.

   b) Determine the correct solution to $(x - 4)(x - 7) = 6$.

82. The radicand in the quadratic formula, $b^2 - 4ac$, is called the **discriminant**. How many real number solutions will the quadratic equation have if the discriminant is (a) greater than 0, (b) equal to 0, or (c) less than zero? Explain your answer.

83. Write an equation that has solutions -1 and 3.

INTERNET/RESEARCH ACTIVITIES

84. Italian mathematician Girolamo Cardano (1501–1576) is recognized for his skill in solving equations. Write a paper about his life and his contributions to mathematics, in particular his contribution to solving equations.

85. Chinese mathematician Foo Ling Awong, who lived during the Pong dynasty, developed a technique, other than trial and error, to factor trinomials of the form $ax^2 + bx + c$, $a \neq 1$. Write a paper about his life and his contributions to mathematics, in particular his technique for factoring trinomials in the form $ax^2 + bx + c$, $a \neq 1$. (References include history of mathematics books, encyclopedias, and the Internet.)

6.10   FUNCTIONS AND THEIR GRAPHS

▲ The more oranges you purchase, the greater the cost.

At a supermarket, oranges cost $0.20 each. Thus, one orange costs $1 \times \$0.20 = \$0.20$, two oranges cost $2 \times \$0.20 = \$0.40$, three oranges cost $3 \times \$0.20 = \$0.60$, and so on. In this section, we will discuss relationships between two quantities such as the number of oranges purchased and the cost of the oranges purchased.

Relations and *functions* are both extremely important mathematical concepts that we will introduce in this section. A *relation* is any set of ordered pairs. Every graph is a set of ordered pairs. Therefore, any graph will be a relation. We can indicate the relation between the number of oranges purchased and the cost of the oranges discussed in the section opening paragraph in a table of values.

Number of Oranges	Cost
0	0.00
1	0.20
2	0.40
3	0.60
⋮	⋮
10	2.00
⋮	⋮

In general, the cost for purchasing n oranges will be 20 cents times the number of oranges, or $0.20n$ dollars. We can represent the cost, c, of n items by the equation $c = 0.20n$. Since the value of c depends on the value of n, we refer to c as the *dependent variable* and n as the *independent variable*. Note that for each value of the independent variable, n, there is one and only one value of the dependent variable, c.

An equation such as $c = 0.20n$ is called a *function*. In the equation $c = 0.20n$, the value of c depends on the value of n, so we say that "c is a function of n."

> A **function** is a special type of relation where each value of the independent variable corresponds to a unique value of the dependent variable.

The set of values that can be used for the independent variable is called the *domain* of the function, and the resulting set of values obtained for the dependent variable is called the *range*. The domain and range for the function $c = 0.20n$ are illustrated in Fig. 6.31.

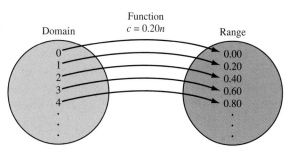

Figure 6.31

When we graphed equations of the form $ax + by = c$ in Section 6.7, we found that they were straight lines. For example, the graph of $y = 2x - 1$ is illustrated in Fig. 6.32.

Is the equation $y = 2x - 1$ a function? To answer this question, we must ask, Does each value of x correspond to a unique value of y? The answer is yes; therefore, this equation is a function.

For the equation $y = 2x - 1$, we say that "y is a function of x" and write $y = f(x)$. The notation $f(x)$ is read "f of x." When we are given an equation that is a function, we may replace the y in the equation with $f(x)$ because $f(x)$ represents y. Thus, $y = 2x - 1$ may be written $f(x) = 2x - 1$.

To evaluate a function for a specific value of x, replace each x in the function with the given value, then evaluate. For example, to evaluate $f(x) = 2x - 1$ when $x = 8$, we do the following.

$$f(x) = 2x - 1$$
$$f(8) = 2(8) - 1 = 16 - 1 = 15$$

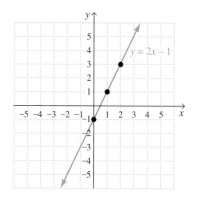

Figure 6.32

Thus, $f(8) = 15$. Since $f(x) = y$, when $x = 8$, $y = 15$. What is the domain and range of $f(x) = 2x - 1$? Because x can be any real number, the domain is the set of real numbers, symbolized $\mathbb{R}$. The range is also $\mathbb{R}$.

We can determine whether a graph represents a function by using the *vertical line test*: If a vertical line can be drawn so that it intersects the graph at more than one point, then each value of x does not have a unique value of y and the graph does not represent a function. If a vertical line cannot be made to intersect the graph in at least two different places, then the graph represents a function.

EXAMPLE ❶ *Using the Vertical Line Test*

Use the vertical line test to determine which of the graphs in Figure 6.33 represent functions.

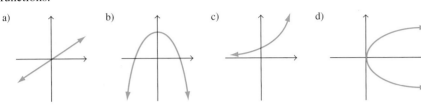

a)   b)   c)   d)

Figure 6.33

SOLUTION (a), (b), and (c) represent functions, but (d) does not.

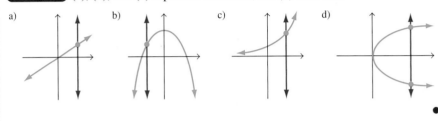

a)   b)   c)   d)

There are many real-life applications of functions. In fact, all the applications illustrated in Sections 6.2 through 6.4 are functions.

In this section, we will discuss three types of functions: linear functions, quadratic functions, and exponential functions.

Linear Functions

In Section 6.7, we graphed linear equations. The graph of any linear equation of the form $y = ax + b$ will pass the vertical line test, and so equations of the form $y = ax + b$ are *linear functions*. If we wished, we could write the linear function as $f(x) = ax + b$ since $f(x)$ means the same as y.

EXAMPLE ❷ *Cost as a Linear Function*

Adam Finiteri's weekly cost of operating a taxi, c, is given by the function $c(m) = 52 + 0.23m$, where m is the number of miles driven per week. What is his weekly cost if he drives 200 miles in a week?

SOLUTION Substitute 200 for m in the function.

$$c(m) = 52 + 0.23m$$
$$c(200) = 52 + 0.23(200)$$
$$c(200) = 52 + 46 = 98$$

Thus, if Adam drives 200 miles in a week, his weekly cost will be $98.

Graphs of Linear Functions

The graphs of linear functions are straight lines that will pass the vertical line test. In Section 6.7, we discussed how to graph linear equations. Linear functions can be graphed by plotting points, by using intercepts, or by using the slope and y-intercept.

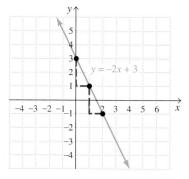

Figure 6.34

EXAMPLE ❸ *Graphing a Linear Function*

Graph $f(x) = -2x + 3$ by using the slope and y-intercept.

SOLUTION Since $f(x)$ means the same as y, we can rewrite this function as $y = -2x + 3$. From Section 6.7, we know that the slope is -2 and the y-intercept is (0, 3). Plot (0, 3) on the y-axis. Then plot the next point by moving *down* 2 units and to the *right* 1 unit (see Fig. 6.34). A third point has been plotted in the same way. The graph of $f(x) = -2x + 3$ is the line drawn through these three points. ●

Quadratic Functions

The standard form of a quadratic equation is $y = ax^2 + bx + c, a \neq 0$. We will learn shortly that graphs of equations of this form always pass the vertical line test and are functions. Therefore, equations of the form $y = ax^2 + bx + c, a \neq 0$, may be referred to as *quadratic functions*. We may express quadratic functions using function notation as $f(x) = ax^2 + bx + c$. Two examples of quadratic functions are $y = 2x^2 + 5x - 7$ and $y = -\frac{1}{2}x^2 + 4$.

EXAMPLE ❹ *Landing on the Moon*

On July 20, 1969, Neil Armstrong became the first person to walk on the moon. The velocity, v, of his spacecraft, the *Eagle*, in meters per second, was a function of time before touchdown, t, given by

$$v = f(t) = 3.2t + 0.45$$

The height of the spacecraft, h, above the moon's surface, in meters, was also a function of time before touchdown, given by

$$h = g(t) = 1.6t^2 + 0.45t$$

What was the velocity of the spacecraft and its distance from the surface of the moon

a) at 3 seconds before touchdown?   b) at touchdown (0 seconds)?

SOLUTION

a) $v = f(t) = 3.2t + 0.45,$ $h = g(t) = 1.6t^2 + 0.45t$
$\quad\quad\ f(3) = 3.2(3) + 0.45 \quad\quad\ g(3) = 1.6(3)^2 + 0.45(3)$
$\quad\quad\quad\quad\ = 9.6 + 0.45 \quad\quad\quad\quad\quad = 1.6(9) + 1.35$
$\quad\quad\quad\quad\ = 10.05 \quad\quad\quad\quad\quad\quad\ = 14.4 + 1.35$
$\quad\quad\quad\quad\quad\quad\quad\quad\quad\quad\quad\quad\quad\quad\ = 15.75$

The velocity 3 seconds before touchdown was 10.05 meters per second, and the height 3 seconds before touchdown was 15.75 meters.

b) $v = f(t) = 3.2t + 0.45,$ $h = g(t) = 1.6t^2 + 0.45t$
$\quad\quad\ f(0) = 3.2(0) + 0.45 \quad\quad\ g(0) = 1.6(0)^2 + 0.45(0)$
$\quad\quad\quad\quad\ = 0 + 0.45 \quad\quad\quad\quad\quad\ = 0 + 0$
$\quad\quad\quad\quad\ = 0.45 \quad\quad\quad\quad\quad\quad\ = 0$

The touchdown velocity was 0.45 meter per second. At touchdown, the *Eagle* is on the moon, and the distance from the surface of the moon is therefore 0 meter. ●

Graphs of Quadratic Functions

The graph of every quadratic function is a parabola. Two parabolas are illustrated in Fig. 6.35. Note that both graphs represent functions since they pass the vertical line test. A parabola opens upward when the coefficient of the squared term, a, is greater than 0, as shown in Figure 6.35(a). A parabola opens downward when the coefficient of the squared term, a, is less than 0, as shown in Fig. 6.35(b).

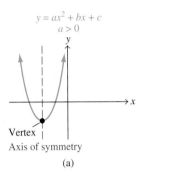

Figure 6.35

The *vertex* of a parabola is the lowest point on a parabola that opens upward and the highest point on a parabola that opens downward. Every parabola is *symmetric* with respect to a vertical line through its vertex. This line is called the *axis of symmetry* of the parabola. The x-coordinate of the vertex and the equation of the axis of symmetry can be found by using the following formula.

> **AXIS OF SYMMETRY OF A PARABOLA**
>
> $$x = \frac{-b}{2a}$$

This formula also gives the x-coordinate of the vertex of a parabola.

Once the x-coordinate of the vertex has been determined, the y-coordinate can be found by substituting the value found for the x-coordinate into the quadratic equation and evaluating the equation. This procedure is illustrated in Example 5.

EXAMPLE ❺ *Describing the Graph of a Quadratic Equation*

Consider the equation $y = -2x^2 + 12x - 10$.

a) Determine whether the graph will be a parabola that opens upward or downward.

b) Determine the equation of the axis of symmetry of the parabola.

c) Determine the vertex of the parabola.

SOLUTION

a) Since $a = -2$, which is less than 0, the parabola opens downward.

b) To find the axis of symmetry, we use the equation $x = -\frac{b}{2a}$. In the equation $y = -2x^2 + 12x - 10$, $a = -2$, $b = 12$, and $c = -10$, so

$$x = \frac{-b}{2a} = \frac{-(12)}{2(-2)} = \frac{-12}{-4} = 3$$

The equation of the axis of symmetry is $x = 3$.

c) The x-coordinate of the vertex is 3 from part (b). To find the y-coordinate, we substitute 3 for x in the equation and then evaluate.

$$y = -2x^2 + 12x - 10$$
$$= -2(3)^2 + 12(3) - 10$$
$$= -2(9) + 36 - 10$$
$$= -18 + 36 - 10$$
$$= 8$$

Therefore, the vertex of the parabola is located at the ordered pair (3, 8) on the graph. ●

GENERAL PROCEDURE TO SKETCH THE GRAPH OF A QUADRATIC EQUATION

1. Determine whether the parabola opens upward or downward.
2. Determine the equation of the axis of symmetry.
3. Determine the vertex of the parabola.
4. Determine the y-intercept by substituting $x = 0$ into the equation.
5. Determine the x-intercepts (if they exist) by substituting $y = 0$ into the equation and solving for x.
6. Draw the graph, making use of the information gained in steps 1 through 5. Remember that the parabola will be symmetric with respect to the axis of symmetry.

In step 5, you may use either factoring or the quadratic formula to determine the x-intercepts.

EXAMPLE ❻ *Graphing a Quadratic Equation*

Sketch the graph of the equation $y = x^2 - 6x + 8$.

SOLUTION We follow the steps outlined in the general procedure.
1. Since $a = 1$, which is greater than 0, the parabola opens upward.
2. Axis of symmetry: $x = \dfrac{-b}{2a} = \dfrac{-(-6)}{2(1)} = \dfrac{6}{2} = 3$
   Thus, the axis of symmetry is $x = 3$.
3. y-coordinate of vertex:
   $y = x^2 - 6x + 8$
   $= (3)^2 - 6(3) + 8 = 9 - 18 + 8 = -1$
   Thus, the vertex is at $(3, -1)$.
4. y-intercept: $y = x^2 - 6x + 8$
   $y = 0^2 - 6(0) + 8 = 8$
   Thus, the y-intercept is at (0, 8).
5. x-intercepts: $0 = x^2 - 6x + 8$, or $x^2 - 6x + 8 = 0$
   We can solve this equation by factoring.
   $$x^2 - 6x + 8 = 0$$
   $$(x - 4)(x - 2) = 0$$
   $$x - 4 = 0 \quad \text{or} \quad x - 2 = 0$$
   $$x = 4 \qquad\qquad x = 2$$
   Thus, the x-intercepts are (4, 0) and (2, 0).

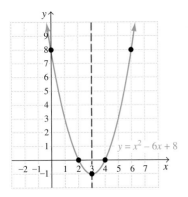

Figure 6.36

6. Plot the vertex $(3, -1)$, the y-intercept $(0, 8)$, and the x-intercepts $(4, 0)$ and $(2, 0)$. Then sketch the graph (Fig. 6.36). ●

Note that the domain of the graph in Example 6, the possible x values, is the set of all real numbers, $\mathbb{R}$. The range, the possible y values, is the set of all real numbers greater than or equal to -1. When graphing parabolas, if you think you need additional points to graph the equation, you can always substitute values for x and find the corresponding values of y and plot those points. For example, if you substituted 1 for x, the corresponding value of y is 3. Thus, you could plot the point $(1, 3)$.

EXAMPLE ❼ *Domain and Range of a Quadratic Function*

a) Sketch the graph of the function $f(x) = -2x^2 + 3x + 4$.

b) Determine the domain and range of the function.

SOLUTION

a) Since $f(x)$ means y, we can replace $f(x)$ with y to obtain $y = -2x^2 + 3x + 4$. Now graph $y = -2x^2 + 3x + 4$ using the steps outlined in the general procedure.

1. Since $a = -2$, which is less than 0, the parabola opens downward.

2. Axis of symmetry: $x = \dfrac{-b}{2a} = \dfrac{-(3)}{2(-2)} = \dfrac{-3}{-4} = \dfrac{3}{4}$

   Thus, the axis of symmetry is $x = \frac{3}{4}$.

3. y-coordinate of vertex:

$$y = -2x^2 + 3x + 4$$

$$= -2\left(\frac{3}{4}\right)^2 + 3\left(\frac{3}{4}\right) + 4$$

$$= -2\left(\frac{9}{16}\right) + \frac{9}{4} + 4$$

$$= -\frac{9}{8} + \frac{9}{4} + 4$$

$$= -\frac{9}{8} + \frac{18}{8} + \frac{32}{8} = \frac{41}{8} \quad \text{or} \quad 5\frac{1}{8}$$

   Thus, the vertex is at $\left(\frac{3}{4}, 5\frac{1}{8}\right)$.

4. y-intercept: $y = -2x^2 + 3x + 4$

$$= -2(0)^2 + 3(0) + 4 = 4$$

   Thus, the y-intercept is $(0, 4)$.

5. x-intercepts: $y = -2x^2 + 3x + 4$

$$0 = -2x^2 + 3x + 4 \quad \text{or} \quad -2x^2 + 3x + 4 = 0$$

This equation cannot be factored, so we will use the quadratic formula to solve it.

$$a = -2, \qquad b = 3, \qquad c = 4$$

$$x = \frac{-b \pm \sqrt{b^2 - 4ac}}{2a}$$

$$= \frac{-3 \pm \sqrt{3^2 - 4(-2)(4)}}{2(-2)}$$

$$= \frac{-3 \pm \sqrt{9 + 32}}{-4}$$

$$= \frac{-3 \pm \sqrt{41}}{-4}$$

Since $\sqrt{41} \approx 6.4$,

$$x \approx \frac{-3 + 6.4}{-4} \approx \frac{3.4}{-4} \approx -0.85 \qquad \text{or} \qquad x \approx \frac{-3 - 6.4}{-4} \approx \frac{-9.4}{-4} \approx 2.35$$

Thus, the x-intercepts are about $(-0.85, 0)$ and $(2.35, 0)$.

6. Plot the vertex $\left(\frac{3}{4}, 5\frac{1}{8}\right)$, the y-intercept $(0, 4)$, and the x-intercepts $(-0.85, 0)$ and $(2.35, 0)$. Then sketch the graph (Fig. 6.37).

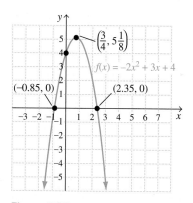

Figure 6.37

b) The domain, the values that can be used for x, is the set of all real numbers, $\mathbb{R}$. The range, the values of y, is $y \leq 5\frac{1}{8}$.

When we use the quadratic formula to find the x-intercepts of a graph, if the radicand, $b^2 - 4ac$, is a negative number, the graph has no x-intercepts. The graph will lie totally above or below the x-axis.

Exponential Functions

In Section 6.3, we discussed exponential equations. Recall that exponential equations are of the form $y = a^x, a > 0, a \neq 1$. The graph of every exponential equation will pass the vertical line test, and so every exponential equation is also an exponential function. Exponential functions may be written as $f(x) = a^x, a > 0, a \neq 1$. When the base a is replaced with the base e, which was discussed on page 333, we get the *natural exponential function* $f(x) = e^x$.

In Section 6.3, we also introduced the natural exponential growth or decay formula $P = P_0 e^{kt}$. We can write this formula in function notation as $P(t) = P_0 e^{kt}$. This expression is referred to as the *natural exponential growth or decay function*. In Examples 8 and 9, we use the natural exponential growth or decay function.

EXAMPLE ❽ *Evaluating an Exponential Decay Function*

Plutonium, a radioactive material used in most nuclear reactors, decays exponentially at a rate of 0.003% per year. If there are originally 2000 grams of plutonium, the amount of plutonium, P, remaining after t years is $P(t) = 2000e^{-0.00003t}$. How much plutonium will remain after 50 years?

SOLUTION Substitute 50 years for t in the function, then evaluate using a calculator as described in Section 6.3.

$$P(t) = 2000e^{-0.00003t}$$
$$P(50) = 2000e^{-0.00003(50)}$$
$$= 2000e^{-0.0015}$$
$$\approx 2000(0.9985011244)$$
$$\approx 1997.0 \text{ grams}$$

Thus, after 50 years, the amount of plutonium remaining will be about 1997 grams. ●

EXAMPLE ❾ *Evaluating an Exponential Decay Function*

Carbon-14 is used by scientists to find the age of fossils and other artifacts. If an object originally had 25 grams of carbon-14, the amount present after t years is $f(t) = 25e^{-0.00012010t}$. How much carbon-14 will be found after 350 years?

SOLUTION Substitute 350 for t in the function, then evaluate using a calculator as described in Section 6.3.

$$f(t) = 25e^{-0.00012010t}$$
$$f(350) = 25e^{-0.00012010(350)}$$
$$= 25e^{-0.042035}$$
$$\approx 25(0.9588362207)$$
$$\approx 23.97090552$$
$$\approx 24 \text{ grams}$$

Thus, after 350 years, about 24 grams of carbon-14 will remain. ●

Graphs of Exponential Functions

What does the graph of an *exponential function* of the form $y = a^x, a > 0, a \neq 1$, look like? Examples 10 and 11 illustrate graphs of exponential functions.

DID YOU KNOW?

Population Growth

P opulation growth during certain time periods can be described by an exponential function. Whether it is a population of bacteria, fish, flowers, or people, the same general trend emerges: a period of rapid (exponential) growth, which is then followed by a leveling-off period.

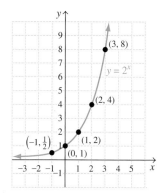

Figure 6.38

EXAMPLE ⑩ *Graphing an Exponential Function with a Base Greater Than 1*

a) Graph $y = 2^x$.

b) Determine the domain and range of the function.

SOLUTION

a) Substitute values for x and find the corresponding values of y. The graph is shown in Fig. 6.38.

$$y = 2^x$$

				x	y
$x = -3,$	$y = 2^{-3} = \dfrac{1}{2^3} = \dfrac{1}{8}$			-3	$\dfrac{1}{8}$
$x = -2,$	$y = 2^{-2} = \dfrac{1}{2^2} = \dfrac{1}{4}$			-2	$\dfrac{1}{4}$
$x = -1,$	$y = 2^{-1} = \dfrac{1}{2^1} = \dfrac{1}{2}$			-1	$\dfrac{1}{2}$
$x = 0,$	$y = 2^0 = 1$			0	1
$x = 1,$	$y = 2^1 = 2$			1	2
$x = 2,$	$y = 2^2 = 4$			2	4
$x = 3,$	$y = 2^3 = 8$			3	8

b) The domain is all real numbers, $\mathbb{R}$. The range is $y > 0$. Note that y can never have a value of 0.   ●

All exponential functions of the form $y = a^x$, $a > 1$, will have the general shape of the graph illustrated in Fig. 6.38. Since $f(x)$ is the same as y, the graphs of functions of the form $f(x) = a^x$, $a > 1$, will also have the general shape of the graph illustrated in Fig. 6.38. Can you now predict the shape of the graph of $y = e^x$? Remember: e has a value of about 2.7183.

EXAMPLE ⑪ *Graphing an Exponential Function with a Base Between 0 and 1*

a) Graph $y = \left(\frac{1}{2}\right)^x$.

b) Determine the domain and range of the function.

SOLUTION

a) We begin by substituting values for x and calculating values for y. We then plot the ordered pairs and use these points to sketch the graph. To evaluate a fraction with a negative exponent, we use the fact that

$$\left(\frac{a}{b}\right)^{-x} = \left(\frac{b}{a}\right)^x$$

For example,

$$\left(\frac{1}{2}\right)^{-3} = \left(\frac{2}{1}\right)^3 = 8$$

Then

$$y = \left(\frac{1}{2}\right)^x$$

				x	y
$x = -3,$	$y = \left(\frac{1}{2}\right)^{-3} = 2^3 = 8$			-3	8
				-2	4
$x = -2,$	$y = \left(\frac{1}{2}\right)^{-2} = 2^2 = 4$			-1	2
$x = -1,$	$y = \left(\frac{1}{2}\right)^{-1} = 2^1 = 2$			0	1
$x = 0,$	$y = \left(\frac{1}{2}\right)^{0} = 1$			1	$\frac{1}{2}$
$x = 1,$	$y = \left(\frac{1}{2}\right)^{1} = \frac{1}{2}$			2	$\frac{1}{4}$
$x = 2,$	$y = \left(\frac{1}{2}\right)^{2} = \frac{1}{4}$			3	$\frac{1}{8}$
$x = 3,$	$y = \left(\frac{1}{2}\right)^{3} = \frac{1}{8}$				

The graph is illustrated in Fig. 6.39.

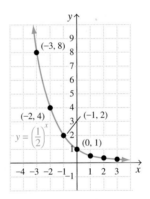

Figure 6.39

b) The domain is the set of all real numbers, $\mathbb{R}$. The range is $y > 0$.

All exponential functions of the form $y = a^x$ or $f(x) = a^x, 0 < a < 1$, will have the general shape of the graph illustrated in Fig. 6.39.

EXAMPLE ⓬ *Is the Growth Exponential?*

The number of new stores opened by Starbucks in the United States has increased greatly since 1994. The graph on the next page shows the number of Starbucks stores in the United States for selected years and projected to 2010.

a) Does the graph approximate the graph of an exponential function?

b) Estimate the number of Starbucks stores in the United States in 2006.

SOLUTION

a) Yes, the graph has the approximate shape of an exponential function. A function that increases rapidly with this general shape is, or approximates, an exponential function.

b) From the graph, we see that Starbucks had about 7500 U.S. stores in 2006.

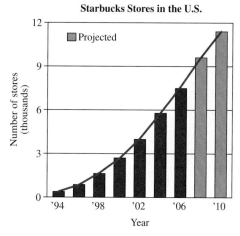

Starbucks Stores in the U.S.

Source: Starbucks, CIBC World Markets

SECTION 6.10 EXERCISES

CONCEPT/WRITING EXERCISES

1. What is a function?

2. What is a relation?

3. What is the domain of a function?

4. What is the range of a function?

5. Explain how and why the vertical line test can be used to determine whether a graph is that of a function.

6. Give three examples of one quantity being a function of another quantity.

7. What is the formula used to determine the equation for the axis of symmetry of a parabola?

8. Explain how to determine whether the graph of a quadratic equation of the form $y = ax^2 + bx + c, a \neq 0$ opens upward or downward.

PRACTICE THE SKILLS

In Exercises 9–26, determine whether the graph represents a function. If it does represent a function, give its domain and range.

9.   10.

11.   12.

13.

14.

23.

24.

15.

16.

25.

17.

26.

18.

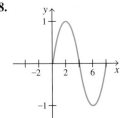

In Exercises 27–32, determine whether the set of ordered pairs is a function.

27. $\{(5, 9), (3, 2), (8, 3), (9, 15)\}$

28. $\{(2, -3), (4, -5), (5, -7), (6, -9)\}$

29. $\{(6, 4), (6, 1), (4, 0)\}$

30. $\{(4, 1), (4, 3), (4, 5)\}$

31. $\{(8, 3), (2, 3), (-1, 3)\}$

32. $\{(1, -1), (1, 0), (1, 1)\}$

19.

20.

In Exercises 33–46, evaluate the function for the given value of x.

33. $f(x) = x + 5, \quad x = 3$

34. $f(x) = 2x + 4, \quad x = 1$

35. $f(x) = -2x - 7, \quad x = -4$

36. $f(x) = -5x + 3, \quad x = -1$

37. $f(x) = 10x - 6, \quad x = 0$

21.

22.

38. $f(x) = 7x - 6, \quad x = 4$

39. $f(x) = x^2 - 3x + 1, \quad x = 4$

40. $f(x) = x^2 - 5, \quad x = 7$

41. $f(x) = -x^2 - 2x + 1, \quad x = 4$

42. $f(x) = 2x^2 + 3x - 1, \quad x = 3$

43. $f(x) = 4x^2 - 6x - 9, \quad x = -3$

44. $f(x) = 3x^2 + 2x - 5, \quad x = -4$

45. $f(x) = -5x^2 + 3x - 9, \quad x = -1$

46. $f(x) = -3x^2 - 6x + 10, \quad x = -2$

In Exercises 47–52, graph the function by using the slope and y-intercept.

47. $f(x) = 3x - 2$

48. $f(x) = -x - 1$

49. $f(x) = -4x + 2$

50. $f(x) = 2x + 5$

51. $f(x) = \frac{3}{2}x - 1$

52. $f(x) = -\frac{1}{2}x + 3$

In Exercises 53–68,

a) *determine whether the parabola will open upward or downward.*

b) *find the equation of the axis of symmetry.*

c) *find the vertex.*

d) *find the y-intercept.*

e) *find the x-intercepts if they exist.*

f) *sketch the graph.*

g) *find the domain and range of the function.*

53. $y = x^2 - 9$

54. $y = x^2 - 16$

55. $y = -x^2 + 4$

56. $y = -x^2 + 16$

57. $y = -2x^2 - 8$

58. $f(x) = -x^2 - 4$

59. $y = 2x^2 - 3$

60. $f(x) = -3x^2 - 6$

61. $f(x) = x^2 + 2x + 6$

62. $y = x^2 - 8x + 1$

63. $y = x^2 + 5x + 6$

64. $y = x^2 - 7x - 8$

65. $y = -x^2 + 4x - 6$

66. $y = -x^2 + 8x - 8$

67. $y = -2x^2 + 3x - 2$

68. $y = -4x^2 - 6x + 4$

In Exercises 69–80, draw the graph of the function and state the domain and range.

69. $y = 3^x$

70. $f(x) = 4^x$

71. $y = \left(\frac{1}{3}\right)^x$

72. $y = \left(\frac{1}{4}\right)^x$

73. $f(x) = 2^x + 1$

74. $y = 3^x - 1$

75. $y = 4^x + 1$

76. $y = 2^x - 1$

77. $y = 3^{x-1}$

78. $y = 3^{x+1}$

79. $f(x) = 4^{x+1}$

80. $y = 4^{x-1}$

PROBLEM SOLVING

81. *Yearly Profit* Marta Rivera is a part owner of a newly opened bagel company. Marta's yearly profit, in dollars, is given by the function $p(x) = 0.3x - 4000$, where x is the number of bagels sold per year. If Marta sells 150,000 bagels a year, determine her yearly profit.

82. *Finding Distances* The distance a car travels, $d(t)$, at a constant 60 mph is given by the function $d(t) = 60t$, where t is the time in hours. Find the distance traveled in

a) 3 hours.

b) 7 hours.

83. *U.S. Foreign-Born Population* The following graph indicates the percentage of the U.S. population that was foreign-born for each decade between 1930 and 2000, and projected percentage for 2010. The function $f(x) = 0.005x^2 - 0.36x + 11.8$ can be used to estimate the percentage of the U.S. population that was foreign-born, where x is the number of years since 1930 and $0 \le x \le 80$.

a) Use the function $f(x)$ to estimate the percentage of the U.S. population that was foreign-born in 2004. Round your answer to the nearest percent.

b) Of the years illustrated on the graph, determine the year in which the percentage of the U.S. population that was foreign-born was a minimum.

c) Determine the x-coordinate of the vertex of the graph of the function $f(x)$. Then use this value in the function $f(x)$ to estimate the minimum percentage of the U.S. population that was foreign-born. Round your answer for the vertex to the nearest whole number. Round your answer for the population to the nearest percent.

Percentage of U.S. Population That is Foreign-Born

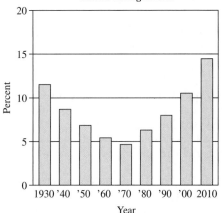

Source: Bureau of the Census, U.S. Dept. of Commerce

84. *Hourly Temperature* The following graph indicates the hourly temperature in Rochester, New York, from 11 P.M. on Thursday, February 16, 2006, through 7 A.M. on Friday, February 17, 2006. The function $t(x) = -1.11x^2 + 6.60x + 51.28$, can be used to estimate the hourly temperature in Rochester where x is the number of hours after midnight on February 16, 2006, and $-1 \le x \le 7$.

a) Use the function $t(x)$ to estimate the temperature in Rochester at 6 A.M. on February 17, 2006. Round your answer to the nearest degree.

b) Use the graph to determine the hour(s) that the temperature was a maximum.

c) Determine the x-coordinate of the vertex of the graph of the function $t(x)$. Then use this value in the function $t(x)$ to estimate the maximum temperature. Round your answer for the vertex to the nearest hour. Round your answer for the temperature to the nearest degree.

Hourly Temperature in Rochester, N.Y. February 16, 2006 – February 17, 2006

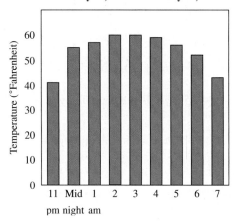

Source: 13 WHAM News Weather Team

85. *Expected Growth* The town of Lockport currently has 4000 residents. The expected future population can be approximated by the function $P(x) = 4000(1.3)^{0.1x}$, where x is the number of years in the future. Find the expected population of Lockport in

a) 10 years.

b) 50 years.

86. *Radioactive Decay* Strontium-90 is a radioactive isotope that decays exponentially at a rate of 2.8% per year. The amount of strontium-90, S, remaining after t years can be found by the function $S(t) = S_0 e^{-0.028t}$, where S_0 is the original amount present. If there are originally 1000 grams of strontium-90, determine the amount of strontium-90 remaining after 40 years. Round your answer to the nearest gram.

87. *Medicare Premiums* The graph at the top of page 405 shows the monthly Medicare Part B premiums for Medicare beneficiaries for the years 2000 through 2005.

a) Does the graph approximate the graph of an exponential function?

b) Estimate the monthly Medicare Part B premium for Medicare beneficiaries in 2003.

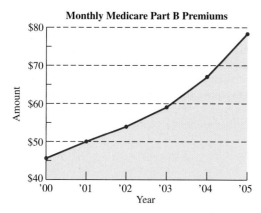

Monthly Medicare Part B Premiums

Source: Centers for Medicare and Medicaid Services

88. U.S. Cellular Telephone Subscribers The following graph shows the number of U.S. cellular subscribers, in thousands, for the years 1985 through 2004.

a) Does the graph approximate the graph of an exponential function?

b) Estimate the number of U.S. cellular subscribers in 2004.

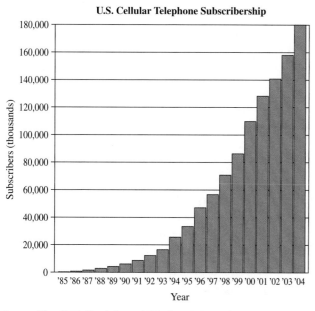

U.S. Cellular Telephone Subscribership

Source: The CTIA Semi-Annual Wireless Industry Survey

89. The spacing of the frets on the neck of a classical guitar is determined from the equation $d = (21.9)(2)^{(20-x)/12}$, where x = the fret number and d = the distance in centimeters of the xth fret from the bridge.

a) Determine how far the 19th fret should be from the bridge (rounded to one decimal place).

b) Determine how far the 4th fret should be from the bridge (rounded to one decimal place).

c) The distance of the nut from the bridge can be found by letting $x = 0$ in the given exponential equation. Find the distance from the nut to the bridge (rounded to one decimal place).

CHALLENGE PROBLEMS/GROUP ACTIVITIES

90. Appreciation of a House A house initially cost $150,000. The value, V, of the house after n years if it appreciates at a constant rate of 4% per year can be determined by the function $V = f(n) = \$150{,}000(1.04)^n$.

a) Determine $f(8)$ and explain its meaning.

b) After how many years is the value of the house greater than $250,000? (Find by trial and error.)

91. Target Heart Rate While exercising, a person's recommended target heart rate is a function of age. The recommended number of beats per minute, y, is given by the function $y = f(x) = -0.85x + 187$, where x represents a person's age in years. Determine the number of recommended heart beats per minute for the following ages and explain the results.

a) 20

b) 30

c) 50

d) 60

e) What is the age of a person with a recommended target heart rate of 85?

92. *Speed of Light* Light travels at about 186,000 miles per second through space. The distance, d, in miles that light travels in t seconds can be determined by the function $d(t) = 186{,}000t$.

a) Light reaches the moon from Earth in about 1.3 sec. Determine the approximate distance from Earth to the moon.

b) Express the distance in miles, d, traveled by light in t minutes as a function of time, t.

c) Light travels from the sun to Earth in about 8.3 min. Determine the approximate distance from the sun to Earth.

INTERNET/RESEARCH ACTIVITY

93. The idea of using variables in algebraic equations was introduced by the French mathematician François Viète (1540–1603). Write a paper about his life and his contributions to mathematics. In particular, discuss his work with algebra equations. References include history of mathematics books, encyclopedias, and the Internet.

CHAPTER ⑥ SUMMARY

IMPORTANT FACTS

PROPERTIES USED TO SOLVE EQUATIONS

Addition property of equality
  If $a = b$, then $a + c = b + c$.

Subtraction property of equality
  If $a = b$, then $a - c = b - c$.

Multiplication property of equality
  If $a = b$, then $ac = bc$.

Division property of equality
  If $a = b$, then $a/c = b/c$, $c \neq 0$.

VARIATION

Direct: $y = kx$

Inverse: $y = \dfrac{k}{x}$

Joint: $y = kxz$

INEQUALITY SYMBOLS

$a < b$ means that a is less than b.

$a \leq b$ means that a is less than or equal to b.

$a > b$ means that a is greater then b.

$a \geq b$ means that a is greater than or equal to b.

INTERCEPTS

To find the x-intercept, set $y = 0$ and solve the resulting equation for x.

To find the y-intercept, set $x = 0$ and solve the resulting equation for y.

SLOPE

Slope (m): $m = \dfrac{y_2 - y_1}{x_2 - x_1}$

EQUATIONS AND FORMULAS

Linear equation in two variables:
$$ax + by = c, \qquad a \neq 0 \qquad \text{and} \qquad b \neq 0$$

Exponential growth or decay formula:
$$P = P_0 a^{kt}, \qquad a \neq 1, \qquad a > 0$$

Slope–intercept form of a line:
$$y = mx + b$$

Quadratic equation in one variable:
$$ax^2 + bx + c = 0, \qquad a \neq 0$$

Quadratic formula:
$$x = \frac{-b \pm \sqrt{b^2 - 4ac}}{2a}$$

Quadratic equation (or function) in two variables:
$$y = ax^2 + bx + c, \qquad a \neq 0$$

Axis of symmetry of a parabola:
$$x = \frac{-b}{2a}$$

Exponential function:
$$f(x) = a^x, \qquad a \neq 1, \qquad a > 0$$

Natural exponential function:
$$f(x) = e^x$$

Natural exponential growth or decay function:
$$P(t) = P_0 e^{kt}$$

ZERO-FACTOR PROPERTY

If $a \cdot b = 0$, then $a = 0$ or $b = 0$.

CHAPTER 6 REVIEW EXERCISES

6.1

In Exercises 1–6, evaluate the expression for the given value(s) of the variable.

1. $x^2 + 15$, $\quad x = 2$

2. $-x^2 - 7$, $\quad x = -1$

3. $4x^2 - 2x + 5$, $\quad x = 2$

4. $-x^2 + 7x - 3$, $\quad x = \frac{1}{2}$

5. $4x^3 - 7x^2 + 3x + 1$, $\quad x = -1$

6. $3x^2 - xy + 2y^2$, $\quad x = 1$, $\quad y = -2$

6.2

In Exercises 7–9, combine like terms.

7. $3x - 2 + x + 7$ 8. $2x + 5(x - 2) + 8x$

9. $4(x - 1) + \frac{1}{3}(9x + 3)$

In Exercises 10–14, solve the equation for the given variable.

10. $4s + 10 = -30$ 11. $3t + 14 = -6t - 13$

12. $\dfrac{x + 5}{2} = \dfrac{x - 3}{4}$ 13. $4(x - 2) = 3 + 5(x + 4)$

14. $\dfrac{x}{4} + \dfrac{3}{5} = 7$

15. **Making Oatmeal** A recipe for Hot Oats Cereal calls for 2 cups of water and for $\frac{1}{3}$ cup of dry oats. How many cups of dry oats would be used with 3 cups of water?

16. **Laying Blocks** A mason lays 120 blocks in 1 hr 40 min. How long will it take her to lay 300 blocks?

6.3

In Exercises 17–20, use the formula to find the value of the indicated variable for the values given.

17. $A = bh$
    Find A when $b = 12$ and $h = 4$ (geometry).

18. $V = 2\pi R^2 r^2$
    Find V when $R = 3$, $r = 1\frac{3}{4}$, and $\pi = 3.14$ (geometry).

19. $z = \dfrac{\bar{x} - \mu}{\dfrac{\sigma}{\sqrt{n}}}$
    Find $\bar{x}$ when $z = 2$, $\mu = 100$, $\sigma = 3$, and $n = 16$ (statistics).

20. $E = mc^2$
    Find m when $E = 400$ and $c = 4$ (physics).

In Exercises 21–24, solve for y.

21. $8x - 4y = 24$ 22. $2x + 7y = 15$

23. $2x - 3y + 52 = 30$ 24. $-3x - 4y + 5z = 4$

In Exercises 25–28, solve for the variable indicated.

25. $A = lw$,   for w

26. $P = 2l + 2w$,   for w

27. $L = 2(wh + lh)$,   for l

28. $a_n = a_1 + (n - 1)d$,   for d

6.4

In Exercises 29–32, write the phrase in mathematical terms.

29. 7 decreased by 4 times x

30. 2 times y increased by 7

31. 10 increased by 3 times r

32. The difference between 9 divided by q and 15

In Exercises 33–36, write an equation that can be used to solve the problems. Solve the equation and find the desired value(s).

33. Three increased by 7 times a number is 17.

34. The product of 3 and a number, increased by 8, is 6 less than the number.

35. Five times the difference of a number and 4 is 45.

36. Fourteen more than 10 times a number is 8 times the sum of the number and 12.

In Exercises 37–40, write the equation and then find the solution.

37. MODELING - *Investing* Jim Lawton received an inheritance of $15,000. If he wants to invest twice as much money in mutual funds as in bonds, how much should he invest in mutual funds?

38. MODELING - *Lawn Chairs* Larry's Lawn Chair Company has fixed costs of $15,000 per month and variable costs of $9.50 per lawn chair manufactured. The company has $95,000 available to meet its total monthly expenditures. What is the maximum number of lawn chairs the company can manufacture in a month? (Fixed costs, such as rent and insurance, are those that occur regardless of the level of production. Variable costs, such as those for materials, depend on the level of production.)

39. MODELING - *Endangered and Threatened Mammals* According to the U.S. Fish and Wildlife Service, the number of species of endangered mammals in the United States in 2005 was 2 more than 6 times the number of species of threatened mammals. The sum of the number of species of endangered mammals and the number of species of threatened mammals in the United States in 2005 was 79. Determine the number of species of endangered mammals and the number of species of threatened mammals.

▲ The Florida panther is an endangered mammal.

40. MODELING - *Restaurant Profit* John Smith owns two restaurants. His profit for a year at restaurant A is $12,000 greater than his profit at restaurant B. The total profit from both restaurants is $68,000. Determine the profit at each restaurant.

6.5

In Exercises 41–44, find the quantity indicated.

41. s is directly proportional to t. If $s = 40$ when $t = 5$, find s when $t = 8$.

42. J is inversely proportional to the square of A. If $J = 8$ when $A = 3$, find J when $A = 6$.

43. W is directly proportional to L and inversely proportional to A. If $W = 80$ when $L = 100$ and $A = 20$, find W when $L = 50$ and $A = 40$.

44. z is jointly proportional to x and y and inversely proportional to the square of r. If $z = 12$ when $x = 20$, $y = 8$, and $r = 8$, find z when $x = 10$, $y = 80$, and $r = 3$.

45. *Buying Fertilizer*

a) A 30 lb bag of fertilizer will cover an area of 2500 ft^2. How many pounds of fertilizer are needed to cover an area of 12,500 ft^2?

b) How many bags of fertilizer are needed?

46. *Map Reading* The scale of a map is 1 in. to 30 mi. What distance on the map represents 120 mi?

47. *Electric Bill* An electric company charges $0.162 per kilowatt-hour (kWh). What is the electric bill if 740 kWh are used in a month?

48. *Stretching a Spring* The length that a spring will stretch, S, varies directly with the weight, w, attached to the spring. If a spring stretches 4.2 in. when a 60 lb weight is attached, how far will the spring stretch when a 25 lb weight is attached?

6.6

In Exercises 49–52, graph the solution set for the set of real numbers.

49. $8 + 7x \leq 5x - 4$

50. $2x + 8 \geq 5x + 11$

51. $3(x + 9) \leq 4x + 11$

52. $-3 \leq x + 1 < 7$

In Exercises 53–56, graph the solution set for the set of integers.

53. $2 + 4x > -10$

54. $5x + 13 \geq -22$

55. $-1 < x \leq 9$

56. $-8 \leq x + 2 \leq 7$

6.7

In Exercises 57–60, graph the ordered pair in the Cartesian coordinate system.

57. $(1, -3)$

58. $(-2, 2)$

59. $(-4, 3)$

60. $(5, 3)$

In Exercises 61 and 62, points A, B, and C are vertices of a rectangle. Plot the points. Find the coordinates of the fourth point, D, to complete the rectangle. Find the area of the rectangle.

61. $A(-3, 3), \quad B(2, 3), \quad C(2, -1)$

62. $A(-3, 1), \quad B(-3, -2), \quad C(4, -2)$

In Exercises 63–66, graph the equation by plotting points.

63. $x - y = 2$

64. $2x + 3y = 12$

65. $x = y$

66. $x = -3$

In Exercises 67–70, graph the equation, using the x- and y-intercepts.

67. $x - 2y = 6$

68. $x + 3y = 6$

69. $4x - 3y = 12$

70. $2x + 3y = 9$

In Exercises 71–74, find the slope of the line through the given points.

71. $(1, 3), (6, 5)$

72. $(3, -1), (5, -4)$

73. $(-1, -4), (2, 3)$

74. $(6, 2), (6, -2)$

In Exercises 75–78, graph the equation by plotting the y-intercept and then plotting a second point by making use of the slope.

75. $y = 2x - 5$

76. $2y - 4 = 3x$

77. $2y + x = 8$

78. $y = -x - 1$

In Exercises 79 and 80, determine the equation of the graph.

79.

80.

81. *Car Wash Profit* The monthly profit, p, in dollars, at Delta Sonic Car Wash can be estimated by $p = 20n - 400$, where n is the number of cars washed in 1 week.

a) Draw a graph of weekly profit versus number of cars washed for up to and including 100 cars.

b) Use the function to determine the weekly profit if 50 cars are washed.

c) If the weekly profit was $1000, use the function to determine how many cars were washed.

82. *Business Space Rental* The monthly rental cost, C, in dollars, for space in the Galleria Mall can be approximated by the equation $C = 1.70A + 3000$, where A is the area, in square feet, of space rented.

a) Draw a graph of monthly rental cost versus square feet for up to and including 12,000 ft^2.

b) Determine the monthly rental cost if 2000 ft^2 are rented.

c) If the rental cost is $10,000 per month, how many square feet are rented?

6.8

In Exercises 83–86, graph the inequality.

83. $2x + 3y \leq 12$

84. $4x + 2y \geq 12$

85. $2x - 3y > 12$

86. $-7x - 2y < 14$

6.9

In Exercises 87–92, factor the trinomial. If the trinomial cannot be factored, so state.

87. $x^2 + 11x + 18$

88. $x^2 + x - 30$

89. $x^2 - 10x + 24$

90. $x^2 - 9x + 20$

91. $6x^2 + 7x - 3$

92. $2x^2 + 13x - 7$

In Exercises 93–96, solve the equation by factoring.

93. $x^2 + 4x + 3 = 0$

94. $x^2 + 3x = 18$

95. $3x^2 - 17x + 10 = 0$

96. $3x^2 = -7x - 2$

In Exercises 97–100, solve the equation using the quadratic formula. If the equation has no real solution, so state.

97. $x^2 - 4x - 1 = 0$

98. $x^2 - 6x - 16 = 0$

99. $2x^2 - 3x + 4 = 0$

100. $2x^2 - x - 3 = 0$

6.10

In Exercises 101–104, determine whether the graph represents a function. If it does represent a function, give its domain and range.

101.

102.

103.

104.

In Exercises 105–108, evaluate f(x) for the given value of x.

105. $f(x) = 3x - 7, \quad x = 4$

106. $f(x) = -2x + 9, \quad x = -3$

107. $f(x) = 2x^2 - 3x + 4, \quad x = 5$

108. $f(x) = -4x^2 + 7x + 9, \quad x = -2$

In Exercises 109 and 110, for each function

   a) *determine whether the parabola will open upward or downward.*

   b) *determine the equation of the axis of symmetry.*

   c) *determine the vertex.*

   d) *determine the y-intercept.*

   e) *determine the x-intercepts if they exist.*

   f) *sketch the graph.*

   g) *determine the domain and range.*

109. $y = -x^2 - 4x + 21$

110. $f(x) = 2x^2 + 8x + 6$

In Exercises 111 and 112, draw the graph of the function and state the domain and range.

111. $y = 2^{2x}$

112. $y = \left(\frac{1}{2}\right)^x$

6.2, 6.3, 6.10

113. *Gas Mileage* The gas mileage, m, of a specific car can be estimated by the equation (or function)

$$m = 30 - 0.002n^2, \qquad 20 \le n \le 80$$

where n is the speed of the car in miles per hour. Estimate the gas mileage when the car travels at 60 mph.

114. *Auto Accidents* The approximate number of accidents in one month, n, involving drivers between 16 and 30 years of age inclusive can be approximated by the equation

$$n = 2a^2 - 80a + 5000, \qquad 16 \le a \le 30$$

where a is the age of the driver. Approximate the number of accidents in one month that involved

  a) 18-year-olds

  b) 25-year-olds.

115. *Filtered Light* The percent of light filtering through Swan Lake, P, can be approximated by the function $P(x) = 100(0.92)^x$, where x is the depth in feet. Find the percent of light filtering through at a depth of 4.5 ft.

CHAPTER ⑥ TEST

1. Evaluate $4x^2 + 3x - 1$, when $x = -2$.

In Exercises 2 and 3, solve the equation.

2. $3x + 5 = 2(4x - 7)$

3. $-2(x - 3) + 6x = 2x + 3(x - 4)$

In Exercises 4 and 5, write an equation to represent the problem. Then solve the equation.

4. The product of 3 and a number, increased by 5 is 17.

5. *Salary* Mary Gilligan's salary is $350 per week plus 6% commission of sales. How much in sales must Mary make to earn a total of $710 per week?

6. Evaluate $L = ah + bh + ch$ when $a = 2, b = 5, c = 4$, and $h = 7$.

7. Solve $3x + 5y = 11$ for y.

8. L varies jointly as M and N and inversely as P. If $L = 12$ when $M = 8, N = 3$, and $P = 2$, find L when $M = 10, N = 5$, and $P = 15$.

9. For a constant area, the length, l, of a rectangle varies inversely as the width, w. If $l = 15$ ft when $w = 9$ ft, find the length of a rectangle with the same area if the width is 20 ft.

10. Graph the solution set of $-3x + 11 \le 5x + 35$ on the real number line.

11. Determine the slope of the line through the points $(-2, 8)$ and $(1, 14)$.

In Exercises 12 and 13, graph the equation.

12. $y = 2x - 4$

13. $2x - 3y = 15$

14. Graph the inequality $3y \ge 5x - 12$.

15. Solve the equation $x^2 + 5x = -4$ by factoring.

16. Solve the equation $3x^2 + 2x = 8$ by using the quadratic formula.

17. Determine whether the graph is a function. Explain your answer.

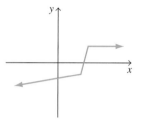

18. Evaluate $f(x) = -3x^2 - 12x + 5$ when $x = -2$.

19. For the equation $y = x^2 - 2x + 4$,

a) determine whether the parabola will open upward or downward.

b) determine the equation of the axis of symmetry.

c) determine the vertex.

d) determine the y-intercept.

e) determine the x-intercepts if they exist.

f) sketch the graph.

g) determine the domain and range of the function.

G R O U P   P R O J E C T S

ARCHEOLOGY: GAINING INFORMATION FROM BONES

1. Archeologists have developed formulas to predict the height and, in some cases, the age at death of the deceased by knowing the lengths of certain bones in the body. The long bones of the body grow at approximately the same rate. Thus, a linear relationship exists between the length of the bones and the person's height. If the length of one of these major bones—the femur (F), the tibia (T), the humerus (H), and the radius (R)— is known, the height, h, of a person can be calculated with one of the following formulas. The relationship between bone length and height is different for males and females.

Male	Female
$h = 2.24F + 69.09$	$h = 2.23F + 61.41$
$h = 2.39T + 81.68$	$h = 2.53T + 72.57$
$h = 2.97H + 73.57$	$h = 3.14H + 64.98$
$h = 3.65R + 80.41$	$h = 3.88R + 73.51$

All measurements are in centimeters.

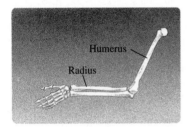

Humerus
Radius

a) Measure your humerus and use the appropriate formula to predict your height in centimeters. How close is this predicted height to your actual height? (The result is an approximation because measuring a bone covered with flesh and muscle is difficult.)

b) Determine and describe where the femur and tibia bones are located.

c) Dr. Juarez, an archeologist, had one female humerus that was 29.42 cm in length. He concluded that the height of the entire skeleton would have been 157.36 cm. Was his conclusion correct?

d) If a 21-year-old woman is 167.64 cm tall, about how long should her tibia be?

e) Sometimes the age of a person may be determined by using the fact that the height of a person, and the length of his or her long bones, decreases at the rate of 0.06 cm per year after the age of 30.

i. At age 30, Jolene is 168 cm tall. Estimate the length of her humerus.

ii. Estimate the length of Jolene's humerus when she is 60 years old.

f) Select six people of the same gender and measure their height and one of the bones for which an equation is given (the same bone on each person). Each measurement should be made to the nearest 0.5 cm. For each person, you will have two measurements, which can be considered an ordered pair (bone length, height). Plot the ordered pairs on a piece of graph paper, with the bone length on the horizontal axis and the height on the vertical axis. Start the scale on both axes at zero. Draw a straight line that you feel is the best approximation, or best fit, through these points. Determine where the line crosses the y-axis and the slope of the line. Your y-intercept and slope should be close to the values in the given equation for that bone. (Reference: M. Trotter and G. C. Gleser, "Estimation of Stature from Long Bones of American Whites and Negroes," *American Journal of Physical Anthropology,* 1952, 10:463–514.)

GRAPHING CALCULATOR

2. The functions that we graphed in this chapter can be easily graphed with a graphing calculator (or grapher). If you do not have a graphing calculator, borrow one from your instructor or a friend.

a) Explain how you would set the domain and range. The calculator key to set the domain and range may be labeled *range* or *window*. Set the grapher with the following range or window settings:

Xmin $= -12$, Xmax $= 12$, Xscl $= 1$,
Ymin $= -13$, Ymax $= 6$, and Yscl $= 1$.

b) Explain how to enter a function in the graphing calculator. Enter the function $y = 3x^2 - 7x - 8$ in the calculator.

c) Graph the function you entered in part (b).

d) Learn how to use the *trace feature.* Then use it to estimate the x-intercepts. Record the estimated values for the x-intercepts.

e) Learn how to use the *zoom feature* to obtain a better approximation of the x-intercepts. Use the zoom feature twice and record the x-intercepts each time.

CHAPTER 7

Systems of Linear Equations and Inequalities

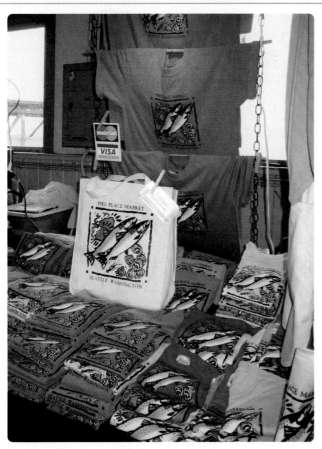

▲ Business owners may solve systems of equations to determine under what conditions their business will be profitable.

WHAT YOU WILL LEARN

- Solving systems of linear equations by graphing, by using the substitution method, by using the addition method, and by using matrices
- Solving systems of linear inequalities
- Adding, subtracting, and multiplying matrices
- Solving application problems using linear programming

WHY IT IS IMPORTANT

Imagine that you and a friend enter the T-shirt business. To make a profit in your business, you need to keep track of the cost of your materials, your overhead, the quantity of T-shirts sold, the price at which you sell the T-shirts, and many other items. To do so, you may need to develop and solve systems of equations.

7.1   SYSTEMS OF LINEAR EQUATIONS

▲ Solving a system of equations can help determine which of two landscaping companies offers the least expensive service.

Suppose you needed to have some landscaping work done at your home. In your local newspaper, two different landscaping services each have advertised specials. One company charges a lower consultation fee than the other company but charges a higher hourly rate for labor. How do you determine the number of hours of service needed for both services to have the same cost? How do you decide which company is less expensive for the services you need? In this section, we will introduce one way to use algebra to answer these questions.

In algebra, it is often necessary to find the common solution to two or more equations. We refer to the equations in this type of problem as a *system of linear equations* or as *simultaneous linear equations*. The solution to a system of linear equations may be found by a number of techniques. In this section, we illustrate how to solve a system of linear equations by graphing.

Solutions to Systems of Linear Equations

A *solution to a system of equations* is the ordered pair or ordered pairs that satisfy *all* equations in the system. A system of linear equations may have exactly one solution, no solution, or infinitely many solutions.

┌─ **EXAMPLE ❶** *Is the Ordered Pair a Solution?*

Determine which of the ordered pairs is a solution to the following system of linear equations.

$$x + 2y = 8$$
$$2x - 3y = 2$$

a) $(6, 1)$ b) $(4, 2)$ c) $(1, 0)$

SOLUTION   For the ordered pair to be a solution to the system, it must satisfy each equation in the system. Substitute the values of x and y into each equation.

a) $x + 2y = 8$ $2x - 3y = 2$
    $6 + 2(1) = 8$ $2(6) - 3(1) = 2$
       $8 = 8$ True $9 = 2$ False

Since $(6, 1)$ does not satisfy *both* equations, it is not a solution to the system.

b) $x + 2y = 8$ $2x - 3y = 2$
    $4 + 2(2) = 8$ $2(4) - 3(2) = 2$
       $8 = 8$ True $2 = 2$ True

Since $(4, 2)$ satisfies *both* equations, it is a solution to the system.

c) $x + 2y = 8$ $2x - 3y = 2$
    $1 + 2(0) = 8$ $2(1) - 3(0) = 2$
       $1 = 8$ False $2 = 2$ True

Since $(1, 0)$ does not satisfy *both* equations, it is not a solution to the system.   ●

Solving a System of Linear Equations by Graphing

To find the solution to a system of linear equations graphically, we graph both of the equations on the same axes. The coordinates of the point or points of intersection of the graphs are the solution or solutions to the system of equations.

$x + y = 4$		$2x - y = -1$	
x	y	x	y
0	4	0	1
1	3	1	3
4	0	-2	-3

PROCEDURE FOR SOLVING A SYSTEM OF EQUATIONS BY GRAPHING

1. Determine three ordered pairs that satisfy each equation.
2. Plot the points that correspond to the ordered pairs and sketch the graphs of *both* equations on the same axes.
3. The coordinates of the point or points of intersection of the graphs are the solution or solutions to the system of equations.

When two linear equations are graphed, three situations are possible. The two lines may intersect at one point, as in Example 2; or the two lines may be parallel and not intersect, as in Example 3; or the two equations may represent the same line, as in Example 4.

Since the solution to a system of equations may not be integer values, you may not be able to obtain the exact solution by graphing.

Figure 7.1

Figure 7.2

EXAMPLE ❷ *A System with One Solution*

Determine the solution to the following system of equations graphically.

$$x + y = 4$$
$$2x - y = -1$$

SOLUTION To determine the solution, graph both $x + y = 4$ and $2x - y = -1$ on the same axes (Fig. 7.1). Three points that satisfy each equation are shown in the tables above Fig. 7.1. Figure 7.2 shows the system $x + y = 4$ and $2x - y = -1$ graphed on a Texas Instrument TI-84 Plus graphing calculator.

The graphs intersect at (1, 3), which is the solution to the system of equations. This point is the only point that satisfies *both* equations.

CHECK:

$x + y = 4$	$2x - y = -1$
$1 + 3 = 4$	$2(1) - 3 = -1$
$4 = 4$ True	$2 - 3 = -1$
	$-1 = -1$ True

TIMELY TIP When checking the solution to a system of equations, it is important to check the solution in both equations. The ordered pair solution must satisfy *both* equations of the system.

The system of equations in Example 2 is an example of a *consistent system of equations*. A consistent system of equations is one that has a solution.

Figure 7.3

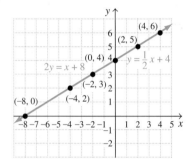

Figure 7.4

EXAMPLE ❸ *A System with No Solution*

Determine the solution to the following system of equations graphically.

$$x + y = 3$$
$$2x + 2y = -4$$

SOLUTION Three ordered pairs that satisfy the equation $x + y = 3$ are $(0, 3)$, $(3, 0)$, and $(1, 2)$. Three ordered pairs that satisfy the equation $2x + 2y = -4$ are $(-2, 0)$, $(0, -2)$, and $(1, -3)$. The graphs of both equations are given in Fig. 7.3. Since the two lines are parallel, they do not intersect; therefore, the system has *no solution*. ●

The system of equations in Example 3 has no solution. A system of equations that has no solution is called an *inconsistent system*.

EXAMPLE ❹ *A System with an Infinite Number of Solutions*

Determine the solution to the following system of equations graphically.

$$y = \frac{1}{2}x + 4$$
$$2y = x + 8$$

SOLUTION Three ordered pairs that satisfy the equation $y = \frac{1}{2}x + 4$ are $(0, 4)$, $(2, 5)$, and $(-2, 3)$. Three ordered pairs that satisfy the equation $2y = x + 8$ are $(-8, 0)$, $(4, 6)$, and $(-4, 2)$. Graph the equations on the same axes (Fig. 7.4). Because all six points are on the same line, the two equations represent the same line. Therefore, every ordered pair that is a solution to one equation is also a solution to the other equation. Every point on the line satisfies both equations; thus, this system has an *infinite number of solutions*. Solving the second equation for y reveals that the equations are equivalent. ●

When a system of equations has an infinite number of solutions, as in Example 4, it is called a *dependent system*. Note that because it has a solution, a dependent system is also a consistent system.

Figure 7.5 summarizes the three possibilities for a system of linear equations.

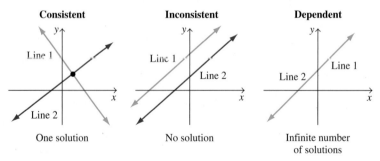

Figure 7.5

In Chapter 6, we introduced *modeling*. Recall that a *mathematical model* is an equation or system of equations that represents a real-life situation. In Examples 5 and 6, we develop equations that model a real-life situation.

EXAMPLE ❺ MODELING - *A Landscape Service Application*

Tom's Tree and Landscape Service charges a consultation fee of $200 plus $50 per hour for labor for landscaping. Lawn Perfect Landscape Service charges a consultation fee of $300 plus $25 per hour for labor for landscaping.

a) Write a system of equations to represent the cost, C, of the two landscaping services, each with h hours of labor.

b) Graph both equations on the same axes and determine the number of hours needed for both services to have the same cost.

c) If the Johnsons need 7 hours of landscaping service done at their home, which service is less expensive?

SOLUTION Let $h =$ the number of hours of labor. The total cost of each service is the consultation fee plus the cost of the labor.

a) Tom's Tree and Landscape Service: $C = 200 + 50h$
   Lawn Perfect Landscape Service: $C = 300 + 25h$

b) We graphed the cost, C, versus the number of hours of labor, h, for 0 to 10 hours (Fig. 7.6). On the graph, the lines intersect at the point (4, 400). Thus, for 4 hours of service, both services would have the same cost, $400.

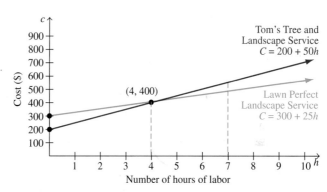

Figure 7.6

c) The graph shows that for more than 4 hours, Lawn Perfect is the least expensive service. Thus, for 7 hours, Lawn Perfect is less expensive than Tom's Tree and Landscape Service.   ●

Break-Even Analysis

Manufacturers use a technique called *break-even analysis* to determine how many units of an item must be sold for the business to "break even," that is, for its total revenue to equal its total cost. Suppose we let the horizontal axis represent the number of units manufactured and sold and the vertical axis represent dollars. Then linear equations for cost, C, and revenue, R, can both be sketched on the same axes (Fig. 7.7). Both C and R are expressed in dollars, and both are a function of the number of units. Profit, P, is the difference between revenue, R, and cost, C. Thus, $P = R - C$. If revenue is greater than cost, the company makes a profit. If cost is greater than revenue, the company has a loss.

Figure 7.7

Initially, the cost graph is higher than the revenue graph because of fixed (overhead) costs such as rent and utilities. During low levels of sales, the manufacturer suffers a loss (the cost graph is greater). During higher levels of sales, the manufacturer realizes a profit (the revenue graph is greater). The point at which the two graphs intersect is called the *break-even point*. At that number of units sold, revenue equals cost and the manufacturer breaks even.

EXAMPLE ⑥ MODELING - *Profit and Loss in Business*

At a collectibles show, Richard Lane can sell model cars for $25. The costs for making the cars are a fixed cost of $150 and a production cost of $10 apiece.

a) Write an equation that represents Richard's revenue. Write an equation that represents Richard's cost.

b) How many model cars must Richard sell to break even?

c) Write an equation for the profit formula. Use the formula to determine Richard's profit if he sells 14 model cars.

d) How many model cars must Richard sell to make a profit of $450?

SOLUTION

a) Let x denote the number of model cars made and sold. The revenue is given by the equation

$$R = 25x \quad \text{($25 times the number of units)}$$

and the cost is given by the equation

$$C = 150 + 10x \quad \text{($150 plus $10 times the number of units)}$$

b) The break-even point is the point at which the revenue and cost graphs intersect. In Fig. 7.8, the graphs intersect at the point (10, 250), which is the break-even point. Thus, for Richard to break even, he must sell 10 model cars. When 10 model cars are made and sold, the cost and revenue are both $250.

Figure 7.8

c) Profit is equal to the revenue minus the cost. Therefore, the profit formula is

$$P = R - C$$
$$= 25x - (150 + 10x)$$
$$= 25x - 150 - 10x$$
$$= 15x - 150$$

For 14 cars, the profit is found as follows.

$$P = 15x - 150$$
$$= 15(14) - 150 = 60$$

Richard has a profit of $60 if he sells 14 model cars. By observing the graph, we can see that if Richard sells 14 model cars, he will have a profit since at 14 cars the revenue line is greater than the cost line.

d) We can determine the number of model cars that Richard must sell to have a profit of $450 by using the profit formula. Substituting 450 for P we have

$$P = 15x - 150$$
$$450 = 15x - 150$$
$$600 = 15x$$
$$40 = x$$

Thus, Richard must sell 40 model cars to make a profit of $450.   ●

TIMELY TIP Following is a summary of the different types of systems of linear equations.

- A *consistent system of equations* is one that has a solution.
- An *inconsistent system of equations* is one that has no solution.
- A *dependent system of equations* is one that has an infinite number of solutions.

SECTION 7.1 EXERCISES

CONCEPT/WRITING EXERCISES

1. What is a system of linear equations?

2. What is the solution to a system of linear equations?

3. Define an *inconsistent system of equations.*

4. Define a *consistent system of equations.*

5. Define a *dependent system of equations.*

6. **a)** Outline the procedure for solving a system of linear equations by graphing.

   b) What is a disadvantage of solving a system of linear equations by graphing?

7. If a system of linear equations has no solution, what does that mean about the graphs of the equations in the system?

8. If a system of linear equations has one solution, what does that mean about the graphs of the equations in the system?

9. If a system of linear equations has an infinite number of solutions, what does that mean about the graphs of the equations in the system?

10. Can a system of linear equations have exactly two solutions? Explain.

PRACTICE THE SKILLS

In Exercises 11 and 12, determine which ordered pairs are solutions to the given system.

11. $y = 3x - 4$ $(3, 5)$ $(2, 2)$ $(1, 7)$
    $y = -x + 8$

12. $x + 2y = 6$ $(-2, 4)$ $(2, 2)$ $(3, -9)$
    $x - y = -6$

In Exercises 13–16, solve the system of equations graphically.

13. $x = 2$
   $y = 4$

14. $x = -1$
   $y = 3$

15. $x = 4$
   $y = -3$

16. $x = -5$
   $y = -3$

In Exercises 17–32, solve the system of equations graphically. If the system does not have a single ordered pair as a solution, state whether the system is inconsistent or dependent.

17. $x = 3$
   $y = -x - 2$

18. $y = 2$
   $y = x - 1$

19. $y = 4x - 8$
   $y = -x + 7$

20. $x + y = 6$
   $-x + y = 4$

21. $x + 2y = 0$
   $2x - 3y = -14$

22. $3x - y = 1$
   $4x - 3y = 3$

23. $2x + y = 3$
   $2y = 6 - 4x$

24. $y = 2x - 4$
   $2x + y = 0$

25. $y = x + 3$
   $y = -1$

26. $x = 1$
   $x + y + 3 = 0$

27. $2x - y = -3$
   $2x + y = -9$

28. $3x + 2y = 6$
   $6x + 4y = 12$

29. $2x - 3y = 12$
   $3y - 2x = 9$

30. $y = \frac{1}{3}x - 4$
   $3y - x = 4$

31. $y = \frac{4}{3}x - 2$
   $2x + 2y = 10$

32. $2(x - 1) + 2y = 0$
   $3x + 2(y + 2) = 0$

33. a) If the two lines in a system of equations have different slopes, how many solutions will the system have? Explain your answer.

   b) If the two lines in a system of equations have the same slope but different y-intercepts, how many solutions will the system have? Explain.

   c) If the two lines in a system of equations have the same slope and the same y-intercept, how many solutions will the system have? Explain.

34. Indicate whether the graph shown represents a consistent, inconsistent, or dependent system. Explain your answer.

a)

b)

c)

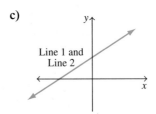

In Exercises 35–46, determine without graphing whether the system of equations has exactly one solution, no solution, or an infinite number of solutions. (Consider your answers to Exercise 33.)

35. $3x + y = 9$
   $y = -3x + 9$

36. $4x + 3y = 8$
   $6y = -8x + 4$

37. $3x + y = -6$
   $4x - 2y = -8$

38. $x - 3y = 8$
   $3x - y = 6$

39. $3x + y = 7$
   $y = -3x + 9$

40. $x + 4y = 12$
   $x = 4y + 3$

41. $2x - 3y = 6$
   $x - \frac{3}{2}y = 3$

42. $x - 2y = 6$
   $x + 2y = 4$

43. $3x = 6y + 5$
   $y = \frac{1}{2}x - 3$

44. $3y = 6x + 4$
   $-2x + y = \frac{4}{3}$

45. $4x + 7y = 2$
   $4x = 6 + 7y$

46. $12x - 5y = 4$
   $3x + 4y = 6$

PROBLEM SOLVING

Two lines are perpendicular *to each other when they meet at a right angle (a 90° angle). Two lines are perpendicular when their slopes are* negative reciprocals. *The negative reciprocal of 2 is* $-\frac{1}{2}$, *the negative reciprocal of* $\frac{3}{5}$ *is* $-1/(\frac{3}{5})$ *or* $-\frac{5}{3}$, *and so on. If a represents any real number, except 0, its negative reciprocal is* $-1/a$. *Note that the product of a number and its negative reciprocal is* -1. *In Exercises 47–50, determine, by finding the slope of each line, whether the lines will be perpendicular to each other when graphed.*

47. $4y - 2x = 15$
$3y - 5x = 9$

48. $4y - x = 6$
$y = x + 8$

49. $2x + y = 3$
$2y - x = 5$

50. $6x + 5y = 3$
$-10x = 2 + 12y$

In Exercises 51–55, part of the question involves determining a system of equations that models the situation.

51. MODELING - *Hot-Water Heater Repair* Phillip Hoffman's plumber says it will cost Phillip $250 to repair his water heater. Phillip can purchase a new, more efficient heater for $700. His current heater costs $375 per year for energy costs, and the new heater would cost $225 per year.

a) Write a system of equations with one equation representing the total cost of repairing the heater plus annual energy costs and the other equation representing the total cost of replacing the heater plus annual energy costs.

b) Graph both equations (for up to and including 6 years) on the same axes.

c) Determine the number of years it would take for the total cost of repair to equal the total cost of replacement.

52. MODELING - *Landscaping Revisited* In Example 5, assume that Tom's Tree and Landscape Service charges $200 for a consultation fee plus $60 per hour for labor and that Lawn Perfect Landscape Service charges $305 for a consultation fee plus $25 per hour for labor.

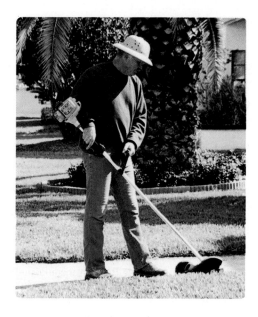

a) Write the system of equations to represent the cost of the two landscaping services.

b) Graph both equations for 0 to 10 hours on the same axes.

c) Determine the number of hours of landscaping that must be used for both services to have the same cost.

53. MODELING - *Selling Backpacks* Benjamin's Backpacks can sell backpacks for $25 per backpack. The costs for making the backpacks are a fixed cost of $400 and a production cost of $15 per backpack (see Example 6 for an example of cost and revenue equations).

a) Write the cost and revenue equations.

b) Graph both equations, for 0 to 50 backpacks, on the same axes.

c) Use the graph to determine the number of backpacks Benjamin's Backpacks must sell to break even.

d) Write the profit formula.

e) Use the profit formula to determine whether Benjamin's Backpacks makes a profit or loss if it sells 30 backpacks. What is the profit or loss?

f) How many backpacks must Benjamin's Backpacks sell to realize a profit of $1000?

54. MODELING - *Purchasing Stocks* When buying or selling stock for a customer, the Mark Demo Agency charges $40 plus 8 cents per share of stock purchased or sold. Andy Harris and Associates charges $15 plus 18 cents per share of stock purchased or sold.

   a) Write a system of equations to represent the cost of purchasing or selling stock with each company.

   b) Graph both equations (for up to and including 350 shares of stock) on the same axes.

   c) Determine the number of shares of stock that must be purchased or sold for the total cost to be the same.

   d) If 300 shares of stock are to be purchased, which firm would be less expensive?

55. MODELING - *Manufacturing PDAs* A manufacturer sells a certain personal digital assistant, or PDA, for $300 per unit. Manufacturing costs consist of a fixed cost of $8400 and a production cost of $230 per unit.

   a) Write the cost and revenue equations.

   b) Graph both equations (for up to and including 150 units) on the same axes.

   c) Use the graph to determine the number of units the manufacturer must sell to break even.

   d) Write the profit formula.

   e) Use the profit formula to determine the manufacturer's profit or loss if 100 units are sold.

   f) How many units must the manufacturer sell to make a profit of $1260?

56. Explain how you can determine whether a system of two linear equations will be consistent, dependent, or inconsistent without graphing the equations.

CHALLENGE PROBLEMS/GROUP ACTIVITIES

57. MODELING - *Job Offers* Hubert Hotchkiss had two job offers for sales positions. One pays a salary of $500 per week plus a 15% commission on his dollar sales volume. The second position pays a salary of $650 per week with no commission.

   a) For each offer, write an equation that expresses the weekly pay.

   b) Graph the system of equations and determine the solution.

   c) For what dollar sales volume will the two offers result in the same pay?

58. MODELING - *Long-Distance Calling*

   a) In September 2006, an AT&T One Rate Plan charged 7 cents per minute for long-distance calls with a monthly fee of $3.95. The Sprint Nickel Anytime Plan charged 5 cents per minute for long-distance calls with a monthly fee of $8.95. Write an equation to determine the monthly cost for long-distance service with the AT&T One Rate Plan and write an equation to determine the monthly cost for long-distance service with the Sprint Nickel Anytime Plan.

   b) Graph the system of equations and determine the solution.

   c) After how many minutes will the cost for the two long-distance service plans be the same?

59. *Points of Intersection* **a)** If two lines have different slopes, what is the maximum possible number of points of intersection?

   b) If three lines all have different slopes, what is the maximum possible number of points of intersection of any two lines?

   c) If four lines all have different slopes, what is the maximum possible number of points of intersection of any two lines?

   d) If five lines all have different slopes, what is the maximum possible number of points of intersection of any two lines?

   e) Is there a pattern in the number of points of intersection? If so, explain the pattern. Use the pattern to determine the maximum possible number of points of intersection for six lines.

RECREATIONAL MATHEMATICS

60. Connect all the following points using exactly four straight-line segments. Do not lift your pencil off the paper.

   •   •   •

   •   •   •

   •   •   •

INTERNET/RESEARCH ACTIVITY

61. *The Rhind Papyrus* The Rhind Papyrus indicates that the early Egyptians used linear equations. Do research and write a paper on the symbols used in linear equations and the use of the linear equations by the early Egyptians. (References include history of mathematics books, encyclopedias, and the Internet.)

7.2 SOLVING SYSTEMS OF LINEAR EQUATIONS BY THE SUBSTITUTION AND ADDITION METHODS

▲ Solving a system of equations may help determine which of two cell phone plans offers the least expensive service.

Bill Ramierez is considering two cell phone plans. Both plans offer 300 free minutes each month. One plan charges $30 per month plus 45 cents for each additional minute after 300 minutes. The other plan charges $35 per month plus 20 cents for each additional minute after 300 minutes. How would Bill determine how long he would have to talk on the phone, in 1 month, for the two plans to have the same cost? In this section, we will illustrate two different algebraic methods, the substitution method and the addition method, for answering this question.

We first discuss the substitution method.

Substitution Method

PROCEDURE FOR SOLVING A SYSTEM OF EQUATIONS USING THE SUBSTITUTION METHOD

1. Solve one of the equations for one of the variables. If possible, solve for a variable with a numerical coefficient of 1. By doing so, you may avoid working with fractions.

2. Substitute the expression found in step 1 into the other equation. This step yields an equation in terms of a single variable.

3. Solve the equation found in step 2 for the variable.

4. Substitute the value found in step 3 into the equation you rewrote in step 1 and solve for the remaining variable.

Examples 1, 2, and 3 illustrate the *substitution method*. These systems of equations are the same as in Examples 2, 3, and 4 in Section 7.1.

EXAMPLE ❶ *A Single Solution, by the Substitution Method*

Solve the following system of equations by substitution.

$$x + y = 4$$
$$2x - y = -1$$

SOLUTION The numerical coefficients of the x and y terms in the equation $x + y = 4$ are both 1. Thus, we can solve this equation for either x or y. Let's solve for x in the first equation.

STEP 1.

$$x + y = 4$$
$$x + y - y = 4 - y \qquad \text{Subtract } y \text{ from both sides of the equation.}$$
$$x = 4 - y$$

STEP 2. Substitute $4 - y$ for x in the second equation.

$$2x - y = -1$$
$$2(4 - y) - y = -1$$

STEP 3. Now solve the equation for y.

$$8 - 2y - y = -1 \qquad \text{Distributive property}$$
$$8 - 3y = -1$$
$$8 - 8 - 3y = -1 - 8 \qquad \text{Subtract 8 from both sides of the equation.}$$
$$-3y = -9$$
$$\frac{-3y}{-3} = \frac{-9}{-3} \qquad \text{Divide both sides of the equation by } -3.$$
$$y = 3$$

STEP 4. Substitute $y = 3$ in the equation solved for x and determine the value of x.

$$x = 4 - y$$
$$x = 4 - 3$$
$$x = 1$$

Thus, the solution is the ordered pair $(1, 3)$. This answer checks with the solution obtained graphically in Section 7.1, Example 2. ●

TIMELY TIP When solving a system of equations, once you successfully solve for one of the variables, make sure you solve for the other variable. Remember that a solution to a system of equations must contain a numerical value for each variable in the system.

EXAMPLE ❷ *No Solution, by the Substitution Method*

Solve the following system of equations by substitution.

$$x + y = 3$$
$$2x + 2y = -4$$

SOLUTION The numerical coefficients of the x and y terms in the equation $x + y = 3$ are both 1. Thus, we can solve this equation for either x or y. Let us solve for y in the first equation.

$$x + y = 3$$
$$x - x + y = 3 - x \qquad \text{Subtract } x \text{ from both sides of the equation.}$$
$$y = 3 - x$$

Now substitute $3 - x$ for y in the second equation.

$$2x + 2y = -4$$
$$2x + 2(3 - x) = -4$$
$$2x + 6 - 2x = -4 \qquad \text{Distributive property}$$
$$6 = -4 \qquad \text{False}$$

Since 6 cannot be equal to -4, there is no solution to the system of equations. Thus, the system of equations is inconsistent. This answer checks with the solution obtained graphically in Section 7.1, Example 3. ●

When solving the system in Example 2, we obtained $6 = -4$ and indicated that the system was inconsistent and that there was no solution. When solving a system of equations, if you obtain a false statement, such as $4 = 0$ or $-2 = 0$, the system is *inconsistent* and has *no solution.*

EXAMPLE ❸ An Infinite Number of Solutions, by the Substitution Method

Solve the following system of equations by substitution.

$$y = \frac{1}{2}x + 4$$
$$2y = x + 8$$

SOLUTION The first equation $y = \frac{1}{2}x + 4$ is already solved for y, so we will substitute $\frac{1}{2}x + 4$ for y in the second equation.

$$2y = x + 8$$
$$2\left(\frac{1}{2}x + 4\right) = x + 8$$
$$x + 8 = x + 8 \qquad \text{Distributive property}$$
$$x - x + 8 = x - x + 8 \qquad \text{Subtract } x \text{ from both sides of the equation.}$$
$$8 = 8 \qquad \text{True}$$

Since 8 equals 8, the system has an infinite number of solutions. Thus, the system of equations is dependent. This answer checks with the solution obtained in Section 7.1, Example 4. ●

When solving Example 3, we obtained $8 = 8$ and indicated that the system was dependent and had an infinite number of solutions. When solving a system of equations, if you obtain a true statement, such as $0 = 0$ or $8 = 8$, the system is *dependent* and has an *infinite number of solutions.*

Addition Method

If neither of the equations in a system of linear equations has a variable with a coefficient of 1, it is generally easier to solve the system by using the *addition* (or *elimination*) *method*.

To solve a system of linear equations by the addition method, it is necessary to obtain two equations whose sum will be a single equation containing only one variable. To achieve this goal, we rewrite the system of equations as two equations where the coefficients of one of the variables are opposites (or additive inverses) of each other. For example, if one equation has a term of $2x$, we might rewrite the other equation so that its x term will be $-2x$. To obtain the desired equations, it might be necessary to multiply one or both equations in the original system by a number. When an equation is to be multiplied by a number, we will place brackets around the equation and place the number that is to multiply the equation before the brackets. For example, $4[2x + 3y = 6]$ means that each term on both sides of the equal sign in the equation $2x + 3y = 6$ is to be multiplied by 4:

$$4[2x + 3y = 6] \qquad \text{gives} \qquad 8x + 12y = 24$$

This notation will make our explanations much more efficient and easier for you to follow.

PROCEDURE FOR SOLVING A SYSTEM OF EQUATIONS BY THE ADDITION METHOD

1. If necessary, rewrite the equations so that the terms containing the variables appear on one side of the equal sign and the constants appear on the other side of the equal sign.

2. If necessary, multiply one or both equations by a constant(s) so that when you add the equations, the sum will be an equation containing only one variable.

3. Add the equations to obtain a single equation in one variable.

4. Solve for the variable in the equation obtained in step 3.

5. Substitute the value found in step 4 into either of the original equations and solve for the other variable.

EXAMPLE ❹ *Eliminating a Variable by the Addition Method*

Solve the following system of equations by the addition method.

$$x + y = 5$$
$$2x - y = 7$$

SOLUTION Since the coefficients of the y terms, 1 and -1, are additive inverses, the sum of the y terms will be zero when the equations are added. Thus, the sum of the two equations will contain only one variable, x. Add the two equations to obtain one equation in one variable. Then solve for the remaining variable.

$$
\begin{aligned}
x + y &= 5 \\
\underline{2x - y} &= \underline{7} \\
3x &= 12 \\
x &= 4
\end{aligned}
$$

Now substitute 4 for x in either of the original equations to find the value of y.

$$x + y = 5$$
$$4 + y = 5$$
$$y = 1$$

The solution to the system is (4, 1). ●

EXAMPLE ⑤ Multiplying by −1 in the Addition Method

Solve the following system of equations by the addition method.

$$x + 4y = 10$$
$$x + 2y = 6$$

SOLUTION We want the sum of the two equations to have only one variable. We can eliminate the variable x by multiplying either equation by −1 and then adding the two equations. We will multiply the first equation by −1.

$$-1[x + 4y = 10] \quad \text{gives} \quad -x - 4y = -10$$
$$x + 2y = 6 \qquad\qquad\qquad x + 2y = 6$$

We now have a system of equations equivalent to the original system.
   Now add the two equations.

$$-x - 4y = -10$$
$$\underline{x + 2y = 6}$$
$$-2y = -4$$
$$y = 2$$

Now we solve for x by substituting 2 for y in either of the original equations.

$$x + 4y = 10$$
$$x + 4(2) = 10$$
$$x + 8 = 10$$
$$x = 2$$

The solution is (2, 2). ●

EXAMPLE ⑥ Multiplying One Equation in the Addition Method

Solve the following system of equations by the addition method.

$$4x + y = 6$$
$$3x + 2y = 7$$

SOLUTION We can multiply the top equation by -2 and then add the two equations to eliminate the variable y.

$$-2[4x + y = 6] \quad \text{gives} \quad -8x - 2y = -12$$
$$3x + 2y = 7 \qquad\qquad 3x + 2y = 7$$

$$\begin{aligned} -8x - 2y &= -12 \\ \underline{3x + 2y} &= \underline{7} \\ -5x &= -5 \\ x &= 1 \end{aligned}$$

Now we find y by substituting 1 for x in either of the original equations.

$$4x + y = 6$$
$$4(1) + y = 6$$
$$4 + y = 6$$
$$y = 2$$

The solution is $(1, 2)$. ●

Note that in Example 6 we could have eliminated the variable x by multiplying the top equation by 3 and the bottom equation by -4, then adding the two equations. Try this method now.

EXAMPLE ❼ *Multiplying Both Equations*

Solve the following system of equations by the addition method.

$$3x - 4y = 8$$
$$2x + 3y = 9$$

SOLUTION In this system, we cannot eliminate a variable by multiplying only one equation by an integer value and then adding. To eliminate a variable, we can multiply each equation by a different number. To eliminate the variable x, we can multiply the top equation by 2 and the bottom by -3 (or the top by -2 and the bottom by 3) and then add the two equations. If we want, we can instead eliminate the variable y by multiplying the top equation by 3 and the bottom by 4 and then adding the two equations. Let's eliminate the variable x.

$$2[3x - 4y = 8] \quad \text{gives} \quad 6x - 8y = 16$$
$$-3[2x + 3y = 9] \quad \text{gives} \quad -6x - 9y = -27$$

$$\begin{aligned} 6x - 8y &= 16 \\ \underline{-6x - 9y} &= \underline{-27} \\ -17y &= -11 \\ y &= \frac{11}{17} \end{aligned}$$

We could now find x by substituting $\frac{11}{17}$ for y in either of the original equations. Although it can be done, it gets messy. Instead, let's solve for x by eliminating the variable y from the two original equations. To do so, we multiply the first equation by 3 and the second equation by 4.

$$3[3x - 4y = 8] \quad \text{gives} \quad 9x - 12y = 24$$
$$4[2x + 3y = 9] \quad \text{gives} \quad 8x + 12y = 36$$

$$\begin{aligned} 9x - 12y &= 24 \\ \underline{8x + 12y} &= \underline{36} \\ 17x \quad\;\; &= 60 \\ x &= \frac{60}{17} \end{aligned}$$

The solution to the system is $\left(\frac{60}{17}, \frac{11}{17}\right)$.

●

> **TIMELY TIP** If you obtain an equation such as $0 = 6$, or any other equation that is false when solving a system of linear equations, the system is *inconsistent* (the two equations represent parallel lines; see Fig. 7.3 on page 416) and there is no solution.
> If you obtain the equation $0 = 0$ when solving a system of linear equations by either the substitution or the addition method, the system is *dependent* (both equations represent the same line; see Fig. 7.4 on page 416) and there are an infinite number of solutions.

EXAMPLE ❽ MODELING - *When Are Repair Costs the Same?*

Melinda Melendez needs to purchase a new radiator for her car and have it installed by a mechanic. She is considering two garages: Steve's Repair and Greg's Garage. At Steve's Repair, the parts cost $200 and the labor cost is $50 per hour. At Greg's Garage, the parts cost $375 and the labor cost is $25 per hour. How many hours would the repair need to take for the total cost at each garage to be the same?

SOLUTION We are asked to find the number of hours the repair would need to take for each garage to have the same total cost, C. First write a system of equations to represent the total cost for each of the garages. The total cost consists of the cost of the parts and the labor cost. The labor cost depends on the number of hours of labor.
Let $x =$ the number of hours of labor.

$$\text{Total cost} = \text{cost of parts} + \text{labor cost}$$
$$\text{Steve's Repair: } C = 200 + 50x$$
$$\text{Greg's Garage: } C = 375 + 25x$$

We want to determine when the cost will be the same, so we set the two costs equal to each other (substitution method) and solve the resulting equation.

$$200 + 50x = 375 + 25x$$
$$200 - 200 + 50x = 375 - 200 + 25x \qquad \text{Subtract 200 from both sides of the equation.}$$
$$50x = 175 + 25x$$
$$50x - 25x = 175 + 25x - 25x \qquad \text{Subtract } 25x \text{ from both sides of the equation.}$$
$$25x = 175$$
$$\frac{25x}{25} = \frac{175}{25} \qquad \text{Divide both sides of the equation by 25.}$$
$$x = 7$$

Thus, for 7 hours of labor, the cost at both garages would be the same. If we construct a graph (Fig. 7.9) of the two cost equations, the point of intersection is (7, 550). If the repair were to require 7 hours of labor, the total cost at either garage would be $550.

Figure 7.9

EXAMPLE 9 **MODELING** - *A Mixture Problem*

Pat Kuby, a pharmacist, needs 500 milliliters (mℓ) of a 10% phenobarbital solution. She has only a 5% phenobarbital solution and a 25% phenobarbital solution available. How many milliliters of each solution should she mix to obtain the desired solution?

SOLUTION First we set up a system of equations. The unknown quantities are the amount of 5% solution and the amount of the 25% solution that must be used. Let

$$x = \text{number of m}\ell \text{ of 5\% solution}$$
$$y = \text{number of m}\ell \text{ of 25\% solution}$$

We know that 500 mℓ of solution are needed. Thus,

$$x + y = 500$$

The total amount of phenobarbital in a solution is determined by multiplying the percent of phenobarbital by the number of milliliters of solution. The second equation comes from the fact that

$$\begin{pmatrix} \text{Total amount of} \\ \text{phenobarbital in} \\ \text{5\% solution} \end{pmatrix} + \begin{pmatrix} \text{total amount of} \\ \text{phenobarbital in} \\ \text{25\% solution} \end{pmatrix} = \begin{pmatrix} \text{total amount of} \\ \text{phenobarbital} \\ \text{in 10\% mixture} \end{pmatrix}$$

$$0.05x \qquad + \qquad 0.25y \qquad = \qquad 0.10(500)$$

or $0.05x + 0.25y = 50$

The system of equations is

$$x + y = 500$$
$$0.05x + 0.25y = 50$$

Let's solve this system of equations by using the addition method. There are various ways of eliminating one variable. To obtain integer values in the second equation, we can multiply both sides of the equation by 100. The result will be an x term of $5x$. If we multiply both sides of the first equation by -5, that will result in an x term of $-5x$. By following this process, we can eliminate the x terms from the system.

$$-5[x + y = 500] \quad \text{gives} \quad -5x - 5y = -2500$$
$$100[0.05x + 0.25y = 50] \quad \text{gives} \quad 5x + 25y = 5000$$

$$
\begin{array}{r}
-5x - 5y = -2500 \\
\underline{5x + 25y = 5000} \\
20y = 2500
\end{array}
$$

$$\frac{20y}{20} = \frac{2500}{20}$$

$$y = 125$$

Now we determine x.

$$x + y = 500$$
$$x + 125 = 500$$
$$x = 375$$

Therefore, 375 mℓ of a 5% phenobarbital solution must be mixed with 125 mℓ of a 25% phenobarbital solution to obtain 500 mℓ of a 10% phenobarbital solution.   ●

Example 9 can also be solved by using substitution. Try to do so now.

SECTION 7.2 EXERCISES

CONCEPT/WRITING EXERCISES

1. In your own words, explain how to solve a system of linear equations by using the substitution method.

2. In your own words, explain how to solve a system of linear equations by using the addition method.

3. How will you know, when solving a system of linear equations by either the substitution or the addition method, whether the system is dependent?

4. How will you know, when solving a system of linear equations by either the substitution or the addition method, whether the system is inconsistent?

5. When solving the following system of equations by the substitution method, which variable, in which equation, would you choose to solve for in order to make the solution easier to obtain? Explain your answer. Do not solve the system.

$$x + 3y = 3$$
$$3x + 4y = -1$$

6. When solving the following system of equations by the addition method, what will your first step be in solving the system? Explain your answer. Do not solve the system.

$$2x + y = 6$$
$$3x + 3y = 9$$

PRACTICE THE SKILLS

In Exercises 7–24, solve the system of equations by the substitution method. If the system does not have a single ordered pair as a solution, state whether the system is inconsistent or dependent.

7. $y = x + 8$
   $y = -x + 4$

8. $y = 4x - 3$
   $y = 3x - 1$

9. $6x + 5y = 1$
   $x - 3y = 4$

10. $4x - y = 3$
    $3x - y = 1$

11. $y - x = 4$
    $x - y = 3$

12. $x + y = 3$
    $y + x = 5$

13. $3y + 2x = 4$
$3y = 6 - x$

14. $x = 5y - 12$
$x - y = 0$

15. $y - 2x = 3$
$2y = 4x + 6$

16. $y = 2$
$y + x + 3 = 0$

17. $x = y + 3$
$x = -3$

18. $x + 2y = 6$
$y = 2x + 3$

19. $y + 3x - 4 = 0$
$2x - y = 7$

20. $x + 4y = 7$
$2x + 3y = 5$

21. $x = 2y + 3$
$y = 3x - 1$

22. $x + 4y = 9$
$2x - y - 6 = 0$

23. $y = -2x + 3$
$4x + 2y = 12$

24. $2x + y = 12$
$x = -\frac{1}{2}y + 6$

In Exercises 25–40, solve the system of equations by the addition method. If the system does not have a single ordered pair as a solution, state whether the system is inconsistent or dependent.

25. $3x + y = 9$
$2x - y = 6$

26. $x + 3y = 9$
$x - 3y = -3$

27. $x + y = 12$
$x - 2y = -3$

28. $2x + y = 10$
$-2x + 2y = -16$

29. $2x - y = -4$
$-3x - y = 6$

30. $x + y = 6$
$-2x + y = -3$

31. $4x + 3y = -1$
$2x - y = -13$

32. $2x + y = 6$
$3x + y = 5$

33. $2x + y = 11$
$x + 3y = 18$

34. $5x - 2y = 11$
$-3x + 2y = 1$

35. $3x - 4y = 11$
$3x + 5y = -7$

36. $4x - 2y = 6$
$4y = 8x - 12$

37. $4x + y = 6$
$-8x - 2y = 13$

38. $2x + 3y = 6$
$5x - 4y = -8$

39. $3x - 4y = 2$
$4x + 3y = 11$

40. $6x + 6y = 1$
$4x + 9y = 4$

PROBLEM SOLVING

In Exercises 41–52, write a system of equations that can be used to solve the problem. Then solve the system and determine the answer.

41. MODELING - *Owning a Business* Sosena Milion can join a small business as a full partner and receive a salary of $12,000 per year plus 15% of the year's profit, or she can join as a sales manager with a salary of $27,000 per year plus 5% of the year's profit. What must the year's profit be

for her total earnings to be the same whether she joins as a full partner or as a sales manager?

42. MODELING - *Investments* David Stewart invested $25,000 in two different corporate bonds for 1 year. One bond pays a 4.5% simple interest rate, and the other pays a 6% simple interest rate. The total annual interest David received from both bonds was $1380. Find the amount he invested in each bond. For 1 year, the simple interest on a specific corporate bond is found by multiplying the amount invested by the simple interest rate.

43. MODELING - *Pizza Orders* Pizza Corner sells medium and large specialty pizzas. A medium Meat Lovers pizza costs $10.95, and a large Meat Lovers pizza costs $14.95. One Saturday a total of 50 Meat Lovers pizzas were sold, and the receipts from the Meat Lovers pizzas were $663.50. How many medium and how many large Meat Lovers pizzas were sold?

44. MODELING - *Basketball Game* The University of Maryland women's basketball team made 45 field goals in a recent game; some were 2-pointers and some were 3-pointers. How many 2-point baskets were made and how many 3-point baskets were made if Maryland scored 101 points?

45. MODELING - *Chemical Mixture* Antonio Gonzalez is a chemist and needs 10 liters (ℓ) of a 40% hydrochloric acid solution. He discovers he is out of the 40% hydrochloric acid solution and does not have sufficient time to reorder. He checks his supply shelf and finds he has a large supply of both 25% and 50% hydrochloric acid solutions. He decides to use the 25% and 50% solutions to make 10 ℓ of a 40% solution. How many liters of the 25% solution and of the 50% solution should he mix?

46. MODELING - *Sets of Dishes* A restaurant manager purchased 100 sets of dishes. One design cost $30 per set, and another design cost $40 per set. If the manager spent $3200 on the dishes, how many sets of each design did she purchase?

47. MODELING - *Choosing a Copy Service* Lori Lanier recently purchased a high-speed copier for her home office and wants to purchase a service contract on the copier. She is considering two sources for the contract. The Economy Sales and Service Company charges $18 a month plus 2 cents per copy. Office Superstore charges $24 a month but only 1.5 cents per copy. How many copies would Lori need to make for the monthly costs of both plans to be the same?

48. MODELING - *Hardwood Floor Installation* The cost to purchase a particular type of hardwood flooring at Home Depot is $2.65 per square foot. In addition, the installation cost is $468.75. The cost to purchase the same flooring at Hardwood Guys is $3.10 per square foot. In addition, the installation cost is $412.50.

a) Determine the number of square feet of this flooring that Roberto Cruz must purchase for the total cost of the flooring and installation to be the same from both stores.

b) If Roberto needs to purchase and have installed 196 square feet of this flooring, which store would be less expensive?

49. MODELING - *Nut and Pretzel Mix* Dave Chwalik wants to purchase 20 pounds of party mix for a total of $30. To obtain the mixture, he will mix nuts that cost $3 per pound with pretzels that cost $1 per pound. How many pounds of each type of mix should he use?

50. MODELING - *Laboratory Research* Animals in an experiment are to be kept on a strict diet. Each animal is to receive, among other things, 20 g of protein and 6 g of carbohydrates. The scientist has only two food mixes of the following compositions available.

	Protein (%)	Carbohydrates (%)
Mix A	10	6
Mix B	20	2

How many grams of each mix should she use to obtain the right diet for a single animal?

51. MODELING - *Concert Ticket Prices* Finger Lakes Performing Arts Center sold 4600 tickets for a summer jazz concert. Covered amphitheatre tickets cost $27 and lawn seats cost $14. If $104,700 in ticket sales was collected for the concert, how many tickets of each type were sold?

52. MODELING - *Golf Club Membership* Membership in Oakwood Country Club costs $3000 per year and entitles a member to play a round of golf for a greens fee of $18. At Pinecrest Country Club, membership costs $2500 per year and the greens fee is $20.

a) How many rounds must a golfer play in a year for the costs at the two clubs to be the same?

b) If Tamika Johnson planned to play 30 rounds of golf in a year, which club would be the least expensive?

53. MODELING - *Car Sales* The following graph shows that General Motors' (GM's) share of the U.S. car and truck market has declined from 2001 through 2005 and is expected to continue to decline through 2010. The graph also shows that Toyota cars and trucks sales are increasing and are expected to continue to increase through 2010. GM's percentage of the U.S. market share can be modeled by the equation $y = -0.83x + 28.33$, and Toyota's market share can be modeled by the equation $y = 0.89x + 9.11$, where x represents the number of years since 2000 and $x \geq 1$. Assuming this trend continues, use the substitution method to approximate when GM's percentage of market share will equal Toyota's.

Source: Center for Automotive Research

54. MODELING - *Printing Photos* The graph on the top of page 434 shows that the number of traditional photos printed at retail photography labs has declined from 2000 through 2005. The graph also shows that the number of digital photos printed at retail photography labs has increased over this time period. The number of traditional photos printed at labs can be modeled by the equation $y = -2.3x + 29.9$, and the number of digital photos printed at labs can be modeled by the equation $y = 1.5x + 0.4$, where x represents the number of years since 2000. Assuming this trend continues, use the substitution method to approximate when the number of traditional photos printed will equal the number of digital photos printed at retail photography labs.

Prints Made at Labs

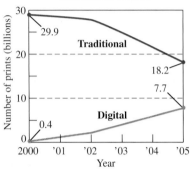

Source: Data from Photo Marketing Association International

CHALLENGE PROBLEMS/GROUP ACTIVITIES

55. Solve the following system of equations for u and v by first substituting x for $\frac{1}{u}$ and y for $\frac{1}{v}$.

$$\frac{1}{u} + \frac{2}{v} = 8$$

$$\frac{3}{u} - \frac{1}{v} = 3$$

56. Develop a system of equations that has (6, 5) as its solution. Explain how you developed your system of equations.

57. The substitution or addition methods can also be used to solve a system of three equations in three variables. Consider the following system.

$$x + y + z = 7$$
$$x - y + 2z = 9$$
$$-x + 2y + z = 4$$

The *ordered triple* (x, y, z) is the solution to the system if it satisfies all three equations.

a) Show that the ordered triple (2, 1, 4) is a solution to the system.

b) Use the substitution or addition method to determine the solution to the system. (*Hint:* Eliminate one variable by using two equations. Then eliminate the same variable by using two different equations.)

58. Construct a system of two equations that has no solution. Explain how you know the system has no solution.

59. Construct a system of two equations that has an infinite number of solutions. Explain how you know the system has an infinite number of solutions.

60. When solving a system of equations by the substitution method, a student obtained the equation $0 = 0$ and gave the solution as (0, 0). What is the student's error?

61. In parts (a)–(d), make up a system of linear equations whose solution will be the ordered pair given. *Hint:* It may be helpful to visualize possible graphs that have the given solution. There are many possible answers for each part.

a) (0, 0)   b) (1, 0)   c) (0, 1)   d) (1, 1)

7.3   MATRICES

▲ In this section, we introduce a method to display information from a survey such as opinions of college students.

Five hundred students at the University of Delaware were asked if they were in favor of or opposed to an increase in their student fees to pay for building a new meeting room for student clubs. Each student was also asked to indicate whether he or she was a freshman, sophomore, junior, or senior. How can the responses from the survey be displayed? In this section, we will introduce a method used to display information such as responses from a survey.

A *matrix* is a rectangular array of elements. An array is a systematic arrangement of numbers or symbols in rows and columns. Matrices (the plural of matrix) may be used

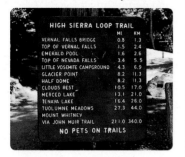
to display information and to solve systems of linear equations. The following matrix displays the responses from the survey of 500 students at the University of Delaware regarding an increase in their student fees.

	Freshmen	Sophomores	Juniors	Seniors
In favor	102	93	22	35
Opposed	82	94	23	49

(Columns across the top, Rows down the side)

The numbers in the rows and columns of a matrix are called the *elements* of the matrix. The matrix given above contains eight elements. The *dimensions* of a matrix may be indicated with the notation $r \times s$, where r is the number of rows and s is the number of columns in the matrix. Because the matrix given above has 2 rows and 4 columns, it is a 2 by 4, written 2×4, matrix. In this text, from this point onward, we use brackets, [ ], to indicate a matrix. Consider the two matrices below. A matrix that contains the same number of rows and columns is called a *square matrix*. Following is an example of a 2×2 square matrix and a 3×3 square matrix.

$$\begin{bmatrix} 2 & 3 \\ 5 & 2 \end{bmatrix} \qquad \begin{bmatrix} 4 & 6 & -1 \\ 2 & 3 & 0 \\ 5 & 2 & 1 \end{bmatrix}$$

Two matrices are equal if and only if they have the same elements in the same relative positions.

EXAMPLE ❶ Equal Matrices

Given $A = B$, determine x and y.

$$A = \begin{bmatrix} 3 & 7 \\ 4 & 9 \end{bmatrix}, \qquad B = \begin{bmatrix} x & 7 \\ 4 & y \end{bmatrix}$$

SOLUTION Since the matrices are equal, the corresponding elements must be the same, so $x = 3$ and $y = 9$. •

Addition of Matrices

Two matrices can be added only if they have the same dimensions (same number of rows and same number of columns). To obtain the sum of two matrices with the same dimensions, add the corresponding elements of the two matrices.

EXAMPLE ❷ Adding Matrices

Determine $A + B$ if

$$A = \begin{bmatrix} 3 & 4 \\ -1 & 7 \end{bmatrix} \qquad \text{and} \qquad B = \begin{bmatrix} 2 & 8 \\ 4 & 0 \end{bmatrix}$$

SOLUTION
$$A + B = \begin{bmatrix} 3 & 4 \\ -1 & 7 \end{bmatrix} + \begin{bmatrix} 2 & 8 \\ 4 & 0 \end{bmatrix}$$
$$= \begin{bmatrix} 3 + 2 & 4 + 8 \\ -1 + 4 & 7 + 0 \end{bmatrix} = \begin{bmatrix} 5 & 12 \\ 3 & 7 \end{bmatrix}$$
•

EXAMPLE ❸ MODELING - *Sales of Bicycles*

Peddler's Bicycle Corporation owns and operates two stores, one in Pennsylvania and one in New Jersey. The number of mountain bicycles (MB) and racing bicycles (RB) sold in each store during January through June and during July through December are indicated in the matrices that follow. We will call the matrices A and B.

$$
\begin{array}{cc}
& \text{Pennsylvania} \\
& \begin{array}{cc} MB & RB \end{array} \\
\begin{array}{c} \text{Jan.–June} \\ \text{July–Dec.} \end{array} & \begin{bmatrix} 515 & 425 \\ 290 & 250 \end{bmatrix} = A
\end{array}
\qquad
\begin{array}{cc}
& \text{New Jersey} \\
& \begin{array}{cc} MB & RB \end{array} \\
& \begin{bmatrix} 520 & 350 \\ 180 & 271 \end{bmatrix} = B
\end{array}
$$

Determine the total number of each type of bicycle sold by the corporation during each time period.

SOLUTION To solve the problem, we add matrices A and B.

$$
\begin{array}{c}
\begin{array}{ccc} & MB & \quad RB \end{array} \\
\begin{array}{c} \text{Jan.–June} \\ \text{July–Dec.} \end{array}
\begin{bmatrix} 515 + 520 & 425 + 350 \\ 290 + 180 & 250 + 271 \end{bmatrix} =
\begin{bmatrix} 1035 & 775 \\ 470 & 521 \end{bmatrix}
\end{array}
$$

We can see from the sum matrix that during the period from January through June, a total of 1035 mountain bicycles and 775 racing bicycles were sold. During the period from July through December, a total of 470 mountain bicycles and 521 racing bicycles were sold. ●

Subtraction of Matrices

Only matrices with the same dimension may be subtracted. To do so, we subtract each entry in one matrix from the corresponding entry in the other matrix.

EXAMPLE ❹ *Subtracting Matrices*

Determine $A - B$ if

$$
A = \begin{bmatrix} 3 & 6 \\ 5 & -1 \end{bmatrix} \quad \text{and} \quad B = \begin{bmatrix} 2 & -4 \\ 8 & -3 \end{bmatrix}
$$

SOLUTION

$$
\begin{aligned}
A - B &= \begin{bmatrix} 3 & 6 \\ 5 & -1 \end{bmatrix} - \begin{bmatrix} 2 & -4 \\ 8 & -3 \end{bmatrix} \\
&= \begin{bmatrix} 3 - 2 & 6 - (-4) \\ 5 - 8 & -1 - (-3) \end{bmatrix} = \begin{bmatrix} 1 & 10 \\ -3 & 2 \end{bmatrix}
\end{aligned}
$$

●

Multiplying a Matrix by a Real Number

A matrix may be multiplied by a real number by multiplying each entry in the matrix by the real number. Sometimes when we multiply a matrix by a real number, we call that real number a *scalar*.

EXAMPLE 5 *Multiplying a Matrix by a Scalar*

For matrices A and B, determine (a) $3A$ and (b) $3A - 2B$.

$$A = \begin{bmatrix} 1 & 4 \\ -3 & 5 \end{bmatrix}, \qquad B = \begin{bmatrix} -1 & 3 \\ 5 & 6 \end{bmatrix}$$

SOLUTION

a) $3A = 3\begin{bmatrix} 1 & 4 \\ -3 & 5 \end{bmatrix} = \begin{bmatrix} 3(1) & 3(4) \\ 3(-3) & 3(5) \end{bmatrix} = \begin{bmatrix} 3 & 12 \\ -9 & 15 \end{bmatrix}$

b) We determined $3A$ in part (a). Now we find $2B$.

$$2B = 2\begin{bmatrix} -1 & 3 \\ 5 & 6 \end{bmatrix} = \begin{bmatrix} 2(-1) & 2(3) \\ 2(5) & 2(6) \end{bmatrix} = \begin{bmatrix} -2 & 6 \\ 10 & 12 \end{bmatrix}$$

$$3A - 2B = \begin{bmatrix} 3 & 12 \\ -9 & 15 \end{bmatrix} - \begin{bmatrix} -2 & 6 \\ 10 & 12 \end{bmatrix}$$

$$= \begin{bmatrix} 3 - (-2) & 12 - 6 \\ -9 - 10 & 15 - 12 \end{bmatrix} = \begin{bmatrix} 5 & 6 \\ -19 & 3 \end{bmatrix} \qquad \bullet$$

Multiplication of Matrices

Multiplication of matrices is slightly more difficult than addition of matrices. Multiplication of matrices is possible only when the number of *columns* of the first matrix, A, is the same as the number of *rows* of the second matrix, B. We use the notation

$$\begin{array}{c} A \\ 3 \times 4 \end{array}$$

to indicate that matrix A has three rows and four columns. Suppose matrix A is a 3×4 matrix and matrix B is a 4×5 matrix. Then

$$\begin{array}{cc} A & B \\ 3 \times 4 & 4 \times 5 \end{array}$$
$$\underbrace{\qquad\text{Same}\qquad}$$
Product matrix 3×5

This notation indicates that matrix A has four columns and matrix B has four rows. Therefore, we can multiply these two matrices. The product matrix will have the same number of rows as matrix A and the same number of columns as matrix B. Thus, the dimensions of the product matrix are 3×5.

EXAMPLE 6 *Can These Matrices Be Multiplied?*

Determine which of the following pairs of matrices can be multiplied.

a) $A = \begin{bmatrix} 3 & 2 \\ 5 & 7 \end{bmatrix}, \qquad B = \begin{bmatrix} 0 & 6 \\ 4 & 1 \end{bmatrix}$

b) $A = \begin{bmatrix} 2 & 3 \\ 5 & 6 \end{bmatrix}, \qquad B = \begin{bmatrix} 2 & 4 & -1 \\ 6 & 8 & 0 \end{bmatrix}$

c) $A = \begin{bmatrix} 2 & 1 & 4 \\ 3 & 2 & 8 \end{bmatrix}, \qquad B = \begin{bmatrix} 2 & 1 & 3 \\ 1 & 0 & -2 \end{bmatrix}$

SOLUTION

a)

$$
\begin{array}{cc}
A & B \\
2 \times 2 & 2 \times 2
\end{array}
$$

Same

Because matrix A has two columns and matrix B has two rows, the two matrices can be multiplied. The product is a 2×2 matrix.

b)

$$
\begin{array}{cc}
A & B \\
2 \times 2 & 2 \times 3
\end{array}
$$

Same

Because matrix A has two columns and matrix B has two rows, the two matrices can be multiplied. The product is a 2×3 matrix.

c)

$$
\begin{array}{cc}
A & B \\
2 \times 3 & 2 \times 3
\end{array}
$$

Not Same

Because matrix A has three columns and matrix B has two rows, the two matrices cannot be multiplied. ●

To explain matrix multiplication, let's use matrices A and B that follow.

$$
A = \begin{bmatrix} 3 & 2 \\ 5 & 7 \end{bmatrix} \quad \text{and} \quad B = \begin{bmatrix} 0 & 6 \\ 4 & 1 \end{bmatrix}
$$

Since A contains two rows and B contains two columns, the product matrix will contain two rows and two columns. To multiply two matrices, we use a row–column scheme of multiplying. The numbers in the *first row* of matrix A are multiplied by the numbers in the *first column* of matrix B. These products are then added to determine the entry in the product matrix.

$$
A \times B = \begin{bmatrix} 3 & 2 \\ 5 & 7 \end{bmatrix} \begin{bmatrix} 0 & 6 \\ 4 & 1 \end{bmatrix}
$$

First row   First column

$$
\begin{bmatrix} 3 & 2 \\ 5 & 7 \end{bmatrix} \qquad \begin{bmatrix} 0 & 6 \\ 4 & 1 \end{bmatrix}
$$

$$
(3 \times 0) + (2 \times 4) = 0 + 8
$$
$$
= 8
$$

The 8 is placed in the first-row, first-column position of the product matrix. The other numbers in the product matrix are obtained similarly, as illustrated in the matrix that follows.

First row First column
$$\begin{bmatrix}3 & 2\\5 & 7\end{bmatrix}\begin{bmatrix}0 & 6\\4 & 1\end{bmatrix}$$
$(3 \times 0) + (2 \times 4) = 8$

First row Second column
$$\begin{bmatrix}3 & 2\\5 & 7\end{bmatrix}\begin{bmatrix}0 & 6\\4 & 1\end{bmatrix}$$
$(3 \times 6) + (2 \times 1) = 20$

$$A \times B = \begin{bmatrix}8 & 20\\28 & 37\end{bmatrix}$$

Second row First column
$$\begin{bmatrix}3 & 2\\5 & 7\end{bmatrix}\begin{bmatrix}0 & 6\\4 & 1\end{bmatrix}$$
$(5 \times 0) + (7 \times 4) = 28$

Second row Second column
$$\begin{bmatrix}3 & 2\\5 & 7\end{bmatrix}\begin{bmatrix}0 & 6\\4 & 1\end{bmatrix}$$
$(5 \times 6) + (7 \times 1) = 37$

We can shorten the procedure as follows.

$$A \times B = \begin{bmatrix}3 & 2\\5 & 7\end{bmatrix}\begin{bmatrix}0 & 6\\4 & 1\end{bmatrix}$$
$$= \begin{bmatrix}3(0)+2(4) & 3(6)+2(1)\\5(0)+7(4) & 5(6)+7(1)\end{bmatrix}$$
$$= \begin{bmatrix}8 & 20\\28 & 37\end{bmatrix}$$

In general, if

$$A = \begin{bmatrix}a & b\\c & d\end{bmatrix} \quad \text{and} \quad B = \begin{bmatrix}e & f\\g & h\end{bmatrix}$$

then

$$A \times B = \begin{bmatrix}a & b\\c & d\end{bmatrix}\begin{bmatrix}e & f\\g & h\end{bmatrix} = \begin{bmatrix}ae+bg & af+bh\\ce+dg & cf+dh\end{bmatrix}$$

Let's do one more multiplication of matrices.

EXAMPLE 7 Multiplying Matrices

Determine $A \times B$ if

$$A = \begin{bmatrix}1 & 5\\4 & 7\end{bmatrix} \quad \text{and} \quad B = \begin{bmatrix}5 & -1 & 3\\2 & 8 & 0\end{bmatrix}$$

SOLUTION Matrix A contains two columns, and matrix B contains two rows. Thus, the matrices can be multiplied. Since matrix A contains two rows and matrix B contains three columns, the product matrix will contain two rows and three columns.

$$A \times B = \begin{bmatrix}1 & 5\\4 & 7\end{bmatrix}\begin{bmatrix}5 & -1 & 3\\2 & 8 & 0\end{bmatrix}$$
$$= \begin{bmatrix}1(5)+5(2) & 1(-1)+5(8) & 1(3)+5(0)\\4(5)+7(2) & 4(-1)+7(8) & 4(3)+7(0)\end{bmatrix}$$
$$= \begin{bmatrix}15 & 39 & 3\\34 & 52 & 12\end{bmatrix}$$

It should be noted that multiplication of matrices *is not* commutative; that is, $A \times B \neq B \times A$, except in special instances.

Square matrices have a *multiplicative identity matrix*. The multiplicative identity matrices for a 2×2 and a 3×3 matrix, denoted I, follow. Note that in any multiplicative identity matrix, 1's go diagonally from top left to bottom right and all other elements in the matrix are 0's.

$$I = \begin{bmatrix} 1 & 0 \\ 0 & 1 \end{bmatrix} \qquad I = \begin{bmatrix} 1 & 0 & 0 \\ 0 & 1 & 0 \\ 0 & 0 & 1 \end{bmatrix}$$

For any square matrix, A, $A \times I = I \times A = A$.

EXAMPLE 8   *Using the Identity Matrix in Multiplication*

Use the multiplicative identity matrix for a 2×2 matrix and matrix A to show that $A \times I = A$.

$$A = \begin{bmatrix} 4 & 3 \\ 2 & 1 \end{bmatrix}$$

SOLUTION The identity matrix is $I = \begin{bmatrix} 1 & 0 \\ 0 & 1 \end{bmatrix}$.

$$\begin{aligned} A \times I &= \begin{bmatrix} 4 & 3 \\ 2 & 1 \end{bmatrix}\begin{bmatrix} 1 & 0 \\ 0 & 1 \end{bmatrix} \\ &= \begin{bmatrix} 4(1) + 3(0) & 4(0) + 3(1) \\ 2(1) + 1(0) & 2(0) + 1(1) \end{bmatrix} \\ &= \begin{bmatrix} 4 & 3 \\ 2 & 1 \end{bmatrix} = A \end{aligned}$$

Verify for yourself that $I \times A = A$ as well.   ●

Example 9 illustrates an application of multiplication of matrices.

EXAMPLE 9   *A Manufacturing Application*

The Fancy Frock Company manufactures three types of women's outfits: a dress, a two-piece suit (skirt and jacket), and a three-piece suit (skirt, jacket, and a vest). On a particular day, the firm produces 20 dresses, 30 two-piece suits, and 50 three-piece suits. Each dress requires 4 units of material and 1 hour of work to produce, each two-piece suit requires 5 units of material and 2 hours of work to produce, and each three-piece suit requires 6 units of material and 3 hours to produce. Use matrix multiplication to determine the total number of units of material and the total number of hours needed for that day's production.

SOLUTION Let matrix A represent the number of each type of women's outfits produced.

	Dress	Two piece	Three piece
$A =$	[20	30	50]

The units of material and time requirements for each type are indicated in matrix B.

$$B = \begin{bmatrix} 4 & 1 \\ 5 & 2 \\ 6 & 3 \end{bmatrix} \begin{matrix} \text{Dress} \\ \text{Two piece} \\ \text{Three piece} \end{matrix}$$

with column headers Material Hours above.

The product of A and B, or $A \times B$, will give the total number of units of material and the total number of hours of work needed for that day's production.

$$A \times B = [20 \quad 30 \quad 50] \begin{bmatrix} 4 & 1 \\ 5 & 2 \\ 6 & 3 \end{bmatrix}$$

$$= [20(4) + 30(5) + 50(6) \quad 20(1) + 30(2) + 50(3)]$$

$$= [530 \quad 230]$$

Thus, a total of 530 units of material and a total of 230 hours of work are needed that day.

TIMELY TIP Matrices can only be added or subtracted if they have the same dimensions.

Matrices can only be multiplied if the number of *columns* in the first matrix is the same as the number of *rows* in the second matrix.

SECTION 7.3 EXERCISES

CONCEPT/WRITING EXERCISES

1. What is a matrix?

2. Explain how to determine the dimensions of a matrix.

3. What is a square matrix?

4. How many columns does a 3×2 matrix have?

5. How many rows does a 4×3 matrix have?

6. To add or subtract two matrices, what must be true about the dimensions of those matrices?

7. **a)** In your own words, explain the procedure used to add matrices.

   b) Use the procedure given in part (a) to add

   $$\begin{bmatrix} 5 & 4 & -1 \\ 3 & 9 & 5 \end{bmatrix} \quad \text{and} \quad \begin{bmatrix} 2 & 5 & -6 \\ -1 & 7 & 4 \end{bmatrix}$$

8. **a)** In your own words, explain the procedure used to subtract matrices.

   b) Use the procedure given in part (a) to subtract

   $$\begin{bmatrix} 8 & 4 & 2 \\ 0 & -2 & 4 \end{bmatrix} \quad \text{from} \quad \begin{bmatrix} 3 & -5 & 6 \\ -2 & 3 & 4 \end{bmatrix}$$

9. **a)** To multiply two matrices, what must be true about the dimensions of those matrices?

   b) What will be the dimensions of the product matrix when multiplying a 2×2 matrix with a 2×3 matrix?

10. **a)** In your own words, explain the procedure used to multiply matrices.

    b) Use the procedure given in part (a) to multiply

    $$\begin{bmatrix} 8 & -1 \\ 6 & 0 \end{bmatrix} \quad \text{by} \quad \begin{bmatrix} 2 & -3 \\ 5 & -4 \end{bmatrix}$$

11. a) What is the multiplicative identity matrix for a 2×2 matrix?

b) What is the multiplicative identity matrix for a 3×3 matrix?

12. A company has three offices: East, West, and Central. Each office has five divisions. The number of employees in each division of the three offices is as follows:
East: 110, 232, 103, 190, 212
West: 107, 250, 135, 203, 189
Central: 115, 218, 122, 192, 210

Express this information in the form of a 3×5 matrix.

PRACTICE THE SKILLS

In Exercises 13–16, determine $A + B$.

13. $A = \begin{bmatrix} 1 & 8 \\ 2 & 7 \end{bmatrix}$, $B = \begin{bmatrix} -4 & 1 \\ 7 & 2 \end{bmatrix}$

14. $A = \begin{bmatrix} 5 & 6 & -7 \\ 0 & 1 & -1 \end{bmatrix}$, $B = \begin{bmatrix} -4 & 2 & -8 \\ 7 & -3 & 0 \end{bmatrix}$

15. $A = \begin{bmatrix} 2 & 1 \\ -1 & 4 \\ 6 & 0 \end{bmatrix}$, $B = \begin{bmatrix} -3 & 3 \\ -4 & 0 \\ 1 & 6 \end{bmatrix}$

16. $A = \begin{bmatrix} 2 & 6 & 3 \\ -1 & -6 & 4 \\ 3 & 0 & 5 \end{bmatrix}$, $B = \begin{bmatrix} -1 & 3 & 1 \\ 7 & -2 & 1 \\ 2 & 3 & 8 \end{bmatrix}$

In Exercises 17–20, determine $A - B$.

17. $A = \begin{bmatrix} 4 & -2 \\ -3 & 5 \end{bmatrix}$, $B = \begin{bmatrix} -2 & 5 \\ 9 & 1 \end{bmatrix}$

18. $A = \begin{bmatrix} 10 & 1 \\ 12 & 2 \\ -3 & -9 \end{bmatrix}$, $B = \begin{bmatrix} -3 & 3 \\ 4 & 7 \\ -2 & 6 \end{bmatrix}$

19. $A = \begin{bmatrix} -5 & 1 \\ 8 & 6 \\ 1 & -5 \end{bmatrix}$, $B = \begin{bmatrix} -6 & -8 \\ -10 & -11 \\ 3 & -7 \end{bmatrix}$

20. $A = \begin{bmatrix} 5 & 3 & -1 \\ 7 & 4 & 2 \\ 6 & -1 & -5 \end{bmatrix}$, $B = \begin{bmatrix} 4 & 3 & 6 \\ -2 & -4 & 9 \\ 0 & -2 & 4 \end{bmatrix}$

In Exercises 21–26,

$A = \begin{bmatrix} 1 & 2 \\ 0 & 5 \end{bmatrix}$, $B = \begin{bmatrix} 3 & 2 \\ 5 & 0 \end{bmatrix}$, and $C = \begin{bmatrix} -2 & 3 \\ 4 & 0 \end{bmatrix}$.

Determine the following.

21. $2B$

22. $-3B$

23. $2B + 4C$

24. $2B + 3A$

25. $4B - 2C$

26. $3C - 2A$

In Exercises 27–32, determine $A \times B$.

27. $A = \begin{bmatrix} 1 & 3 \\ 0 & 6 \end{bmatrix}$, $B = \begin{bmatrix} 2 & 6 \\ 8 & 4 \end{bmatrix}$

28. $A = \begin{bmatrix} 1 & -1 \\ 2 & 6 \end{bmatrix}$, $B = \begin{bmatrix} 4 & -2 \\ -3 & -2 \end{bmatrix}$

29. $A = \begin{bmatrix} 2 & 3 & -1 \\ 0 & 4 & 6 \end{bmatrix}$, $B = \begin{bmatrix} 2 \\ 4 \\ 1 \end{bmatrix}$

30. $A = \begin{bmatrix} 1 & 1 \\ 1 & 1 \end{bmatrix}$, $B = \begin{bmatrix} 1 & -1 \\ -1 & 2 \end{bmatrix}$

31. $A = \begin{bmatrix} 4 & 7 & 6 \\ -2 & 3 & 1 \\ 5 & 1 & 2 \end{bmatrix}$, $B = \begin{bmatrix} 1 & 0 & 0 \\ 0 & 1 & 0 \\ 0 & 0 & 1 \end{bmatrix}$

32. $A = \begin{bmatrix} 1 & -3 \\ 7 & 2 \end{bmatrix}$, $B = \begin{bmatrix} 0 & 4 \\ 6 & 1 \end{bmatrix}$

In Exercises 33–38, determine $A + B$ and $A \times B$. If an operation cannot be performed, explain why.

33. $A = \begin{bmatrix} 2 & 3 & 5 \\ 4 & 0 & 3 \end{bmatrix}$, $B = \begin{bmatrix} 7 & -2 & 3 \\ 2 & -1 & 1 \end{bmatrix}$

34. $A = \begin{bmatrix} 6 & 4 & -1 \\ 2 & 3 & 4 \end{bmatrix}$, $B = \begin{bmatrix} 1 & 0 \\ 4 & -1 \end{bmatrix}$

35. $A = \begin{bmatrix} 4 & 5 & 3 \\ 6 & 2 & 1 \end{bmatrix}$, $B = \begin{bmatrix} 3 & 2 \\ 4 & 6 \\ -2 & 0 \end{bmatrix}$

36. $A = \begin{bmatrix} 6 & 5 \\ 4 & 3 \\ 2 & 1 \end{bmatrix}$, $B = \begin{bmatrix} 6 & 5 \\ 4 & 3 \\ 2 & 1 \end{bmatrix}$

37. $A = \begin{bmatrix} 1 & 2 \\ 3 & 4 \end{bmatrix}$, $B = \begin{bmatrix} -3 \\ 2 \end{bmatrix}$

38. $A = \begin{bmatrix} 5 & -1 \\ 6 & -2 \end{bmatrix}$, $B = \begin{bmatrix} 1 & 2 \\ 3 & 4 \end{bmatrix}$

In Exercises 39–41, show the commutative property of addition, $A + B = B + A$, holds for matrices A and B.

39. $A = \begin{bmatrix} 3 & 5 \\ -2 & -3 \end{bmatrix}, \qquad B = \begin{bmatrix} 4 & 5 \\ 6 & 7 \end{bmatrix}$

40. $A = \begin{bmatrix} 9 & 4 \\ 1 & 7 \end{bmatrix}, \qquad B = \begin{bmatrix} 2 & 0 \\ -1 & 6 \end{bmatrix}$

41. $A = \begin{bmatrix} 0 & -1 \\ 3 & -4 \end{bmatrix}, \qquad B = \begin{bmatrix} 8 & 1 \\ 3 & -4 \end{bmatrix}$

42. Create two matrices with the same dimensions, A and B, and show that $A + B = B + A$.

In Exercises 43–45, show that the associative property of addition, $(A + B) + C = A + (B + C)$, holds for the matrices given.

43. $A = \begin{bmatrix} 5 & 2 \\ 3 & 6 \end{bmatrix}, \quad B = \begin{bmatrix} 3 & 4 \\ -2 & 7 \end{bmatrix}, \quad C = \begin{bmatrix} -1 & 4 \\ 5 & 0 \end{bmatrix}$

44. $A = \begin{bmatrix} 4 & 1 \\ 6 & 7 \end{bmatrix}, \quad B = \begin{bmatrix} -9 & 1 \\ -7 & 2 \end{bmatrix}, \quad C = \begin{bmatrix} -6 & -3 \\ 3 & 6 \end{bmatrix}$

45. $A = \begin{bmatrix} 7 & 4 \\ 9 & -36 \end{bmatrix}, \quad B = \begin{bmatrix} 5 & 6 \\ -1 & -4 \end{bmatrix}, \quad C = \begin{bmatrix} -7 & -5 \\ -1 & 3 \end{bmatrix}$

46. Create three matrices with the same dimensions, A, B, and C, and show that $(A + B) + C = A + (B + C)$.

In Exercises 47–51, determine whether the commutative property of multiplication, $A \times B = B \times A$, holds for the matrices given.

47. $A = \begin{bmatrix} 1 & -2 \\ 4 & -3 \end{bmatrix}, \qquad B = \begin{bmatrix} -1 & -3 \\ 2 & 4 \end{bmatrix}$

48. $A = \begin{bmatrix} 3 & 1 \\ 6 & 6 \end{bmatrix}, \qquad B = \begin{bmatrix} 1 & 0 \\ 0 & 1 \end{bmatrix}$

49. $A = \begin{bmatrix} 4 & 2 \\ 1 & -3 \end{bmatrix}, \qquad B = \begin{bmatrix} 2 & 4 \\ -3 & 1 \end{bmatrix}$

50. $A = \begin{bmatrix} -3 & 2 \\ 6 & -5 \end{bmatrix}, \qquad B = \begin{bmatrix} -\frac{5}{3} & -\frac{2}{3} \\ -2 & -1 \end{bmatrix}$

51. $A = \begin{bmatrix} 3 & 2 & 1 \\ 4 & 2 & 0 \\ 0 & -2 & 5 \end{bmatrix}, \qquad B = \begin{bmatrix} 1 & 0 & 0 \\ 0 & 1 & 0 \\ 0 & 0 & 1 \end{bmatrix}$

52. Create two square matrices A and B with the same dimensions, and determine whether $A \times B = B \times A$.

In Exercises 53–57, show that the associative property of multiplication, $(A \times B) \times C = A \times (B \times C)$, holds for the matrices given.

53. $A = \begin{bmatrix} 1 & 3 \\ 4 & 0 \end{bmatrix}, \qquad B = \begin{bmatrix} 4 & 2 \\ 3 & 1 \end{bmatrix}, \qquad C = \begin{bmatrix} 2 & 1 \\ 3 & 0 \end{bmatrix}$

54. $A = \begin{bmatrix} -2 & 3 \\ 0 & 4 \end{bmatrix}, \qquad B = \begin{bmatrix} 4 & 0 \\ 3 & 5 \end{bmatrix}, \qquad C = \begin{bmatrix} 3 & 4 \\ -2 & 5 \end{bmatrix}$

55. $A = \begin{bmatrix} 4 & 3 \\ -6 & 2 \end{bmatrix}, \qquad B = \begin{bmatrix} 1 & 2 \\ 0 & 1 \end{bmatrix}, \qquad C = \begin{bmatrix} 4 & 3 \\ 0 & -2 \end{bmatrix}$

56. $A = \begin{bmatrix} -1 & -2 \\ -3 & -4 \end{bmatrix}, \qquad B = \begin{bmatrix} 1 & 0 \\ 0 & 1 \end{bmatrix}, \qquad C = \begin{bmatrix} 0 & 0 \\ 0 & 0 \end{bmatrix}$

57. $A = \begin{bmatrix} 3 & 4 \\ -1 & -2 \end{bmatrix}, \qquad B = \begin{bmatrix} 0 & 1 \\ 1 & 0 \end{bmatrix}, \qquad C = \begin{bmatrix} 2 & 0 \\ 3 & 0 \end{bmatrix}$

58. Create three matrices, A, B, and C, and show that $(A \times B) \times C = A \times (B \times C)$.

PROBLEM SOLVING

59. MODELING - *Purchasing Produce* Matrix A represents the weight, in pounds, of tomatoes, onions, and carrots that Wegmans Food Markets purchased from two different local farmers in 1 week. Matrix B represents the weight, in pounds, of tomatoes, onions, and carrots that Wegmans Food Markets purchased from the same farmers the following week.

$$A = \begin{array}{c} \\ \begin{bmatrix} 40 & 22 & 31 \\ 38 & 25 & 34 \end{bmatrix} \end{array} \begin{array}{l} \\ \text{Chase's Farm} \\ \text{Gro-More Farms} \end{array}$$

Tomatoes Onions Carrots

$$B = \begin{array}{c} \\ \begin{bmatrix} 48 & 36 & 39 \\ 40 & 29 & 37 \end{bmatrix} \end{array} \begin{array}{l} \\ \text{Chase's Farm} \\ \text{Gro-More Farms} \end{array}$$

Tomatoes Onions Carrots

Use matrix addition to determine the total weight, in pounds, Wegmans purchased over the 2-week period for each item from each farm.

60. MODELING - *Sweatshirt Inventory* Dick's Sporting Goods sells sweatshirts for adults and youth. Each type of sweatshirt comes in four sizes: small, medium, large, and extra large. Matrix *A* represents the stock on hand at the beginning of a given week for each type of sweatshirt. Matrix *B* represents the stock on hand at the end of the same week for the same three types of sweatshirts.

$$A = \begin{array}{c} \\ \text{Small} \\ \text{Medium} \\ \text{Large} \\ \text{Extra large} \end{array} \quad \overset{\text{Adult Youth}}{\begin{bmatrix} 31 & 18 \\ 39 & 16 \\ 41 & 22 \\ 34 & 21 \end{bmatrix}} \quad B = \overset{\text{Adult Youth}}{\begin{bmatrix} 14 & 9 \\ 18 & 9 \\ 19 & 15 \\ 15 & 9 \end{bmatrix}} \begin{array}{c} \text{Small} \\ \text{Medium} \\ \text{Large} \\ \text{Extra large} \end{array}$$

Use matrix subtraction to determine the number of sweatshirts sold in the given week for each size of each type.

61. MODELING - *Supplying Muffins* A bakery supplies chocolate chip, blueberry, and coffee cake muffins to Java's Coffee Shop and to Spot Coffee Shop. Matrix *A* shows the number of each type of muffin, in dozens, supplied to each coffee shop in 1 week. The bakery's cost for ingredients, per dozen, for chocolate chip, blueberry, and coffee cake muffins is $3, $2, and $1.5, respectively. Matrix *B* represents the bakery's cost for ingredients, per dozen, for each type of muffin.

$$A = \begin{bmatrix} 7 & 8.5 & 10 \\ 7.5 & 8 & 11 \end{bmatrix} \begin{array}{l} \text{Java's Coffee Shop} \\ \text{Spot Coffee Shop} \end{array}$$

Chocolate Blue- Coffee
chip berry cake

$$B = \begin{bmatrix} 3 \\ 2 \\ 1.5 \end{bmatrix} \begin{array}{l} \text{Chocolate chip} \\ \text{Blueberry} \\ \text{Coffee cake} \end{array}$$

Use matrix multiplication to determine the total weekly cost of ingredients for each shop.

62. MODELING - *Menu Choices* The cafeteria manager at the University of Virginia recently added three new meals that each consist of a meat, a vegetable, a salad, and a dessert to the lunch menu. The first day these new meals were offered, 35 students purchased meal 1, 28 students purchased meal 2, and 23 students purchased meal 3. Matrix *A* represents the number of students who purchased each type of meal. Matrix *B* represents the cost per serving, in cents, for each item from the three meals.

Meal Meal Meal
1   2   3

$$A = [35 \quad 28 \quad 23]$$

$$B = \begin{bmatrix} 45 & 12 & 8 & 10 \\ 52 & 8 & 6 & 15 \\ 49 & 7 & 9 & 11 \end{bmatrix} \begin{array}{l} \text{Meal 1} \\ \text{Meal 2} \\ \text{Meal 3} \end{array}$$

Meat Vegetable Salad Dessert

Use matrix multiplication to determine the total cost for each item.

63. MODELING - *Cookie Company Costs* The Original Cookie Factory bakes and sells four types of cookies: chocolate chip, sugar, molasses, and peanut butter. Matrix *A* shows the number of units of various ingredients used in baking a dozen of each type of cookie.

$$A = \begin{bmatrix} 2 & 2 & \frac{1}{2} & 1 \\ 3 & 2 & 1 & 2 \\ 0 & 1 & 0 & 3 \\ \frac{1}{2} & 1 & 0 & 0 \end{bmatrix} \begin{array}{l} \text{Chocolate chip} \\ \text{Sugar} \\ \text{Molasses} \\ \text{Peanut butter} \end{array}$$

Sugar Flour Milk Eggs

The cost, in cents per cup or per egg, for each ingredient when purchased in small quantities and in large quantities is given in matrix *B*.

$$B = \begin{bmatrix} 10 & 12 \\ 5 & 8 \\ 8 & 8 \\ 4 & 6 \end{bmatrix} \begin{array}{l} \text{Sugar} \\ \text{Flour} \\ \text{Milk} \\ \text{Eggs} \end{array}$$

Large   Small
quantities quantities

Use matrix multiplication to find a matrix representing the comparative cost per item for small and large quantities purchased.

In Exercises 64 and 65, use the information given in Exercise 63. Suppose a typical day's order consists of 40 dozen chocolate chip cookies, 30 dozen sugar cookies, 12 dozen molasses cookies, and 20 dozen peanut butter cookies.

64. a) Express these orders as a 1×4 matrix.

b) Use matrix multiplication to determine the amount of each ingredient needed to fill the day's order.

65. Use matrix multiplication to determine the cost under the two purchase options (small and large quantities) to fill the day's order.

66. MODELING - *Food Prices* To raise money for a local charity, the Spanish Club at Montclair High School sold hot dogs, soft drinks, and candy bars for 3 days in the student lounge. The sales for the 3 days are summarized in matrix A.

$$
A = \begin{array}{c c c c} & \text{Hot} & \text{Soft} & \text{Candy} \\ & \text{dogs} & \text{drinks} & \text{bars} \\ \begin{bmatrix} 52 & 50 & 75 \\ 48 & 43 & 60 \\ 62 & 57 & 81 \end{bmatrix} & & & \begin{array}{c} \text{Day 1} \\ \text{Day 2} \\ \text{Day 3} \end{array} \end{array}
$$

The cost and revenue (in dollars) for hot dogs, soft drinks, and candy are summarized in matrix B.

$$
B = \begin{array}{c c c} & \text{Cost} & \text{Revenue} \\ \begin{bmatrix} 0.30 & 0.75 \\ 0.25 & 0.50 \\ 0.15 & 0.45 \end{bmatrix} & & \begin{array}{c} \text{Hot dogs} \\ \text{Soft drinks} \\ \text{Candy bars} \end{array} \end{array}
$$

Multiply the two matrices to form a 3×2 matrix that shows the total cost and revenue for each item.

In Exercises 67 and 68, there are many acceptable answers.

67. a) Construct two matrices A and B whose product is a 3×1 matrix. Explain how you determined your answer.

b) For your matrices, determine $A \times B$.

68. a) Construct two matrices A and B whose product is a 4×1 matrix. Explain how you determined your answer.

b) For your matrices, determine $A \times B$.

Two matrices whose product is the multiplicative identity matrix are said to be multiplicative inverses. *That is, if* $A \times B = B \times A = I$, *where I is the multiplicative identity matrix, then A and B are multiplicative inverses.*

In Exercises 69 and 70, determine whether A and B are multiplicative inverses.

69. $A = \begin{bmatrix} 5 & -2 \\ -2 & 1 \end{bmatrix}$, $B = \begin{bmatrix} 1 & 2 \\ 2 & 5 \end{bmatrix}$

70. $A = \begin{bmatrix} 7 & 3 \\ 2 & 1 \end{bmatrix}$, $B = \begin{bmatrix} 1 & -3 \\ -2 & 7 \end{bmatrix}$

CHALLENGE PROBLEMS/GROUP ACTIVITIES

In Exercises 71 and 72, determine whether the statement is true or false. Give an example to support your answer.

71. $A - B = B - A$, where A and B are any matrices.

72. For scalar a and matrices B and C, $a(B + C) = aB + aC$.

73. MODELING - *Sofa Manufacturing Costs* The number of hours of labor required to manufacture one sofa of various sizes is summarized in matrix L.

$$
L = \begin{array}{c c c c} & \multicolumn{3}{c}{\text{Department}} \\ & \text{Cutting} & \text{Assembly} & \text{Packing} \\ \begin{bmatrix} 1.4 \text{ hr} & 0.7 \text{ hr} & 0.3 \text{ hr} \\ 1.8 \text{ hr} & 1.4 \text{ hr} & 0.3 \text{ hr} \\ 2.7 \text{ hr} & 2.8 \text{ hr} & 0.5 \text{ hr} \end{bmatrix} & & & \begin{array}{c} \text{Small} \\ \text{Medium} \\ \text{Large} \end{array} \end{array}
$$

Sofa size

The hourly labor rates for cutting, assembly, and packing at the Ames City Plant and at the Bay City Plant are given in matrix C.

$$
C = \begin{array}{c c c} & \multicolumn{2}{c}{\text{Plant}} \\ & \text{Ames} & \text{Bay} \\ & \text{City} & \text{City} \\ \begin{bmatrix} \$14 & \$12 \\ \$10 & \$9 \\ \$7 & \$5 \end{bmatrix} & & \begin{array}{c} \text{Cutting} \\ \text{Assembly} \\ \text{Packaging} \end{array} \end{array}
$$

Department

a) What is the total labor cost for manufacturing a small-sized sofa at the Ames City plant?

b) What is the total cost for manufacturing a large-sized sofa at the Bay City plant?

c) Determine the product $L \times C$ and explain the meaning of the results.

74. Is it possible that two matrices could be added but not multiplied? If so, give an example.

75. Is it possible that two matrices could be multiplied but not added? If so, give an example.

RECREATIONAL MATHEMATICS

76. Make up two matrices A and B such that

$$A + B = \begin{bmatrix} 1 & 0 \\ 0 & 1 \end{bmatrix} \text{ and } A \times B = \begin{bmatrix} 0 & 0 \\ 0 & 0 \end{bmatrix}.$$

INTERNET/RESEARCH ACTIVITIES

77. Find an article that shows information illustrated in matrix form. Write a short paper explaining how to interpret the information provided by the matrix. Include the article with your report.

78. *Messages* The study of encoding and decoding messages is called *cryptography*. Do research on current real-life uses of cryptography and write a paper on how matrix multiplication is used to encode and decode messages. In your paper, include current real-life uses of cryptography.

7.4 SOLVING SYSTEMS OF LINEAR EQUATIONS BY USING MATRICES

▲ If you know the total cost of two different types of candy in a box and the weight of each type of candy, you can use a system of equations to determine the cost of 1 pound of each item.

Consider the following situation. One package of chocolate-covered mints and chocolate-covered cherries sells for $23. A different package of chocolate-covered mints and chocolate-covered cherries sells for $14. If you know the weight of the chocolate-covered mints and the weight of the chocolate-covered cherries in each package, would you be able to determine the cost of 1 pound of the chocolate-covered mints and the cost of 1 pound of the chocolate-covered cherries? There are several ways to determine these costs. In this section, we will illustrate a method for answering this question by using matrices.

In Section 7.3, we introduced matrices. Now we will discuss the procedure to solve a system of linear equations using matrices. We will illustrate how to solve a system of two equations and two unknowns. Systems of equations containing three equations and three unknowns (called third-order systems) and higher-order systems can also be solved by using matrices, but we will not discuss third or higher order systems in this book.

The first step in solving a system of equations using matrices is to represent the system of equations with an *augmented matrix*. An augmented matrix consists of two smaller matrices, one for the coefficients of the variables in the equations and one for the constants in the equations. To determine the augmented matrix, first write each equation in standard form, $ax + by = c$. For the system of equations below, its augmented matrix is shown to its right.

System of equations	**Augmented matrix**
$a_1 x + b_1 y = c_1$ $a_2 x + b_2 y = c_2$	$\begin{bmatrix} a_1 & b_1 & \vert & c_1 \\ a_2 & b_2 & \vert & c_2 \end{bmatrix}$

Following is another example.

System of equations	Augmented matrix

$$x + 2y = 8$$
$$3x - y = 7$$

$$\begin{bmatrix} 1 & 2 & | & 8 \\ 3 & -1 & | & 7 \end{bmatrix}$$

Note that the vertical bar in the augmented matrix separates the numerical coefficients from the constants. The matrix is just a shortened way of writing the system of equations. Thus, we can solve a system of equations by using matrices in a manner very similar to solving a system of equations with the addition method.

To solve a system of equations by using matrices, we use *row transformations* to obtain new matrices that have the same solution as the original system. We will discuss three row transformation procedures.

PROCEDURES FOR ROW TRANSFORMATIONS

1. Any two rows of a matrix may be interchanged (which is the same as interchanging any two equations in the system of equations).

2. All the numbers in any row may be multiplied by any nonzero real number (which is the same as multiplying both sides of an equation by any nonzero real number).

3. All the numbers in any row may be multiplied by any nonzero real number, and these products may be added to the corresponding numbers in any other row of numbers.

We use row transformations to obtain an augmented matrix whose numbers to the left of the vertical bar are the same as in the *multiplicative identity matrix*. From this type of augmented matrix, we can determine the solution to the system of equations. For example, if we get

$$\begin{bmatrix} 1 & 0 & | & 3 \\ 0 & 1 & | & -2 \end{bmatrix}$$

it tells us that $1x + 0y = 3$ or $x = 3$, and $0x + 1y = -2$ or $y = -2$. Thus, the solution to the system of equations that yielded this augmented matrix is $(3, -2)$. Now let's work an example.

EXAMPLE ❶ *Using Row Transformations*

Solve the following system of equations by using matrices.

$$x + 2y = 5$$
$$3x - y = 8$$

SOLUTION First we write the augmented matrix.

$$\begin{bmatrix} 1 & 2 & | & 5 \\ 3 & -1 & | & 8 \end{bmatrix}$$

Our goal is to obtain a matrix of the form

$$\begin{bmatrix} 1 & 0 & | & c_1 \\ 0 & 1 & | & c_2 \end{bmatrix}$$

where c_1 and c_2 may represent any real numbers. It is generally easier to work by columns. Therefore, we will try to get the first column of the augmented matrix to be $\begin{smallmatrix}1\\0\end{smallmatrix}$ and the second column to be $\begin{smallmatrix}0\\1\end{smallmatrix}$. Since the element in the top left position is already a 1, we must work to change the 3 in the first column, second row, into a 0. We use row transformation procedure 3 to change the 3 into a 0. If we multiply the top row of numbers by -3 and add these products to the second row of numbers, the element in the first column, second row will become a 0:

$$\begin{bmatrix} 1 & 2 & | & 5 \\ 3 & -1 & | & 8 \end{bmatrix} \quad \text{Original augmented matrix}$$

The top row of numbers multiplied by -3 gives

$$1(-3), \qquad 2(-3), \qquad \text{and} \qquad 5(-3)$$

Now add these products to their respective numbers in row 2.

$$\begin{bmatrix} 1 & 2 & | & 5 \\ 3 + 1(-3) & -1 + 2(-3) & | & 8 + 5(-3) \end{bmatrix} = \begin{bmatrix} 1 & 2 & | & 5 \\ 0 & -7 & | & -7 \end{bmatrix}$$

The next step is to obtain a 1 in the second column, second row. At present, -7 is in that position. To change the -7 to a 1, we use row transformation procedure 2. If we multiply -7 by $-\frac{1}{7}$, the product will be 1. Therefore, we multiply all the numbers in the second row by $-\frac{1}{7}$ to get

$$\begin{bmatrix} 1 & 2 & | & 5 \\ 0(-\frac{1}{7}) & -7(-\frac{1}{7}) & | & -7(-\frac{1}{7}) \end{bmatrix} = \begin{bmatrix} 1 & 2 & | & 5 \\ 0 & 1 & | & 1 \end{bmatrix}$$

The next step is to obtain a 0 in the second column, first row. At present, a 2 is in that position. Multiplying the numbers in the second row by -2 and adding the products to the corresponding numbers in the first row gives a 0 in the desired position.

$$\begin{bmatrix} 1 + 0(-2) & 2 + 1(-2) & | & 5 + 1(-2) \\ 0 & 1 & | & 1 \end{bmatrix} = \begin{bmatrix} 1 & 0 & | & 3 \\ 0 & 1 & | & 1 \end{bmatrix}$$

We now have the desired augmented matrix:

$$\begin{bmatrix} 1 & 0 & | & 3 \\ 0 & 1 & | & 1 \end{bmatrix}$$

With this matrix, we see that $1x + 0y = 3$, or $x = 3$, and $0x + 1y = 1$, or $y = 1$. The solution to the system is $(3, 1)$.

CHECK:

$$x + 2y = 5 \qquad\qquad 3x - y = 8$$
$$3 + 2(1) = 5 \qquad\qquad 3(3) - 1 = 8$$
$$5 = 5 \quad \text{True} \qquad\qquad 8 = 8 \quad \text{True}$$

Now we give a general procedure to change an augmented matrix to the desired form.

TO CHANGE AN AUGMENTED MATRIX TO THE FORM

$$\begin{bmatrix} 1 & 0 & c_1 \\ 0 & 1 & c_2 \end{bmatrix}$$

Use row transformations to:

1. Change the element in the first column, first row, to a 1.
2. Change the element in the first column, second row, to a 0.
3. Change the element in the second column, second row, to a 1.
4. Change the element in the second column, first row, to a 0.

Generally, when changing an element in the augmented matrix to a 1, we use step 2 in the row transformation box on page 447. When changing an element to a 0, we use step 3 in the row transformation box.

EXAMPLE 2 *Using Matrices to Solve a System of Equations*

Solve the following system of equations using matrices.

$$2x + 4y = 6$$
$$4x - 2y = -8$$

SOLUTION First write the augmented matrix.

$$\begin{bmatrix} 2 & 4 & 6 \\ 4 & -2 & -8 \end{bmatrix}$$

To obtain a 1 in the first column, first row, multiply the numbers in the first row by $\frac{1}{2}$.

$$\begin{bmatrix} 1 & 2 & 3 \\ 4 & -2 & -8 \end{bmatrix}$$

To obtain a 0 in the first column, second row, multiply the numbers in the first row by -4 and add the products to the corresponding numbers in the second row.

$$\begin{bmatrix} 1 & 2 & 3 \\ 4 + 1(-4) & -2 + 2(-4) & -8 + 3(-4) \end{bmatrix} = \begin{bmatrix} 1 & 2 & 3 \\ 0 & -10 & -20 \end{bmatrix}$$

To obtain a 1 in the second column, second row, multiply the numbers in the second row by $-\frac{1}{10}$.

$$\begin{bmatrix} 1 & 2 & \bigm| & 3 \\ 0(-\frac{1}{10}) & -10(-\frac{1}{10}) & \bigm| & -20(-\frac{1}{10}) \end{bmatrix} = \begin{bmatrix} 1 & 2 & \bigm| & 3 \\ 0 & 1 & \bigm| & 2 \end{bmatrix}$$

To obtain a 0 in the second column, first row, multiply the numbers in the second row by -2 and add the products to the corresponding numbers in the first row.

$$\begin{bmatrix} 1 + 0(-2) & 2 + 1(-2) & \bigm| & 3 + 2(-2) \\ 0 & 1 & \bigm| & 2 \end{bmatrix} = \begin{bmatrix} 1 & 0 & \bigm| & -1 \\ 0 & 1 & \bigm| & 2 \end{bmatrix}$$

The solution to the system of equations is $(-1, 2)$. ●

Inconsistent and Dependent Systems

Assume that you solve a system of two equations and obtain an augmented matrix in which one row of numbers on the left side of the vertical line are all zeroes but a zero does not appear in the same row on the right side of the vertical line. This situation indicates that the system is inconsistent and has no solution. For example, a system of equations that yields the following augmented matrix is an inconsistent system.

$$\begin{bmatrix} 1 & 2 & \bigm| & 5 \\ 0 & 0 & \bigm| & 4 \end{bmatrix} \quad \text{Inconsistent system}$$

The second row of the matrix represents the equation

$$0x + 0y = 4 \qquad \text{or} \qquad 0 = 4$$

which is never true. This matrix represents a system of equations that has no solution.

If you obtain a matrix in which a 0 appears across an entire row, the system of equations is dependent. For example, a system of equations that yields the following matrix is a dependent system.

$$\begin{bmatrix} 1 & 5 & \bigm| & -6 \\ 0 & 0 & \bigm| & 0 \end{bmatrix} \quad \text{Dependent system}$$

The second row of the matrix represents the equation

$$0x + 0y = 0 \qquad \text{or} \qquad 0 = 0$$

which is always true. This matrix represents a system of equations that has an infinite number of solutions.

Triangularization Method

Another procedure to solve a system of two equations is to use row transformation procedures to obtain an augmented matrix of the form

$$\begin{bmatrix} 1 & a & \bigm| & b \\ 0 & 1 & \bigm| & c \end{bmatrix}$$

where a, b, and c represent real numbers. This procedure is called the *triangulariza-tion method* because the ones and zeroes form a triangle.

When the matrix is in this form, we can write the following system of equations.

$$\begin{array}{ll} 1x + ay = b & \quad x + ay = b \\ 0x + 1y = c & \quad y = c \end{array}$$

or

Using substitution, we can solve the system.

┌─EXAMPLE ❸ *Solving a System of Equations Using the Triangularization Method*

Solve the following system of equations using the triangularization method.

$$2x + 4y = 6$$
$$4x - 2y = -8$$

SOLUTION In Example 2, in the process of solving the system we obtained the augmented matrix

$$\begin{bmatrix} 1 & 2 & | & 3 \\ 0 & 1 & | & 2 \end{bmatrix}$$

This matrix represents the following system of equations.

$$x + 2y = 3$$
$$y = 2$$

To solve for x, we substitute 2 for y in the equation

$$x + 2y = 3$$
$$x + 2(2) = 3$$
$$x + 4 = 3$$
$$x = -1$$

Thus, the solution to the system is $(-1, 2)$, as was obtained in Example 2. You may use either method when solving a system of equations with matrices unless your instructor specifies otherwise.   ●

SECTION 7.4   EXERCISES

CONCEPT/WRITING EXERCISES

1. a) What is an augmented matrix?

   b) Determine the augmented matrix for the following system.
   $$x + 3y = 7$$
   $$2x - y = 4$$

2. In your own words, write the three row transformation procedures.

3. How will you know, when solving a system of equations by using matrices, whether the system is dependent?

4. How will you know, when solving a system of equations by using matrices, whether the system is inconsistent?

5. If you obtained the following augmented matrix when solving a system of equations, what would be your next step in completing the process? Explain your answer.
   $$\begin{bmatrix} 1 & 3 & | & 5 \\ 0 & -2 & | & 1 \end{bmatrix}$$

6. If you obtained the following augmented matrix when solving a system of equations, what would be your next step in completing the process? Explain your answer.
   $$\begin{bmatrix} 1 & -2 & | & 1 \\ 0 & 1 & | & 3 \end{bmatrix}$$

PRACTICE THE SKILLS

In Exercises 7–20, use matrices to solve the system of equations.

7. $x + 3y = 7$
   $-x + y = 1$

8. $x - y = 2$
   $2x - y = 4$

9. $x - 2y = -1$
   $2x + y = 8$

10. $x + y = -1$
    $2x + 3y = -5$

11. $2x - 5y = -6$
    $-4x + 10y = 12$

12. $x + y = 5$
    $3x - y = 3$

13. $2x - 3y = 10$
    $2x + 2y = 5$

14. $x + 3y = 1$
    $-2x + y = 5$

15. $4x + 2y = 6$
    $5x + 4y = 9$

16. $4x + 2y = -10$
    $-2x + y = -7$

17. $-3x + 6y = 5$
    $2x - 4y = 8$

18. $2x - 5y = 10$
    $3x + y = 15$

19. $3x + y = 13$
    $x + 3y = 15$

20. $4x - 3y = 7$
    $-2x + 5y = 14$

PROBLEM SOLVING

In Exercises 21–24, use matrices to solve the problem.

21. **MODELING - *Selling Baseball Caps*** Lids sells fitted baseball caps for $35 and stretch-fit baseball caps for $25. Recently, the company sold a total of 32 baseball caps in a single day. If the baseball cap receipts for the day totaled $980, how many of each type of baseball cap were sold?

22. **MODELING - *TV Dimensions*** Vita Gunta just purchased a new high-definition television. She noticed that the perimeter of the screen is 124 in. The width of the screen is 8 in. greater than its height. Find the dimensions of the screen.

23. **MODELING - *On the Job*** Peoplepower, Inc., a daily employment agency, charges $10 per hour for a truck driver and $8 per hour for a laborer. On a certain job, the laborer worked two more hours than the truck driver, and together they cost $144. How many hours did each work?

24. **MODELING - *Snack Mix*** If Andy DaLora buys 2 lb of caramel corn and 3 lb of mixed nuts, his total cost would be $27. If he buys 1 lb of caramel corn and 2 lb of mixed nuts, his total cost would be $17. Find the cost of 1 lb of caramel corn and 1 lb of mixed nuts.

25. MODELING - *Fill in the Missing Information* Hammer-Mill sells two types of laser printer paper to office supply stores. The premium laser paper sells for $6 a ream, and the paper for color laser printers sells for $7.50 a ream. HammerMill received an order for 200 reams of paper and a check for $1275. When placing the order, the office supply store clerk failed to specify the number of reams of each type of paper being ordered. Can HammerMill fill the order with the information given? If so, determine the number of reams of premium laser paper and the number of reams of paper for the color laser printer the clerk ordered.

INTERNET/RESEARCH ACTIVITY

26. Do research and write a paper on the development of matrices. In your paper, cover the contributions of James Joseph Sylvester, Arthur Cayley, and William Rowan Hamilton. References include history of mathematics books, encyclopedias, and the Internet.

7.5 SYSTEMS OF LINEAR INEQUALITIES

▲ A system of inequalities may be used to determine how many camcorders should be stocked to satisfy store requirements.

Assume that Circuit City sells only two different camcorders, one made by Panasonic and one made by Sony. Based on demand, the store must stock at least twice as many Panasonic units as Sony units. The costs to the store for the two camcorders are $300 and $600, respectively. The management wants at least 10 Panasonic camcorders and 5 Sony camcorders in inventory at all times and does not want more than $18,000 in camcorder inventory at any one time. To determine how many camcorders of each type should be stocked to satisfy the store's requirements, we could set up and solve a system of linear inequalities. In this section, we will explore the techniques of finding the solution set to a system of linear inequalities.

The solution set of a system of linear inequalities is the set of points that satisfy all inequalities in the system. The solution set of a system of linear inequalities may consist of infinitely many ordered pairs. To determine the solution set to a system of linear inequalities, graph each inequality on the same axes. The ordered pairs common to all the inequalities are the solution set to the system.

PROCEDURE FOR SOLVING A SYSTEM OF LINEAR INEQUALITIES

1. Select one of the inequalities. Replace the inequality symbol with an equal sign and draw the graph of the equation. Draw the graph with a dashed line if the inequality is $<$ or $>$ and with a solid line if the inequality is $\leq$ or $\geq$.

2. Select a test point on one side of the line and determine whether the point is a solution to the inequality. If so, shade the area on the side of the line containing the point. If the point is not a solution, shade the area on the other side of the line.

3. Repeat steps 1 and 2 for the other inequality.

4. The intersection of the two shaded areas and any solid line common to both inequalities form the solution set to the system of inequalities.

┌─ **EXAMPLE ❶** *Solving a System of Inequalities*

Graph the following system of inequalities and indicate the solution set.

$$x + y < 5$$
$$x - y < 3$$

SOLUTION Graph both inequalities on the same axes. First draw the graph of $x + y < 5$. When drawing the graph, remember to use a dashed line, since the inequality is "less than" (see Fig. 7.10a). If you have forgotten how to graph inequalities, review Section 6.8. Shade the half plane that satisfies the inequality $x + y < 5$.

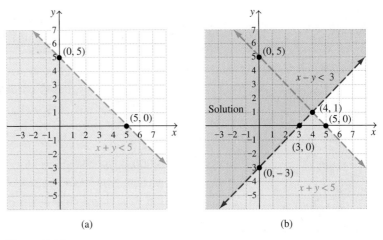

(a)   (b)

Figure 7.10

Now, on the same axes, shade the half plane that satisfies the inequality $x - y < 3$ (see Fig. 7.10b). The solution set consists of all the points common to the two shaded half planes. These are the points in the region on the graph containing both color shadings. In Figure 7.10(b), we have indicated this region in green. Figure 7.10(b) shows that the two lines intersect at $(4, 1)$. This ordered pair can also be found by any of the algebraic methods discussed in Sections 7.2 and 7.3. ●

┌─ **EXAMPLE ❷** *Solving a System of Linear Inequalities*

Graph the following system of inequalities and indicate the solution set.

$$4x - 2y \geq 8$$
$$2x + 3y < 6$$

SOLUTION Graph the inequality $4x - 2y \geq 8$. Remember to use a solid line because the inequality is "greater than or equal to"; see Fig. 7.11(a).

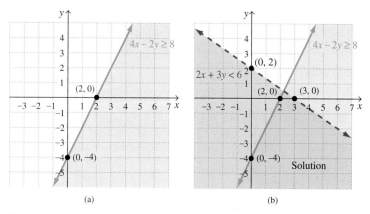

Figure 7.11

On the same set of axes, draw the graph of $2x + 3y < 6$. Use a dashed line since the inequality is "less than"; see Fig. 7.11(b). The solution is the region of the graph that contains both color shadings and the part of the solid line that satisfies the inequality $2x + 3y < 6$. Note that the point of intersection of the two lines is not a part of the solution set.

EXAMPLE ❸ *Another System of Inequalities*

Graph the following system of inequalities and indicate the solution set.

$$x \geq -2$$
$$y < 3$$

SOLUTION Graph the inequality $x \geq -2$; see Fig. 7.12(a). On the same axes, graph the inequality $y < 3$; see Fig. 7.12(b). The solution set is that region of the graph that is shaded in both colors and the part of the solid line that satisfies the inequality $y < 3$. The point of intersection of the two lines, $(-2, 3)$, is not part of the solution because it does not satisfy the inequality $y < 3$.

Figure 7.12

SECTION 7.5   EXERCISES

CONCEPT/WRITING EXERCISES

1. What is the solution set of a system of linear inequalities?

2. If in a system of inequalities one inequality contains $\leq$ and the other inequality contains $>$, is the point of intersection of the two boundary lines of the inequalities in the solution set? Explain.

3. If in a system of inequalities one inequality contains $\leq$ and the other inequality contains $\geq$, is the point of intersection of the two boundary lines of the inequalities in the solution set? Explain.

4. If in a system of inequalities one inequality contains $<$ and the other inequality contains $>$, is the point of intersection of the two boundary lines of the inequalities in the solution set? Explain.

PRACTICE THE SKILLS

In Exercises 5–20, graph the system of linear inequalities and indicate the solution set.

5. $y > x + 4$
   $y > 3x$

6. $y > -3x$
   $y \leq -2x + 1$

7. $x + y \geq 1$
   $x - y > 3$

8. $x + y < 4$
   $3x + 2y \geq 6$

9. $x - y < 4$
   $x + y < 5$

10. $3x - y \leq 6$
    $x - y > 4$

11. $x + 2y \geq 4$
    $3x - y \geq -6$

12. $x - 3y \leq 3$
    $x + 2y \geq 4$

13. $y \leq 3x$
    $x \geq 3y$

14. $y \leq 4$
    $x - y < 1$

15. $x \leq 0$
    $y \leq 0$

16. $x \geq 1$
    $y \leq 1$

17. $4x + 2y > 8$
    $x \geq y - 1$

18. $5y > 3x + 10$
    $3y < -2x - 3$

19. $3x + 2y > 8$
    $x < 5y - 5$

20. $3x \geq 2y + 10$
    $x \leq y + 8$

CHALLENGE PROBLEMS/GROUP ACTIVITIES

21. **MODELING** - *Craft Sales* Julie Gratien makes small and large decorated bowls. It takes Julie 20 minutes to decorate a small bowl and 30 minutes to decorate a large bowl. She can spend no more than a total of 600 minutes decorating the bowls. The number of small bowls made needs to be at least twice the number of large bowls made. Julie must also make at least 10 small bowls and 5 large bowls.

    a) Using x to represent the number of small bowls made and y to represent the number of large bowls made, translate the problem into a system of linear inequalities.

    b) Solve the system graphically. Graph sales for small bowls on the horizontal axis and sales for large bowls on the vertical axis.

    c) Select a point in the solution set. Determine the sales for the two types of bowls that corresponds to the point selected.

22. **MODELING** - *Special Diet* Ruben Gonzalez is on a special diet. He must consume fewer than 500 calories at a meal that consists of one serving of chicken and one serving of rice. The meal must contain at least 150 calories from each source.

    a) Using x to represent the number of calories from chicken and y to represent the number of calories from rice, translate the problem into a system of linear inequalities.

    b) Solve the system graphically. Graph calories from chicken on the horizontal axis and calories from rice on the vertical axis.

    c) There are about 180 calories in 3 oz of chicken and about 200 calories in 8 oz of rice. Select a point in the solution set. For the point selected, determine the number of ounces of chicken and the number of ounces of rice to be served.

23. Write a system of linear inequalities whose solution is the second quadrant, including the axes.

24. a) Do all systems of linear inequalities have solutions? Explain.

   b) Write a system of inequalities that has no solution.

25. Can a system of linear inequalities have a solution set consisting of a single point? Explain.

26. Can a system of linear inequalities have as its solution set all the points on the coordinate plane? Explain your answer, giving an example to support it.

27. Write a system of linear inequalities that has the ordered pair (0, 0) as its only solution. There are many possible answers.

28. Write a system of linear inequalities that has the following ordered pairs as some of its solutions. There are many possible answers.

$$\ldots (-3, -3), (-2, -2), (-1, -1), (0, 0), (1, 1), (2, 2), (3, 3), \ldots$$

7.6   LINEAR PROGRAMMING

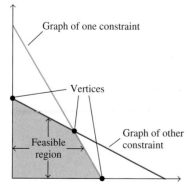

Consider a small company that manufactures skateboards and in-line skates. The maximum number of skateboards and pairs of in-line skates that can be produced per day is 20. To meet demand, the company must make at least 2 pairs of in-line skates per day. The company wants to make at least 3 skateboards but no more than 6 skateboards per day. The company makes a profit of $25 on a skateboard and a profit of $20 on a pair of in-line skates. How many skateboards and pairs of in-line skates should be made to maximize profit while satisfying all requirements? In this section, we will discuss how to solve questions such as this one by using a method called linear programming.

▲ Linear programming can be used to determine how many skateboards and in-line skates a company should make in order to maximize profit.

Government, business, and industry often require decision makers to find cost-effective solutions to a variety of problems. Linear programming provides businesses and governments with a mathematical form of decision making that makes the most efficient use of time and resources. Linear programming often serves as a method of expressing the relationships in many of these problems and uses systems of linear inequalities.

The typical linear programming problem has many variables and is generally so lengthy that it is solved on a computer by a technique called the *simplex method*. The simplex method was developed in the 1940s by George B. Dantzig (see the Profile in Mathematics on page 459). Linear programming is used to solve problems in the social sciences, health care, land development, nutrition, military, and many other fields.

We will not discuss the simplex method in this textbook. We will merely give a brief introduction to how linear programming works. You can find a detailed explanation in books on finite mathematics.

In a linear programming problem, there are restrictions called *constraints*. Each constraint is represented as a linear inequality. The list of constraints forms a system of linear inequalities. When the system of inequalities is graphed, we often obtain a region bounded on all sides by line segments (Fig. 7.13). This region is called the *feasible region*. The points where two or more boundaries intersect are called the *vertices* of the feasible region. The points on the boundary of the region and the points inside the feasible region are the solution set for the system of inequalities.

Figure 7.13

For each linear programming problem, we will obtain a formula of the form $K = Ax + By$, called the *objective function*. The objective function is the formula for the quantity K (or some other variable) that we want to maximize or minimize. The values we substitute for x and y determine the value of K. From the information given in the problem, we determine the real number constants A and B. In a particular linear programming problem, a typical equation that might be used to find the maximum profit, P, is $P = 3x + 7y$. We would find the maximum profit by substituting the ordered pairs (x, y) of the vertices of the feasible region into the formula $P = 3x + 7y$ to see which ordered pair yields the greatest value of P and therefore the maximum profit. The ordered pair that yields the smallest value of P determines the minimum profit.

Linear programming is used to determine which ordered pair will yield the maximum (or minimum) value of the variable that is being maximized (or minimized). The fundamental principle of linear programming provides a rule for finding those maximum and minimum values.

FUNDAMENTAL PRINCIPLE OF LINEAR PROGRAMMING

If the objective function, $K = Ax + By$, is evaluated at each point in a feasible region, the maximum and minimum values of the equation occur at vertices of the region.

Linear programming is a powerful tool for finding the maximum and minimum values of an objective function. Using the fundamental principle of linear programming, we are quickly able to determine the maximum and minimum values of an objective function by using just a few of the infinitely many points in the feasible region.

Example 1 illustrates how the fundamental principle is used to solve a linear programming problem.

EXAMPLE ❶   MODELING - *Using the Fundamental Principle of Linear Programming*

The Ric Shaw Chair company makes two types of rocking chairs, a plain chair and a fancy chair. Each rocking chair must be assembled and then finished. The plain chair takes 4 hours to assemble and 4 hours to finish. The fancy chair takes 8 hours to assemble and 12 hours to finish. The company can provide at most 160 worker-hours of assembling and 180 worker-hours of finishing a day. If the profit on a plain chair is $40 and the profit on a fancy chair is $65, how many rocking chairs of each type should the company make per day to maximize profits? What is the maximum profit?

SOLUTION From the information given, we know the following facts.

	Assembly Time (hr)	Finishing Time (hr)	Profit ($)
Plain chair	4	4	40.00
Fancy chair	8	12	65.00

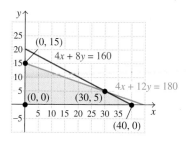

Figure 7.14

Let

$$x = \text{the number of plain chairs made per day}$$
$$y = \text{the number of fancy chairs made per day}$$
$$40x = \text{profit on the plain chairs}$$
$$65y = \text{profit on the fancy chairs}$$
$$P = \text{the total profit}$$

The total profit is the sum of the profit on the plain chairs and the profit on the fancy chairs. Since $40x$ is the profit on the plain chairs and $65y$ is the profit on the fancy chairs, the profit formula is $P = 40x + 65y$.

The maximum profit, P, is dependent on several conditions, called *constraints*. The number of chairs manufactured each day cannot be a negative amount. This condition gives us the constraints $x \geq 0$ and $y \geq 0$. Another constraint is determined by the total number of hours allocated for assembling. Four hours are needed to assemble the plain chair, so the total number of hours per day to assemble x plain chairs is $4x$. Eight hours are required to assemble a fancy chair, so the total number of hours needed to assemble y fancy chairs is $8y$. The maximum number of hours allocated for assembling is 160 per day. Thus, the third constraint is $4x + 8y \leq 160$. The final constraint is determined by the number of hours allotted for finishing. Finishing a plain chair takes 4 hours, or $4x$ hours to finish x plain chairs. Finishing a fancy chair takes 12 hours, or $12y$ hours to finish y fancy chairs. The total number of hours allotted for finishing is 180 per day. Therefore, the fourth constraint is $4x + 12y \leq 180$. Thus, the four constraints are

$$x \geq 0$$
$$y \geq 0$$
$$4x + 8y \leq 160$$
$$4x + 12y \leq 180$$

The list of constraints is a system of linear inequalities in two variables. The solution to the system of inequalities is the set of ordered pairs that satisfy all the constraints. These points are plotted in Fig. 7.14. Note that the solution to the system consists of the colored region and the solid boundaries. The points $(0, 0)$, $(0, 15)$, $(30, 5)$, and $(40, 0)$ are the points at which the boundaries intersect. These points can also be found by the addition or substitution method described in Section 7.2.

The goal in this example is to maximize the profit. The objective function is given by the profit formula $P = 40x + 65y$. According to the fundamental principle, the maximum profit will be found at one of the vertices of the feasible region.

Calculate P for each one of the vertices.

$$P = 40x + 65y$$

At $(0, 0)$,	$P = 40(0) + 65(0) = 0$
At $(0, 15)$,	$P = 40(0) + 65(15) = 975$
At $(30, 5)$,	$P = 40(30) + 65(5) = 1525$
At $(40, 0)$,	$P = 40(40) + 65(0) = 1600$

The maximum profit is at (40, 0), which means that the company should manufacture 40 plain rocking chairs and no fancy rocking chairs. The maximum profit would be $1600. The minimum profit would be at (0, 0), when no rocking chairs of either style were manufactured. ●

A variation of the problem in Example 1 could be that the company knows that it cannot make more than 15 plain rocking chairs per day. With this additional constraint, we now have the following set of constraints.

$$x \geq 0$$
$$x \leq 15$$
$$y \geq 0$$
$$4x + 8y \leq 160$$
$$4x + 12y \leq 180$$

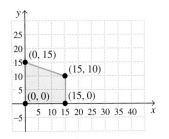

Figure 7.15

The graph of the feasible region of these constraints is shown in Fig. 7.15.

The vertices of the feasible region are (0, 0), (0, 15), (15, 10), and (15, 0). To determine the maximum profit, we calculate P for each of these vertices:

$$P = 40x + 65y$$

At (0, 0),	$P = 40(0) + 65(0) = 0$
At (0, 15),	$P = 40(0) + 65(15) = 975$
At (15, 10),	$P = 40(15) + 65(10) = 1250$
At (15, 0),	$P = 40(15) + 65(0) = 600$

This set of constraints gives the maximum profit of $1250 when the company manufactures 15 plain rocking chairs and 10 fancy rocking chairs.

EXAMPLE ❷ MODELING - *Washers and Dryers, Maximizing Profit*

The Admiral Appliance Company makes washers and dryers. The company must manufacture at least one washer per day to ship to one of its customers. No more than 6 washers can be manufactured due to production restrictions. The number of dryers manufactured cannot exceed 7 per day. Also, the number of washers manufactured cannot exceed the number of dryers manufactured per day. If the profit on each washer is $20 and the profit on each dryer is $30, how many of each appliance should the company make per day to maximize profits? What is the maximum profit?

SOLUTION Let

$$x = \text{the number of washers manufactured per day}$$
$$y = \text{the number of dryers manufactured per day}$$
$$20x = \text{the profit on washers}$$
$$30y = \text{the profit on dryers}$$
$$P = \text{the total profit}$$

The maximum profit is dependent on several constraints. The number of appliances manufactured each day cannot be a negative amount. This condition gives us the constraints $x \geq 0$ and $y \geq 0$. The company must manufacture at least one washer

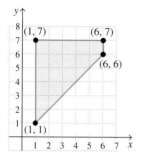

Figure 7.16

per day; therefore, $x \geq 1$. No more than 6 washers can be manufactured per day; therefore, $x \leq 6$. No more than 7 dryers can be manufactured per day; therefore, $y \leq 7$. The number of washers cannot exceed the number of dryers manufactured per day; therefore, $x \leq y$. Thus, the six constraints are

$$x \geq 0, y \geq 0, x \geq 1, x \leq 6, y \leq 7, x \leq y$$

In this example, the objective function is the profit formula. Since $20x$ is the profit on x washers and $30y$ is the profit on y dryers, the profit formula is $P = 20x + 30y$. Figure 7.16 shows the feasible region. The feasible region consists of the shaded region and the boundaries. The vertices of the feasible region are at $(1, 1)$, $(1, 7)$, $(6, 7)$, and $(6, 6)$.

Next we calculate the value of the objective function, P, at each one of the vertices.

$$P = 20x + 30y$$

At $(1, 1)$,	$P = 20(1) + 30(1) = 50$
At $(1, 7)$,	$P = 20(1) + 30(7) = 230$
At $(6, 7)$,	$P = 20(6) + 30(7) = 330$
At $(6, 6)$,	$P = 20(6) + 30(6) = 300$

The maximum profit is at $(6, 7)$. Therefore, the company should manufacture 6 washers and 7 dryers to maximize its profit. The maximum profit is $330. ●

Use the following steps to solve a linear programming problem.

SOLVING A LINEAR PROGRAMMING PROBLEM

1. Determine all necessary constraints.
2. Determine the objective function.
3. Graph the constraints and determine the feasible region.
4. Determine the vertices of the feasible region.
5. Determine the value of the objective function at each vertex.

The solution is determined by the values in the ordered pair of the vertex that yields the maximum or minimum value of the objective function.

SECTION 7.6   EXERCISES

CONCEPT/WRITING EXERCISES

1. What are constraints in a linear programming problem? How are they represented?

2. In a linear programming problem, how is a feasible region formed?

3. What are the points of intersection of the boundaries of the feasible region called?

4. a) What is the general form of the objective function?

   b) What is the purpose of the objective function in a linear programming problem?

5. In your own words, state the fundamental principle of linear programming.

6. A profit function is $P = 4x + 6y$ and the vertices of the feasible region are $(1, 1)$, $(1, 4)$, $(5, 1)$, and $(7, 1)$. Determine the maximum profit. Explain how you determined your answer.

PRACTICE THE SKILLS

Exercises 7–10 show a feasible region and its vertices. Find the maximum and minimum values of the given objective function.

7. $K = 6x + 4y$

8. $K = 10x + 8y$

9. $K = 2x + 3y$

10. $K = 40x + 50y$

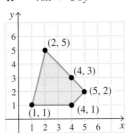

In Exercises 11–16, a set of constraints and a profit formula are given.

a) *Draw the graph of the constraints and find the vertices of the feasible region.*

b) *Use the vertices as obtained in part (a) to determine the maximum and minimum profit.*

11. $x + y \leq 6$
    $2x + y \leq 8$
    $x \geq 0$
    $y \geq 0$
    $P = 4x + 5y$

12. $2x + 3y \leq 8$
    $4x + 2y \leq 8$
    $x \geq 0$
    $y \geq 0$
    $P = 2x + 6y$

13. $x + y \leq 4$
    $x + 3y \leq 6$
    $x \geq 0$
    $y \geq 0$
    $P = 7x + 6y$

14. $x + y \leq 50$
    $x + 3y \leq 90$
    $x \geq 0$
    $y \geq 0$
    $P = 20x + 40y$

15. $2x + 3y \geq 18$
    $4x + 2y \leq 20$
    $x \geq 1$
    $y \geq 4$
    $P = 2.20x + 1.65y$

16. $x + 2y \leq 14$
    $7x + 4y \geq 28$
    $x \geq 2$
    $x \leq 10$
    $y \geq 1$
    $P = 15.13x + 9.35y$

PROBLEM SOLVING

17. **MODELING - *Stocking Cameras*** A small electronics store stocks two brands of digital cameras, Kodak and Canon. The manager does not want to keep more than 24 cameras on hand. The number of Kodak cameras stocked needs to be at least twice the number of Canon cameras stocked. She also wants to stock at least 4 Canon cameras. Assume that the store makes a profit of $40 on a Kodak camera and a profit of $55 on a Canon camera.

a) List the constraints.

b) Determine the objective function for maximizing profit.

c) Graph the set of constraints.

d) Determine the vertices of the feasible region.

e) How many cameras of each brand should the manager stock to maximize the store's profit?

f) Determine the maximum profit.

18. **MODELING - *On Wheels*** The Boards and Blades Company manufactures skateboards and in-line skates. The company can produce a maximum of 20 skateboards and pairs of in-line skates per day. It makes a profit of $25 on a skateboard and a profit of $20 on a pair of in-line skates. The company's planners want to make at least 3 skateboards but not more than 6 skateboards per day. To keep customers happy, they must make at least 2 pairs of in-line skates per day.

a) List the constraints.

b) Determine the objective function.

c) Graph the set of constraints.

d) Determine the vertices of the feasible region.

e) How many skateboards and pairs of in-line skates should be made to maximize the profit?

f) Find the maximum profit.

▲ Central Park, New York City, see Exercise 18

19. MODELING - *Paint Production* A paint supplier has two machines that produce both indoor paint and outdoor paint. To meet one of its contractual obligations, the company must produce at least 60 gal of indoor paint and 100 gal of outdoor paint. Machine I makes 3 gal of indoor paint and 10 gal of outdoor paint per hour. Machine II makes 4 gal of indoor paint and 5 gal of outdoor paint per hour. It costs $28 per hour to run machine I and $33 per hour to run machine II.

a) List the constraints.

b) Determine the objective function.

c) Graph the set of constraints.

d) Determine the vertices of the feasible region.

e) How many hours should each machine be operated to fulfill the contract at a minimum cost?

f) Determine the minimum cost.

CHALLENGE PROBLEMS/GROUP ACTIVITIES

20. MODELING - *Hot Dog Profits* To make one package of all-beef hot dogs, a manufacturer uses 1 lb of beef; to make one package of regular hot dogs, the manufacturer uses $\frac{1}{2}$ lb each of beef and pork. The profit on the all-beef hot dogs is 40 cents per pack and the profit on regular hot dogs is 30 cents per pack. If there are 200 lb of beef and 150 lb of pork available, how many packs of all-beef and regular hot dogs should the manufacturer make to maximize the profit? What is the profit?

21. MODELING - *Car Seats and Strollers* A company makes car seats and strollers. Each car seat and stroller passes through three processes: assembly, safety testing, and packaging. A car seat requires 1 hr in assembly, 2 hr in safety testing, and 1 hr in packaging. A stroller requires 3 hr in assembly, 1 hr in safety testing, and 1 hr in packaging. Employee work schedules allow for 24 hr per day for assembly, 16 hr per day for safety testing, and 10 hr per day for packaging. The profit for each car seat is $25, and the profit for each stroller is $35. How many units of each type should the company make per day to maximize the profit? What is the maximum profit?

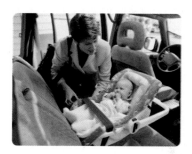

22. MODELING - *Special Diet* A dietitian prepares a special diet using two food groups, *A* and *B*. Each ounce of food group *A* contains 3 units of vitamin C and 1 unit of vitamin D. Each ounce of food group *B* contains 1 unit of vitamin C and 2 units of vitamin D. The minimum daily requirements with this diet are at least 9 units of vitamin C and at least 8 units of vitamin D. Each ounce of food group *A* costs 50 cents, and each ounce of food group *B* costs 30 cents.

a) List the constraints.

b) Determine the objective function for minimizing cost.

c) Graph the set of constraints.

d) Determine the vertices of the feasible region.

e) How many ounces of each food group should be used to meet the daily requirements and minimize the cost?

f) Determine the minimum cost.

INTERNET/RESEARCH ACTIVITY

23. *Operations research* draws on several disciplines, including mathematics, probability theory, statistics, and economics. George B. Dantzig (see the Profile in Mathematics on page 459) was one of the key people in developing operations research. Write a paper on Dantzig and his contributions to operations research and linear programming.

CHAPTER **7** SUMMARY

IMPORTANT FACTS

SYSTEMS OF EQUATIONS

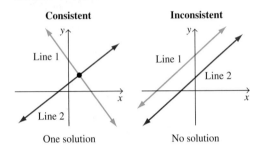

Consistent   Inconsistent

One solution   No solution

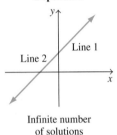

Dependent

Infinite number
of solutions

METHODS OF SOLVING SYSTEMS OF EQUATIONS

1. Graphing
2. Substitution
3. Addition (or elimination) method

MULTIPLICATIVE IDENTITY MATRIX

$$\begin{bmatrix} 1 & 0 \\ 0 & 1 \end{bmatrix}, \begin{bmatrix} 1 & 0 & 0 \\ 0 & 1 & 0 \\ 0 & 0 & 1 \end{bmatrix}, \ldots$$

FUNDAMENTAL PRINCIPLE OF LINEAR PROGRAMMING
If the objective function $K = Ax + By$ is evaluated at
each point in a feasible region, the maximum and minimum values of the equation occur at vertices of the region.

CHAPTER **7** REVIEW EXERCISES

7.1

*In Exercises 1–4, solve the system of equations graphically.
If the system does not have a single ordered pair as a solution, state whether the system is inconsistent or dependent.*

1. $x = 2$
  $y = -1$

2. $x + 2y = 4$
  $2x - 2y = 2$

3. $x = 3$
  $x + y = 5$

4. $x + 2y = 5$
  $2x + 4y = 4$

*In Exercises 5–8, determine without graphing whether the
system of equations has exactly one solution, no solution,
or an infinite number of solutions.*

5. $y = \frac{1}{3}x + 5$
  $3y - x = 15$

6. $2x + y = 4$
  $y = -2x + 6$

7. $6y - 2x = 20$
  $4y + 2x = 10$

8. $2x - 4y = 8$
  $-2x + y = 6$

7.2

*In Exercises 9–12, solve the system of equations by the
substitution method. If the system does not have a single
ordered pair as a solution, state whether the system is
inconsistent or dependent.*

9. $-x + y = -2$
  $x + 2y = 5$

10. $x - 2y = 9$
  $y = 2x - 3$

11. $2x - y = 4$
  $3x - y = 2$

12. $3x + y = 1$
  $3y = -9x - 4$

*In Exercises 13–18, solve the system of equations by the ad-
dition method. If the system does not have a single ordered
pair as a solution, state whether the system is inconsistent
or dependent.*

13. $x - 2y = 8$
  $2x + y = 6$

14. $2x + y = 2$
  $-3x - y = 5$

15. $x + y = 2$
$x + 3y = -2$

16. $4x - 8y = 16$
$x - 2y = 4$

17. $3x - 4y = 10$
$5x + 3y = 7$

18. $3x + 4y = 6$
$2x - 3y = 4$

7.3

Given $A = \begin{bmatrix} 1 & -3 \\ 2 & 4 \end{bmatrix}$ and $B = \begin{bmatrix} -2 & -5 \\ 6 & 3 \end{bmatrix}$, determine the following.

19. $A + B$

20. $A - B$

21. $2A$

22. $2A - 3B$

23. $A \times B$

24. $B \times A$

7.4

In Exercises 25–30, use an augmented matrix to solve the system of equations.

25. $x + 3y = 8$
$x + y = 4$

26. $-x + y = 4$
$x + 3y = 4$

27. $2x + y = 3$
$3x - y = 12$

28. $2x + 3y = 2$
$4x - 9y = 4$

29. $x + 3y = 3$
$3x - 2y = 2$

30. $3x - 6y = -9$
$4x + 5y = 14$

7.1–7.4

31. MODELING - *Borrowing Money* A company borrows $600,000 for 1 year to expand its product line. Some of the money was borrowed at a 4% simple interest rate, and the rest of the money was borrowed at a 6% simple interest rate. How much money was borrowed at each rate if the total annual interest was $29,000?

32. MODELING - *Chemistry* In chemistry class, Tom Northrup has an 80% acid solution and a 50% acid solution. How much of each solution should he mix to get 100 liters of a 75% acid solution?

33. MODELING - *Landscaping* The Garden Factory purchased 4 tons of topsoil and 3 tons of mulch for $1529. The next week the company purchased 2 tons of topsoil and 5 tons of mulch for $1405. Determine the price per ton for topsoil and the price per ton for mulch.

34. MODELING - *Cool Air* Emily Richelieu needs to purchase a new air conditioner for the office. Model 1600A costs $950 to purchase and $32 per month to operate. Model 6070B, a more efficient unit, costs $1275 to purchase and $22 per month to operate.

a) After how many months will the total cost of both units be equal?

b) Which model will be the more cost effective if the life of both units is guaranteed for 10 years?

35. MODELING - *Minimizing Parking Costs* The cost of parking in All-Day parking lot is $5 for the first hour and $0.50 for each additional hour. Sav-a-Lot parking lot costs $4.25 for the first hour and $0.75 for each additional hour.

a) In how many hours after the first hour would the total cost of parking at All-Day and Sav-a-Lot be the same?

b) If Mark McMahon needed to park his car for 5 hr, which parking lot would be less expensive?

7.5

In Exercises 36–39, graph the system of linear inequalities and indicate the solution set.

36. $y \leq 3x - 1$
$y > -2x + 1$

37. $2x + y < 8$
$y \geq 2x - 1$

38. $x + 3y \leq 6$
$2x - 7y \geq 14$

39. $x - y > 5$
$6x + 5y \leq 30$

7.6

40. The set of constraints and profit formula for a linear programming problem are

$$2x + 3y \leq 12$$
$$2x + y \leq 8$$
$$x \geq 0$$
$$y \geq 0$$
$$P = 5x + 3y$$

a) Draw the graph of the constraints and determine the vertices of the feasible region.

b) Use the vertices to determine the maximum and minimum profit.

CHAPTER **7** TEST

1. From a graph, explain how you would identify a consistent system of equations, an inconsistent system of equations, and a dependent system of equations.

2. Solve the system of equations graphically.
$$y = 2x - 12$$
$$2x + 2y = -6$$

3. Determine without graphing whether the system of equations has exactly one solution, no solution, or an infinite number of solutions.
$$4x + 5y = 6$$
$$-3x + 5y = 13$$

Solve the system of equations by the method indicated.

4. $x + y = -1$
   $2x + 3y = -5$
   (substitution)

5. $y = 3x - 7$
   $y = 5x - 3$
   (substitution)

6. $x - y = 4$
   $2x + y = -10$
   (addition)

7. $4x + 3y = 5$
   $2x + 4y = 10$
   (addition)

8. $3x + 4y = 6$
   $2x - 3y = 4$
   (addition)

9. $x + 3y = 4$
   $5x + 7y = 4$
   (matrices)

In Exercises 10–12, for $A = \begin{bmatrix} 2 & -5 \\ 1 & 3 \end{bmatrix}$ and
$B = \begin{bmatrix} -1 & -3 \\ 5 & 2 \end{bmatrix}$, *determine the following.*

10. $A + B$ 11. $A - 2B$ 12. $A \times B$

13. Graph the system of linear inequalities and indicate the solution set.
$$y < -2x + 2$$
$$y > 3x + 2$$

Solve Exercises 14 and 15 by using a system of equations.

14. **MODELING - *Truck Rental*** U-Haul charges a daily fee plus a mileage charge to rent a truck. Dorothy DiMento rented a truck with U-Haul and was charged $132 for 3 days rental and 150 miles driven. Elena

Dilai rented the same truck and was charged $142 for 2 days rental and 400 miles driven. Determine the daily fee and the mileage charge for renting this truck.

15. **MODELING - *Checking Accounts*** The charge for maintaining a checking account at Union Bank is $6 per month plus 10 cents for each check that is written. The charge at Citrus Bank is $2 per month and 20 cents per check.

a) How many checks would a customer have to write in a month for the total charges to be the same at both banks?

b) If Brent Pickett planned to write 14 checks per month, which bank would be the least expensive?

16. The set of constraints and profit formula for a linear programming problem are
$$x + 3y \leq 6$$
$$4x + 3y \leq 15$$
$$x \geq 0$$
$$y \geq 0$$
$$P = 6x + 4y$$

a) Draw the graph of the constraints and determine the vertices of the feasible region.

b) Use the vertices to determine the maximum and minimum profit.

9.1   POINTS, LINES, PLANES, AND ANGLES

▲ Billiard games involve many geometric concepts.

Playing billiards involves many geometric concepts. In this section, we will introduce the geometric terms *points, lines, planes*, and *angles*. In a typical billiard game, each of these terms is evident. The billiard balls rest on specific points on the table, which is part of a plane. After being struck with a cue, the ball travels along a path that is part of a line. Angles are involved in the path the ball takes when the ball hits a table bumper or another ball. The concepts we will discuss in this section form an important basis for the study of geometry.

Human beings recognized shapes, sizes, and physical forms long before geometry was developed. Geometry as a science is said to have begun in the Nile Valley of ancient Egypt. The Egyptians used geometry to measure land and to build pyramids and other structures.

The word *geometry* is derived from two Greek words, *ge*, meaning earth, and *metron*, meaning measure. Thus, geometry means "earth measure" or "measurement of the earth."

Unlike the Egyptians, the Greeks were interested in more than just the applied aspects of geometry. The Greeks attempted to apply their knowledge of logic to geometry. In about 600 B.C., Thales of Miletus was the first to be credited with using deductive methods to develop geometric concepts. Another outstanding Greek mathematician, Pythagoras, continued the systematic development of geometry that Thales had begun.

In about 300 B.C., Euclid (see Profile in Mathematics on page 516) collected and summarized much of the Greek mathematics of his time. In a set of 13 books called *Elements*, Euclid laid the foundation for plane geometry, which is also called *Euclidean geometry*.

Euclid is credited with being the first mathematician to use the *axiomatic method* in developing a branch of mathematics. First, Euclid introduced *undefined terms* such as point, line, plane, and angle. He related these to physical space by such statements as "A line is length without breadth" so that we may intuitively understand them. Because such statements play no further role in his system, they constitute primitive or undefined terms.

Second, Euclid introduced certain *definitions*. The definitions are introduced when needed and are often based on the undefined terms. Some terms that Euclid introduced and defined include triangle, right angle, and hypotenuse.

Third, Euclid stated certain primitive propositions called *postulates* (now called *axioms**) about the undefined terms and definitions. The reader is asked to accept these statements as true on the basis of their "obviousness" and their relationship with the physical world. For example, the Greeks accepted all right angles as being equal, which is Euclid's fourth postulate.

Fourth, Euclid proved, using deductive reasoning (see Section 1.1), other propositions called *theorems*. One theorem that Euclid proved is known as the Pythagorean theorem: "The sum of the areas of the squares constructed on the arms of a right triangle

---

*The concept of the axiom has changed significantly since Euclid's time. Now any statement may be designated as an axiom, whether it is self-evident or not. All axioms are *accepted* as true. A set of axioms forms the foundation for a mathematical system.

is equal to the area of the square constructed on the hypotenuse." He also proved that the sum of the angles of a triangle is 180°.

Using only 10 axioms, Euclid deduced 465 propositions (or theorems) in plane and solid geometry, number theory, and Greek geometric algebra.

Point and Line

Three basic terms in geometry are *point*, *line*, and *plane*. These three terms are not given a formal definition, but we recognize points, lines, and planes when we see them.

Let's consider some properties of a line. Assume that a line means a straight line unless otherwise stated.

1. A line is a set of points. Each point is on the line and the line passes through each point. When we wish to refer to a specific point, we will label it with a single capital letter. For example, in Figure 9.1(a) three points are labled A, B, and C, respectively.

2. Any two distinct points determine a unique line. Figure 9.1(a) illustrates a line. The arrows at both ends of the line indicate that the line continues in each direction. The line in Fig. 9.1(a) may be symbolized with any two points on the line by placing a line with a double-sided arrow above the letters that correspond to the points, such as $\overleftrightarrow{AB}$, $\overleftrightarrow{BA}$, $\overleftrightarrow{AC}$, $\overleftrightarrow{CA}$, $\overleftrightarrow{BC}$, or $\overleftrightarrow{CB}$.

3. Any point on a line separates the line into three parts: the point itself and two *half lines* (neither of which includes the point). For example, in Fig. 9.1(a) point B separates the line into the point B and two half lines. Half line BA, symbolized $\overset{\circ}{BA}$, is illustrated in Fig. 9.1(b). The open circle above the B indicates that point B is not included in the half line. Figure 9.1(c) illustrates half line BC, symbolized $\overset{\circ}{BC}$.

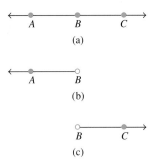

Figure 9.1

Look at the half line $\overset{\circ}{AB}$ in Fig. 9.2(b) on page 517. If the *end point*, A, is included with the set of points on the half line, the result is called a *ray*. Ray AB, symbolized $\overrightarrow{AB}$, is illustrated in Fig. 9.2(c). Ray BA, symbolized $\overrightarrow{BA}$, is illustrated in Fig. 9.2(d).

A *line segment* is that part of a line between two points, including the end points. Line segment AB, symbolized $\overline{AB}$, is illustrated in Fig. 9.2(e).

An open line segment is the set of points on a line between two points, excluding the end points. Open line segment AB, symbolized $\overset{\circ\circ}{AB}$, is illustrated in Fig. 9.2(f).

Figure 9.2(g) illustrates two half open line segments, symbolized $\overset{\;\;\circ}{AB}$ and $\overset{\circ}{AB}$.

Compass and Straightedge Constructions

Straightedge   Compass

Geometric constructions were central to ancient Greek mathematics. Although these constructions are often referred to as *Euclidean constructions*, they were used centuries before Euclid wrote his classic work, *Elements*. The tools *allowed* in geometric constructions are a pencil, an unmarked straightedge, and a drawing compass. The straightedge is used to draw line segments, and the compass is used to draw circles and arcs. The Internet has many sites devoted to classic geometric constructions. Exercise 105 shows you how to construct a triangle with sides of equal length (an equilateral triangle).

Description	Diagram	Symbol
(a) Line AB		$\overleftrightarrow{AB}$
(b) Half line AB		$\overrightarrow{AB}$
(c) Ray AB		$\overrightarrow{AB}$
(d) Ray BA		$\overleftarrow{BA}$
(e) Line segment AB		$\overline{AB}$
(f) Open line segment AB		$\overset{\circ\,\circ}{AB}$
(g) Half open line segments AB		$\overset{\,\circ}{AB}$ $\overset{\circ}{AB}$

Figure 9.2

In Chapter 2, we discussed intersection of sets. Recall that the intersection (symbolized $\cap$) of two sets is the set of elements (points in this case) common to both sets.

Consider the rays $\overrightarrow{AB}$ and $\overrightarrow{BA}$ in Fig. 9.3(a). The intersection of $\overrightarrow{AB}$ and $\overrightarrow{BA}$ is $\overline{AB}$. Thus, $\overrightarrow{AB} \cap \overrightarrow{BA} = \overline{AB}$.

We also discussed the union of two sets in Chapter 2. The union (symbolized $\cup$) of two sets is the set of elements (points in this case) that belong to either of the sets or both sets. The union of $\overrightarrow{AB}$ and $\overrightarrow{BA}$ is $\overleftrightarrow{AB}$ (Fig. 9.3b). Thus, $\overrightarrow{AB} \cup \overrightarrow{BA} = \overleftrightarrow{AB}$.

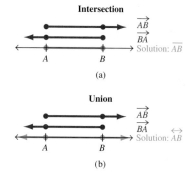

Figure 9.3

EXAMPLE ❶ *Unions and Intersections of Parts of a Line*

Using line AD, determine the solution to each part.

a) $\overrightarrow{AB} \cap \overrightarrow{DC}$ b) $\overrightarrow{AB} \cup \overrightarrow{DC}$ c) $\overrightarrow{AB} \cap \overrightarrow{CD}$ d) $\overline{AD} \cup \overset{\smile}{CA}$

SOLUTION

a) $\overrightarrow{AB} \cap \overrightarrow{DC}$

Ray AB and ray DC are shown below. The intersection of these two rays is that part of line AD that is a part of *both* ray AB and ray DC. The intersection of ray AB and ray DC is line segment AD.

$$\overrightarrow{AB} \cap \overrightarrow{DC} = \overline{AD}$$

b) $\overrightarrow{AB} \cup \overrightarrow{DC}$

Once again ray AB and ray DC are shown below. The union of these two rays is that part of line AD that is part of *either* ray AB or ray DC. The union of ray AB and ray DC is the entire line AD.

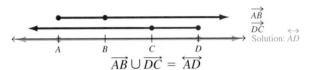

$$\overrightarrow{AB} \cup \overrightarrow{DC} = \overleftrightarrow{AD}$$

c) $\overline{AB} \cap \overrightarrow{CD}$

Line segment AB and ray CD have no points in common, so their intersection is empty.

$$\overline{AB} \cap \overrightarrow{CD} = \varnothing$$

d) $\overline{AD} \cup \overset{\circ}{CA}$

The union of line segment AD and half line CA is ray DA (or equivalently, $\overrightarrow{DB}$ or $\overrightarrow{DC}$).

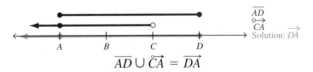

$$\overline{AD} \cup \overset{\circ}{CA} = \overrightarrow{DA}$$

Plane

The term *plane* is one of Euclid's undefined terms. For our purposes, we can think of a plane as a two-dimensional surface that extends infinitely in both directions, like an infinitely large blackboard. Euclidean geometry is called *plane geometry* because it is the study of two-dimensional figures in a plane.

Two lines in the same plane that do not intersect are called *parallel lines*. Figure 9.4(a) on page 519 illustrates two parallel lines in a plane ($\overleftrightarrow{AB}$ is parallel to $\overleftrightarrow{CD}$). Properties of planes include the following:

1. Any three points that are not on the same line (noncollinear points) determine a unique plane (Fig. 9.4b).
2. A line in a plane divides the plane into three parts, the line and two half planes (Fig. 9.4c).

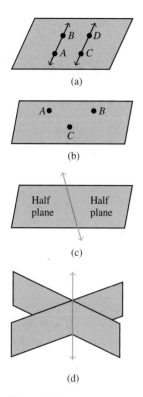

Figure 9.4

3. Any line and a point not on the line determine a unique plane.
4. The intersection of two distinct planes is a line (Fig. 9.4d).

Two planes that do not intersect are said to be *parallel planes*. For example, in Fig. 9.5 plane *ABE* is parallel to plane *GHF*.

Two lines that do not lie in the same plane and do not intersect are called *skew lines*. Figure 9.5 illustrates many skew lines (for example, $\overleftrightarrow{AB}$ and $\overleftrightarrow{CD}$).

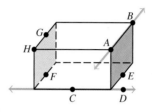

Figure 9.5

Angles

An *angle*, denoted ⌊, is the union of two rays with a common end point (Fig. 9.6):

$$\overrightarrow{BA} \cup \overrightarrow{BC} = \angle ABC \text{ (or } \angle CBA)$$

An angle can be formed by the rotation of a ray about a point. An angle has an initial side and a terminal side. The initial side indicates the position of the ray prior to rotation; the terminal side indicates the position of the ray after rotation. The point common to both rays is called the *vertex* of the angle. The letter designating the vertex is always the middle one of the three letters designating an angle. The rays that make up the angle are called its *sides*.

There are several ways to name an angle. The angle in Fig. 9.6 may be denoted

$$\angle ABC, \qquad \angle CBA, \qquad \text{or} \qquad \angle B$$

An angle divides a plane into three distinct parts: the angle itself, its interior, and its exterior. In Fig. 9.6, the angle is represented by the blue lines, the interior of the angle is shaded pink, and the exterior is shaded green.

The *measure of an angle*, symbolized *m*, is the amount of rotation from its initial side to its terminal side. In Fig. 9.6, the letter *x* represents the measure of ⌊*ABC*; therefore, we may write $m\angle ABC = x$.

Angles can be measured in *degrees*, radians, or gradients. In this text, we will discuss only the degree unit of measurement. The symbol for degrees is the same as the symbol for temperature degrees. An angle of 45 degrees is written 45°. A *protractor* is used to measure angles. The angle shown being measured by the protractor in Fig. 9.7 is 50°.

Figure 9.6

Figure 9.7

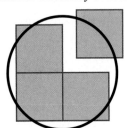
EXAMPLE ❷ *Union and Intersection*

Refer to Fig. 9.8. Determine the following.

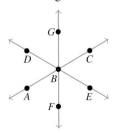

Figure 9.8

a) $\overleftrightarrow{AC} \cap \overleftrightarrow{DE}$ b) $\overrightarrow{AB} \cup \overrightarrow{CB}$ c) $\overrightarrow{BE} \cup \overrightarrow{BF}$ d) $\angle DBG \cap \angle CBG$

SOLUTION

a) $\overleftrightarrow{AC} \cap \overleftrightarrow{DE} = \{B\}$ b) $\overrightarrow{AB} \cup \overrightarrow{CB} = \overleftrightarrow{AC}$
c) $\overrightarrow{BE} \cup \overrightarrow{BF} = \angle EBF$ d) $\angle DBG \cap \angle CBG = \overrightarrow{BG}$ ●

Consider a circle whose circumference is divided into 360 equal parts. If we draw a line from each mark on the circumference to the center of the circle, we get 360 wedge-shaped pieces. The measure of an angle formed by the straight sides of each wedge-shaped piece is defined to be 1°.

Angles are classified by their degree measurement, as shown in the following summary. A *right angle* has a measure of 90°, an *acute angle* has a measure less than 90°, an *obtuse angle* has a measure greater than 90° but less than 180°, and a *straight angle* has a measure of 180°.

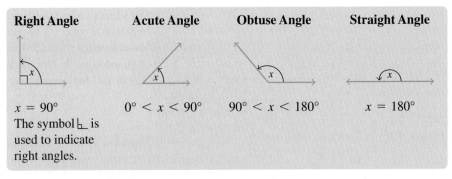

Right Angle	Acute Angle	Obtuse Angle	Straight Angle
$x = 90°$ The symbol ∟ is used to indicate right angles.	$0° < x < 90°$	$90° < x < 180°$	$x = 180°$

Two angles in the same plane are *adjacent angles* when they have a common vertex and a common side but no common interior points. In Fig. 9.9, $\angle DBC$ and $\angle CBA$ are adjacent angles, but $\angle DBA$ and $\angle CBA$ are not adjacent angles.

Two angles are called *complementary angles* if the sum of their measures is 90°. Two angles are called *supplementary angles* if the sum of their measures is 180°.

Figure 9.9

Figure 9.10

Figure 9.11

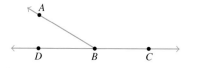

Figure 9.12

EXAMPLE ③ *Determining Complementary and Supplementary Angles*

In Fig. 9.10, we see that $m\angle ABC = 25°$.

a) $\angle ABC$ and $\angle CBD$ are complementary angles. Determine $m\angle CBD$.

b) $\angle ABC$ and $\angle CBE$ are supplementary angles. Determine $m\angle CBE$.

SOLUTION

a) The sum of two complementary angles must be 90°, so

$$m\angle ABC + m\angle CBD = 90°$$
$$25° + m\angle CBD = 90°$$
$$m\angle CBD = 90° - 25° = 65° \quad \text{Subtract 25° from each side of the equation.}$$

b) The sum of two supplementary angles must be 180°, so

$$m\angle ABC + m\angle CBE = 180°$$
$$25° + m\angle CBE = 180°$$
$$m\angle CBE = 180° - 25° = 155° \quad \text{Subtract 25° from each side of the equation.}$$

EXAMPLE ④ *Determining Complementary Angles*

If $\angle ABC$ and $\angle CBD$ are complementary angles and $m\angle ABC$ is 26° less than $m\angle CBD$, determine the measure of each angle (Fig. 9.11).

SOLUTION Let $m\angle CBD = x$. Then $m\angle ABC = x - 26$ since it is 26° less than $m\angle CBD$. Because these angles are complementary, we have

$$m\angle CBD + m\angle ABC = 90°$$
$$x + (x - 26) = 90°$$
$$2x - 26 = 90°$$
$$2x = 116°$$
$$x = 58°$$

Therefore, $m\angle CBD = 58°$ and $m\angle ABC = 58° - 26°$, or 32°. Note that $58° + 32° = 90°$, which is what we expected.

EXAMPLE ⑤ *Determining Supplementary Angles*

If $\angle ABC$ and $\angle ABD$ are supplementary and $m\angle ABC$ is five times larger than $m\angle ABD$, determine $m\angle ABC$ and $m\angle ABD$ (Fig. 9.12).

SOLUTION Let $m\angle ABD = x$, then $m\angle ABC = 5x$. Since these angles are supplementary, we have

$$m\angle ABD + m\angle ABC = 180°$$
$$x + 5x = 180°$$
$$6x = 180°$$
$$x = 30°$$

Thus, $m\angle ABD = 30°$ and $m\angle ABC = 5(30°) = 150°$. Note that $30° + 150° = 180°$, which is what we expected.

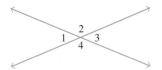

Figure 9.13

When two straight lines intersect, the nonadjacent angles formed are called *vertical angles*. In Fig. 9.13, $\angle 1$ and $\angle 3$ are vertical angles, and $\angle 2$ and $\angle 4$ are vertical angles. We can show that vertical angles have the same measure, that is, they are equal. For example, Fig. 9.13 shows that

$$m\angle 1 + m\angle 2 = 180°. \qquad \text{Why?}$$
$$m\angle 2 + m\angle 3 = 180°. \qquad \text{Why?}$$

Since $\angle 2$ has the same measure in both cases, $m\angle 1$ must equal $m\angle 3$.

> Vertical angles have the same measure.

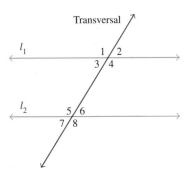

Figure 9.14

A line that intersects two different lines, l_1 and l_2, at two different points is called a *transversal*. Figure 9.14 illustrates that when two parallel lines are cut by a transversal, eight angles are formed. Angles 3, 4, 5, and 6 are called *interior angles*, and angles 1, 2, 7, and 8 are called *exterior angles*. Eight pairs of supplementary angles are formed. Can you list them?

Special names are given to the angles formed by a transversal crossing two parallel lines. We describe these angles below.

Name	Description	Illustration	Pairs of Angles Meeting Criteria
Alternate interior angles	Interior angles on opposite sides of the transversal		$\angle 3$ and $\angle 6$ $\angle 4$ and $\angle 5$
Alternate exterior angles	Exterior angles on opposite sides of the transversal		$\angle 1$ and $\angle 8$ $\angle 2$ and $\angle 7$
Corresponding angles	One interior and one exterior angle on the same side of the transversal		$\angle 1$ and $\angle 5$ $\angle 2$ and $\angle 6$ $\angle 3$ and $\angle 7$ $\angle 4$ and $\angle 8$

> When two parallel lines are cut by a transversal,
> 1. alternate interior angles have the same measure.
> 2. alternate exterior angles have the same measure.
> 3. corresponding angles have the same measure.

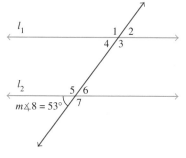

Figure 9.15

EXAMPLE ➏ *Determining Angle Measures*

Figure 9.15 shows two parallel lines cut by a transversal. Determine the measure of ∡1 through ∡7.

SOLUTION

$m∡6 = 53°$	∡8 and ∡6 are vertical angles.
$m∡5 = 127°$	∡8 and ∡5 are supplementary angles.
$m∡7 = 127°$	∡5 and ∡7 are vertical angles.
$m∡1 = 127°$	∡1 and ∡7 are alternate exterior angles.
$m∡4 = 53°$	∡4 and ∡6 are alternate interior angles.
$m∡2 = 53°$	∡6 and ∡2 are corresponding angles.
$m∡3 = 127°$	∡3 and ∡1 are vertical angles.

In Example 6, the angles could have been determined in alternate ways. For example, we mentioned $m∡1 = 127°$ because ∡1 and ∡7 are alternate exterior angles. We could have also stated that $m∡1 = 127°$ because ∡1 and ∡5 are corresponding angles.

SECTION 9.1 EXERCISES

CONCEPT/WRITING EXERCISES

1. a) What are the four key parts in the axiomatic method used by Euclid?

    b) Discuss each of the four parts.

2. What is the difference between an axiom and a theorem?

3. What are skew lines?

4. What are parallel lines?

5. What are adjacent angles?

6. What are complementary angles?

7. What are supplementary angles?

8. What is a straight angle?

9. What is an obtuse angle?

10. What is a right angle?

11. What is an acute angle?

12. Draw two intersecting lines. Identify the two pairs of vertical angles.

PRACTICE THE SKILLS

In Exercises 13–20, identify the figure as a line, half line, ray, line segment, open line segment, or half open line segment. Denote the figure by its appropriate symbol.

13.

14.

15.

16.

17.

18.

19.

20.

In Exercises 21–32, use the figure to find the following:

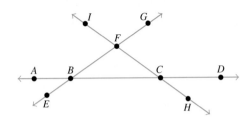

21. $\angle ABF \cap \angle GBC$

22. $\overleftrightarrow{FE} \cup \overrightarrow{FG}$

23. $\overleftrightarrow{BC} \cup \overleftrightarrow{CD}$

24. $\overline{AB} \cup \overline{BD}$

25. $\angle ICA \cap \overleftrightarrow{EG}$

26. $\angle IFG \cap \angle EFH$

27. $\overleftrightarrow{AB} \cap \overline{HC}$

28. $\overrightarrow{BD} \cap \overrightarrow{CB}$

29. $\overrightarrow{FG} \cup \overrightarrow{FC}$

30. $\overline{BC} \cup \overline{CF} \cup \overline{FB}$

31. $\overrightarrow{BD} \cup \overrightarrow{CB}$

32. $\{C\} \cap \overleftrightarrow{CH}$

In Exercises 33–44, use the figure to find each of the following.

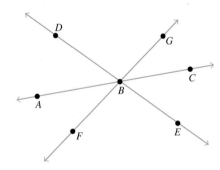

33. $\overleftrightarrow{DE} \cap \overleftrightarrow{FG}$

34. $\overleftrightarrow{BD} \cup \overrightarrow{BE}$

35. $\angle GBC \cap \angle CBE$

36. $\overleftrightarrow{DE} \cup \overrightarrow{BE}$

37. $\angle ABE \cup \overleftrightarrow{AB}$

38. $\overrightarrow{BF} \cup \overrightarrow{BE}$

39. $\overrightarrow{FG} \cap \overrightarrow{BF}$

40. $\overrightarrow{BE} \cup \overline{BD}$

41. $\overleftrightarrow{AC} \cap \overline{AC}$

42. $\overleftrightarrow{AC} \cap \overrightarrow{BE}$

43. $\overrightarrow{EB} \cap \overrightarrow{BE}$

44. $\overline{FB} \cup \overline{BG}$

In Exercises 45–52, classify the angle as acute, right, straight, obtuse, or none of these angles.

45.

46.

47.

48.

49.

50.

51.

52.

In Exercises 53–58, find the complementary angle of the given angle.

53. $26°$

54. $89°$

55. $32\frac{3}{4}°$

56. $31\frac{2}{5}°$

57. $64.7°$

58. $0.01°$

In Exercises 59–64, find the supplementary angle of the given angle.

59. $89°$

60. $8°$

61. $20.5°$

62. $148.7°$

63. $43\frac{5}{7}°$

64. $64\frac{7}{16}°$

In Exercises 65–70, match the names of the angles with the corresponding figure in parts (a)–(f).

65. Vertical angles

66. Corresponding angles

67. Complementary angles

68. Supplementary angles

69. Alternate exterior angles

70. Alternate interior angles

a)

b)

c)

d)

e)

f)

81.

82.

PROBLEM SOLVING

71. MODELING - *Complementary Angles* If ∡1 and ∡2 are complementary angles and if the measure of ∡1 is eight more than the measure of ∡2, determine the measures of ∡1 and ∡2.

72. MODELING - *Complementary Angles* The difference between the measures of two complementary angles is 16°. Determine the measures of the two angles.

73. MODELING - *Supplementary Angles* The difference between the measures of two supplementary angles is 88°. Determine the measures of the two angles.

74. MODELING - *Supplementary Angles* If ∡1 and ∡2 are supplementary angles and if the measure of ∡2 is 17 times the measure of ∡1, determine the measures of the two angles.

In Exercises 75–78, parallel lines are cut by the transversal shown. Determine the measures of ∡1 through ∡7.

75.

76.

77.

78.

In Exercises 79–82, the angles are complementary angles. Determine the measures of ∡1 and ∡2.

79.

80.

In Exercises 83–86, the angles are supplementary angles. Determine the measures of ∡1 and ∡2.

83.

84.

85.

86.

87. a) How many lines can be drawn through a given point?

b) How many planes can be drawn through a given point?

88. What is the intersection of two distinct nonparallel planes?

89. How many planes can be drawn through a given line?

90. a) Will three noncollinear points *A*, *B*, and *C* always determine a plane? Explain.

b) Is it possible to determine more than one plane with three noncollinear points? Explain.

c) How many planes can be constructed through three collinear points?

The figure suggests a number of lines and planes. The lines may be described by naming two points, and the planes may be described by naming three points. In Exercises 91–98, use the figure to name the following.

91. Two parallel planes

92. Two parallel lines

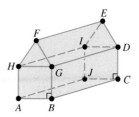

93. Two lines that intersect at right angles

94. Two planes that intersect at right angles

95. Three planes whose intersection is a single point

96. Three planes whose intersection is a line

97. A line and a plane whose intersection is a point

98. A line and a plane whose intersection is a line

In Exercises 99–104, determine whether the statement is always true, sometimes true, or never true. Explain your answer.

99. Two lines that are both parallel to a third line must be parallel to each other.

100. A triangle contains exactly two acute angles.

101. Vertical angles are complementary angles.

102. Alternate exterior angles are supplementary angles.

103. Alternate interior angles are complementary angles.

104. A triangle contains two obtuse angles.

CHALLENGE PROBLEMS/GROUP ACTIVITIES

105. Use a straightedge and a compass to construct a triangle with sides of equal length (an equilateral triangle) by doing the following:

    a) Use the straightedge to draw a line segment of any length and label the end points A and B (Fig. a).

    b) Place one end of the compass at point A and the other end at point B and draw an arc as shown (Fig. b).

    c) Now turn the compass around and draw another arc as shown. Label the point of intersection of the two arcs C (Fig. c).

    d) Draw line segments AC and BC. This completes the construction of equilateral triangle ABC (Fig. d).

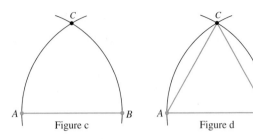

Figure c   Figure d

106. If lines l and m are parallel lines and if lines l and n are skew lines, is it true that lines m and n must also be skew? (*Hint:* Look at Fig. 9.5 on page 519.) Explain your answer and include a sketch to support your answer.

107. Two lines are *perpendicular* if they intersect at right angles. If lines l and m are perpendicular and if lines m and n are perpendicular, is it true that lines l and n must also be perpendicular? Explain your answer and include a sketch to support your answer.

108. Suppose you have three distinct lines, all lying in the same plane. Find all the possible ways in which the three lines can be related. Sketch each case (four cases).

RECREATIONAL MATHEMATICS

109. If two straight lines intersect at a point, determine the sum of the measures of the 4 angles formed.

110. $\angle ABC$ and $\angle CBD$ are complementary and $m\angle CBD$ is twice the $m\angle ABC$. $\angle ABD$ and $\angle DBE$ are supplementary angles.

    a) Draw a sketch illustrating $\angle ABC$, $\angle CBD$, and $\angle DBE$.

    b) Determine $m\angle ABC$.

    c) Determine $m\angle CBD$.

    d) Determine $m\angle DBE$.

INTERNET/RESEARCH ACTIVITIES

111. Using the Internet and other sources, write a research paper on Euclid's contributions to geometry.

112. Using the Internet and other sources, write a research paper on the three classic geometry problems of Greek antiquity (see the Did You Know? on page 520).

113. Search the Internet or other sources such as a geometry textbook to study the geometric constructions that use a straightedge and a compass only. Prepare a poster demonstrating five of these basic constructions.

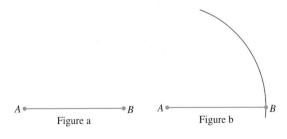

Figure a   Figure b

9.2 POLYGONS

What shape would you use to best describe the following road signs: a stop sign, a yield sign, a speed limit sign? In this section, we will study the shapes of these and other geometric figures that can be classified as *polygons*.

▲ The shapes of these road signs are examples of polygons.

A *polygon* is a closed figure in a plane determined by three or more straight line segments. Examples of polygons are given in Fig. 9.16.

The straight line segments that form the polygon are called its *sides*, and a point where two sides meet is called a *vertex* (plural *vertices*). The union of the sides of a polygon and its interior is called a *polygonal region*. A *regular polygon* is one whose sides are all the same length and whose interior angles all have the same measure. Figures 9.16(b) and (d) are regular polygons.

(a)   (b)   (c)   (d)

Figure 9.16

Polygons are named according to their number of sides. The names of some polygons are given in Table 9.1.

Table 9.1

Number of Sides	Name	Number of Sides	Name
3	Triangle	8	Octagon
4	Quadrilateral	9	Nonagon
5	Pentagon	10	Decagon
6	Hexagon	12	Dodecagon
7	Heptagon	20	Icosagon

One of the most important polygons is the triangle. The sum of the measures of the interior angles of a triangle is 180°. To illustrate, consider triangle *ABC* given in Fig. 9.17. The triangle is formed by drawing two transversals through two parallel lines l_1 and l_2 with the two transversals intersecting at a point on l_1.

Figure 9.17

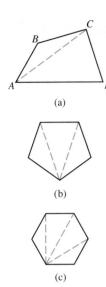

(a)

(b)

(c)

Figure 9.18

In Fig. 9.17, notice that $\angle A$ and $\angle A'$ are corresponding angles. Recall from Section 9.1 that corresponding angles are equal, so $m\angle A = m\angle A'$. Also, $\angle C$ and $\angle C'$ are corresponding angles; therefore, $m\angle C = m\angle C'$. Next, we notice that $\angle B$ and $\angle B'$ are vertical angles. In Section 9.1, we learned that vertical angles are equal; therefore, $m\angle B = m\angle B'$. Figure 9.17 shows that $\angle A'$, $\angle B'$, and $\angle C'$ form a straight angle; therefore, $m\angle A' + m\angle B' + m\angle C' = 180°$. Since $m\angle A = m\angle A'$, $m\angle B = m\angle B'$, and $m\angle C = m\angle C'$, we can reason that $m\angle A + m\angle B + m\angle C = 180°$. This example illustrates that the sum of the interior angles of a triangle is $180°$.

Consider the quadrilateral $ABCD$ (Fig. 9.18a). Drawing a straight line segment between any two vertices forms two triangles. Since the sum of the measures of the angles of a triangle is $180°$, the sum of the measures of the interior angles of a quadrilateral is $2 \cdot 180°$, or $360°$.

Now let's examine a pentagon (Fig. 9.18b). We can draw two straight line segments to form three triangles. Thus, the sum of the measures of the interior angles of a five-sided figure is $3 \cdot 180°$, or $540°$. Figure 9.18(c) shows that four triangles can be drawn in a six-sided figure. Table 9.2 summarizes this information.

Table 9.2

Sides	Triangles	Sum of the Measures of the Interior Angles
3	1	$1(180°) = 180°$
4	2	$2(180°) = 360°$
5	3	$3(180°) = 540°$
6	4	$4(180°) = 720°$

If we continue this procedure, we can see that for an n-sided polygon the sum of the measures of the interior angles is $(n - 2)180°$.

The **sum** of the measures of the interior angles of an n-sided polygon is $(n - 2)180°$.

EXAMPLE ❶ Angles of a Hexagon

The surfaces of the heads of many bolts are in the shape of regular hexagons. A regular hexagon is a six-sided figure with all the sides the same length and all interior angles with the same measure. See Fig. 9.19. Determine

a) the measure of an interior angle.

b) the measure of exterior $\angle 1$.

Figure 9.19

SOLUTION

a) Using the formula $(n - 2)180°$, we can determine the sum of the measures of the interior angles of a hexagon as follows.

$$\text{Sum} = (6 - 2)180°$$
$$= 4(180°)$$
$$= 720°$$

The measure of an interior angle of a regular polygon can be determined by dividing the sum of the interior angles by the number of angles.

The measure of an interior angle of a regular hexagon is determined as follows:

$$\text{Measure} = \frac{720°}{6} = 120°$$

b) Since $\angle 1$ is the supplement of an interior angle,

$$m\angle 1 = 180° - 120° = 60°$$ ●

To discuss area in the next section, we must be able to identify various types of triangles and quadrilaterals. The following is a summary of certain types of triangles and their characteristics.

TRIANGLES

Acute Triangle	**Obtuse Triangle**	**Right Triangle**
All angles are acute angles.	One angle is obtuse.	hypotenuse One angle is a right angle.

Isosceles Triangle	**Equilateral Triangle**	**Scalene Triangle**
Two equal sides Two equal angles	Three equal sides Three equal angles (60° each)	No two sides are equal in length.

Similar Figures

In everyday living, we often have to deal with geometric figures that have the "same shape" but are of different sizes. For example, an architect will make a small-scale drawing of a floor plan or a photographer will make an enlargement of a photograph. Figures that have the same shape but may be of different sizes are called *similar figures*. Two similar figures are illustrated in Fig. 9.20.

Similar figures have *corresponding angles* and *corresponding sides*. In Fig. 9.20, triangle *ABC* has angles *A*, *B*, and *C*. Their respective corresponding angles in triangle *DEF* are angles *D*, *E*, and *F*. Sides $\overline{AB}$, $\overline{BC}$, and $\overline{AC}$ in triangle *ABC* have corresponding sides $\overline{DE}$, $\overline{EF}$, and $\overline{DF}$, respectively, in triangle *DEF*.

> Two polygons are **similar** if their corresponding angles have the same measure and the lengths of their corresponding sides are in proportion.

In Figure 9.20, $\angle A$ and $\angle D$ have the same measure, $\angle B$ and $\angle E$ have the same measure, and $\angle C$ and $\angle F$ have the same measure. Also, the lengths of corresponding sides of similar triangles are in proportion.

When we refer to the line segment *AB*, we place a line over the *AB* and write $\overline{AB}$. **When we refer to the *length* of a line segment, we do not place a bar above the two letters**. Thus, when we refer to the *length* of line segment $\overline{AB}$, we write *AB* (without the bar on top of the two letters). For example, if we write *AB* = 12, we are indicating the

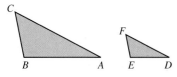

Figure 9.20

length of line segment $\overline{AB}$ is 12. When setting up proportions, as is shown below, we will be comparing the lengths of corresponding sides of similar figures, and so we will not be using the bar above the letters of the line segments. The proportion below shows that the lengths of the corresponding sides of the similar triangles in Figure 9.20 are in proportion.

$$\frac{AB}{DE} = \frac{BC}{EF} = \frac{AC}{DF}$$

EXAMPLE ❷   *Similar Figures*

Consider the similar figures in Fig. 9.21.

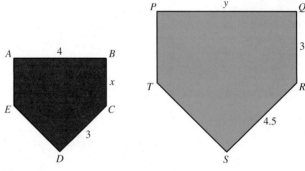

Figure 9.21

Determine

a) the length of side $\overline{BC}$.   b) the length of side $\overline{PQ}$.

SOLUTION

a) We will represent the length of side $\overline{BC}$ with the variable x. Because the corresponding sides of similar figures must be in proportion, we can write a proportion (as explained in Section 6.2) to determine the length of side $\overline{BC}$. Corresponding sides $\overline{CD}$ and $\overline{RS}$ are known, so we use them as one ratio in the proportion. The side corresponding to $\overline{BC}$ is $\overline{QR}$.

$$\frac{BC}{QR} = \frac{CD}{RS}$$
$$\frac{x}{3} = \frac{3}{4.5}$$

Now we solve for x.

$$x \cdot 4.5 = 3 \cdot 3$$
$$4.5x = 9$$
$$x = 2$$

Thus, the length of side $\overline{BC}$ is 2 units.

b) We will represent the length of side $\overline{PQ}$ with the variable y. The side corresponding to $\overline{PQ}$ is $\overline{AB}$. We will work part (b) in a manner similar to part (a).

$$\frac{PQ}{AB} = \frac{RS}{CD}$$
$$\frac{y}{4} = \frac{4.5}{3}$$

$$y \cdot 3 = 4 \cdot 4.5$$
$$3y = 18$$
$$y = 6$$

Thus, the length of side $\overline{PQ}$ is 6 units.

EXAMPLE ❸ *Using Similar Triangles to Find the Height of a Tree*

Saraniti Walker plans to remove a tree from her back yard. She needs to know the height of the tree. Saraniti is 5 ft tall and determines that when her shadow is 8 ft long, the shadow of the tree is 50 ft long (see Fig. 9.22). How tall is the tree?

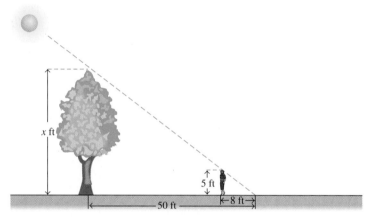

Figure 9.22

SOLUTION We will let x represent the height of the tree. From Fig. 9.22, we can see that the triangle formed by the sun's rays, Saraniti, and her shadow is similar to the triangle formed by the sun's rays, the tree, and its shadow. To find the height of the tree, we will set up and solve the following proportion:

$$\frac{\text{Height of the tree}}{\text{Height of Saraniti}} = \frac{\text{length of tree's shadow}}{\text{length of Saraniti's shadow}}$$

$$\frac{x}{5} = \frac{50}{8}$$

$$8x = 250$$

$$x = 31.25$$

Therefore, the tree is 31.25 ft tall.

Congruent Figures

If the corresponding sides of two similar figures are the same length, the figures are called *congruent figures*. Corresponding angles of congruent figures have the same measure, and the corresponding sides are equal in length. Two congruent figures coincide when placed one upon the other.

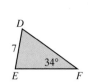

Figure 9.23

┌─ **EXAMPLE ❹** *Congruent Triangles*

Triangles *ABC* and *DEF* in Fig. 9.23 are congruent. Determine

a) the length of side $\overline{DF}$.   b) the length of side $\overline{AB}$.

c) $m\angle FDE$.   d) $m\angle ACB$.

e) $m\angle ABC$.

SOLUTION Because $\triangle ABC$ is congruent to $\triangle DEF$, we know that the corresponding side lengths are equal and corresponding angle measures are equal.

a) $DF = AC = 12$

b) $AB = DE = 7$

c) $m\angle FDE = m\angle CAB = 65°$

d) $m\angle ACB = m\angle DFE = 34°$

e) The sum of the angles of a triangle is 180°. Since $m\angle BAC = 65°$ and
   $m\angle ACB = 34°$, $m\angle ABC = 180° - 65° - 34° = 81°$.   ●

Earlier we learned that *quadrilaterals* are four-sided polygons, the sum of whose interior angles is 360°. Quadrilaterals may be classified according to their characteristics, as illustrated in the summary box below.

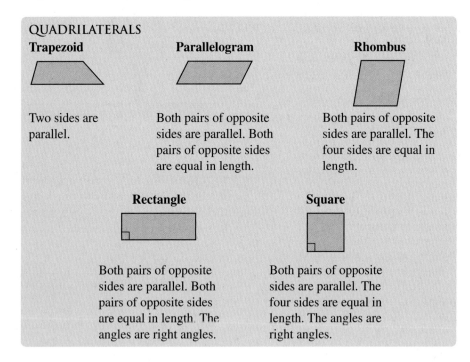

QUADRILATERALS

Trapezoid

Two sides are parallel.

Parallelogram

Both pairs of opposite sides are parallel. Both pairs of opposite sides are equal in length.

Rhombus

Both pairs of opposite sides are parallel. The four sides are equal in length.

Rectangle

Both pairs of opposite sides are parallel. Both pairs of opposite sides are equal in length. The angles are right angles.

Square

Both pairs of opposite sides are parallel. The four sides are equal in length. The angles are right angles.

┌─ **EXAMPLE ❺** *Angles of a Trapezoid*

Trapezoid *ABCD* is shown in Fig. 9.24.

a) Determine the measure of the interior angle, *x*.

b) Determine the measure of the exterior angle, *y*.

SOLUTION

a) We know that each of the two right angles in trapezoid *ABCD* has a measure of 90°. We also know that the sum of the interior angles in any quadrilateral is 360°. Therefore, we have

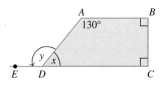

Figure 9.24

$$m\angle DAB + m\angle ABC + m\angle BCD + m\angle x = 360°$$
$$130° + 90° + 90° + m\angle x = 360°$$
$$310° + m\angle x = 360°$$
$$m\angle x = 50°$$

Thus, the measure of the interior angle, x, is 50°.

b) Angle x and angle y are supplementary angles. Therefore, $m\angle x + m\angle y = 180°$ and $m\angle y = 180° - m\angle x = 180° - 50° = 130°$. Thus, the measure of the exterior angle, y, is 130°. ●

Note that in Example 5 part (b) we could also have determined the measure of angle y as follows. By the definition of a trapezoid, sides $\overline{AB}$ and $\overline{CD}$ must be parallel. Therefore, side $\overline{AD}$ may be considered a transversal and $\angle BAD$ and $\angle ADE$ are alternate interior angles. Recall from Section 9.1 that alternate interior angles are equal. Thus, $m\angle BAD = m\angle ADE$ and $m\angle y = 130°$.

SECTION 9.2 EXERCISES

CONCEPT/WRITING EXERCISES

1. What is a polygon?

2. What distinguishes regular polygons from other polygons?

3. List six different types of triangles and in your own words describe the characteristics of each.

4. List five different types of quadrilaterals and in your own words describe the characteristics of each.

5. What are congruent figures?

6. What are similar figures?

In Exercises 7–14, (a) name the polygon. If the polygon is a quadrilateral, give its specific name. (b) State whether or not the polygon is a regular polygon.

7.

8.

9.

10.

11.

12.

13.

14.

In Exercises 15–22, identify the triangle as (a) scalene, isosceles, or equilateral and as (b) acute, obtuse, or right. The parallel markings (the two small parallel lines) on two or more sides indicate that the marked sides are of equal length.

15.

16.

17.

18.

19.

20.

21.

22.

In Exercises 23–28, identify the quadrilateral.

23.

24.

25.

26.

27.

28.

In Exercises 29–32, find the measure of $\angle x$.

29.

30.

31.

32.

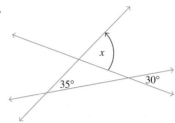

In Exercises 33–34, lines l_1 and l_2 are parallel. Determine the measures of $\angle 1$ through $\angle 12$.

33.

34.

In Exercises 35–40, determine the sum of the measures of the interior angles of the indicated polygon.

35. Hexagon

36. Heptagon

37. Octagon

38. Decagon

39. Icosagon

40. Dodecagon

In Exercises 41–46, (a) determine the measure of an interior angle of the named regular polygon. (b) If a side of the polygon is extended, determine the supplementary angle of an interior angle. See Example 1.

41. Triangle   **42.** Quadrilateral

43. Pentagon   **44.** Nonagon

45. Decagon   **46.** Icosagon

In Exercises 47–52, the figures are similar. Find the length of side x and side y.

47.

48.

49.

50.

51.

52.
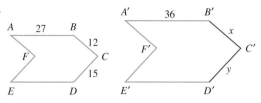

In Exercises 53–56, triangles ABC and DEC are similar figures. Determine the length of

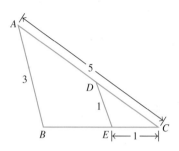

53. side $\overline{BC}$.   **54.** side $\overline{DC}$.

55. side $\overline{AD}$.   **56.** side $\overline{BE}$.

In Exercises 57–62, find the length of the sides and the measures of the angles for the congruent triangles ABC and A'B'C'.

57. The length of side $\overline{AC}$

58. The length of side $\overline{A'B'}$

59. The length of side $\overline{B'C'}$

60. $\angle B'A'C'$

61. $\angle ACB$

62. $\angle ABC$

In Exercises 63–68, determine the length of the sides and the measures of the angles for the congruent quadrilaterals ABCD and A′B′C′D′.

63. The length of side $\overline{AD}$

64. The length of side $\overline{B'C'}$

65. The length of side $\overline{A'B'}$

66. $\measuredangle BCD$ **67.** $\measuredangle A'D'C'$

68. $\measuredangle DAB$

PROBLEM SOLVING

In Exercises 69–72, determine the measure of the angle. In the figure, $\measuredangle ABC$ makes an angle of 125° with the floor and l_1 and l_2 are parallel.

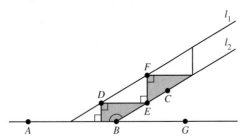

69. $\measuredangle GBC$ **70.** $\measuredangle EDF$

71. $\measuredangle DFE$ **72.** $\measuredangle DEC$

73. *Height of a Silo* Steve Runde is buying a farm and needs to determine the height of a silo on the farm. Steve, who is 6 ft tall, notices that when his shadow is 9 ft long, the shadow of the silo is 105 ft long (see diagram). How tall is the silo? Note that the diagram is not to scale.

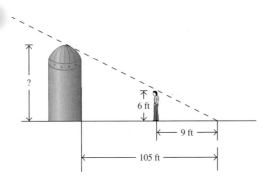

74. *Angles on a Picnic Table* The legs of a picnic table form an isosceles triangle as indicated in the figure. If $\measuredangle ABC = 80°$, determine $m\measuredangle x$ and $m\measuredangle y$ so that the top of the table will be parallel to the ground.

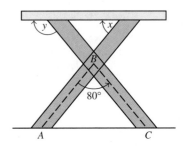

75. *Distances in Minnesota* A triangle can be formed by drawing line segments on a map of Minnesota connecting the cities of Austin, Rochester, and St. Paul (see figure). If the actual distance from Austin to Rochester is approximately 44 miles, use the lengths of the line segments indicated in the figure along with similar triangles to approximate

a) the actual distance from St. Paul to Austin.

b) the actual distance from St. Paul to Rochester.

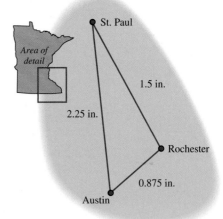

76. *Distances in Illinois* A triangle can be formed by drawing line segments on a map of Illinois connecting the cities of Rockford, Chicago, and Bloomington (see figure on top of page 537). If the actual distance from Chicago to Rockford is approximately 90 miles, use the lengths of the line segments indicated in the figure along with similar triangles to approximate

a) the actual distance from Chicago to Bloomington.

b) the actual distance from Bloomington to Rockford.

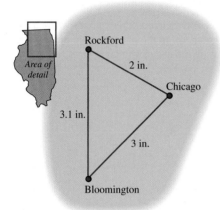

CHALLENGE PROBLEMS/GROUP ACTIVITIES

Scaling Factor Examine the similar triangles ABC and
A′B′C′ in the figure below.

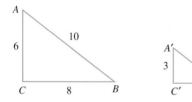

If we calculate the ratios $\dfrac{AB}{A'B'}$, $\dfrac{BC}{B'C'}$, *and* $\dfrac{CA}{C'A'}$, *we see*
that each of these ratios is equal to 2. We call this common
ratio the **scaling factor** *of* $\triangle ABC$ *with respect to* $\triangle A'B'C'$.
If we calculate the reciprocal ratios $\dfrac{A'B'}{AB}$, $\dfrac{B'C'}{BC}$, *and* $\dfrac{C'A'}{CA}$,
we see that each of these ratios is equal to $\frac{1}{2}$. *We call this*
common ratio the scaling factor of $\triangle A'B'C'$ *with respect to*
$\triangle ABC$. *Every pair of similar figures has two scaling fac-*
tors that show the relationship between the corresponding
side lengths. Notice that the length of each side of $\triangle ABC$ *is*
two times the length of the corresponding side in $\triangle A'B'C'$.
We can also state that the length of each side of $\triangle A'B'C'$ *is*
one-half the length of the corresponding side of $\triangle ABC$.

77. In the figure, $\triangle DEF$ is similar to $\triangle D'E'F'$. The length
of the sides of $\triangle DEF$ is shown in the figure. If the scaling

factor of $\triangle DEF$ with respect to $\triangle D'E'F'$ is 3, determine
the length of the sides of triangle $\triangle D'E'F'$.

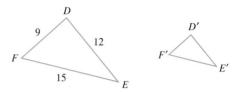

78. In the figure, quadrilateral $EFGH$ is similar to quadrilateral
$E'F'G'H'$. The length of the sides of quadrilateral $EFGH$
is shown in the figure. If the scaling factor of quadrilateral
$E'F'G'H'$ with respect to quadrilateral $EFGH$ is $\frac{1}{3}$, determine the length of the sides of quadrilateral $E'F'G'H'$.

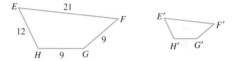

79. **Height of a Wall** You are asked to measure the height of
an inside wall of a warehouse. No ladder tall enough to
measure the height is available. You borrow a mirror from
a salesclerk and place it on the floor. You then move away
from the mirror until you can see the reflection of the top
of the wall in it, as shown in the figure.

a) Explain why triangle HFM is similar to triangle TBM.
(*Hint:* In the reflection of light the angle of incidence
equals the angle of reflection. Thus, $\angle HMF = \angle TMB$.)

b) If your eyes are $5\frac{1}{2}$ ft above the floor and you are $2\frac{1}{2}$ ft
from the mirror and the mirror is 20 ft from the wall,
how high is the wall?

RECREATIONAL MATHEMATICS

80. *Distance Across a Lake*

   a) In the figure, $m\angle CED = m\angle ABC$. Explain why triangles *ABC* and *DEC* must be similar.

   b) Determine the distance across the lake, *DE*.

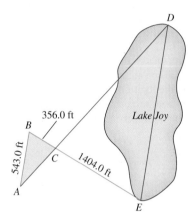

INTERNET/RESEARCH ACTIVITIES

81. Using the Internet and history of mathematics texts, write a paper on the history and use of the theodolite, a surveying instrument.

82. Using the Internet and other sources, write a paper on the use of geometry in the photographic process. Include discussions on the use of similar figures.

9.3   PERIMETER AND AREA

▲ We will study the perimeter and area of rectangles, such as those on which basketball games are played.

If you were to walk around the outside edge of a basketball court, how far would you walk? If you had to place floor tiles that each measured 1 foot by 1 foot on a basketball court, how many tiles would you need? In this section, we will study the geometric concepts of perimeter and area that are used to answer these and other questions.

Perimeter and Area

The *perimeter*, *P*, of a two-dimensional figure is the sum of the lengths of the sides of the figure. In Figs. 9.25 and 9.26, the sums of the lengths of the red line segments are the perimeters. Perimeters are measured in the same units as the sides. For example, if the sides of a figure are measured in feet, the perimeter will be measured in feet.

Figure 9.25

Figure 9.26

The *area*, A, is the region within the boundaries of the figure. The blue color in Figs. 9.25 and 9.26 indicates the areas of the figures. Area is measured in square units. For example, if the sides of a figure are measured in inches, the area of the figure will be measured in square inches (in.^2). (See Table 8.7 on page 501 for common units of area in the U.S. customary and metric systems.)

Consider the rectangle in Fig. 9.26. Two sides of the rectangle have length l, and two sides of the rectangle have width w. Thus, if we add the lengths of the four sides to get the perimeter, we find $P = l + w + l + w = 2l + 2w$.

PERIMETER OF A RECTANGLE

$$P = 2l + 2w$$

Consider a rectangle of length 5 units and width 3 units (Fig. 9.27). Counting the number of 1-unit by 1-unit squares within the figure we obtain the area of the rectangle, 15 square units. The area can also be obtained by multiplying the number of units of length by the number of units of width, or 5 units × 3 units = 15 square units. We can find the area of a rectangle by the formula area = length × width.

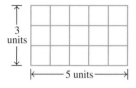

3 units

5 units

Figure 9.27

AREA OF A RECTANGLE

$$A = l \times w$$

Using the formula for the area of a rectangle, we can determine the formulas for the areas of other figures.

A square (Fig. 9.28) is a rectangle that contains four equal sides. Therefore, the length equals the width. If we call both the length and the width of the square s, then

$$A = l \times w, \qquad \text{so} \qquad A = s \times s = s^2$$

s

s s

s

Figure 9.28

AREA OF A SQUARE

$$A = s^2$$

A parallelogram with height h and base b is shown in Fig. 9.29(a).

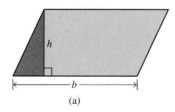

h h

b b

(a)   (b)

Figure 9.29

If we were to cut off the red portion of the parallelogram on the left, Fig. 9.29(a), and attach it to the right side of the figure, the resulting figure would be a rectangle, Fig. 9.29(b). Since the area of the rectangle is $b \times h$, the area of the parallelogram is also $b \times h$.

AREA OF A PARALLELOGRAM

$$A = b \times h$$

Consider the triangle with height, h, and base, b, shown in Fig. 9.30(a). Using this triangle and a second identical triangle, we can construct a parallelogram, Fig. 9.30(b). The area of the parallelogram is bh. The area of the triangle is one-half that of the parallelogram. Therefore, the area of the triangle is $\frac{1}{2}$(base)(height).

(a)

(b)

Figure 9.30

AREA OF A TRIANGLE

$$A = \tfrac{1}{2}bh$$

Now consider the trapezoid shown in Fig. 9.31(a). We can partition the trapezoid into two triangles by drawing diagonal $\overline{DB}$, as in Fig. 9.31(b). One triangle has base $\overline{AB}$ (called b_2) with height $\overline{DE}$, and the other triangle has base $\overline{DC}$ (called b_1) with height $\overline{FB}$. Note that the line used to measure the height of the triangle need not be inside the triangle. Because heights $\overline{DE}$ and $\overline{FB}$ are equal, both triangles have the same height, h. The area of triangles DCB and ADB are $\frac{1}{2}b_1h$ and $\frac{1}{2}b_2h$, respectively. The area of the trapezoid is the sum of the areas of the triangles, $\frac{1}{2}b_1h + \frac{1}{2}b_2h$, which can be written $\frac{1}{2}h(b_1 + b_2)$.

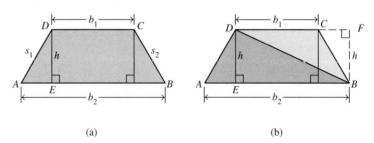

(a)   (b)

Figure 9.31

AREA OF A TRAPEZOID

$$A = \tfrac{1}{2}h(b_1 + b_2)$$

Following is a summary of the perimeters and areas of selected figures.

PERIMETERS AND AREAS

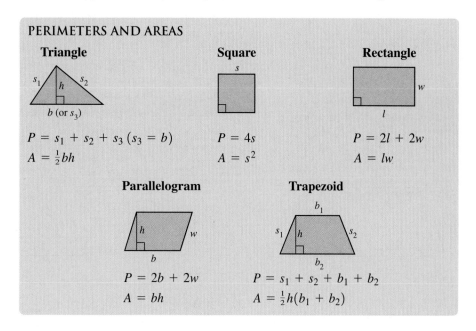

Triangle

$P = s_1 + s_2 + s_3 \ (s_3 = b)$
$A = \frac{1}{2}bh$

Square

$P = 4s$
$A = s^2$

Rectangle

$P = 2l + 2w$
$A = lw$

Parallelogram

$P = 2b + 2w$
$A = bh$

Trapezoid

$P = s_1 + s_2 + b_1 + b_2$
$A = \frac{1}{2}h(b_1 + b_2)$

EXAMPLE ❶ *Sodding a Backyard*

Kevin Geis wishes to replace the grass (sod) in his backyard. One pallet of Milberger's sod costs $130 and covers 450 square feet. If Kevin's backyard is rectangular in shape with a length of 90 feet and width of 60 feet, determine

a) the area of the backyard.

b) how many pallets of sod Kevin needs to purchase.

c) the cost of the sod purchased.

SOLUTION

a) The area of the backyard is

$$A = l \cdot w = 90 \cdot 60 = 5400 \text{ ft}^2$$

The area of the backyard is in square feet because both the length and width are measured in feet.

b) To determine the number of pallets of sod Kevin needs, divide the area of the backyard by the area covered by one pallet of sod.

$$\frac{\text{Area of backyard}}{\text{Area covered by one pallet}} = \frac{5400}{450} = 12$$

Kevin needs to purchase 12 pallets of sod.

c) The cost of 12 pallets of sod is 12 × $130, or $1560. ●

Pythagorean Theorem

We introduced the Pythagorean theorem in Chapter 5. Because this theorem is an important tool for finding the perimeter and area of triangles, we restate it here.

PYTHAGOREAN THEOREM

The sum of the squares of the lengths of the legs of a right triangle equals the square of the length of the hypotenuse.

$$\text{leg}^2 + \text{leg}^2 = \text{hypotenuse}^2$$

Symbolically, if a and b represent the lengths of the legs and c represents the length of the hypotenuse (the side opposite the right angle), then

$$a^2 + b^2 = c^2$$

EXAMPLE ❷   *Crossing a Moat*

The moat surrounding a castle is 18 ft wide and the wall by the moat of the castle is 24 ft high (see Fig. 9.32). If an invading army wishes to use a ladder to cross the moat and reach the top of the wall, how long must the ladder be?

Figure 9.32

SOLUTION   The moat, the castle wall, and the ladder form a right triangle. The moat and the castle wall form the legs of the triangle (sides a and b), and the ladder forms the hypotenuse (side c). By the Pythagorean theorem,

$$c^2 = a^2 + b^2$$
$$c^2 = (18)^2 + (24)^2$$
$$c^2 = 324 + 576$$
$$c^2 = 900$$
$$\sqrt{c^2} = \sqrt{900} \qquad \text{Take the square root of both sides of the equation.}$$
$$c = 30$$

Therefore, the ladder would need to be at least 30 ft long.

Circumference
$C = 2\pi r$

Radius

Area
$A = \pi r^2$

Diameter

Figure 9.33

Circles

A commonly used plane figure that is not a polygon is a *circle*. A *circle* is a set of points equidistant from a fixed point called the center. A *radius*, *r*, of a circle is a line segment from the center of the circle to any point on the circle (Fig. 9.33). A *diameter*, *d*, of a circle is a line segment through the center of a circle with both end points on the circle. Note that the diameter of the circle is twice its radius. The *circumference* is the length of the simple closed curve that forms the circle. The formulas for the area and circumference of a circle are given in Fig. 9.33. The symbol pi, π, was introduced in Chapter 5. Recall that π is approximately 3.14. If your calculator contains a $\boxed{\pi}$ key, you should use that key when working calculations involving pi.

⌐**EXAMPLE ❸** *Comparing Pizzas*

Victoria Montoya wishes to order a large cheese pizza. She can choose among three pizza parlors in town: Antonio's, Brett's, and Dorsey's. Antonio's large cheese pizza is a round 16-in.-diameter pizza that sells for $15. Brett's large cheese pizza is a round 14-in.-diameter pizza that sells for $12. Dorsey's large cheese pizza is a square 12-in. by 12-in. pizza that sells for $10. All three pizzas have the same thickness. To get the most for her money, from which pizza parlor should Victoria order her pizza?

SOLUTION To determine the best value, we will calculate the cost per square inch of pizza for each of the three pizzas. To do so, we will divide the cost of each pizza by its area. The areas of the two round pizzas can be determined using the formula for the area of a circle, $A = \pi r^2$. Since the radius is half the diameter, we will use $r = 8$ and $r = 7$ for Antonio's and Brett's large pizzas, respectively. The area for the square pizza can be determined using the formula for the area of a square, $A = s^2$. We will use $s = 12$.

$$\text{Area of Antonio's pizza} = \pi r^2 \approx (3.14)(8)^2 \approx 3.14(64) \approx 200.96 \text{ in.}^{2*}$$
$$\text{Area of Brett's pizza} = \pi r^2 \approx (3.14)(7)^2 \approx 3.14(49) \approx 153.86 \text{ in.}^{2*}$$
$$\text{Area of Dorsey's pizza} = s^2 = (12)^2 = 144 \text{ in.}^2$$

Now, to find the cost per square inch of pizza, we will divide the cost of the pizza by the area of the pizza.

$$\text{Cost per square inch of Antonio's pizza} \approx \frac{\$15}{200.96 \text{ in.}^2} \approx \$0.0746$$

Thus, Antonio's pizza costs about $0.0746, or about 7.5 cents, per square inch.

$$\text{Cost per square inch of Brett's pizza} \approx \frac{\$12}{153.86 \text{ in.}^2} \approx \$0.0780$$

Thus, Brett's pizza costs about $0.0780, or about 7.8 cents, per square inch.

$$\text{Cost per square inch of Dorsey's pizza} = \frac{\$10}{144 \text{ in.}^2} \approx \$0.0694$$

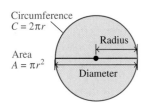
_____

*If you use the $\boxed{\pi}$ key on your calculator, your answers will be slightly more accurate.

Thus, Dorsey's pizza costs about $0.0694, or about 6.9 cents, per square inch.

Since the cost per square inch of pizza is the lowest for Dorsey's pizza, Victoria would get the most pizza for her money by ordering her pizza from Dorsey's.   ●

EXAMPLE ④   *Applying Lawn Fertilizer*

Steve May plans to fertilize his lawn. The shapes and dimensions of his lot, house, driveway, pool, and rose garden are shown in Fig. 9.34. One bag of fertilizer costs $29.95 and covers 5000 ft². Determine how many bags of fertilizer Steve needs and the total cost of the fertilizer.

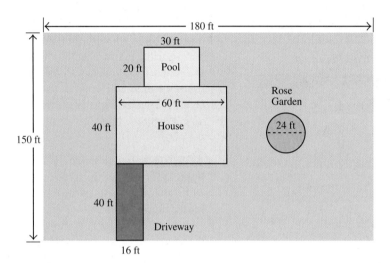

Figure 9.34

SOLUTION The total area of the lot is $150 \cdot 180$, or 27,000 ft². To determine the area to be fertilized, subtract the area of house, driveway, pool, and rose garden from the total area.

$$\text{Area of house} = 60 \cdot 40 = 2400 \text{ ft}^2$$
$$\text{Area of driveway} = 40 \cdot 16 = 640 \text{ ft}^2$$
$$\text{Area of pool} = 20 \cdot 30 = 600 \text{ ft}^2$$

The diameter of the rose garden is 24 ft, so its radius is 12 ft.

$$\text{Area of rose garden} = \pi r^2 = \pi(12)^2 \approx 3.14(144) \approx 452.16 \text{ ft}^2$$

The total area of the house, driveway, pool, and rose garden is approximately $2400 + 640 + 600 + 452.16$, or 4092.16 ft². The area to be fertilized is $27,000 - 4092.16$ ft², or 22,907.84 ft². The number of bags of fertilizer is found by dividing the total area to be fertilized by the number of square feet covered per bag.

The number of bags of fertilizer is $\dfrac{22,907.84}{5000}$, or about 4.58 bags. Therefore, Steve needs five bags. At $29.95 per bag, the total cost is $5 \times \$29.95$, or $149.75.   ●

EXAMPLE ❺ *Converting Between Square Feet and Square Inches*

a) Convert 1 ft^2 to square inches.

b) Convert 37 ft^2 to square inches.

c) Convert 432 in.^{2} to square feet.

d) Convert 2196 in.^{2} to square feet.

SOLUTION

a) 1 ft = 12 in. Therefore, 1 ft^2 = 12 in. × 12 in. = 144 in.^{2}.

b) From part (a), we know that 1 ft^2 = 144 in.^{2}. Therefore, 37 ft^2 = 37 × 144 in.^{2} = 5328 in.^{2}.

c) In part (b), we converted from square feet to square inches by *multiplying* the number of square feet by 144. Now, to convert from square inches to square feet we will *divide* the number of square inches by 144. Therefore, 432 in.^{2} = $\frac{432}{144}$ ft^2 = 3 ft^2.

d) As in part (c), we will divide the number of square inches by 144. Therefore, 2196 in.^{2} = $\frac{2196}{144}$ ft^2 = 15.25 ft^2. ●

EXAMPLE ❻ *Installing Ceramic Tile*

Debra Levy wishes to purchase ceramic tile for her family room, which measures 30 ft × 27 ft. The cost of the tile, including installation, is $21 per square yard.

a) Find the area of Debra's family room in square *yards*.

b) Determine Debra's cost of the ceramic tile for her family room.

SOLUTION

a) The area of the family room in square feet is 30 · 27 = 810 ft^2. Since 1 yd = 3 ft, 1 yd^2 = 3 ft × 3 ft = 9 ft^2. To find the area of the family room in square yards, divide the area in square feet by 9 ft^2.

$$\text{Area in square yards} = \frac{810}{9} = 90$$

Therefore, the area is 90 yd^2.

b) The cost of 90 yd^2 of ceramic tile, including installation, is 90 · $21 = $1890. ●

When multiplying units of length, be sure that the units are the same. You can multiply feet by feet to get square feet or yards by yards to get square yards. However, you cannot get a valid answer if you multiply numbers expressed in feet by numbers expressed in yards.

SECTION 9.3 EXERCISES

CONCEPT/WRITING EXERCISES

1. **a)** Describe in your own words how to determine the *perimeter* of a two-dimensional figure.

   b) Describe in your own words how to determine the *area* of a two-dimensional figure.

   c) Draw a rectangle with a length of 6 units and a width of 2 units. Determine the area and perimeter of this rectangle.

2. What is the relationship between the *radius* and the *diameter* of a circle?

3. a) How do you convert an area from square yards into square feet?

b) How do you convert an area from square feet into square yards?

4. a) How do you convert an area from square feet into square inches?

b) How do you convert an area from square inches into square feet?

PRACTICE THE SKILLS

In Exercises 5–8, find the area of the triangle.

5.

5 in.
7 in.

6.

6 yd
2 ft

7.

5 cm
7 cm

8.

√3 m
2 m

In Exercises 9–14, determine (a) the area and (b) the perimeter of the quadrilateral.

9.

10 ft
21 ft

10.

7 in.
9 in.
14 in.

11.

20 cm
27 cm
3 m

12.

2 yd
6 ft

13.

5 in.
25 in.   2 ft   25 in.
19 in.

14.

16 in.
13 in.   12 in.   13 in.
6 in.

In Exercises 15–18, determine (a) the area and (b) the circumference of the circle. Use the $\boxed{\pi}$ key on your calculator and round your answer to the nearest hundredth.

15.

4 m

16.

28 yd

17.

13 ft

18.

13 mm

In Exercises 19–22, (a) use the Pythagorean theorem to determine the length of the unknown side of the triangle, (b) determine the perimeter of the triangle, and (c) determine the area of the triangle.

19.

c   15 yd
8 yd

20.

15 in.   12 in.
a

21.

5 km   13 km
b

22.

10 cm   24 cm
c

PROBLEM SOLVING

In Exercises 23–32, find the shaded area. When appropriate, use the $\boxed{\pi}$ key on your calculator and round your answer to the nearest hundredth.

23.

5 m

24.

3 cm
4 cm

25.

4 in.

4 in.

26.

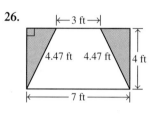

3 ft
4.47 ft   4.47 ft
4 ft
7 ft

27.

9 yd   10 yd   11 yd

18 yd

28.

5 in.
10 in.

29.

20 in.
4 in.
9 in.

30.

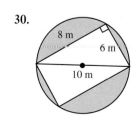

8 m
6 m
10 m

31.

28 cm

32.

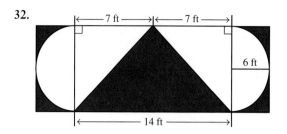

7 ft   7 ft
6 ft
14 ft

One square yard equals 9 ft². Use this information to convert the following.

33. 207 ft² to square yards

34. 15.2 ft² to square yards

35. 14.7 yd² to square feet

36. 15.2 yd² to square feet

One square meter equals 10,000 cm². Use this information to convert the following.

37. 23.4 m² to square centimeters

38. 0.375 m² to square centimeters

39. 8625 cm² to square meters

40. 608 cm² to square meters

Nancy Wallin has just purchased a new house that is in need of new flooring. In Exercises 41–46, use the measurements given on the floor plans of Nancy's house to obtain the answer.

20′

Bedroom
10′ × 20′

28′

Bath
8′ × 14′

Bedroom
10′ × 14′

Stairs

Second floor

42′
20′

Laundry

Kitchen
12′ × 14′

Living/Dining
25′ × 22′

25′

28′

Bath
6′ × 10′

Bedroom
10′ × 14′

Stairs

First floor

41. *Cost of Hardwood Flooring* The cost of Mannington Chestnut hardwood flooring is $10.86 per square foot if Nancy installs the flooring herself or $13.86 per square foot if she has the flooring installed by the flooring company. Determine the cost for hardwood flooring in the living/dining room if

a) Nancy installs it herself.

b) Nancy has it installed by the flooring company.

42. *Cost of Laminate Flooring* The cost of Pergo Select Helsinki Birch laminate flooring is $5.89 per square foot if Nancy installs the flooring herself or $8.89 per square foot if she has the flooring installed by the flooring company. Determine the cost for the flooring in the living/dining room if

a) Nancy installs it herself.

b) Nancy has it installed by the flooring company.

43. *Cost of Ceramic Tile* The cost of Mohawk Porcelain ceramic tile is $8.50 per square foot. This price includes the cost of installation. Determine the cost for Nancy to have this ceramic tile installed in the kitchen and in both bathrooms.

44. *Cost of Linoleum* The cost of Armstrong Solarian Woodcut linoleum is $5.00 per square foot. This price includes the cost of installation. Determine the cost for Nancy to have this linoleum installed in the kitchen and in both bathrooms.

45. *Cost of Berber Carpeting* The cost of Bigelow Commodore Berber carpeting is $6.06 per square foot. This price includes the cost of installation. Determine the cost for Nancy to have this carpeting installed in all three bedrooms.

46. *Cost of Saxony Carpeting* The cost of DuPont Stainmaster Saxony carpeting is $5.56 per square foot. This price includes the cost of installation. Determine the cost for Nancy to have this carpeting installed in all three bedrooms.

47. *Cost of a Lawn Service* Clarence and Rose Cohen's home lot is illustrated here. Clarence and Rose wish to hire Picture Perfect Lawn Service to cut their lawn. How much will it cost Clarence and Rose to have their lawn cut if Picture Perfect charges $0.02 per square yard?

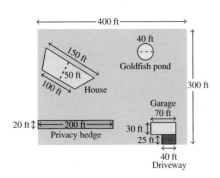

48. *Cost of a Lawn Service* Jim and Wendy Scott's home lot is illustrated here. The Scotts wish to hire a lawn service to cut their lawn. M&M Lawn Service charges $0.02 per square yard of lawn. How much will it cost the Scotts to have their lawn cut?

49. *Area of a Basketball Court* A National Basketball Association (NBA) basketball court is a rectangle that is 94 ft long and 50 ft wide.

a) If you were to walk around the outside edge of a basketball court, how far would you walk?

b) If you had to place floor tiles that each measured 1 foot by 1 foot on a basketball court, how many tiles would you need?

50. *Quartz Countertops* Larry Shedden wishes to have three Cambria Windsor quartz countertops installed in his new kitchen. The countertops are rectangular and have the following dimensions: $3\frac{1}{2}$ ft × 6 ft, $2\frac{1}{2}$ ft × 8 ft, and 3 ft × $11\frac{1}{2}$ ft. The cost of the countertops is $87 per square foot, which includes the cost of installation.

a) Determine the total area of the three countertops.

b) Determine the total cost to have all three countertops installed.

51. *Ladder on a Wall* Lorrie Morgan places a 29 ft ladder against the side of a building with the bottom of the ladder 20 ft away from the building (see figure). How high up on the wall does the ladder reach?

52. *Docking a Boat* Brian Murphy is bringing his boat into a dock that is 9 ft above the water level (see figure). If a 41 ft rope is attached to the dock on one side and to the boat on the other side, determine the horizontal distance from the dock to the boat.

9 ft
41 ft rope

53. *The Green Monster* In Fenway Park, home of baseball's Boston Red Sox, the left field wall is known as the *Green Monster*. The distance from home plate down the third baseline to the bottom of the wall is 310 feet (see photo). In left field, at the end of the baseline, the Green Monster is perpendicular to the ground and is 37 feet tall. Determine the distance from home plate to the top of the Green Monster along the third baseline. Round your answer to the nearest foot.

CHALLENGE PROBLEMS/GROUP ACTIVITIES

54. *Plasma Television* The screen of a plasma television is in the shape of a rectangle with a diagonal of length 43 in. If the height of the screen is 21 in., determine the width of the screen.

55. *Doubling the Sides of a Square* In the figure below, an original square with sides of length s is shown. Also shown is a larger square with sides double in length, or $2s$.

a) Express the area of the original square in terms of s.

b) Express the area of the larger square in terms of s.

c) How many times larger is the area of the square in part (b) than the area of the square in part (a)?

56. *Doubling the Sides of a Parallelogram* In the figure below, an original parallelogram with base b and height h is shown. Also shown is a larger parallelogram with base and height double in length, or $2b$ and $2h$, respectively.

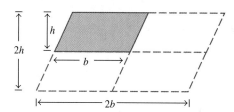

a) Express the area of the original parallelogram in terms of b and h.

b) Express the area of the larger parallelogram in terms of b and h.

c) How many times larger is the area of the parallelogram in part (b) than the area of the parallelogram in part (a)?

57. *Heron's Formula* A second formula for determining the area of a triangle (called Heron's formula) is

$$A = \sqrt{s(s-a)(s-b)(s-c)}$$

where $s = \frac{1}{2}(a+b+c)$ and a, b, and c are the lengths of the sides of the triangle. Use Heron's formula to determine the area of right triangle ABC and check your answer using the formula $A = \frac{1}{2}ab$.

58. **Expansion of (a + b)²** In the figure, one side of the largest square has length $a + b$. Therefore, the area of the largest square is $(a + b)^2$. Answer the following questions to find a formula for the expansion of $(a + b)^2$.

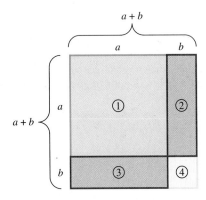

a) What is the area of the square marked ① ?

b) What is the area of the rectangle marked ② ?

c) What is the area of the rectangle marked ③ ?

d) What is the area of the square marked ④ ?

e) Add the four areas found in parts (a) through (d) to write a formula for the expansion of $(a + b)^2$.

RECREATIONAL MATHEMATICS

59. **Sports Areas** Use the Internet to research the official dimensions of the playing surface of your favorite sport and determine the area of the playing surface. Choices may include a basketball court, volleyball court, tennis court, racquetball court, hockey rink, football field, soccer field, lacrosse field, and a baseball diamond.

60. **Scarecrow's Error** In the movie *The Wizard of Oz*, once the scarecrow gets his diploma he states the following: "In an isosceles triangle, the sum of the square roots of the two equal sides is equal to the square root of the third side." Discuss why this statement is incorrect.

INTERNET/RESEARCH ACTIVITIES

For Exercises 61–63, references include the Internet, history of mathematics textbooks, and encyclopedias.

61. Research the proof of the Pythagorean theorem provided by President James Garfield. Write a brief paper and make a poster of this proof and the associated diagrams.

62. The early Babylonians and Egyptians did not know about π and had to devise techniques to approximate the area of a circle. Do research and write a paper on the techniques these societies used to approximate the area of a circle.

63. Write a paper on the contributions of Heron of Alexandria to geometry.

9.4 VOLUME AND SURFACE AREA

On the label of a can of Color Place paint is the following sentence: "One gallon will cover about 400 sq. ft." This sentence refers to the two main geometric topics that we will cover in this section. *One gallon* refers to volume of the paint in the can, and *400 sq. ft* refers to the surface area that the paint will cover.

▲ The amount of paint in a can refers to volume, and the amount of wall space that the paint covers refers to surface area.

When discussing a one-dimensional figure such as a line, we can find its length. When discussing a two-dimensional figure such as a rectangle, we can find its area and its perimeter. When discussing a three-dimensional figure such as a cube, we can find its volume and its surface area. *Volume* is a measure of the capacity of a three-dimensional figure. *Surface area* is the sum of the areas of the surfaces of a three-dimensional figure. Volume refers to the amount of material that you can put *inside* a three-dimensional figure, and surface area refers to the total area that is on the *outside* surface of the figure.

Solid geometry is the study of three-dimensional solid figures, also called *space figures*. Volumes of three-dimensional figures are measured in cubic units such as cubic feet or cubic meters. Surface areas of three-dimensional figures are measured in square units such as square feet or square meters.

Rectangular Solids, Cylinders, Cones, and Spheres

We will begin our discussion with the *rectangular solid*. If the length of the solid is 5 units, the width is 2 units, and the height is 3 units, the total number of cubes is 30 (Fig. 9.35). Thus, the volume is 30 cubic units. The volume of a rectangular solid can also be found by multiplying its length times width times height; in this case, 5 units × 2 units × 3 units = 30 cubic units. In general, the volume of any rectangular solid is $V = l \times w \times h$.

The surface area of the rectangular solid in Fig. 9.35 is the sum of the area of the surfaces of the rectangular solid. Notice that each surface of the rectangular solid is a rectangle. The left and right side of the rectangular solid each has an area of 5 units × 3 units, or 15 square units. The front and back sides of the rectangular solid each has an area of 2 units × 3 units, or 6 square units. The top and bottom sides of the rectangular solid each has an area of 5 units × 2 units, or 10 square units. Therefore, the surface area of the rectangular solid is $2(5 \times 3) + 2(2 \times 3) + 2(5 \times 2) = 2(15) + 2(6) + 2(10) = 30 + 12 + 20 = 62$ square units. In general, the surface area of any rectangular solid is $SA = 2lw + 2wh + 2lh$.

Figure 9.35

VOLUME AND SURFACE AREA OF A RECTANGULAR SOLID

$$V = lwh \qquad SA = 2lw + 2wh + 2lh$$

A *cube* is a rectangular solid with the same length, width, and height. If we call the length of the side of the cube s and use the volume and surface area formulas for a rectangular solid, substituting s in for l, w, and h, we obtain $V = s \cdot s \cdot s = s^3$ and $SA = 2 \cdot s \cdot s + 2 \cdot s \cdot s + 2 \cdot s \cdot s = 2s^2 + 2s^2 + 2s^2 = 6s^2$.

VOLUME AND SURFACE AREA OF A CUBE

$$V = s^3 \qquad SA = 6s^2$$

Now consider the right circular cylinder shown in Fig. 9.36(a) on page 552. When we use the term *cylinder* in this book, we mean a right circular cylinder. The volume of the cylinder is found by multiplying the area of the circular base, πr^2, by the height, h, to get $V = \pi r^2 h$.

The surface area of the cylinder is the sum of the area of the top, the area of the bottom, and the area of the side of the cylinder. Both the top and bottom of the cylinder are circles with an area of πr^2. To determine the area of the side of the cylinder, examine Fig. 9.36(b) and (c).

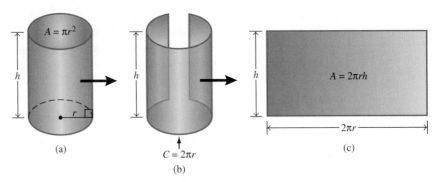

(a)

$C = 2\pi r$

(b)

(c)

Figure 9.36

Notice how when flattened, the side of the cylinder is a rectangle whose length is the circumference of the base of the cylinder and whose width is the height of the cylinder. Thus, the side of the cylinder has an area of $2\pi r \cdot h$, or $2\pi rh$. Therefore, the surface area of the cylinder is $\pi r^2 + \pi r^2 + 2\pi rh$, or $SA = 2\pi rh + 2\pi r^2$.

VOLUME AND SURFACE AREA OF A CYLINDER

$$V = \pi r^2 h \qquad SA = 2\pi rh + 2\pi r^2$$

Now consider the right circular cone illustrated in Fig. 9.37. When we use the term *cone* in this book, we mean a right circular cone. The volume of a cone is less than the volume of a cylinder that has the same base and the same height. In fact, the volume of the cone is one-third the volume of the cylinder. The formula for the surface area of a cone is the sum of the area of the circular base of the cone, πr^2, and the area of the side of the cone, $\pi r\sqrt{r^2 + h^2}$, or $SA = \pi r^2 + \pi r\sqrt{r^2 + h^2}$. The derivation of the area of the side of the cone is beyond the scope of this book.

Figure 9.37

VOLUME AND SURFACE AREA OF A CONE

$$V = \tfrac{1}{3}\pi r^2 h \qquad SA = \pi r^2 + \pi r\sqrt{r^2 + h^2}$$

The next shape we will discuss in this section is the *sphere*. Baseballs, tennis balls, and so on have the shape of a sphere. The formulas for the volume and surface area of a sphere are as follows. The derivation of the volume and surface area of a sphere are beyond the scope of this book.

VOLUME AND SURFACE AREA OF A SPHERE

$$V = \tfrac{4}{3}\pi r^3 \qquad SA = 4\pi r^2$$

Following is a summary of the formulas for the volumes and surface areas of the three-dimensional figures we have discussed thus far.

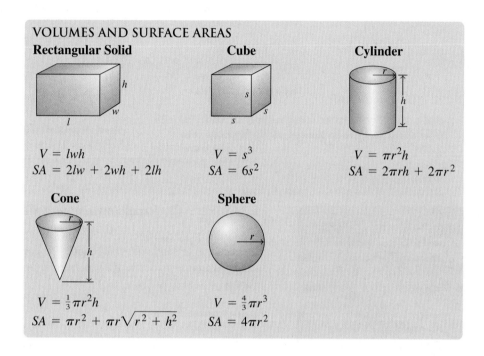

VOLUMES AND SURFACE AREAS

Rectangular Solid

$V = lwh$
$SA = 2lw + 2wh + 2lh$

Cube

$V = s^3$
$SA = 6s^2$

Cylinder

$V = \pi r^2 h$
$SA = 2\pi rh + 2\pi r^2$

Cone

$V = \frac{1}{3}\pi r^2 h$
$SA = \pi r^2 + \pi r\sqrt{r^2 + h^2}$

Sphere

$V = \frac{4}{3}\pi r^3$
$SA = 4\pi r^2$

EXAMPLE ❶ *Volume and Surface Area*

Determine the volume and surface area of each of the following three-dimensional figures. When appropriate, use the $\boxed{\pi}$ key on your calculator and round your answer to the nearest hundredths.

(a)

(b)

(c)

(d)

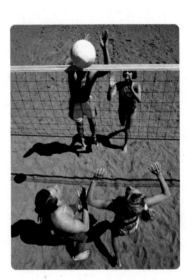

SOLUTION

a) $V = lwh = 8 \cdot 3 \cdot 4 = 96 \text{ ft}^3$
   $SA = 2lw + 2wh + 2lh = 2 \cdot 8 \cdot 3 + 2 \cdot 3 \cdot 4 + 2 \cdot 8 \cdot 4$
   $\qquad\qquad = 48 + 24 + 64 = 136 \text{ ft}^2$

b) $V = \pi r^2 h = \pi \cdot 5^2 \cdot 9 = 225\pi \approx 706.86 \text{ in.}^3$
   $SA = 2\pi rh + 2\pi r^2 = 2\pi \cdot 5 \cdot 9 + 2\pi \cdot 5^2$
   $\qquad\qquad = 90\pi + 50\pi = 140\pi \approx 439.82 \text{ in.}^2$

c) $V = \frac{1}{3}\pi r^2 h = \frac{1}{3}\pi \cdot 3^2 \cdot 4 = \frac{1}{3}\pi \cdot 9 \cdot 4 = 12\pi \text{ m}^3 \approx 37.70 \text{ m}^3$
   $SA = \pi r^2 + \pi r\sqrt{r^2 + h^2} = \pi \cdot 3^2 + \pi \cdot 3 \cdot \sqrt{3^2 + 4^2}$
   $\qquad\qquad = \pi \cdot 9 + \pi \cdot 3 \cdot \sqrt{25}$
   $\qquad\qquad = \pi \cdot 9 + \pi \cdot 3 \cdot 5 = 9\pi + 15\pi$
   $\qquad\qquad = 24\pi \approx 75.40 \text{ m}^2$

d) $V = \frac{4}{3}\pi r^3 = \frac{4}{3} \cdot \pi \cdot 6^3 = \frac{4}{3} \cdot \pi \cdot 216 = 288\pi \approx 904.78 \text{ cm}^3$
   $SA = 4\pi r^2 = 4 \cdot \pi \cdot 6^2 = 4 \cdot \pi \cdot 36 = 144\pi \approx 452.39 \text{ cm}^2$

EXAMPLE ❷   *Replacing a Sand Volleyball Court*

Joan Lind is the manager at the Colony Apartments and needs to replace the sand in the rectangular sand volleyball court. The court is 30 ft wide by 60 ft long, and the sand has a uniform depth of 18 in. (see figure below). Sand sells for $15 per cubic yard.

18 in.

60 ft

30 ft

a) How many cubic yards of sand does Joan need?
b) How much will the sand cost?

SOLUTION

a) Since we are asked to find the volume in cubic yards, we will convert each measurement to yards. There are 3 ft in a yard. Thus, 30 ft equals $\frac{30}{3}$ or 10 yd, and 60 ft equals $\frac{60}{3}$ or 20 yd. There are 36 in. in a yard, so 18 in. equals $\frac{18}{36}$, or $\frac{1}{2}$ yd. The amount of sand needed is determined using the formula for the volume of a rectangular solid, $V = l \cdot w \cdot h$. In this case, the height of the rectangular solid can be considered the depth of the sand.

$$V = l \cdot w \cdot h = 10 \cdot 20 \cdot \tfrac{1}{2} = 100 \text{ yd}^3$$

Note that since the measurements for length, width, and height are each in terms of yards, the answer is in terms of cubic yards.

b) One cubic yard of sand costs $15, so 100 yd^3 will cost 100 × $15, or $1500.

┌─ **EXAMPLE ❸** *Homecoming Float*

The basketball team at Southwestern High School is building a float for the home-coming parade. On the float is a large papier-maché basketball that has a radius of 4.5 ft. Team members need to know the surface area of the basketball so that they can determine how much paint they will need to buy to paint the basketball.

a) Determine the surface area of the basketball.

b) If 1 quart of paint covers approximately 100 sq ft, how many quarts of paint will team members need to buy?

SOLUTION

a) The basketball has the shape of a sphere. We will use the formula for the surface area of a sphere: $SA = 4\pi r^2 = 4 \cdot \pi \cdot (4.5)^2 = 81\pi \approx 254.47$ sq ft.

b) Since each quart of paint will cover about 100 sq ft, they will need about $\frac{254.47}{100} = 2.5447$ quarts. Since you cannot buy a portion of a quart of paint, the team will need to buy 3 quarts of paint to paint the basketball.   ●

┌─ **EXAMPLE ❹** *Silage Storage*

Gordon Langeneger has three silos on his farm. The silos are each in the shape of a right circular cylinder (see Fig. 9.38). One silo has a 12 ft diameter and is 40 ft tall. The second silo has a 14 ft diameter and is 50 ft tall. The third silo has an 18 ft diameter and is 60 ft tall.

Figure 9.38

a) What is the total capacity of the three silos in cubic feet?

b) If Gordon fills all three of his silos and then feeds his cattle 150 ft^3 of silage per day, in how many days will all three silos be empty?

SOLUTION

a) The capacity of each silo can be determined using the formula for the volume of a right circular cylinder, $V = \pi r^2 h$. Since the radius is half the diameter, the radii for the three silos are 6 ft, 7 ft, and 9 ft, respectively. Now let's determine the volumes.

$$\text{Volume of the first silo} = \pi r^2 h = \pi \cdot 6^2 \cdot 40$$
$$\approx 3.14 \cdot 36 \cdot 40 \approx 4521.6 \text{ ft}^3$$

$$\text{Volume of the second silo} = \pi r^2 h = \pi \cdot 7^2 \cdot 50$$
$$\approx 3.14 \cdot 49 \cdot 50 \approx 7693.0 \text{ ft}^3$$
$$\text{Volume of the third silo} = \pi r^2 h = \pi \cdot 9^2 \cdot 60$$
$$\approx 3.14 \cdot 81 \cdot 60 \approx 15,260.4 \text{ ft}^3$$

Therefore, the total capacity of all three silos is about

$$4521.6 + 7693.0 + 15,260.4 \approx 27,475.0 \text{ ft}^3.$$

b) To find how long it takes to empty all three silos, we will divide the total capacity by 150 ft^3, the amount fed to Gordon's cattle every day.

$$\frac{27,475}{150} \approx 183.17$$

Thus, the silos will be empty in about 183 days. •

Polyhedra, Prisms, and Pyramids

Now let's discuss polyhedra. A *polyhedron* (plural is *polyhedra*) is a closed surface formed by the union of polygonal regions. Figure 9.39 illustrates some polyhedra.

Polyhedra

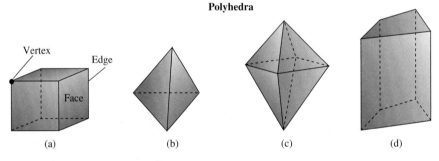

Figure 9.39

Each polygonal region is called a *face* of the polyhedron. The line segment formed by the intersection of two faces is called an *edge*. The point at which two or more edges intersect is called a *vertex*. In Fig. 9.39(a), there are 6 faces, 12 edges, and 8 vertices. Note that

$$\text{Number of vertices} - \text{number of edges} + \text{number of faces} = 2$$
$$8 \qquad - \qquad 12 \qquad + \qquad 6 \qquad = 2$$

This formula, credited to Leonhard Euler, is true for any polyhedron.

EULER'S POLYHEDRON FORMULA

Number of vertices − number of edges + number of faces = 2

We suggest that you verify that this formula holds for Fig. 9.39(b), (c), and (d).

EXAMPLE ⑤ *Using Euler's Polyhedron Formula*

A certain polyhedron has 13 vertices and 8 faces. Determine the numb of edges on the polyhedron.

SOLUTION Since we are seeking the number of edges, we will let x represent the number of edges on the polyhedron. Next, we will use Euler's polyhedron formula to set up an equation:

$$\text{Number of vertices} - \text{number of edges} + \text{number of faces} = 2$$
$$13 - x + 8 = $$
$$21 - x = $$
$$-x = 19$$
$$x = 1$$

Therefore, the polyhedron has 19 edges.   •

A *platonic solid*, also known as a *regular polyhedron*, is a polyhedron whose faces are all regular polygons of the same size and shape. There are exactly five platonic solids. All five platonic solids are illustrated in the Did You Know? at left.

A *prism* is a special type of polyhedron whose bases are congruent polygons and whose sides are parallelograms. These parallelogram regions are called the *lateral faces* of the prism. If all the lateral faces are rectangles, the prism is said to be a *right prism*. The prisms illustrated in Fig. 9.40 are all right prisms. When we use the word *prism* in this book, we are referring to a right prism.

Prisms

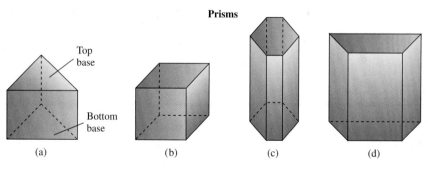

Top base / Bottom base

(a)   (b)   (c)   (d)

Figure 9.40

The volume of any prism can be found by multiplying the area of the base, B, by the height, h, of the prism.

> **VOLUME OF A PRISM**
>
> $$V = Bh$$
>
> where B is the area of a base and h is the height.

16 in.

24 in.

8 in.

8 in.

8 in.

Figle 9.41

EXAMPLE 6 *Volume of a Hexagonal Prism Fish Tank*

Frank Nicolzaao's fish tank is in the shape of a hexagonal prism as shown in Fig. 9.41. Use the dimensions shown in the figure and the fact that 1 gal $= 231$ in.^{3} to

a) determine the volume of the fish tank in cubic inches.

b) determine the volume of the fish tank in gallons (round your answer to the nearest gallon).

SOLUTION

a) First we will need to calculate the area of the hexagonal base of the fish tank. Notice from Fig. 9.41 that by drawing a diagonal as indicated, the base can be divided into two identical trapezoids. To find the area of the hexagonal base, we will calculate the area of one of these trapezoids and then multiply by 2.

$$\text{Area of one trapezoid} = \tfrac{1}{2}h(b_1 + b_2)$$
$$= \tfrac{1}{2}(8)(16 + 8) = 96 \text{ in.}^2$$
$$\text{Area of the hexagonal base} = 2(96) = 192 \text{ in.}^2$$

Now to determine the volume of the fish tank, we will use the formula for the volume of a prism, $V = Bh$. We already determined that the area of the base, B, is 192 in.^{2}

$$V = B \cdot h = 192 \cdot 24 = 4608 \text{ in.}^3$$

In the above calculation, the area of the base, B, was measured in square inches and the height was measured in inches. The product of square inches and inches is cubic inches, or in.^{3}.

b) To determine the volume of the fish tank in gallons, we will divide the volume of the fish tank in cubic inches by 231.

$$V = \frac{4608}{231} \approx 19.95 \text{ gal}$$

Thus, the volume of the fish tank is approximately 20 gal.   ●

EXAMPLE 7 *Volumes Involving Prisms*

Determine the volume of the remaining solid after the cylinder, triangular prism, and square prism have been cut from the solid (Fig. 9.42).

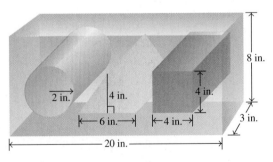

8 in.

2 in.

4 in.

4 in.

6 in.

4 in.

3 in.

20 in.

Figure 9.42

SOLUTION To determine the volume of the remaining solid, first determine e volume of the rectangular solid. Then subtract the volume of the two prisms a the cylinder that were cut out.

$$\text{Volume of rectangular solid} = l \cdot w \cdot h$$
$$= 20 \cdot 3 \cdot 8 = 480 \text{ in.}^3$$
$$\text{Volume of circular cylinder} = \pi r^2 h$$
$$\approx (3.14)(2^2)(3)$$
$$\approx (3.14)(4)(3) \approx 37.68 \text{ in.}^3$$
$$\text{Volume of triangular prism} = \text{area of the base} \cdot \text{height}$$
$$= \tfrac{1}{2}(6)(4)(3) = 36 \text{ in.}^3$$
$$\text{Volume of square prism} = s^2 \cdot h$$
$$= 4^2 \cdot 3 = 48 \text{ in.}^3$$
$$\text{Volume of solid} \approx 480 - 37.68 - 36 - 48$$
$$\approx 358.32 \text{ in.}^3$$

Another special category of polyhedra is the *pyramid*. Unlike prisms, pyramids have only one base. The figures illustrated in Fig. 9.43 are pyramids. Note that all but one face of a pyramid intersect at a common vertex.

Pyramids

(a)   (b)   (c)   (d)

Figure 9.43

If a pyramid is drawn inside a prism, as shown in Fig. 9.44, the volume of the pyramid is less than that of the prism. In fact, the volume of the pyramid is one-third the volume of the prism.

Figure 9.44

VOLUME OF A PYRAMID

$$V = \tfrac{1}{3} Bh$$

where B is the area of the base and h is the height.

EXAMPLE ❽ *Volume of a Pyramid*

Find the volume of the pyramid shown in Fig. 9.45.

SOLUTION First find the area of the base of the pyramid. Since the base of the pyramid is a square,

$$\text{Area of base} = s^2 = 8^2 = 64 \text{ m}^2$$

Figure 9.45

Now use this information to find the volume of the pyramid.

$$V = \frac{1}{3} \cdot B \cdot h$$
$$= \frac{1}{3} \cdot 64 \cdot 12$$
$$= 256 \text{ m}^3$$

Thus, the volume of the pyramid is 256 m^3.

Cubic Unit Conversions

In certain situations, converting volume from one cubic unit to a different cubic unit might be necessary. For example, when purchasing topsoil you might have to change the amount of topsoil from cubic feet to cubic yards prior to placing your order. Example 9 shows how that may be done.

EXAMPLE 9   *Cubic Yards and Cubic Feet*

a) Convert 1 yd^3 to cubic feet. (See Fig. 9.46.)

b) Convert 18.5 yd^3 to cubic feet.

c) Convert 1302.75 ft^3 to cubic yards.

SOLUTION

a) We know that $1 \text{ yd} = 3\text{ft}$. Thus, $1 \text{ yd}^3 = 3 \text{ ft} \times 3 \text{ ft} \times 3 \text{ ft} = 27 \text{ ft}^3$.

b) In part (a), we learned that $1 \text{ yd}^3 = 27\text{ft}^3$. Thus, $18.5 \text{ yd}^3 = 18.5 \times 27 = 499.5 \text{ ft}^3$.

c) In part (b), we converted from cubic yards to cubic feet by *multiplying* the number of cubic yards by 27. Now, to convert from cubic feet to cubic yards we will *divide* the number of cubic feet by 27. Therefore, $1302.75 \text{ ft}^3 = \frac{1302.75}{27} \text{ yd}^3 = 48.25 \text{ yd}^3$.

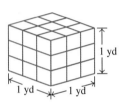

Figure 9.46

EXAMPLE 10   *Filling in a Swimming Pool*

Julianne Peterson recently purchased a home with a rectangular swimming pool. The pool is 30 ft long and 15 ft wide, and it has a uniform depth of 4.5 ft. Julianne lives in a cold climate, so she plans to fill the pool in with dirt to make a flower garden. How many cubic yards of dirt will Julianne have to purchase to fill in the swimming pool?

SOLUTION To find the amount of dirt, we will use the formula for the volume of a rectangular solid:

$$V = lwh$$
$$= (30)(15)(4.5)$$
$$= 2025 \text{ ft}^3$$

Now we must convert this volume from cubic feet to cubic yards. In Example 9, we learned that $1 \text{ yd}^3 = 27 \text{ ft}^3$. Therefore, $2025 \text{ ft}^3 = \frac{2025}{27} = 75 \text{ yd}^3$. Thus, Julianne needs to purchase 75 yd^3 of dirt to fill in her swimming pool.

SECTION 9.4 EXERCISES

CONCEPT/WRITING EXERCISES

1. **a)** Define *volume*.

   b) Define *surface area*.

2. What is solid geometry?

3. What is the difference between a polyhedron and a regular polyhedron?

4. What is the difference between a prism and a right prism?

5. In your own words, explain the difference between a prism and a pyramid.

6. In your own words, state Euler's polyhedron formula.

PRACTICE THE SKILLS

In Exercises 7–16, determine (a) the volume and (b) the surface area of the three-dimensional figure. When appropriate, use the $\boxed{\pi}$ key on your calculator and round your answer to the nearest hundredth.

7.

8.

9.

10.

11.

12.

13.

14.

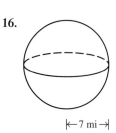

15.

16.

3.

In Exercises 17–20, determine the volume of the three-dimensional figure. When appropriate, round your answer to the nearest hundredth.

17.

18.

19.

20.

23.

24.

25.

PROBLEM SOLVING

In Exercises 21–28, determine the volume of the shaded region. When appropriate, use the $\boxed{\pi}$ *key on your calculator and round your answer to the nearest hundredth.*

21.

26.

27.

22.

28.

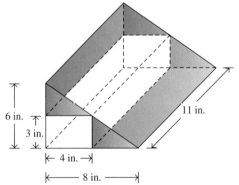

In Exercises 29–32, use the fact that 1 yd³ equals 27 ft³ to make the conversion.

29. 9 yd³ to cubic feet

30. 7.25 yd³ to cubic feet

31. 153 ft³ to cubic yards

32. 2457 ft³ to cubic yards

In Exercises 33–36, use the fact that 1 m³ equals 1,000,000 cm³ to make the conversion.

33. 3.7 m³ to cubic centimeters

34. 17.6 m³ to cubic centimeters

35. 7,500,000 cm³ to cubic meters

36. 7,300,000 cm³ to cubic meters

37. *Ice-Cream Comparison* The Louisburg Creamery packages its homemade ice cream in tubs and in boxes. The tubs are in the shape of a right cylinder with a radius of 3 in. and height of 5 in. The boxes are in the shape of a cube with each side measuring 5 in. Determine the volume of each container.

38. *Volume of a Freezer* The dimensions of the interior of an upright freezer are height 46 in., width 25 in., and depth 25 in. Determine the volume of the freezer

a) in cubic inches.   **b)** in cubic feet.

39. *CD Case* A compact disc case is a rectangular solid that is 142 mm long, 125 mm wide, and 10 mm high. Determine its surface area.

40. *Globe Surface Area* The Everest model globe has a diameter of 20 in. Determine the surface area of this globe. Round your answer to the nearest hundredth.

41. *Volume of a Bread Pan* A bread pan is 12 in. × 4 in. × 3 in. How many quarts does it hold, if 1 in.³ ≈ 0.01736 qt?

42. *Gasoline Containers* Mark Russo has two right cylindrical containers for storing gasoline. One has a diameter of 10 in. and a height of 12 in. The other has a diameter of 12 in. and a height of 10 in.

a) Which container holds the greater amount of gasoline, the taller one or the one with the greater diameter?

b) What is the difference in volume?

43. *A Fish Tank*

a) How many cubic centimeters of water will a rectangular fish tank hold if the tank is 80 cm long, 50 cm wide, and 30 cm high?

b) If 1 cm³ holds 1 mℓ of liquid, how many milliliters will the tank hold?

c) If 1ℓ = 1000 mℓ, how many liters will the tank hold?

44. *The Pyramid of Cheops* The Pyramid of Cheops in Egypt has a square base measuring 720 ft on a side. Its height is 480 ft. What is its volume?

45. *Engine Capacity* The engine in a 1957 Chevrolet Corvette has eight cylinders. Each cylinder is a right cylinder with a bore (diameter) of 3.875 in. and a stroke (height) of 3 in. Determine the total displacement (volume) of this engine.

46. *Rose Garden Topsoil* Marisa Raffaele wishes to plant a rose garden in her backyard. The rose garden will be in the shape of a 9 ft by 18 ft rectangle. Marisa wishes to add a 4 in. layer of organic topsoil on top of the rectangular area. The topsoil sells for $32.95 per cubic yard. Determine

   a) how many cubic yards of topsoil Marisa will need.

   b) how much the topsoil will cost.

47. *Pool Toys* A Wacky Noodle Pool Toy, frequently referred to as a "noodle," is a cylindrical flotation device made from cell foam (see photo). One style of noodle is a cylinder that has a diameter of 2.5 in. and a length of 5.5 ft. Determine the volume of this style of noodle in

   a) cubic inches.

   b) cubic feet.

48. *Comparing Cake Pans* When baking a cake, you can choose between a round pan with a 9 in. diameter and a 7 in. × 9 in. rectangular pan.

   a) Determine the area of the base of each pan.

   b) If both pans are 2 in. deep, determine the volume of each pan.

   c) Which pan has the larger volume?

49. *Cake Icing* A bag used to apply icing to a cake is in the shape of a cone with a diameter of 3 in. and a height of 6 in. How much icing will this bag hold when full?

50. *Flower Box* The flower box shown at the top of the right-hand column is 4 ft long, and its ends are in the shape of a trapezoid. The upper and lower bases of the trapezoid measure 12 in. and 8 in., respectively, and the height is 9 in. Find the volume of the flower box

   a) in cubic inches.

   b) in cubic feet.

In Exercises 51–56, find the missing value indicated by the question mark. Use the following formula.

$$\left(\begin{array}{c}\text{Number of}\\\text{vertices}\end{array}\right) - \left(\begin{array}{c}\text{number of}\\\text{edges}\end{array}\right) + \left(\begin{array}{c}\text{number}\\\text{of faces}\end{array}\right) = 2$$

	Number of Vertices	Number of Edges	Number of Faces
51.	8	?	4
52.	12	16	?
53.	?	8	4
54.	7	12	?
55.	11	?	5
56.	?	10	4

CHALLENGE PROBLEMS/GROUP ACTIVITIES

57. *Earth and Moon Comparisons* The diameter of Earth is approximately 12,756.3 km. The diameter of the moon is approximately 3474.8 km. Assume that both Earth and the moon are spheres.

   a) Determine the surface area of Earth.

   b) Determine the surface area of the moon.

   c) How many times larger is the surface area of Earth than the surface area of the moon?

   d) Determine the volume of Earth.

   e) Determine the volume of the moon.

   f) How many times larger is the volume of Earth than the volume of the moon?

58. *Packing Orange Juice* A box is packed with six cans of orange juice. The cans are touching each other and the sides of the box, as shown. What percent of the volume of the interior of the box is not occupied by the cans?

59. *Doubling the Edges of a Cube* In this exercise, we will explore what happens to the volume of a cube if we double the length of each edge of the cube.

a) Choose a number between 1 and 10 and call this number *s*.

b) Calculate the volume of a cube with the length of each edge equal to *s*.

c) Now double *s* and call this number *t*.

d) Calculate the volume of a cube with the length of each edge equal to *t*.

e) Repeat parts (a) through (d) for a different value of *s*.

f) Compare the results from part (b) with the results from part (d) and explain what happens to the volume of a cube if we double the length of each edge.

60. *Doubling the Radius of a Sphere* In this exercise, we will explore what happens to the volume of a sphere if we double the radius of the sphere.

a) Choose a number between 1 and 10 and call this number *r*.

b) Calculate the volume of a sphere with radius *r* (use the $\boxed{\pi}$ key on your calculator).

c) Now double *r* and call this number *t*.

d) Calculate the volume of a sphere with radius *t*.

e) Repeat parts (a) through (d) for a different value of *r*.

f) Compare the results from part (b) with the results from part (d) and explain what happens to the volume of a sphere if we double the radius.

61. a) Explain how to demonstrate, using the cube shown below, that
$$(a + b)^3 = a^3 + 3a^2b + 3ab^2 + b^3$$

b) What is the volume in terms of *a* and *b* of each numbered piece in the figure?

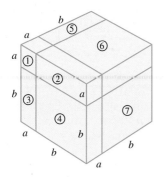

c) An eighth piece is not illustrated. What is its volume?

RECREATIONAL MATHEMATICS

62. *More Pool Toys* Wacky Noodle Pool Toys (see Exercise 47) come in many different shapes and sizes.

Base for part (a)   Base for part (b)

a) Determine the volume, in cubic inches, of a noodle that is in the shape of a 5.5-ft-long solid octagonal prism whose base has an area of 5 in.^{2}.

b) Determine the volume, in cubic inches, of a hollow noodle that has the same shape as the noodle described in part (a) except that a right circular cylinder of diameter 0.75 in. has been removed from the center.

INTERNET/RESEARCH ACTIVITIES

63. *Air-Conditioner Selection* Calculate the volume of the room in which you sleep or study. Go to a store that sells room air conditioners and find out how many cubic feet can be cooled by the different models available. Describe the model that would be the proper size for your room. What is the initial cost? How much does that model cost to operate? If you moved to a room that had twice the amount of floor space and the same height, would the air conditioner you selected still be adequate? Explain.

64. Pappus of Alexandria (ca. A.D. 350) was the last of the well-known ancient Greek mathematicians. Write a paper on his life and his contributions to mathematics.

65. *Platonic Solids* Construct cardboard models of one or more of the platonic solids. Visit the web site www. mathsnet.net/geometry/solid/platonic.html or similar web sites for patterns to follow.

9.5 TRANSFORMATIONAL GEOMETRY, SYMMETRY, AND TESSELLATIONS

Consider the capital letters of the alphabet. Now consider a kindergarten-age child who is practicing to write capital letters. Usually, children will write some of the letters "backwards." However, some letters cannot be made backwards. For example, the capital letter A would look the same backwards as it does forwards, but the capital letters B and C would look different backwards than they do forwards. The capital letters that appear the same forwards as they do backwards have a property called *symmetry* that we will define and study in this section. Symmetry is part of another branch of geometry known as *transformational geometry*.

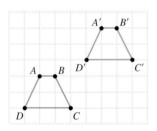

▲ Capital letters can be used to display symmetry.

In our study of geometry, we have thus far focused on definitions, axioms, and theorems that are used in the study of *Euclidean geometry*. We will now introduce a second type of geometry called *transformational geometry*. In *transformational geometry*, we study various ways to move a geometric figure without altering the shape or size of the figure. When discussing transformational geometry, we often use the term *rigid motion*.

> The act of moving a geometric figure from some starting position to some ending position without altering its shape or size is called a **rigid motion** (or **transformation**).

Figure 9.47

Consider trapezoid *ABCD* in Figure 9.47. If we move each point on this trapezoid 4 units to the right and 3 units up, the trapezoid is in the location specified by trapezoid *A′B′C′D′*. This figure illustrates one type of rigid motion. When studying rigid motions, we are only concerned about the starting and ending positions of the figure and not what happens in between. When discussing rigid motions of two-dimensional figures, there are four basic types of rigid motions: reflections, rotations, translations, and glide reflections. We call these four types of rigid motions the *basic rigid motions in a plane*. After we discuss the four rigid motions we will discuss symmetry of geometric figures and tessellations.

Reflections

The first rigid motion we will study is *reflection*. In our everyday life, we are quite familiar with the concept of reflection. In transformational geometry, a reflection is an image of a geometric figure that appears on the opposite side of a designated line.

> A **reflection** is a rigid motion that moves a geometric figure to a new position such that the figure in the new position is a mirror image of the figure in the starting position. In two dimensions, the figure and its mirror image are equidistant from a line called the **reflection line** or the **axis of reflection**.

Figure 9.48 shows trapezoid *ABCD*, a reflection line *l*, and the reflected trapezoid *A′B′C′D′*. Notice that vertex *A* is 6 units to the *left* of reflection line *l* and that vertex *A′* is 6 units to the *right* of reflection line *l*. Next notice that vertex *B* is 2 units to the *left* of *l* and that vertex *B′* is 2 units to the *right* of *l*. A similar relationship holds true for vertices *C* and *C′* and for vertices *D* and *D′*. It is important to see that the trapezoid is not simply *moved* to the other side of the reflection line, but instead it is *reflected*. Notice in the trapezoid *ABCD* that the longer base $\overline{BC}$ is on the *right* side of the trapezoid, but in the reflected trapezoid *A′B′C′D′* the longer base $\overline{B′C′}$ is on the *left* side of the trapezoid. Finally, notice the colors of the sides of the two trapezoids. Side $\overline{AB}$ in trapezoid *ABCD* and side $\overline{A′B′}$ in the reflected trapezoid are both blue. Side $\overline{BC}$ and side $\overline{B′C′}$ are both red, sides $\overline{CD}$ and $\overline{C′D′}$ are both gold, and sides $\overline{DA}$ and $\overline{D′A′}$ are both green. In this section, we will occasionally use such color coding to help you visualize the effect of a rigid transformation on a figure.

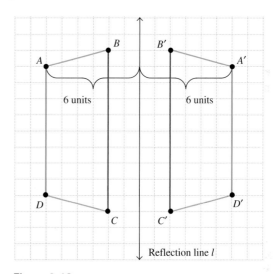

Figure 9.48

┌─EXAMPLE ❶ *Reflection of a Square*

Construct the reflection of square *ABCD*, shown in Fig. 9.49, about reflection line *l*.

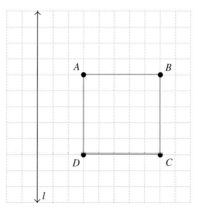

Figure 9.49

SOLUTION The reflection of square $ABCD$ will be called $A'B'C'D'$. To determine the position of the reflection, we first examine vertex A in Fig. 9.49. Notice that vertex A is 3 units to the *right* of line l. Thus, in the reflected square $A'B'C'D'$, vertex A' must also be 3 units away from, but to the *left* of, reflection line l. Next, notice that vertex B is 8 units to the right of l. Thus, in the reflected square $A'B'C'D'$, vertex B' must also be 8 units away from, but to the *left* of, reflection line l. Next, notice that vertices C and D are 8 units and 3 units, respectively, to the *right* of l. Thus, vertices C' and D' must be 8 units and 3 units, respectively, to the *left* of l. Figure 9.50 shows vertices A', B', C', and D'. Finally, we draw line segments between vertices A' and B', B' and C', C' and D', and D' and A' to form the sides of the reflection, square $A'B'C'D'$, as illustrated in Fig. 9.50.

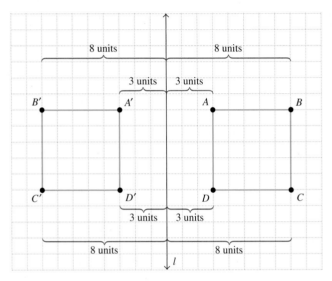

Figure 9.50

In Example 1, the reflection line did not intersect the figure being reflected. We will now study an example in which the reflection line goes directly through the figure to be reflected.

EXAMPLE ❷ *Reflection of a Hexagon*

Construct the reflection of hexagon $ABCDEF$, shown in Fig. 9.51, about reflection line l.

SOLUTION From Fig. 9.51, we see that vertex A in hexagon $ABCDEF$ is 2 units to the left of reflection line l. Thus, vertex A' in the reflected hexagon will be 2 units to the right of l (see Fig. 9.52). Notice that vertex A' of the reflected hexagon is in the same location as vertex B of hexagon $ABCDEF$ in Fig. 9.51.

We next see that vertex B in hexagon $ABCDEF$ is 2 units to the right of l. Thus, vertex B' in the reflected hexagon will be 2 units to the left of l. Notice that vertex B' of the reflected hexagon is in the same location as vertex A of hexagon $ABCDEF$. We continue this process to determine the locations of vertices C', D', E', and F' of the reflected hexagon. Notice once again that each vertex of

Figure 9.51

Figure 9.52

Figure 9.53

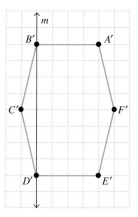

Figure 9.54

the reflected hexagon is in the same location as a vertex of hexagon *ABCDEF*. Finally, we draw the line segments to complete the reflected hexagon *A′B′C′D′E′F′* (see Fig. 9.52). For this example, we see that other than the vertex labels, the positions of the hexagon before and after the reflection are identical.   ●

In Example 2, the reflection line was in the center of the hexagon in the original position. As a result, the reflection line was also in the center of the reflected hexagon. In this particular case the reflected hexagon lies directly on top of the hexagon in its original position. We will revisit reflections such as that in Example 2 again when we discuss *reflective symmetry* later in this section.

Now consider hexagon *ABCDEF* in Fig. 9.53 and its reflection about line *m*, hexagon *A′B′C′D′E′F′* in Fig. 9.54. Notice that the positions of the hexagon before and after the reflection, relative to line *m*, are not the same. Furthermore, if we line up reflection line *m* in Fig. 9.53 and Fig. 9.54, we would see that hexagon *ABCDEF* and hexagon *A′B′C′D′E′F′* are in different positions.

Translations

The next rigid motion we will discuss is the *translation*. In a translation, we simply move a figure along a straight line to a new position.

> A **translation** (or **glide**) is a rigid motion that moves a geometric figure by sliding it along a straight line segment in the plane. The direction and length of the line segment completely determine the translation.

After conducting a translation, we say the figure was *translated* to a new position.

A concise way to indicate the direction and the distance that a figure is moved during a translation is with a *translation vector.* In mathematics, vectors are typically represented with boldface letters. For example, in Fig. 9.55 we see trapezoid *ABCD* and a translation vector, **v**, which is pointing to the right and upward. This translation vector indicates a translation of 9 units to the right and 4 units upward. Notice that in Fig. 9.55 the translated vector appears on the right side of the polygon. The placement of the translation vector does not matter. Therefore, the translation vector could have been placed to the left, above, or below the polygon, and the translation would not change. When trapezoid *ABCD* is translated using **v**, every point on trapezoid *ABCD* is moved 9 units to the right and 4 units upward. This movement is demonstrated for vertex *A* in Fig. 9.56(a) on page 570. Figure 9.56(b) shows trapezoid *ABCD* and the translated trapezoid *A′B′C′D′*. Notice in Fig. 9.56(b) that every point on trapezoid *A′B′C′D′* is 9 units to the right and 4 units up from its corresponding point on trapezoid *ABCD*.

Figure 9.55

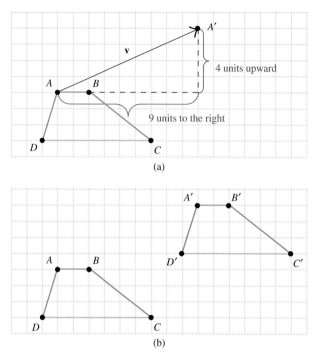

Figure 9.56

EXAMPLE ❸ *A Translated Parallelogram*

Given parallelogram *ABCD* and translation vector, **v**, shown in Fig. 9.57, construct the translated parallelogram *A′B′C′D′*.

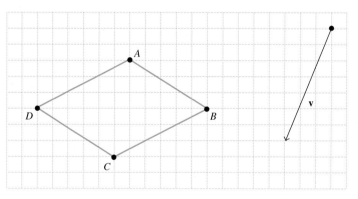

Figure 9.57

SOLUTION The translated figure will be a parallelogram of the same size and shape as parallelogram *ABCD*. We notice that the translation vector, **v**, points 7 units downward and 3 units to the left. We next examine vertex *A*. To determine the location of vertex *A′* of the translated parallelogram, start at vertex *A* of parallelogram *ABCD* and move down 7 units and to the left 3 units. We label this vertex *A′* (see Fig. 9.58a on page 571). We determine vertices *B′*, *C′*, and *D′* in a similar manner by moving down 7 units and to the left 3 units from vertices *B*, *C*, and *D*, respectively.

Figure 9.58(b) shows parallelogram *ABCD* and the translated parallelogram *A'B'C'D'*. Notice in Fig. 9.58(b) that every point on parallelogram *A'B'C'D'* is 7 units down and 3 units to the left of its corresponding point on parallelogram *ABCD*.

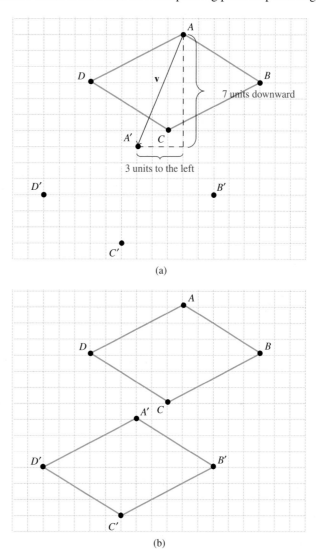

(a)

(b)

Figure 9.58

Rotations

The next rigid motion we will discuss is *rotation*. To help visualize a rotation, examine Fig. 9.59, which shows right triangle *ABC* and point *P* about which right triangle *ABC* is to be rotated.

Imagine that this page was removed from this book and attached to a bulletin board with a single pin through point *P*. Next imagine rotating the page 90° in the *counterclockwise* direction. The triangle would now appear as triangle *A'B'C'* shown in Fig. 9.60. Next, imagine rotating the original triangle 180° in a counterclockwise direction. The triangle would now appear as triangle *A"B"C"* shown in Fig. 9.61.

Figure 9.59

Figure 9.60

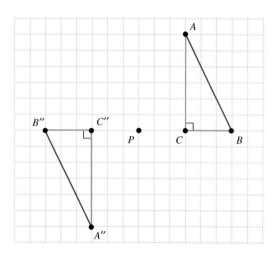

Figure 9.61

Now that we have an intuitive idea of how to determine a rotation, we give the definition of rotation.

A **rotation** is a rigid motion performed by rotating a geometric figure in the plane about a specific point, called the **rotation point** or the **center of rotation**. The angle through which the object is rotated is called the **angle of rotation**.

We will measure angles of rotation using degrees. In mathematics, generally, *counterclockwise angles have positive degree measures and clockwise angles have negative degree measures.*

EXAMPLE ❹ A Rotated Rectangle

Given rectangle *ABCD* and rotation point *P*, shown in Fig. 9.62, construct rectangles that result from rotations through

a) 90°.   b) 180°.   c) 270°.

SOLUTION

a) First, since 90 is a *positive* number, we will rotate the figure in a counterclockwise direction. We also note that the rotated rectangle will be the same size and shape as rectangle *ABCD*. To get an idea of what the rotated rectangle will look like, pick up this book and rotate it counterclockwise 90°. Figure 9.63 on page 573 shows rectangle *ABCD* and rectangle *A′B′C′D′* which is rectangle *ABCD* rotated 90° about point *P*. Notice how line segment *AB* in rectangle *ABCD* is horizontal, but in the rotated rectangle in Fig. 9.63 line segment *A′B′* is vertical. Also notice that in rectangle *ABCD* vertex *D* is 3 units to the *right* and 1 unit *above* rotation point *P*, but in the rotated rectangle, vertex *D′* is 3 units *above* and 1 unit to the *left* of rotation point *P*.

Figure 9.62

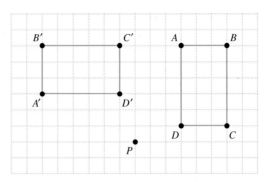

Figure 9.63

b) To gain some perspective on a 180° rotation, again pick up this book, but this time rotate the book 180° in the counterclockwise direction. The rotated rectangle $A''B''C''D''$ is shown along with the rectangle $ABCD$ in Fig. 9.64.

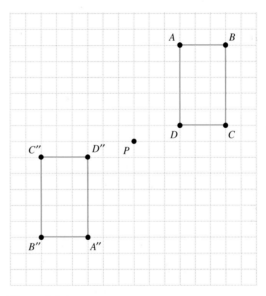

Figure 9.64

c) To gain some perspective on a 270° rotation, rotate this book 270° in the counterclockwise direction. The rotated rectangle $A'''B'''C'''D'''$ is shown along with rectangle $ABCD$ in Fig. 9.65.   ●

Thus far in our examples of rotations, the rotation point was outside the figure being rotated. We now will study an example where the rotation point is inside the figure to be rotated.

Figure 9.65

EXAMPLE ❺ *A Rotation Point Inside a Polygon*

Given polygon *ABCDEFGH* and rotation point *P*, shown in Fig. 9.66, construct polygons that result from rotations through

a) 90°.   b) 180°.

Figure 9.66

Figure 9.67

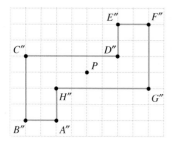

Figure 9.68

SOLUTION

a) We will rotate the polygon 90° in a counterclockwise direction. The resulting polygon will be the same size and shape as polygon *ABCDEFGH*. To visualize what the rotated polygon will look like, pick up this book and rotate it counterclockwise 90°. Figure 9.67 shows the polygon *ABCDEFGH*, in pale blue, and the rotated polygon *A'B'C'D'E'F'G'H'*, in deeper blue. Notice how line segments $\overline{AB}$, $\overline{CD}$, $\overline{EF}$, and $\overline{GH}$ in polygon *ABCDEFGH* are *horizontal*, but in the rotated polygon *A'B'C'D'E'F'G'H'*, line segments $\overline{A'B'}$, $\overline{C'D'}$, $\overline{E'F'}$, and $\overline{G'H'}$ are *vertical*. Also notice in polygon *ABCDEFGH* that line segment $\overline{GH}$ is 1 unit *above* rotation point *P*, but in the rotated polygon, line segment $\overline{G'H'}$ is 1 unit to the *left* of rotation point *P*.

b) To visualize the polygon obtained through a 180° rotation, we can pick up this book and rotate it 180° in the counterclockwise direction. Notice from Fig. 9.68 that vertex *A''* of the rotated polygon is in the same position as vertex *E* of polygon *ABCDEFGH*. Also notice from Fig. 9.68 that vertex *B''* of the rotated polygon is in the same position as vertex *F* of polygon *ABCDEFGH*. In fact, each of the vertices in the rotated polygon is in the same position as a different vertex in polygon *ABCDEFGH*. From Fig. 9.68 we see that, other than vertex labels, the position of rotated polygon *A''B''C''D''E''F''G''H''* is the same as the position of polygon *ABCDEFGH*. ●

The polygon used in Example 5 will be discussed again later when we discuss *rotational symmetry*. The three rigid motions we have discussed thus far are reflection, translation, and rotation. Now we will discuss the fourth rigid motion, *glide reflection*.

Glide Reflections

A **glide reflection** is a rigid motion formed by performing a *translation* (or *glide*) followed by a *reflection*.

A glide reflection, as its name suggests, is a translation (or glide) followed by a reflection. Both translations and reflections were discussed earlier in this section. Consider triangle *ABC* (shown in blue), translation vector **v**, and reflection line *l* in Fig. 9.69. The translation of triangle *ABC*, obtained using translation vector **v**, is triangle *A'B'C'* (shown in red). The reflection of triangle *A'B'C'* about reflection line *l* is triangle *A''B''C''* (shown in green). Thus, triangle *A''B''C''* is the glide reflection of triangle *ABC* using translation vector **v** and reflection line *l*.

Figure 9.69

Figure 9.70

(a)

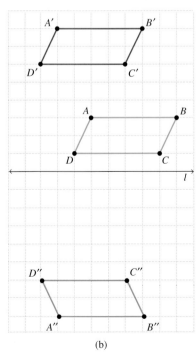

(b)

Figure 9.71

┌─ **EXAMPLE ❻** *A Glide Reflection of a Parallelogram*

Construct a glide reflection of parallelogram *ABCD*, shown in Fig. 9.70, using translation vector **v** and reflection line *l*.

SOLUTION To construct the glide reflection of parallelogram *ABCD*, first translate the parallelogram 2 units to the left and 5 units up as indicated by translation vector **v**. This translated parallelogram is labeled *A'B'C'D'*, shown in red in Fig. 9.71(a). Next, we will reflect parallelogram *A'B'C'D'* about reflection line *l*. Parallelogram *A'B'C'D'*, shown in red, and the reflected parallelogram, labeled *A"B"C"D"*, shown in green, are shown in Fig. 9.71(b). The glide reflection of the parallelogram *ABCD* is parallelogram *A"B"C"D"*.   ●

Symmetry

We are now ready to discuss symmetry. Our discussion of symmetry involves a rigid motion of an object.

> A **symmetry** of a geometric figure is a rigid motion that moves the figure back onto itself. That is, the beginning position and ending position of the figure must be identical.

Suppose we start with a figure in a specific position and perform a rigid motion on this figure. If the position of the figure after the rigid motion is identical to the position of the figure before the rigid motion (if the beginning and ending positions of the figure coincide), then the rigid motion is a symmetry and we say that the figure has symmetry. For a two-dimensional figure, there are four types of symmetries: reflective symmetry, rotational symmetry, translational symmetry, and glide reflective symmetry. In this textbook, however, we will only discuss reflective symmetry and rotational symmetry.

Consider the polygon and reflection line *l* shown in Fig. 9.72(a). If we use the rigid motion of reflection and reflect the polygon *ABCDEFGH* about line *l*, we get polygon *A'B'C'D'E'F'G'H'*. Notice that the ending position of the polygon is identical to the starting position, as shown in Fig. 9.72(b). Compare Fig. 9.72(a) with Fig. 9.72(b). Although the vertex labels are different, the reflected polygon is in the same position as the polygon in the original position. Thus, we say that the polygon has *reflective symmetry* about line *l*. We refer to line *l* as a *line of symmetry*.

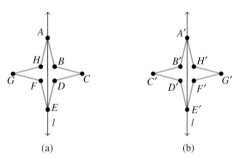

(a)   (b)

Figure 9.72

Recall Example 2 on page 568 in which hexagon *ABCDEF* was reflected about reflection line *l*. Examine the hexagon in the original position (Fig. 9.51) and the hexagon in the final position after being reflected about line *l* (Fig. 9.52). Other than the labels of the vertices, the beginning and ending positions of the hexagon are identical. Therefore, hexagon *ABCDEF* has reflective symmetry about line *l*.

Figure 9.73

(a)

(b)

Figure 9.74

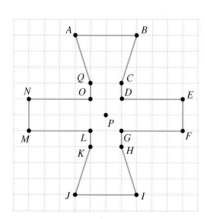

Figure 9.76

EXAMPLE ❼   *Reflective Symmetries of Polygons*

Determine whether the polygon shown in Fig. 9.73 has reflective symmetry about each of the following lines.

a) Line *l*   b) Line *m*

SOLUTION

a) Examine the reflection of the polygon about line *l* as seen in Fig. 9.74(a). Notice that other than the vertex labels, the beginning and ending positions of the polygon are identical. Thus, the polygon has reflective symmetry about line *l*.

b) Examine the reflection of the polygon about line *m* as seen in Fig. 9.74(b). Notice that the position of the reflected polygon is different from the original position of the polygon. Thus, the polygon does not have reflective symmetry about line *m*.   ●

We will now discuss a second type of symmetry, rotational symmetry. Consider the polygon and rotation point *P* shown in Fig. 9.75(a). The rigid motion of rotation of polygon *ABCDEFGH* through a 90° angle about point *P* gives polygon *A'B'C'D'E'F'G'H'* shown in Fig. 9.75(b). Compare Fig. 9.75(a) with Fig. 9.75(b). Although the vertex labels are different, the position of the polygon before and after the rotation is identical. Thus, we say that the polygon has 90° *rotational symmetry* about point *P*. We refer to point *P* as the *point of symmetry*.

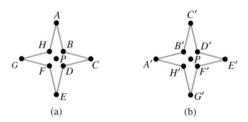

Figure 9.75

Recall Example 5 on page 573 in which polygon *ABCDEFGH* was rotated 90° about point *P* in part (a) and 180° in part (b). First examine the polygon in the original position in Fig. 9.66 on page 573 and the 90° rotated polygon in Fig. 9.67 on page 574. Notice the position of the polygon after the 90° rotation is different from the original position of the polygon. Therefore, polygon *ABCDEFGH* in Fig. 9.67 does not have 90° rotational symmetry about point *P*. Now examine the 180° rotated polygon in Fig. 9.68 on page 574. Notice that other than the vertex labels, the position of the two polygons *ABCDEFGH* and *A'B'C'D'E'F'G'H'* is identical with respect to rotation about point *P*. Therefore, polygon *ABCDEFGH* in Fig. 9.66 has 180° rotational symmetry about point *P*.

EXAMPLE ❽   *Rotational Symmetries*

Determine whether the polygon shown in Fig. 9.76 has rotational symmetry about point *P* for rotations through each of the following angles.

a) 90°   b) 180°

SOLUTION

a) To determine whether the polygon has 90° counterclockwise rotational symmetry about point *P* we rotate the polygon 90° as shown in Fig. 9.77(a). Compare Fig. 9.77(a) with Fig. 9.76. Notice that the position of the polygon after

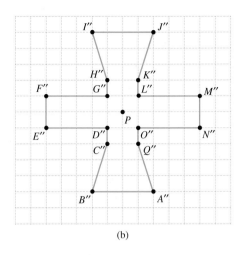

(a)   (b)

Figure 9.77

the rotation in Fig. 9.77(a) is different than the original position of the polygon (Fig. 9.76). Therefore, the polygon does not have 90° rotational symmetry.

b) To determine whether the polygon has 180° counterclockwise rotational symmetry about the point P, we rotate the polygon 180° as shown in Fig. 9.77(b). Compare Fig. 9.77(b) with Fig. 9.76. Notice that other than vertex labels, the position of the polygon after the rotation in Fig. 9.77(b) is identical to the position of the polygon before the rotation (Fig. 9.76). Therefore, the polygon has 180° rotational symmetry.   ●

Tessellations

A fascinating application of transformational geometry is the creation of *tessellations*.

> A **tessellation** (or **tiling**) is a pattern consisting of the repeated use of the same geometric figures to entirely cover a plane, leaving no gaps. The geometric figures used are called the **tessellating shapes** of the tessellation.

Figure 9.78 shows an example of a tessellation from ancient Egypt. Perhaps the most famous person to incorporate tessellations into his work is M. C. Escher (see Profiles in Mathematics at left).

Figure 9.78

Figure 9.79

Figure 9.80

The simplest tessellations use one single regular polygon as the tessellating shape. Recall that a *regular polygon* is one whose sides are all the same length and whose interior angles all have the same measure. A tessellation that uses one single regular polygon as the tessellating shape is called a *regular tessellation*. It can be shown that only three regular tessellations exist: those that use an equilateral triangle, a square, or a regular hexagon as the tessellating shape. Figure 9.79 shows each of these regular tessellations. Notice that each tessellation can be obtained from a single tessellating shape through the use of reflections, translations, or rotations.

We will now learn how to create unique tessellations. We will do so by constructing a unique tessellating shape from a square. We could also construct other tessellating shapes using an equilateral triangle or a regular hexagon. If you wish to follow along with our construction, you will need some lightweight cardboard, a ruler, cellophane tape, and a pair of scissors. We will start by measuring and cutting out a square 2 in. by 2 in. from the cardboard. We next cut the square into two parts by cutting it from top to bottom using any kind of cut. One example is shown in Fig. 9.80. We then rearrange the pieces and tape the two vertical edges together as shown in Fig. 9.81. Next we cut this new shape into two parts by cutting it from left to right using any kind of cut as shown in Fig. 9.82. We then rearrange the pieces and tape the two horizontal edges together as shown in Fig. 9.83. This completes our tessellating shape.

Figure 9.81

Figure 9.82

Figure 9.83

Figure 9.84

We now set the cardboard tessellating shape in the middle of a blank piece of paper (the tessellating shape can be rotated to any position as a starting point) and trace the outline of the shape onto the paper. Next move the tessellating shape so that it lines up with the figure already drawn and trace the outline again. Continue to do that until the page is completely covered. Once the page is covered with the tessellation, we can add some interesting colors or even some unique sketches to the tessellation. Figure 9.84 shows one tessellation created using the tessellation shape in Fig. 9.83. In Fig. 9.84, the tessellation shape was rotated about 45° counterclockwise.

An infinite number of different tessellations can be created using the method described by altering the cuts made. We could also create different tessellations using an equilateral triangle, a regular hexagon, or other types of polygons. There are also other, more complicated ways to create the tessellating shape. The Internet has many sites devoted to the creation of tessellations by hand. Many computer programs that generate tessellations are also available.

SECTION 9.5 EXERCISES

CONCEPT/WRITING EXERCISES

1. In the study of transformational geometry, what is a rigid motion? List the four rigid motions studied in this section.

2. What is transformational geometry?

3. In terms of transformational geometry, describe a reflection.

4. Describe how to construct a reflection of a given figure about a given line.

5. In terms of transformational geometry, describe a rotation.

6. Describe how to construct a rotation of a given figure, about a given point, through a given angle.

7. In terms of transformational geometry, describe a translation.

8. Describe how to construct a translation of a given figure using a translation vector.

9. In terms of transformational geometry, describe a glide reflection.

10. Describe how to construct a glide reflection of a given figure using a given translation vector and a given reflection line.

11. Describe what it means for a figure to have reflective symmetry about a given line.

12. Describe what it means for a figure to have rotational symmetry about a given point.

13. What is a tessellation?

14. Describe one way to make a unique tessellation from a 2-in. by 2-in. cardboard square.

PRACTICE THE SKILLS/PROBLEM SOLVING

In Exercises 15–22, use the given figure and lines of reflection to construct the indicated reflections. Show the figure in the positions both before and after the reflection.

In Exercises 15 and 16, use the following figure. Construct

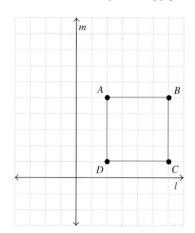

15. the reflection of square *ABCD* about line *m*.

16. the reflection of square *ABCD* about line *l*.

In Exercises 17 and 18, use the following figure. Construct

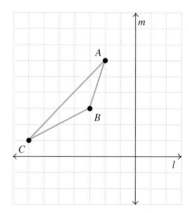

17. the reflection of triangle *ABC* about line *l*.

18. the reflection of triangle *ABC* about line *m*.

In Exercises 19 and 20, use the following figure. Construct

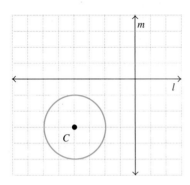

19. the reflection of circle *C* about line *l*.

20. the reflection of circle *C* about line *m*.

In Exercises 21 and 22, use the following figure. Construct

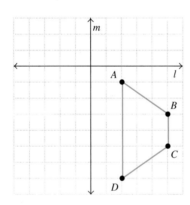

21. the reflection of trapezoid *ABCD* about line *m*.

22. the reflection of the trapezoid *ABCD* about line *l*.

*In Exercises 23–30, use the translation vectors, **v** and **w** shown below, to construct the translations indicated in the exercises. Show the figure in the positions both before and after the translation.*

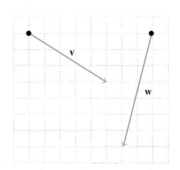

In Exercises 23 and 24, use the following figure. Construct

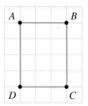

23. the translation of rectangle *ABCD* using translation vector **v**.

24. the translation of rectangle *ABCD* using translation vector **w**.

In Exercises 25 and 26, use the following figure. Construct

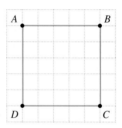

25. the translation of square *ABCD* using translation vector **w**.

26. the translation of square *ABCD* using translation vector **v**.

In Exercises 27 and 28, use the following figure. Construct

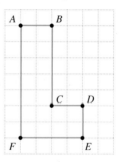

27. the translation of polygon *ABCDEF* using translation vector **v**.

28. the translation of polygon *ABCDEF* using translation vector **w**.

In Exercises 29 and 30, use the following figure. Construct

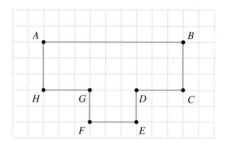

29. the translation of polygon *ABCDEFGH* using translation vector **w** (shown on page 580).

30. the translation of polygon *ABCDEFGH* using translation vector **v** (shown on page 580).

In Exercises 31–38, use the given figure and rotation point P to construct the indicated rotations. Show the figure in the positions both before and after the rotation.

In Exercises 31 and 32, use the following figure. Construct

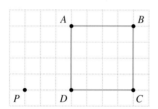

31. a 90° rotation of square *ABCD* about point *P*.

32. a 180° rotation of square *ABCD* about point *P*.

In Exercises 33 and 34, use the following figure. Construct

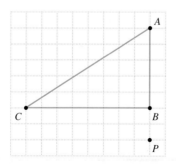

33. a 180° rotation of triangle *ABC* about point *P*.

34. a 270° rotation of triangle *ABC* about point *P*.

In Exercises 35 and 36, use the following figure. Construct

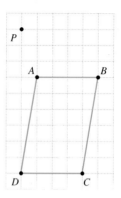

35. a 270° rotation of parallelogram *ABCD* about point *P*.

36. a 180° rotation of parallelogram *ABCD* about point *P*.

In Exercises 37 and 38, use the following figure. Construct

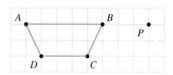

37. a 90° rotation of trapezoid *ABCD* about point *P*.

38. a 270° rotation of trapezoid *ABCD* about point *P*.

In Exercises 39–46, use the given figure, translation vectors **v** *and* **w**, *and reflection lines l and m to construct the indicated glide reflections. Show the figure in the positions before and after the glide reflection.*

In Exercises 39 and 40, use the following figure. Construct

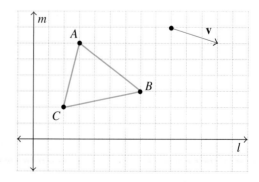

39. a glide reflection of triangle *ABC* using vector **v** and reflection line *l*.

40. a glide reflection of triangle *ABC* using vector **v** (shown on page 581) and reflection line *m*.

In Exercises 41 and 42, use the following figure. Construct

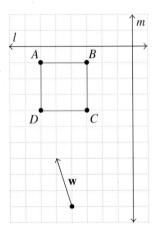

41. a glide reflection of square *ABCD* using vector **w** and reflection line *l*.

42. a glide reflection of square *ABCD* using vector **w** and reflection line *m*.

In Exercises 43 and 44, use the following figure. Construct

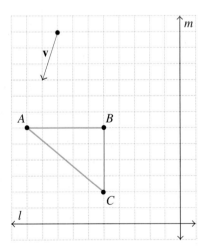

43. a glide reflection of triangle *ABC* using vector **v** and reflection line *l*.

44. a glide reflection of triangle *ABC* using vector **v** and reflection line *m*.

In Exercises 45 and 46, use the following figure. Construct

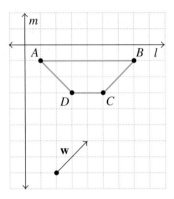

45. a glide reflection of trapezoid *ABCD* using vector **w** and reflection line *l*.

46. a glide reflection of trapezoid *ABCD* using vector **w** and reflection line *m*.

47. a) Reflect triangle *ABC*, shown below, about line *l*. Label the reflected triangle *A'B'C'*.

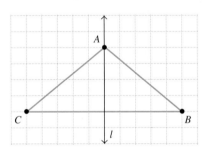

b) Other than vertex labels, is the position of triangle *A'B'C'* identical to the position of triangle *ABC*?

c) Does triangle *ABC* have reflective symmetry about line *l*?

48. a) Reflect rectangle *ABCD*, shown below, about line *l*. Label the reflected rectangle *A'B'C'D'*.

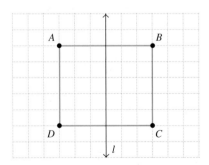

b) Other than vertex labels, is the position of rectangle $A'B'C'D'$ identical to the position of rectangle $ABCD$?

c) Does rectangle $ABCD$ have reflective symmetry about line l?

49. a) Reflect parallelogram $ABCD$, shown below, about line l. Label the reflected parallelogram $A'B'C'D'$.

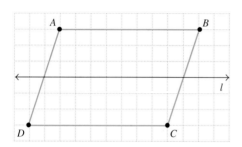

b) Other than vertex labels, is the position of parallelogram $A'B'C'D'$ identical to the position of parallelogram $ABCD$?

c) Does parallelogram $ABCD$ have reflective symmetry about line l?

50. a) Reflect square $ABCD$, shown below, about line l. Label the reflected square $A'B'C'D'$.

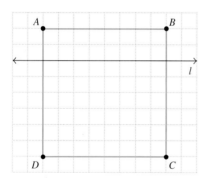

b) Other than vertex labels, is the position of square $A'B'C'D'$ identical to the position of square $ABCD$?

c) Does square $ABCD$ have reflective symmetry about line l?

51. a) Rotate rectangle $ABCD$, shown below, 90° about point P. Label the rotated rectangle $A'B'C'D'$.

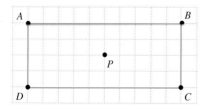

b) Other than vertex labels, is the position of rectangle $A'B'C'D'$ identical to the position of rectangle $ABCD$?

c) Does rectangle $ABCD$ have 90° rotational symmetry about point P?

d) Now rotate the rectangle in the original position, rectangle $ABCD$, 180° about point P. Label the rotated rectangle $A''B''C''D''$.

e) Other than vertex labels, is the position of rectangle $A''B''C''D''$ identical to the position of rectangle $ABCD$?

f) Does rectangle $ABCD$ have 180° rotational symmetry about point P?

52. a) Rotate parallelogram $ABCD$, shown below, 90° about point P. Label the rotated parallelogram $A'B'C'D'$.

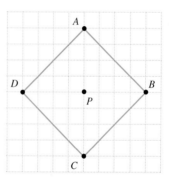

b) Other than vertex labels, is the position of parallelogram $A'B'C'D'$ identical to the position of parallelogram $ABCD$?

c) Does parallelogram $ABCD$ have 90° rotational symmetry about point P?

d) Now rotate the parallelogram in the original position, parallelogram $ABCD$, shown above, 180° about point P. Label the rotated parallelogram $A''B''C''D''$.

e) Other than vertex labels, is the position of parallelogram $A''B''C''D''$ identical to the position of parallelogram $ABCD$?

f) Does parallelogram $ABCD$ have 180° rotational symmetry about point P?

53. Consider the following figure.

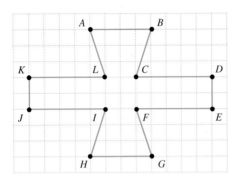

a) Insert a vertical line *m* through the figure so the figure has reflective symmetry about line *m*.

b) Insert a horizontal line *l* through the figure so the figure has reflective symmetry about line *l*.

c) Insert a point *P* within the figure so the figure has 180° rotational symmetry about point *P*.

d) Is it possible to insert a point *P* within the figure so the figure has 90° rotational symmetry about point *P*? Explain your answer.

54. Consider the following figure.

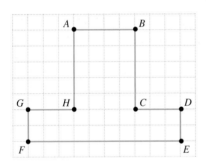

a) Insert a vertical line *m* through the figure so the figure has reflective symmetry about line *m*.

b) Is it possible to insert a horizontal line *l* through the figure so the figure has reflective symmetry about line *l*? Explain your answer.

c) Is it possible to insert a point *P* within the figure so the figure has 90° rotational symmetry about point *P*? Explain your answer.

d) Is it possible to insert a point *P* within the figure so the figure has 180° rotational symmetry about point *P*? Explain your answer.

CHALLENGE PROBLEMS/GROUP ACTIVITIES

55. *Glide Reflection, Order* Examine the figure below and then do the following:

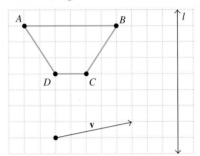

a) Determine a glide reflection of trapezoid *ABCD* by first applying translation vector **v** and then reflecting about the line *l*. Label the glide reflection *A′B′C′D′*.

b) In this step, we will reverse the order of the translation and the reflection. First reflect trapezoid *ABCD* about the line *l* and then translate the reflection using vector **v**. Label the resulting figure *A″B″C″D″*.

c) Is figure *A′B′C′D′* in the same position as figure *A″B″C″D″*?

d) What can be said about the order of the translation and the reflection used in a glide reflection? Is the figure obtained in part (a) or part (b) the glide reflection?

56. *Tessellation with a Square* Create a unique tessellation from a square piece of cardboard by using the method described on page 578. Be creative using color and sketches to complete your tessellation.

57. *Tessellation with a Hexagon* Using the method described on page 578, create a unique tessellation using a regular hexagon like the one shown below. Be creative using color and sketches to complete your tessellation.

58. *Tessellation with an Octagon?* **a)** Trace the regular octagon, shown below, onto a separate piece of paper.

b) Try to create a regular tessellation by tracing this octagon repeatedly. Attempt to cover the entire piece of paper where no two octagons overlap each other. What conclusion can you draw about using a regular octagon as a tessellating shape?

59. *Tessellation with a Pentagon?* Repeat Exercise 58 using the regular pentagon below instead of a regular octagon.

RECREATIONAL MATHEMATICS

60. Examine each capital letter in the alphabet and determine which letters have reflective symmetry about a horizontal line through the center of the letter.

61. Examine each capital letter in the alphabet and determine which letters have reflective symmetry about a vertical line through the center of the letter.

62. Examine each capital letter in the alphabet and determine which letters have 180° rotational symmetry about a point in the center of the letter.

INTERNET/RESEARCH ACTIVITIES

63. In the study of biology, reflective symmetry is called *bilateral symmetry* and rotational symmetry is called *radial symmetry*. Do research and write a report on the role symmetry plays in the study of biology.

64. Write a paper on the mathematics displayed in the artwork of M. C. Escher. Include such topics as tessellations, optical illusions, perspective, and non-Euclidean geometry.

9.6 TOPOLOGY

Examine the outline of the map of the continental United States shown below. Now suppose you were given four crayons, each of a different color. Could you color this map with the four crayons in a way so that no two bordering states have the same color? In this section, we will discuss this question and many other questions that are relevant to the branch of mathematics known as *topology*.

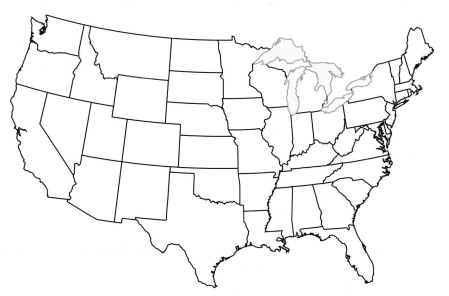

▲ How many different colors are needed so that no two bordering states share the same color?

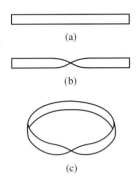

(a)

(b)

(c)

Figure 9.85

Figure 9.86

Figure 9.87

Figure 9.88

Figure 9.89

The branch of mathematics called *topology* is sometimes referred to as "rubber sheet geometry" because it deals with bending and stretching of geometric figures.

One of the first pioneers of topology was the German astronomer and mathematician August Ferdinand Möbius (1790–1866). A student of Gauss, Möbius was the director of the University of Leipzig's observatory. He spent a great deal of time studying geometry and he played an essential part in the systematic development of projective geometry. He is best known for his studies of the properties of one-sided surfaces, including the one called the Möbius strip.

Möbius Strip

If you place a pencil on one surface of a sheet of paper and do not remove it from the sheet, you must cross the edge to get to the other surface. Thus, a sheet of paper has one edge and two surfaces. The sheet retains these properties even when crumpled into a ball. The *Möbius strip*, also called a *Möbius band*, is a one-sided, one-edged surface. You can construct one, as shown in Fig. 9.85, by (a) taking a strip of paper, (b) giving one end a half twist, and (c) taping the ends together.

The Möbius strip has some very interesting properties. To better understand these properties, perform the following experiments.

Experiment 1 Make a Möbius strip using a strip of paper and tape as illustrated in Fig. 9.85. Place the point of a felt-tip pen on the edge of the strip (Fig. 9.86). Pull the strip slowly so that the pen marks the edge; do not remove the pen from the edge. Continue pulling the strip and observe what happens.

Experiment 2 Make a Möbius strip. Place the tip of a felt-tip pen on the surface of the strip (Fig. 9.87). Pull the strip slowly so that the pen marks the surface. Continue and observe what happens.

Experiment 3 Make a Möbius strip. Use scissors to make a small slit in the middle of the strip. Starting at the slit, cut along the strip, keeping the scissors in the middle of the strip (Fig. 9.88). Continue cutting and observe what happens.

Experiment 4 Make a Möbius strip. Make a small slit at a point about one-third of the width of the strip. Cut along the strip, keeping the scissors the same distance from the edge (Fig. 9.89). Continue cutting and observe what happens.

If you give a strip of paper several twists, you get variations on the Möbius strip. To a topologist, the important distinction is between an odd number of twists, which leads to a one-sided surface, and an even number of twists, which leads to a two-sided surface. All strips with an odd number of twists are topologically the same as a Möbius strip, and all strips with an even number of twists are topologically the same as an ordinary cylinder, which has no twists.

Klein Bottle

Another topological object is the punctured *Klein bottle*; see Fig. 9.90 on page 587. This object, named after Felix Klein (1849–1925), resembles a bottle but only has one side.

A punctured Klein bottle can be made by stretching a hollow piece of glass tubing. The neck is then passed through a hole and joined to the base.

Look closely at the model of the Klein bottle shown in Fig. 9.90. The punctured Klein bottle has only one edge and no outside or inside because it has just one side. Figure 9.91 shows a Klein bottle blown in glass by Alan Bennett of Bedford, England.

Limericks from unknown writers:

"A mathematician confided
That a Möbius band is one-sided,
And you'll get quite a laugh
If you cut one in half
For it stays in one piece when divided."

"A mathematician named Klein
Thought the Möbius band was divine.
He said, 'If you glue
the edges of two
You'll get a weird bottle like mine.' "

Figure 9.90

Figure 9.91 *Klein bottle*, a one-sided surface, blown in glass by Alan Bennett.

Imagine trying to paint a Klein bottle. You start on the "outside" of the large part and work your way down the narrowing neck. When you cross the self-intersection, you have to pretend temporarily that it is not there, so you continue to follow the neck, which is now inside the bulb. As the neck opens up, to rejoin the bulb, you find that you are now painting the inside of the bulb! What appear to be the inside and outside of a Klein bottle connect together seamlessly since it is one-sided.

If a Klein bottle is cut along a curve, the results are two (one-twist) Möbius strips, see Fig. 9.92. Thus, a Klein bottle could also be made by gluing together two Möbius strips along the edges.

Figure 9.92 Two Möbius strips result from cutting a Klein bottle along a curve.

Maps

Maps have fascinated topologists for years because of the many challenging problems they present. Mapmakers have known for a long time that regardless of the complexity of the map and whether it is drawn on a flat surface or a sphere, only four colors are needed to differentiate each country (or state) from its immediate neighbors. Thus, every map can be drawn by using only four colors, and no two countries with a common border will have the same color. Regions that meet at only one point (such as the states of Arizona, Colorado, Utah, and New Mexico) are not considered to have a common border. In Fig. 9.93(a), no two states with a common border are marked with the same color.

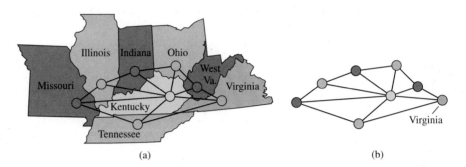

Figure 9.93

The "four-color" problem was first suggested by a student of Augustus DeMorgan in 1852. In 1976, Kenneth Appel and Wolfgang Haken of the University of Illinois—using their ingenuity, logic, and 1200 hours of computer time—succeeded in proving that only four colors are needed to draw a map. They solved the four-color map problem by reducing any map to a series of points and connecting line segments. They replaced each country with a point. They connected two countries having a common border with a straight line; see Fig. 9.93(b). They then showed that the points of any graph in the plane could be colored by using only four colors in such a way that no two points connected by the same line were the same color.

Mathematicians have shown that, on different surfaces, more than four colors may be needed to draw a map. For example, a map drawn on a Möbius strip requires a maximum of six colors, as in Fig. 9.94(a). A map drawn on a torus (the shape of a doughnut) requires a maximum of seven colors, as in Fig. 9.94(b).

Figure 9.94

Jordan Curves

A *Jordan curve* is a topological object that can be thought of as a circle twisted out of shape; see Fig. 9.95 (a)–(d). Like a circle, it has an inside and an outside. To get from one side to the other, at least one line must be crossed. Consider the Jordan curve in Fig. 9.95(d). Are points *A* and *B* inside or outside the curve?

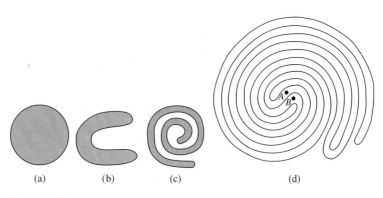

(a)   (b)   (c)   (d)

Figure 9.95

A quick way to tell whether the two dots are inside or outside the curve is to draw a straight line from each dot to a point that is clearly outside the curve. If the straight line crosses the curve an even number of times, the dot is outside. If the straight line crosses the curve an odd number of times, the dot is inside the curve. Can you explain why this procedure works? Determine whether point *A* and point *B* are inside or outside the curve (see Exercises 21 and 22 at the end of this section).

Topological Equivalence

Figure 9.96

Someone once said that a topologist is a person who does not know the difference between a doughnut and a coffee cup. Two geometric figures are said to be *topologically equivalent* if one figure can be elastically twisted, stretched, bent, or shrunk into the other figure without puncturing or ripping the original figure. If a doughnut is made of elastic material, it can be stretched, twisted, bent, shrunk, and distorted until it resembles a coffee cup with a handle, as shown in Fig. 9.96. Thus, the doughnut and the coffee cup are topologically equivalent.

In topology, figures are classified according to their *genus*. The *genus* of an object is determined by the number of holes that go *through* the object. A cup and a doughnut each have one hole and are of genus 1 (and are therefore topologically equivalent). Notice that the cup's handle is considered a hole, whereas the opening at the rim of the cup is not considered a hole. For our purposes, we will consider an object's opening a hole if you could pour liquid *through* the opening. For example, a typical bowling ball has three openings in the surface into which you can put your fingers when preparing to roll the ball, but liquid cannot be poured *through* any of these openings. Therefore, a bowling ball has genus 0 and is topologically equivalent to a marble. Figure 9.97 on page 590 illustrates the genus of several objects.

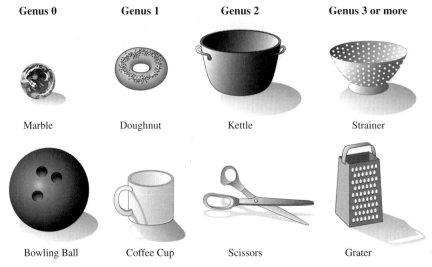

Genus 0   **Genus 1**   **Genus 2**   **Genus 3 or more**

Marble   Doughnut   Kettle   Strainer

Bowling Ball   Coffee Cup   Scissors   Grater

Figure 9.97

SECTION 9.6 EXERCISES

CONCEPT/WRITING EXERCISES

1. Explain why topology is sometimes referred to as "rubber sheet geometry."

2. What is a Möbius strip?

3. Explain how to make a Möbius strip.

4. What is a Klein bottle?

5. What is the maximum number of colors needed to create a map on a flat surface if no two regions colored the same are to share a common border?

6. What is the maximum number of colors needed to create a map if no two regions colored the same are to share a common border if the surface is a

  a) Möbius strip?

  b) torus?

7. What is a Jordan curve?

8. When testing to determine whether a point is inside or outside a Jordan curve, explain why if you count an odd number of lines, the point is inside the curve, and if you count an even number of lines, the point is outside the curve.

9. How is the genus of a figure determined?

10. When are two figures topologically equivalent?

PRACTICE THE SKILLS

In Exercises 11–16, color the map by using a maximum of four colors so that no two regions with a common border have the same color.

11.

1		2
	3	
4		5

12.

13.

14.

15.

16.

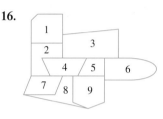

Using the Four-Color Theorem In Exercises 17–20, maps show certain areas of the United States, Canada, and Mexico. Shade in the states (or provinces) using a maximum of four colors so that no two states (or provinces) with a common border have the same color.

17.

18.

19.

20.

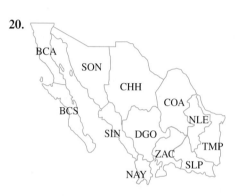

21. Determine whether point *A* in Fig. 9.95(d) on page 589 is inside or outside the Jordan curve.

22. Determine whether point *B* in Fig. 9.95(d) is inside or outside the Jordan curve.

At right is a Jordan curve. In Exercises 23–28, determine if the point is inside or outside the curve.

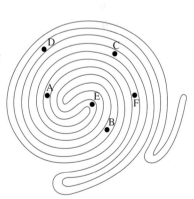

23. Point *F*

24. Point *E*

25. Point *D*

26. Point *C*

27. Point *B*

28. Point *A*

In Exercises 29–40, give the genus of the object. If the object has a genus larger than 5, write "larger than 5."

29.

30.

31.

32.

33.

34.

35.

36.

37.

38.

39.

40.

41. Name at least three objects not mentioned in this section that have

   a) genus 0.

   b) genus 1.

   c) genus 2.

   d) genus 3 or more.

42. Use the result of Experiment 1 on page 586 to find the number of edges on a Möbius strip.

43. Use the result of Experiment 2 on page 586 to find the number of surfaces on a Möbius strip.

44. How many separate strips are obtained in Experiment 3 on page 586?

45. How many separate strips are obtained in Experiment 4 on page 586?

46. Make a Möbius strip. Cut it one-third of the way from the edge, as in Experiment 4. You should get two loops, one going through the other. Determine whether either (or both) of these loops is itself a Möbius strip.

47. a) Take a strip of paper, give it one full twist, and connect the ends. Is the result a Möbius strip with only one side? Explain.

   b) Determine the number of edges, as in Experiment 1.

   c) Determine the number of surfaces, as in Experiment 2.

   d) Cut the strip down the middle. What is the result?

48. Take a strip of paper, make one whole twist and another half twist, and then tape the ends together. Test by a method of your choice to determine whether this has the same properties as a Möbius strip.

CHALLENGE PROBLEMS/GROUP ACTIVITIES

49. Using clay (or glazing compound), make a doughnut. Without puncturing or tearing the doughnut, reshape it into a topologically equivalent figure, a cup with a handle.

50. Using at most four colors, color the map of South America. Do not use the same color for any two countries that share a common border.

Source: www.mapsofworld.com
<http://www.mapsofworld.com>
used with permission.

51. Using at most four colors, color the following map of the counties of New Mexico. Do not use the same color for any two counties that share a common border.

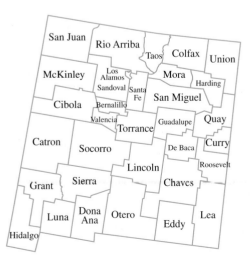

RECREATIONAL MATHEMATICS

52. *An Interesting Surface* Construct the following surface using two strips of paper, scissors, and tape. Then answer the following questions.

a) How many sides does this surface have?

b) How many edges does this surface have?

c) Attempt to cut the surface "in half" by making a small slit in the middle of the paper surface. Then cut along the surface (see dashed line in the figure), keeping the scissors the same distance from the edge. In your own words, describe what happens.

Although the surface shown in the figure above shares some of the same traits as a Möbius strip, this surface is not topologically equivalent to the Möbius strip.

INTERNET/RESEARCH ACTIVITIES

53. Use the Internet to find a map of your state that shows the outline of all the counties within your state. Print this map and, using at most four colors, color it. Do not use the same color for any two counties that share a common border.

54. The short story *Paul Bunyan versus the Conveyor Belt* (1947) by William Hazlett Upson focuses on a conveyor belt in the shape of a Möbius strip. The story can be found in several books that include mathematical essays. Read Upson's short story and write a 200-word description of what Paul Bunyan does to the conveyor belt. Confirm the outcome of the story by repeating Paul's actions with a paper Möbius strip.

9.7 NON-EUCLIDEAN GEOMETRY AND FRACTAL GEOMETRY

▲ Many branches of geometry are needed to accurately represent space.

Ponder the following question: Given a line *l* and a point *P* not on the line *l*, how many lines can you draw through *P* that are parallel to *l*?

You may answer that only one line may be drawn through *P* parallel to *l*. This answer would be correct *provided* the setting of the problem is in a plane and not on the surface of a curved object. The study of this question led to the development of several new branches of geometry. It is now believed all of these branches of geometry, taken together, can be used to accurately represent space. In this section, we will study the geometry of surfaces other than the geometry of the plane.

Non-Euclidean Geometry

In Section 9.1, we stated that postulates or axioms are statements to be accepted as true. In his book *Elements*, Euclid's fifth postulate was, "If a straight line falling on two straight lines makes the interior angles on the same side less than two right angles, the two straight lines, if produced indefinitely, meet on that side on which the angles are less than the two right angles."

Euclid's fifth axiom may be better understood by observing Fig. 9.98. The sum of angles *A* and *B* is less than the sum of two right angles (180°). Therefore, the two lines will meet if extended.

John Playfair (1748–1819), a Scottish physicist and mathematician, wrote a geometry book that was published in 1795. In his book, Playfair gave a logically equivalent interpretation of Euclid's fifth postulate. This version is often referred to as Playfair's postulate or the Euclidean parallel postulate.

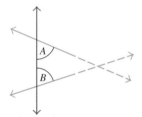

Figure 9.98

THE EUCLIDEAN PARALLEL POSTULATE

Given a line and a point not on the line, one and only one line can be drawn through the given point parallel to the given line (Fig. 9.99)

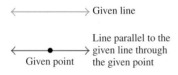

Figure 9.99

The Euclidean parallel postulate may be better understood by looking at Fig. 9.99. Many mathematicians after Euclid believed that this postulate was not as self-evident as the other nine postulates given by Euclid. Others believed that this postulate could be proved from the other nine postulates and therefore was not needed at all. Of the many attempts to prove that the fifth postulate was not needed, the most noteworthy one was presented by Girolamo Saccheri (1667–1733), a Jesuit priest in Italy. In the course of his elaborate chain of deductions, Saccheri proved many of the theorems of what is now called hyperbolic geometry. However, Saccheri did not realize what he had done. He believed that Euclid's geometry was the only "true" geometry and concluded that his own work was in error. Thus, Saccheri narrowly missed receiving credit for a great achievement: the founding of *non-Euclidean geometry*.

Over time, geometers became more and more frustrated at their inability to prove Euclid's fifth postulate. One of them, a Hungarian named Farkos Bolyai, in a letter to his son, Janos Bolyai, wrote, "I entreat you leave the science of parallels alone. . . . I have traveled past all reefs of this infernal dead sea and have always come back with a broken mast and torn sail." The son, refusing to heed his father's advice, continued to think about parallels until, in 1823, he saw the whole truth and enthusiastically declared, "I have created a new universe from nothing." He recognized that geometry branches in two directions, depending on whether Euclid's fifth postulate is applied. He recognized two different geometries and published his discovery as a 24-page appendix to a textbook written by his father. The famous mathematician George Bruce Halsted called it "the most extraordinary two dozen pages in the whole history of thought." Farkos Bolyai proudly presented a copy of his son's work to his friend Carl Friedrich Gauss, then Germany's greatest mathematician, whose reply to the father had a devastating effect on the son. Gauss wrote, "I am unable to praise this work. . . . To praise it would be to praise myself. Indeed, the whole content of the work, the path taken by your son, the results to which he is led, coincides almost entirely with my meditations which occupied my mind partly for the last thirty or thirty-five years." We now know from his earlier correspondence that Gauss had indeed been familiar with *hyperbolic geometry* even before Janos was born. In his letter, Gauss also indicated that it was his intention not to let his theory be published during his lifetime, but to record it so that the theory would not perish with him. It is believed that the reason Gauss did not publish his work was that he feared being ridiculed by other prominent mathematicians of his time.

At about the same time as Bolyai's publication, Nikolay Ivanovich Lobachevsky, a Russian, published a paper that was remarkably like Bolyai's, although it was quite independent of it. Lobachevsky made a deeper investigation and wrote several books. In marked contrast to Bolyai, who received no recognition during his lifetime, Lobachevsky received great praise and became a professor at the University of Kazan.

After the initial discovery, little attention was paid to the subject until 1854, when G. F. Bernhard Riemann (1826–1866), a student of Gauss, suggested a second type of non-Euclidean geometry, which is now called *spherical*, *elliptical*, or *Riemannian geometry*. The hyperbolic geometry of his predecessors was synthetic; that is, it was not based on or related to any concrete model when it was developed. Riemann's geometry was closely related to the theory of surfaces. A *model* may be considered a physical interpretation of the undefined terms that satisfies the axioms. A model may be a picture or an actual physical object.

The two types of non-Euclidean geometries we have mentioned are elliptical geometry and hyperbolic geometry. The major difference among the three geometries lies in the fifth axiom. The fifth axiom of the three geometries is summarized here.

THE FIFTH AXIOM OF GEOMETRY

Euclidean	Elliptical	Hyperbolic
Given a line and a point not on the line, one and only one line can be drawn parallel to the given line through the given point.	Given a line and a point not on the line, no line can be drawn through the given point parallel to the given line.	Given a line and a point not on the line, two or more lines can be drawn through the given point parallel to the given line.

To understand the fifth axiom of the two non-Euclidean geometries, remember that the term *line* is undefined. Thus, a line can be interpreted differently in different geometries. A model for Euclidean geometry is a plane, such as a blackboard (Fig. 9.100a). A model for elliptical geometry is a sphere (Fig. 9.100b). A model for hyperbolic geometry is a pseudosphere (Fig. 9.100c). A pseudosphere is similar to two trumpets placed bell to bell. Obviously, a line on a plane cannot be the same as a line on either of the other two figures.

(a) Plane   (b) Sphere   (c) Pseudosphere

Figure 9.100

Elliptical Geometry

A circle on the surface of a sphere is called a great circle if it divides the sphere into two equal parts. If we were to cut through a sphere along a great circle, we would have two identical pieces. If we interpret a line to be a great circle, then the two red curves in Fig. 9.101(a) are lines. Figure 9.101(a) shows that the fifth axiom of elliptical geometry is true. Two great circles on a sphere must intersect; hence, there can be no parallel lines (Fig. 9.101a).

If we were to construct a triangle on a sphere, the sum of its angles would be greater than 180° (Fig. 9.101b). The theorem "The sum of the measures of the angles of a triangle is greater than 180°" has been proven by means of the axioms of elliptical geometry. The sum of the measures of the angles varies with the area of the triangle and gets closer to 180° as the area decreases.

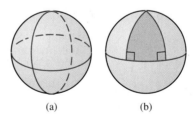

(a)   (b)

Figure 9.101

Hyperbolic Geometry

Lines in hyperbolic geometry* are represented by geodesics on the surface of a pseudosphere. A *geodesic* is the shortest and least-curved arc between two points on a surface. Figure 9.102 shows two different lines represented by geodesics on the surface of a pseudosphere. For simplicity of the diagrams, we only show one of the "bells" of the pseudosphere.

Figure 9.103(a) on page 597 illustrates the fifth axiom of hyperbolic geometry. The diagram illustrates one way that, through the given point, two lines are drawn parallel to the given line. If we were to construct a triangle on a pseudosphere, the sum of

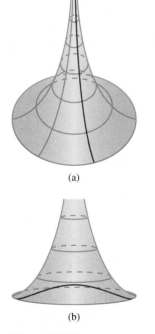

(a)

(b)

Figure 9.102

*A formal discussion of hyperbolic geometry is beyond the scope of this text.

"The Great Architect of the universe now appears to be a great mathematician."
    British physicist Sir James Jeans

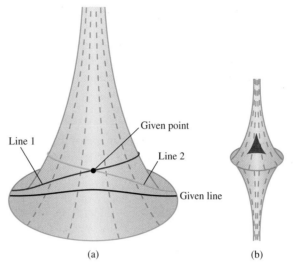

Figure 9.103

the measures of the angles would be less than 180° (Fig. 9.103b). The theorem "The sum of the measures of the angles of a triangle is less than 180°" has been proven by means of the axioms of hyperbolic geometry.

We have stated that the sum of the measures of the angles of a triangle is 180°, is greater than 180°, and is less than 180°. Which statement is correct? Each statement is correct *in its own geometry*. Many theorems hold true for all three geometries; vertical angles still have the same measure, we can uniquely bisect a line segment with a straightedge and compass alone, and so on.

The many theorems based on the fifth postulate may differ in each geometry. It is important for you to realize that each theorem proved is true *in its own geometry* because each is logically deduced from the given set of axioms of the geometry. No one system is the "best" system. Euclidean geometry may appear to be the one to use in the classroom, where the blackboard is flat. In discussions involving Earth as a whole, however, elliptical geometry may be the most useful since Earth is a sphere. If the object under consideration has the shape of a saddle or pseudosphere, hyperbolic geometry may be the most useful.

Fractal Geometry

We are familiar with one-, two-, and three-dimensional figures. Many objects, however, are difficult to categorize as one-, two-, or three-dimensional. For example, how would you classify the irregular shapes we see in nature, such as a coastline, or the bark on a tree, or a mountain, or a path followed by lightning? For a long time mathematicians assumed that making realistic geometric models of natural shapes and figures was almost impossible, but the development of *fractal geometry* now makes it possible. Both color photos on the next page were made by using fractal geometry. The discovery and study of fractal geometry has been one of the most popular mathematical topics in recent times.

The word *fractal* (from the Latin word *fractus*, "broken up, fragmented") was first used in the mid-1970s by mathematician Benoit Mandelbrot to describe shapes that had several common characteristics, including some form of "self-similarity," as will be seen shortly in the Koch snowflake.

▲ Fractal images

Typical fractals are extremely irregular curves or surfaces that "wiggle" enough so that they are not considered one-dimensional. Fractals do not have integer dimensions; their dimensions are between 1 and 2. For example, a fractal may have a dimension of 1.26. Fractals are developed by applying the same rule over and over again, with the end point of each simple step becoming the starting point for the next step, in a process called *recursion*.

Using the recursive process, we will develop a famous fractal called the *Koch snowflake* named after Helga von Koch, a Swedish mathematician who first discovered its remarkable characteristics. The Koch snowflake illustrates a property of all fractals called *self-similarity*; that is, each smaller piece of the curve resembles the whole curve.

To develop the Koch snowflake:

1. Start with an equilateral triangle (step 1, Fig. 9.104).
2. Whenever you see an edge —— replace it with ⎺⋀⎺ (steps 2–4).

What is the perimeter of the snowflake in Fig. 9.104, and what is its area? A portion of the boundary of the Koch snowflake known as the Koch curve or the snowflake curve is represented in Fig. 9.105.

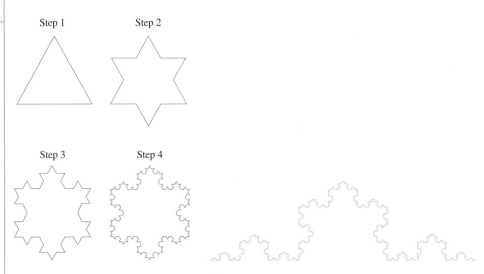

Step 1　　Step 2

Step 3　　Step 4

Figure 9.104　　　**Figure 9.105**

The Koch curve consists of infinitely many pieces of the form ⎺⋀⎺. Notice that after each step, the perimeter is $\frac{4}{3}$ times the perimeter of the previous step. Therefore, the Koch snowflake has an infinite perimeter. It can be shown that the area of the snowflake is 1.6 times the area of the starting equilateral triangle. Thus, the area of the snowflake is finite. The Koch snowflake has a finite area enclosed by an infinite boundary! This fact may seem difficult to accept, but it is true. However, the Koch snowflake, like other fractals, is not an everyday run-of-the-mill geometric shape.

Let us look at a few more fractals made using the recursive process. We will now construct what is known as a *fractal tree*. Start with a tree trunk (Fig. 9.106a on page 599). Draw two branches, each one a bit smaller than the trunk (Fig. 9.106b). Draw two branches from each of those branches, and continue; see Fig. 9.106(c) and (d). Ideally, we continue the process forever.

Figure 9.108 Sierpinski carpet

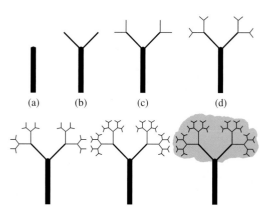

Figure 9.106 The fractal tree

If you take a little piece of any branch and zoom in on it, it will look exactly like the original tree. Fractals are *scale independent*, which means that you cannot really tell whether you are looking at something very big or something very small because the fractal looks the same whether you are close to it or far from it.

In Figs. 9.107 and 9.108, we develop two other fractals through the process of recursion. Figure 9.107 shows a fractal called the Sierpinski triangle, and Fig. 9.108 shows a fractal called the Sierpinski carpet. Both fractals are named after Waclaw Sierpinski, a Polish mathematician who is best known for his work with fractals and space-filling curves.

Figure 9.107 Sierpinski triangle

Fractals provide a way to study natural forms such as coastlines, trees, mountains, galaxies, polymers, rivers, weather patterns, brains, lungs, and blood supply. Fractals also help explain that which appears chaotic. The blood supply in the body is one example. The branching of arteries and veins appears chaotic, but closer inspection reveals that the same type of branching occurs for smaller and smaller blood vessels, down to the capillaries. Thus, fractal geometry provides a geometric structure for chaotic processes in nature. The study of chaotic processes is called *chaos theory*.

Fractals nowadays have a potentially important role to play in characterizing weather systems and in providing insight into various physical processes such as the occurrence of earthquakes or the formation of deposits that shorten battery life. Some scientists view fractal statistics as a doorway to unifying theories of medicine, offering a powerful glimpse of what it means to be healthy.

Fractals lie at the heart of current efforts to understand complex natural phenomena. Unraveling their intricacies could reveal the basic design principles at work in our world. Until recently, there was no way to describe fractals. Today, we are beginning to see such features everywhere. Tomorrow, we may look at the entire universe through a fractal lens.

SECTION 9.7 EXERCISES

CONCEPT/WRITING EXERCISES

In Exercises 1–6, describe the accomplishments of the mathematician.

1. Benoit Mandelbrot

2. G. F. Bernhard Riemann

3. Nikolay Ivanovich Lobachevsky

4. Carl Friedrich Gauss

5. Janos Bolyai

6. Girolamo Saccheri

7. State the fifth axiom of

   a) Euclidean geometry.

   b) elliptical geometry.

   c) hyperbolic geometry.

8. State the theorem concerning the sum of the measures of the angles of a triangle in

   a) Euclidean geometry.

   b) hyperbolic geometry.

   c) elliptical geometry.

9. What model is often used in describing and explaining Euclidean geometry?

10. What model is often used in describing and explaining elliptical geometry?

11. What model is often used in describing and explaining hyperbolic geometry?

12. What do we mean when we say that no one axiomatic system of geometry is "best"?

13. List the three types of curvature of space and the types of geometry that correspond to them.

14. List at least five natural forms that appear chaotic that we can study using fractals.

PRACTICE THE SKILLS

In the following, we show a fractal-like figure made using a recursive process with the letter "M." In Exercises 15–18, use this fractal-like figure as a guide in constructing fractal-like figures with the letter given. Show three steps, as is done here.

15. I   **16.** V   **17.** H   **18.** W

19. **a)** Develop a fractal by beginning with a square and replacing each side —— with a ⌐L . Repeat this process twice.

   b) If you continue this process, will the fractal's perimeter be finite or infinite? Explain.

   c) Will the fractal's area be finite or infinite? Explain.

PROBLEM SOLVING/GROUP ACTIVITY

20. In forming the Koch snowflake in Figure 9.104 on page 598, the perimeter becomes greater at each step in the process. If each side of the original triangle is 1 unit, a general formula for the perimeter, L, of the snowflake at any step, n, may be found by the formula

$$L = 3\left(\frac{4}{3}\right)^{n-1}$$

For example, at the first step when $n = 1$, the perimeter is 3 units, which can be verified by the formula as follows:

$$L = 3\left(\frac{4}{3}\right)^{1-1} = 3\left(\frac{4}{3}\right)^{0} = 3 \cdot 1 = 3$$

At the second step, when $n = 2$, we find the perimeter as follows:

$$L = 3\left(\frac{4}{3}\right)^{2-1} = 3\left(\frac{4}{3}\right) = 4$$

Thus, at the second step the perimeter of the snowflake is 4 units.

a) Use the formula to complete the following table.

Step	Perimeter
1	
2	
3	
4	
5	
6	

b) Use the results of your calculations to explain why the perimeter of the Koch snowflake is infinite.

c) Explain how the Koch snowflake can have an infinite perimeter, but a finite area.

INTERNET/RESEARCH ACTIVITIES

In Exercises 21–23, references include the Internet, books on art, encyclopedias, and history of mathematics books.

21. To complete his masterpiece *Circle Limit III*, M. C. Escher studied a model of hyperbolic geometry called the *Poincaré disk*. Write a paper on the Poincaré disk and how it was used in Escher's art. Include representations of *infinity* and the concepts of *point* and *line* in hyperbolic geometry.

▲ Escher's *Circle Limit III*

22. To transfer his two-dimensional tiling known as *Symmetry Work 45* to a sphere, M. C. Escher used the spherical geometry of Bernhard Riemann. Write a paper on Escher's use of geometry to complete this masterpiece.

23. Go to the web site *Fantastic Fractals* at www.fantastic-fractals.com and study the information about fractals given there. Print copies, in color if a color printer is available, of the Mandlebrot set and the Julia set.

CHAPTER ⑨ SUMMARY

IMPORTANT FACTS

The sum of the measures of the angles of a triangle is 180°.

The sum of the measures of the angles of a quadrilateral is 360°.

The sum of the measures of the interior angles of an n-sided polygon is $(n - 2)180°$.

TRIANGLE

b (or s_3)

$P = s_1 + s_2 + s_3(s_3 = b)$
$A = \frac{1}{2}bh$

SQUARE

s

$P = 4s$
$A = s^2$

RECTANGLE

w

l

$P = 2l + 2w$
$A = lw$

PARALLELOGRAM

h w

b

$P = 2b + 2w$
$A = bh$

TRAPEZOID

b_1

s_1 h s_2

b_2

$P = s_1 + s_2 + b_1 + b_2$
$A = \frac{1}{2}h(b_1 + b_2)$

PYTHAGOREAN THEOREM

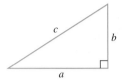

$$a^2 + b^2 = c^2$$

CIRCLE

$$A = \pi r^2; C = 2\pi r \text{ or } C = \pi d$$

CUBE

$$V = s^3$$
$$SA = 6s^2$$

RECTANGULAR SOLID

$$V = lwh$$
$$SA = 2lw + 2wh + 2lh$$

CYLINDER

$$V = \pi r^2 h$$
$$SA = 2\pi rh + 2\pi r^2$$

CONE

$$V = \frac{1}{3}\pi r^2 h$$
$$SA = \pi r^2 + \pi r\sqrt{r^2 + h^2}$$

SPHERE

$$V = \frac{4}{3}\pi r^3$$
$$SA = 4\pi r^2$$

PRISM

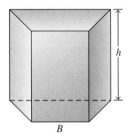

$V = Bh$, where B is the area of the base

PYRAMID

$V = \frac{1}{3}Bh$, where B is the area of the base

FIFTH POSTULATE IN EUCLIDEAN GEOMETRY
Given a line and a point not on the line, only one line can be drawn through the given point parallel to the given line.

FIFTH POSTULATE IN ELLIPTICAL GEOMETRY
Given a line and a point not on the line, no line can be drawn through the given point parallel to the given line.

FIFTH POSTULATE IN HYPERBOLIC GEOMETRY
Given a line and a point not on the line, two or more lines can be drawn through the given point parallel to the given line.

CHAPTER ⑨ REVIEW EXERCISES

9.1

In Exercises 1–6, use the figure shown to determine the following.

1. $\angle ABF \cap \angle CBH$

2. $\overrightarrow{AB} \cap \overrightarrow{DC}$

3. $\overline{BF} \cup \overline{FC} \cup \overline{BC}$

4. $\overrightarrow{BH} \cup \overrightarrow{HB}$

5. $\overleftrightarrow{HI} \cap \overleftrightarrow{EG}$

6. $\overleftrightarrow{CF} \cap \overleftrightarrow{CG}$

7. $m\angle A = 23.7°$. Determine the measure of the complement of $\angle A$.

8. $m\angle B = 124.7°$. Determine the measure of the supplement $\angle B$.

9.2

In Exercises 9–12, use the similar triangles ABC and A'B'C shown to determine the following.

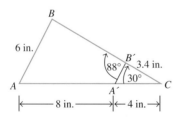

9. The length of $\overline{BC}$

10. The length of $\overline{A'B'}$

11. $m\angle BAC$

12. $m\angle ABC$

13. In the following figure, l_1 and l_2 are parallel lines. Determine $m\angle 1$ through $m\angle 6$.

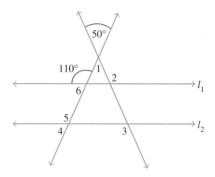

14. Determine the sum of the measures of the interior angles of a pentagon.

9.3

In Exercises 15–18, determine (a) the area and (b) the perimeter of the figure.

15.

[square: 8 cm height, 10 cm base]

16.

[trapezoid: 4 in. top, 3.2 in. left side, 2 in. height, 3.2 in. right side, 9 in. bottom]

17.

[parallelogram: 9 in. slant side, 7 in. height, 12 in. base]

18.

[triangle: 3 km, 4 km]

19. Determine (a) the area and (b) the circumference of the circle. Use the $\boxed{\pi}$ key on a calculator and round your answer to the nearest hundredth.

[circle with radius 13 cm]

In Exercises 20 and 21, determine the shaded area. When appropriate, use the $\boxed{\pi}$ key on your calculator and round your answer to the nearest hundredth.

20.

8 in.

21.

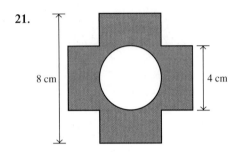

8 cm   4 cm

22. *Cost of Kitchen Tile* Determine the total cost of covering a 14 ft by 16 ft kitchen floor with ceramic tile. The cost of the tile selected is $5.25 per square foot.

9.4

In Exercises 23–26, determine (a) the volume and (b) the surface area of the figure. When appropriate, use the $\boxed{\pi}$ key on your calculator and round your answer to the nearest hundredth.

23.

4 cm

4 cm

10 cm

3 cm

24.

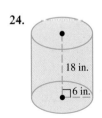

18 in.

6 in.

25.

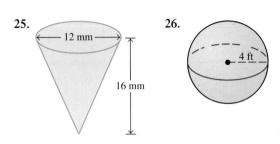

12 mm

16 mm

26.

4 ft

In Exercises 27 and 28, determine the volume of the figure.

27.

9 m   12 m

8 m

15 m

28.

5 ft   7 ft

6 ft   5 ft

In Exercises 29 and 30, determine the volume of the shaded area. When appropriate, use the $\boxed{\pi}$ key and round your answer to the nearest hundredth.

29.

9 cm

4 cm

30.

6 cm

12 cm

31. *Water Trough* Steven Dale has a water trough whose ends are trapezoids and whose sides are rectangles, as illustrated. He is afraid that the base it is sitting on will not support the weight of the trough when it is filled with water. He knows that the base will support 4800 lb.

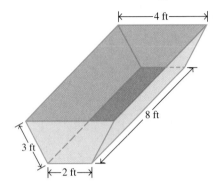

a) If the trough is filled with water, determine the number of cubic feet of water contained in the trough.

b) Determine the total weight, assuming that the trough weighs 375 lb and the water weighs 62.4 lb per cubic foot. Is the base strong enough to support the trough filled with water?

c) If 1 gal of water weighs 8.3 lb, how many gallons of water will the trough hold?

9.5

In Exercises 32 and 33, use the given triangle and reflection lines to construct the indicated reflections. Show the triangle in the positions both before and after the reflection.

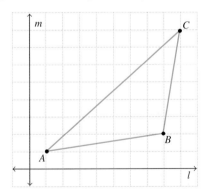

32. Construct the reflection of triangle *ABC* about line *l*.

33. Construct the reflection of triangle *ABC* about line *m*.

*In Exercises 34 and 35, use translation vectors **v** and **w** to construct the indicated translations. Show the parallelogram in the positions both before and after the translation.*

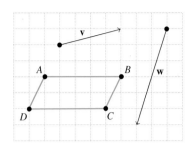

34. Construct the translation of parallelogram *ABCD* using translation vector **v**.

35. Construct the translation of parallelogram *ABCD* using translation vector **w**.

In Exercises 36–38, use the given figure and rotation point P to construct the indicated rotations. Show the trapezoid in the positions both before and after the rotation.

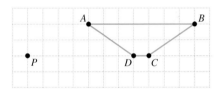

36. Construct a 90° rotation of trapezoid *ABCD* about point *P*.

37. Construct a 180° rotation of trapezoid *ABCD* about point *P*.

38. Construct a 270° rotation of trapezoid *ABCD* about point *P*.

*In Exercises 39 and 40, use the given figure, translation vector **v**, and reflection lines l and m to construct the indicated glide reflections. Show the triangle in the positions both before and after the glide reflection.*

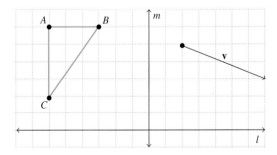

39. Construct a glide reflection of triangle *ABC* using vector **v** and reflection line *l*.

40. Construct a glide reflection of triangle *ABC* using vector **v** and reflection line *m*.

In Exercises 41 and 42, use the following figure to answer the following questions.

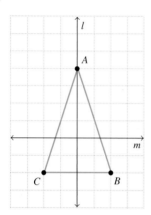

41. Does triangle *ABC* have reflective symmetry about line *l*? Explain.

42. Does triangle *ABC* have reflective symmetry about line *m*? Explain.

In Exercises 43 and 44, use the following figure to answer the following questions.

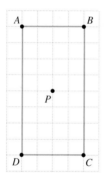

43. Does rectangle *ABCD* have 90° rotational symmetry about point *P*? Explain.

44. Does rectangle *ABCD* have 180° rotational symmetry about point *P*? Explain.

9.6

45. Give an example of an object that has

    a) genus 0.

    b) genus 1.

    c) genus 2.

    d) genus 3 or more.

46. The map shows the states of Germany. Shade in the states using a maximum of four colors so that no two states with a common border have the same color.

47. Determine whether point *A* is inside or outside the Jordan curve.

9.7

48. State the fifth axiom of Euclidean, elliptical, and hyperbolic geometry.

49. Develop a fractal by beginning with a square and replacing each side ⎯ with a ⌐⌐. Repeat this process twice.

50. Construct a Koch snowflake by beginning with an equilateral triangle and replacing each side with a ⌃. Repeat this process twice.

CHAPTER ❾ TEST

In Exercises 1–4, use the figure to describe the following sets of points.

1. $\overleftrightarrow{AF} \cap \overleftrightarrow{EF}$

2. $\overline{BC} \cup \overline{CD} \cup \overline{BD}$

3. $\angle EDF \cap \angle BDC$

4. $\overrightarrow{AC} \cup \overrightarrow{BA}$

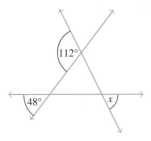

5. $m\angle A = 12.4°$. Determine the measure of the complement of $\angle A$.

6. $m\angle B = 51.7°$. Determine the measure of the supplement of $\angle B$.

7. In the figure, determine the measure of $\angle x$.

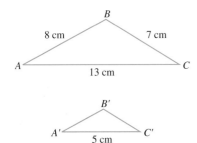

8. Determine the sum of the measures of the interior angles of a decagon.

9. Triangles ABC and $A'B'C'$ are similar figures. Determine the length of side $\overline{B'C'}$.

10. Right triangle ABC (see top of right-hand column) has one leg of length 5 in. and a hypotenuse of length 13 in.

   a) Determine the length of the other leg.

   b) Determine the perimeter of the triangle.

   c) Determine the area of the triangle.

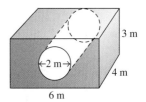

11. Determine (a) the volume and (b) the surface area of a sphere of diameter 14 cm.

12. Determine the volume of the shaded area. Use the $\boxed{\pi}$ key on your calculator and round your answer to the nearest hundredth.

13. Determine the volume of the pyramid.

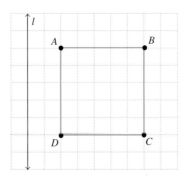

14. Construct a reflection of square $ABCD$, shown below, about line l. Show the square in the positions both before and after the reflection.

15. Construct a translation of quadrilateral *ABCD*, shown below, using translation vector **v**. Show the quadrilateral in the positions both before and after the translation.

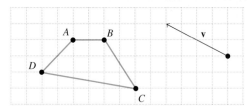

16. Construct a 180° rotation of triangle *ABC*, shown below, about rotation point *P*. Show the triangle in the positions both before and after the rotation.

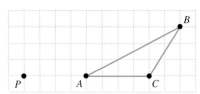

17. Construct a glide reflection of rectangle *ABCD*, shown below, using translation vector **v** and reflection line *l*. Show the rectangle in the positions both before and after the glide reflection.

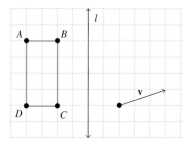

18. Use the figure below to answer the following questions.

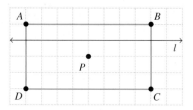

a) Does rectangle *ABCD* have reflective symmetry about line *l*? Explain.

b) Does rectangle *ABCD* have 180° rotational symmetry about point *P*? Explain.

19. What is a Möbius strip?

20. a) Sketch an object of genus 1.

b) Sketch an object of genus 2.

21. Explain how the fifth axiom in Euclidean geometry, elliptical geometry, and hyperbolic geometry differ.

G R O U P   P R O J E C T S

1. Samantha Saraniti is thinking of buying a circular hot tub 12 ft in diameter, 4 ft deep, and weighing 475 lb. She wants to place the hot tub on a deck built to support 30,000 lb.

a) Determine the volume of the water in the hot tub in cubic feet.

b) Determine the number of gallons of water the hot tub will hold. (*Note:* 1 ft^3 ≈ 7.5 gal.)

c) Determine the weight of the water in the hot tub. (*Hint:* Fresh water weighs about 62.4 lb/ft^3.)

d) Will the deck support the weight of the hot tub and water?

e) Will the deck support the weight of the hot tub, water, and four people, whose average weight is 115 lb?

DESIGNING A RAMP

2. David and Sandra Jessee are planning to build a ramp so that the front entrance of their home is wheelchair accessible. The ramp will be 3 feet wide. It will rise 2 in. for each foot of length of horizontal distance. Where the ramp meets the porch, the ramp must be 2 ft high. To provide stability for the ramp, the Jessees will install a slab of concrete 4 in. thick and 6 in. longer and wider than the ramp (see accompanying figure). The top of the slab will be level with the ground. The ramp may be constructed of concrete or pressure-treated lumber. You are to estimate the cost of materials for constructing the slab, the ramp of concrete, and the ramp of pressure-treated lumber.

Slab

a) Determine the length of the base of the ramp.

b) Determine the dimensions of the concrete slab on which the ramp will set.

c) Determine the volume of the concrete in cubic yards needed to construct the slab.

d) If ready-mix concrete costs $45 per cubic yard, determine the cost of the concrete needed to construct the slab.

Concrete Ramp

e) To build the ramp of concrete, a form in the shape of the ramp must be framed. The two sides of the form are triangular, and the shape of the end, which is against the porch, is rectangular. The form will be framed from $\frac{3}{4}$ in. plywood, which comes in 4 ft × 8 ft sheets. Determine the number of sheets of plywood needed. Assume that the entire sheet(s) will be used to make the sides and the end of the form and that there is no waste.

f) If the plywood costs $18.95 for a 4 ft × 8 ft sheet, determine the cost of the plywood.

g) To brace the form, the Jessees will need two boards 2 in. × 4 in. × 8 ft (referred to as 8 ft 2 × 4's) and six pieces of lumber 2 in. × 4 in. × 3 ft. These six pieces of lumber will be cut from 8 ft 2 × 4 boards. Determine the number of 8 ft 2 × 4 boards needed.

h) Determine the cost of the 8 ft 2 × 4 boards needed in part (g) if one board costs $2.14.

i) Determine the volume, in cubic yards, of concrete needed to fill the form.

j) Determine the cost of the concrete needed to fill the form.

k) Determine the total cost of materials for building the ramp of concrete by adding the results in parts (d), (f), (h), and (j).

Wooden Ramp

l) Determine the length of the top of the ramp.

m) The top of the ramp will be constructed of $\frac{5}{4}$ in. × 6 in. × 10 ft pressure-treated lumber. The boards will be butted end to end to make the necessary length and will be supported from underneath by a wooden frame. Determine the number of boards needed to cover the top of the ramp. The boards are laid lengthwise on the ramp.

n) Determine the cost of the boards to cover the top of the ramp if the price of a 10 ft length is $6.47.

o) To support the top of the ramp, the Jessees will need 10 pieces of 8 ft 2 × 4's. The price of a pressure-treated 8 ft 2 × 4 is $2.44. Determine the cost of the supports.

p) Determine the cost of the materials for building a wooden ramp by adding the amounts from parts (d), (n), and (o).

q) Are the materials for constructing a concrete ramp or a wooden ramp less expensive?

CHAPTER 10

Mathematical Systems

▲ Group theory is used in the field of medical image analysis including magnetic resonance imaging.

WHAT YOU WILL LEARN

- Mathematical systems
- Commutative and associative properties
- Closure, identity element, and inverses
- Groups and commutative groups
- Clock arithmetic
- Mathematical systems without numbers
- Modular arithmetic

WHY IT IS IMPORTANT

Mathematics plays a crucial role in the study of the interaction between many objects of interest to scientists. One branch of mathematics of particular importance to this study is group theory. Physicists and chemists rely on group theory to study the behavior of the tiniest particles that make up an atom in a scientific field known as *quantum mechanics*. Group theory can be used to predict the interactive behavior of subatomic particles within an atom, atoms within a molecule, and molecules within a chemical reaction. Group theory also plays an important role in robotics, computer graphics, weather forecasting, musical theory, and medical image analysis.

10.1 GROUPS

▲ The big bang theory uses groups to address questions relating to the size, structure, and future of the universe.

How big is the universe? Is there an end to the universe? Does the universe continue to expand? Such questions are addressed by the scientific theory known as the *big bang theory*. In this section, we will introduce a mathematical structure known as a *group*. Many modern scientific theories, including the big bang theory, make use of groups and other mathematical topics discussed in this book.

We begin our discussion by introducing a mathematical system. As you will learn shortly, you already know and use many mathematical systems.

> A **mathematical system** consists of a set of elements and at least one binary operation.

In the above definition, we mention binary operation. A *binary operation* is an operation, or rule, that can be performed on two and only two elements of a set. The result is a single element. When we add *two* integers, the sum is *one* integer. When we multiply *two* integers, the product is *one* integer. Thus, addition and multiplication are both binary operations. Is finding the reciprocal of a number a binary operation? No, it is an operation on a single element of a set.

> **TIMELY TIP** Throughout this section, we will work with many different sets of numbers, such as natural numbers, whole numbers, integers, rational numbers, and real numbers. A review of the numbers that make up each of these sets can be found on page 270 in Section 5.5.

When you learned how to add integers, you were introduced to a mathematical system. The set of elements is the set of integers, and the binary operation is addition. When you learned how to multiply integers, you became familiar with a second mathematical system. The set of integers with the operation of subtraction and the set of integers with the operation of division are two other examples of mathematical systems since subtraction and division are also binary operations.

Some mathematical systems are used in solving everyday problems, such as planning work schedules. Others are more abstract and are used primarily in research, chemistry, physical structure, matter, the nature of genes, and other scientific fields.

Commutative and Associative Properties

Once a mathematical system is defined, its structure may display certain properties. Consider the set of integers

$$I = \{\ldots, -3, -2, -1, 0, 1, 2, 3, \ldots\}$$

Recall that the ellipsis, the three dots at each end of the set, indicates that the set continues in the same manner.

The set of integers can be studied with the operations of addition, subtraction, multiplication, and division as separate mathematical systems. For example, when we study the set of integers under the operations of addition or multiplication, we see that the commutative and associative properties hold. The general forms of the properties are shown here.

FOR ANY ELEMENTS a, b, AND c		
	Addition	**Multiplication**
Commutative property	$a + b = b + a$	$a \cdot b = b \cdot a$
Associative property	$(a + b) + c = a + (b + c)$	$(a \cdot b) \cdot c = a \cdot (b \cdot c)$

The integers *are commutative* under the operations of *addition and multiplication*. For example,

$$\begin{array}{ccc} \textbf{Addition} & & \textbf{Multiplication} \\ 2 + 4 = 4 + 2 & \text{and} & 2 \cdot 4 = 4 \cdot 2 \\ 6 = 6 & & 8 = 8 \end{array}$$

The integers, however, *are not commutative* under the operations of *subtraction and division*. For example,

$$\begin{array}{ccc} \textbf{Subtraction} & & \textbf{Division} \\ 4 - 2 \neq 2 - 4 & \text{and} & 4 \div 2 \neq 2 \div 4 \\ 2 \neq -2 & & 2 \neq \frac{1}{2} \end{array}$$

The integers *are associative* under the operations of *addition and multiplication*. For example,

$$\begin{array}{ccc} \textbf{Addition} & & \textbf{Multiplication} \\ (1 + 2) + 3 = 1 + (2 + 3) & \text{and} & (1 \cdot 2) \cdot 3 = 1 \cdot (2 \cdot 3) \\ 3 + 3 = 1 + 5 & & 2 \cdot 3 = 1 \cdot 6 \\ 6 = 6 & & 6 = 6 \end{array}$$

The integers, however, *are not associative* under the operations of *subtraction and division*. See Exercises 21 and 22 at the end of this section.

To say that a set of elements is commutative under a given operation means that the commutative property holds for *any* elements a and b in the set. Similarly, to say that a set of elements is associative under a given operation means that the associative property holds for *any* elements a, b, and c in the set.

Consider the mathematical system consisting of the set of integers under the operation of addition. Because the set of integers is infinite, this mathematical system is an example of an *infinite mathematical system*. We will study certain properties of this mathematical system. The first property we will examine is closure.

Closure

The sum of any two integers is an integer. Therefore, the set of integers is said to be *closed*, or to satisfy the *closure property*, under the operation of addition.

If a binary operation is performed on any two elements of a set and the result is an element of the set, then that set is **closed** (or has **closure**) under the given binary operation.

Is the set of integers closed under the operation of multiplication? The answer is yes. When any two integers are multiplied, the product will be an integer.

Is the set of integers closed under the operation of subtraction? Again, the answer is yes. The difference of any two integers is an integer.

Is the set of integers closed under the operation of division? The answer is no because two integers may have a quotient that is not an integer. For example, if we select the integers 2 and 3, the quotient of 2 divided by 3 is $\frac{2}{3}$, which is not an integer. Thus, the integers are not closed under the operation of division.

We showed that the set of integers was not closed under the operation of division by finding two integers whose quotient was not an integer. A specific example illustrating that a specific property is not true is called a *counterexample*. Mathematicians and scientists often try to find a counterexample to confirm that a specific property is not always true.

Identity Element

Now we will discuss the identity element for the set of integers under the operation of addition. Is there an element in the set that, when added to any given integer, results in a sum that is the given integer? The answer is yes. The sum of 0 and any integer is the given integer. For example, $1 + 0 = 0 + 1 = 1$, $-4 + 0 = 0 + (-4) = -4$, and so on. For this reason, we call 0 the *additive identity element* for the set of integers. Note that for any integer a, $a + 0 = 0 + a = a$.

An **identity element** is an element in a set such that when a binary operation is performed on it and any given element in the set, the result is the given element.

Is there an identity element for the set of integers under the operation of multiplication? The answer is yes; it is the number 1. Note that $2 \cdot 1 = 1 \cdot 2 = 2$, $3 \cdot 1 = 1 \cdot 3 = 3$, and so on. For any integer a, $a \cdot 1 = 1 \cdot a = a$. For this reason, 1 is called the *multiplicative identity element* for the set of integers.

Inverses

What integer, when added to 4, gives a sum of 0; that is, $4 + \boxed{} = 0$? The shaded area is to be filled in with the integer -4: $4 + (-4) = 0$. We say that -4 is the additive inverse of 4 and that 4 is the additive inverse of -4. Note that the sum of the element and its additive inverse gives the additive identity element 0. What is the additive inverse of 12? Since $12 + (-12) = 0$, -12 is the additive inverse of 12.

Other examples of integers and their additive inverses are

Element	+	Additive Inverse	=	Identity Element
0	+	0	=	0
2	+	(−2)	=	0
−5	+	5	=	0

Note that for the operation of addition, every integer a has a unique inverse, $-a$, such that $a + (-a) = -a + a = 0$.

> When a binary operation is performed on two elements in a set and the result is the identity element for the binary operation, each element is said to be the **inverse** of the other.

Does every integer have an inverse under the operation of multiplication? For multiplication, the product of an integer and its inverse must yield the multiplicative identity element, 1. What is the multiplicative inverse of 2? That is, 2 times what number gives 1?

$$2 \cdot \boxed{} = 1 \qquad 2 \cdot \tfrac{1}{2} = 1$$

However, since $\tfrac{1}{2}$ is not an integer, 2 does not have a multiplicative inverse in the set of integers.

Group

Let's review what we have learned about the mathematical system consisting of the set of integers under the operation of addition.

1. The set of integers is *closed* under the operation of addition.
2. The set of integers has an *identity element* under the operation of addition.
3. Each element in the set of integers has an *inverse* under the operation of addition.
4. The *associative property* holds for the set of integers under the operation of addition.

The set of integers under the operation of addition is an example of a *group*. The properties of a group can be summarized as follows.

> ### PROPERTIES OF A GROUP
> Any mathematical system that meets the following four requirements is called a **group**.
> 1. The set of elements is *closed* under the given operation.
> 2. An *identity element* exists for the set under the given operation.
> 3. Every element in the set has an *inverse* under the given operation.
> 4. The set of elements is *associative* under the given operation.

> **TIMELY TIP** Note that the set of elements need *not* be commutative for the mathematical system to be a group. Also note that *every* element in the set must have an inverse for the mathematical system to be a group.

It is often very time consuming to show that the associative property holds for all cases. In many of the examples that follow, we will state that the associative property holds for the given set of elements under the given operation.

Commutative Group

The commutative property does not need to hold for a mathematical system to be a group. However, if a mathematical system meets the four requirements of a group and is also commutative under the given operation, the mathematical system is a *commutative* (or *abelian*) *group.* The abelian group is named after Niels Abel (see the Profiles in Mathematics on page 614).

A group that satisfies the commutative property is called a **commutative group** (or **abelian group**).

Because the commutative property holds for the set of integers under the operation of addition, the set of integers under the operation of addition is not only a group, but it is a commutative group.

PROPERTIES OF A COMMUTATIVE GROUP
A mathematical system is a **commutative group** if all five of the following conditions hold.
1. The set of elements is *closed* under the given operation.
2. An *identity element* exists for the set under the given operation.
3. Every element in the set has an *inverse* under the given operation.
4. The set of elements is *associative* under the given operation.
5. The set of elements is *commutative* under the given operation.

To determine whether a mathematical system is a group under a given operation, *check, in the following order,* whether (1) the system is closed under the given operation, (2) there is an identity element in the set for the given operation, (3) every element in the set has an inverse under the given operation, and (4) the associative property holds under the given operation. If *any* of these four requirements is *not* met, stop and state that the mathematical system is not a group. If asked to determine whether the mathematical system is a commutative group, you also need to determine whether the commutative property holds for the given operation.

EXAMPLE ❶ *Is It a Group?*

Determine whether the set of rational numbers under the operation of multiplication forms a group.

SOLUTION Recall from Chapter 5 that the rational numbers are the set of numbers of the form $\frac{p}{q}$ where p and q are integers and $q \neq 0$. The set of rational numbers includes all fractions and integers.

1. *Closure*: The product of any two rational numbers is a rational number. Therefore, the rational numbers are closed under the operation of multiplication.
2. *Identity element*: The multiplicative identity element for the set of rational numbers is 1. Note, for example, that $3 \cdot 1 = 1 \cdot 3 = 3$, and $\frac{3}{8} \cdot 1 = 1 \cdot \frac{3}{8} = \frac{3}{8}$. For any rational number a, $a \cdot 1 = 1 \cdot a = a$.
3. *Inverse elements*: For the mathematical system to be a group under the operation of multiplication, *each and every* rational number must have a multiplicative inverse in the set of rational numbers. Remember that for the operation of

multiplication, the product of a number and its inverse must give the multiplicative identity element, 1. Let's check a few rational numbers:

$$\text{Rational Number} \cdot \text{Inverse} = \text{Identity Element}$$

Rational Number	·	Inverse	=	Identity Element
3	·	$\dfrac{1}{3}$	=	1
$\dfrac{2}{3}$	·	$\dfrac{3}{2}$	=	1
$-\dfrac{1}{5}$	·	-5	=	1

Looking at these examples you might deduce that each rational number does have an inverse. However, one rational number, 0, does not have an inverse.

$$0 \cdot ? = 1$$

Because there is no rational number that when multiplied by 0 gives 1, 0 does not have a multiplicative inverse. Since *every* rational number does not have an inverse, this mathematical system is not a group.

There is no need at this point to check the associative property because we have already shown that the mathematical system of rational numbers under the operation of multiplication is not a group.

EXAMPLE ❷ *Real Numbers under Addition*

Determine whether the set of real numbers under the operation of addition forms a commutative group.

SOLUTION Recall from Chapter 5 that the real numbers can be thought of as the numbers that correspond to *each* point on the real number line. The set of real numbers include all rational numbers and all irrational numbers.

1. *Closure*: The sum of any two real numbers is a real number. Therefore, the real numbers are closed under the operation of addition.
2. *Identity element*: The additive identity element for the set of real numbers is 0. For example, $\frac{2}{3} + 0 = 0 + \frac{2}{3} = \frac{2}{3}$ and $\sqrt{5} + 0 = 0 + \sqrt{5} = \sqrt{5}$. For any real number a, $a + 0 = 0 + a = a$.
3. *Inverse elements*: For the set of real numbers under the operation of addition to be a commutative group, each and every real number must have an additive inverse in the set of real numbers. For the operation of addition, the sum of a real number and its additive inverse, or *opposite*, must be the additive identity element, 0. From Chapter 5, we saw that every real number a has an opposite $-a$ such that $a + (-a) = -a + a = 0$. Recall also that $0 + 0 = 0$; therefore, 0 is its own additive inverse. Thus, all real numbers have an additive inverse.
4. *Associative property*: From Chapter 5, we know that the real numbers are associative under the operation of addition. That is, for any real numbers a, b, and c, $(a + b) + c = a + (b + c)$.
5. *Commutative property*: From Chapter 5, we know that the real numbers are commutative under the operation of addition. That is, for any real numbers a and b, $a + b = b + a$.

Since all five properties (closure, identity element, inverse elements, associative property, and commutative property) hold, the set of real numbers under the operation of addition form a commutative group.

SECTION 10.1 EXERCISES

CONCEPT/WRITING EXERCISES

1. What is a binary operation?

2. What does a mathematical system consist of?

3. Explain why each of the following is a binary operation. Give an example to illustrate each binary operation.

   a) Addition   **b)** Subtraction

   c) Multiplication   **d)** Division

4. What properties are required for a mathematical system to be a group?

5. What properties are required for a mathematical system to be a commutative group?

6. What is another name for a commutative group?

7. What is an identity element? Give the additive and multiplicative identity elements for the set of integers.

8. Explain the closure property. Give an example of the property.

9. What is an inverse element? Give the additive and multiplicative inverse of the number 2 for the set of rational numbers.

10. What is a counterexample?

11. Is it possible that a mathematical system is a group but not a commutative group? Explain.

12. Is it possible that a mathematical system is a commutative group but not a group? Explain.

13. Which of the following properties is not required for a mathematical system to be a group?
   a) Closure
   b) An identity element
   c) Every element must have an inverse.
   d) The commutative property must apply.
   e) The associative property must apply.

14. For the set of integers, list two operations that are not binary. Explain.

PRACTICE THE SKILLS

15. Give the commutative property of addition and illustrate the property with an example.

16. Give the commutative property of multiplication and illustrate the property with an example.

17. Give the associative property of multiplication and illustrate the property with an example.

18. Give the associative property of addition and illustrate the property with an example.

19. Give an example to show that the commutative property does not hold for the set of integers under the operation of division.

20. Give an example to show that the commutative property does not hold for the set of integers under the operation of subtraction.

21. Give an example to show that the associative property does not hold for the set of integers under the operation of subtraction.

22. Give an example to show that the associative property does not hold for the set of integers under the operation of division.

PROBLEM SOLVING

In Exercises 23–38, explain your answer.

23. Is the set of integers a commutative group under the operation of addition?

24. Is the set of integers a group under the operation of addition?

25. Is the set of positive integers a group under the operation of addition?

26. Is the set of positive integers a commutative group under the operation of addition?

27. Is the set of positive integers a group under the operation of subtraction?

28. Is the set of negative integers a group under the operation of division?

29. Is the set of negative integers a commutative group under the operation of addition?

30. Is the set of integers a group under the operation of multiplication?

31. Is the set of positive integers a commutative group under the operation of multiplication?

32. Is the set of negative integers a group under the operation of multiplication?

33. Is the set of rational numbers a group under the operation of addition?

34. Is the set of rational numbers a commutative group under the operation of multiplication?

35. Is the set of rational numbers a commutative group under the operation of division?

36. Is the set of rational numbers a group under the operation of subtraction?

37. Is the set of rational numbers a commutative group under the operation of subtraction?

38. Is the set of positive integers a commutative group under the operation of division?

CHALLENGE PROBLEMS/GROUP ACTIVITIES

In Exercises 39–42, explain your answer.

39. Is the set of real numbers a group under the operation of addition?

40. Is the set of real numbers a group under the operation of multiplication?

41. Is the set of irrational numbers a group under the operation of addition?

42. Is the set of irrational numbers a group under the operation of multiplication?

43. Create a mathematical system with two binary operations. Select a set of elements and two binary operations so that one binary operation with the set of elements meets the requirements for a group and the other binary operation does not. Explain why the one binary operation with the set of elements is a group. For the other binary operation and the set of elements, find counterexamples to show that it is not a group.

INTERNET/RESEARCH ACTIVITY

44. There are other classifications of mathematical systems besides groups. For example, there are *rings* and *fields*. Do research to determine the requirements that must be met for a mathematical system to be (a) a ring and (b) a field. (c) Is the set of real numbers, under the operations of addition and multiplication, a field? Ask your instructor for references to use.

10.2 FINITE MATHEMATICAL SYSTEMS

Suppose you decide to use your slow cooker to cook a roast for your family's dinner. You would like to eat at 6:30 P.M., and it will take 10 hours for the roast to cook. What time should you begin to cook the roast in the slow cooker? Or, suppose you need to take medicine every eight hours. You last took the medicine at 7:45 A.M. When should you take the medicine again? Questions such as these involve the use of mathematical systems that we will study in this section.

In the preceding section, we presented infinite mathematical systems. In this section, we present some finite mathematical systems. A *finite mathematical system* is one whose set contains a finite number of elements.

Clock Arithmetic

▲ Timekeeping makes use of mathematical systems.

Let's develop a finite mathematical system called *clock 12 arithmetic*. The set of elements in this system will be the hours on a clock: $\{1, 2, 3, 4, 5, 6, 7, 8, 9, 10, 11, 12\}$.

Figure 10.1

The binary operation that we will use is addition, which we define as movement of the hour hand in a clockwise direction. Assume that it is 4 o'clock. What time will it be in 9 hours? (See Fig. 10.1.) If we add 9 hours to 4 o'clock, the clock will read 1 o'clock. Thus, $4 + 9 = 1$ in clock arithmetic. Would $9 + 4$ be the same as $4 + 9$? Yes it will because $4 + 9 = 9 + 4 = 1$.

Table 10.1 is the addition table for clock arithmetic. Its elements are based on the definition of addition as previously illustrated. For example, the sum of 4 and 9 is 1, so we put a 1 in the table where the row to the right of the 4 intersects the column below the 9. Likewise, the sum of 11 and 10 is 9, so we put a 9 in the table where the row to the right of the 11 intersects the column below the 10.

The binary operation of this system is defined by the table. It is denoted by the symbol $+$. To determine the value of $a + b$, where a and b are any two numbers in the set, find a in the left-hand column and find b along the top row. Assume that there is a horizontal line through a and a vertical line through b; the point of intersection of these two lines is where you find the value of $a + b$. For example, $10 + 4 = 2$ has been circled in Table 10.1. Note that $4 + 10$ also equals 2, but this result will not necessarily hold for all examples in this chapter.

Table 10.1 Clock 12 Arithmetic

+	1	2	3	4	5	6	7	8	9	10	11	12
1	2	3	4	5	6	7	8	9	10	11	12	1
2	3	4	5	6	7	8	9	10	11	12	1	2
3	4	5	6	7	8	9	10	11	12	1	2	3
4	5	6	7	8	9	10	11	12	1	2	3	4
5	6	7	8	9	10	11	12	1	2	3	4	5
6	7	8	9	10	11	12	1	2	3	4	5	6
7	8	9	10	11	12	1	2	3	4	5	6	7
8	9	10	11	12	1	2	3	4	5	6	7	8
9	10	11	12	1	2	3	4	5	6	7	8	9
10	11	12	1	②	3	4	5	6	7	8	9	10
11	12	1	2	3	4	5	6	7	8	9	10	11
12	1	2	3	4	5	6	7	8	9	10	11	12

EXAMPLE ❶ A Commutative Group?

Determine whether the clock arithmetic system under the operation of addition is a commutative group.

SOLUTION Check the five requirements that must be satisfied for a commutative group.

1. *Closure*: Is the set of elements in clock arithmetic closed under the operation of addition? Yes it is since Table 10.1 contains only the elements in the set $\{1, 2, 3, 4, 5, 6, 7, 8, 9, 10, 11, 12\}$. If Table 10.1 had contained an element other than the numbers 1 through 12, the set would not have been closed under addition.
2. *Identity element*: Is there an identity element for clock arithmetic? If the time is currently 4 o'clock, how many hours have to pass before it is 4 o'clock again? Twelve hours: $4 + 12 = 12 + 4 = 4$. In fact, given any hour, in 12 hours the clock will return to the starting point. Therefore, 12 is the additive identity element in clock arithmetic.

In examining Table 10.1, we see that the row of numbers next to the 12 in the left-hand column is identical to the row of numbers along the top. We also see that the column of numbers under the 12 in the top row is identical to the column of numbers on the left. The search for such a column and row is one technique for determining whether an identity element exists for a system defined by a table.

3. *Inverse elements*: Is there an inverse for the number 4 in clock arithmetic for the operation of addition? Recall that the identity element in clock arithmetic is 12. What number when added to 4 gives 12; that is, $4 + \underline{\quad} = 12$? Table 10.1 shows that $4 + 8 = 12$ and also that $8 + 4 = 12$. Thus, 8 is the additive inverse of 4 and 4 is the additive inverse of 8.

    To find the additive inverse of 7, find 7 in the left-hand column of Table 10.1. Look to the right of the 7 until you come to the identity element 12. Then determine the number at the top of this column. The number is 5. Since $7 + 5 = 5 + 7 = 12$, 5 is the inverse of 7 and 7 is the inverse of 5. The other inverses can be found in the same way. Table 10.2 shows each element in clock 12 arithmetic and its inverse. Note that each element in the set has an *inverse*.

Table 10.2 Clock 12 Inverses

Element	+	Inverse	=	Identity Element
1	+	11	=	12
2	+	10	=	12
3	+	9	=	12
4	+	8	=	12
5	+	7	=	12
6	+	6	=	12
7	+	5	=	12
8	+	4	=	12
9	+	3	=	12
10	+	2	=	12
11	+	1	=	12
12	+	12	=	12

4. *Associative property*: Now consider the associative property. Does $(a + b) + c = a + (b + c)$ for all values a, b, and c of the set? Remember to always evaluate the values within the parentheses first. Let's select some values for a, b, and c. Let $a = 2$, $b = 6$, and $c = 8$. Then

$$(2 + 6) + 8 = 2 + (6 + 8)$$
$$8 + 8 = 2 + 2$$
$$4 = 4 \quad \text{True}$$

Let $a = 5$, $b = 12$, and $c = 9$. Then

$$(5 + 12) + 9 = 5 + (12 + 9)$$
$$5 + 9 = 5 + 9$$
$$2 = 2 \quad \text{True}$$

Randomly selecting *any* elements a, b, and c of the set reveals $(a + b) + c = a + (b + c)$. Thus, the system of clock arithmetic is associative under the operation of addition. Note that if there is just one set of values a, b, and c such that $(a + b) + c \neq a + (b + c)$, the system is not associative. Usually, you will not be asked to check every case to determine whether the associative property holds. *If every element in the set does not appear in every row and column of the table, however, you need to check the associative property carefully.*

5. *Commutative property*: Does the commutative property hold under the given operation? Does $a + b = b + a$ for all elements a and b of the set? Let's randomly select some values for a and b to determine whether the commutative property appears to hold. Let $a = 5$ and $b = 8$; then Table 10.1 shows that

$$5 + 8 = 8 + 5$$
$$1 = 1 \quad \text{True}$$

Let $a = 9$ and $b = 6$; then

$$9 + 6 = 6 + 9$$
$$3 = 3 \quad \text{True}$$

The commutative property holds for these two specific cases. In fact, if we were to select *any* values for a and b, we would find that $a + b = b + a$. Thus, the commutative property of addition is true in clock arithmetic. Note that if there is just one set of values a and b such that $a + b \neq b + a$, the system is not commutative.

This system satisfies the five properties required for a mathematical system to be a commutative group. Thus, clock arithmetic under the operation of addition is a commutative or abelian group. ●

Table 10.3 Symmetry about the Main Diagonal

+	0	1	2	3	4
0	0	1	2	3	4
1	1	2	3	4	0
2	2	3	4	0	1
3	3	4	0	1	2
4	4	0	1	2	3

One method that can be used to determine whether a system defined by a table is commutative under the given operation is to determine whether the elements in the table are symmetric about the *main diagonal*. The main diagonal is the diagonal from the upper left-hand corner to the lower right-hand corner of the table. In Table 10.3, the main diagonal is shaded in color.

If the elements are symmetric about the main diagonal, then the system is commutative. If the elements are not symmetric about the main diagonal, then the system is not commutative. If you examine the system in Table 10.3, you see that its elements are symmetric about the main diagonal because the same numbers appear in the same relative positions on opposite sides of the main diagonal. Therefore, this mathematical system is commutative.

It is possible to have groups that are not commutative. Such groups are called *noncommutative* or *nonabelian groups*. However, a *noncommutative group defined by a table must be at least a six-element by six-element table*. Nonabelian groups are illustrated in Exercises 83 through 85 at the end of this section.

Now we will look at another finite mathematical system.

Table 10.4 Four-Element System

☼	6	7	8	9
6	8	9	6	7
7	9	6	7	8
8	6	7	8	9
9	7	8	9	6

EXAMPLE ❷ *A Finite System*

Consider the mathematical system defined by Table 10.4. Assume that the associative property holds for the given operation.

a) List the elements in the set of this mathematical system.

b) Identify the binary operation.

c) Determine whether this mathematical system is a commutative group.

SOLUTION

a) The set of elements for this mathematical system consists of the elements found on the top (or left-hand side) of the table: $\{6, 7, 8, 9\}$.

b) The binary operation is ✿.

c) We must determine whether the five requirements for a commutative group are satisfied.

1. *Closure*: All the elements in the table are in the original set of elements $\{6, 7, 8, 9\}$, so the mathematical system is closed.

2. *Identity element*: The identity element is 8. Note that the row of elements to the right of the 8 is identical to the top row *and* that the column of elements under the 8 is identical to the left-hand column.

3. *Inverse elements*: When an element operates on its inverse element, the result is the identity element. For this example, the identity element is 8. To determine the inverse of 6, determine which element (6, 7, 8, or 9) replaces the shaded area in the following equation and gives a true statement:

$$6 \; ✿ \; \boxed{} = 8$$

From Table 10.4, we see that $6 \; ✿ \; 6 = 8$. Thus, 6 is its own inverse.

To determine the inverse of 7, determine which element replaces the shaded area in the following equation and gives a true statement:

$$7 \; ✿ \; \boxed{} = 8$$

From Table 10.4, we see that $7 \; ✿ \; 9 = 9 \; ✿ \; 7 = 8$. Therefore, 7 is the inverse of 9 and 9 is the inverse of 7. The elements and their inverses are shown in Table 10.5. Every element has an inverse.

4. *Associative property*: It is given that the associative property holds for this binary operation, ✿. One example of the associative property is

$$(7 \; ✿ \; 6) \; ✿ \; 9 = 7 \; ✿ \; (6 \; ✿ \; 9)$$
$$9 \; ✿ \; 9 = 7 \; ✿ \; 7$$
$$6 = 6 \quad \text{True}$$

5. *Commutative property*: The elements in Table 10.4 are symmetric about the main diagonal, so the commutative property holds for the operation of ✿. One example of the commutative property is

$$6 \; ✿ \; 7 = 7 \; ✿ \; 6$$
$$9 = 9 \quad \text{True}$$

The five necessary properties hold. Thus, the mathematical system is a commutative group.   ●

Table 10.5 Inverses under ✿

Element	✿	Inverse	=	Identity Element
6	✿	6	=	8
7	✿	9	=	8
8	✿	8	=	8
9	✿	7	=	8

Mathematical Systems Without Numbers

Thus far, all the systems we have discussed have been based on sets of numbers. Example 3 illustrates a mathematical system of symbols rather than numbers.

EXAMPLE ❸ *Investigating a System of Symbols*

Use the mathematical system defined by Table 10.6 and determine

Table 10.6 A System of Symbols

$\cdot$	@	*P*	*W*
@	*P*	*W*	@
P	*W*	@	*P*
W	@	*P*	*W*

a) the set of elements.

b) the binary operation.

c) closure or nonclosure of the system.

d) the identity element.

e) the inverse of @.

f) $(@ \cdot P) \cdot P$ and $@ \cdot (P \cdot P)$.

g) $W \cdot P$ and $P \cdot W$.

SOLUTION

a) The set of elements of this mathematical system is $\{@, P, W\}$.

b) The binary operation is $\cdot$.

c) Because the table does not contain any symbols other than @, *P*, and *W*, the system is closed under $\cdot$.

d) The identity element is *W*. Note that the row next to *W* in the left-hand column is the same as the top row and that the column under *W* is identical to the left-hand column. We see that

$$@ \cdot W = W \cdot @ = @$$
$$P \cdot W = W \cdot P = P$$
$$W \cdot W = W$$

e) We know that

$$\text{element} \cdot \text{inverse element} = \text{identity element,}$$

and since *W* is the identity element, to find the inverse of @ we write

$$@ \cdot \blacksquare = W$$

To find the inverse of @, we must determine the element to replace the shaded area. Since $@ \cdot P = W$ and $P \cdot @ = W$, *P* is the inverse of @.

f) We first evaluate the information within parentheses.

$$(@ \cdot P) \cdot P = W \cdot P \qquad \text{and} \qquad @ \cdot (P \cdot P) = @ \cdot @$$
$$= P \qquad\qquad\qquad = P$$

g) $W \cdot P = P \qquad$ and $\qquad P \cdot W = P$

●

Table 10.7

♫	α	β	γ	δ
α	δ	α	β	γ
β	α	β	γ	δ
γ	β	γ	δ	α
δ	γ	δ	α	β

Table 10.8 Inverses under ♫

Element	♫	Inverse	=	Identity Element
α	♫	γ	=	β
β	♫	β	=	β
γ	♫	α	=	β
δ	♫	δ	=	β

EXAMPLE ❹ *Is the System a Commutative Group?*

Determine whether the mathematical system defined by Table 10.7 is a commutative group. Assume that the associative property holds for the given operation.

SOLUTION

1. *Closure*: The system is closed.
2. *Identity element*: The identity element is β.
3. *Inverse elements*: Each element has an inverse as illustrated in Table 10.8.
4. *Associative property*: It is given that the associative property holds. An example illustrating the associative property is

$$(\gamma \text{ ♫ } \delta) \text{ ♫ } \alpha = \gamma \text{ ♫ } (\delta \text{ ♫ } \alpha)$$
$$\alpha \text{ ♫ } \alpha = \gamma \text{ ♫ } \gamma$$
$$\delta = \delta \quad \text{True}$$

5. *Commutative property*: By examining Table 10.7, we can see that the elements are symmetric about the main diagonal. Thus, the system is commutative under the given operation, ♫. One example of the commutative property is

$$\alpha \text{ ♫ } \delta = \delta \text{ ♫ } \alpha$$
$$\gamma = \gamma \quad \text{True}$$

All five properties are satisfied. Thus, the system is a commutative group. ●

EXAMPLE ❺ *Another System to Study*

Determine whether the mathematical system in Table 10.9 is a commutative group under the operation of ☺.

Table 10.9

☺	x	y	z
x	x	z	y
y	z	y	x
z	y	x	z

SOLUTION

1. The system is closed.
2. No row is identical to the top row, so there is no identity element. Therefore, this mathematical system is *not a group*. There is no need to go any further, but for practice, let's look at a few more items.
3. Since there is no identity element, there can be no inverses.
4. The associative property does not hold. The following counterexample illustrates the associative property does not hold for every case.

$$(x \text{ ☺ } y) \text{ ☺ } z \neq x \text{ ☺ } (y \text{ ☺ } z)$$
$$z \text{ ☺ } z \neq x \text{ ☺ } x$$
$$z \neq x$$

5. The table is symmetric about the main diagonal. Therefore, the commutative property does hold for the operation of ☺.

Ⓦ hat makes group theory such a powerful tool is that it can be used to reveal the underlying structure of just about any physical phenomenon that involves symmetry and patterning, such as wallpaper or quilt patterns. Interest in the formal study of symmetry in design came out of the Industrial Revolution in the late nineteenth century. The new machines of the Industrial Revolution could vary any given pattern almost indefinitely. Designers needed a way to describe and manipulate patterns systematically. Shown here are the four geometric motions that generate all two-dimensional patterns: (a) reflection, (b) translation, (c) rotation, and (d) glide reflection. How these motions are applied, or not applied, is the basis of pattern analysis. The geometric motions, called *rigid motions*, were discussed in Section 9.5.

Note that the associative property does not hold even though the commutative property does hold. This outcome can occur when there is no identity element and every element does not have an inverse, as in this example. ●

EXAMPLE ❻ *Is the System a Commutative Group?*

Determine whether the mathematical system in Table 10.10 is a commutative group under the operation of $*$.

Table 10.10

$*$	□	b	c
□	□	b	c
b	b	b	□
c	c	□	c

SOLUTION

1. The system is closed.
2. There is an identity element, □.
3. Each element has an inverse; □ is the inverse of □, b is the inverse of c, and c is the inverse of b.
4. Every element in the set does not appear in every row and every column of the table, so we need to check the associative property carefully. There are many specific cases in which the associative property does hold. However, the following counterexample illustrates that the associative property does not hold for every case.

$$(b*b)*c \neq b*(b*c)$$
$$b*c \neq b*□$$
$$□ \neq b$$

5. The commutative property holds because there is symmetry about the main diagonal.

   Since we have shown that the associative property does not hold under the operation of $*$, this system is not a group. Therefore, it cannot be a commutative group. ●

SECTION ⑩.② EXERCISES

CONCEPT/WRITING EXERCISES

1. Explain how the clock 12 addition table is formed.

2. What is $12 + 12$ in clock 12 arithmetic? Explain how you obtained your answer.

3. **a)** Explain how to add the numbers $(2 + 11) + 5$ in clock 12 arithmetic using the addition table.

   b) What is $(2 + 11) + 5$ in clock 12 arithmetic?

4. **a)** Explain how to determine a difference of two numbers in clock arithmetic by using the face of a clock.

   b) Determine $3 - 5$ in clock 12 arithmetic using the method explained in part (a).

5. **a)** Explain how to find $2 - 6$ in clock 12 arithmetic by adding the number 12 to one of the numbers.

   b) Determine $2 - 6$ in clock 12 arithmetic using the method explained in part (a).

   c) Explain why the procedure you give in part (a) works.

6. Explain one method of determining whether a system defined by a table is commutative under the given operation.

7. Is there an identity element for addition in clock 12 arithmetic? If so, what is it?

8. Is clock 12 arithmetic closed under the operation of addition? Explain.

9. Does each element in clock 12 arithmetic have an inverse? If so, give each element and its corresponding inverse.

10. Give an example to illustrate the associative property of addition in clock 12 arithmetic.

11. Is clock 12 arithmetic commutative? Give an example to verify your answer.

12. Is clock 12 arithmetic under the operation of addition a commutative group? Explain.

13. Consider clock 7 arithmetic under the operation of addition. The set of elements in such a mathematical system is $\{1, 2, 3, 4, 5, 6, 7\}$.

    a) Which of these elements is the additive identity element?

    b) What is the additive inverse of 2? Explain your answer.

14. Consider clock 6 arithmetic under the operation of addition. The set of elements in such a mathematical system is $\{1, 2, 3, 4, 5, 6\}$.

    a) Which of these elements is the additive identity element?

    b) What is the additive inverse of 3? Explain your answer.

In Exercises 15 and 16, determine if the system is commutative. Explain how you determined your answer.

15.

I	P	A	L
P	L	P	A
A	P	L	A
L	A	L	P

16.

$\boxminus$	A	$\otimes$	W
A	$\otimes$	W	A
$\otimes$	W	A	$\otimes$
W	A	$\otimes$	$\otimes$

In Exercises 17 and 18, determine if the system has an identity element. If so, list the identity element. Explain how you determined your answer.

17.

W	A	B	C
A	C	B	A
B	B	C	B
C	A	B	C

18.

$\ominus$	$\square$	$\odot$	$\triangle$
$\square$	$\triangle$	$\square$	$\odot$
$\odot$	$\odot$	$\triangle$	$\square$
$\triangle$	$\square$	$\odot$	$\triangle$

In Exercises 19 and 20, the identity element is C. Determine the inverse of (a) A, (b) B, and (c) C.

19.

$\oslash$	A	B	C
A	B	C	A
B	C	A	B
C	A	B	C

20.

$\frac{\perp}{\top}$	A	B	C
A	C	B	A
B	B	C	B
C	A	B	C

PRACTICE THE SKILLS

In Exercises 21–32, use Table 10.1 on page 619 to determine the sum in clock 12 arithmetic.

21. $2 + 5$

22. $6 + 10$

23. $8 + 8$

24. $11 + 7$

25. $4 + 12$

26. $12 + 12$

27. $3 + (8 + 9)$

28. $(8 + 7) + 6$

29. $(6 + 4) + 8$

30. $(6 + 10) + 12$

31. $(7 + 8) + (9 + 6)$

32. $(7 + 11) + (9 + 5)$

In Exercises 33–44, determine the difference in clock 12 arithmetic by starting at the first number and counting counterclockwise on the clock the number of units given by the second number.

33. $11 - 6$

34. $12 - 5$

35. $6 - 7$

36. $8 - 11$

37. $11 - 8$

38. $4 - 10$

39. $1 - 12$

40. $6 - 10$

41. $5 - 5$

42. $8 - 8$

43. $12 - 12$

44. $5 - 8$

45. Use the following figure to develop an addition table for clock 6 arithmetic. The figure will also be used in Exercises 46–54.

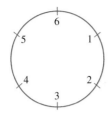

In Exercises 46–54, determine the sum or difference in clock 6 arithmetic.

46. 3 + 4   **47.** 4 + 5   **48.** 5 + 6

49. 5 − 2   **50.** 4 − 5   **51.** 2 − 6

52. 3 − 4   **53.** (3 − 5) − 6   **54.** 2 + (1 − 3)

55. Use the following figure to develop an addition table for clock 7 arithmetic. The figure will also be used in Exercises 56–64.

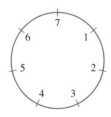

In Exercises 56–64, determine the sum or difference in clock 7 arithmetic.

56. 6 + 6   **57.** 4 + 7

58. 3 + 6   **59.** 7 + 6

60. 2 − 3   **61.** 3 − 6

62. 2 − 4   **63.** (4 − 5) − 6

64. 3 − (2 − 6)

65. Determine whether clock 7 arithmetic under the operation of addition is a commutative group. Explain.

66. A mathematical system is defined by a three-element by three-element table where every element in the set appears in each row and each column. Must the mathematical system be a commutative group? Explain.

67. Consider the mathematical system defined by the following table. Assume that the associative property holds for the given operation.

£	0	2	4	6
0	0	2	4	6
2	2	4	6	0
4	4	6	0	2
6	6	0	2	4

a) What are the elements of the set in this mathematical system?

b) What is the binary operation?

c) Is the system closed? Explain.

d) Is there an identity element for the system under the given operation? If so, what is it?

e) Does every element in the system have an inverse? If so, give each element and its corresponding inverse.

f) Give an example to illustrate the associative property.

g) Is the system commutative? Give an example to verify your answer.

h) Is the mathematical system a commutative group? Explain.

In Exercises 68–70, repeat parts (a)–(h) of Exercise 67 for the mathematical system defined by the following table. Assume that the associative property holds for the given operation.

68.

®	♣	♦	♥	♠
♣	♥	♠	♣	♦
♦	♠	♣	♦	♥
♥	♣	♦	♥	♠
♠	♦	♥	♠	♣

69.

🐎	*	5	L
*	5	L	*
5	L	*	5
L	*	5	L

70.

☿	3	5	8	4
3	5	8	4	3
5	8	4	3	5
8	4	3	5	8
4	3	5	8	4

71. For the mathematical system

🐕	*f*	*r*	*o*	*m*
f	*f*	*r*	*o*	*m*
r	*r*	*o*	*m*	*f*
o	*o*	*m*	*f*	*r*
m	*m*	*f*	*r*	*o*

determine

a) the elements in the set.

b) the binary operation.

c) closure or nonclosure of the system.

d) $(r \, \mathbb{A} \, o) \, \mathbb{A} \, f$

e) $(f \, \mathbb{A} \, r) \, \mathbb{A} \, m$

f) the identity element.

g) the inverse of *r*.

h) the inverse of *m*.

72. a) Is the following mathematical system a group? Explain your answer.

b) Find an example showing that the associative property does not hold for the given set of elements.

〜	1	2	3	4
1	2	3	4	1
2	3	4	1	2
3	4	1	2	3
4	1	4	3	2

In Exercises 73–78, for the mathematical system given, determine which of the five properties of a commutative group do not hold.

73.

‡	Δ	Φ	Γ
Δ	Δ	Φ	Γ
Φ	Φ	Γ	Δ
Γ	Γ	Δ	Π

74.

⊗	⊡	*M*	🔔
⊡	⊡	*M*	🔔
M	*M*	⊡	🔔
🔔	🔔	*M*	⊡

75.

·	⊙	*	?	*T*	*P*
⊙	*T*	*P*	⊙	*	*
*	*P*	⊙	*	⊙	*T*
?	⊙	*	?	*T*	*P*
T	*	⊙	*T*	*P*	?
P	*	*T*	*P*	?	⊙

76.

☺	*a*	*b*	π	0	Δ
a	Δ	*a*	*b*	π	0
b	*a*	*b*	π	0	Δ
π	*b*	π	0	Δ	*a*
0	π	0	Δ	*a*	*b*
Δ	0	Δ	π	*b*	*a*

77.

⇔	*a*	*b*	*c*	*d*	*e*
a	*c*	*d*	*e*	*a*	*b*
b	*d*	*e*	*a*	*b*	*c*
c	*e*	*a*	*b*	*c*	*d*
d	*a*	*b*	*c*	*e*	*d*
e	*b*	*c*	*d*	*e*	*a*

78.

▽	0	1	2	3	4	5
0	0	0	0	0	0	0
1	0	1	2	3	4	5
2	0	2	4	0	2	4
3	0	3	0	3	0	3
4	0	4	2	0	4	2
5	0	5	4	3	2	1

PROBLEM SOLVING

79. a) Consider the set consisting of two elements $\{E, O\}$, where *E* stands for an even number and *O* stands for an odd number. For the operation of addition, complete the table.

+	*E*	*O*
E		
O		

b) Determine whether this mathematical system forms a commutative group under the operation of addition. Explain your answer.

80. a) Let *E* and *O* represent even numbers and odd numbers, respectively, as in Exercise 79. Complete the table for the operation of multiplication.

×	*E*	*O*
E		
O		

b) Determine whether this mathematical system forms a commutative group under the operation of multiplication. Explain your answer.

In Exercises 81 and 82, make up your own mathematical system that is a group. List the identity element and the inverses of each element. Do so with sets containing

81. three elements.

82. four elements.

In Exercises 83 and 84 the tables shown are examples of noncommutative or nonabelian groups. For each exercise, do the following.

a) *Show that the system under the given operation is a group. (It would be very time consuming to prove that the associative property holds, but you can give some examples to show that it appears to hold.)*

b) *Determine a counterexample to show that the commutative property does not hold.*

83.

∞	1	2	3	4	5	6
1	5	3	4	2	6	1
2	4	6	5	1	3	2
3	2	1	6	5	4	3
4	3	5	1	6	2	4
5	6	4	2	3	1	5
6	1	2	3	4	5	6

84.

$\ominus$	A	B	C	D	E	F
A	E	C	D	B	F	A
B	D	F	E	A	C	B
C	B	A	F	E	D	C
D	C	E	A	F	B	D
E	F	D	B	C	A	E
F	A	B	C	D	E	F

CHALLENGE PROBLEMS/GROUP ACTIVITIES

85. *Book Arrangements* Suppose that three books numbered 1, 2, and 3 are placed next to one another on a shelf. If we remove volume 3 and place it before volume 1, the new order of books is 3, 1, 2. Let's call this replacement R. We can write

$$R = \begin{pmatrix} 1 & 2 & 3 \\ 3 & 1 & 2 \end{pmatrix}$$

which indicates the books were switched in order from 1, 2, 3 to 3, 1, 2. Other possible replacements are $S, T, U, V,$ and I, as indicated.

$$S = \begin{pmatrix} 1 & 2 & 3 \\ 2 & 1 & 3 \end{pmatrix} \quad T = \begin{pmatrix} 1 & 2 & 3 \\ 3 & 2 & 1 \end{pmatrix} \quad U = \begin{pmatrix} 1 & 2 & 3 \\ 1 & 3 & 2 \end{pmatrix}$$

$$V = \begin{pmatrix} 1 & 2 & 3 \\ 2 & 3 & 1 \end{pmatrix} \quad I = \begin{pmatrix} 1 & 2 & 3 \\ 1 & 2 & 3 \end{pmatrix}$$

Replacement set I indicates that the books were removed from the shelves and placed back in their original order. Consider the mathematical system with the set of elements, $\{R, S, T, U, V, I\}$ with the operation of $*$.

To evaluate $R * S$, write

$$R*S = \begin{pmatrix} 1 & 2 & 3 \\ 3 & 1 & 2 \end{pmatrix} * \begin{pmatrix} 1 & 2 & 3 \\ 2 & 1 & 3 \end{pmatrix}.$$

As shown in Fig. 10.2, R replaces 1 with 3 and S replaces 3 with 3 (no change), so $R * S$ replaces 1 with 3. R replaces 2 with 1 and S replaces 1 with 2, so $R * S$ replaces 2 with 2 (no change). R replaces 3 with 2 and S replaces 2 with 1, so $R * S$ replaces 3 with 1. $R * S$ replaces 1 with 3, 2 with 2, and 3 with 1.

$$R*S = \begin{pmatrix} 1 & 2 & 3 \\ 3 & 2 & 1 \end{pmatrix} = T$$

Figure 10.2

Since this result is the same as replacement set T, we write $R * S = T$.

a) Complete the table for the operation using the procedure outlined.

$*$	R	S	T	U	V	I
R		T				
S						
T						
U						
V						
I						

b) Is this mathematical system a group? Explain.

c) Is this mathematical system a commutative group? Explain.

86. If a mathematical system is defined by a four-element by four-element table, how many specific cases must be illustrated to prove the set of elements is associative under the given operation?

Exercises 87–89 will help prepare you for the next section.

87. A table is shown below.

+	0	1	2	3	4
0	0		2		4
1	1		3		
2			4		1
3	3		0	1	2
4		0	1		

Fill in the blank areas of the table by performing the following operations.

1. Add the number in the left-hand column and the number in the top row.

2. Divide this sum by 5.

3. Place the *remainder* in the cell where the number in the left-hand column and the number in the top row would meet.

For example, in the table, the number 2 is in red. This 2 is found as follows:

$$(3 + 4) \div 5 = 1, \text{remainder } 2$$

88. Using Exercise 87 as a guide, complete the following table by dividing the sum of the numbers by 6. Place the remainders in the table as explained in Exercise 87.

+	0	1	2	3	4	5
0						
1						
2						
3						
4						
5						

89. Consider the mathematical system defined by the following table.

+	0	1	2	3
0	0	1	2	3
1	1	2	3	0
2	2	3	0	1
3	3	0	1	2

After working Exercises 87 and 88, can you explain how this table was formed?

RECREATIONAL MATHEMATICS

90. *A Tiny Group* Consider the mathematical system with the single element {1} under the operation of multiplication.

a) Is this system closed?

b) Does the system have an identity element?

c) Does 1 have an inverse?

d) Does the associative property hold?

e) Does the commutative property hold?

f) Is {1} under the operation of multiplication a commutative group?

INTERNET/RESEARCH ACTIVITIES

91. a) *Matrices* In Section 7.3, we introduced matrices. Show that 2×2 matrices under the operation of addition form a commutative group.

b) Show that 2×2 matrices under the operation of multiplication do not form a commutative group.

10.3 MODULAR ARITHMETIC

▲ When will your birthday next land on a Saturday?

If your birthday is on a Monday this year, on what day of the week will it fall next year? What if next year is a leap year? When will your birthday next fall on a Saturday? These questions can be addressed with the type of arithmetic that we will study in this section.

The clock arithmetic we discussed in the previous section is similar to modular arithmetic. The set of elements $\{0, 1, 2, 3, 4, 5, 6, 7, 8, 9, 10, 11\}$ together with the operation of addition is called a modulo 12 or mod 12 system. There is one difference in notation between clock 12 arithmetic and modulo 12 arithmetic. In the modulo 12 system, the symbol 12 is replaced with the symbol 0.

Modulo *m* Systems

A *modulo m system* consists of *m* elements, 0 through $m - 1$, and a binary operation. In this section, we will discuss modular arithmetic systems and their properties.

If today is Sunday, what day of the week will it be in 23 days? The answer, Tuesday, is arrived at by dividing 23 by 7. The quotient is 3 and the remainder is 2. Twenty-three days represent 3 weeks plus 2 days. Since we are interested only in the day of the week on which the twenty-third day will fall, the 3-week segment is unimportant to the answer. The remainder of 2 indicates the answer will be 2 days later than Sunday, which is Tuesday.

If we place the days of the week on a clock face as shown in Fig. 10.3, then in 23 days the hand would have made three complete revolutions and end on Tuesday. If we replace the days of the week with numbers, then a modulo 7 arithmetic system will result. See Fig. 10.4: Sunday = 0, Monday = 1, Tuesday = 2, and so on. If we start at 0 and move the hand 23 places, we will end at 2. Table 10.11 shows a modulo 7 addition table.

Table 10.11 Modular 7 Addition

+	0	1	2	3	4	5	6
0	0	1	2	3	4	5	6
1	1	2	3	4	5	6	0
2	2	3	4	5	6	0	1
3	3	4	5	6	0	1	2
4	4	5	6	0	1	2	③
5	5	6	0	1	2	3	4
6	6	0	1	2	3	4	5

Figure 10.3

Figure 10.4

Figure 10.5

If we start at 4 and add 6, we end at 3 on the clock in Fig. 10.5. This number is circled in Table 10.11. The other numbers can be obtained in the same way.

A second method of determining the sum of $4 + 6$ in modulo 7 arithmetic is to divide the sum, 10, by 7 and observe the remainder.

$$10 \div 7 = 1, \quad \text{remainder } 3$$

The remainder, 3, is the sum of $4 + 6$ in a modulo 7 arithmetic system.

The concept of congruence is important in modular arithmetic.

a is **congruent** to b modulo m, written, $a \equiv b \pmod{m}$, if a and b have the same remainder when divided by m.

We can show, for example, that $10 \equiv 3 \pmod{7}$ by dividing both 10 and 3 by 7 and observing that we obtain the same remainder in each case.

$$10 \div 7 = 1, \quad \text{remainder } 3 \quad \text{and} \quad 3 \div 7 = 0, \quad \text{remainder } 3$$

Since the remainders are the same, 3 in each case, 10 is congruent to 3 modulo 7, and we may write $10 \equiv 3 \pmod{7}$.

Now consider $37 \equiv 5 \pmod{8}$. If we divide both 37 and 5 by 8, each has the same remainder, 5.

In any modulo system, we can develop a set of *modulo classes* by placing all numbers with the same remainder in the appropriate modulo class. In a modulo 7 system, every number must have a remainder of either 0, 1, 2, 3, 4, 5, or 6. Thus, a modulo 7 system has seven modulo classes. The seven classes are presented in Table 10.12.

Every number is congruent to a number from 0 to 6 in modulo 7. For example, $24 \equiv 3 \pmod{7}$ because 24 is in the same modulo class as 3.

The solution to a problem in modular arithmetic, if it exists, will always be a number from 0 through $m - 1$, where m is the *modulus* of the system. For example, in a modulo 7 system, because 7 is the modulus, the solution will be a number from 0 through 6.

Table 10.12 Modulo 7 Classes

0	1	2	3	4	5	6
0	1	2	3	4	5	6
7	8	9	10	11	12	13
14	15	16	17	18	19	20
21	22	23	24	25	26	27
28	29	30	31	32	33	34
⋮	⋮	⋮	⋮	⋮	⋮	⋮

EXAMPLE ❶ Congruence Modulo 7

Determine which number, from 0 through 6, the following numbers are congruent to in modulo 7.

a) 62   b) 53   c) 105

SOLUTION We could determine the answer by listing more entries in Table 10.12. Another method of finding the answer is to divide the given number by 7 and observe the remainder. In the solutions, we will use a question mark, ?, as a place holder. The ? will represent a number from 0 through 6. When we determine the answer, we will replace the ? with the answer.

a) $62 \equiv ? \pmod{7}$

To determine the value that 62 is congruent to in mod 7, divide 62 by 7 and find the remainder.

$$\begin{array}{r} 8 \\ 7\overline{)62} \\ \underline{56} \\ 6 \end{array} \leftarrow \text{remainder}$$

$$62 \div 7 = 8, \quad \text{remainder } 6$$

Thus, $62 \equiv 6 \pmod{7}$.

b) $53 \equiv ? \pmod{7}$

$$53 \div 7 = 7, \quad \text{remainder } 4$$

Thus, $53 \equiv 4 \pmod{7}$.

c) $105 \equiv ? \pmod 7$

$$105 \div 7 = 15, \quad \text{remainder } 0$$

Thus, $105 \equiv 0 \pmod 7$. ●

EXAMPLE ❷ *Congruence Modulo 6*

Evaluate each of the following in modulo 6.

a) $3 + 5$ b) $5 - 2$ c) $3 \cdot 5$

SOLUTION In each part, because we are working in modulo 6, the answer will be a number from 0 through 5.

a)
$$3 + 5 \equiv ? \pmod 6$$
$$8 \equiv ? \pmod 6$$
$$8 \div 6 = 1, \quad \text{remainder } 2$$

Therefore, $3 + 5 \equiv 2 \pmod 6$.

b)
$$5 - 2 \equiv ? \pmod 6$$
$$3 \equiv ? \pmod 6$$
$$3 \equiv 3 \pmod 6$$

Remember that we want to replace the question mark with a number between 0 and 5, inclusive. Thus, $5 - 2 \equiv 3 \pmod 6$.

c)
$$3 \cdot 5 \equiv ? \pmod 6$$
$$15 \equiv ? \pmod 6$$

Since $15 \div 6 = 2$, remainder 3, $15 \equiv 3 \pmod 6$. Thus, $3 \cdot 5 \equiv 3 \pmod 6$. ●

Note in Table 10.12 on page 632 that every number in the same modulo class differs by a multiple of the modulus, in this case a multiple of 7. Adding (or subtracting) a multiple of the modulus to (or from) a given number does not change the modulo class or congruence of the given number. For example, $3, 3 + 1(7), 3 + 2(7), 3 + 3(7), \ldots,$ $3 + n(7)$ are all in the same modulo class, namely, 3. We use this fact in the solution to Example 3.

EXAMPLE ❸ *Using Modulo Classes in Subtraction*

Determine the positive number replacement (less than the modulus) for the question mark that makes the statement true.

a) $3 - 5 \equiv ? \pmod 7$ b) $? - 4 \equiv 3 \pmod 5$ c) $5 - ? \equiv 7 \pmod 8$

SOLUTION In each part, we wish to replace the question mark with a positive number less than the modulus. Therefore, in part (a) we wish to replace the ? with a positive number less than 7. In part (b), we wish to replace the ? with a positive number less than 5. In part (c), we wish to replace the ? with a positive number less than 8.

a) In mod 7, adding 7, or a multiple of 7, to a number results in a sum that is in the same modulo class. Thus, if we add 7, 14, 21, ... to 3, the result will be a number in the same modulo class. We want to replace 3 with an equivalent mod 7 number that is greater than 5. Adding 7 to 3 yields a sum of 10, which is greater than 5.

$$3 - 5 \equiv ? \,(\text{mod } 7)$$
$$(3 + 7) - 5 \equiv ? \,(\text{mod } 7)$$
$$10 - 5 \equiv ? \,(\text{mod } 7)$$
$$5 \equiv ? \,(\text{mod } 7)$$
$$5 \equiv 5 \,(\text{mod } 7)$$

Therefore, $? = 5$ and $3 - 5 \equiv 5 \,(\text{mod } 7)$.

b) We wish to replace the $?$ with a number less than 5. We know that $7 - 4 \equiv 3 \bmod 5$ because $3 \equiv 3 \bmod 5$. Therefore, we need to determine what number, less than 5, the number 7 is congruent to in mod 5. If we subtract the modulus, 5, from 7, we obtain 2. Thus, 2 and 7 are in the same modular class. Therefore, $? = 2$.

$$? - 4 \equiv 3 \,(\text{mod } 5)$$
$$7 - 4 \equiv 3 \,(\text{mod } 5)$$
$$2 - 4 \equiv 3 \,(\text{mod } 5)$$

Notice in the last equivalence that if we add 5 to 2, we get $7 - 4 \equiv 3 \,(\text{mod } 5)$, which is a true statement.

c) $5 - ? \equiv 7 \,(\text{mod } 8)$

In mod 8, adding 8, or a multiple of 8, to a number results in a sum that is in the same modulo class. Thus, we can add 8 to 5 so that the statement becomes

$$(8 + 5) - ? \equiv 7 \,(\text{mod } 8)$$
$$13 - ? \equiv 7 \,(\text{mod } 8)$$

We can see that $13 - 6 = 7$. Therefore, $? = 6$ and $5 - 6 \equiv 7 \,(\text{mod } 8)$.   ●

EXAMPLE ❹ *Using Modulo Classes in Multiplication*

Determine all positive number replacements (less than the modulus) for the question mark that make the statement true.

a) $4 \cdot ? \equiv 2 \,(\text{mod } 5)$ b) $3 \cdot ? \equiv 3 \,(\text{mod } 6)$ c) $3 \cdot ? \equiv 4 \,(\text{mod } 6)$

SOLUTION

a) One method of determining the solution is to replace the question mark with the numbers 0–4 and then determine the equivalent modulo class of the product. We use the numbers 0–4 because we are working in modulo 5.

$$4 \cdot ? \equiv 2 \,(\text{mod } 5)$$
$$4 \cdot 0 \equiv 0 \,(\text{mod } 5)$$
$$4 \cdot 1 \equiv 4 \,(\text{mod } 5)$$
$$4 \cdot 2 \equiv 3 \,(\text{mod } 5)$$
$$4 \cdot 3 \equiv 2 \,(\text{mod } 5)$$
$$4 \cdot 4 \equiv 1 \,(\text{mod } 5)$$

Therefore, $? = 3$ since $4 \cdot 3 \equiv 2 \,(\text{mod } 5)$.

b) Since we are working in modulo 6, replace the question mark with the numbers 0–5 and follow the procedure used in part (a).

$$3 \cdot ? \equiv 3 \ (\text{mod } 6)$$
$$3 \cdot 0 \equiv 0 \ (\text{mod } 6)$$
$$3 \cdot 1 \equiv 3 \ (\text{mod } 6)$$
$$3 \cdot 2 \equiv 0 \ (\text{mod } 6)$$
$$3 \cdot 3 \equiv 3 \ (\text{mod } 6)$$
$$3 \cdot 4 \equiv 0 \ (\text{mod } 6)$$
$$3 \cdot 5 \equiv 3 \ (\text{mod } 6)$$

Therefore, replacing the question mark with 1, 3, or 5 results in true statements. The answers are 1, 3, and 5.

c) $3 \cdot ? \equiv 4 \ (\text{mod } 6)$

Examining the products in part (b) shows there are no values that satisfy the statement. The answer is "no solution."   ●

Whenever a process is repetitive, modular arithmetic may be helpful in answering some questions about the process. Now let's look at an application of modular arithmetic.

EXAMPLE 5 *Work Schedule*

Ellis drives a bus for a living. His working schedule is to drive for 6 days, and then he gets 2 days off. If today is the third day that Ellis has been driving, determine the following.

a) Will he be driving 60 days from today?
b) Will he be driving 82 days from today?
c) Was he driving 124 days ago?

SOLUTION

a) Since Ellis drives for 6 days and then gets 2 days off, his working schedule may be considered a modular 8 system. That is, 8, 16, 24, ... days from today will be just like today, the third day of the 8-day cycle.

   If we divide 60 by 8, we obtain

$$
\begin{array}{r}
7 \\
8 \overline{)60} \\
56 \\
\hline
4 \leftarrow \text{remainder}
\end{array}
$$

Therefore, in 60 days Ellis will go through 7 complete cycles and be 4 days further into the next cycle. If we let D represent a driving day and N represent a not driving day, then Ellis's cycle may be represented as follows:

$$D \ D \ \overset{\frown}{D} \ D \ D \ D \ \overset{\frown}{N} \ N$$

$$\underset{\text{today}}{\uparrow} \qquad \underset{\substack{\text{4 days} \\ \text{from today}}}{\uparrow}$$

Notice that 4 days from today will be his first not driving day. Therefore, Ellis will not be driving 60 days from today.

b) We work this part in the same way we worked part (a). Divide 82 by 8 and determine the remainder.

$$
\begin{array}{r}
10 \\
8\overline{)82} \\
\underline{80} \\
2 \leftarrow \text{remainder}
\end{array}
$$

Thus, in 82 days it will be 2 days later in the cycle than it is today. Because he is currently in day 3 of his cycle, Ellis will be in day 5 of his cycle and will be driving the bus.

c) This part is worked in the same way as parts (a) and (b), but once we find the remainder we must move backward in the cycle.

$$
\begin{array}{r}
15 \\
8\overline{)124} \\
\underline{120} \\
4 \leftarrow \text{remainder}
\end{array}
$$

Thus, 124 days ago was 4 days earlier in the cycle. Marking day 3 of the cycle (indicated by the word *today*) and then moving 4 days backwards brings us to the first nondriving day. The two N's shown at the beginning of the letters below are actually the end days of the previous cycle.

$$
N\ N\ D\ D\ D\ D\ D\ D\ N\ N
$$

$$\uparrow \qquad\qquad\qquad \uparrow$$

4 days   today
earlier
than today

Therefore, 124 days ago Ellis was not driving.   ●

Modular Arithmetic and Groups

Modular arithmetic systems under the operation of addition are commutative groups, as illustrated in Example 6.

EXAMPLE ❻ *A Commutative Group*

Construct a mod 5 addition table and show that the mathematical system is a commutative group. Assume that the associative property holds for the given operation.

SOLUTION The set of elements in modulo 5 arithmetic is $\{0, 1, 2, 3, 4\}$; the binary operation is $+$.

+	0	1	2	3	4
0	0	1	2	3	4
1	1	2	3	4	0
2	2	3	4	0	1
3	3	4	0	1	2
4	4	0	1	2	3

For this system to be a commutative group, it must satisfy the five properties of a commutative group.

1. *Closure*: Every entry in the table is a member of the set $\{0, 1, 2, 3, 4\}$, so the system is closed under addition.
2. *Identity element*: An easy way to determine whether there is an identity element is to look for a row in the table that is identical to the elements at the top of the table. Note that the row next to 0 is identical to the top of the table, which indicates that 0 *might be* the identity element. Now look at the column under the 0 at the top of the table. If this column is identical to the left-hand column, then 0 is the identity element. Since the column under 0 is the same as the left-hand column, 0 is the additive identity element in modulo 5 arithmetic.

Element	+	Identity	=	Element
0	+	0	=	0
1	+	0	=	1
2	+	0	=	2
3	+	0	=	3
4	+	0	=	4

3. *Inverse elements*: Does every element have an inverse? Recall that an element plus its inverse must equal the identity element. In this example, the identity element is 0. Therefore, for each of the given elements 0, 1, 2, 3, and 4, we must find the element that when added to it results in a sum of zero. These elements will be the inverses.

Element	+	Inverse	=	Identity	
0	+	?	=	0	Since $0 + 0 = 0$, 0 is its own inverse.
1	+	?	=	0	Since $1 + 4 = 0$, 4 is the inverse of 1.
2	+	?	=	0	Since $2 + 3 = 0$, 3 is the inverse of 2.
3	+	?	=	0	Since $3 + 2 = 0$, 2 is the inverse of 3.
4	+	?	=	0	Since $4 + 1 = 0$, 1 is the inverse of 4.

Note that each element has an inverse.

4. *Associative property*: It is given that the associative property holds. One example that illustrates the associative property is

$$(2 + 3) + 4 = 2 + (3 + 4)$$
$$0 + 4 = 2 + 2$$
$$4 = 4 \quad \text{True}$$

5. *Commutative property*: Is $a + b = b + a$ for *all* elements a and b of the given set? The table shows that the system is commutative because the elements are symmetric about the main diagonal. We will give one example to illustrate the commutative property.

$$4 + 2 = 2 + 4$$
$$1 = 1 \quad \text{True}$$

All five properties are satisfied. Thus, modulo 5 arithmetic under the operation of addition is a commutative group.

SECTION 10.3 EXERCISES

CONCEPT/WRITING EXERCISES

1. What does a *modulo m* system consist of?

2. **a)** Explain the meaning of the statement, "*a* is congruent to *b* modulo *m*."

   b) Explain the meaning of $27 \equiv 2 \pmod 5$.

3. In a modulo 3 system, how many modulo classes will there be? Construct a table similar to Table 10.12 on page 632 showing elements from each class.

4. In general, for a modulo *m* system, how are modulo classes developed?

5. In a modulo 16 system, how many modulo classes will there be? Explain.

6. In a modulo *n* system, how many modulo classes will there be? Explain.

7. Consider $29 \equiv ? \pmod 5$. Which of the following values could replace the question mark and result in a true statement? Explain.

   a) 20   **b)** 4   **c)** 9   **d)** 204

8. Consider $50 \equiv ? \pmod 7$. Which of the following values could replace the question mark and result in a true statement? Explain.

   a) 78   **b)** 71   **c)** 7   **d)** 17

PRACTICE THE SKILLS

In Exercises 9–16, assume that Sunday is represented as day 0, Monday is represented by day 1, and so on. If today is Thursday (day 4), determine the day of the week it will be in the specified number of days. Assume no leap years.

9. 10 days

10. 161 days

11. 365 days

12. 3 years

13. 3 years, 34 days

14. 463 days

15. 400 days

16. 3 years, 27 days

In Exercises 17–24, consider 12 months to be a modulo 12 system. If it is currently October, determine the month it will be in the specified number of months.

17. 16 months

18. 48 months

19. 2 years, 7 months

20. 4 years, 8 months

21. 83 months

22. 7 years

23. 105 months

24. 5 years, 9 months

In Exercises 25–36, determine what number the sum, difference, or product is congruent to in mod 5.

25. $4 + 7$

26. $5 + 10$

27. $2 + 5 + 5$

28. $9 - 3$

29. $5 - 12$

30. $8 \cdot 7$

31. $8 \cdot 9$

32. $1 - 4$

33. $4 - 8$

34. $3 - 7$

35. $(15 \cdot 4) - 8$

36. $(4 - 9) \cdot 7$

In Exercises 37–50, find the modulo class to which each number belongs for the indicated modulo system.

37. 10, mod 5

38. 23, mod 7

39. 48, mod 12

40. 43, mod 6

41. 41, mod 9

42. 75, mod 8

43. 30, mod 7

44. 53, mod 4

45. −1, mod 7

46. −7, mod 4

47. −13, mod 11

48. −11, mod 13

49. 205, mod 10

50. −12, mod 4

In Exercises 51–66, determine all positive number replacements (less than the modulus) for the question mark that make the statement true.

51. $2 + 3 \equiv ? \pmod 6$

52. $? + 5 \equiv 3 \pmod 8$

53. $1 + ? \equiv 4 \pmod 5$

54. $4 + ? \equiv 3 \pmod 6$

55. $4 - ? \equiv 5 \pmod 6$

56. $2 \cdot 5 \equiv ? \pmod 7$

57. $5 \cdot ? \equiv 7 \pmod 9$

58. $3 \cdot ? \equiv 5 \pmod 6$

59. $2 \cdot ? \equiv 1 \pmod 6$

60. $3 \cdot ? \equiv 3 \pmod{12}$

61. $4 \cdot ? \equiv 4 \pmod{10}$

62. $? - 3 \equiv 4 \pmod 8$

63. $? - 8 \equiv 9 \pmod{12}$

64. $6 - ? \equiv 8 \pmod 9$

65. $3 \cdot ? \equiv 0 \pmod{10}$

66. $4 \cdot ? \equiv 5 \pmod 8$

PROBLEM SOLVING

67. *Presidential Elections* The upcoming presidential election years are 2008, 2012, 2016, ...

a) List the next five presidential election years after 2016.

b) What will be the first election year after the year 3000?

c) List the election years between the years 2550 and 2575.

68. *Flight Schedules* A pilot is scheduled to fly for 5 consecutive days and rest for 3 consecutive days. If today is the second day of her rest shift, determine whether she will be flying or resting

a) 60 days from today.

b) 90 days from today.

c) 240 days from today.

d) Was she flying 6 days ago?

e) Was she flying 20 days ago?

69. *Workout Schedule* A tennis pro's workout schedule is to give both morning and afternoon lessons for 3 days, rest for 1 day, give only morning lessons for 2 days, rest for 2 days, and then start the cycle again. If the tennis pro is on his rest for 1 day part of the schedule, determine what he will be doing

a) 28 days from today.

b) 60 days from today.

c) 127 days from today.

d) Will the tennis pro have a day off 82 days from today?

70. *Physical Therapy* A man has an Achilles' tendon injury and is receiving physical therapy. He must have physical therapy twice a day for 5 days, physical therapy once a day for 3 days, and then 2 days off; then the cycle begins again. If he is in his second day of his twice-a-day therapy cycle, determine what he will be doing

a) 20 days from today.

b) 49 days from today.

c) 103 days from today.

d) Will the man have a day off 78 days from today?

71. *The Weekend Off* A manager of a theater has both Saturday and Sunday off every 7 weeks. It is week 2 of the 7 weeks.

a) Determine the number of weeks before he will have both Saturday and Sunday off.

b) Will he have both Saturday and Sunday off 25 weeks from this week?

c) What is the first week, after 50 weeks from this week, that he will have both Saturday and Sunday off?

72. *Nursing Shifts* A nurse's work pattern at Community Hospital consists of working the 7 A.M.–3 P.M. shift for 3 weeks and then the 3 P.M.–11 P.M. shift for 2 weeks.

a) If it is the third week of the pattern, what shift will the nurse be working 6 weeks from now?

b) If it is the fourth week of the pattern, what shift will the nurse be working 7 weeks from now?

c) If it is the first week of the pattern, what shift will the nurse be working 11 weeks from now?

73. *Restaurant Rotation* A waiter at a restaurant works both daytime and evening shifts. He works daytime for 5 consecutive days, then evenings for 3 consecutive days, then daytime for 4 consecutive days, then evenings for 2 consecutive days. Then the rotation starts again. If it is day 2 of the 5-day consecutive daytime shift, determine whether he will be working the daytime or evening shift

a) 20 days from today.

b) 52 days from today.

c) 365 days from today.

74. *A Truck Driver's Schedule* A truck driver's routine is as follows: Drive 3 days from New York to Chicago, rest 1 day in Chicago, drive 3 days from Chicago to Los Angeles, rest 2 days in Los Angeles, drive 5 days to return to New York, rest 3 days in New York. Then the cycle begins again. If the truck driver is starting her trip to Chicago today, what will she be doing

a) 30 days from today?

b) 70 days from today?

c) 2 years from today?

75. a) Construct a modulo 3 addition table.

b) Is the system closed? Explain.

c) Is there an identity element for the system? If so, what is it?

d) Does every element in the system have an inverse? If so, list the elements and their inverses.

e) The associative property holds for the system. Give an example.

f) Does the commutative property hold for the system? Give an example.

g) Is the system a commutative group?

h) Will every modulo system under the operation of addition be a commutative group? Explain.

76. Construct a modulo 8 addition table. Repeat parts (b)–(h) in Exercise 75.

77. a) Construct a modulo 4 multiplication table.

b) Is the system closed under the operation of multiplication?

c) Is there an identity element in the system? If so, what is it?

d) Does every element in the system have an inverse? Make a list showing the elements that have a multiplicative inverse and list the inverses.

e) The associative property holds for the system. Give an example.

f) Does the commutative property hold for the system? Give an example.

g) Is this mathematical system a commutative group? Explain.

78. Construct a modulo 7 multiplication table. Repeat parts (b)–(g) in Exercise 77.

CHALLENGE PROBLEMS/GROUP ACTIVITIES

We have not discussed division in modular arithmetic. With what number or numbers, if any, can you replace the question marks to make the statement true? The question mark must be a number less than the modulus. Hint: Use the fact that $\frac{a}{b} \equiv c \ (mod \ m)$ means $a \equiv b \cdot c \ (mod \ m)$ and use trial and error to obtain your answer.

79. $? \div 5 \equiv 5 \ (\text{mod } 9)$

80. $5 \div 7 \equiv ? \ (\text{mod } 9)$

81. $? \div ? \equiv 1 \ (\text{mod } 4)$

82. $1 \div 2 \equiv ? \ (\text{mod } 5)$

In Exercises 83–85, solve for x where k is any counting number.

83. $5k \equiv x \ (\text{mod } 5)$

84. $5k + 4 \equiv x \ (\text{mod } 5)$

85. $4k - 2 \equiv x \ (\text{mod } 4)$

86. Find the smallest positive number divisible by 5 to which 2 is congruent in modulo 6.

87. *Birthday Question* During a certain year, Clarence Dee's birthday is on Monday, April 18.

a) If next year is *not* a leap year, on what day of the week will Clarence's birthday fall next year?

b) If next year *is* a leap year, on what day of the week will Clarence's birthday fall next year?

88. *More Birthday Questions* During a certain leap year, Josephine Emme's birthday is on Tuesday, May 18. In how many years will Josephine's birthday fall on a

a) Friday?

b) Saturday?

RECREATIONAL MATHEMATICS

89. *Rolling Wheel* The wheel shown below is to be rolled. Before the wheel is rolled, it is resting on number 0. The wheel will be rolled at a uniform rate of one complete roll every 4 minutes. In exactly 1 year (not a leap year), what number will be at the bottom of the wheel?

90. *Deciphering a Code* One important use of modular arithmetic is in coding. One type of coding circle is given in Fig. 10.6. To use it, the person you are sending the message to must know the code key to decipher the code. The code key to this message is *j*. Can you decipher this code? (*Hint*: Subtract the code key from the code numbers.)

23   11   3   18   10   19   2   10   16   4   24

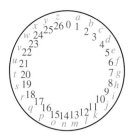

Figure 10.6

INTERNET/RESEARCH ACTIVITY

91. *The Enigma Cipher Machine* Do research on the Enigma ciphering machine (see the Did You Know? on page 634). Write a paper on the role the machine played prior to and during World War II.

92. The concept and notation for modular systems were introduced by Carl Friedrich Gauss in 1801. Write a paper on Gauss's contribution to modular systems.

CHAPTER ⑩ SUMMARY

IMPORTANT FACTS

If the elements in a table are symmetric about the main diagonal, then the system is commutative.

	Addition	Multiplication
Commutative property	$a + b = b + a$	$a \cdot b = b \cdot a$
Associative property	$(a + b) + c$ $= a + (b + c)$	$(a \cdot b) \cdot c = a \cdot (b \cdot c)$

a is **congruent** to *b* modulo *m*, written $a \equiv b \pmod{m}$, if *a* and *b* have the same remainder when divided by *m*.

PROPERTIES OF A GROUP AND A COMMUTATIVE GROUP

A mathematical system is a group if the first four conditions hold and a commutative group if all five conditions hold.

Commutative Group / Group

1. The set of elements is *closed* under the given operation.
2. An *identity element* exists for the set.
3. Every element in the set has an *inverse*.
4. The set of elements is *associative* under the given operation.
5. The set of elements is *commutative* under the given operation.

CHAPTER ⑩ REVIEW EXERCISES

10.1, 10.2

1. List the parts of a mathematical system.

2. What is a binary operation?

3. Are the whole numbers closed under the operation of subtraction? Explain.

4. Are the real numbers closed under the operation of subtraction? Explain.

Determine the sum or difference in clock 12 arithmetic.

5. $8 + 7$　　　**6.** $5 + 12$　　　**7.** $4 - 7$

8. $4 + 7 + 9$　　**9.** $7 - 4 + 6$　　**10.** $2 - 8 - 7$

11. List the properties of a group and explain what each property means.

12. What is another name for an abelian group?

In Exercises 13 and 14, explain your answer.

13. Determine whether the set of integers under the operation of addition forms a group.

14. Determine whether the set of integers under the operation of multiplication forms a group.

15. Determine whether the set of natural numbers under the operation of addition forms a group.

16. Determine whether the set of rational numbers under the operation of multiplication forms a group.

In Exercises 17–19, for the mathematical system, determine which of the five properties of a commutative group do not hold.

17.

●	*R*	*S*	⊗
R	*R*	⊗	*S*
S	⊗	*S*	*R*
⊗	*S*	*R*	⊗

18.

?	**4**	**#**	**L**	**P**
4	P	4	#	L
#	4	#	L	P
L	#	L	P	4
P	L	P	4	L

19.

□	**!**	**?**	**Δ**	***p***
!	?	Δ	!	*p*
?	Δ	*p*	?	!
Δ	!	?	Δ	*p*
p	*p*	!	*p*	Δ

20. Consider the following mathematical system. Assume the associative property holds for the given operation.

⌂	☺	●	♀	♂
☺	☺	●	♀	♂
●	●	♀	♂	☺
♀	♀	♂	☺	●
♂	♂	☺	●	♀

a) What are the elements of the set in this mathematical system?

b) What is the binary operation? ⌂

c) Is the system closed? Explain.

d) Is there an identity element for the system under the given operation?

e) Does every element in the system have an inverse? If so, give each element and its corresponding inverse.

f) Give an example to illustrate the associative property. ⌂　⌂　　⌂　⌂

g) Is the system commutative? Give an example.　⌂　　⌂

h) Is this mathematical system a commutative group? Explain.

10.3

In Exercises 21–30, find the modulo class to which the number belongs for the indicated modulo system.

21. 15, mod 5　　　　　　**22.** 26, mod 7

23. 31, mod 6　　　　　　**24.** 59, mod 8

25. 71, mod 12　　　　　**26.** 54, mod 4

27. 52, mod 12　　　　　**28.** 54, mod 14

29. 97, mod 11　　　　　**30.** 31, mod 20

In Exercises 31–40, determine all positive number replacements (less than the modulus) for the question mark that make the statement true.

31. $3 + 7 \equiv ? \pmod 9$　　**32.** $? - 4 \equiv 2 \pmod 5$

33. $4 \cdot ? \equiv 3 \pmod 7$　　**34.** $6 - ? \equiv 5 \pmod 7$

35. $? \cdot 4 \equiv 0 \pmod 8$　　**36.** $10 \cdot 7 \equiv ? \pmod{11}$

37. $3 - 5 \equiv ? \pmod 7$　　**38.** $? \cdot 7 \equiv 3 \pmod{10}$

39. $5 \cdot ? \equiv 3 \pmod 8$　　**40.** $7 \cdot ? \equiv 2 \pmod 9$

41. Construct a modulo 6 addition table. Then determine whether the modulo 6 system forms a commutative group under the operation of addition.

42. Construct a modulo 4 multiplication table. Then determine whether the modulo 4 system forms a commutative group under the operation of multiplication.

43. *Firefighter Shifts* Linda Zientek, a firefighter, has the following work pattern. She works 3 days, has 4 days off, then works 4 days, then has 3 days off, and then the pattern repeats. Assume that today is the first day of the work pattern.

a) Will Linda be working 30 days from today?

b) Linda's sister is getting married 45 days from today. Will Linda have the day off?

CHAPTER ⑩ TEST

1. What is a mathematical system?

2. List the requirements needed for a mathematical system to be a commutative group.

3. Is the set of integers a commutative group under the operation of addition? Explain your answer completely.

4. Develop a clock 5 arithmetic addition table.

5. Is clock 5 arithmetic under the operation of addition a commutative group? Assume that the associative property holds. Explain your answer completely.

Determine the following in clock 5 arithmetic.

6. $1 + 3 + 4$

7. $2 - 5$

8. Consider the mathematical system

□	*W*	*S*	*T*	*R*
W	*T*	*R*	*W*	*S*
S	*R*	*W*	*S*	*T*
T	*W*	*S*	*T*	*R*
R	*S*	*T*	*R*	*W*

a) What is the binary operation?

b) Is this system closed? Explain.

c) Is there an identity element for this system under the given operation? Explain.

d) What is the inverse of the element *R*?

e) What is $(T \,\square\, R) \,\square\, W$?

In Exercises 9 and 10, determine whether the mathematical system is a commutative group. Explain your answer completely.

9.
*	*a*	*b*	*c*
a	*a*	*b*	*c*
b	*b*	*b*	*a*
c	*c*	*a*	*d*

10.
?	**1**	**2**	**3**
1	3	1	2
2	1	2	3
3	2	3	1

11. Determine whether the mathematical system is a commutative group. Assume that the associative property holds. Explain your answer.

○	@	$	&	%
@	%	@	$	&
$	@	$	&	%
&	$	&	%	@
%	&	%	@	$

In Exercises 12 and 13, determine the modulo class to which the number belongs for the indicated modulo system.

12. 27, mod 5

13. 91, mod 13

In Exercises 14–18, determine all positive number replacements for the question mark, less than the modulus, that make the statement true.

14. $7 + 7 \equiv ? \pmod 8$

15. $? - 3 \equiv 4 \pmod 5$

16. $3 - ? \equiv 7 \pmod 9$

17. $4 \cdot 2 \equiv ? \pmod 6$

18. $3 \cdot ? \equiv 2 \pmod 6$

19. To what number is 239 congruent to in modulo 8?

20. a) Construct a modulo 5 multiplication table.

    b) Is this mathematical system a commutative group? Explain your answer completely.

G R O U P   P R O J E C T S

1. **Rotating a Square** The square *ABCD* below has a pin through it at point *P* so that the square can be rotated in a clockwise direction. The mathematical system consists of the set $\{A, B, C, D\}$ and the operation ♣. The elements of the set represent different rotations of the square clockwise, as follows.

A = rotate about the point *P* clockwise 90°

B = rotate about the point *P* clockwise 180°

C = rotate about the point *P* clockwise 270°

D = rotate about the point *P* clockwise 360°

The operation ♣ means *followed by*. For example, A ♣ B means a clockwise rotation of 90° followed by a clockwise rotation of 180° or a clockwise rotation of 270°. Since C represents a clockwise rotation of 270°, we write A ♣ $B = C$. This result and two other results are shown in the following table.

♣	A	B	C	D
A		C		
B			A	
C	D			
D				

Complete the table and determine if the mathematical system is a commutative group. Explain.

2. **Product of Zero** In arithmetic and algebra, the statement, "If $a \cdot b = 0$, then $a = 0$ or $b = 0$" is true. That is, for the product of two numbers to be 0, at least one of the factors must be 0. Can the product of two nonzero numbers equal 0 in a specific modulo system? If so, in what type of modulo systems can this result occur?

    a) Construct multiplication tables for modulo systems 3–9.

    b) Which, if any, of the multiplication tables in part (a) have products equal to 0 when neither factor is 0?

    c) Which, if any, of the multiplication tables in part (a) have products equal to 0 only when at least one factor is 0?

    d) Using the results in parts (b) and (c), can you write a conjecture as to which modulo systems have a product of 0 when neither factor is 0?

3. **Conjecture about Multiplicative Inverses** Are there certain modulo systems where all numbers have multiplicative inverses?

    a) If you have not worked Group Projects Exercise 2, construct multiplication tables for modulo systems 3–9.

    b) Which of the multiplication tables in part (a) contain multiplicative inverses for all nonzero numbers?

    c) Which, if any, of the multiplication tables in part (a) do not contain multiplicative inverses for all nonzero numbers?

    d) Using the results in parts (b) and (c), can you write a conjecture as to which modulo systems contain multiplicative inverses for all nonzero numbers?

CHAPTER 11

Consumer Mathematics

▲ Knowledge of consumer mathematics may help you reach your long-term goals.

WHY IT IS IMPORTANT

Managing your money takes much thought and planning. All your daily needs must be met, but at the same time you should consider your long-term goals such as purchasing a house, saving for college expenses, and investing for retirement. This chapter will provide information to help you reach these important goals.

645

11.1   PERCENT

▲ Daily weather reports include percents.

Your clock radio turns on at 6 A.M. and you hear a weather report that says there is a 75% chance of rain. Based on this information, you decide to carry your umbrella to your class. You used your knowledge of percents to make an everyday decision. In this section, we will investigate the meaning of percent and learn how percents relate to mathematics and to daily life.

Percent, Fractions, and Decimals

The study of mathematics is crucial to understanding how to make better financial decisions. A basic topic necessary for understanding the material in this chapter is percent.

The word *percent* comes from the Latin *per centum*, meaning "per hundred." A *percent* is simply a ratio of some number to 100. Thus, $\frac{15}{100} = 15\%$, and $\frac{x}{100} = x\%$.

Percents are useful in making comparisons. Consider Ross, who took two psychology tests. On the first test Ross answered 18 of the 20 questions correctly, and on the second test he answered 23 of 25 questions correctly. On which test did he have the higher score? One way to compare the results is to write ratios of the number of correct answers to the number of questions on the test and then convert the ratios to percents. We can find the grades in percent for each test by (a) writing a ratio of the number of correct answers to the total number of questions, (b) rewriting these ratios with a denominator of 100, and (c) expressing the ratios as percents.

$$\text{Test 1} \qquad\qquad \overset{(a)}{} \qquad \overset{(b)}{} \qquad \overset{(c)}{}$$

$$\frac{\text{Number of correct answers}}{\text{Number of questions on the test}} = \frac{18}{20} = \frac{18 \times 5}{20 \times 5} = \frac{90}{100} = 90\%$$

$$\text{Test 2} \qquad\qquad \overset{(a)}{} \qquad \overset{(b)}{} \qquad \overset{(c)}{}$$

$$\frac{\text{Number of correct answers}}{\text{Number of questions on the test}} = \frac{23}{25} = \frac{23 \times 4}{25 \times 4} = \frac{92}{100} = 92\%$$

By changing the results of both tests to percents, we have a common standard for comparison. The results show that Ross scored 90% on the first test and 92% on the second test. Thus, he had a higher score on the second test.

Another procedure to change a fraction to a percent follows.

PROCEDURE TO CHANGE A FRACTION TO A PERCENT

1. Divide the numerator by the denominator to obtain a decimal number.
2. Multiply the decimal number by 100 (which has the effect of moving the decimal point two places to the right).
3. Add a percent sign.

The above procedure has the effect of moving the decimal point in the original number two places to the right and adding a percent sign.

Note that steps 2 and 3 together are the equivalent of multiplying by 100%. Since 100% = 100/100 = 1, we are not changing the *value* of the number, we are simply changing the number to a percent.

─EXAMPLE ❶ *Converting Fractions to Percents*

Change each of the following fractions to a percent.

a) $\dfrac{67}{100}$ b) $\dfrac{11}{20}$ c) $\dfrac{15}{16}$

SOLUTION For each fraction, follow the three steps in the procedure box on page 646.

a) 1. $67 \div 100 = 0.67$ 2. $0.67 \times 100 = 67$ 3. 67%

   Thus, $\dfrac{67}{100} = 67\%$.

b) 1. $11 \div 20 = 0.55$ 2. $0.55 \times 100 = 55$ 3. 55%

   Thus, $\dfrac{11}{20} = 55\%$.

c) 1. $15 \div 16 = 0.9375$ 2. $0.9375 \times 100 = 93.75$ 3. 93.75%

   Thus, $\dfrac{15}{16} = 93.75\%$.   ●

In future examples, when changing a fraction to a percent, we will follow the three-step procedure given on page 646 without listing the individual steps. We will do this in Examples 4 and 5.

> ## PROCEDURE TO CHANGE A DECIMAL NUMBER TO A PERCENT
>
> 1. Multiply the decimal number by 100.
> 2. Add a percent sign.

The procedure for changing a decimal number to a percent is equivalent to moving the decimal point two places to the right and adding a percent sign.

─EXAMPLE ❷ *Converting Decimal Numbers to Percents*

Change each of the following decimal numbers to a percent.

a) 0.14   b) 0.893   c) 0.7625

SOLUTION

a) $0.14 = (0.14 \times 100)\% = 14\%$. Thus, $0.14 = 14\%$.
b) $0.893 = (0.893 \times 100)\% = 89.3\%$. Thus, $0.893 = 89.3\%$.
c) $0.7625 = (0.7625 \times 100)\% = 76.25\%$. Thus, $0.7625 = 76.25\%$.   ●

To change a number given as a percent to a decimal number, follow this procedure.

> ## PROCEDURE TO CHANGE A PERCENT TO A DECIMAL NUMBER
>
> 1. Divide the number by 100.
> 2. Remove the percent sign.

The procedure for changing a number given as a percent to a decimal number is equivalent to moving the decimal point two places to the left and removing the percent sign. Another way to remember that is: percent means per hundred (or to *divide by 100*).

EXAMPLE ❸ *Converting a Percent to a Decimal Number*

a) Change 35% to a decimal number.

b) Change 69.8% to a decimal number.

c) Change $\frac{1}{2}$% to a decimal number.

SOLUTION

a) $35\% = \dfrac{35}{100} = 0.35$. Thus, $35\% = 0.35$.

b) $69.8\% = \dfrac{69.8}{100} = 0.698$. Thus, $69.8\% = 0.698$.

c) $\dfrac{1}{2}\% = 0.5\% = \dfrac{0.5}{100} = 0.005$. Thus, $\dfrac{1}{2}\% = 0.005$. ●

EXAMPLE ❹ *How Old Is Old?*

A Yahoo! Question of the Week asked, "At what age do you consider someone old?" Out of the 3496 people who responded, 1017 said age 80 is old. What percent of respondents believe that age 80 is old?

SOLUTION To find the percent that believe age 80 is old, divide the number of those who responded that age 80 is old by the total number of respondents. Then move the decimal point two places to the right and add a percent sign:

$$\text{Percent who believe age 80 is old} = \frac{1017}{3496} \approx 0.2909 \approx 29.09\%$$

Thus, about 29.1% (to the nearest tenth of a percent) believe that age 80 is old. ●

All answers in this section will be given to the nearest tenth of a percent. When rounding answers to the nearest tenth of a percent, we carry the division to four places after the decimal point (ten-thousandths) and then round to the nearest thousandths position.

EXAMPLE ❺ *Cherry Production*

According to the U.S. Department of Agriculture, in 2005 the United States produced about 244 million pounds of cherries. Of these cherries, 190 million pounds were produced in Michigan, 19 million pounds were produced in Utah, 18 million pounds were produced in Washington, and 17 million pounds were produced in other states. Determine the percent of cherries produced in each of the following: Michigan, Utah, Washington, and other states.

SOLUTION To determine each percent, divide the number of cherries produced in each state by the total number of cherries produced in the United States. Because all the numbers given are in millions of pounds, we will calculate the percents using the number of millions of pounds.

$$\text{Percent of cherries produced in Michigan} = \frac{190}{244} \approx 0.7786\ldots \approx 77.9\%$$

$$\text{Percent of cherries produced in Utah} = \frac{19}{244} \approx 0.0778\ldots \approx 7.8\%$$

$$\text{Percent of cherries produced in Washington} = \frac{18}{244} \approx 0.0737\ldots \approx 7.4\%$$

$$\text{Percent of cherries produced in other states} = \frac{17}{244} \approx 0.0696\ldots \approx 7.0\%$$

The sum of the percents, 77.9% + 7.8% + 7.4% + 7.0%, equals 100.1%. Since we rounded each percent, the sum is slightly different from 100%. ●

Percent Change

The percent increase or decrease, or percent change, over a period of time is determined by the following formula:

$$\textbf{Percent change} = \frac{\left(\substack{\text{amount in}\\\text{latest period}}\right) - \left(\substack{\text{amount in}\\\text{previous period}}\right)}{\text{amount in previous period}} \times 100$$

If the amount in the latest period is greater than the amount in the previous period, the answer will be positive and will indicate a percent increase. If the amount in the latest period is smaller than the amount in the previous period, the answer will be negative and will indicate a percent decrease.

┌─ **EXAMPLE ❻** *Baseball Record*

In 2005, the major league baseball team with the most improved record for winning games was the Arizona Diamondbacks. In 2004, the Diamondbacks won 51 games. In 2005, the Diamondbacks won 77 games. Find the percent increase in the number of games won by the Diamondbacks from 2004 to 2005.

SOLUTION The previous period is 2004, and the latest period is 2005.

$$\text{Percent change} = \frac{\left(\substack{\text{amount in}\\\text{latest period}}\right) - \left(\substack{\text{amount in}\\\text{previous period}}\right)}{\text{amount in previous period}} \times 100$$

$$= \frac{77 - 51}{51} \times 100$$

$$= \frac{26}{51} \times 100$$

$$\approx 0.5098\ldots \times 100$$

$$\approx 51.0$$

Therefore, there was about a 51.0% increase in the number of games won by the Diamondbacks from 2004 to 2005. ●

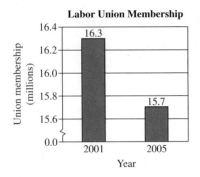

Labor Union Membership

EXAMPLE ❼ *Labor Union Membership*

There were approximately 15.7 million labor union members in the United States in 2005. This number is down from approximately 16.3 million labor union members in 2001 (see the bar graph in the margin). Determine the percent change in labor union membership from 2001 to 2005.

SOLUTION The previous period is 2001, and the latest period is 2005.

$$\text{Percent change} = \frac{\left(\begin{array}{c}\text{amount in}\\\text{latest period}\end{array}\right) - \left(\begin{array}{c}\text{amount in}\\\text{previous period}\end{array}\right)}{\text{amount in previous period}} \times 100$$

$$= \frac{15.7 - 16.3}{16.3} \times 100$$

$$= \frac{-0.6}{16.3} \times 100$$

$$\approx -0.0368 \times 100$$

$$\approx -3.7\%$$

Thus, union membership decreased by about 3.7% from 2001 to 2005.   ●

Percent Markup and Markdown

A similar formula is used to calculate percent markup or markdown on cost. A positive answer indicates a markup, and a negative answer indicates a markdown.

$$\textbf{Percent markup (or markdown) on cost} = \frac{\text{selling price} - \text{dealer's cost}}{\text{dealer's cost}} \times 100$$

EXAMPLE ❽ *Determining Percent Markup*

Holdren Hardware stores pay $48.76 for glass fireplace screens. They regularly sell them for $79.88. At a sale they sell them for $69.99. Determine

a) the percent markup on the regular price.

b) the percent markup on the sale price.

SOLUTION

a) We determine the percent markup on the regular price as follows.

$$\text{Percent markup} = \frac{\text{selling price} - \text{dealer's cost}}{\text{dealer's cost}} \times 100$$

$$= \frac{79.88 - 48.76}{48.76} \times 100$$

$$\approx 0.6382 \times 100$$

$$\approx 63.8$$

Thus, the percent markup on the regular price was about 63.8%.

b) We determine the percent markup on the sale price as follows.

$$\text{Percent markup} = \frac{69.99 - 48.76}{48.76} \times 100$$

$$\approx 0.4354 \times 100$$

$$\approx 43.5$$

Thus, the percent markup on the sale price was about 43.5%.

EXAMPLE 9 *Determining Percent Markdown*

To help get rid of its remaining 2008 calendars, Cardin's Pharmacy reduces the price to $4.00. If Cardin's cost per calendar was $5.00, determine the percent markdown on the calendars.

SOLUTION We will use the same formula that we used for percent markup in Example 8.

$$\text{Percent markdown} = \frac{\text{selling price} - \text{dealer's cost}}{\text{dealer's cost}} \times 100$$

$$= \frac{4.00 - 5.00}{5.00} \times 100$$

$$= -0.20 \times 100$$

$$= -20$$

Thus, the percent markdown on the calendars was 20%.

Other Percent Problems

In daily life, we may need to know how to solve any one of the following three types of problems involving percent:

1. What is a 15% tip on a restaurant bill of $24.66? The problem can be stated as

    15% of $24.66 is what number?

2. If Nancy Johnson made a sale of $500 and received a commission of $25, what percent of the sale is the commission? The problem can be stated as

    What percent of $500 is $25?

3. If the price of a jacket was reduced by 25% or $12.50, what was the original price of the jacket? The problem can be stated as

    25% of what number is 12.50?

To answer these questions, we will write each problem as an equation. The word *is* means "is equal to," or =. In each problem, we will represent the unknown quantity with the letter *x*. Therefore, the preceding problems can be represented as

1. 15% of $24.66 = *x*   2. *x*% of $500 = $25   3. 25% of *x* = $12.50

The word *of* in such problems indicates multiplication. To solve each problem, change the percent to a decimal number and express the problem as an equation; then solve the equation for the variable *x*. The solutions follow.

1.   15% of $24.66 = x$

        $0.15(24.66) = x$ 15% is written as 0.15 in decimal form.

              $3.699 = x$

Since 15% of $24.66 is $3.699, the tip would be $3.70.

2. $x\%$ of $500 = 25

      $(0.01x)500 = 25$ $x\%$ is written as 0.01x in decimal form.

           $5x = 25$

$$\frac{5x}{5} = \frac{25}{5}$$

            $x = 5$

Since 5% of $500 is $25, the commission is 5% of the sale.

3.   25% of $x = 12.50

      $0.25(x) = 12.50$ 25% is written as 0.25 in decimal form.

$$\frac{0.25x}{0.25} = \frac{12.50}{0.25}$$

         $x = 50$

Since 25% of $50 is $12.50, the original price of the jacket was $50.00.

EXAMPLE ❿ *Down Payment on a House*

Melissa Bell wishes to buy a house for $287,000. To obtain a mortgage, she needs to pay 20% of the selling price as a down payment. Determine the amount of Melissa's down payment.

SOLUTION We want to find the amount of the down payment. Let $x =$ the down payment. Then

$$
\begin{aligned}
x &= 20\% \text{ of the selling price} \\
&= 20\% \text{ of } \$287,000 \\
&= 0.20(287,000) \\
&= 57,400.
\end{aligned}
$$

Melissa will have a down payment of $57,400.

EXAMPLE ⓫ *Chess Membership*

In 2006, the U.S. Chess Federation (USCF) had about 84,000 members. Of these members, about 11,000 were considered *youth* members (under the age of 18). What percent of USCF members were youth members?

SOLUTION We need to determine what percent of the 84,000 is 11,000. Let $x =$ percent of USCF members who are youth members. Then

$$
\begin{aligned}
x\% \text{ of } 84,000 &= 11,000 \\
0.01x(84,000) &= 11,000 \\
840x &= 11,000 \\
x &= \frac{11,000}{840} \\
x &\approx 13.1
\end{aligned}
$$

Therefore, about 13.1% of the USCF members are youth members.

▲ The Taj Mahal in Agra, India

┌─ **EXAMPLE ⑫** *Population of India*

About 337,567,000, or about 31%, of India's population is younger than 15 years old. What is the approximate population of India?

SOLUTION This problem can be stated as

31% of what number is 337,567,000?

Let $x =$ the population of India. Then

$$31\% \text{ of } x = 337{,}567{,}000$$
$$0.31x = 337{,}567{,}000$$
$$x = \frac{337{,}567{,}000}{0.31}$$
$$x \approx 1{,}088{,}925{,}806$$

Thus, the population of India is about 1,088,925,806 people.

SECTION 11.1 EXERCISES

CONCEPT/WRITING EXERCISES

1. What is a percent?

2. Explain how to change a percent to a decimal number.

3. Explain how to change a fraction to a percent.

4. Explain how to change a decimal number to a percent.

5. Explain how to determine percent change.

6. Explain how to determine percent markup on cost.

PRACTICE THE SKILLS

In Exercises 7–14, change the number to a percent. Express your answer to the nearest tenth of a percent.

7. $\frac{3}{4}$

8. $\frac{3}{5}$

9. $\frac{11}{20}$

10. $\frac{7}{8}$

11. 0.007654

12. 0.5688

13. 3.78

14. 13.678

In Exercises 15–24, change the percent to a decimal number.

15. 5%

16. 3.2%

17. 5.15%

18. 0.75%

19. $\frac{1}{4}\%$

20. $\frac{3}{8}\%$

21. $\frac{1}{5}\%$

22. 135.9%

23. 1%

24. 0.50%

PROBLEM SOLVING

For Exercises 25–67, where appropriate, round answers to the nearest tenth of a percent.

25. *Niacin* The U.S. Department of Agriculture's recommended daily allowance (USRDA) of niacin for adults is 20 mg. One serving of Cheerios provides 5 mg of niacin. What percent of the USRDA of niacin does one serving of Cheerios provide?

26. *Potassium* The U.S. Department of Agriculture's recommended daily allowance (USRDA) of potassium for adults is 3500 mg. One serving of Cheerios provides 95 mg of potassium. What percent of the USRDA of potassium does one serving of Cheerios provide?

27. *Reduced Fat Milk* One serving of whole milk contains 8 g of fat. Reduced-fat milk contains 41.25% less fat per serving than whole milk. How many grams of fat does one serving of reduced-fat milk contain?

28. *River Pollution* In a study of U.S. rivers, 693,905 miles of river were studied. According to the U.S. Environmental Protection Agency, it was found that 36% of the total miles of rivers studied had impaired water quality. How many miles of rivers had impaired quality?

Florida Lottery In Exercises 29–32, use the circle graph to answer the questions. In April 2006, the Florida Lottery had total education appropriations of $2,604,264. The areas where this money was appropriated along with the corresponding percents are represented in the circle graph below.

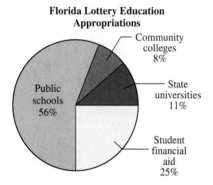

Florida Lottery Education Appropriations

Source: Florida Department of Education

29. Determine the amount of money appropriated to community colleges.

30. Determine the amount of money appropriated to state universities.

31. Determine the amount of money appropriated to student financial aid.

32. Determine the amount of money appropriated to public schools.

Gasoline Cost In Exercises 33–36, use the circle graph to answer the questions. The circle graph shows the breakdown of the cost, for a gallon of gasoline in the United States, in 2006.

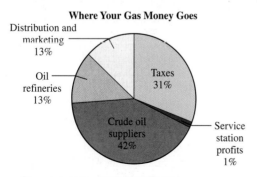

Where Your Gas Money Goes

Source: American Petroleum Institute

Suppose in 2006 your family spent $5000 on gasoline.

33. Determine the amount of money that went to taxes.

34. Determine the amount of money that went to service station profits.

35. Determine the amount of money that went to crude oil suppliers.

36. Determine the amount of money that went to oil refineries.

Super Bowl Advertisers In Exercises 37–40, use the circle graph to answer the questions. During the years 1986–2005, the top five Super Bowl advertisers spent about $559 million on advertising.

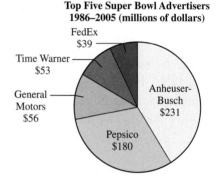

Top Five Super Bowl Advertisers 1986–2005 (millions of dollars)

37. What percent of the total did Anheuser-Busch spend?

38. What percent of the total did Pepsico spend?

39. What percent of the total did General Motors spend?

40. What percent of the total did Time Warner spend?

41. *Decreasing Population* In 2000, the population of North Dakota was 642,200. In 2005, the population of North Dakota had decreased to 636,677. Determine the percent decrease in North Dakota's population from 2000 to 2005.

42. *Increasing Population* The population of the United States rose from approximately 248.7 million in 1990 to approximately 298.3 million in 2006. Determine the percent increase in the U.S. population from 1990 to 2006.

43. *U.S. Debt* The graph on top of page 655 shows the federal U.S. debt, in trillions of U.S. dollars, for the years 1966, 1976, 1986, 1996, and 2006.

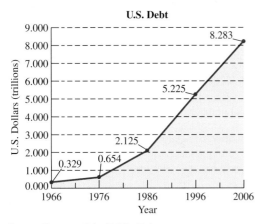

U.S. Debt

Source: Bureau of the Public Debt

a) Determine the percent increase in the U.S. debt from 1966 to 1976.

b) Determine the percent increase in the U.S. debt from 1976 to 1986.

c) Determine the percent increase in the U.S. debt from 1986 to 1996.

d) Determine the percent increase in the U.S. debt from 1996 to 2006.

44. *American Smokers* The graph shows the number of cigarette smokers, in millions, in the United States for the years 1990, 1994, 1998, 2002, and 2006.

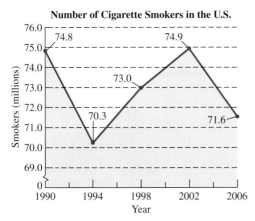

Number of Cigarette Smokers in the U.S.

Source: Action on Smoking and Health

a) Determine the percent decrease in the number of smokers from 1990 to 1994.

b) Determine the percent increase in the number of smokers from 1994 to 1998.

c) Determine the percent increase in the number of smokers from 1998 to 2002.

d) Determine the percent decrease in the number of smokers from 2002 to 2006.

45. *Dow Jones* The following graph shows the Dow Jones Industrial Average (DJIA) for 1996–2005 at closing on the last trading day of the year.

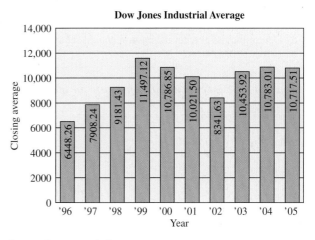

Dow Jones Industrial Average

Source: Dow Jones indexes

a) Determine the percent increase in the DJIA from 1996 to 1999.

b) Determine the percent decrease in the DJIA from 1999 to 2002.

c) Determine the percent increase in the DJIA from 2002 to 2005.

d) Determine the percent increase in the DJIA from 1996 to 2005.

46. *Presidential Salary* The following graph shows the annual salary for the president of the United States for selected years from 1789 to 2006.

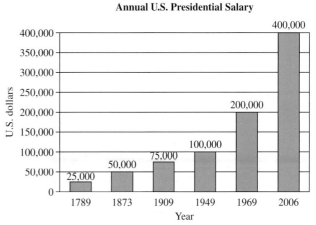

Annual U.S. Presidential Salary

Source: *Congressional Quarterly*

a) Determine the percent increase in annual salary from 1789 to 1873.

b) Determine the percent increase in annual salary from 1909 to 1949.

c) Determine the percent increase in annual salary from 1969 to 2006.

d) Determine the percent increase in annual salary from 1789 to 2006.

In Exercises 47–52, determine the answer to the question.

47. What is 15% of $45.00?

48. What is 6.5% of $150.00?

49. What percent of 96 is 24?

50. What percent of 75 is 15?

51. Five percent of what number is 15?

52. Ten percent of what number is 75?

53. *Tax and Tip* According to the Original Tipping Page (*www.tipping.org*), it is appropriate to tip waiters and waitresses 15–20% of the total restaurant bill, *including the tax.* Mary and Keith's dinner costs $43.50 before tax, and the tax rate is 6%.

a) What is the tax on Mary and Keith's dinner?

b) What is the total bill, including tax, before the tip?

c) If Mary and Keith decide to tip 15% of the total bill, how much is the tip?

d) What is the total cost of the dinner including tax and tip?

54. *Fishing* The Gordon and Stallard Charter Fishing Boat Company has recently increased the number of crewmembers by 25%, or 10 crewmembers. What was the original number of crewmembers?

55. *Percentage of A's* In a mathematics class, eighteen students received an A on the third test, which is 150% of the students who received an A on the second test. How many students received an A on the second test?

56. *Employee Increase* The Fastlock Company hired 57 new employees, which increased its staff by 30%. What was the original number of employees?

57. *Salary Increase* Qami Brown's present salary is $36,500. He is getting an increase of 7% in his salary next year. What will his new salary be?

58. *Salad Dressing Preference* In a survey of 300 people, 17% prefer ranch dressing on their salad. How many people in the surveyed group prefer ranch dressing?

59. *Vacuum Cleaner Sales* A Kirby vacuum cleaner dealership sold 430 units in 2006 and 407 units in 2007. Find the percent increase or decrease in the number of units sold.

60. *Vacuum Cleaner Markup* If the Kirby dealership in Exercise 59 pays $320 for each unit and sells it for $699, what is the percent markup?

61. *U.S. Poverty* In the year 2001, the United States had 32.9 million people living in poverty. In the year 2004, there were 37.0 million people living in poverty. Determine the percent increase in the number of people in the United States living in poverty from 2001 to 2004.

62. *Weight Loss* On January 1, Max Saindoux weighed 235 pounds and decided to diet and exercise. On June 30, Max weighed 210 pounds. Determine the percent decrease in Max's weight from January 1 to June 30.

63. *Television Sale* The regular price of a Phillips color TV is $539.62. During a sale, Hill TV is selling the TV for $439. Find the percent decrease in the price of this TV.

64. *Restaurant Markup* The cost of a fish dinner to the owner of the Golden Wharf restaurant is $7.95. The fish dinner is sold for $11.95. Find the percent markup.

65. *Truck Sale Profit* Bonnie James sold a truck and made a profit of $675. Her profit was 18% of the sale price. What was the sale price?

66. *Furniture Sale* Kane's Furniture Store advertised a table at a 15% discount. The original price was $115, and the sale price was $100. Was the sale price consistent with the ad? Explain.

67. *Reselling a Car* Quincy Carter purchased a used car for $1000. He decided to sell the car for 10% above his purchase price. Quincy could not sell the car so he reduced his asking price by 10%. If he sells the car at the reduced price, will he have a profit or a loss or will he break even? Explain how you arrived at your answer.

CHALLENGE PROBLEMS/GROUP ACTIVITIES

68. *Comparing Markdowns*

a) A coat is marked down 10%, and the customer is given a second discount of 15%. Is that the same as a single discount of 25%? Explain.

b) The regular price of a chair is $189.99. Determine the sale price of the chair if the regular price is reduced by 10% and this price is then reduced another 15%.

c) Determine the sale price of the chair if the regular price of $189.99 is reduced by 25%.

d) Examine the answers obtained in parts (b) and (c). Does your answer to part (a) appear to be correct? Explain.

69. *Selling Ties* The Tie Shoppe paid $5901.79 for a shipment of 500 ties and wants to make a profit of 40% of the cost on the whole shipment. The store is having two special sales. At the first sale it plans to sell 100 ties for $9.00 each, and at the second sale it plans to sell 150 ties for $12.50 each. What should be the selling price of the other 250 ties for the Tie Shoppe to make a 40% profit on the whole shipment?

RECREATIONAL MATHEMATICS

70. In parts (a) through (c), determine which is the greater amount and by how much.

a) $100 increased by 25% or $200 decreased by 25%.

b) $100 increased by 50% or $200 decreased by 50%.

c) $100 increased by 100% or $200 decreased by 100%.

INTERNET/RESEARCH ACTIVITY

71. Find two circle graphs in newspapers, magazines, or on the Internet whose data are not given in percents. Redraw the graphs and label them with percents.

11.2 PERSONAL LOANS AND SIMPLE INTEREST

▲ A personal loan may be used to pay for emergency expenses.

Consider the following dilemma. Your car breaks down and the mechanic tells you that the repairs will cost $450. You currently do not have the $450, but your family desperately needs to have a car. One option for you would be to borrow the money from a friend, a family member, a bank, or other lending institution. In this section, we will discuss personal loans and the cost of obtaining such loans.

The money a bank or other lender is willing to lend you is called the amount of *credit* extended or the *principal of the loan*. The amount of credit and the interest rate that you may obtain depend on the assurance you can give the lender that you will be able to repay the loan. Your credit is determined by your business reputation for honesty, by your earning power, and by what you can pledge as security to cover the loan. *Security* (or *collateral*) is anything of value pledged by the borrower that the lender may sell or keep if the borrower does not repay the loan. Acceptable security may be a business, a mortgage on a property, the title to an automobile, savings accounts, or stocks or bonds. The more marketable the security, the easier it is to obtain the loan, and in some cases marketability may help in getting a lower interest rate.

Bankers sometimes grant loans without security, but they require the signature of one or more other persons, called *cosigners*, who guarantee the loan will be repaid. For either of the two types of loans, the secured loan or the cosigner loan, the borrower (and cosigner, if there is one) must sign an agreement called a *personal note*. This document states the terms and conditions of the loan.

The most common way for individuals to borrow money is through an installment loan or using a credit card. (Installment loans and credit cards are discussed in Section 11.4.)

The concept of simple interest is essential to the understanding of installment buying. *Interest* is the money the borrower pays for the use of the lender's money. One type of interest is called simple interest. *Simple interest* is based on the entire amount of the loan for the total period of the loan. The formula used to find simple interest follows.

SIMPLE INTEREST FORMULA

$$\text{Interest} = \text{principal} \times \text{rate} \times \text{time}$$
$$i = prt$$

In the simple interest formula, the *principal*, p, is the amount of money lent, the *rate*, r, is the rate of interest expressed as a percent, and the *time*, t, is the number of days, months, or years for which the money will be lent. *Time is expressed in the same period as the rate.* For example, if the rate is 2% per month, the time must be expressed in months. Typically, rate means the annual rate unless otherwise stated. Principal and interest are expressed in dollars in the United States.

Ordinary Interest

The most common type of simple interest is called *ordinary interest*. For computing ordinary interest, each month has 30 days and a year has 12 months or 360 days. On the *due date* of a *simple interest note* the borrower must repay the principal plus the interest. (*Note*: Simple interest will mean ordinary interest unless stated otherwise.)

EXAMPLE ❶ *Eye Surgery Loan*

Chad Cohen needs to borrow $2200 to have corrective eye surgery. From his credit union, he obtains an 18-month loan with an annual simple interest rate of 7.5%.

a) Calculate the simple interest on the loan.

b) Determine the amount (principal + interest) Chad will pay the credit union at the end of the 18 months.

SOLUTION

a) To determine the interest on the loan, we use the formula $i = prt$. We know that $p = \$2200$, $r = 7.5\%$, which when written as a decimal is 0.075, and t in years $= \frac{18}{12} = 1.5$. We substitute the appropriate values in the simple interest formula

$$i = p \times r \times t$$
$$= \$2200 \times 0.075 \times 1.5$$
$$= \$247.5$$

The simple interest on $2200 at 7.5% for 18 months is $247.50.

b) The amount to be repaid is equal to the principal plus interest, or

$$A = p + i$$
$$= \$2200 + \$247.50$$
$$= \$2447.50$$

To pay off his loan, Chad will pay the credit union $2447.50 at the end of 18 months. ●

EXAMPLE ❷ *Determining the Annual Rate of Interest*

Philip Mahler agrees to lend $850 to his friend Joe Mahoney to help Joe travel to Cincinnati to attend a family wedding. Nine months later, Joe repaid the original $850 plus $51 interest. What annual rate of interest did Joe pay to Philip?

SOLUTION We need to solve for the interest rate, r. Since the time is 9 months, the time in years is $\frac{9}{12}$, or 0.75. Using the formula $i = prt$, we get

$$\$51 = \$850 \times r \times 0.75$$
$$51 = 637.5r$$
$$\frac{51}{637.5} = \frac{637.5r}{637.5}$$

Divide both sides of the equation by 637.5.

$$0.08 = r$$

Thus, the annual rate of interest paid by Joe to Philip is 0.08, or 8%. ●

EXAMPLE ❸ *A Pawn Loan*

To obtain money for new eyeglasses, Gilbert French decides to pawn his trumpet. Gilbert borrows $240 and after 30 days gets his trumpet back by paying the pawnbroker $288. What annual rate of interest did Gilbert pay?

SOLUTION Gilbert paid $288 − $240 = $48 in interest, and the length of the loan is for one month, or $\frac{1}{12}$ of a year.
Using the formula $i = prt$, we get

$$\$48 = \$240 \times r \times \frac{1}{12}$$
$$48 = 20r$$
$$\frac{48}{20} = r$$
$$2.4 = r$$

The annual rate of interest as a decimal number is 2.4. To change this number to a percent, multiply the number by 100 and add a percent sign. Thus, the annual rate of interest paid is 240%. •

In Examples 1, 2, and 3, we illustrated a simple interest loan, for which the interest and principal are paid on the due date of the note. There is another type of loan, the *discount note*, for which the interest is paid at the time the borrower receives the loan. The interest charged in advance is called the *bank discount*. A Federal Reserve Treasury bill is a bank discount note issued by the U.S. government. Example 4 illustrates a discount note.

EXAMPLE ❹ *True Interest Rate of a Discount Note*

Siegrid Cook took out a $500 loan using a 10% discount note for a period of 3 months. Determine

a) the interest she must pay to the bank on the date she receives the loan.

b) the net amount of money she receives from the bank.

c) the actual rate of interest for the loan.

SOLUTION

a) To determine the interest, use the simple interest formula.

$$i = prt$$
$$= \$500 \times 0.10 \times \frac{3}{12}$$
$$= \$12.50$$

b) Since Siegrid must pay $12.50 interest when she first receives the loan, the net amount she receives is $500 − $12.50, or $487.50.

c) We calculate the actual rate of interest charged using the simple interest formula. In the formula for the principal, we use the amount Siegrid actually received from the bank, $487.50. For the interest, we use the interest calculated in part (a).

$$i = prt$$
$$\$12.50 = \$487.50 \times r \times \frac{3}{12}$$
$$12.50 = 121.875 \times r$$
$$\frac{12.50}{121.875} = r$$
$$0.1026 \approx r$$

Thus, the actual rate of interest is about 10.3% rather than the quoted 10%. •

EXAMPLE ❺ *Partial Payments*

Ken Hurley wishes to purchase a new racing bicycle but does not have the $2000 purchase price. Luckily, the bike shop has two payment options. With option 1, Ken can pay $1000 as a down payment and then pay $1150 in 6 months. With option 2, Ken can pay $500 as a down payment and then pay $1700 in 6 months. Which payment option has a higher annual simple interest rate?

SOLUTION Option 1: To determine the principal of the loan, subtract the down payment from the purchase price of the bicycle. Therefore, $p = \$2000 - \$1000 = \$1000$.

To determine the interest charged, subtract the purchase price of the bicycle from the total amount paid. Therefore, $i = (\$1000 + \$1150) - \$2000 = \$150$. Then

$$i = p \times r \times t$$
$$150 = 1000 \times r \times \frac{1}{2}$$
$$150 = 500r$$
$$\frac{150}{500} = r$$
$$0.3 = r$$

Option 1 has a 30% annual simple interest rate.

Option 2: The principal is $p = \$2000 - \$500 = \$1500$, and interest is $i = (\$500 + \$1700) - \$2000 = \$200$. Then

$$i = p \times r \times t$$
$$200 = 1500 \times r \times \frac{1}{2}$$
$$200 = 750r$$
$$\frac{200}{750} = r$$
$$0.2667 \approx r$$

Option 2 has about a 26.7% annual simple interest rate. Therefore, option 1 charges a higher annual simple interest rate than option 2. ●

The United States Rule

A loan has a date of maturity, at which time the principal and interest are due. It is possible to make payments on a loan before the date of maturity. A payment that is less than the full amount owed and made prior to the due date is known as a *partial payment*. A Supreme Court decision specified the method by which these payments are credited. The procedure is called the *United States rule*.

The United States rule states that if a partial payment is made on the loan, interest is computed on the principal from the first day of the loan until the date of the partial payment. The partial payment is used to pay the interest first; then the rest of the payment is used to reduce the principal. The next time a partial payment is made, interest is calculated on the unpaid principal from the date of the previous date of payment. Again, the payment goes first to pay the interest, with the rest of the payment used to reduce the principal. An individual can make as many partial payments as he or she wishes; the procedure is repeated for each payment. The balance due on the date of maturity is found by computing interest due since the last partial payment and adding this interest to the unpaid principal.

The *Banker's rule* is used to calculate simple interest when applying the United States rule. The Banker's rule considers a year to have 360 days, and any fractional part of a year is the exact number of days of the loan.

To determine the exact number of days in a period, we can use Table 11.1 on page 662. Example 6 illustrates how to use the table.

Table 11.1

					Days in Each Month							
	31	28	31	30	31	30	31	31	30	31	30	31
Day of Month	**Jan.**	**Feb.**	**Mar.**	**Apr.**	**May**	**June**	**July**	**Aug.**	**Sept.**	**Oct.**	**Nov.**	**Dec.**
Day 1	1	32	60	91	121	152	182	213	244	274	305	335
Day 2	2	33	61	92	122	153	183	214	245	275	306	336
Day 3	3	34	62	93	123	154	184	215	246	276	307	337
Day 4	4	35	63	94	124	155	185	216	247	277	308	338
Day 5	5	36	64	95	125	156	186	217	248	278	309	339
Day 6	6	37	65	96	126	157	187	218	249	279	310	340
Day 7	7	38	66	97	127	158	188	219	250	280	311	341
Day 8	8	39	67	98	128	159	189	220	251	281	312	342
Day 9	9	40	68	99	129	160	190	221	252	282	313	343
Day 10	10	41	69	100	130	161	191	222	253	283	314	344
Day 11	11	42	70	101	131	162	192	223	254	284	315	345
Day 12	12	43	71	102	132	163	193	224	255	285	316	346
Day 13	13	44	72	103	133	164	(194)	225	256	286	317	347
Day 14	14	45	73	104	134	165	195	226	257	287	318	348
Day 15	15	46	(74)	105	135	166	196	227	258	288	319	349
Day 16	16	47	75	106	136	167	197	228	259	289	320	350
Day 17	17	48	76	107	137	168	198	229	260	290	321	351
Day 18	18	49	77	108	138	169	199	230	261	291	322	352
Day 19	19	50	78	109	139	170	200	231	262	292	323	353
Day 20	20	51	79	110	140	171	201	232	263	293	324	354
Day 21	21	52	80	111	141	172	202	233	264	294	325	355
Day 22	22	53	81	112	142	173	203	234	265	295	326	356
Day 23	23	54	82	113	143	174	204	235	266	296	327	357
Day 24	24	55	83	114	144	175	205	236	267	297	328	358
Day 25	25	56	84	115	145	176	206	237	268	298	329	359
Day 26	26	57	85	116	146	177	207	238	269	299	330	360
Day 27	27	58	86	117	147	178	208	239	270	300	331	361
Day 28	28	59	87	118	148	179	209	240	271	301	332	362
Day 29	29		88	119	149	180	210	241	272	302	333	363
Day 30	30		89	120	150	181	211	242	273	303	334	364
Day 31	31		90		151		212	243		304		365

Add 1 day for leap year if February 29 falls between the two dates under consideration.

EXAMPLE ❻ *Determining the Due Date of a Note*

Use Table 11.1 to determine (a) the due date of a loan made on March 15 for 120 days and (b) the number of days from April 18 to July 31.

SOLUTION

a) To determine the due date of the loan, do the following. In Table 11.1, find Day 15 in the left column (with heading Day of Month) and then move three columns to the right (heading at the top of the column is March) to find the number 74 (circled in red). Thus, March 15 is the 74th day of the year. Add 120 to 74 since the loan will be due 120 days after March 15:

$$74 + 120 = 194$$

Thus, the due date of the note is the 194th day of the year. Find 194 (circled in blue) in Table 11.1 in the column headed July. The number in the same row as 194 in the left column is Day 13. Thus, the due date of the note is July 13.

b) To determine the number of days from April 18 to July 31, use Table 11.1 to find that April 18 is the 108th day of the year and that July 31 is the 212th day of the year. Then find the difference: $212 - 108 = 104$. Thus, the number of days from April 18 to July 31 is 104 days.   ●

EXAMPLE ❼ *Using the Banker's Rule*

Determine the simple interest that will be paid on a $300 loan at an interest rate of 5% for the period March 3 to May 3 using the Banker's rule.

SOLUTION The exact number of days from March 3 to May 3 is 61. The period of time in years is 61/360. Substituting in the simple interest formula gives

$$i = prt$$
$$= \$300 \times 0.05 \times \frac{61}{360}$$
$$\approx \$2.54$$

The interest is $2.54.   ●

A borrower may wish to pay back part of the loan prior to the due date. The next example illustrates how a partial payment is credited under the United States rule. Making partial payments reduces the amount of interest paid and thus the cost of the loan.

EXAMPLE ❽ *Using the United States Rule*

Cathy Panik is a mathematics teacher, and she plans to attend a national conference. To pay for her airfare, on November 1, 2006, Cathy takes out a 120-day loan for $400 at an interest rate of 12.5%. Cathy uses some birthday gift money to make a partial payment of $150 on January 5, 2007. She makes a second partial payment of $100 on February 2, 2007.

a) Determine the due date of the loan.

b) Determine the interest and the amount credited to the principal on January 5.

c) Determine the interest and the amount credited to the principal on February 2.

d) Determine the amount that Cathy must pay on the due date.

SOLUTION

a) Using Table 11.1, we see that November 1 is the 305th day of the year. Next, we note that the sum of 305 and 120 is 425. Since this due date will extend into the next year, we subtract 365 from 425 to get 60. From Table 11.1, we see that the 60th day of the year is March 1. Therefore, the loan due date is March 1, 2007. Had 2007 been a leap year, the due date would have been February 29, 2007.

b) Using Table 11.1, January 5 is the 5th day of the year and November 1 is the 305th day of the year. The number of days from November 1 to January 5 can be computed as follows: $(365 - 305) + 5 = 65$. Then, using $i = prt$ and the Banker's rule, we get

$$i = \$400 \times 0.125 \times \frac{65}{360}$$
$$\approx \$9.03$$

The interest of $9.03 that is due January 5, 2007, is deducted from the payment of $150. The remaining payment of $150 − $9.03, or $140.97, is then credited to the principal. Therefore, the adjusted principal is now $400 − $140.97, or $259.03.

Note that if that payment had been the only partial payment made, then the balance due on March 1, 2007, would be calculated by determining the interest on the balance, $259.03, for the remainder of the loan, 55 days, and adding this interest to the principal of $259.03. If this payment had been the only partial payment made, then the balance due on March 1 would be

$$\text{Balance due} = \text{principal} + \text{interest}$$
$$\approx 259.03 + \left(259.03 \times 0.125 \times \frac{55}{360} \right)$$
$$\approx 259.03 + 4.95$$
$$\approx 263.98$$

c) Since there was a second partial payment made, we use the Banker's rule to calculate the interest on the unpaid principal for the period from January 5 to February 2. According to Table 11.1, the number of days from January 5 to February 2 is $33 - 5$, or 28 days.

$$i = \$259.03 \times 0.125 \times \frac{28}{360}$$
$$\approx \$2.52$$

The interest of $2.52 that is due February 2, 2007, is deducted from the payment of $100. The remaining payment of $100 − $2.52, or $97.48, is then credited to the principal. Therefore, the new adjusted principal is now $259.03 − $97.48, or $161.55.

d) The due date of the loan is March 1. Using Table 11.1, we see that there are $60 - 33$, or 27, days from February 2 to March 1. The interest is computed on the remaining balance of $161.51 by using the simple interest formula.

$$i = \$161.55 \times 0.125 \times \frac{27}{360}$$
$$\approx \$1.51$$

Therefore, the balance due on the maturity date of the loan is the sum of the principal and the interest, $161.55 + $1.51, or $163.06. *Note*: The sum of the days in the three calculations, $65 + 28 + 27$, equals the total number of days in the loan, 120.

SECTION 11.2 EXERCISES

CONCEPT/WRITING EXERCISES

1. What is interest?

2. What is credit?

3. What is security (or collateral)?

4. What is a cosigner?

5. Explain what each letter in the simple interest formula, $i = prt$, represents.

6. What is a personal note?

7. Explain the United States rule.

8. Explain the difference between ordinary interest and interest under the Banker's rule.

In Exercises 9–18, determine the simple interest. (The rate is an annual rate unless otherwise noted. Assume 360 days in a year.)

PRACTICE THE SKILLS

9. $p = \$250, r = 3\%, t = 2$ years

10. $p = \$375, r = 4.5\%, t = 5$ years

11. $p = \$1100, r = 8.75\%, t = 90$ days

12. $p = \$2500, r = 3\frac{7}{8}\%, t = 9$ months

13. $p = \$587, r = 0.045\%$ per day, $t = 2$ months

14. $p = \$6742.75, r = 6.05\%, t = 90$ days

15. $p = \$2756.78, r = 10.15\%, t = 103$ days

16. $p = \$550.31, r = 8.9\%, t = 67$ days

17. $p = \$1372.11, r = 1\frac{3}{8}\%$ per month, $t = 6$ months

18. $p = \$41,864, r = 0.0375\%$ per day, $t = 60$ days

In Exercises 19–24, use the simple interest formula to determine the missing value.

19. $p = \$2000, r = ?, t = 4$ years, $i = \$400$

20. $p = ?, r = 6\%, t = 30$ days, $i = \$26.90$

21. $p = ?, r = 8\%, t = 3$ months, $i = \$12.00$

22. $p = \$800.00, r = 6\%, t = ?, i = \$64.00$

23. $p = \$957.62, r = 6.5\%, t = ?, i = \$124.49$

24. $p = \$1650.00, r = ?, t = 6.5$ years, $i = \$343.20$

PROBLEM SOLVING

25. *Business Loan* The city of Bradenton is offering simple interest loans to start-up businesses at a rate of 3.5%. Gaetano Cannata obtains one of these loans of $6000 for 3 years to help pay for start-up costs of his restaurant, Ortygia. Determine the amount of money Gaetano must repay the city after 3 years.

26. *Credit Union Loan* Steve and Laurie Carah borrowed $4500 from their credit union to remodel their kitchen. The simple interest rate was 4.75%, and the length of the loan was 3 years.

   a) How much did the Carahs pay for the use of the money?

   b) Determine the amount the Carahs paid to the credit union on the date of maturity of the loan.

27. *Bank Personal Note* Kelly Droessler borrowed $3500 from a bank for 6 months. Her friend Ms. Harris was cosigner of Kelly's personal note. The bank collected $7\frac{1}{2}\%$ simple interest on the date of maturity.

   a) How much did Kelly pay for the use of the money?

   b) Determine the amount she repaid to the bank on the due date of the note.

28. *Bank Discount Note* Kwame Adebele borrowed $2500 for 5 months from a bank, using U.S. government bonds as security. The bank discounted the loan at 8%.

   a) How much interest did Kwame pay the bank for the use of its money?

b) How much did he receive from the bank?

c) What was the actual rate of interest he paid?

29. *Bank Discount Note* Julie Jansen borrowed $3650 from her bank for 8 months. The bank discounted the loan at 7.5%.

a) How much interest did Julie pay the bank for the use of its money?

b) How much did she receive from the bank?

c) What was the actual rate of interest she paid?

30. *Credit Union Loan* Enrico Montoyo wants to borrow $350 for 6 months from his credit union, using his savings account as security. The credit union's policy is that the maximum amount a person can borrow is 80% of the amount in the person's savings account. The interest rate is 2 percentage points higher than the interest rate being paid on the savings account. The current rate on the savings account is $3\frac{1}{4}\%$.

a) How much money must Enrico have in his account to borrow $350?

b) What is the rate of interest the credit union will charge for the loan?

c) Find the amount Enrico must repay in 6 months.

31. *Investing Tuition Payments* Sand Ridge School is requiring parents to pay half of the yearly tuition at the time of registration and half on the date classes begin. Registration is held 5 months prior to the beginning of school, and administrators expect 470 students to register. If annual tuition is $4500 and if the money paid at the time of registration is placed in an account paying 5.4% simple interest, how much interest will Sand Ridge School earn by the time school begins?

32. *A Pawn Loan* Jeffrey Kowalski wants to take his mother out for dinner on her birthday, but he doesn't get paid until the following week. To borrow money, Jeffrey pawns his watch. Based on the value of the watch, the pawnbroker loans Jeffrey $75. Fourteen days later, Jeffrey gets his watch back by paying the pawnbroker $80.25. What annual simple interest rate did the pawnbroker charge Jeffrey?

In Exercises 33–38, determine the exact time from the first date to the second date. Use Table 11.1. Assume the year is not a leap year unless otherwise indicated.

33. February 14 to November 25

34. January 17 to October 10

35. February 2 to October 31 (the loan is due in a leap year)

36. June 14 to January 24

37. August 24 to May 15

38. December 21 to April 28

In Exercises 39–42, determine the due date of the loan, using the exact time, if the loan is made on the given date for the given number of days.

39. March 15 for 90 days

40. May 11 for 120 days

41. November 25 for 120 days (the loan is due in a leap year)

42. July 5 for 210 days

In Exercises 43–52, a partial payment is made on the date(s) indicated. Use the United States rule to determine the balance due on the note at the date of maturity. (The Effective Date is the date the note was written.) Assume the year is not a leap year.

	Principal	Rate	Effective Date	Partial Payment(s) Amount	Date(s)	Maturity Date
43.	$3000	4%	May 1	$1000	June 1	July 1
44.	$6000	7%	Jan. 15	$2500	Mar. 1	Apr. 15
45.	$2400	5.5%	Feb. 1	$1000	May 1	Aug. 31
46.	$8500	11.5%	Sep. 1	$4250	Oct. 31	Dec. 31
47.	$9000	6%	July 15	$4000	Dec. 27	Feb. 1
48.	$1000	12.5%	Jan. 1	$300	Jan. 15	Feb. 15
49.	$1800	15%	Aug. 1	$500	Sept. 1	Nov. 1
				$500	Oct. 1	
50.	$5000	14%	Oct. 15	$800	Nov. 15	Jan. 1
				$800	Dec. 15	
51.	$11,600	6%	Mar. 1	$2000	Aug. 1	Dec. 1
				$4000	Nov. 15	
52.	$21,000	$4\frac{3}{8}\%$	July 12	$8000	Oct. 10	Jan. 30
				$6000	Dec. 8	

667 PERSONAL LOANS AND SIMPLE INTEREST

53. *Company Loan* On March 1, the Zwick Balloon Company signed a $6500 note with simple interest of $10\frac{1}{2}\%$ for 180 days. The company made payments of $1750 on May 1 and $2350 on July 1. How much will the company owe on the date of maturity?

54. *Restaurant Loan* The Sweet Tooth Restaurant borrowed $3000 on a note dated May 15 with simple interest of 11%. The maturity date of the loan is September 1. The restaurant made partial payments of $875 on June 15 and $940 on August 1. Find the amount due on the maturity date of the loan.

55. *U.S. Treasury Bills* The U.S. government borrows money by selling Treasury bills. Treasury bills are discounted notes issued by the U.S. government. On May 5, 2007, Kris Greenhalgh purchased a 182-day, $1000 U.S. Treasury bill at a 4.34% discount. On the date of maturity, Kris will receive $1000.

a) What is the date of maturity of the Treasury bill?

b) How much did Kris actually pay for the Treasury bill?

c) How much interest did the U.S. goverment pay Kris on the date of maturity?

d) What is the actual rate of interest of the Treasury bill? (Round the answer to the nearest hundredth of a percent.)

▲ United States Treasury building, Washington, D.C.

56. *U.S. Treasury Bills* On August 31, 2007, Trinity Lopez purchased a 364-day, $6000 U.S. Treasury bill at a 4.4% discount. (See Exercise 55.)

a) What is the date of maturity of the Treasury bill (2008 is a leap year)?

b) How much did Trinity actually pay for the Treasury bill?

c) How much interest did the U.S. government pay Trinity on the date of maturity?

d) What is the actual rate of interest of the Treasury bill?

57. *Tax Preparation Loan* Many tax preparation organizations will prepay customers' tax refunds if they pay a one-time finance charge. In essence, the customer is borrowing the money (the refund minus the finance charge) from the tax preparer, prepaying the interest (as in a discount note), and then repaying the loan with the tax refund. This procedure allows customers access to their tax refund money without having to wait. Joy Stallard had a tax refund of $743.21 due. She was able to get her tax refund immediately by paying a finance charge of $39.95. What annual simple interest rate is Joy paying for this loan assuming

a) the tax refund check would be available in 5 days?

b) the tax refund check would be available in 10 days?

c) the tax refund check would be available in 20 days?

58. *Prime Interest Rate* Nick St. Louis borrowed $600 for 3 months. The banker said that Nick must repay the loan at the rate of $200 per month plus interest. The bank was charging a rate of 2% above the prime interest rate. The *prime interest rate* is the rate charged to preferred customers of the bank. During the first month the prime interest rate was 4.75%, during the second month it was 5%, and during the third month it was 5.25%.

a) Find the amount Nick paid the bank at the end of the first month, at the end of the second month, and at the end of the third month.

b) What was the total amount of interest Nick paid the bank?

CHALLENGE PROBLEM/GROUP ACTIVITY

59. *U.S. Treasury Bills* Mark Beiley purchased a 52-week U.S. Treasury bill (see Exercises 55 and 56) for $93,337. The par value (the value of the bill upon maturity) was $100,000.

a) What was the discount rate?

b) What was the actual interest rate Mark received?

c) Since U.S. Treasury bills are sold through auctions at Federal Reserve Banks, Mark did not know the purchase price of his Treasury bill until after he was notified by mail. If Mark had sent the Federal Reserve

Bank a check for $100,000 to purchase the Treasury bill, how much would the Federal Reserve Bank have to rebate him upon notice of his purchase?

d) On the day Mark received the rebate that was discussed in part (c), he invested it in a 1-year certificate of deposit yielding 5% interest. What is the total amount of interest he will receive from both investments?

RECREATIONAL MATHEMATICS

60. *Columbus Investment* On August 3, 1492, Christopher Columbus set sail on a voyage that would eventually lead him to the Americas. If on this day Columbus had invested $1 in a 5% simple interest account, determine the amount of interest the account would have earned by the following dates. Use a scientific calculator and disregard leap years in your calculations.

a) December 11, 1620 (Pilgrims land on Plymouth Rock)

b) July 4, 1776 (Declaration of Independence)

c) December 7, 1941 (U.S. enters World War II)

d) Today's date

INTERNET/RESEARCH ACTIVITY

61. *Loan Sources* Consider the following places where a loan may be obtained: banks, savings and loans, credit unions, and pawnshops. Write a report that includes the following information:

a) Describe the ownership of each.

b) Historically, what need is each fulfilling?

c) What are the advantages and disadvantages of obtaining a loan from each of those places listed?

11.3 COMPOUND INTEREST

▲ Albert Einstein

Albert Einstein said of compound interest, "The most powerful force in the universe is compound interest." What is this marvelous concept that warrants such a strong statement from one of the most intelligent human beings of all time? Another very intelligent person, Benjamin Franklin, described compound interest quite well when he said, "Money makes money and the money that money makes makes more money." In this section, we will introduce compound interest and show how it can be used to help you with investing for some of your long-term goals.

Compound Interest

An *investment* is the use of money or capital for income or profit. We can divide investments into two classes: fixed investments and variable investments. In a *fixed investment*, the amount invested as principal is guaranteed and the interest is computed at a fixed rate. *Guaranteed* means that the exact amount invested will be paid back together with any accumulated interest. Examples of a fixed investment are savings accounts, money market deposit accounts, and certificates of deposit. Another fixed investment is a government savings bond. In a *variable investment*, neither the principal nor the interest is guaranteed. Examples of variable investments are stocks, mutual funds, and commercial bonds.

Simple interest, introduced earlier in the chapter, is calculated once for the period of a loan or investment using the formula $i = prt$. The interest paid on savings accounts at most banks is compound interest. A bank computes the interest periodically (for example, daily or quarterly) and adds this interest to the original principal. The interest for the following period is computed by using the new principal (original principal plus interest). In effect, the bank is computing interest on interest, which is called compound interest.

Interest that is computed on the principal and any accumulated interest is called **compound interest**.

EXAMPLE ❶ *Computing Compound Interest*

Marjorie Thrall recently won the $1000 first prize in a raffle contest. Marjorie deposits the $1000 in a 1-year certificate of deposit paying 2.0% compounded quarterly. Find the amount, A, to which the $1000 will grow in 1 year.

SOLUTION Compute the interest for the first quarter using the simple interest formula. Add this interest to the principal to find the amount at the end of the first quarter. In our calculation, time is $\frac{1}{4}$ of a year, or $t = 0.25$.

$$i = prt$$
$$= \$1000 \times 0.02 \times 0.25 = \$5.00$$
$$A = \$1000 + \$5 = \$1005$$

Now repeat the process for the second quarter, this time using a principal of $1005.

$$i = \$1005 \times 0.02 \times 0.25 \approx \$5.03$$
$$A = \$1005 + \$5.03 = \$1010.03$$

For the third quarter, use a principal of $1010.03.

$$i = \$1010.03 \times 0.02 \times 0.25 \approx \$5.05$$
$$A = \$1010.03 + \$5.05 = \$1015.08$$

For the fourth quarter, use a principal of $1015.08.

$$i = \$1015.08 \times 0.02 \times 0.25 \approx \$5.08$$
$$A = \$1015.08 + \$5.08 = \$1020.16$$

Hence, the $1000 grows to a final value of $1020.16 over the 1-year period. ●

This example shows the effect of earning interest on interest, or compounding interest. In 1 year, the amount of $1000 has grown to $1020.16, compared with $1020 that would have been obtained with a simple interest rate of 2%. Thus, in 1 year alone the gain was $0.16 more with compound interest than with simple interest.

A simpler and less time-consuming way to calculate compound interest is to use the compound interest formula and a calculator.

COMPOUND INTEREST FORMULA

$$A = p\left(1 + \frac{r}{n}\right)^{nt}$$

where A is the amount that accumulates in the account, p is the principal, r is the annual interest rate as a decimal, n is the number of compounding periods per year, and t is the time in years.

We will now use this formula to show how an investment can grow using compound interest.

EXAMPLE ❷   Using the Compound Interest Formula

Kathy Mowers invested $5600 in a savings account with an interest rate of 7.5% compounded monthly. If Kathy makes no other deposits into this account, determine the amount in the account after 10 years.

SOLUTION We will use the formula for compound interest, $A = p\left(1 + \dfrac{r}{n}\right)^{nt}$. The principal, p, is the amount of money invested, so $p = 5600$. The interest rate, r, is 7.5%, so $r = 0.075$. Because the interest is compounded monthly, there are 12 periods per year and $n = 12$. Because the money is invested for 10 years, $t = 10$.

$$
\begin{aligned}
A &= 5600\left(1 + \frac{0.075}{12}\right)^{12\cdot 10} \\
&= 5600(1 + 0.00625)^{120} \\
&= 5600(1.00625)^{120} \\
&\approx 5600(2.1120646) \\
&\approx 11{,}827.56
\end{aligned}
$$

Thus, the amount in the account after 10 years would be $11,827.56.   ●

TECHNOLOGY TIP

Scientific Calculator

There are various ways to determine compound interest using a scientific calculator. In Example 2, we had to evaluate

$$5600\left(1 + \frac{0.075}{12}\right)^{12\cdot 10}$$

To evaluate this expression on a scientific calculator, enter the following keys.

5600 $\boxed{\times}$ $\boxed{(}$ 1 $\boxed{+}$ 0.075 $\boxed{\div}$ 12 $\boxed{)}$ $\boxed{y^x}$ $\boxed{(}$ 12 $\boxed{\times}$ 10 $\boxed{)}$ $\boxed{=}$

After the $\boxed{=}$ key is pressed, the answer 11827.56197 is displayed. This answer checks with the answer obtained in Example 2.

Microsoft Excel

The software program Excel can also be used to determine compound interest. Excel is a spreadsheet, so once you insert a formula, you can change the value of the variables and the new answer will be automatically displayed. Before you read the following procedure, read Appendix A on Excel at the back of the book.

To use Excel to determine the compound interest, set up the following columns.

	A	B	C	D	E
1	p	r	n	t	A
2	5600	0.075	12	10	
3					

Sheet1 / Sheet2 / Sheet3 /

Ready

Then, to set up the formula in cell E2, go to the formula bar on top and enter

$$= A2*(1+B2/C2)\wedge(C2*D2)$$

Then press the Enter key. The answer, 11827.56, will appear in cell E2.

Suppose you now wanted to determine the compound interest for $p = 6000$, $r = 0.05$, $n = 24$, and $t = 6$. If you were to change the values in cells A2 through D2 to these values respectively, the new answer, 8096.626, will now be displayed in cell E2.

EXAMPLE ❸ *Calculating Compound Interest*

Calculate the interest on $650 at 8% compounded semiannually for 3 years, using the compound interest formula.

SOLUTION Since interest is compounded semiannually, there are two periods per year. Thus, $n = 2$, $r = 0.08$, and $t = 3$. Substituting into the formula, we find the amount, A.

$$A = p\left(1 + \frac{r}{n}\right)^{nt}$$

$$= 650\left(1 + \frac{0.08}{2}\right)^{(2)(3)}$$

$$= 650(1.04)^6$$

$$\approx 650(1.2653190)$$

$$\approx \$822.46$$

Since the total amount is $822.46 and the original principal is $650, the interest must be $822.46 − $650, or $172.46. ●

In Example 3, the interest rate is stated as an annual rate of 8%, but the number of compounding periods per year is two. In applying the compound interest formula, the rate for one period is $r \div n$, which was 8% ÷ 2, or 4% per period.

Now we will calculate the amount, and interest, on $1 invested at 8% compounded semiannually for 1 year. The result is

$$A = 1\left(1 + \frac{0.08}{2}\right)^{(2)(1)} = 1.0816$$

$$\text{Interest} = \text{amount} - \text{principal}$$

$$i = 1.0816 - 1$$

$$= 0.0816$$

The interest for 1 year is 0.0816. This amount written as a percent, 8.16%, is called the *effective annual yield*. *Most financial institutions refer to the effective annual yield as the annual percentage yield (APY)*. If $1 was invested at a simple interest rate of 8.16% and $1 was invested at 8% interest compounded semiannually (equivalent to an effective yield of 8.16%), the interest from both investments would be the same.

The **effective annual yield** or **annual percentage yield (APY)** is the simple interest rate that gives the same amount of interest as a compound rate over the same period of time.

Suncoast Schools Credit Union

CD Rates

Type	Rate	APY*
12 mo	4.50%	4.60%
24 mo	4.88%	5.00%
36 mo	4.97%	5.10%
48 mo	5.12%	5.25%

* Annual Percentage Yield

Many banks compound interest daily. When computing the effective annual yield, they use 360 for the number of periods in a year. To determine the effective annual yield for any interest rate, calculate the amount using the compound interest formula where p is $1. Then subtract $1 from that amount. The difference, written as a percent, is the effective annual yield, as illustrated in Example 4.

The sign in the margin shows interest rates and the corresponding annual percentage yields (APY) that were available for certificates of deposit (CDs) from Suncoast Schools Credit Union on April 7, 2006. To determine the effective annual yield (or the APY) for the 48-month CD, calculate the amount of interest earned on $1 for 1 year.

$$A = 1\left(1 + \frac{0.0512}{360}\right)^{360} \approx 1.0525295 \approx 1.0525$$

From this result, we subtract 1 to get $1.0525 - 1 = 0.0525$, or 5.25%. Confirm that the other annual percentage yields shown on the sign are correct.

EXAMPLE ❹ *Determining Annual Percentage Yield*

Determine the annual percentage yield or the effective annual yield for $1 invested for 1 year at

a) 8% compounded daily.

b) 6% compounded quarterly.

SOLUTION

a) With daily compounding, $n = 360$.

$$A = p\left(1 + \frac{r}{n}\right)^{nt}$$

$$= 1\left(1 + \frac{0.08}{360}\right)^{(360)(1)}$$

$$\approx 1.0832774 \approx 1.0833$$

$$i = A - 1$$

$$\approx 1.0833 - 1$$

$$\approx 0.0833$$

Thus, when the interest is 8% compounded daily, the annual percentage yield or the effective annual yield is about 8.33%.

b) With quarterly compounding, $n = 4$.

$$A = p\left(1 + \frac{r}{n}\right)^{nt}$$

$$= 1\left(1 + \frac{0.06}{4}\right)^{(4)(1)}$$

$$\approx 1.0613636 \approx 1.0614$$

$$i = A - 1$$

$$\approx 1.0614 - 1$$

$$\approx 0.0614$$

Thus, when the interest is 6% compounded quarterly, the annual percentage yield or the effective annual yield is about 6.14%. ●

There are numerous types of savings accounts. Many savings institutions compound interest daily. Some pay interest from the day of deposit to the day of withdrawal, and others pay interest from the first of the month on all deposits made before the tenth of the month. In each of these accounts in which interest is compounded daily, interest is entered into the depositor's account only once each quarter. Some savings institutions will not pay any interest on a day-to-day account if the balance falls below a set amount.

Present Value

You may wonder about what amount of money you must deposit in an account today to have a certain amount of money in the future. For example, how much must you deposit in an account today at a given rate of interest so that it will accumulate to $25,000 to pay your child's college costs in 4 years? The principal, p, that would have to be invested now is called the *present value*. Following is a formula for determining the present value.

PRESENT VALUE FORMULA

$$p = \frac{A}{\left(1 + \dfrac{r}{n}\right)^{nt}}$$

where p is the present value, or the principal to invest now, A is the amount to be accumulated in the account, r is the annual interest rate as a decimal, n is the number of compounding periods per year, and t is the time in years.

EXAMPLE ❺ *Savings for College*

Will Hunting would like his daughter to attend college in 6 years when she finishes high school. Will would like to invest enough money in a certificate of deposit (CD) now to pay for his daughter's college expenses. If Will estimates that he will need $30,000 in 6 years, how much should he invest now in a CD that has a rate of 4.72% compounded quarterly?

SOLUTION To answer this question, we will use the present value formula with $A = \$30{,}000$, $r = 0.0472$, $n = 4$, and $t = 6$.

$$p = \frac{A}{\left(1 + \dfrac{r}{n}\right)^{nt}}$$

$$= \frac{30{,}000}{\left(1 + \dfrac{0.0472}{4}\right)^{4 \cdot 6}}$$

$$= \frac{30{,}000}{(1.0118)^{24}}$$

$$\approx \frac{30{,}000}{1.3251718}$$

$$\approx 22{,}638.57$$

Will Hunting needs to invest approximately \$22,638.57 now to have \$30,000 in 6 years. ●

TECHNOLOGY TIP

Scientific Calculator

To determine the present value in Example 5, we had to evaluate the expression

$$\frac{30{,}000}{\left(1 + \dfrac{0.0472}{4}\right)^{4 \cdot 6}}$$

This expression can be evaluated on a scientific calculator as follows.

30000 ÷ (1 + 0.0472 ÷ 4) y^x (4 × 6) =

After the = key is pressed, the answer 22638.57326 is displayed. This answer agrees with the answer obtained in Example 5.

Excel

Read the Technology Tip on compound interest on page 670. To determine the present value using Excel, enter A, r, n, t, and p in cells A1 through E1, respectively. In cells A2 through D2, enter the values of A, r, n, and t, respectively. (See the table in the Technology Tip on page 670 for an example of the row and column setup.) When cell E2 is highlighted, in the formula box at the top, enter

= A2/(1 + B2/C2)^(C2*D2)

After the Enter key is pressed, the answer, 22638.57, is displayed in cell E2.

SECTION 11.3 EXERCISES

CONCEPT/WRITING EXERCISES

1. What is an investment?

2. What is a fixed investment?

3. What is a variable investment?

4. What is compound interest?

5. **a)** What is effective annual yield?

   b) What is another name for effective annual yield?

6. What is meant by present value in the present value formula?

PRACTICE THE SKILLS

In Exercises 7–14, use the compound interest formula to compute the total amount accumulated and the interest earned.

7. $1000 for 5 years at 4% compounded annually

8. $3000 for 2 years at 5% compounded semiannually

9. $2000 for 4 years at 3% compounded quarterly

10. $5000 for 3 years at 7% compounded quarterly

11. $7000 for 3 years at 5.5% compounded monthly

12. $4000 for 2 years at 6% compounded semiannually

13. $8000 for 2 years at 4% compounded daily (use $n = 360$)

14. $8500 for 5 years at 4.5% compounded monthly

In Exercises 15–18, use the present value formula to determine the amount to be invested now, or the present value needed.

15. The desired accumulated amount is $50,000 after 10 years invested in an account with 5% interest compounded annually.

16. The desired accumulated amount is $75,000 after 5 years invested in an account with 4% interest compounded semiannually.

17. The desired accumulated amount is $100,000 after 4 years invested in an account with 4% interest compounded quarterly.

18. The desired accumulated amount is $25,000 after 20 years invested in an account with 12% interest compounded quarterly.

PROBLEM SOLVING

19. *Little League* Braden River Little League receives a $50,000 donation for building a new snack bar and office building. The league decides to invest this money in a money market account that pays 4% interest compounded quarterly. How much will the league have in this account after 2 years?

20. *Contest Winnings* Mary Robinson wins $2500 in a singing contest and invests the money in a 4-year CD that pays 3% interest compounded monthly. How much money will Mary receive when she redeems the CD at the end of the 4 years?

21. *Class Trip* To help pay for a class trip at the end of their senior year, the sophomore class at Cortez High School invests $1500 from fund-raisers in a 30-month CD paying 3.9% interest compounded monthly. Determine the amount the class will receive when it cashes in the CD after 30 months.

22. *Investing Prize Winnings* Marcella Laddon wins third prize in the Clearinghouse Sweepstakes and receives a check for $250,000. After spending $10,000 on a vacation, she decides to invest the rest in a money market account that pays 1.5% interest compounded monthly. How much money will be in the account after 10 years?

23. *Investing Gifts and Scholarships* Cliff Morris just graduated from high school and has received $800 in gifts of cash from friends and relatives. In addition, Cliff received three scholarships in the amounts of $150, $300, and $1000. If Cliff takes all his gift and scholarship money and invests it in a 24-month CD paying 2% interest compounded daily, how much will he have when he cashes in the CD at the end of the 24 months?

24. *Investing a Signing Bonus* Joe Gallegos just started a new job and has received a $5000 signing bonus. Joe decides to invest this money now so that he can buy a new car in 5 years. If Joe invests in a 5-year CD paying 3.35% interest compounded quarterly, how much money will he receive from his CD in 5 years?

25. *Savings Account Investment* When Richard Zucker was born, his father deposited $2000 in his name in a savings account. The account was paying 5% interest compounded semiannually.

   a) If the rate did not change, what was the value of the account after 15 years?

   b) If the money had been invested at 5% compounded quarterly, what would the value of the account have been after 15 years?

26. *Savings and Loan Investment* When Lois Martin was born, her father deposited $2000 in a savings account in her name at a savings and loan association. At the time, the savings and loan was paying 6% interest compounded semiannually on savings accounts. After 10 years, the savings and loan association changed to an interest rate of 6% compounded quarterly. How much had the $2000 amounted to after 18 years when the money was withdrawn for Lois to use to help pay her college expenses?

27. *Personal Loan* Brent Pickett borrowed $3000 from his brother Dave. He agreed to repay the money at the end of 2 years, giving Dave the same amount of interest that he would have received if the money had been invested at 1.75% compounded quarterly. How much money did Brent repay his brother?

28. *Forgoing Interest* Rikki Blair borrowed $6000 from her daughter, Lynette. She repaid the $6000 at the end of 2 years. If Lynette had left the money in a bank account that paid an interest rate of $5\frac{1}{4}$% compounded monthly, how much interest would she have accumulated?

29. *Saving for a Down Payment* Jean Woody invested $6000 at 8% compounded quarterly. Three years later she withdrew the full amount and used it for the down payment on a house. How much money did she put down on the house?

30. *Investing a Salary* After Karen Estes began her job as a waitress, she invested in a money market account paying 5.6% interest compounded daily. What was the effective annual yield of this account?

▲ See Exercise 30

31. *Determining Effective Annual Yield* Determine the effective annual yield for $1 invested for 1 year at 3.5% compounded semiannually.

32. *Determining Effective Annual Yield* Determine the effective annual yield for $1 invested for 1 year at 4.75% compounded monthly.

33. *Verifying APY* Suppose you saw a sign at your local bank that said, "2.4% rate compounded monthly—2.6% Annual Percentage Yield (APY)." Is there anything wrong with the sign? Explain.

34. *Verifying APY* Suppose you saw an advertisement in the newspaper for a financial planner who was recommending a certificate of deposit that paid 4.5% interest compounded quarterly. In the fine print at the bottom of the advertisement, it stated that the APY on the CD was 4.58%. Was this advertisement accurate? Explain.

35. *Comparing Investments* Dave Dudley won a photography contest and received a $1000 cash prize. Will he earn more interest in 1 year if he invests his winnings in a simple interest account that pays 5% or in an account that pays 4.75% interest compounded monthly?

36. *Comparing Loan Sources* Tom Angelo needs to borrow $1500 to expand his farm implement maintenance business. He learns that the local bank will lend him the money for 2 years at a rate of 10% compounded quarterly. After hearing of this rate, Tom's grandfather offers to lend him the money for 2 years with a simple interest rate of 7%. How much money will Tom save by borrowing the money from his grandfather?

37. *A New Water Tower* The village of Kieler recently completed the construction of a new water tower. The entire cost of the water tower was $925,000, and the state paid $370,000 of the total cost through the awarding of a grant.

In addition, the village can delay paying the balance of the cost for 30 years (without paying any interest during the 30 years). To finance the balance, the village board will at this time assess its 598 homeowners a one-time flat fee surcharge and then invest this money in a 30-year CD paying 7.5% interest compounded monthly.

a) What is the balance due on the water tower?

b) How much will the village of Kieler need to invest at this time in the CD to raise the balance due in 30 years?

c) What amount should each homeowner pay as a surcharge?

38. After seeing its neighboring village obtain a new water tower (see Exercise 37), the city board of East Dubuque begins planning to replace its water towers. The board estimates that it will need $1,750,000 to build the new water towers in 20 years. At this time, the city board plans to assess its 2753 homeowners with a one-time flat fee surcharge and then invest the money received in a CD paying 9% interest compounded daily (use $n = 360$) for 20 years.

a) How much money will the board need to raise at this time to meet the city's water tower needs at the end of the 20 years?

b) Before applying the surcharge, the city board receives a federal grant of $100,000 toward the water tower investment. Taking this grant into account, how much should the surcharge be on each homeowner?

39. *Saving for a Tractor* Jim Roznowski wants to invest some money now to buy a new tractor in the future. If he wants to have $275,000 available in 5 years, how much does he need to invest now in a CD paying 5.15% interest compounded monthly?

▲ See Exercise 39

40. *Investing for Retirement* Carl and Kathy Minieri are planning to retire in 20 years and believe that they will need $200,000 in addition to income from their retirement plans. How much must they invest today at 7.5% compounded quarterly to accomplish their goal?

41. *Investment for a Newborn* How much money should parents invest at the birth of their child to provide their child with $50,000 at age 18? Assume that the money earns interest at 8% compounded quarterly.

42. *Future Value* How much money must Harry Kim invest today to have $20,000 in 15 years? Assume that the money earns interest at 7% compounded quarterly.

43. *Doubling the Rate* Determine the total amount and the interest paid on $1000 with interest compounded semiannually for 2 years at

a) 2%.   **b)** 4%.   **c)** 8%.

d) Is there a predictable outcome in either the amount or the interest when the rate is doubled? Explain.

44. *Doubling the Principal* Compute the total amount and the interest paid at 12% compounded monthly for 2 years for the following principals.

a) $100   **b)** $200   **c)** $400

d) Is there a predictable outcome in the interest when the principal is doubled? Explain.

CHALLENGE PROBLEMS/GROUP ACTIVITIES

45. *A Loaf of Bread* If the cost of a loaf of bread was $1.35 in 2007 and the annual average inflation rate is $2\frac{1}{2}\%$, what will be the cost of a loaf of bread in 2012?

46. *Determining the Interest Rate* For a total accumulated amount of $3586.58, a principal of $2000, and a time period of 5 years, use the compound interest formula to find r if interest is compounded monthly.

47. *Rule of 72* A simple formula can help you estimate the number of years required to double your money. It's called the *rule of 72*. You simply divide 72 by the interest rate (without the percent sign). For example, with an interest rate of 4%, your money would double in approximately $72 \div 4$ or 18 years. In (a)–(d), determine the approximate number of years it will take for $1000 to double at the given interest rate.

   a) 3%   **b)** 6%   **c)** 8%   **d)** 12%

   e) If $120 doubles in approximately 22 years, estimate the rate of interest.

48. *Determining the Interest Rate* Richard Maruszewski borrowed $2000 from Linda Tonolli. The terms of the loan are as follows: The period of the loan is 3 years, and the rate of interest is 8% compounded semiannually. What rate of simple interest would be equivalent to the rate Linda charged Richard?

RECREATIONAL MATHEMATICS

49. *Interest Comparison* You are given a choice of taking the simple interest on $100,000 invested for 4 years at a rate of 5% or the interest on $100,000 invested for 4 years at an interest rate of 5% compounded daily. Which would you select? Explain your answer and give the difference in the two investments.

INTERNET/RESEARCH ACTIVITIES

50. Imagine you have $4000 to invest and that you need this money to grow to $5000 by investing in a CD. Contact a local bank, a savings and loan, and a credit union to obtain CD information: the interest rate, the length of the term, and the number of times per year the CD is compounded. Find out how long it would take you to reach your goal with the institution selected. Write a report summarizing your findings.

51. Write a paper on the history of simple interest and compound interest. Answer the questions: When was simple interest first charged on loans? When was compound interest first given on investments?

11.4   INSTALLMENT BUYING

▲ Financial decisions may involve deciding whether to use a credit card when making a purchase.

While you are with some friends at the mall, you spot a new pair of shoes that you really like. Although you don't have money to pay for the shoes, you do have a credit card with you. You now must make a very important decision. Do you charge the shoes to your credit card and worry about paying for them later, or do you decide that you can't afford the shoes and leave the store without them? In this section, we will look at the real cost of using a credit card to make purchases. We will also study other common types of loans frequently used to purchase big-ticket items such as cars, appliances, and home improvements.

In Section 11.2, we discussed personal notes and discounted notes. When borrowing money by either of these methods, the borrower normally repays the loan as a single payment at the end of the specified time period. There may be circumstances under which it is more convenient for the borrower to repay the loan on a weekly or monthly basis or to use some other convenient time period. One method of doing so is to borrow money on an *installment plan.*

There are two types of installment loans: fixed payment and open-end. A *fixed installment loan* is one on which you pay a fixed amount of money for a set number of payments. Examples of items purchased with fixed-payment installment loans are college tuition loans and loans for cars, boats, appliances, and furniture. These loans are generally repaid in 24, 36, 48, or 60 equal monthly payments. An *open-end installment loan* is a loan on which you can make variable payments each month. Credit cards, such as MasterCard, Visa, and Discover, are actually open-end installment loans, used to purchase items such as clothing, textbooks, and meals.

Lenders give any individual wishing to borrow money or purchase goods or services on an installment plan a credit rating to determine if the borrower is likely to repay the loan. The lending institution determines whether the applicant is a good "credit risk" by examining the individual's income, assets, liabilities, and history of repaying debts.

The advantage of installment buying is that the buyer has the use of an article while paying for it. If the article is essential, installment buying may serve a real need. A disadvantage is that some people buy more on the installment plan than they can afford. Another disadvantage is the interest the borrower pays for the loan. The method of determining the interest charged on an installment plan may vary with different lenders.

To provide the borrower with a way to compare interest charged, Congress passed the Truth in Lending Act in 1969. The law requires that the lending institution tell the borrower two things: the annual percentage rate and the finance charge. The *annual percentage rate (APR)* is the true rate of interest charged for the loan. The APR is calculated by using a complex formula, so we use a table to determine the APR. The technique of using a table to find the APR is illustrated in Example 2. The total *finance charge* is the total amount of money the borrower must pay for borrowing the money. The finance charge includes the interest plus any additional fees charged. The additional fees may include service charges, credit investigation fees, mandatory insurance premiums, and so on.

The finance charge a consumer pays when purchasing goods or services on an installment plan is the difference between the total installment price and the cash price. The *total installment price* is the sum of all the monthly payments and the down payment, if any.

Fixed Installment Loan

In Example 1, we learn how to determine the finance charge and the monthly payment on a fixed installment loan.

EXAMPLE ❶ *Window Blinds*

Kristin Aiken wishes to purchase new window blinds for her house at a cost of $1500. The home improvement store has an advertised finance option of no down payment and 7% APR for 24 months.

a) Determine the finance charge.

b) Determine Kristin's monthly payment.

SOLUTION

a) Table 11.2 on page 680 gives the finance charge per $100 of the amount financed. The table shows that the finance charge per $100 for 24 months at 7% is $7.45 (circled in red). Because Kristin is financing $1500, the number of hundreds of dollars financed is $\frac{1500}{100} = 15$. To determine the total finance charge, multiply the finance charge per $100 by the number of hundreds of dollars financed.

$$\text{Total finance charge} = \$7.45 \times 15 = \$111.75$$

Therefore, Kristin will pay a total finance charge of $111.75.

b) To determine the monthly payments, first calculate the total installment price by adding the finance charge to the purchase price.

$$\text{Total installment price} = \$1500 + \$111.75 = \$1611.75$$

Next divide the total installment price by the number of payments.

$$\text{Monthly payment} = \frac{\$1611.75}{24} \approx \$67.16$$

Kristin will have 24 monthly payments of $67.16. ●

Table 11.2 Annual Percentage Rate Table for Monthly Payment Plans

Number of Payments	Annual Percentage Rate												
	4.0%	4.5%	5.0%	5.5%	6.0%	6.5%	7.0%	7.5%	8.0%	8.5%	9.0%	9.5%	10.0%
	(Finance charge per $100 of amount financed)												
6	1.17	1.32	1.46	1.61	1.76	1.90	2.05	2.20	2.35	2.49	2.64	2.79	2.93
12	2.18	2.45	2.73	3.00	3.28	3.56	3.83	4.11	4.39	4.66	4.94	5.22	5.50
18	3.20	3.60	4.00	4.41	4.82	5.22	5.63	6.04	6.45	6.86	7.28	7.69	8.10
24	4.22	4.75	5.29	5.83	6.37	6.91	(7.45)	8.00	8.54	9.09	9.64	10.19	10.75
30	5.25	5.92	6.59	7.26	7.94	8.61	9.30	9.98	10.66	11.35	12.04	12.74	13.43
36	6.29	7.09	7.90	8.71	9.52	10.34	11.16	11.98	12.81	13.64	14.48	15.32	16.16
48	8.38	9.46	10.54	11.63	12.73	13.83	14.94	(16.06)	17.18	18.31	19.45	20.59	21.74
60	10.50	11.86	13.23	14.61	16.00	17.40	18.81	20.23	21.66	(23.10)	24.55	26.01	27.48

Example 2 shows how to determine the annual percentage rate on a loan that has a schedule for repaying a fixed amount each period.

┌EXAMPLE ❷ *Determining the APR*

Jan Ford is purchasing a new boat for $25,000, including taxes. Jan decides to make a $5000 down payment and finance the balance, $20,000, through her bank. The loan officer informs her that her monthly payment will be $410.33 for 60 months.

a) Determine the finance charge.   b) Determine the APR.

SOLUTION

a) The total installment price is the down payment plus the total monthly installment payments.

$$\text{Total installment price} = \$5000 + (60 \times \$410.33)$$
$$= \$5000 + \$24{,}619.80 = \$29{,}619.80$$

The finance charge is the total installment price minus the cash price.

$$\text{Finance charge} = \$29{,}619.80 - \$25{,}000 = \$4619.80$$

Therefore, Jan will pay a finance charge of $4619.80.

b) To determine the annual percentage rate, use Table 11.2 on page 680. First divide the finance charge by the amount financed and multiply the quotient by 100. The result is the finance charge per $100 of the amount financed.

$$\frac{\text{Finance charge}}{\text{Amount financed}} \times 100 = \frac{4619.80}{20{,}000} \times 100 = 0.23099 \times 100 \approx 23.10$$

Thus, Jan pays about $23.10 for each $100 being financed. To use Table 11.2, look for 60 in the far left-hand column under the heading Number of Payments. Then move across to the right until you find the value closest to $23.10. In this case, $23.10 is in the table (circled in blue). At the top of this column is the value of 8.5%. Therefore, the annual percentage rate is approximately 8.5%. ●

Much more complete APR tables similar to Table 11.2 are available at your local lending institution or on the Internet. Also available on the Internet are loan "calculators" that will calculate your payment for you if you provide the loan amount, the length of the loan, and the APR. One such loan calculator can be found at www.bankrate.com.

EXAMPLE ❸ *Financing a Restored Car*

Tino Garcia borrowed $9800 to purchase a classic 1966 Ford Mustang. He does not recall the APR of the loan but remembers that there are 48 payments of $237. If he did not make a down payment on the car, determine the APR.

SOLUTION First determine the finance charge by subtracting the cash price from the total amount paid.

$$\begin{aligned}
\text{Finance charge} &= (237 \times 48) - 9800 \\
&= 11{,}376 - 9800 \\
&= \$1576
\end{aligned}$$

Next divide the finance charge by the amount of the loan and multiply this quotient by 100.

$$\frac{1576}{9800} \times 100 \approx 16.08$$

Next find 48 payments in the left column of Table 11.2. Move to the right until you find the value that is closest to 16.08. The value closest to 16.08 is 16.06 (circled in green). At the top of the column is the APR of 7.5%. Thus, the APR for Tino's loan is approximately 7.5%. ●

In Example 3, if Tino made all 48 payments, his finance charge would be $1576. If he decides to repay the loan after making 30 payments, must he pay the total finance charge? The answer is no. By paying off his loan early, Tino is not obligated to pay the entire finance charge. The amount of the reduction of the finance charge from

DID YOU KNOW?

Actuarial Method versus Rule of 78s

When making automobile or other installment loans, consumers should be aware of the "fine print" described in the loan agreement. If a consumer repays a loan early, the lender is required by law to rebate a portion of the original finance charge or interest. Two methods exist for determining this rebate. The actuarial method is the primary method used by financial institutions today. This method provides the consumer with an accurate interest rebate for repaying the loan early. The second method, the rule of 78s, however, is still used by some lenders. The rule of 78s charges the consumer more interest during the first few months of the loan. The result is a lower interest rebate for the consumer. The rule of 78s originated as an easy way for lenders to compute interest rebates. It has been phased out gradually as calculators and computers become more available. Today the rule of 78s is most likely to be applied by used car dealers who specialize in offering loans to borrowers with a poor credit history.

In 1992, the U.S. Congress outlawed the use of the rule of 78s for loans longer than 61 months. In 2001, federal legislation was introduced to outlaw it altogether. As of 2006, 17 states outlaw the use of the rule of 78s for all loans.

paying off a loan early is called the *unearned interest*. We will learn about the most common way of calculating unearned interest, the *actuarial method*. In the past, another method, called the *rule of 78s*, was also used. The rule of 78s, however, is less favorable to the consumer and is outlawed in much of the country (See the Did You Know? at left and Exercises 39 and 40). Now we give the formula for finding the unearned interest using the actuarial method.

ACTUARIAL METHOD FOR UNEARNED INTEREST

$$u = \frac{n \cdot P \cdot V}{100 + V}$$

where u is the unearned interest, n is the number of remaining monthly payments (excluding the current payment), P is the monthly payment, and V is the value from the APR table that corresponds to the annual percentage rate for the number of remaining payments (excluding the current payment).

Example 4 illustrates the actuarial method for calculating unearned interest when an installment loan is paid off early.

EXAMPLE ❹   *Using the Actuarial Method*

In Example 3, we determined the APR of Tino's loan to be 7.5%. Instead of making his 30th payment of his 48-payment loan, Tino wishes to pay his remaining balance and terminate the loan.

a) Use the actuarial method to determine how much interest Tino will save (the unearned interest, u) by repaying the loan early.

b) What is the total amount due to pay off the loan early on the day he makes his final payment?

SOLUTION

a) Recall from Example 3 that Tino's monthly payments are $237.00. After 30 payments have been made, 18 payments remain. Thus, $n = 18$ and $P = \$237$. To determine V, use the APR table (Table 11.2). In the Number of Payments column, find the number of remaining payments, 18, and then look to the right until you reach the column headed by 7.5%, the APR. This row and column intersect at 6.04. Thus, $V = 6.04$. Now use the actuarial method formula to determine the unearned interest, u.

$$u = \frac{n \cdot P \cdot V}{100 + V}$$

$$= \frac{(18)(237)(6.04)}{100 + 6.04}$$

$$\approx 242.99$$

Tino will save $242.99 in interest by the actuarial method.

b) Because the remaining payments total $18(\$237) = \$4266$, Tino's remaining balance excluding his 30th monthly payment, is

$4266.00	Total of remaining payments (which includes interest)
− 242.99	Interest saved (unearned interest)
$4023.01	Balance due (excluding the 30th monthly payment)

A payment of $4023.01 plus the 30th monthly payment of $237 will terminate Tino's installment loan.

$4023.01	Balance due (excluding the 30th monthly payment)
+ 237.00	30th monthly payment
$4260.01	Total amount due

The total amount due is $4260.01.   •

Open-End Installment Loan

A credit card is a popular way of making purchases or borrowing money. Use of a credit card is an example of an open-end installment loan. A typical credit card charge account with a bank or store may have the terms in Table 11.3.

Table 11.3 Credit Card Terms

Type of Charge	Daily Periodic Rate*	Annual Percentage Rate*
Purchases	0.04792%	17.49%
Cash advances	0.05477%	19.99%

*These rates vary with different credit card accounts and localities.

Typically, credit card monthly statements contain the following information: balance at the beginning of the period, balance at the end of the period (or new balance), the transactions for the period, statement closing date (or billing date), payment due date, and the minimum payment due. For *purchases*, there is no finance or interest charge if there is no previous balance due and you pay the entire new balance by the payment due date. The period between when a purchase is made and when the credit card company begins charging interest is called the *grace period* and is usually 20 to 25 days. However, if you use a credit card to borrow money, called a *cash advance*, there generally is no grace period and a finance charge is applied from the date you borrowed the money until the date you repay the money. When you make purchases or obtain cash advances, the minimum monthly payment will generally be any new fees and interest plus at least 1% of the outstanding principal (see Mathematics Today at left). This sum is then rounded up to the nearest whole dollar. Credit card companies often have $20 as the lowest possible minimum monthly payment. These guidelines will vary among the different credit card companies. If you currently have a credit card, you can find out how your minimum monthly payment is calculated by reading the back of your monthly statement or by reading the literature given to you when you obtained the credit card.

EXAMPLE ❺ *Calculating Minimum Monthly Payments*

Mary Beth Orrange's credit card company determines her minimum monthly payment by adding any new interest owed to 1.5% of the outstanding principal. The credit card company charges an interest rate of 0.04792% per day. For Mother's Day, May 17, Mary Beth used her credit card to purchase a $1500 plasma television set for her mother. She made no other purchases in May.

a) Assuming Mary Beth owed no interest, determine Mary Beth's minimum payment due on June 1, her billing date.

b) On June 1, instead of making the minimum payment, Mary Beth makes a payment of $200. Assuming there are no additional charges or cash advances, determine Mary Beth's minimum payment due on July 1.

SOLUTION

a) Since there is no new interest due, Mary Beth's minimum monthly payment is determined by taking 1.5% of the outstanding principal, or $0.015 \times \$1500 = \$22.50$. Rounding up to the nearest whole dollar, we determine that Mary Beth's minimum monthly payment due on June 1 is $23.

b) Mary Beth's minimum monthly payment will be the sum of the new interest charges plus 1.5% of her outstanding principal. To calculate her new interest charges, we will use the simple interest formula, $i = prt$. The principal p is $\$1500 - \$200 = \$1300$, the rate r is 0.04792% per day or 0.0004792 per day, and the time t is 30 days because June has 30 days.

$$
\begin{aligned}
i &= prt \\
&= (\$1300)(0.0004792)(30) \\
&\approx \$18.69
\end{aligned}
$$

To determine 1.5% of the outstanding principal, multiply 0.015 by $1300 to get $19.50. Now we add the new interest to 1.5% of the outstanding principal to get $\$19.50 + \$18.69 = \$38.19$. Rounding up to the nearest whole dollar, we determine that Mary Beth's minimum monthly payment due on July 1 is $39.   ●

In Example 5, Mary Beth made no additional charges during the month. When additional charges are made during the month, the finance charges on open-end installment loans or credit cards are generally calculated in one of two ways: the *unpaid balance method* or the *average daily balance method*. Example 6 illustrates the unpaid balance method, and Example 7 illustrates the average daily balance method.

With the *unpaid balance method*, the borrower is charged interest or a finance charge on the unpaid balance from the previous charge period.

EXAMPLE ❻ *Finance Charges Using the Unpaid Balance Method*

In October, Ed Laughbaum charged all the supplies for his Halloween party to his Visa card. On November 5, the billing date, Ed had a balance due of $275. From November 5 through December 4, he did some shopping and charged items totaling $320, and he also made a payment of $145.

a) Determine the finance charge due on December 5, Ed's next billing date, using the unpaid balance method. Assume that the interest rate charged is 1.3% per month.

b) Determine the new account balance on December 5.

SOLUTION

a) The finance charge is based on the $275 balance due on November 5. To find the finance charge due on December 5, we used the simple interest formula with a time of 1 month.

$$ i = \$275 \times 0.013 \times 1 \approx \$3.58 $$

The finance charge due on December 5 is $3.58.

b) The balance due on December 5 is found by adding the costs of the new purchases and the calculated interest to the balance due on November 5, and then subtracting the payment made from this sum.

$$\$275 + \$320 + \$3.58 - \$145 = \$453.58$$

The balance due on December 5 is $453.58. The finance charge on January 5 is based on the December 5 balance of $453.58. •

Many lending institutions use the *average daily balance method* of calculating the finance charge because they believe that it is fairer to the customer. With the average daily balance method, a balance is determined each day of the billing period for which there is a transaction in the account. The average daily balance method is illustrated in Example 7.

EXAMPLE ❼ Finance Charges Using the Average Daily Balance Method

The balance on Min Zeng's credit card account on July 1, the billing date, was $375.80. The following transactions occurred during the month of July.

July 5	Payment	$150.00
July 10	Charge: Toy store	74.35
July 18	Charge: Garage	123.50
July 28	Charge: Restaurant	42.50

a) Determine the average daily balance for the billing period.
b) Determine the finance charge to be paid on August 1, Min's next billing date. Assume that the interest rate is 1.3% per month.
c) Determine the balance due on August 1.

SOLUTION

a) To determine the average daily balance, we do the following. (i) Find the balance due for each transaction date.

July 1	$375.80					
July 5	$375.80	−	$150	=	$225.80
July 10	$225.80	+	$74.35	=	$300.15
July 18	$300.15	+	$123.50	=	$423.65
July 28	$423.65	+	$42.50	=	$466.15

(ii) Find the number of days that the balance did not change between each transaction. Count the first day in the period but not the last day. Note that the time period from July 28 through August 1, the beginning of the next billing cycle, is 4 days. (iii) Multiply the balance due by the number of days the balance did not change. (iv) Find the sum of the products.

Date	(i) Balance Due	(ii) Number of Days Balance Did Not Change	(iii) (Balance)(Days)
July 1	$375.80	4	($375.80)(4) = $1503.20
July 5	$225.80	5	($225.80)(5) = $1129.00
July 10	$300.15	8	($300.15)(8) = $2401.20
July 18	$423.65	10	($423.65)(10) = $4236.50
July 28	$466.15	4	($466.15)(4) = $1864.60
		31	(iv) Sum = $11,134.50

(v) Divide this sum by the number of days in the billing cycle (in the month). The number of days may be found by adding the days in column (ii).

$$\frac{\$11,134.50}{31} = \$359.18$$

Thus, the average daily balance is $359.18.

b) The finance charge for the month is found using the simple interest formula with the average daily balance as the principal.

$$i = \$359.18 \times 0.013 \times 1 = \$4.67$$

c) Since the finance charge for the month is $4.67, the balance owed on August 1 is $466.15 + $4.67, or $470.82. ●

The calculations in Example 7 are tedious. These calculations, however, are made almost instantaneously with computers.

Example 8 illustrates how a credit card may be used to borrow money.

EXAMPLE ❽ *Using a Credit Card for a Cash Advance*

To obtain money to buy a stereo system, Bobby Bueker obtained a cash advance of $1500 from his credit card. He borrowed the money on July 10 and repaid it on July 31. If Bobby is charged an interest rate of 0.05477% per day, how much did Bobby pay the credit card company on July 31?

SOLUTION The amount Bobby pays is the original principal plus any accrued interest. Interest on cash advances is generally calculated for the exact number of days of the loan, starting with the day the money is obtained. The time of the loan in this case is 21 days. Using the simple interest formula, we get the following:

$$
\begin{aligned}
i &= prt \\
&= \$1500 \times 0.0005477 \times 21 \\
&= \$17.25
\end{aligned}
$$

Therefore, on July 31, Bobby must repay the credit card company $1500 + $17.25, or $1517.25. ●

Anyone purchasing a car or other costly items should consider a number of different sources for a loan. Example 9 illustrates one method of making a comparison.

EXAMPLE ❾ *Comparing Loan Sources*

Franz Helfenstein purchased carpeting costing $2400 with his credit card. When the bill comes due on February 1, Franz realizes that he can pay $350 per month until the debt is paid off. His credit card charges 1.5% interest per month.

a) Assuming Franz makes no other purchases with this credit card, how many payments are necessary to retire this debt?

b) What is the total interest Franz will pay?

c) How much money could Franz have saved by obtaining a fixed installment loan of $2400 with an annual percentage rate of 6% interest with six equal monthly payments?

SOLUTION

a) Franz would make his first monthly payment of $350 on February 1, resulting in a new balance of $2400 − $350 = $2050. His next bill reflects the $2050 balance plus the monthly interest. He continues to make payments until the debt is retired. For each date indicated, the amount on the far right represents the amount due on that date.

February 1	$2400 − $350 = $2050
March 1	$2050 + 0.015($2050) = $2080.75; $2080.75 − $350 = $1730.75
April 1	$1730.75 + 0.015($1730.75) = $1756.71; $1756.71 − $350 = $1406.71
May 1	$1406.71 + 0.015($1406.71) = $1427.81; $1427.81 − $350 = $1077.81
June 1	$1077.81 + 0.015($1077.81) = $1093.98; $1093.98 − $350 = $743.98
July 1	$743.98 + 0.015($743.98) = $755.14; $755.14 − $350 = $405.14
August 1	$405.14 + 0.015($405.14) = $411.22; $411.22 − $350 = $61.22
September 1	$61.22 + 0.015($61.22) = $62.14

After eight payments—seven for $350 and one for $62.14—Franz has paid off his credit card bill for his carpeting.

b) To calculate the total interest paid, we add up all his payments and then subtract the cost of the carpeting:

$$\text{Total of all payments} = 7(\$350) + \$62.14$$
$$= \$2512.14$$
$$\text{Total interest} = \$2512.14 - \$2400$$
$$= \$112.14$$

c) To determine the interest that Franz pays with the installment loan, we will use Table 11.2 on page 680. From Table 11.2, we see that a fixed installment loan with an APR of 6% for 6 months corresponds to a finance charge of $1.76 per $100 of the amount financed. So the interest, or finance charge, is

$$\text{Finance charge} = 1.76\left(\frac{2400}{100}\right)$$
$$= 1.76(24)$$
$$= \$42.24$$

Therefore, Franz would save $112.14 − $42.24 = $69.90 in interest by using an installment loan instead of a credit card.   ●

SECTION 11.4   EXERCISES

CONCEPT/WRITING EXERCISES

1. Explain the difference between an open-end installment loan and a fixed installment loan.

2. Explain how an installment plan differs from a personal note.

3. What is an annual percentage rate (APR)?

4. What is a finance charge?

5. What is a total installment price?

6. a) What is unearned interest?

   b) What is the most common method used to calculate unearned interest?

7. Name the two methods used to determine the finance charge on an open-end installment loan.

8. What is a cash advance?

PRACTICE THE SKILLS/PROBLEM SOLVING

9. *A New Roof* Nancy Sattler paid $10,000 for a new roof for her house. She paid 15% as a down payment and financed the balance with a 60-month fixed installment loan with an APR of 7.5%.

   a) Determine Nancy's finance charge.

   b) Determine Nancy's monthly payment.

10. *A New Air Conditioner* Juan Avalos paid $7000 for a new central air-conditioning unit for his house. He paid 20% as a down payment and financed the balance with a 36-month fixed installment loan with an APR of 5%.

    a) Determine Juan's finance charge.

    b) Determine Juan's monthly payment.

11. *New Appliances* Becky Kubiac wishes to purchase all new kitchen appliances for her home for $9900. To finance the purchase, Becky makes a 10% down payment and finances the balance with a 48-month fixed installment loan with an APR of 9.5%.

    a) Determine Becky's total finance charge.

    b) Determine Becky's monthly payment.

12. *PT Cruiser GT* Sam Bazzi purchased a 2006 PT Cruiser GT for $25,000. Sam obtained a 60-month, no-money-down installment loan with an APR of 8%.

    a) Determine Sam's total finance charge.

    b) Determine Sam's monthly payment.

13. *Financing a New Business* Cheryl Sisson is a hair designer and wishes to convert her garage into a hair salon to use for her own business. The entire project would have a cash price of $3200. She decides to finance the project by paying 20% down, with the balance paid in 60 monthly payments of $53.14.

    a) What finance charge will Cheryl pay?

    b) What is the APR to the nearest half percent?

14. *Financing a Computer* Ilga Ross purchased a new laptop computer on a monthly purchase plan. The computer sold for $1495. Ilga paid 5% down and $64 a month for 24 months.

    a) What finance charge did Ilga pay?

    b) What is the APR to the nearest half percent?

15. *Financing a Used Car* Jack Fitzgerald wishes to purchase a used car that has a cash price of $12,000. The installment terms include a down payment of $3000 and 48 monthly payments of $224.

    a) What finance charge will Jack pay?

    b) What is the APR to the nearest half percent?

16. *Financing Furniture* Mr. and Mrs. Chan want to buy furniture that has a cash price of $3450. On the installment plan, they must pay 25% of the cash price as a down payment and make six monthly payments of $437.

a) What finance charge will the Chans pay?

b) What is the APR to the nearest half percent?

17. *Early Repayment of a Loan* Ray Flagg took out a 60-month fixed installment loan of $12,000 to open a new pet store. He paid no money down and began making monthly payments of $232. Ray's business does better than expected and instead of making his 24th payment, Ray wishes to repay his loan in full.

a) Determine the APR of the installment loan.

b) How much interest will Ray save by paying off his loan early?

c) What is the total amount due to pay off the loan?

18. *Early Repayment of a Loan* Jeslie Ann Hernandez has a 48-month installment loan with a fixed monthly payment of $83.81. The amount she borrowed was $3500. Instead of making her 18th payment, Jeslie Ann is paying the remaining balance on the loan.

a) Determine the APR of the installment loan.

b) How much interest will Jeslie Ann save by paying off the loan early?

c) What is the total amount due to pay off the loan?

19. *Early Repayment of a Loan* Nina Abu buys a new sport utility vehicle for $32,000. She trades in her old truck and receives $10,000, which she uses as a down payment. She finances the balance at 8% APR over 36 months. Before making her 24th payment, she decides to pay off the loan.

a) Use Table 11.2 to determine the total interest Nina would pay if all 36 payments were made.

b) What were Nina's monthly payments?

c) How much interest will Nina save by paying off the loan early?

d) What is the total amount due to pay off the loan?

20. *Early Repayment of a Loan* The cash price for a new washer and dryer for Toshio Nakamura's new apartment was $1250. Toshio made a $100 down payment and financed the balance on a 24-month fixed payment installment loan. The monthly payments are $50.71. Instead of making his 12th payment, Toshio decides to pay off the loan.

a) Determine the APR on the installment loan.

b) How much interest will Toshio save by paying off the loan early?

c) What is the total amount due to pay off the loan?

21. *Credit Card Minimum Monthly Payment* Shiing Shen Chern's credit card company determines his minimum monthly payment by adding all new interest to 1% of the outstanding principal. The credit card company charges an interest rate of 0.039698% per day. On March 17, Shiing uses his credit card to purchase airline tickets for his family for $2600. He makes no other purchases during March.

a) Assuming Shiing had no new interest, determine Shiing's minimum payment due on April 1, his billing date.

b) On April 1, instead of making the minimum payment, Shiing makes a payment of $500. Assuming there are no additional charges or cash advances, determine Shiing's minimum payment due on May 1.

22. *Credit Card Minimum Monthly Payment* Evelyn Boyd Granville's credit card company determines her minimum monthly payment by adding all new interest to 2% of the outstanding principal. The credit card company charges an interest rate of 0.04238% per day. On September 15, Evelyn uses her credit card to purchase a new picture window for her house for $3200. She makes no other purchases during September.

a) Assuming Evelyn had no new interest, determine Evelyn's minimum payment due on October 1, her billing date.

b) On October 1, instead of making the minimum payment, Evelyn makes a payment of $1200. Assuming there are no additional charges or cash advances, determine Evelyn's minimum payment due on November 1.

23. *Business Expenses* Kevin Devlin's credit card company determines his minimum monthly payment by adding all new interest to 1.5% of the outstanding principal. The credit card company charges an interest rate of 0.05163% per day. On November 12, Kevin used his credit card to pay for the following business expenses: van repairs ($677), equipment maintenance ($452), office supplies ($139), and dinner with clients ($141).

a) Assuming Kevin had no new interest, determine his minimum payment due on December 1, his billing date.

b) On December 1, instead of making his minimum payment, Kevin makes a payment of $300. Assuming there are no additional charges or cash advances, determine Kevin's minimum payment on January 1.

24. *Vacation Expenses* Harry Waldman's credit card company determines his minimum monthly payment by adding all new interest to 2.5% of the outstanding principal. The credit card company charges an interest rate of 0.03164% per day. On August 21, while on vacation, Harry used his credit card to pay for the following expenses: airfare ($359), car rental ($273), hotel ($653), meals ($315), and surfboard rental ($225).

   a) Assuming Harry had no new interest, determine Harry's minimum payment due on September 1, his billing date.

   b) On September 1, instead of making his minimum payment, Harry makes a payment of $750. Assuming there are no additional charges or cash advances, determine Harry's minimum payment on October 1.

25. *Unpaid Balance Method* On the April 5 billing date, Michaelle Chappell had a balance due of $1097.86 on her credit card. From April 5 through May 4, Michaelle charged an additional $425.79 and made a payment of $800.

   a) Find the finance charge on May 5, using the unpaid balance method. Assume that the interest rate is 1.8% per month.

   b) Find the new balance on May 5.

26. *Unpaid Balance Method* On September 5, the billing date, Verna Brown had a balance due of $567.20 on her credit card. The transactions during the following month were

September 8	Payment	$275.00
September 21	Charge: Airline ticket	330.00
September 27	Charge: Hotel bill	190.80
October 2	Charge: Clothing	84.75

   a) Find the finance charge on October 5, using the unpaid balance method. Assume that the interest rate is 1.1% per month.

   b) Find the new balance on October 5.

27. *Unpaid Balance Method* On February 3, the billing date, Carol Ann Bluesky had a balance due of $124.78 on her credit card. Her bank charges an interest rate of 1.25% per month. She made the following transactions during the month.

February 8	Charge: Art supplies	$25.64
February 12	Payment	100.00
February 14	Charge: Flowers delivered	67.23
February 25	Charge: Music CD	13.90

   a) Find the finance charge on March 3, using the unpaid balance method.

   b) Find the new balance on March 3.

28. *Unpaid Balance Method* On April 15, the billing date, Gabrielle Michaelis had a balance due of $57.88 on her credit card. She is redecorating her apartment and has the following transactions.

April 16	Charge: Paint	$64.75
April 20	Payment	45.00
May 3	Charge: Curtains	72.85
May 10	Charge: Chair	135.50

   a) Find the finance charge on May 15, using the unpaid balance method. Assume that the interest rate is 1.35% per month.

   b) Find the new balance on May 15.

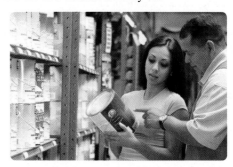

29. *Average Daily Balance Method* The balance on the Razazadas' credit card on May 12, their billing date, was $378.50. For the period ending June 12, they had the following transactions.

May 13	Charge: Toys	$129.79
May 15	Payment	50.00
June 1	Charge: Clothing	135.85
June 8	Charge: Housewares	37.63

   a) Find the average daily balance for the billing period.

   b) Find the finance charge to be paid on June 12. Assume an interest rate of 1.3% per month.

   c) Find the balance due on June 12.

30. *Average Daily Balance Method* The Levys' credit card statement shows a balance due of $1578.25 on March 23, the billing date. For the period ending April 23, they had the following transactions.

March 26	Charge: Party supplies	$79.98
March 30	Charge: Restaurant meal	52.76
April 3	Payment	250.00
April 15	Charge: Clothing	190.52
April 22	Charge: Car repairs	190.85

   a) Find the average daily balance for the billing period.

b) Find the finance charge to be paid on April 23. Assume an interest rate of 1.3% per month.

c) Find the balance due on April 23.

31. *Average Daily Balance Method* Refer to Exercise 27. Instead of the unpaid balance method, suppose that Carol Ann's bank uses the average daily balance method.

  a) Find Carol Ann's average daily balance for the billing period from February 3 to March 3. Assume it is not a leap year.

  b) Find the finance charge to be paid on March 3.

  c) Find the balance due on March 3.

  d) Compare these answers with those in Exercise 27.

32. *Average Daily Balance Method* Refer to Exercise 26. Instead of the unpaid balance method, suppose Verna's bank uses the average daily balance method.

  a) Find Verna's average daily balance for the billing period from September 5 to October 5.

  b) Find the finance charge to be paid on October 5.

  c) Find the balance due on October 5.

  d) Compare these answers with those in Exercise 26.

33. *A Cash Advance* John Richards borrowed $875 against his charge account on September 12 and repaid the loan on October 14 (32 days later). Assume that the interest rate is 0.04273% per day.

  a) How much interest did John pay on the loan?

  b) What amount did he pay the bank when he repaid the loan?

34. *A Cash Advance* Travis Thompson uses his credit card to obtain a cash advance of $600 to pay for his textbooks in medical school. The interest rate charged for the loan is 0.05477% per day. Travis repays the money plus the interest after 27 days.

  a) Determine the interest charged for the cash advance.

  b) When he repaid the loan, how much did he pay the credit card company?

35. *Comparing Loan Sources* Grisha Stewart needs to borrow $1000 for an automobile repair. She finds that State National Bank charges 5% simple interest on the amount borrowed for the duration of the loan and requires the loan to be repaid in six equal monthly payments. Consumer's Credit Union offers loans of $1000 to be repaid in 12 monthly payments of $86.30.

a) How much interest is charged by the State National Bank?

b) How much interest is charged by the Consumer's Credit Union?

c) What is the APR, to the nearest half percent, on the State National Bank loan?

d) What is the APR, to the nearest half percent, on the Consumer Credit Union loan?

36. *Comparing Loan Options* Sara Lin wants to purchase a new television set. The purchase price is $890. If she purchases the set today and pays cash, she must take money out of her savings account. Another option is to charge the TV on her credit card, take the set home today, and pay next month. Next month she will have cash and can pay her credit card balance without paying any interest. The simple interest rate on her savings account is $5\frac{1}{4}\%$. How much is she saving by using the credit card instead of taking the money out of her savings account?

CHALLENGE PROBLEMS/GROUP ACTIVITY

37. *Determining Purchase Price* Ken Tucker bought a new car, but now he cannot remember the original purchase price. His payments are $379.50 per month for 36 months. He remembers that the salesperson said the simple interest rate for the period of the loan was 6%. He also recalls he was allowed $2500 on his old car. Find the original purchase price.

38. *Comparing Loans* Suppose the Chans in Exercise 16 use a credit card rather than an installment plan. Assume that they make the same down payment, have no finance charge the first month, make no additional purchases on their credit card, and pay $432 per month plus the finance charge starting with the second month. The interest rate is 1.3% per month.

  a) How many months will it take them to repay the loan?

  b) How much interest will they pay on the loan?

  c) Which method of borrowing will cost the Chans the least amount of interest, the installment loan in Exercise 16 or the credit card?

39. *Rule of 78s* When an installment loan is repaid early, there is a second method for calculating unearned interest called the *rule of 78s*. Although rarely used today, the following formula can be used to calculate unearned interest.

$$u = \frac{f \cdot k(k + 1)}{n(n + 1)}$$

where u is the unearned interest, f is the original finance charge, k is the number of remaining monthly payments (excluding the current payment), and n is the original number of payments. Joscelyn Jarrett obtained a new sport utility vehicle that had a cash price of $35,000 by paying 15% down and financing the balance with a 60-month fixed installment loan. The APR on the loan was 8.5%. Before making the 24th payment, Joscelyn decides to pay off the loan.

a) Determine the original finance charge on the 60-month loan.

b) Determine Joscelyn's monthly payment.

c) If the actuarial method is used, determine the amount of interest Joscelyn will save by paying the loan off early.

d) If the rule of 78s is used, determine the amount of interest Joscelyn will save by paying the loan off early.

40. *Repayment Comparisons* Christine Biko obtained a new speedboat that had a cash price of $23,000 by paying 10% down and financing the balance with a 48-month fixed installment loan. The APR on the loan was 6.0%. Before making the 12th payment, Christine decides to pay off the loan.

a) Determine the original finance charge on the 48-month loan.

b) Determine Christine's monthly payment.

c) If the actuarial method is used, determine the amount of interest Christine will save by paying the loan off early.

d) If the rule of 78s (see Exercise 39) is used, determine the amount of interest Christine will save by paying the loan off early.

41. *Installment Payment Formula* The following formula can be used to calculate installment payments rather than using Table 11.2.

$$m = \frac{P\left(\dfrac{r}{n}\right)}{1 - \left(1 + \dfrac{r}{n}\right)^{-n \cdot t}}$$

In the formula, m is the installment payment, P is the amount financed, r is the annual percentage rate, n is the number of payments per year, and t is the time in years. Use the installment payment formula to calculate the monthly installment payment on a loan of $4500 with an APR of 7% for 2 years.

INTERNET/RESEARCH ACTIVITIES

42. *Paying Off Your Credit Card Debt* Suppose you purchase a new wardrobe for $1000 with a credit card. You decide to pay for your new wardrobe by making the minimum monthly payment each month. You do not charge anything else to this credit card until you have the wardrobe paid off. Your credit card company determines your minimum monthly payment by adding all new interest to 1% of the outstanding principal. The credit card company charges an annual percentage rate of 18%. Go to an Internet credit card payment calculator web site such as www.bankrate.com to answer the following.

a) How long will it take you to pay off the entire credit card debt if you make the minimum monthly payments?

b) How much interest will you pay?

c) Adding the interest paid to the cash price, determine the total cost of your wardrobe.

43. Write a brief report giving the advantages and disadvantages of leasing a car. Determine all the individual costs involved with leasing a car. Indicate why you would prefer to lease or purchase a car at the present time.

44. Assume that you are married and have a child. You don't own a washer and dryer and have no money to buy the appliances. Would it be cheaper to borrow money on an installment loan and buy the appliances or to continue to go to the local coin-operated laundry for 5 years until you have saved enough to pay cash for a washer and dryer?

With the aid of parents or friends, establish how many loads of laundry you would be doing each week. Then determine the cost of doing that number of loads at a coin-operated laundry. (Don't forget the cost of transportation to and from the laundry.) Shop around for a washer and dryer, and determine the total cost on an installment plan. Don't forget to include the cost of gas, electricity, and water. This information can be obtained from a local gas and electric company. With this information, you should be able to make a decision about whether to buy now or wait for 5 years.

11.5 BUYING A HOUSE WITH A MORTGAGE

▲ Owning your own home is part of the American dream.

Part of the American dream is to own your own home. Owning your own home means that you don't pay rent to someone else, but rather, in a way, you pay rent to yourself. Often you are also rewarded with several financial benefits. In this section, we will discuss the many advantages of owning your own home. We will also discuss the many financial obligations that must be met prior to obtaining a home loan as well as the obligations that must be met after you obtain the loan.

When purchasing a home, buyers usually seek a *mortgage* from a bank or other lending institution. Before approving a mortgage, which is a long-term loan, the bank will require the buyer to have a specified minimum amount for the down payment. The *down payment* is the amount of cash the buyer must pay to the seller before the lending institution will grant the buyer a mortgage. If the buyer has the down payment and meets the other criteria for the mortgage, the lending institution prepares a written agreement called the mortgage, stating the terms of the loan. The loan specifies the repayment schedule, the duration of the loan, whether the loan can be assumed by another party, and the penalty if payments are late. The party borrowing the money accepts the terms of this agreement and gives the lending institution the title or deed to the property as security.

HOMEOWNER'S MORTGAGE
A long-term loan in which the property is pledged as security for payment of the difference between the down payment and the sale price.

The two most popular types of mortgage loans available today are the *conventional loan* and the *adjustable-rate loan* (or *variable-rate loan*). The major difference between the two is that the interest rate for a conventional loan is fixed for the duration of the loan, whereas the interest rate for the variable-rate loan may change every period, as specified in the loan. We will first discuss the requirements that are the same for both types of loans.

The size of the down payment required depends on who is lending the money, how old the property is, and whether or not it is easy to borrow money at that particular time. The down payment required by the lending institution can vary from 5% to 50% of the purchase price. A larger down payment is required when money is "tight," that is, when it is difficult to borrow money. Furthermore, most lending institutions tend to require larger down payments on older homes and smaller down payments on newer homes.

Lending institutions may require the buyer to pay one or more *points* for a loan at the time of the *closing* (the final step in the sale process). According to the Internal Revenue Service, *points* are interest prepaid by the buyer and may be used to reduce the stated interest rate the lender charges. One point is equal to 1% of the loan amount. This reduction in the interest rate allows the lender to reduce the size of the monthly mortgage payment, which enables more people to purchase homes. However, because points are considered interest, the rate of interest that lenders state when you are applying for a mortgage is not the annual percentage rate (APR) for the loan. Determining the

APR involves a number of steps, including adding the amount paid for points to the total interest paid. The APR can then be determined using an APR table, a calculator, or an Internet web site. Often, when shopping for a mortgage, a buyer may have two choices: pay points and get a lower interest rate, or not pay points and get a higher interest rate. Web sites such as www.hsh.com allow potential buyers to research whether paying points is to their advantage.

Conventional Loans

Example 1 illustrates purchasing a house with a conventional mortgage loan.

EXAMPLE ❶ *Down Payment and Points*

Patricia and Marshall Martin wish to purchase a house selling for $249,000. They plan to obtain a loan from their bank. The bank requires a 15% down payment, payable to the seller, and a payment of 2 points, payable to the bank, at the time of closing.

a) Determine the Martins' down payment.

b) Determine the amount of the Martins' mortgage.

c) Determine the cost of the 2 points paid by the Martins on their mortgage.

SOLUTION

a) The down payment is 15% of $249,000, or

$$0.15 \times \$249,000 = \$37,350$$

b) The mortgage on the Martin's new home is the selling price minus the down payment.

$$\$249,000 - \$37,350 = \$211,650$$

c) Each point equals 1% of the mortgage amount, so 2 points equals 2% of the mortgage amount.

$$0.02 \times \$211,650 = \$4233$$

At the closing, the Martins will pay the down payment of $37,350 to the seller and the 2 points, or $4233, to the bank.   ●

Banks use a formula to determine the maximum monthly payment that they believe is within the purchaser's ability to pay. A mortgage loan officer first determines the buyer's *adjusted monthly income* by subtracting from the gross monthly income (total income before any deductions) any fixed monthly payments with more than 10 months remaining (such as for a student loan, a car, furniture, or a television). The loan officer then multiplies the adjusted monthly income by 28%. (This percent, and the maximum number of payments remaining on other fixed loans, may vary in different locations.) In general, this product is the maximum monthly house payment the lending institution believes that the purchaser can afford to pay. This payment must cover principal, interest, property taxes, and insurance. Taxes and insurance are not necessarily paid to the bank; they may be paid directly to the tax collector and the insurance company. Example 2 shows how a bank uses the formula to determine whether a prospective buyer qualifies for a mortgage.

DID YOU KNOW?

The Benefits of Home Ownership

 ost of us are aware that a big part of the American dream is to own your own home. What we may not realize is that in addition to having a place to call your own, several important financial benefits occur when you own your own home. First, instead of paying rent to someone else, you make a mortgage payment that builds the equity in your home. *Equity* is the difference between the appraised value of your home and your loan balance, and it usually increases with each payment you make. As years go by, this equity may also help you qualify for other loans such as college and car loans. Second, the interest and real estate taxes you pay (in most cases) are deductible on your federal income tax returns. These deductions can add up to significant savings each year and may result in a larger tax refund. Finally, over time you can typically expect your home to increase in value. Thus, your home not only becomes your place of dwelling; in most cases, it also serves as a wise financial investment.

EXAMPLE ❷ *Qualifying for a Mortgage*

Suppose the Martins' (see Example 1) gross monthly income is $7250 and they have 23 remaining monthly payments of $225 on their car loan, 17 remaining monthly payments of $175 on their daughter's orthodontic braces, and 11 remaining monthly payments of $45 on a loan used to purchase new furniture. The property taxes and homeowners' insurance on the house they wish to buy are $165 and $115 per month, respectively. Their bank will approve a loan that has a total monthly mortgage payment of principal, interest, property taxes, and homeowners' insurance that is less than or equal to 28% of their adjusted monthly income.

a) Determine 28% of the Martins' adjusted monthly income.

b) The Martins want a 30-year, $211,650 mortgage. If the interest rate is 7.0%, determine the total monthly mortgage payment (including principal, interest, property taxes, and homeowners' insurance) for this mortgage.

c) Determine whether the Martins qualify for this mortgage.

SOLUTION

a) To determine the Martins' adjusted monthly income, subtract the sum of their monthly payments, $225 + $175 + $45 = $445, from their gross monthly income, $7250.

$7250	Gross monthly income
− 445	Monthly payments
$6805	Adjusted monthly income

Next, take 28% of the adjusted monthly income.

$$0.28 \times \$6805 = \$1905.40$$

Thus, 28% of the Martins' adjusted monthly income is $1905.40.

b) To determine the total monthly mortgage payment, we first need to determine the monthly principal and interest payments. Then we add the monthly property taxes and homeowners' insurance. We will use a table to calculate the monthly principal and interest payments the Martins will need to pay. Table 11.4 on page 696 gives monthly principal and interest payments per $1000 of mortgage. With an interest rate of 7.0% and a 30-year mortgage, the Martins would have a monthly principal and interest payment of $6.65 (circled in blue) per thousand dollars of mortgage.

To determine the Martins' monthly principal and interest payment on their mortgage, first divide the mortgage amount by $1000, which will give the number of thousands of dollars of the mortgage.

$$\frac{\$211,650}{\$1000} = 211.65$$

Then, to determine the monthly principal and interest payment, multiply the number of thousands of dollars of mortgage, 211.65, by the value found in Table 11.4, $6.65.

$$211.65 \times \$6.65 \approx \$1407.47$$

Thus, the monthly principal and interest payment is $1407.47. To this amount, we add the monthly property taxes, $165, and the monthly homeowner's insurance, $115.

$$\$1407.47 + \$165 + \$115 = \$1687.47$$

Therefore, the Martins' total monthly mortgage payment is $1687.47.

c) In part (a), we determined that 28% of the Martins' adjusted monthly income is $1905.40. In part (b), we determined the Martins' total monthly mortgage payment is $1687.47. Because their total monthly mortgage payment is less than or equal to 28% of their adjusted monthly income, the Martins would most likely qualify for the mortgage.   ●

Table 11.4 Monthly Principal and Interest Payment per $1000 of Mortgage

	Number of Years				
Rate %	**10**	**15**	**20**	**25**	**30**
4.0	$10.12	$7.40	$6.06	$5.28	$4.77
4.5	10.36	7.65	6.33	5.56	5.07
5.0	10.61	7.91	6.60	5.85	5.37
5.5	10.85	8.17	6.88	6.14	5.68
6.0	11.10	8.44	7.16	6.44	6.00
6.5	11.35	8.71	7.46	6.75	6.32
7.0	11.61	8.99	7.75	7.07	(6.65)
7.5	11.87	9.27	8.06	7.39	6.99
8.0	12.13	9.56	8.36	7.72	7.34
8.5	12.40	9.85	8.68	8.05	7.69
9.0	12.67	10.14	9.00	8.40	8.05
9.5	12.94	10.44	9.32	8.74	8.41
10.0	13.22	10.75	9.65	9.09	8.78
10.5	13.49	11.05	9.98	9.44	9.15
11.0	13.78	11.37	10.32	9.80	9.52
11.5	14.06	11.68	10.66	10.16	9.90
12.0	14.35	12.00	11.01	10.53	10.29

In everyday practice, lending institutions, real estate professionals, and lawyers use computer programs or calculators to determine the monthly principal and interest payment on a mortgage. When the amount of the loan, the interest rate, and the length of the mortgage are known, the monthly principal and interest payments can also be determined at many Internet web sites such as www.bankrate.com. The formula used in these calculations is discussed in Exercise 26.

What is the effect on the total monthly mortgage payments when only the period of time of the mortgage has been changed? The total monthly mortgage payments for a 7.0% mortgage for the Martins in Example 2 would be about $2737.26 for 10 years, $2182.73 for 15 years, $1920.29 for 20 years, $1776.37 for 25 years, and $1687.47 for 30 years. (You should verify these numbers for yourself.) Increasing the length of time decreases the monthly payment but increases the total amount of interest paid because the borrower is paying for a longer period of time (see Exercise 25). The longer the term of the mortgage, the more expensive the total cost of the house. Although the total cost of a house with the shorter-term mortgage is less expensive overall, it is not always possible for the buyer to obtain the shorter-term mortgage. Note that based on their current adjusted monthly income, the Martins most likely would not have qualified for the 10-year, 15-year, or 20-year mortgage.

MATHEMATICS TODAY

Affordable and Overvalued Housing

H ow much does a house cost? Many factors affect the price of a house, including the city in which the house is located. The tables below show median house prices in the five most *affordable* cities and median house prices in the most *overvalued* cities as of August 2006. The ranking is determined by the percentage of the median household income needed to pay the median mortgage payment.

Most Affordable Cities

City	Median House Price
Danville, Illinois	$58,900
Odessa, Texas	$65,600
Wichita Falls, Texas	$80,500
Youngstown, Ohio	$83,500
South Bend, Indiana	$88,200

Most Overvalued Cities

City	Median House Price
Salinas, California	$608,600
Naples, Florida	$383,000
Merced, California	$291,300
Bend, Oregon	$276,100
Port St. Lucie, Florida	$240,800

Source: *U.S. News & World Report*

EXAMPLE ❸ *The Total Cost of a House*

Patricia and Marshall Martin of Examples 1 and 2 purchased a house selling for $249,000. They made a 15% down payment of $37,350 and obtained a 30-year conventional mortgage for $211,650 at 7.0%. They also paid 2 points at closing. Their monthly principal and interest payment on their mortgage is $1407.47. Recall that points are considered prepaid interest.

a) Determine the total amount including principal, interest, down payment, and points the Martins will pay for their house over 30 years.

b) How much of the cost in part (a) is interest?

c) How much of the first mortgage payment is applied to the principal?

SOLUTION

a) To determine the total amount the Martins will pay for their house, we first note that on a 30-year mortgage, there will be $30 \times 12 = 360$ monthly payments. We then perform the following computations.

$1407.47	Monthly principal and interest payment
$\times$ 360	Number of monthly payments
$506,689.20	Total principal and interest paid
+ 37,350.00	Down payment, from Example 1, part (a)
+ 4,233.00	2 points, from Example 1, part (c)
$548,272.20	Total cost of the house

Note that the result might not be the *exact* cost of the house because the final mortgage payment might be slightly more or less than the rest of the monthly mortgage payments.

b) To determine the amount of the interest paid over 30 years, subtract the purchase price of the house from the total price of the house.

$548,272.20	Total cost of the house
$-249,000.00$	Purchase price
$299,272.20	Total interest (including the 2 points)

With this mortgage, the Martins will pay $299,272.20 in interest, including the 2 points.

c) To determine the amount of the first payment that is applied to the principal, subtract the amount of interest on the first payment from the monthly principal and interest payment. We will use the simple interest formula, $i = prt$, to determine the interest on the first payment.

$$i = prt$$
$$= \$211{,}650 \times 0.07 \times \frac{1}{12}$$
$$\approx \$1234.63$$

Now subtract the interest for the first month from the monthly principal and interest payment. The difference will be the amount paid on the principal for the first month.

$1407.47	Monthly principal and interest payment
-1234.63	Interest paid for the first month
$ 172.84	Principal paid for the first month

Thus, the first monthly principal and interest payment consists of $1234.63 in interest and $172.84 in principal. The $172.84 is applied to reduce the outstanding balance due on the loan. Thus, the balance due on the loan after the first monthly payment is made is $211,650.00 − $172.84, or $211,477.16.   ●

By repeatedly using the simple interest formula month to month on the unpaid balance, you could calculate the principal and the interest for all the payments, which is a tedious task. However, a list containing the payment number, payment on the interest, payment on the principal, and balance of the loan, can be prepared using a computer. Such a list is called a loan *amortization schedule*. One way to obtain an amortization schedule is by using a computer spreadsheet program. Another way is to access an amortization "calculator" program on the Internet. A part of the amortization schedule for the Martins' loan in Example 3 is given in Table 11.5. This schedule was generated from an Internet site called *Monthly Mortgage Payment Calculator* (*www.hsh.com*). Note that the monthly payment in Table 11.5 ($1408.11) is slightly more than the monthly payment we calculated using Table 11.4 ($1407.47). This difference is due to a rounding error that occurs when estimating the monthly principal and interest payment from Table 11.4.

Table 11.5 Amortization Schedule

Annual % Rate: 7.0			
Loan: $211,650		**Monthly Payment: $1408.11**	
Periods: 360		**Term: Years 30, Months 0**	
Payment Number	**Interest**	**Principal**	**Balance of Loan**
1	$1234.62	$173.49	$211,476.51
2	$1233.61	$174.50	$211,302.01
3	$1232.60	$175.52	$211,126.49
4	$1231.57	$176.54	$210,949.95
11	$1224.24	$183.88	$209,684.99
12	$1223.16	$184.95	$209,500.04
119	$1063.49	$344.62	$181,968.54
120	$1061.48	$346.63	$181,621.91
239	$715.54	$692.57	$121,972.22
240	$711.50	$696.61	$121,275.62
359	$16.29	$1391.83	$1,399.95
360	$8.17	$1399.95	$0.00

Adjustable-Rate Mortgages

Now let's consider *adjustable-rate mortgages*, ARMs (also called *variable-rate mortgages*). The rules for ARMs vary from state to state and from bank to bank, so the material presented on ARMs may not apply in your state or at your local lending institution. Generally, with adjustable-rate mortgages, the monthly mortgage payment remains the same for a 1-, 2-, or 5-year period even though the interest rate of the mortgage may change every 3 months, 6 months, or some other predetermined period. The interest rate for an adjustable-rate mortgage may be based on an index that is determined by the Federal Home Loan Bank Association or on the interest rate of a 3-month, 6-month, or 1-year Treasury bill. The interest rate of a 3-month Treasury bill may change every 3 months, the interest rate on a 6-month Treasury bill may change every 6 months, and so on. When the base is a Treasury bill, the actual interest rate charged for the mortgage is often determined by adding 3% to $3\frac{1}{2}$%, called the *add on rate* or *margin*, to the rate of the Treasury bill. Thus, if the rate of the Treasury bill is 6% and the add on rate is 3%, the interest rate charged is 9%.

EXAMPLE ❹ *An Adjustable-Rate Mortgage*

Tony and Keisha Torrence purchased a house for $115,000 with a down payment of $23,100. They obtained a 30-year adjustable-rate mortgage with the following terms. The interest rate is based on a 6-month Treasury bill. The interest rate charged is 3% above the interest rate of the 6-month Treasury bill (3% is the add on rate). The interest rate is adjusted every 6 months on the date of adjustment. The interest rate will not change more than 1% (up or down) when it is adjusted. The maximum interest rate for the duration of the loan is 12%. There is no lower limit on the interest rate. The initial mortgage interest rate is 5.5%, and the monthly payments (including principal and interest) are adjusted every 5 years.

a) Determine the initial monthly payment.

b) Determine the adjusted interest rate in 6 months if the interest rate on the Treasury bill at that time is 2%.

SOLUTION

a) To determine the initial monthly payment of interest and principal, divide the amount of the loan, $115,000 − $23,100 = $91,900, by $1000. The result is 91.9. Now multiply the number of thousands of dollars of mortgage, 91.9, by the value found in Table 11.4 with r = 5.5% for 30 years. The value found in the table is $5.68.

$$\$5.68 \times 91.9 \approx \$521.99$$

Thus, the initial monthly payment for principal and interest is $521.99. This amount will not change for the first 5 years of the mortgage.

b) The adjusted interest rate in 6 months will be the Treasury bill rate plus the add on rate.

$$2\% + 3\% = 5\% \qquad\qquad \bullet$$

In Example 4(b), note that the rate after 6 months, 5%, is lower than the initial rate of 5.5%. Since the monthly payment remains the same, the additional money paid the bank is applied to reduce the principal. The monthly interest and principal payment of $521.99 would pay off the loan in 30 years if the interest remained constant at 5.5%. What happens if the interest rate drops and stays lower than the initial 5.5% for the length of the loan? In this case, at the end of each 5-year period the bank reduces the monthly payment so that the loan will be paid off in 30 years. What happens if the interest rate increases above the initial 5.5% rate? In that case, part of, or if necessary, all the mortgage payment that would normally go toward repaying

MATHEMATICS TODAY

Credit Scores

When consumers borrow money by obtaining car loans, home mortgages, or other loans or through the use of credit cards, the lender takes a risk by giving money to the borrower in exchange for the borrower's promise to repay the loan plus interest. To minimize their risk, lenders do research on borrowers to determine the likelihood the borrowers will *default*, or fail to repay the loans. Lenders often obtain this research from a *consumer credit bureau*. Credit bureaus assign a credit score to an individual after researching the individual's credit history, income, age, and other factors. A high credit score suggests that the individual is more likely to repay a loan and often enables the borrower to get lower interest rates on loans. In March 2006, the three major credit bureaus in the United States—Equifax, Experian, and TransUnion—announced that they would all use a new system to establish a credit score called a *VantageScore*. You can find your own credit score online at www.creditscoring.com. The following table shows the range of credit scores, grades, and credit outlook for these scores.

Score	Grade	Credit Outlook
901–990	A	Excellent rating; low interest rate
801–900	B	Good rating but short of excellent
701–800	C	Fair rating; moderate risk
601–700	D	Poor rating; higher risk and high interest rate
501–600	F	Lowest rating; highest risk; may not qualify for loan

the principal would be used to meet the interest obligation. At the end of the 5-year period, the bank will increase the monthly payment so that the loan can be repaid by the end of the 30-year period. Or the bank may increase the time period of the loan beyond 30 years so that the monthly payment is affordable.

To prevent rapid increases in interest rates, some banks have a rate cap. A *rate cap* limits the maximum amount the interest rate may change. A *periodic rate cap* limits the amount the interest rate may increase in any one period. For example, your mortgage could provide that even if the index increases by 2% in 1 year, your rate can only go up 1% per year. An *aggregate rate cap* limits the interest rate increase and decrease over the entire life of the loan. If the initial interest rate is 6% and the aggregate rate cap is 2%, the interest rate could go no higher than 8% and no lower than 4% over the life of the mortgage. A *payment cap* limits the amount the monthly payment may change but does not limit changes in interest rates. If interest rates increase rapidly on a loan with a payment cap, the monthly payment may not be large enough to pay the monthly principal and interest on the loan. If that happens, the borrower could end up paying interest on interest.

Other Types of Mortgages

Conventional mortgages and adjustable-rate mortgages are not the only methods of financing the purchase of a house. Next, we briefly describe four other methods and then briefly discuss home equity loans.

FHA Mortgage   A house may be purchased with a smaller down payment than with a conventional mortgage if the buyer qualifies for a Federal Housing Authority (FHA) loan. The loan application is made through a lender such as a bank, credit union, or mortgage broker, but the FHA guarantees the loan against default by the borrower. This guarantee by the FHA, along with the smaller down payment, makes it possible for many people who might not qualify for a conventional mortgage to buy a home. The down payment for an FHA loan may be as low as 3% of the purchase price, rather than the standard 5% or higher. FHA loans also allow the buyer to have slightly higher mortgage payments than with a conventional loan. Typically, lenders will approve an FHA mortgage with a monthly mortgage payment of 29% of the lenders' adjusted monthly income as opposed to 28% with a conventional loan. Furthermore, many of the costs that are paid by the buyer at the closing of a conventional mortgage may be added to the amount of money borrowed, again making it easier for the buyer to obtain a mortgage to buy a house.

Another advantage is that FHA loans can be assumed at the original rate of interest on the loan. For example, if you purchase a home today that already has a 6% FHA mortgage, you, the new buyer, can assume that 6% mortgage regardless of the current interest rates. However, to be able to assume a mortgage, the purchaser must be able to make a down payment equal to the difference between the purchase price and the balance due on the original mortgage. There are some drawbacks to an FHA loan. Each county in the United States has a maximum house purchase price that can be financed through an FHA loan. Also, because the buyer pays such a small down payment, the borrower must pay for loan insurance. FHA borrowers must pay a one-time upfront insurance premium that is 1.5% of the loan amount. The borrower also pays an annual premium of 0.5% of the balance due on the loan amount that can be added to the monthly mortgage amount. Even with these drawbacks, FHA loans are the only mortgage option for many people seeking to buy a home and are very popular with first-time home buyers.

VA Mortgage   A veteran certified by the Department of Veterans Affairs (VA) who wants to purchase a house can apply for a mortgage with a bank or lending institution. The individual must meet the requirements set by the bank or lending institution for a mortgage. The VA guarantees repayment of a certain percentage of the loan obtained by the individual should the individual default on the loan. For example, if the loan is less than $45,000, the VA guarantees 50% of the loan. For loans from $45,000 to $144,000, the VA guarantees the lesser of 40% of the loan or $36,000. Since the bank has this guarantee from the VA, qualified veterans often do not have to make a down payment. A veteran who can make the monthly mortgage payments may therefore obtain a certain mortgage without a down payment.

The government sets the maximum interest rate that the lender may charge. A VA loan is always assumable. With a VA loan, the seller may be asked to pay points, but there is no monthly insurance premium.

Graduated Payment Mortgage (GPM)   A GPM mortgage is designed so that for the first 5 to 10 years the size of the mortgage payment is smaller than the payments for the remaining time of the mortgage. After the first 5- to 10-year period, the mortgage payments are increased. The mortgage payments then remain constant for the duration of the mortgage. Depending on the size of the mortgage, the lender might find that the monthly payments made for the first few years may actually be less than the interest owed on the loan for those few years. The interest not paid during the first few years of the mortgage is then added to the original loan. This type of loan is strictly for those who are confident their annual incomes will increase as rapidly as the mortgage payments do.

Balloon-Payment Mortgage (BPM)   The BPM type of loan could be for the person who needs time to find a permanent loan. It may work this way: The individual pays the interest for 3 to 8 years, the period of the loan. At the end of the 3- to 8-year period, the buyer must repay the entire remaining principal unless the lender agrees to a loan extension. Balloon-payment loans generally offer lower rates than conventional mortgages. This type of loan may be advantageous if the buyer plans to sell the house before the maturity date of the balloon-payment mortgage.

Home Equity Loans   As you make monthly payments and pay off the principal you owe on your home, you are said to be gaining *equity* in your home. **Equity** is the difference between the appraised value of your home and your loan balance. This equity can be used as collateral in obtaining a loan. Such a loan is referred to as a **home equity loan** or a **second mortgage**. One advantage of home equity loans over other types of loans (such as installment loans) is that the interest charged on a home equity loan is often tax deductible on federal income taxes. Home equity loans are commonly used for home improvements, for bill consolidation, or to pay for college education expenses.

For further information about the types of mortgages or loans discussed in this section, consult a loan officer at a local bank, a savings and loan, or a credit union. Additional information on buying a house may be obtained from the U.S. Government Printing Office in Pueblo, CO 81009. This information is also available on the Internet at www.access.gpo.gov.

SECTION 11.5 EXERCISES

CONCEPT/WRITING EXERCISES

1. What is a mortgage?

2. What is a down payment?

3. What is the difference between a variable-rate mortgage and a conventional mortgage?

4. a) What are points in a mortgage agreement?

   b) Explain how to determine the cost of x points.

5. Explain how to determine a buyer's adjusted monthly income.

6. What is an add on rate, or margin?

7. What is an amortization schedule?

8. Who insures an FHA loan? Who provides the money for an FHA loan?

9. What is equity?

10. What is a home equity loan?

PROBLEM SOLVING

11. *Buying a House* Greg Tobin is buying a house selling for $350,000. His bank requires him to make a 20% down payment. The current mortgage rate is 6.5%.

    a) Determine the amount of the required down payment.

    b) Determine the monthly principal and interest payment for a 15-year mortgage with a 20% down payment.

12. *Buying a Condo* Lauren Morse is buying a condominium for $289,000. Her credit union is requiring a down payment of 10%. The current mortgage rate is 8.0%.

    a) Determine the amount of the required down payment.

    b) Determine the monthly principal and interest payment for a 30-year mortgage with the 10% down payment.

13. *Buying a Townhouse* Karen Guardino is purchasing a brownstone townhouse in Brooklyn for $1,750,000. The mortgage broker she is working with is requiring her to make a 15% down payment. The current mortgage rate is 6.0%.

    a) Determine the amount of the required down payment.

    b) Determine the monthly principal and interest payment for a 25-year loan with a 15% down payment.

▲ See Exercise 13

14. *Buying a First Home* Stacie Williams is purchasing her first home for $275,000. She is obtaining an FHA mortgage through her credit union and is required to make a 3% down payment. The current mortgage rate is 7.5%.

    a) Determine the amount of the required down payment.

    b) Determine the monthly principal and interest payment for a 30-year loan with a 3% down payment.

15. *Paying Points* Martha Cutler is buying a house selling for $195,000. The bank is requiring a minimum down payment of 20%. To obtain a 20-year mortgage at 6% interest, she must pay 2 points at the time of closing.

    a) What is the required down payment?

    b) With the 20% down payment, what is the amount of the mortgage?

    c) What is the cost of the 2 points?

16. *Down Payment and Points* The Nicols are buying a house selling for $245,000. They pay a down payment of $45,000 from the sale of their current house. To obtain a 15-year mortgage at 4.5% interest, the Nicols must pay 1.5 points at the time of closing.

    a) What is the amount of the mortgage?

    b) What is the cost of the 1.5 points?

17. *Qualifying for a Mortgage* Pietr and Helga Guenther's gross monthly income is $3200. They have 25 remaining car payments of $335. The Guenthers are applying for a 15-year, $150,000 mortgage at 5% interest to buy a new house. The taxes and insurance on the house are $225 per month. Their credit union will approve a loan that has a total monthly mortgage payment of principal, interest, property taxes, and homeowners' insurance that is less than or equal to 28% of their adjusted monthly income.

a) Determine 28% of the Guenther's adjusted monthly income.

b) Determine the Guenther's total monthly mortgage payment, including principal, interest, taxes, and homeowners' insurance.

c) Do the Guenthers qualify for this mortgage?

18. *Qualifying for a Mortgage* Ting-Fang and Su-hua Zheng's gross monthly income is $4100. They have 18 remaining boat payments of $505. The Zhengs are applying for a 20-year, $275,000 mortgage at 9% interest to buy a new house. The taxes and insurance on the house are $425 per month. Their bank will approve a loan that has a total monthly mortgage payment of principal, interest, property taxes, and homeowners' insurance that is less than or equal to 28% of their adjusted monthly income.

a) Determine 28% of the Zheng's adjusted monthly income.

b) Determine the Zheng's total monthly mortgage payment, including principal, interest, taxes, and insurance.

c) Do the Zhengs qualify for this mortgage?

19. *A 30-Year Conventional Mortgage* Ingrid Holzner obtains a 30-year, $63,750 conventional mortgage at 8.5% on a house selling for $75,000. Her monthly payment, including principal and interest, is $490.24. She pays 0 points.

a) Determine the total amount Ingrid will pay for her house.

b) How much of the cost will be interest?

c) How much of the first payment on the mortgage is applied to the principal?

20. *A 25-Year Conventional Mortgage* Mr. and Mrs. Alan Bell obtain a 25-year, $110,000 conventional mortgage at 10.5% on a house selling for $160,000. Their monthly mortgage payment, including principal and interest, is $1038.40. They also pay 2 points at closing.

a) Determine the total amount the Bells will pay for their house.

b) How much of the cost will be interest (including the 2 points)?

c) How much of the first payment on the mortgage is applied to the principal?

21. *Evaluating a Loan Request* The Rosens found a house selling for $113,500. The taxes on the house are $1200 per year, and insurance is $320 per year. They are requesting a conventional loan from the local bank. The bank is currently requiring a 15% down payment and 3 points, and the interest rate is 10%. The Rosen's gross monthly income is $4750. They have more than 10 monthly payments remaining on a car, a boat, and furniture. The total monthly payments for these items is $420. Their bank will approve a loan that has a total monthly mortgage payment of principal, interest, property taxes, and homeowners' insurance that is less than or equal to 28% of their adjusted monthly income.

a) Determine the required down payment.

b) Determine the cost of the 3 points.

c) Determine 28% of their adjusted monthly income.

d) Determine the monthly payments of principal and interest for a 20-year loan.

e) Determine their total monthly payment, including homeowners' insurance and taxes.

f) Determine whether the Rosens qualify for the 20-year loan.

g) Determine how much of the first payment on the loan is applied to the principal.

22. *Evaluating a Loan Request* Kathy Fields wants to buy a condominium selling for $95,000. The taxes on the property are $1500 per year, and homeowners' insurance is $336 per year. Kathy's gross monthly income is $4000. She has 15 monthly payments of $135 remaining on her van. The bank is requiring 20% down and is charging 9.5% interest with no points. Her bank will approve a loan that has a total monthly mortgage payment of principal, interest, property taxes, and homeowners' insurance that is less than or equal to 28% of her adjusted monthly income.

a) Determine the required down payment.

b) Determine 28% of her adjusted monthly income.

c) Determine the monthly payment of principal and interest for a 25-year loan.

d) Determine her total monthly payment, including homeowners' insurance and taxes.

e) Does Kathy qualify for the loan?

f) Determine how much of the first payment on the mortgage is applied to the principal.

g) Determine the total amount she pays for the condominium with a 25-year conventional loan. (Do not include taxes or homeowners' insurance.)

h) Determine the total interest paid for the 25-year loan.

23. *Comparing Loans* The Riveras are negotiating with two banks for a mortgage to buy a house selling for $105,000. The terms at bank A are a 10% down payment, an interest rate of 10%, a 30-year conventional mortgage, and 3 points to be paid at the time of closing. The terms at bank B are a 20% down payment, an interest rate of 11.5%, a 25-year conventional mortgage, and no points. Which loan should the Riveras select for the total cost of the house to be less?

24. *Comparing Loans* Paul Westerberg is negotiating with two credit unions for a mortgage to buy a condominium selling for $525,000. The terms at Grant County Teacher's Credit Union are a 20% down payment, an interest rate of 7.5%, a 15-year mortgage, and 1 point to be paid at the time of closing. The terms at Sinnipee Consumer's Credit Union are a 15% down payment, an interest rate of 8.5%, a 20-year mortgage, and no points. Which loan should Paul select for the total cost of the down payment, points, and total mortgage payments of the house to be less?

CHALLENGE PROBLEMS/GROUP ACTIVITIES

25. *Changing Lengths of Mortgages* Rose Hulman and her husband, George Mason, are purchasing a home for $450,000. Their bank requires them to pay a down payment of 20%. The current mortgage rate is 10%, and they are required to pay 1 point at the time of closing. Determine the total amount Rose and George will pay for their house, including principal, interest, down payment, and points (do not include taxes and homeowners' insurance) if the length of their mortgage is

a) 10 years.

b) 20 years.

c) 30 years.

26. *Principal and Interest Formula* Rather than using Table 11.4, the following formula can be used to calculate principal and interest payments on a mortgage.

$$m = \frac{P\left(\dfrac{r}{n}\right)}{1 - \left(1 + \dfrac{r}{n}\right)^{-n \cdot t}}$$

In the formula, m is the principal and interest payment, P is the amount of the mortgage, r is mortgage rate in decimal form, n is the number of payments per year, and t is the time in years. This formula allows you to calculate principal and interest payments that could not be calculated using Table 11.4. For a mortgage of $200,000, calculate the principal and interest payment with each of the following conditions.

a) $r = 6.75\%$, monthly payments for 30 years

b) $r = 13\%$, weekly payments for 15 years

c) $r = 9\%$, monthly payments for 40 years

d) $r = 3.9\%$, semimonthly (use $n = 24$) payments for 50 years

27. *An Adjustable-Rate Mortgage* The Simpsons purchased a house for $105,000 with a down payment of $5000. They obtained a 30-year adjustable-rate mortgage. The terms of the mortgage are as follows: The interest rate is based on a 3-month Treasury bill, the interest rate charged is 3.25% above the rate of the Treasury bill on the date of adjustment, the interest rate is adjusted every 3 months, the interest rate will not change more than 1% (up or down) when the interest rate is adjusted, the maximum interest rate that can be charged for the duration of the loan is 16%, there is no lower limit on the interest rate, the initial mortgage interest rate is 9%, and the monthly payment of interest and principal is adjusted semiannually.

a) Determine the initial monthly payment for interest and principal.

b) Determine an amortization schedule for months 1–3.

c) Determine the interest rate for months 4–6 if the interest rate on the Treasury bill at the time is 6.13%.

d) Determine an amortization schedule for months 4–6.

e) Determine the interest rate for months 7–9 if the interest rate on the Treasury bill at the time is 6.21%.

28. *An Adjustable-Rate Mortgage* The Bretz family purchased a house for $95,000 with a down payment of $13,000. They obtained a 30-year adjustable-rate mortgage. The terms of the mortgage are as follows: The interest rate is based on a 3-month Treasury bill, the interest rate charged is 3.25% above the rate of the Treasury bill on the date of adjustment, the interest rate is adjusted every 3 months, the interest rate will not change more than 1% (up or down) when the interest rate is adjusted, the maximum interest rate that can be charged for the duration of the loan is 16%, there is no lower limit on the interest rate, the initial mortgage interest rate is 8.5%, and the monthly payment of interest and principal is adjusted annually.

a) Determine the initial monthly payment for interest and principal.

b) Determine the interest rate in 3 months if the interest rate on the Treasury bill at the time is 5.65%.

c) Determine the interest rate in 6 months if the interest rate on the Treasury bill at the time is 4.85%.

29. *Comparing Mortgages* The Hassads are applying for a $90,000 mortgage. They can choose between a 30-year conventional mortgage and a 30-year variable-rate mortgage. The interest rate on a 30-year conventional mortgage is 9.5%. The terms of the variable-rate mortgage are 6.5% interest rate the first year, an annual cap of 1%, and an aggregate cap of 6%. The interest rates and the mortgage payments are adjusted annually. Assume that the interest rates for the variable-rate mortgage increase by the maximum amount each year. Then the monthly mortgage payments for the variable-rate mortgage for years 1–6 are $568.80, $629.10, $692.10, $756.90, $823.50, and $891.00, respectively.

a) Knowing that they will be in the house for only 6 years, which mortgage, the conventional mortgage or the variable-rate mortgage, will be the least expensive for that period?

b) How much will they save by choosing the less expensive mortgage?

INTERNET/RESEARCH ACTIVITIES

30. *Finding Your Dream Home* Examine a local newspaper to find your "dream home" and note the asking price. Next contact a loan officer from your local bank, savings and loan, or credit union. Assuming you can make a 20% down payment, determine the interest rates for a 15-year and a 30-year mortgage. Use an amortization "calculator" (see page 698) with the data you obtained to print amortizations schedules. Compare the monthly payments with the 15-year mortgage to those of the 30-year mortgage. Compare the total interest costs of the 15-year mortgage with those of the 30-year mortgage. Write a report summarizing your findings.

31. *Closing Costs* An important part of buying a house is the closing. The exact procedures for the closing differ with individual cases and in different parts of the country. In any closing, however, both the buyer and the seller have certain expenses. To determine what is involved in the closing of a property in your community, contact a lawyer, a real estate agent, or a banker. Explain that you are a student and that your objective is to understand the procedure for closing a real estate purchase and the costs to both buyer and seller. Select a specific piece of property that is for sale. Use the asking price to determine the total closing costs to both buyer and seller. The following is a partial list of the most common closing costs. Consider them in your research.

a) Fee for title search and title insurance
b) Credit report on buyer
c) Fees to the lender for services in granting the loan
d) Fee for property survey
e) Fee for recording of the deed
f) Appraisal fee
g) Lawyer's fee
h) Escrow accounts (taxes, insurance)
i) Mortgage assumption fee

11.6 ORDINARY ANNUITIES, SINKING FUNDS, AND RETIREMENT INVESTMENTS

▲ Saving money on a regular basis may lead to a comfortable retirement.

In Sections 11.2 and 11.3, we studied simple and compound interest, respectively. In both sections, we answered questions involving the investment of one lump sum of money. Although such questions may be relevant to our everyday financial matters, more appropriate questions might involve investing smaller amounts of money on a regular basis over a longer period of time. For example, the Brabsons, who are planning for retirement, might ask the question: If we deposit $100 a month at 5% interest compounded monthly, how much money will we accumulate after 30 years when we're ready to retire? Also consider the Weismans, who are saving for their child's college education. They might ask the question: If we know we will need $50,000 to send our child to college in 10 years, how much should we invest each month beginning now in an account paying 6% interest compounded monthly? To answer these and similar questions, we will study two investments, the *annuity* and the *sinking fund*.

Annuities

An **annuity** is an account into which, or out of which, a sequence of scheduled payments is made.

There are many different types of annuities. An annuity may be an investment account that you have with a bank, insurance company, or financial management firm. Annuities may contain investments in stocks, bonds, mutual funds, money market accounts, and other types of investments. Annuities are often used to save for long-term goals such as saving money for college or for retirement. An annuity can also be used to provide long-term regular payments to individuals. Lottery jackpots and professional athletes' salaries are often paid out over time from annuities. Retirees may invest some of their retirement savings into an annuity and then receive monthly payments that come from that annuity. We will focus primarily on two basic types of annuities that are used as investment accounts, *ordinary annuities* and *sinking funds*. Later in this section, we will discuss other types of annuities.

An annuity into which equal payments are made at regular intervals, with the interest compounded at the end of each interval and with a fixed interest rate for each compounding period, is called an **ordinary annuity** or a **fixed annuity**.

For example, the Brabsons, mentioned in the opening paragraph of this section, are asking a question that may involve an ordinary annuity. They plan to make $100 payments each month into an account that pays 5% interest compounded monthly. With an ordinary annuity, the payment period and frequency of the compounding are the same, so the Brabsons make *monthly* payments *and* the interest is compounded *monthly*. They would like to know how much money would accumulate in this annuity after 30 years. The amount of money that is present in an ordinary annuity after *t* years is known as the *accumulated amount* or the *future value* of an annuity.

To determine the accumulated amount of an ordinary annuity, we could make many individual calculations using the compound interest formula discussed in Section 11.3. Such calculations would be quite tedious and time consuming even with the assistance of a scientific calculator. A better way is to use a computer to generate a spreadsheet. A portion of such a spreadsheet is shown below. This spreadsheet was generated using the data from the example involving the Brabsons. Note that the Brabsons would make monthly payments for 30 years for a total of 360 payments. Also note that had the Brabsons simply put $100 in their cookie jar each month, their total amount at the end of 30 years would have been $100 × 360, or $36,000. Instead, by investing in an interest-bearing annuity, they will have accumulated $83,225.86 after 30 years.

Accumulated Amount in an Ordinary Annuity with $100 Monthly Investments with Interest Rate 5% Compounded Monthly

Payment Number	Monthly Investment	Interest	Accumulated Amount
1	$100.00	$ —	$ 100.00
2	$100.00	$ 0.42	$ 200.42
3	$100.00	$ 0.84	$ 301.25
4	$100.00	$ 1.26	$ 402.51
5	$100.00	$ 1.68	$ 504.18
6	$100.00	$ 2.10	$ 606.28
7	$100.00	$ 2.53	$ 708.81
8	$100.00	$ 2.95	$ 811.76
9	$100.00	$ 3.38	$ 915.15
10	$100.00	$ 3.81	$ 1,018.96
⋮	⋮	⋮	⋮
351	$100.00	$328.58	$79,287.40
352	$100.00	$330.36	$79,717.76
353	$100.00	$332.16	$80,149.92
354	$100.00	$333.96	$80,583.88
355	$100.00	$335.77	$81,019.65
356	$100.00	$337.58	$81,457.23
357	$100.00	$339.41	$81,896.63
358	$100.00	$341.24	$82,337.87
359	$100.00	$343.07	$82,780.94
360	$100.00	$344.92	$83,225.86

Another way, and perhaps a more efficient way, to calculate the accumulated amount in an ordinary annuity is to use the ordinary annuity formula.

ORDINARY ANNUITY FORMULA

The accumulated amount, A, of an ordinary annuity with payments of p dollars made n times per year, for t years, at interest rate, r, compounded at the end of each payment period is given by the formula

$$A = \frac{p\left[\left(1 + \frac{r}{n}\right)^{nt} - 1\right]}{\frac{r}{n}}$$

EXAMPLE ❶ *Using the Ordinary Annuity Formula*

Bill and Megan Lutes are depositing $250 each quarter in an ordinary annuity that pays 4% interest compounded quarterly. Determine the accumulated amount in this annuity after 35 years.

SOLUTION We will use the ordinary annuity formula. The payment, p, is $250, the interest rate, r, is 4% or 0.04, the account is compounded quarterly and the payments are made quarterly, so n is 4, and the number of years, t, is 35. We substitute these values into the formula to obtain the following.

$$A = \frac{p\left[\left(1 + \frac{r}{n}\right)^{nt} - 1\right]}{\frac{r}{n}}$$

$$= \frac{250\left[\left(1 + \frac{0.04}{4}\right)^{(4\cdot35)} - 1\right]}{\frac{0.04}{4}}$$

Substitute the given values into the formula.

$$= \frac{250[1.01^{140} - 1]}{0.01}$$

To evaluate 1.01^{140}, we use the y^x or the $\wedge$ key on a scientific calculator.

$$\approx \frac{250[4.027099 - 1]}{0.01}$$

We round 1.01^{140} to six decimal places.

$$\approx \frac{250(3.027099)}{0.01}$$

$$\approx \frac{756.77475}{0.01} \approx 75{,}677.475$$

Thus, there will be about $75,677.48 in Bill and Megan's annuity after 35 years. ●

Example 1 illustrated the power of establishing and maintaining a consistent investment plan and how using an ordinary annuity can help us save money.

TECHNOLOGY TIP

Scientific Calculator

In Example 1, we determined the accumulated amount by evaluating the expression

$$250\left[\dfrac{\left(1 + \dfrac{0.04}{4}\right)^{(4\cdot35)} - 1}{\dfrac{0.04}{4}}\right]$$

There are various ways to evaluate this expression on a scientific calculator. One method is to press the following sequence of keys.

$$(\ \ 250 \ \times \ (\ \ (\ \ 1 \ + \ 0.04 \ \div \ 4 \ ) \ y^x \ (\ \ 4 \ \times$$

$$35 \ ) \ - \ 1 \ ) \ ) \ \div \ (\ \ 0.04 \ \div \ 4 \ ) \ =$$

After the $=$ key is pressed, the answer 75677.48042 is displayed.

Another way to perform the calculations is to use the store and display keys (or memory keys). Read the manual that came with your calculator to learn how to use those keys.

Microsoft Excel

Read the Technology Tip on page 670. To determine the accumulated interest using Excel, place the letters p, r, n, t and A in cells A1 through E1, respectively. In cells A2 through D2, place the values for p, r, n, and t, respectively. In cell E2, use the formula box to add the formula

$$= (A2*((1 + B2/C2)\wedge(C2*D2) - 1))/(B2/C2)$$

After the Enter key is pressed, the result is as follows.

	A	B	C	D	E
1	p	r	n	t	A
2	250	0.04	4	35	75677.48
3					

⏮ ◀ ▶ ⏭ \ Sheet1 / Sheet2 / Sheet3 / |◀

Ready

Next we will study a specific type of annuity called a *sinking fund*.

Sinking Funds

We will return to the second question asked at the beginning of this section. Recall that the Weismans have a goal of saving $50,000 in 10 years for their child's college education and want to know how much they should begin to invest each month in an account paying 6% interest compounded monthly. The Weismans can help reach this goal by investing in a special kind of annuity called a *sinking fund*.

A **sinking fund** is a type of annuity in which the goal is to save a specific amount of money in a specific amount of time.

There are many types of sinking funds that are used for many different purposes by individuals, corporations, and governments. Historically, sinking funds were used by governments to set aside money to pay off bonds that were used to borrow large sums of money for major civic projects such as sewers, roads, or bridges. The term *sinking fund* referred to the *sinking* of the debt that was owed to repay the bonds. Our discussion of sinking funds will focus on answering questions similar to the question facing the Weismans. Because they are interested in saving a specific amount of money, $50,000, in a specific amount of time, 10 years, their investment can be considered a sinking fund.

> **TIMELY TIP** An ordinary annuity is used when you wish to determine the accumulated amount obtained over n years when you contribute a fixed amount each period. A sinking fund is used when you wish to determine how much money an investor must invest each period to reach an accumulated amount at a specific time.

To determine the amount of money to be invested in a sinking fund, we can solve the ordinary annuity formula, given on page 708, for the payment, p. Solving this formula for p gives us the following formula.

SINKING FUND PAYMENT FORMULA

$$p = \frac{A\left(\dfrac{r}{n}\right)}{\left(1 + \dfrac{r}{n}\right)^{nt} - 1}$$

In the formula, p is the payment needed to reach the accumulated amount, A. Payments are made n times per year, for t years, into a sinking fund with interest rate r, compounded n times per year.

⌐EXAMPLE ❷ *Using the Sinking Fund Payment Formula*

Richard Ries would like to have $30,000 in 5 years to buy a new grand piano. Richard decides to invest monthly in a sinking fund that pays 4.5% interest compounded monthly. How much should Richard invest in the sinking fund each month to accumulate $30,000 in 5 years?

SOLUTION Since we are looking for how much Richard should invest each month to obtain an accumulated amount, we will use the sinking fund payment formula. The accumulated amount, A, is $30,000, the interest rate, r, is 4.5%, the payments are made monthly and the interest is compounded monthly, so n is 12, and the number of years, t, is 5.

$$p = \frac{A\left(\dfrac{r}{n}\right)}{\left(1 + \dfrac{r}{n}\right)^{nt} - 1}$$

$$= \frac{30{,}000\left(\dfrac{0.045}{12}\right)}{\left(1 + \dfrac{0.045}{12}\right)^{(12 \cdot 5)} - 1}$$

$$\approx \frac{112.50}{0.2517958}$$

$$\approx 446.79061$$

We often round to the nearest cent, or in this case, to $446.79. However, because $446.79 is actually slightly *less* than the answer we obtained, we will round our answer *up* to $446.80 to ensure that Richard reaches his goal of $30,000. *In all our exercises involving sinking fund calculations, we will round our final answers up to the nearest cent to ensure that we reach the desired amount in the sinking fund.*

TECHNOLOGY TIP

Scientific Calculator

To determine the sinking fund payment in Example 2, we evaluated the expression

$$\frac{30{,}000\left(\dfrac{0.045}{12}\right)}{\left(1 + \dfrac{0.045}{12}\right)^{12 \cdot 5} - 1}$$

This expression can be evaluated on a scientific calculator as follows.

30000 $\boxed{\times}$ $\boxed{(}$ 0.045 $\boxed{\div}$ 12 $\boxed{)}$ $\boxed{\div}$ $\boxed{(}$ $\boxed{(}$ 1 $\boxed{+}$ 0.045 $\boxed{\div}$
12 $\boxed{)}$ $\boxed{y^x}$ $\boxed{(}$ 12 $\boxed{\times}$ 5 $\boxed{)}$ $\boxed{-}$ 1 $\boxed{)}$

After the $\boxed{=}$ key is pressed, the answer 446.7905772 is displayed.

Microsoft Excel

Read the Technology Tip on page 709 and follow a similar procedure using the variables A, r, n, t, and p, respectively. To evaluate the sinking fund payment, in cell E2 enter the formula

$$= \text{A2*(B2/C2)/((1 + B2/C2)\textasciicircum(C2*D2) - 1)}$$

After the Enter key is pressed, the answer 446.7906 is displayed in cell E2.

In Example 2, we can check our answer by substituting $446.80 back into the ordinary annuity formula. Doing this calculation with a scientific calculator gives us an accumulated amount of about $30,000.63. If we substitute $446.79 for A into the ordinary annuity formula, the accumulated amount would be about $29,999.96, which is slightly short of the desired goal of $30,000.

Other Annuities

Ordinary annuities and sinking funds are not the only types of annuities. We will now briefly discuss two other common types of annuities, *variable annuities* and *immediate annuities*.

Variable Annuities A *variable annuity* is an annuity that is invested in stocks, bonds, mutual funds, or other investments that do not provide a guaranteed interest rate. The term *variable* is used because the value of the annuity will vary depending on the performance of the investment options chosen by the investor. Like an ordinary annuity, an investor usually invests in a variable annuity by making regular periodic payments. Unlike an ordinary annuity, however, a variable annuity does not have a fixed rate of interest.

Most variable annuities are investments made to save for retirement. Upon retirement, the variable annuity investor generally receives periodic, usually monthly, payments for the rest of the investor's life. Often, variable annuity investors may name a designated person who will continue receiving the monthly payments after the investor's death. Such a person, usually a spouse or other family member, is called a *beneficiary*. A very attractive feature of variable annuities is that during the investment period, variable annuities are *tax-deferred*, meaning that no taxes are paid on the income and investment gains from the annuity until the investor begins receiving payments. Investors may also transfer money from one investment option to another within a variable annuity without paying taxes at the time of the transfer. Note, however, that payments received from a variable annuity after a person retires are generally subject to income taxes. One major drawback to investing in variable annuities is that the fees charged by the investment company are often higher than those charged for other investments. Another possible drawback of a variable annuity is that it is sometimes possible for an investor to lose money in a variable annuity. For example, if an investor invests in stocks that drop in price, the value of the annuity may decrease.

As mentioned, variable annuities are usually used to save for retirement. Next we will discuss another type of annuity that is commonly used once an investor nears or reaches retirement.

Immediate Annuities An *immediate annuity* is an annuity that is established with a lump sum of money for the purpose of providing the investor with regular, usually monthly, payments for the rest of the investor's life. In exchange for giving the investment company a lump sum of money, the investor is guaranteed to receive a monthly income for the duration of the investor's life.

An investor has many options when investing in an immediate annuity. One option is to receive monthly payments that increase a certain percentage each year. Another option is the ability to name a beneficiary who will continue receiving the monthly payments after the investor dies. A third option is a guarantee by the investment company that the investor's heirs will receive the remaining balance of the annuity upon the deaths of both the investor and the beneficiary. Although most of these options appear attractive to the investor, including any of these options in an immediate annuity contract may decrease the amount of the monthly payments.

Both variable annuities and immediate annuities have an array of options that carry several advantages and disadvantages. All these options are described in an informational brochure from the investment company called a *prospectus*. Furthermore, many Internet web sites such as www.morningstar.com and www.smartmoney.com have independent information about annuities and other investments. Before investing in any annuity, it is very important to do independent research. Thoroughly read the prospectus, and ask any questions you may have.

Other Retirement Savings Options

Annuities can be used to save for retirement, but depending on the investor's circumstances, other investment options for retirement savings may be more advantageous. We will briefly discuss several options here.

▲ It is never too early to start saving for retirement.

Individual Retirement Accounts   *Individual retirement accounts*, also known as *IRAs*, are savings accounts that allow individuals to invest up to a certain amount of money each year for the purpose of saving for retirement. IRAs have distinct tax advantages over ordinary savings accounts. There are several different kinds of IRAs, but most can be classified as either a *traditional IRA* or a *Roth IRA*. A *traditional IRA* is an IRA into which *pretax* money is invested, meaning that any money invested into a traditional IRA is not subject to income taxes. However, when money is withdrawn from a traditional IRA, the money *is* subject to income taxes. A *Roth IRA* is an IRA into which *posttax* money is invested, meaning that the investor has already paid income taxes on the money invested. When money is withdrawn from a Roth IRA after the investor reaches retirement age, the money *is not* subject to income taxes. Also, as of 2007, married couples with an adjusted gross income of more than $166,000 per year and single taxpayers with an adjusted gross income of more than $114,000 cannot invest in Roth IRAs.

Both types of IRAs have annual limits on the amount that can be invested. In 2008, investors age 49 or younger could invest an annual maximum of $5000 into a traditional or Roth IRA, whereas investors age 50 or older could invest an annual maximum of $6000. Both types of IRAs can contain a variety of investments, including, but not limited to, stocks, bonds, mutual funds, and money market funds. Depending on an investor's financial situation, there may be advantages of one type of IRA over the other. Therefore, before investing, it is important to do research or talk to a financial advisor.

401ks and 403bs   A *401k plan* is a retirement savings plan that allows employees of private companies to make contributions of pre–income tax dollars that are then pooled with other employees' money. These funds are then invested in a variety of stocks, bonds, money markets, and mutual funds. Although 401k plans share many of the same tax advantages as IRAs, they also have several distinct advantages. First, employees are generally allowed to invest more money annually in a 401k plan than in an IRA. Often, employees are eligible to invest up to 15% of their annual income into a 401k plan. A second advantage is that some employers will match employee contributions up to a certain limit. A *Roth 401k plan* is a retirement savings plan that differs from a 401k plan in that an investor first pays income tax on the money used to invest in a Roth 401k plan. Like a Roth IRA, money withdrawn from a Roth 401k plan after the individual reaches retirement age is not subject to income tax.

A *403b plan* is a retirement plan similar to a 401k plan, but it is available only to employees of schools, civil governments, hospitals, charities, and other not-for-profit organizations. A 403b plan shares many of the same advantages as 401k plans. One major difference, though, is that a 403b plan is not necessarily established by the employer as is a 401k plan. Rather, a 403b plan is frequently established independently by the employee with a representative of an investment or insurance company.

As you can see from this brief discussion of annuities, IRAs, 401k plans, and 403b plans, many important details can affect your decisions as you begin to invest for retirement. For more information, and for other types of retirement accounts not mentioned here, consult a financial advisor or visit a library or Internet web site. Regardless of the source of your information, it is very important to do your own research and ask many questions prior to investing.

SECTION 11.6 EXERCISES

CONCEPT/WRITING EXERCISES

1. What is an *annuity*?

2. In the ordinary annuity formula

$$A = \frac{p\left[\left(1 + \dfrac{r}{n}\right)^{nt} - 1\right]}{\dfrac{r}{n}}$$

state what each of the following variables represents: A, p, r, n, and t.

3. **a)** What is an *ordinary* annuity?

   b) What is a *sinking fund*?

4. What is a variable annuity?

5. What is an immediate annuity?

6. Explain the difference between a *traditional IRA* and a *Roth IRA*.

7. Explain the difference between a *401k* and a *Roth 401k*.

8. What is a *403b* plan?

PRACTICE THE SKILLS

In Exercises 9–12, use the ordinary annuity formula

$$A = \frac{p\left[\left(1 + \dfrac{r}{n}\right)^{nt} - 1\right]}{\dfrac{r}{n}}$$

to determine the accumulated amount in each annuity. Round all answers to the nearest cent.

9. $3000 invested annually for 30 years at 5% compounded annually

10. $800 invested semiannually for 25 years at 7% compounded semiannually

11. $400 invested quarterly for 35 years at 8% compounded quarterly

12. $200 invested monthly for 40 years at 6% compounded monthly

In Exercises 13–16, use the sinking fund formula

$$p = \frac{A\left(\dfrac{r}{n}\right)}{\left(1 + \dfrac{r}{n}\right)^{nt} - 1}$$

to determine the payment needed to reach the accumulated amount. To ensure that enough is invested each period, round each answer up to the next cent.

13. Semiannual payments with 4% interest compounded semiannually for 15 years to accumulate $20,000

14. Annual payments with 5% interest compounded annually for 20 years to accumulate $35,000

15. Monthly payments with 6% interest compounded monthly for 35 years to accumulate $250,000

16. Quarterly payments with 10% interest compounded quarterly for 40 years to accumulate $1,000,000

PROBLEM SOLVING

In Exercises 17–20, round all answers to the nearest cent.

17. *Retirement Savings* To save for retirement, Jana Bryant invests $50 each month in an ordinary annuity with 3% interest compounded monthly. Determine the accumulated amount in Jana's annuity after 35 years.

18. *Starting a Business* To save money to start a new business, Sandra Jessee invests $200 each quarter in an ordinary annuity with 5% interest compounded quarterly. Determine the accumulated amount in Sandra's annuity after 20 years.

19. *Saving for Graduate School* To save for graduate school, Ena Salter invests $2000 semiannually in an ordinary annuity with 7% interest compounded semiannually. Determine the accumulated amount in Ena's annuity after 10 years.

20. *Grandchild's College Expenses* To save for their grandchildren's college expenses, Ken and Laura Machol invest $500 each month in an ordinary annuity with 6% interest compounded monthly. Determine the accumulated amount in the Machol's annuity after 18 years.

In Exercises 21–24, round all answers up to the next cent.

21. *Saving for a New Car* Erica Atalla would like to have $25,000 to buy a new car in 5 years. To accumulate $25,000 in 5 years, how much should she invest monthly in a sinking fund with 9% interest compounded monthly?

22. *Campaign Fund-Raising* Ashley Smith would like to run for Congress in 10 years and would like to have $500,000 to spend on campaign expenses. How much should she invest monthly in a sinking fund with 6% interest compounded monthly to accumulate $500,000 in 10 years?

▲ See Exercise 22

23. *Becoming a Millionaire* Becky Anderson would like to be a millionaire in 20 years. How much would she need to invest quarterly in a sinking fund paying 8% interest compounded quarterly to accumulate $1,000,000 in 20 years?

24. *Repaying a Debt* The city of Corpus Christi wishes to repay a debt of $25,000,000 in 50 years. How much should the city invest semiannually in a sinking fund paying 7.5% interest compounded semiannually to accumulate $25,000,000 in 50 years?

CHALLENGE PROBLEMS/GROUP ACTIVITIES

25. *Don't Wait to Invest* The purpose of this exercise is to demonstrate that it is much more beneficial to begin to invest for retirement early than it is to wait. We will compare the investments of two investors, Alberto and Zachary. Alberto begins to invest immediately after graduating from college, whereas Zachary waits 10 years to begin investing. For the purposes of this demonstration, we will assume that the interest rate remains the same over the entire period of time.

   a) From age 25 to 35, Alberto invests $100 per month in an annuity that pays 12% interest compounded monthly. Determine the accumulated amount in Alberto's annuity after 10 years.

   b) At age 35, Alberto stops making monthly payments into his annuity, but he takes the money he accumulated in his annuity and invests it in a savings account with an interest rate of 12% compounded monthly. Using the compound interest formula $A = p\left(1 + \dfrac{r}{n}\right)^{n \cdot t}$, determine the amount Alberto has in his savings account 30 years later when he retires at age 65.

   c) From age 35 to 65, Zachary invests $100 per month in an annuity that pays 12% interest compounded monthly. Determine the accumulated amount in Zachary's annuity after 30 years.

   d) Determine the total amount Alberto invested during the 10 years from age 25 to 35.

e) Determine the total amount Zachary invested during the 30 years from age 35 to 65.

f) Which investor has more money from the investments described here upon retirement?

INTERNET/RESEARCH ACTIVITIES

26. *Investing in Stocks* Using the Internet, financial magazines, books, and other resources, write a paper on investing in stocks (see the Did You Know? on page 707). Include in your paper a description of the following terms: *New York Stock Exchange*, *National Association of Securities Dealers Automated Quotations (NASDAQ)*, *dividends*, and *capital gains*.

27. *Investing in Bonds* Using the Internet, financial magazines, books, and other resources, write a paper on investing in bonds (see the Did You Know? on page 708). Include in your paper a description of the following terms: *Treasury bonds*, *corporate bonds*, and *municipal bonds*.

28. *Investing in Mutual Funds* Using the Internet, financial magazines, books, and other resources, write a paper on investing in mutual funds (see the Did You Know? on page 712). Include in your paper a description of the following terms: *net asset value (NAV)*, *portfolio manager*, and *expense ratios*.

29. *Financial Advisor Interview* Annuities, IRAs, 401k plans, 403b plans, SEP accounts, and Keogh accounts are all types of investments that can be used to save for retirement. Interview a financial advisor to discuss each of these investment types. Write a paper in which you explain the differences among the different investment types and describe the advantages of each type of investment over the others.

CHAPTER ⑪ SUMMARY

IMPORTANT FACTS

ORDINARY INTEREST

When computing ordinary interest, each month is considered to have 30 days and a year is considered to have 360 days.

UNITED STATES RULE

If a partial payment is made on a loan, interest is computed on the principal from the first day of the loan until the date of the partial payment. The partial payment is used to pay the interest first; then the rest of the payment is used to reduce the principal.

BANKER'S RULE

When computing interest with the Banker's rule, a year is considered to have 360 days and any fractional part of a year is the exact number of days.

SIMPLE INTEREST FORMULA

Interest = principal × rate × time or $i = prt$

PERCENT CHANGE

$$\text{Percent change} = \frac{\text{amount in latest period} - \text{amount in previous period}}{\text{amount in previous period}} \times 100$$

PERCENT MARKUP

$$\text{Percent markup on cost} = \frac{\text{selling price} - \text{dealer's cost}}{\text{dealer's cost}} \times 100$$

COMPOUND INTEREST FORMULA

$$A = p\left(1 + \frac{r}{n}\right)^{nt}$$

PRESENT VALUE FORMULA

$$p = \frac{A}{\left(1 + \frac{r}{n}\right)^{nt}}$$

ACTUARIAL METHOD

$$u = \frac{n \cdot P \cdot V}{100 + V}$$

ORDINARY ANNUITY FORMULA

$$A = \frac{p\left[\left(1 + \frac{r}{n}\right)^{nt} - 1\right]}{\frac{r}{n}}$$

SINKING FUND PAYMENT FORMULA

$$p = \frac{A\left(\frac{r}{n}\right)}{\left(1 + \frac{r}{n}\right)^{nt} - 1}$$

CHAPTER ⑪ REVIEW EXERCISES

11.1

Change the number to a percent. Express your answer to the nearest tenth of a percent.

1. $\dfrac{3}{4}$ 2. $\dfrac{5}{6}$ 3. $\dfrac{5}{8}$

4. 0.041   5. 0.0098   6. 3.141

Change the percent to a decimal number.

7. 9%   8. 14.1%   9. 123%

10. $\dfrac{1}{4}\%$ 11. $\dfrac{5}{6}\%$ 12. 0.00045%

13. *Lambeau Field* Before undergoing renovations, the seating capacity at Lambeau Field (in Green Bay, Wisconsin) was 60,790. After the renovations were complete the seating capacity was 71,500. Determine the percent increase (to the nearest tenth of a percent) in the seating capacity of Lambeau Field.

▲ Lambeau Field

14. *Salary Increase* Charlotte Newsom had a salary of $46,200 in 2006 and a salary of $51,300 in 2007. Determine the percent increase in Charlotte's salary from 2006 to 2007.

In Exercises 15–17, solve for the unknown quantity.

15. What percent of 80 is 15?

16. Forty-four is 55% of what number?

17. What is 17% of 540?

18. *Tipping* At Empress Garden Restaurant, Vishnu and Krishna's bill comes to $42.79, including tax. If they wish to leave a 15% tip on the total bill, how much should they tip the waiter?

19. *Increased Membership* If the number of people in your chess club increased by 20%, or 8 people, what was the original number of people in the club?

20. *Increased Membership* The Sarasota Wheelers skateboard club had 75 members and increased the number of members to 95. What is the percent increase in the number of members?

11.2

In Exercises 21–24, determine the missing quantity by using the simple interest formula.

21. $p = \$6000$, $r = 2\%$, $t = 30$ days, $i = ?$

22. $p = \$2700$, $r = ?$, $t = 100$ days, $i = \$37.50$

23. $p = ?$, $r = 8\frac{1}{2}\%$, $t = 3$ years, $i = \$114.75$

24. $p = \$5500$, $r = 11\frac{1}{2}\%$, $t = ?$, $i = \$316.25$

25. *Roof Replacement Loan* Chris Sharek borrowed $5300 from his father to replace the roof on his house. The loan was for 36 months and had a simple interest rate of 5.75%. Determine the amount Chris paid his father on the date of maturity of the loan.

26. *A Bank Loan* Lori Holdren borrowed $3000 from her bank for 240 days at a simple interest rate of 8.1%.

a) How much interest did she pay for the use of the money?

b) How much did she pay the bank on the date of maturity?

27. **A Bank Loan** Nikos Pappas borrowed $6000 for 24 months from his bank, using stock as security. The bank discounted the loan at $11\frac{1}{2}\%$.

   a) How much interest did Nikos pay the bank for the use of the money?

   b) How much did he receive from the bank?

   c) What was the actual rate of interest?

28. **Savings as Security** Golda Frankl borrowed $800 for 6 months from her bank, using her savings account as security. A bank rule limits the amount that can be borrowed in this manner to 85% of the amount in the borrower's savings account. The rate of interest is 2% higher than the interest rate being paid on the savings account. The current rate on the savings account is $5\frac{1}{2}\%$.

   a) What rate of interest will the bank charge for the loan?

   b) Determine the amount that Golda must repay in 6 months.

   c) How much money must she have in her account to borrow $800?

11.3

29. **Comparing Compounding Periods** Determine the amount and the interest when $5000 is invested for 5 years at 6%.

   a) compounded annually.

   b) compounded semi-annually.

   c) compounded quarterly.

   d) compounded monthly.

   e) compounded daily (use $n = 360$).

30. **Total Amount** Choi deposited $2500 in a savings account that pays 4.75% interest compounded quarterly. What will be the total amount of money in the account 15 years from the day of deposit?

31. **Effective Annual Yield** Determine the effective annual yield of an investment if the interest is compounded daily at an annual rate of 5.6%.

32. **Present Value** How many dollars must you invest today to have $40,000 in 20 years? Assume that the money earns 5.5% interest compounded quarterly.

11.4

33. **Repaying a Loan Early** Bill Jordan has a 48-month installment loan with a fixed monthly payment of $176.14. The amount borrowed was $7500. Instead of making his 24th payment, Bill is paying the remaining balance on the loan.

   a) Determine the APR of the installment loan.

   b) How much interest will Bill save, computed by the actuarial method?

   c) What is the total amount due to pay off the loan?

34. **Repaying a Loan Early** Carter Fenton is buying a book collection that costs $4000. He is making a down payment of $500 and 24 monthly payments of $155.91.

   a) Determine the APR.

   b) Instead of making his 12th payment, Carter decides to pay the total remaining balance and terminate the loan. How much will Carter save by paying off the loan early?

   c) What is the total amount due to pay off the loan?

35. **Installment Loan** Dara Holliday's cost for a new wardrobe was $3420. She made a down payment of $860 and financed the balance on a 24-month fixed payment installment loan. The monthly payments are $111.73. Instead of making her 12th payment, Dara decides to pay the total remaining balance and terminate the loan.

   a) Determine the APR of the installment loan.

   b) How much interest will Dara save, computed by the actuarial method?

   c) What is the total amount due to pay off the loan?

36. **Finance Charge Comparison** On June 1, the billing date, Tim McGarry had a balance due of $485.75 on his credit card. The transactions during the month of June were

June 4	Payment	$375.00
June 8	Charge: Car repair	370.00
June 21	Charge: Airline ticket	175.80
June 28	Charge: Clothing	184.75

a) Determine the finance charge on July 1 by using the unpaid balance method. Assume that the interest rate is 1.3% per month.

b) Determine the new account balance on July 1 using the finance charge found in part (a).

c) Determine the average daily balance for the period.

d) Determine the finance charge on July 1 by using the average daily balance method. Assume that the interest rate is 1.3% per month.

e) Determine the new account balance on July 1 using the finance charge found in part (d).

37. **Finance Charge Comparison** On August 5, the billing date, Pat Schaefer had a balance due of $185.72 on her credit card. The transactions during the month of August were

August 8	Charge: Shoes	$85.75
August 10	Payment	75.00
August 15	Charge: Dry cleaning	72.85
August 21	Charge: Textbooks	275.00

a) Determine the finance charge on September 5 by using the unpaid balance method. Assume that the interest rate is 1.4% per month.

b) Determine the new account balance on September 5 using the finance charge found in part (a).

c) Determine the average daily balance for the period.

d) Determine the finance charge on September 5 by using the average daily balance method. Assume that the interest rate is 1.4% per month.

e) Determine the new account balance on September 5 using the finance charge found in part (d).

38. **Financing a Corvette** David Snodgress bought a new Chevrolet Corvette for $72,000. He made a 20% down payment and financed the balance with the dealer on a 60-month payment plan. The monthly payments were $1100.26. Determine the

a) down payment.

b) amount to be financed.

c) total finance charge.

d) APR.

39. **Financing a Ski Outfit** Lucille Groenke can buy a cross-country skiing outfit for $275. The store is offering the following terms: $50 down and 12 monthly payments of $19.62.

a) Determine the interest Lucille would pay.

b) Determine the APR.

11.5

40. **Building a House** The Freemans have decided to build a new house. The contractor quoted them a price of $135,700. The taxes on the house will be $3450 per year, and homeowners' insurance will be $350 per year. They have applied for a conventional loan from a local bank. The bank is requiring a 25% down payment, and the interest rate on the loan is 9.5%. The Freemans' annual income is $64,000. They have more than 10 monthly payments remaining on each of the following: $218 on a car, $120 on new furniture, and $190 on a camper. Their bank will approve a loan that has a total monthly mortgage payment of principal, interest, property taxes, and homeowners' insurance that is less than or equal to 28% of their adjusted monthly income. Determine

a) the required down payment.

b) 28% of their adjusted monthly income.

c) the monthly payment of principal and interest for a 30-year loan.

d) their total monthly payment, including insurance and taxes.

e) Do the Freemans qualify for the mortgage?

41. **Thirty-Year Mortgage** James Whitehead purchased a home selling for $89,900 with a 15% down payment. The period of the mortgage is 30 years, and the interest rate is 11.5% with no points. Determine the

a) amount of the down payment.

b) monthly mortgage payment.

c) amount of the first payment applied to the principal.

d) total cost of the house.

e) total interest paid.

42. **Adjustable-Rate Mortgage** The Nguyens purchased a house for $105,000 with a down payment of $26,250. They obtained a 30-year adjustable-rate mortgage. The terms of the mortgage are as follows: The interest rate is based on the 6-month Treasury bill, the interest rate charged is 3.00% above the rate of the Treasury bill on the date of adjustment, the interest rate is adjusted every 6 months, the interest rate will not change more than 1% (up or down) when the interest rate is adjusted, the maximum interest rate that can be charged for the duration of the loan is 16%, there is no lower limit on the interest rate,

the initial mortgage interest rate is 7.5%, and the monthly payment of interest and principal is adjusted annually. Determine the

a) initial monthly payment for principal and interest.

b) interest rate in 6 months if the interest rate on the Treasury bill at the time is 5.00%.

c) interest rate in 6 months if the interest rate on the Treasury bill at the time is 4.75%.

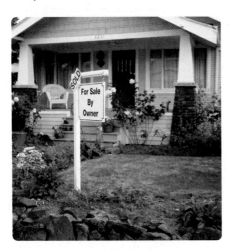

11.6

43. *Vacation Savings* To save money to take an around-the-world vacation, Kim Ghiselin invests $250 monthly in an ordinary annuity with 9% interest compounded monthly. Determine the accumulated amount in Kim's annuity after 10 years.

44. *Saving for a New Combine* Steve Carah would like to save $100,000 to buy a used combine (a farm machine used to harvest crops). To accumulate $100,000 in 15 years, how much should he invest quarterly in a sinking fund with 7% interest compounded quarterly?

CHAPTER ⑪ TEST

1. Find the missing quantity by using the simple interest formula.

   a) $i = ?, p = \$2000, r = 4\%$ per year, $t = 6$ months

   b) $i = \$288, p = \$1200, r = 8\%$ per year, $t = ?$

In Exercises 2 and 3, Rob Wilson borrowed $5700 from a bank for 30 months. The rate of simple interest charged is 5.75%.

2. How much interest did he pay for the use of the money?

3. What is the amount he repaid to the bank on the due date of the loan?

In Exercises 4 and 5, Yolanda Fernandez received a $5400 loan with interest at 12.5% for 90 days on August 1. Yolanda made a payment of $3000 on September 15.

4. How much did she owe the bank on the date of maturity?

5. What total amount of interest did she pay on the loan?

6. Compute the amount and the compound interest.

	Principal	Time	Rate	Compounded
a)	$7500	2 years	3%	Quarterly
b)	$2500	3 years	6.5%	Monthly

A New Bicycle In Exercises 7–9, a new touring bicycle sells for $2350. To finance it through a bank the bank will require a down payment of 15% and monthly payments of $90.79 for 24 months.

7. How much money will the purchaser borrow from the bank?

8. What finance charge will the individual pay the bank?

9. What is the APR?

10. *Actuarial Method* Gino Sedillo borrowed $7500. To repay the loan, he was scheduled to make 36 monthly installment payments of $223.10. Instead of making his 24th payment, Gino decides to pay off the loan.

  a) Determine the APR of the installment loan.

  b) How much interest will Gino save (use the actuarial method)?

  c) What is the total amount due to pay off the loan?

11. *Unpaid Balance Method* Michael Murphy's credit card statement shows a balance due of $878.25 on March 23, the billing date. For the period ending on April 23, he had the following transactions.

March 26	Charge: Groceries	$ 95.89
March 30	Charge: Restaurant bill	68.76
April 3	Payment	450.00
April 15	Charge: Clothing	90.52
April 22	Charge: Eyeglasses	450.85

  a) Determine the finance charge on March 23 by using the unpaid balance method. Assume that the interest rate is 1.4% per month.

  b) Determine the new account balance on April 23 using the finance charge found in part (a).

  c) Determine the average daily balance for the period.

  d) Determine the finance charge on March 23 by using the average daily balance method. Assume that the interest rate is 1.4% per month.

  e) Determine the new account balance on April 23 using the finance charge found in part (d).

Building a House In Exercises 12–18, the Leungs decided to build a new house. The contractor quoted them a price of $144,500, including the lot. The taxes on the house would be $3200 per year, and homeowners' insurance would cost $450 per year. They have applied for a conventional loan from a bank. The bank is requiring a 15% down payment, and the interest rate is $10\frac{1}{2}\%$ with 2 points. The Leung's annual income is $86,500. They have more than 10 monthly payments remaining on each of the following: $220 for a car, $175 for new furniture, and $210 on a college education loan. Their bank will approve a loan that has a total monthly mortgage payment of principal, interest, property taxes, and homeowners' insurance that is less than or equal to 28% of their adjusted monthly income.

12. What is the required down payment?

13. Determine the amount paid for points.

14. Determine 28% of their adjusted monthly income.

15. Determine the monthly payments of principal and interest for a 30-year loan.

16. Determine their total monthly payments, including homeowners' insurance and taxes.

17. Do the Leungs meet the requirements for the mortgage?

18. a) Determine the total cost of the house (excluding homeowners' insurance and taxes) after 30 years.

  b) How much of the total cost is interest, including points?

19. *Opening a Restaurant* To raise money to open her own restaurant, Carol Kuper invests $500 monthly in an ordinary annuity with 6% interest compounded monthly. Determine the accumulated amount in Carol's annuity after 20 years.

20. *Saving for a New Car* Elisa Bolotin is a freshman in high school and would like to have $15,000 in 4 years to buy a car for college. How much should she invest monthly in a sinking fund with 6% interest compounded monthly to accumulate $15,000 in 4 years?

G R O U P   P R O J E C T S

MORTGAGE LOAN

1. The Young family is purchasing a $130,000 house with a VA mortgage. The bank is offering them a 25-year mortgage with an interest rate of 9.5%. They have $20,000 invested that could be used for a down payment. Since they do not need a down payment, Mr. Young wants to keep the money invested. Mrs. Young believes that they should make a down payment of $20,000.

   a) Determine the total cost of the house with no down payment.

   b) Determine the total cost of the house if they make a down payment of $20,000.

   c) Mr. Young believes that the $20,000 investment will have an annual rate of return of 10% compounded quarterly. Assuming that Mr. Young is right, calculate the value of the investment in 25 years. (Use the Compound Interest Formula on page 669.)

   d) If the Youngs use the $20,000 as a down payment, their monthly payments will decrease. Determine the difference of the monthly payments in parts (a) and (b).

   e) Assume that the difference in monthly payments, part (d), is invested each month at a rate of 6% compounded monthly for 25 years. Determine the value of the investment in 25 years. (Use the Ordinary Annuity Formula on page 708.)

   f) Use the information from parts (a)–(e) to analyze the problem. Would you recommend that the Youngs make the down payment of $20,000 and invest the difference in their monthly payments as in part (e) or that they do not make the down payment and keep the $20,000 invested as in part (c)? Explain.

CREDIT CARD TERMS

2. With each credit card comes a credit agreement (or security agreement) the cardholder must sign. Select two members of your group who have a major credit card (MasterCard, Visa, Discover, or American Express). If possible, one of the credit cards should be a gold or platinum card. Obtain a copy of the credit agreement signed by each cardholder and answer questions (a)–(p).

   a) What are the cardholder's responsibilities?

   b) What is the cardholder's maximum line of credit?

   c) What restrictions apply to the use of the credit card?

   d) How many days after the billing date does the cardholder have to make a payment without being charged interest?

   e) What is the minimum monthly payment required, and how is it determined?

   f) What is the interest rate charged on purchases?

   g) How does the bank determine when to start charging interest on purchases?

   h) Is there an annual fee for the credit card? If so, what is it?

   i) What late charge applies if payments are not made on time?

   j) What information is given on the monthly statement?

   k) How is the finance charge computed?

   l) What other fees, if any, may the bank charge you?

   m) If the card is lost, what is the responsibility of the cardholder?

   n) If the card is lost, what is the liability of the cardholder?

   o) What are the advantages of a gold or platinum card over a regular card?

   p) In your opinion, which of the two cards is more desirable? Explain.

17.

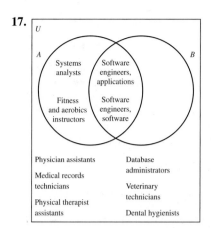

19. The set of animals in U.S. zoos that are not in the San Diego Zoo

21. The set of insurance companies in the United States that do not offer life insurance

23. The set of insurance companies in the United States that offer life insurance or car insurance

25. The set of insurance companies in the United States that offer life insurance and do not offer car insurance

27. The set of furniture stores that sell mattresses and outdoor furniture

29. The set of furniture stores that do not sell outdoor furniture and sell leather furniture

31. The set of furniture stores that sell mattresses or outdoor furniture or leather furniture

33. $\{b, c, t, w, a, h\}$ **35.** $\{a, h\}$

37. $\{c, w, b, t, a, h, f, r, d, g\}$ **39.** $\{p, m, z\}$

41. $\{L, \Delta, @, \$, *\}$

43. $\{L, \Delta, @, *, \$, R, \Box, \alpha, \infty, \Sigma, Z\}$

45. $\{R, \Box, \alpha, *, \$, \infty, Z, \Sigma\}$ **47.** $\{R, \Box, \alpha\}$

49. $\{1, 2, 3, 4, 5, 6, 8\}$ **51.** $\{1, 5, 7, 8\}$ **53.** $\{7\}$

55. $\{\ \}$ **57.** $\{7\}$ **59.** $\{a, e, h, i, j, k\}$

61. $\{a, f, i\}$ **63.** $\{b, c, d, e, g, h, j, k\}$

65. $\{a, c, d, e, f, g, h, i, j, k\}$

67. $\{a, b, c, d, e, f, g, h, i, j, k\}$, or U

69. $\{2, 6, 9\}$ **71.** $\{1, 4\}$ **73.** $\{1, 3, 4, 5, 7, 8, 10\}$

75. $\{4, 9\}$ **77.** $\{(a, 1), (a, 2), (b, 1), (b, 2), (c, 1), (c, 2)\}$

79. No. The ordered pairs are not the same. For example, $(a, 1) \neq (1, a)$.

81. 6   **83.** $\{\ \}$ **85.** $\{2, 4, 6, 8\}$, or B **87.** $\{7, 9\}$

89. $\{1, 3, 5, 6, 7, 8, 9\}$ **91.** $\{6, 8\}$

93. $\{1, 2, 3, 4, 5, 6, 7, 8, 9\}$, or U

95. $\{1, 2, 3, 4, 5\}$, or C

97. A set and its complement will always be disjoint. For example, if $U = \{1, 2, 3\}$ and $A = \{1, 2\}$, then $A' = \{3\}$, and $A \cap A' = \{\ \}$.

99. 49

101. a) $8 = 4 + 6 - 2$ **b)** and **c)** Answers will vary.

103. $\{1, 2, 3, 4, \ldots\}$, or A **105.** $\{4, 8, 12, 16, \ldots\}$, or B

107. $\{2, 4, 6, 8, \ldots\}$, or C **109.** $\{2, 6, 10, 14, 18, \ldots\}$

111. $\{2, 6, 10, 14, 18, \ldots\}$ **113.** U **115.** A

117. U **119.** U **121.** $B \subseteq A$

123. A and B are disjoint sets.   **125.** $A \subseteq B$

SECTION 2.4, PAGE 77

1. 8   **3.** II, IV, VI   **5.** 5

7. a) Yes
b) No, one specific case cannot be used as proof.
c) No

9.

11.

13.

15.

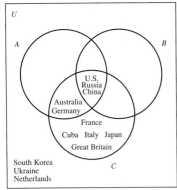

17. V   **19.** I   **21.** III   **23.** II   **25.** VIII   **27.** III
29. VI   **31.** III   **33.** III   **35.** V   **37.** II   **39.** VII
41. I   **43.** VIII   **45.** VI   **47.** $\{1, 3, 4, 5, 7, 9\}$
49. $\{2, 3, 4, 5, 6, 8, 12, 14\}$ **51.** $\{3, 4, 5\}$
53. $\{1, 2, 3, 7, 9, 10, 11, 12, 13, 14\}$
55. $\{1, 2, 3, 4, 5, 6, 7, 8, 9, 12, 14\}$
57. $\{2, 11, 12, 13, 14\}$
59. $\{2, 6, 8, 10, 11, 12, 13, 14\}$
61. Yes   **63.** No   **65.** No   **67.** Yes   **69.** No
71. Yes   **73.** Yes   **75.** Yes   **77.** No
79. $(A \cup B)'$ **81.** $(A \cup B) \cap C'$
83. a) Both equal $\{6, 7\}$.   **b)** Answers will vary.
    c) Both are represented by the regions IV, V, VI.

85.

87. a)

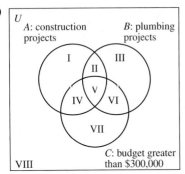

b) V; $A \cap B \cap C$
c) VI; $A' \cap B \cap C$
d) I; $A \cap B' \cap C'$

89. a)

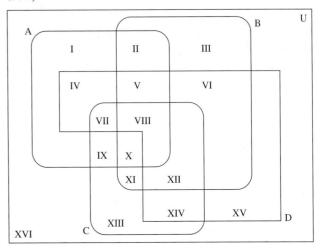

b)

Region	Set	Region	Set
I	$A \cap B' \cap C' \cap D'$	IX	$A \cap B' \cap C \cap D'$
II	$A \cap B \cap C' \cap D'$	X	$A \cap B \cap C \cap D'$
III	$A' \cap B \cap C' \cap D'$	XI	$A' \cap B \cap C \cap D'$
IV	$A \cap B' \cap C' \cap D$	XII	$A' \cap B \cap C \cap D$
V	$A \cap B \cap C' \cap D$	XIII	$A' \cap B' \cap C \cap D'$
VI	$A' \cap B \cap C' \cap D$	XIV	$A' \cap B' \cap C \cap D$
VII	$A \cap B' \cap C \cap D$	XV	$A' \cap B' \cap C' \cap D$
VIII	$A \cap B \cap C \cap D$	XVI	$A' \cap B' \cap C' \cap D'$

SECTION 2.5, PAGE 86

1.

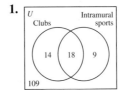

a) 14
b) 9
c) 109

3.

a) 17
b) 12
c) 59

5.

a) 3
b) 6
c) 22
d) 11
e) 12

7.

a) 22
b) 11
c) 64
d) 50
e) 23

9.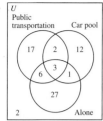

a) 17
b) 27
c) 2
d) 31
e) 2

11.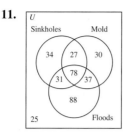

a) 67
b) 237
c) 37
d) 25

13. In a Venn diagram, regions II, IV, and V contain a total of 37 cars driven by women. This total is greater than the 35 cars driven by women, as given in the exercise.

15.

a) 410
b) 35
c) 90
d) 50

SECTION 2.6, PAGE 93

1. An infinite set is a set that can be placed in a one-to-one correspondence with a proper subset of itself.

3. $\{5, 6, 7, 8, 9, \ldots, n + 4, \ldots\}$
$\downarrow \downarrow \downarrow \downarrow \downarrow \quad \downarrow$
$\{6, 7, 8, 9, 10, \ldots, n + 5, \ldots\}$

5. $\{3, 5, 7, 9, 11, \ldots, 2n + 1, \ldots\}$
$\downarrow \downarrow \downarrow \downarrow \downarrow \quad \downarrow$
$\{5, 7, 9, 11, 13, \ldots, 2n + 3, \ldots\}$

7. $\{3, 7, 11, 15, 19, \ldots, 4n - 1, \ldots\}$
$\downarrow \downarrow \downarrow \downarrow \downarrow \quad \downarrow$
$\{7, 11, 15, 19, 23, \ldots, 4n + 3, \ldots\}$

9. $\{6, 11, 16, 21, 26, \ldots, 5n + 1, \ldots\}$
$\downarrow \downarrow \downarrow \downarrow \downarrow \quad \downarrow$
$\{11, 16, 21, 26, 31, \ldots, 5n + 6, \ldots\}$

11. $\left\{\dfrac{1}{2}, \dfrac{1}{4}, \dfrac{1}{6}, \dfrac{1}{8}, \dfrac{1}{10}, \ldots, \dfrac{1}{2n}, \ldots\right\}$
$\downarrow \downarrow \downarrow \downarrow \downarrow \quad \downarrow$
$\left\{\dfrac{1}{4}, \dfrac{1}{6}, \dfrac{1}{8}, \dfrac{1}{10}, \dfrac{1}{12}, \ldots, \dfrac{1}{2n + 2}, \ldots\right\}$

13. $\{1, 2, 3, 4, 5, \ldots, n, \ldots\}$
$\downarrow \downarrow \downarrow \downarrow \downarrow \quad \downarrow$
$\{3, 6, 9, 12, 15, \ldots, 3n, \ldots\}$

15. $\{1, 2, 3, 4, 5, \ldots, n, \ldots\}$
$\downarrow \downarrow \downarrow \downarrow \downarrow \quad \downarrow$
$\{4, 6, 8, 10, 12, \ldots, 2n + 2, \ldots\}$

17. $\{1, 2, 3, 4, 5, \ldots, n, \ldots\}$
$\downarrow \downarrow \downarrow \downarrow \downarrow \quad \downarrow$
$\{2, 5, 8, 11, 14, \ldots, 3n - 1, \ldots\}$

19. $\{1, 2, 3, 4, 5, \ldots, n, \ldots\}$
$\downarrow \downarrow \downarrow \downarrow \downarrow \quad \downarrow$
$\{5, 9, 13, 17, 21, \ldots, 4n + 1, \ldots\}$

21. $\{1, 2, 3, 4, 5, \ldots, n, \ldots\}$
$\downarrow \downarrow \downarrow \downarrow \downarrow \quad \downarrow$
$\left\{\dfrac{1}{3}, \dfrac{1}{4}, \dfrac{1}{5}, \dfrac{1}{6}, \dfrac{1}{7}, \ldots, \dfrac{1}{n + 2}, \ldots\right\}$

23. $\{1, 2, 3, 4, 5, \ldots, n, \ldots\}$
$\downarrow \downarrow \downarrow \downarrow \downarrow \quad \downarrow$
$\{1, 4, 9, 16, 25, \ldots, n^2, \ldots\}$

25. $\{1, 2, 3, 4, 5, \ldots, n, \ldots\}$
$\downarrow \downarrow \downarrow \downarrow \downarrow \quad \downarrow$
$\{3, 9, 27, 81, 243, \ldots, 3^n, \ldots\}$

27. $=$ **29.** $=$ **31.** $=$

REVIEW EXERCISES, PAGE 94

1. True

2. False; the word *best* makes the statement not well defined.

3. True

4. False; no set is a proper subset of itself.

5. False; the elements $6, 12, 18, 24, \ldots$ are members of both sets.

6. True

7. False; both sets do not contain exactly the same elements.

8. True **9.** True **10.** True **11.** True **12.** True

13. True **14.** True **15.** $A = \{7, 9, 11, 13, 15\}$

16. {Colorado, Nebraska, Missouri, Oklahoma}

17. $C = \{1, 2, 3, 4, \ldots, 161\}$

18. $D = \{9, 10, 11, 12, \ldots, 96\}$

19. $A = \{x \mid x \in N \text{ and } 52 < x < 100\}$

20. $B = \{x \mid x \in N \text{ and } x > 42\}$

21. $C = \{x \mid x \in N \text{ and } x < 5\}$

22. $D = \{x \mid x \in N \text{ and } 27 \le x \le 51\}$

23. A is the set of capital letters in the English alphabet from E through M, inclusive.

24. *B* is the set of U.S. coins with a value of less than a dollar.

25. *C* is the set of the last three lowercase letters in the English alphabet.

26. *D* is the set of numbers greater than or equal to 3 and less than 9.

27. $\{5, 7\}$ **28.** $\{1, 2, 3, 4, 5, 6, 7, 8\}$ **29.** $\{9, 10\}$

30. $\{1, 2, 4, 6, 7, 8, 10\}$ **31.** $\{1, 3\}$ **32.** $\{1, 7\}$

33. $\{(1, 1), (1, 7), (1, 10), (3, 1), (3, 7), (3, 10), (5, 1), (5, 7),$ $(5, 10), (7, 1), (7, 7), (7, 10)\}$

34. $\{(5, 1), (5, 3), (5, 5), (5, 7), (7, 1), (7, 3), (7, 5), (7, 7), (9, 1),$ $(9, 3), (9, 5), (9, 7), (10, 1), (10, 3), (10, 5), (10, 7)\}$

36. 15

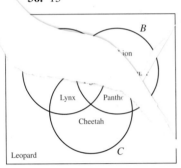

38. $\{a, c, d, f, g, i, k, l\}$ **39.** $\{i, k\}$

40. $\{a, b, c, d, f, g, h, i, k, l\}$ **41.** $\{f\}$ **42.** $\{a, f, i\}$

43. $\{a, b, d, f, h, i, l\}$ **44.** True **45.** True **46.** II

47. III **48.** I **49.** IV **50.** IV **51.** II **52.** $450

53.

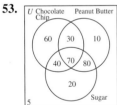

a) 315
b) 10
c) 30
d) 110

54.

a) 38
b) 298
c) 28
d) 236
e) 106

55. $\{2, 4, 6, 8, 10, \ldots, \quad 2n, \ldots\}$
 ↓ ↓ ↓ ↓ ↓   ↓
$\{4, 6, 8, 10, 12, \ldots, 2n + 2, \ldots\}$

56. $\{3, 5, 7, 9, 11, \ldots, 2n + 1, \ldots\}$
 ↓ ↓ ↓ ↓ ↓   ↓
$\{5, 7, 9, 11, 13, \ldots, 2n + 3, \ldots\}$

57. $\{1, 2, 3, 4, 5, \ldots, \quad n, \ldots\}$
 ↓ ↓ ↓ ↓ ↓   ↓
$\{5, 8, 11, 14, 17, \ldots, 3n + 2, \ldots\}$

58. $\{1, 2, 3, 4, 5, \ldots, \quad n, \ldots\}$
 ↓ ↓ ↓ ↓ ↓   ↓
$\{4, 9, 14, 19, 24, \ldots, 5n - 1, \ldots\}$

CHAPTER TEST, PAGE 97

1. True

2. False; the sets do not contain exactly the same elements.

3. True

4. False; the second set has no subset that contains the element 7.

5. False; the empty set is a subset of every set.

6. False; the set has 2^4, or 16 subsets. **7.** True

8. False; for any set A, $A \cup A' = U$, not $\{\ \}$. **9.** True

10. $A = \{1, 2, 3, 4, 5, 6, 7, 8\}$

11. Set *A* is the set of natural numbers less than 9.

12. $\{7, 9\}$ **13.** $\{3, 5, 7, 9, 13\}$

14. $\{3, 5, 7, 9\}$, or A **15.** 2 **16.** $\{3, 5\}$

17. $\{(3, 3), (3, 11), (3, 15), (5, 3), (5, 11), (5, 15), (7, 3),$ $(7, 11), (7, 15), (9, 3), (9, 11), (9, 15)\}$

18.

19. Equal

20.

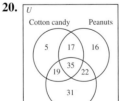

a) 52 **b)** 10 **c)** 93 **d)** 17 **e)** 38 **f)** 31

21. $\{7, 8, 9, 10, 11, \ldots, n + 6, \ldots\}$
 ↓ ↓ ↓ ↓ ↓   ↓
$\{8, 9, 10, 11, 12, \ldots, n + 7, \ldots\}$

22. $\{1, 2, 3, 4, 5, \ldots, \quad n, \ldots\}$
 ↓ ↓ ↓ ↓ ↓   ↓
$\{1, 3, 5, 7, 9, \ldots, 2n - 1, \ldots\}$

CHAPTER 3

SECTION 3.1, PAGE 109

1. **a)** A sentence that can be judged either true or false is called a statement.
   b) A simple statement is a statement that conveys only one idea.
   c) Compound statements are statements consisting of two or more simple statements.

3. *All, none,* and *some* are quantifiers.

5. **a)** Some are. **b)** All are.
   c) Some are not. **d)** None are.

7. **a)** No, the *exclusive or* is used when one or the other of the events can take place, but not both.
   b) Yes, the *inclusive or* is used when one or the other, or both events can take place.
   c) *Inclusive or*

9. Compound; disjunction, $\vee$

11. Compound; biconditional, $\leftrightarrow$

13. Compound; conjunction, $\wedge$

15. Simple statement

17. Compound; negation, $\sim$

19. Compound; conjunction, $\wedge$

21. Compound; negation, $\sim$

23. Some butterflies are not insects.

25. Some aldermen are running for mayor.

27. All turtles have claws.

29. Some bicycles have three wheels.

31. All pine trees produce pinecones.

33. No pedestrians are in the crosswalk.

35. $\sim p$ 37. $\sim q \vee \sim p$ 39. $\sim p \rightarrow \sim q$

41. $\sim p \wedge q$ 43. $\sim q \leftrightarrow p$ 45. $\sim(p \vee q)$

47. Ken Jennings did not win more than $3 million.

49. Ken Jennings won 74 games of *Jeopardy!* and Ken Jennings won more than $3 million.

51. If Ken Jennings did not win 74 games of *Jeopardy!* then Ken Jennings won more than $3 million.

53. Ken Jennings did not win 74 games of *Jeopardy!* or Ken Jennings did not win more than $3 million.

55. It is false that Ken Jennings won 74 games of *Jeopardy!* and Ken Jennings won more than $3 million.

57. $(p \wedge \sim q) \wedge r$ 59. $(p \wedge q) \vee r$ 61. $p \rightarrow (q \vee \sim r)$

63. $(r \leftrightarrow q) \wedge p$ 65. $q \rightarrow (p \leftrightarrow r)$

67. The water is 70° or the sun is shining, and we do not go swimming.

69. The water is not 70°, and the sun is shining or we go swimming.

71. If we do not go swimming, then the sun is shining and the water is 70°.

73. If the sun is shining then we go swimming, and the water is 70°.

75. The sun is shining if and only if the water is 70°, and we go swimming.

77. Not permissible, you cannot have both soup and salad. The *or* used on menus is the *exclusive or*.

79. Not permissible, you cannot have both potatoes and pasta. The *or* used on menus is the *exclusive or*.

81. **a)** $w \wedge \sim p$ **b)** Conjunction

83. **a)** $\sim(b \rightarrow \sim p)$ **b)** Negation

85. **a)** $(f \vee v) \rightarrow h$ **b)** Conditional

87. **a)** $c \leftrightarrow (\sim f \vee p)$ **b)** Biconditional

89. **a)** $(c \leftrightarrow w) \vee s$ **b)** Disjunction

91. **a)** Answers will vary.
    b) Answers will vary.

SECTION 3.2, PAGE 122

1. **a)** 4   **b)**

p	q
T	T
T	F
F	T
F	F

3. **a)**

p	q	$p \wedge q$
T	T	T
T	F	F
F	T	F
F	F	F

b) Only when both p and q are true

c)

p	q	$p \vee q$
T	T	T
T	F	T
F	T	T
F	F	F

d) Only when both p and q are false

5. F   7. T   9. F   11. T

5.	7.	9.	11.
F	F	T	F
	T	T	T
	T	T	T

13. T   15. T   17. F   19. T

13.	15.	17.	19.
F	F	F	T
T	T	T	T
T	T	F	T
F	T	F	F
F	F	F	F
T	T	T	T
T	F	T	F

21. $p \wedge q$

T
F
F
F

23. $p \wedge \sim q$

F
T
F
F

25. $\sim(p \wedge q)$

F
T
T
T

27. $p \vee (q \vee r)$

T
T
T
T
T
T
T
F

29. $p \wedge (q \vee \sim q)$

T
T
F
F

31. a) False **b)** True
33. a) True **b)** True
35. a) True **b)** False
37. a) False **b)** True
39. a) True **b)** True
41. a) True **b)** True
43. True **45.** True **47.** False **49.** False
51. False **53.** True **55.** True **57.** False

59. $p \wedge \sim q$

F
T
F
F

True in case 2

61. $p \vee \sim q$

T
T
F
T

True in cases 1, 2, and 4, when p is true, or when p and q are both false.

63. $(r \vee q) \wedge p$

T
T
T
F
F
F
F
F

True in cases 1, 2, and 3

65. $q \vee (p \wedge \sim r)$

T
T
F
T
T
T
F
F

True in cases 1, 2, 4, 5, and 6. True except when p, q, r have truth values TFT, FFT, or FFF.

67. a) Mr. Duncan and Mrs. Tuttle qualify.
b) Mrs. Rusinek does not qualify, since their combined income is less than $46,000.

69. a) Wing Park qualifies; the other four do not.
b) Gina Vela is returning on April 2. Kara Sharo is returning on a Monday. Christos Supernaw is not staying over on a Saturday. Alex Chang is returning on a Monday.

71. T
T
T
T
F
T
F
T

73. Yes

SECTION 3.3, PAGE 133

1. a)

p	q	$p \to q$
T	T	T
T	F	F
F	T	T
F	F	T

b) The conditional is false only when the antecedent is true and the consequent is false.

c)

p	q	$p \leftrightarrow q$
T	T	T
T	F	F
F	T	F
F	F	T

d) The biconditional is true only when both p and q are true or when p and q are both false.

3. a) Substitute the truth values for the simple statements. Then evaluate the compound statement, using the assigned truth values.
b) True

5. A self-contradiction is a compound statement that is always false.

7. T
T
T
F

9. T
F
F
F

11. F
T
T
F

13. T
T
F
T

15. F
T
T
T

17. T
T
T
T
T
F
F
F

19. T
T
F
F
F
F
F
T

21. F
F
T
F
T
T
F
T

23. T
T
T
T
T
F
F
F

25. T
T
F
T
T
F
F
T

27. $p \to (q \wedge r)$

T
F
F
F
T
T
T
T

29. $(p \leftrightarrow q) \vee r$

T
T
T
F
T
F
T
T

31. $(\sim p \rightarrow q) \vee r$

T
T
T
T
T
T
T
F

33. Neither **35.** Self-contradiction **37.** Tautology

39. Not an implication **41.** Implication **43.** Implication

45. True **47.** False **49.** False **51.** True

53. True **55.** True **57.** True **59.** False

61. True **63.** True **65.** True **67.** False

69. False **71.** True **73.** True **75.** True

77. No, the statement only states what will occur if your sister gets straight A's. If your sister does not get straight A's, your parents may still get her a computer.

79. F
F
T
T
T
F
F
F

81. It is a tautology. The statement may be expressed as $(p \rightarrow q) \vee (\sim p \rightarrow q)$, where p: It is a head and q: I win. This statement is a tautology.

83.

Tiger	Boots	Sam	Sue
Blue	Yellow	Red	Green
Nine Lives	Whiskas	Friskies	Meow Mix

SECTION 3.4, PAGE 146

1. a) Statements that have exactly the same truth values

b) Construct truth tables for each statement. If both have the same truth values in the answer columns of the truth tables, the statements are equivalent.

3. $\sim(p \wedge q) \Leftrightarrow \sim p \vee \sim q$

$\sim(p \vee q) \Leftrightarrow \sim p \wedge \sim q$

5. a and c, b and d **7.** $p \wedge \sim q$ **9.** Not equivalent

11. Not equivalent **13.** Equivalent **15.** Equivalent

17. Equivalent **19.** Equivalent **21.** Equivalent

23. Equivalent **25.** Not equivalent

27. Not equivalent **29.** Equivalent

31. The Rocky Mountains are not in the East or the Appalachian Mountains are not in the West.

33. It is false that the watch was a Swatch or the watch was a Swiss Army watch.

35. It is false that the hotel has a weight room and the conference center has an auditorium.

37. If Ashley Tabai takes the new job, then it is false that she will move and she will not buy a new house in town.

39. Ena Salter does not select a new textbook or she will have to write a new syllabus.

41. If Bob the Tomato didn't visit the nursing home then he did not visit the Cub Scout meeting.

43. The plumbers do not meet in Kansas City or the Rainmakers will provide the entertainment.

45. If Chase is hiding, then the pitcher is broken.

47. We go to Cincinnati and we will not go to the zoo.

49. It is false that if I am cold then the heater is working.

51. Borders has a sale and we will not buy $100 worth of books.

53. It is false that if John Deere will hire new workers then the city of Dubuque will not retrain the workers.

55. *Converse:* If we can finish the quilt in 1 week, then we work every night.
Inverse: If we do not work every night, then we cannot finish the quilt in 1 week.
Contrapositive: If we cannot finish the quilt in 1 week, then we do not work every night.

57. *Converse:* If I buy silver jewelry, then I go to Mexico.
Inverse: If I do not go to Mexico, then I do not buy silver jewelry.
Contrapositive: If I do not buy silver jewelry, then I do not go to Mexico.

59. *Converse:* If I scream, then that annoying paper clip shows up on my computer screen.
Inverse: If that annoying paper clip does not show up on my computer screen, then I will not scream.
Contrapositive: If I do not scream, then that annoying paper clip does not show up on my screen.

61. If a natural number is divisible by 10, then the natural number is divisible by 5. True.

63. If a natural number is not divisible by 6, then the natural number is not divisible by 3. False.

65. If two lines are not parallel, then the two lines intersect in at least one point. True.

67. b) and **c)** are equivalent.

69. a) and **c)** are equivalent.

71. b) and **c)** are equivalent.

73. b) and **c)** are equivalent.

75. None are equivalent.

77. None are equivalent.

79. a) and **c)** are equivalent.

81. a) and **b)** are equivalent.

83. True. If $p \rightarrow q$ is false, it must be of the form $T \rightarrow F$. Therefore, the converse must be of the form $F \rightarrow T$, which is true.

85. False. A conditional statement and its contrapositive always have the same truth values.

87. Answers will vary.

89. Answers will vary.

SECTION 3.5, PAGE 159

1. a) The conclusion necessarily follows from the given set of premises.
   b) The conclusion does not necessarily follow from the given set of premises.

3. Yes, if the conclusion does not necessarily follow from the premises, the argument is invalid, even if the conclusion is a true statement.

5. Yes, if the conclusion does not follow from the set of premises.

7. a) $p \rightarrow q$
   $$\frac{p}{\therefore q}$$
   b) Answers will vary.

9. a) $p \rightarrow q$
   $$\frac{\sim q}{\therefore}$$
   ers will vary.

11. a) $p \rightarrow q$
   $$\frac{q}{\therefore p}$$
   b) Answers will vary.

13. Invalid   **15.** Valid   **17.** alid   **19.** Invalid

21. Valid   **23.** Valid   **25.** Valid   **27.** Invalid

29. Invalid   **31.** Valid

33. a) $p \rightarrow q$
   $$\frac{\sim p}{\therefore \sim q}$$
   b) Invalid

35. a) $p \rightarrow q$
   $$\frac{p}{\therefore q}$$
   b) Valid

37. a) $p \rightarrow q$
   $$\frac{\sim q}{\therefore \sim p}$$
   b) Valid

39. a) $p \rightarrow q$
   $$\frac{q}{\therefore p}$$
   b) Invalid

41. a) $p \vee q$
   $$\frac{\sim p}{\therefore q}$$
   b) Valid

43. a) $p \rightarrow q$
   $$\frac{q \rightarrow r}{\therefore p \rightarrow r}$$
   b) Valid

45. a) $p \wedge q$
   $$\frac{q \rightarrow r}{\therefore r \rightarrow p}$$
   b) Valid

47. a) $s \wedge g$
   $$\frac{g \rightarrow c}{\therefore s \rightarrow c}$$
   b) Valid

49. a) $h \rightarrow b$
   $$\frac{\sim p \rightarrow \sim b}{\therefore h \rightarrow p}$$
   b) Valid

51. a) $p \rightarrow q$
   $$\frac{\sim q}{\therefore \sim p}$$
   b) Valid

53. a) $p \vee q$
   $$\frac{\sim p}{\therefore q}$$
   b) Valid

55. a) $t \wedge g$
   $$\frac{\sim t \vee \sim g}{\therefore \sim t}$$
   b) Valid

57. a) $c \wedge \sim h$
   $$\frac{h \rightarrow c}{\therefore h}$$
   b) Invalid

59. a) $f \rightarrow d$
   $$\frac{d \rightarrow \sim s}{\therefore f \rightarrow s}$$
   b) Invalid

61. Therefore, your face will break out.

63. Therefore, I am stressed out.

65. Therefore, you did not close the deal.

67. Therefore, if you do not pay off your credit card bills, then the bank makes money.

69. No. The conditional statement will always be true, and therefore it will be a tautology, and a valid argument.

SECTION 3.6, PAGE 167

1. a) It is a valid argument.
   b) It is an invalid argument.

3. a)

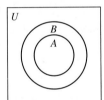

   b)

   c)

5. Yes, if the conclusion necessarily follows from the premises, the argument is valid.

7. Valid   **9.** Valid   **11.** Invalid

13. Valid   **15.** Invalid   **17.** Invalid

19. Invalid   **21.** Valid   **23.** Invalid   **25.** Invalid

27. Valid   **29.** Invalid

SECTION 3.7, PAGE 174

1. a) Answers will vary.
   b) $\wedge$ (and)

3. It is a series circuit; therefore, both switches must be closed for current to flow and the lightbulb to go on. When the p switch is closed, the $\bar{p}$ switch is open and no current will flow through the circuit. When the $\bar{p}$ switch is closed, the p switch is open and no current will flow through the circuit.

5. a) $p \wedge q$
   b) The lightbulb will be on when both p and q are closed.

7. a) $(p \vee q) \wedge \sim q$
   b) The lightbulb will be on when p is closed and q is open.

9. a) $(p \wedge q) \wedge [(p \wedge \sim q) \vee r]$
   b) The lightbulb will be on when p, q, and r are all closed.

11. a) $p \vee q \vee (r \wedge \sim p)$
   b) The lightbulb will be on in all cases except when p, q, and r are all open.

13.

15.

17.

19.

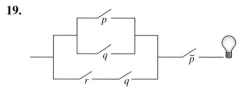

21. $p \vee q$; $\sim p \wedge \sim q$; not equivalent

23. $[(p \wedge q) \vee r] \wedge p$; $(q \vee r) \wedge p$; equivalent

25. $(p \vee \sim p) \wedge q \wedge r$; $p \wedge q \wedge r$; not equivalent

27. a)   **b)**

REVIEW EXERCISES, PAGE 176

1. Some gift cards are exchangeable.

2. Some bears are not mammals.

3. No women are presidents.

4. All pine trees are green.

5. The coffee is Maxwell House or the coffee is hot.

6. The coffee is not hot and the coffee is strong.

7. If the coffee is hot, then the coffee is strong and the coffee is not Maxwell House.

8. The coffee is Maxwell House if and only if the coffee is not strong.

9. The coffee is not Maxwell House, if and only if the coffee is strong and the coffee is not hot.

10. The coffee is Maxwell House or the coffee is not hot, and the coffee is not strong.

11. $r \wedge q$ **12.** $p \rightarrow r$ **13.** $(r \rightarrow q) \vee \sim p$

14. $(q \leftrightarrow p) \wedge \sim r$ **15.** $(r \wedge q) \vee \sim p$ **16.** $\sim(r \wedge q)$

17. F   **18.** T   **19.** T   **20.** T   **21.** F   **22.** F
F   F   T   F   T   T
T   F   T   T   F   T
F   F   T   T   F   T
            T   F   T   T
            F   F   T   T
            F   F   T   T
            T   F   T   T

23. False   **24.** True   **25.** False   **26.** True   **27.** True

28. True   **29.** False   **30.** False   **31.** Not equivalent

32. Equivalent   **33.** Equivalent   **34.** Not equivalent

35. It is false that if Bobby Darin sang *Mack the Knife* then Elvis wrote *Memphis*.

36. If Lynn Swann did not play for the Steelers, then Jack Tatum played for the Raiders.

37. Altec Lansing does not produce only speakers and Harman Kardon does not produce only stereo receivers.

38. It is false that Travis Tritt won an Academy Award or Randy Jackson does commercials for Milk Bone Dog Biscuits.

39. The temperature is above 32° or we will go ice fishing at O'Leary's Lake.

40. a) If you soften your opinion, then you hear a new voice today.
   b) If you do not hear a new voice today, then you do not soften your opinion.
   c) If you do not soften your opinion, then you do not hear a new voice today.

41. a) If we will learn the table's value, then we take the table to *Antiques Roadshow*.
   b) If we do not take the table to *Antiques Roadshow*, then we will not learn the table's value.
   c) If we will not learn the table's value, then we do not take the table to *Antiques Roadshow*.

42. a) If Maureen Gerald is helping at the school, then she is not in attendance.
b) If Maureen Gerald is in attendance, then she is not helping at the school.
c) If Maureen Gerald is not helping at the school, then she is in attendance.

43. a) If we will not buy a desk at Miller's Furniture, then the desk is made by Winner's Only and the desk is in the Rose catalog.
b) If the desk is not made by Winner's Only or the desk is not in the Rose catalog, then we will buy a desk at Miller's Furniture.
c) If we will buy a desk at Miller's Furniture, then the desk is not made by Winner's Only or the desk is not in the Rose catalog.

44. a) If I let you attend the prom, then you will get straight A's on your report card.
b) If you do not get straight A's on your report card, then I will not let you attend the prom.
c) If I will not let you attend the prom, then you did not get straight A's on your report card.

45. a), b), and **c)** are equivalent.   **46.** None are equivalent.

47. a) and **c)** are equivalent.   **48.** None are equivalent.

49. Invalid   **50.** Valid   **51.** Valid   **52.** Invalid

53. Invalid   **54.** Invalid   **55.** Invalid   **56.** Invalid

57. a) $p \wedge [(q \wedge r) \vee \sim p]$

b) The lightbulb will be on when p, q, and r are all closed.

58.

59. Equivalent

CHAPTER TEST, PAGE 179

1. $(p \wedge r) \vee \sim q$ **2.** $(r \rightarrow q) \vee \sim p$ **3.** $\sim(r \leftrightarrow \sim q)$

4. Ann is not the secretary and Elaine is the president, if and only if Dick is not the vice president.

5. If Ann is the secretary or Dick is not the vice president, then Elaine is the president.

6. F   **7.** T
  T   T
  F   T
  F   T
  F   F
  F   T
  F   T
  F   F

8. True   **9.** True   **10.** True   **11.** True   **12.** Equivalent

13. a) and **b)** are equivalent.

14. a) and **b)** are equivalent.

15. $s \rightarrow f$ **16.** Invalid
$\dfrac{f \rightarrow p}{\therefore s \rightarrow p}$
Valid

17. Some highways are not roads.

18. Nick did not play football or Max did not play baseball.

19. *Converse:* If today is Saturday, then the garbage truck comes.
Inverse: If the garbage truck does not come, then today is not Saturday.
Contrapositive: If today is not Saturday, then the garbage truck does not come.

20.

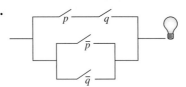

CHAPTER 4

SECTION 4.1, PAGE 188

1. A number is a quantity, and it answers the question "How many?" A numeral is a symbol used to represent the number.

3. A system of numeration consists of a set of numerals and a scheme or rule for combining the numerals to represent numbers.

5. The Hindu–Arabic numeration system

7. In a multiplicative system, there are numerals for each number less than the base and for powers of the base. Each numeral less than the base is multiplied by a numeral for the power of the base, and these products are added to obtain the number.

9. 345   **11.** 2423   **13.** 334,214   **15.** ∩∩∩∩|||

17. 𝄞𝄞∩∩∩||||| **19.** ◁𝍸𝍸𝍸𝍸𝍸𝍸𝍸𝄆𝄆99999999∩∩∩|||||

21. 8   **23.** 43   **25.** 1236   **27.** 2946   **29.** 12,666

31. 9464   **33.** XXVII   **35.** CCCXLI   **37.** MMV

39. $\overline{\text{IV}}$DCCXCIII   **41.** $\overline{\text{IX}}$CMXCIX   **43.** $\overline{\text{XX}}$DCXLIV

45. 74   **47.** 4081   **49.** 8550   **51.** 4003

53. 五十三   **55.** 三百七十八   **57.** 四千二百六十   **59.** 七千零五十六

61. 26   **63.** 279   **65.** 2883

67. $\nu\theta$ **69.** $\psi\kappa\digamma$ **71.** $^\iota\varepsilon\varepsilon$

73. 1021, MXXI, 一, $^\iota\alpha\kappa\alpha$

千
零
二十
一

75. 527, 𓏺𓏺𓏺𓏺𓏺∩∩||||||,
DXXVII, $\phi\kappa\zeta$

77. $\overline{\text{CMXCIX}}$CMXCIX

79. Advantage: You can write some numerals more compactly.

Disadvantage: There are more numerals to memorize.

81. Advantage: You can write some numerals more compactly.

Disadvantage: There are more numerals to memorize.

83. MM

SECTION 4.2, PAGE 196

1. Positional value system

3. The 4 in 40 represents four tens. The 4 in 400 represents 4 hundreds.

5. a) 10   **b)** 0, 1, 2, 3, 4, 5, 6, 7, 8, 9

7. Write the sum of the products of each digit times its corresponding positional value.

9. a) There may be confusion because numerals could be interpreted in different ways. For example, ▌ could be interpreted to be either 1 or 60.

b) ▐▐ ◄▐▐▐ for both numerals.

11. $1, 20, 18 \times 20, 18 \times (20)^2, 18 \times (20)^3$

13. $(2 \times 10) + (3 \times 1)$

15. $(3 \times 100) + (5 \times 10) + (9 \times 1)$

17. $(8 \times 100) + (9 \times 10) + (7 \times 1)$

19. $(4 \times 1000) + (3 \times 100) + (8 \times 10) + (7 \times 1)$

21. $(1 \times 10,000) + (6 \times 1000) + (4 \times 100) + (0 \times 10) + (2 \times 1)$

23. $(3 \times 100,000) + (4 \times 10,000) + (6 \times 1000) + (8 \times 100) + (6 \times 10) + (1 \times 1)$

25. 14   **27.** 784   **29.** 4868   **31.** ◄◄◄▐▐▐▐▐

33. ◄▐▐▐▐ ◄◄◄◄◄▐   **35.** ▐ ▐ ◄◄▐▐▐▐▐   **37.** 57

39. 4682   **41.** 4000   **43.** ⠛   **45.** ⠿   **47.** ⠂

49. 1944, ◄◄◄▐▐ ◄◄▐▐▐▐

51. $\left(\triangle \times \square^2\right) + \left(\square \times \bigcirc\right) + \left(\diamondsuit \times 1\right)$

53. a) No largest numeral

b) (symbols)

55. ▐▐ ◄◄◄◄◄▐▐▐▐   **57.** (symbols)

59. a) Answers will vary.   **b)** Answers will vary.

c) Answers will vary.

SECTION 4.3, PAGE 202

1. a) Answers will vary.   **b)** Answers will vary.

3. a) There is no digit 4 in base 4.

b) There is no digit 3 in base 2.

c) There is no digit L in base 12.

d) There is no digit G in base 16.

5. 1   **7.** 13   **9.** 184   **11.** 441   **13.** 1395   **15.** 53

17. 4012   **19.** 1549   **21.** 50,809   **23.** 110_2 **25.** 110212_3

27. 311_4 **29.** 402_5 **31.** 2112_8 **33.** 1861_9 **35.** 5264_{12}

37. $1ED_{16}$ **39.** $24EF_{16}$ **41.** 2202102_3 **43.** 31014_5

45. 5600_7 **47.** 1567_{11} **49.** $8DE_{15}$ **51.** 23   **53.** 78

55. (symbols)$_5$ **57.** (symbols)$_5$ **59.** 14   **61.** 36   **63.** (symbols)$_4$

65. (symbols)$_4$

67. a) Answers will vary.   **b)** 10213_5 **c)** 1373_8

69. Answers will vary.   **71.** $b = 6$

73. a) 876   **b)** (symbols)$_4$

SECTION 4.4, PAGE 212

1. a) $1, b, b^2, b^3, b^4$ **b)** $1, 2, 2^2, 2^3, 2^4$

3. No, cannot have a 3 in base 3.

5. Answers will vary.   **7.** 111_3 **9.** 2130_5

11. $9B5_{12}$ **13.** 2200_3 **15.** 24001_7 **17.** 10100_2

19. 11_3 **21.** 831_9 **23.** 110_2 **25.** 11_2 **27.** 3616_7

29. 1011_3 **31.** 121_3 **33.** 4103_8 **35.** 21020_6

37. 77676_{12} **39.** 100011_2 **41.** 6031_7 **43.** 101_2

45. $223_6 R1_6$ **47.** 123_4 **49.** $103_4 R1_4$ **51.** $41_5 R1_5$

53. $45_7 R2_7$ **55.** (symbols)$_5$ **57.** (symbols)$_5$ **59.** (symbols)$_4$

61. (symbols)$_4$ **63.** (symbols)$_4$ **65.** (symbols)$_4$

67. 2302_5, 327   **69.** $3EAC_{16}$ **71.** 114_4R3_4

73. a) 21252_8 **b)** 306 and 29   **c)** 8874   **d)** 8874

e) Yes

75. ⬤ = 0,   ⬤ = 1,   ⬤ = 2,   ⬤ = 3,

SECTION 4.5, PAGE 217

1. Duplation and mediation, lattice multiplication, and Napier's rods

3. a) Answers will vary.   **b)** 3718

5. 1539   **7.** 1431   **9.** 8260   **11.** 8649

13. 955   **15.** 3752   **17.** 900   **19.** 204,728

21. 147   **23.** 315   **25.** 625   **27.** 60,678

29. a) 46×253 **b)** 11,638

31. a) 4×382 **b)** 1528

33. 99∩∩∩∩∩∩∩∩∩IIIIII   **35.** 1211_3

37. a) 1776   **b)** Answers will vary.

REVIEW EXERCISES, PAGE 219

1. 2121   **2.** 1214   **3.** 1311   **4.** 2114   **5.** 2314

6. 2312   **7.** *bbba*   **8.** *cbbaaaaa*

9. *ccbbbbbbbbbaaa*   **10.** *ddaaaaaaaa*

11. *dddddccccccccbbbba*   **12.** *ddcccbaaaa*

13. 32   **14.** 85   **15.** 749   **16.** 4068   **17.** 5648

18. 6905   **19.** *cxg*   **20.** *ayexg*   **21.** *hyfxb*

22. *czixd*   **23.** *fzd*   **24.** *bza*   **25.** 93   **26.** 203

27. 568   **28.** 46,883   **29.** 64,481   **30.** 60,529

31. la   **32.** tpd   **33.** vrc   **34.** BArg   **35.** ODvog

36. QFvrf   **37.** ʃ999999999∩∩∩∩∩∩∩IIIIII

38. MDCCLXXVI   **39.** 千七百七十六   **40.** 'αψof

41. <<<↑I <<<IIIIIII   **42.** ••••   **43.** 222,035

44. 8254   **45.** 685   **46.** 1991   **47.** 1277   **48.** 2690   **49.** 39

50. 5   **51.** 28   **52.** 1244   **53.** 1552   **54.** 186

55. 111001111_2 **56.** 122011_3 **57.** 13033_4

58. 717_8 **59.** 327_{12} **60.** $1CF_{16}$ **61.** 140_7 **62.** 101111_2

63. 166_{12} **64.** $70F_{16}$ **65.** 12102_5 **66.** 12423_8

67. 3411_7 **68.** 100_2 **69.** $9A6_{12}$ **70.** 3324_5 **71.** 450_8

72. $CC1_{16}$ **73.** 110111_2 **74.** 21102_3 **75.** 1314_5

76. 13632_8 **77.** 5656_{12} **78.** $1D76_{16}$ **79.** 21_3R1_3

80. 130_4 **81.** 23_5R1_5 **82.** 433_6 **83.** $411_6 R1_6$

84. $664_8 R 2_8$ **85.** 3408   **86.** 3408   **87.** 3408

CHAPTER TEST, PAGE 220

1. A number is a quantity and answers the question "How many?" A numeral is a symbol used to represent a number.

2. 2484   **3.** 1275   **4.** 8090   **5.** 969   **6.** 122,142

7. 2745   **8.** ʃʃ9∩∩IIII   **9.** 'βυof

10. (tally/dot symbol)   **11.** <<IIIIII <<<IIIIII

12. MMMDCCVI

13. In an additive system, the number represented by a particular set of numerals is the sum of the values of the numerals.

14. In a multiplicative system, there are numerals for each numeral less than the base and for powers of the base. Each numeral less than the base is multiplied by a numeral for the power of the base, and these products are added to obtain the number.

15. In a ciphered system, the number represented by a particular set of numerals is the sum of the values of the numerals. There are numerals for each number up to and including the base and multiples of the base.

16. In a place-value system, each numeral is multiplied by a power of the base. The position of the numeral indicates the power of the base by which it is multiplied.

17. 11   **18.** 103   **19.** 45   **20.** 559   **21.** 100100_2

22. 135_8 **23.** 1444_{12} **24.** $B7A_{16}$ **25.** 11000_2

26. 142_6 **27.** 2003_6 **28.** 220_5 **29.** 980   **30.** 8428

CHAPTER 5

SECTION 5.1, PAGE 232

1. Number theory is the study of numbers and their properties.

3. a) *a* divides *b* means that *b* divided by *a* has a remainder of zero.

b) *a* is divisible by *b* means that *a* divided by *b* has a remainder of zero.

5. A composite number is a natural number that is divisible by a number other than itself and 1.

7. a) The LCM of a set of natural numbers is the smallest natural number that is divisible by each number in the set.

b) Answers will vary.   **c)** 80

9. Mersenne primes are prime numbers of the form $2^n - 1$, where *n* is a prime number.

11. Goldbach's conjecture states that every even number greater than or equal to 4 can be represented as the sum of two (not necessarily distinct) prime numbers.

13. The prime numbers between 1 and 100 are 2, 3, 5, 7, 11, 13, 17, 19, 23, 29, 31, 37, 41, 43, 47, 53, 59, 61, 67, 71, 73, 79, 83, 89, and 97.

15. True   **17.** False; 26 is a multiple of 13.

19. False; 56 is divisible by 8.   **21.** True

23. False; if a number is divisible by 3, then the sum of the digits of the number is divisible by 3.

25. True   **27.** 11,115 is divisible by 3, 5, and 9.

29. 474,138 is divisible by 2, 3, 6, and 9.

31. 1,882,320 is divisible by 2, 3, 4, 5, 6, 8, and 10.

33. 60 (other answers are possible)

35. $48 = 2^4 \cdot 3$ **37.** $168 = 2^3 \cdot 3 \cdot 7$

39. $332 = 2^2 \cdot 83$ **41.** $513 = 3^3 \cdot 19$

43. $1336 = 2^3 \cdot 167$ **45.** $2001 = 3 \cdot 23 \cdot 29$

47. a) 3   **b)** 42   **49. a)** 5   **b)** 140   **51. a)** 20   **b)** 1800

53. a) 4   **b)** 5088   **55. a)** 8   **b)** 384

57. 17, 19, and 29, 31

59. a) Yes   **b)** No   **c)** Yes   **d)** Yes

61. 5, 17, and 257 are all prime.

63. 2×60, 3×40, 4×30, 5×24, 6×20, 8×15, 10×12, 12×10, 15×8, 20×6, 24×5, 30×4, 40×3, 60×2

65. 210 days   **67.** 35 cars   **69.** 30 trees   **71.** 30 days

73. A number is divisible by 15 if both 3 and 5 divide the number.

75. 5   **77.** 35   **79.** 30   **81.** No   **83.** No

85. a) 12   **b)** 1, 2, 3, 4, 5, 6, 10, 12, 15, 20, 30, 60

87. For any three consecutive natural numbers, one of the numbers is divisible by 2 and another number is divisible by 3. Therefore, the product of the three numbers would be divisible by 6.

89. Yes

91. $8 = 2 + 3 + 3$, $9 = 3 + 3 + 3$, $10 = 2 + 3 + 5$, $11 = 2 + 2 + 7$, $12 = 2 + 5 + 5$, $13 = 3 + 3 + 7$, $14 = 2 + 5 + 7$, $15 = 3 + 5 + 7$, $16 = 2 + 7 + 7$, $17 = 5 + 5 + 7$, $18 = 2 + 5 + 11$, $19 = 3 + 5 + 11$, $20 = 2 + 7 + 11$

93. The answer most people select is Denmark, Kangaroo, and Orange.

SECTION 5.2, PAGE 242

1. Begin at zero. Represent the first addend with an arrow. Draw the arrow to the right if the addend is positive, to the left if negative. From the tip of the first arrow, represent the second addend with a second arrow. The sum of the two integers is at the tip of the second arrow.

3. To rewrite a subtraction problem as an addition problem, rewrite the minus sign as a plus sign and change the second number to its opposite.

5. The product of two numbers with like signs is positive. The product of two numbers with unlike signs is negative.

7. 3   **9.** -4 **11.** -5 **13.** 2   **15.** -21 **17.** -6

19. -9 **21.** -2 **23.** -6 **25.** -2 **27.** -30 **29.** 64

31. 96   **33.** -60 **35.** -720 **37.** 7   **39.** -1

41. -7 **43.** -15 **45.** -48

In Exercises 47–55, false answers can be modified in a variety of ways. We give one possible answer.

47. True

49. False; the difference of two negative integers may be a positive integer, a negative integer, or zero.

51. True   **53.** True

55. False; the sum of a positive integer and a negative integer may be a positive integer, a negative integer, or zero.

57. 6   **59.** 20   **61.** -12 **63.** -5 **65.** -6

67. $-9, -6, -3, 0, 3, 6$ **69.** $-6, -5, -4, -3, -2, -1$

71. 11,109   **73.** 14,777 ft   **75.** 9 yards; no

77. a) 9 hours   **b)** 2 hours   **79.** -1

81. $0 + 1 - 2 + 3 + 4 - 5 + 6 - 7 - 8 + 9 = 1$

SECTION 5.3, PAGE 255

1. The set of rational numbers is the set of numbers of the form $\frac{p}{q}$, where p and q are integers and $q \neq 0$.

3. a) Divide both the numerator and the denominator by their greatest common factor.
   b) $\frac{4}{5}$

5. For positive mixed numbers, multiply the denominator of the fraction by the integer preceding it. Add this product to the numerator. This sum is the numerator of the improper fraction; the denominator is the same as the denominator in the mixed number. For negative mixed numbers, temporarily ignore the negative sign, perform the conversion described above, and then add the negative sign.

7. a) The reciprocal of a number is 1 divided by the number.
   b) $-\frac{1}{2}$

9. a) To add or subtract two fractions with a common denominator, perform the indicated operation on the numerators. Keep the common denominator. Reduce the new fraction to lowest terms, if possible.
   b) $\frac{7}{12}$ **c)** $\frac{1}{2}$

11. Answers will vary.   **13.** $\frac{1}{2}$ **15.** $\frac{4}{9}$ **17.** $\frac{19}{25}$ **19.** $\frac{7}{11}$

21. $\frac{1}{11}$ **23.** $\frac{29}{8}$ **25.** $-\frac{31}{16}$ **27.** $-\frac{79}{16}$ **29.** $\frac{9}{8}$ **31.** $\frac{15}{8}$

33. $2\frac{3}{5}$ **35.** $-12\frac{1}{6}$ **37.** $-58\frac{8}{15}$ **39.** 0.7 **41.** $0.\overline{2}$

43. 0.375 **45.** $2.1\overline{6}$ **47.** $5.\overline{6}$ **49.** $\frac{75}{100} = \frac{3}{4}$

51. $\frac{45}{1000} = \frac{9}{200}$ **53.** $\frac{2}{10} = \frac{1}{5}$ **55.** $\frac{131}{10,000}$ **57.** $\frac{1}{10,000}$ **59.** $\frac{1}{9}$

61. $\frac{2}{1}$ **63.** $\frac{15}{11}$ **65.** $\frac{37}{18}$ **67.** $\frac{574}{165}$ **69.** $\frac{2}{5}$ **71.** $\frac{2}{5}$ **73.** $\frac{49}{64}$

75. $\frac{36}{35}$ **77.** $\frac{20}{21}$ **79.** $\frac{11}{12}$ **81.** $\frac{9}{22}$ **83.** $\frac{23}{54}$ **85.** $\frac{17}{144}$

87. $-\frac{109}{600}$ **89.** $\frac{19}{24}$ **91.** $-\frac{1}{24}$ **93.** $\frac{19}{24}$ **95.** 1 **97.** $\frac{11}{10}$

99. $\frac{23}{42}$ **101.** $1\frac{3}{4}$ in. **103.** $120\frac{3}{4}$ in. **105.** $9\frac{15}{16}$ in.

107. $\frac{1}{10}$ **109.** $58\frac{7}{8}$ in.

111. **a)** $1\frac{49}{60}, 2\frac{48}{60}, 9\frac{6}{60}, 6\frac{3}{60}, 2\frac{9}{60}, \frac{22}{60}$

   b) $22\frac{17}{60}$; 22 hours 17 minutes

113. $26\frac{5}{32}$ in. **115. a)** $29\frac{3}{16}$ in. **b)** 33 in. **c)** $32\frac{3}{4}$ in.

In Exercises 117–121, an infinite number of answers are possible. We give one answer.

117. 0.105 **119.** -2.1755 **121.** 4.8725

123. $\frac{1}{2}$ **125.** $\frac{11}{200}$ **127.** $\frac{11}{200}$

129. a) $1\frac{3}{8}$ cup water (or milk) and $\frac{3}{4}$ cup oatmeal

   b) $1\frac{1}{2}$ cup water (or milk) and $\frac{3}{4}$ cup oatmeal

131. a) $\frac{1}{8}$ **b)** $\frac{1}{16}$ **c)** 5 **d)** 6

SECTION 5.4, PAGE 266

1. A rational number can be written as a ratio of two integers. Real numbers that cannot be written as a ratio of two integers are irrational numbers.

3. A perfect square is any number that is the square of a natural number.

5. a) To add or subtract two or more square roots with the same radicand, add or subtract their coefficients and then multiply the sum or difference by the common radical.
   b) $5\sqrt{5}$

7. a) Multiply both the numerator and denominator by a radical that will result in the radicand in the denominator becoming a perfect square.
   b) $\dfrac{2\sqrt{3}}{3}$

9. Rational **11.** Rational **13.** Irrational

15. Rational **17.** Irrational **19.** 4 **21.** 10

23. -13 **25.** -9 **27.** -10

29. Rational number, integer, natural number

31. Rational number, integer, natural number

33. Rational number **35.** Rational number

37. Rational number

39. $2\sqrt{3}$ **41.** $4\sqrt{3}$ **43.** $3\sqrt{7}$ **45.** $2\sqrt{21}$ **47.** $9\sqrt{2}$

49. $5\sqrt{5}$ **51.** $\sqrt{2}$ **53.** $-13\sqrt{3}$ **55.** $4\sqrt{3}$ **57.** $23\sqrt{2}$

59. 9 **61.** $2\sqrt{15}$ **63.** $10\sqrt{2}$ **65.** 2 **67.** 3

69. $\dfrac{\sqrt{5}}{5}$ **71.** $\dfrac{\sqrt{21}}{7}$ **73.** $\dfrac{2\sqrt{15}}{3}$ **75.** $\dfrac{\sqrt{15}}{3}$ **77.** $\dfrac{\sqrt{15}}{3}$

79. $\sqrt{5}$ is between 2 and 3 since 5 is between 4 and 9. $\sqrt{5}$ is between 2 and 2.5 since 5 is closer to 4 than to 9. $\sqrt{5} \approx 2.24$.

81. $\sqrt{107}$ is between 10 and 11 since 107 is between 100 and 121. $\sqrt{107}$ is between 10 and 10.5 since 107 is closer to 100 than to 121. $\sqrt{107} \approx 10.34$.

83. $\sqrt{170}$ is between 13 and 14 since 170 is between 169 and 196. $\sqrt{170}$ is between 13 and 13.5 since 170 is closer to 169 than to 196. $\sqrt{170} \approx 13.04$.

In Exercises 85–89, false answers can be modified in a variety of ways. We give one possible answer.

85. False. $\sqrt{c}$ may be a rational number or an irrational number for a composite number c. (For example, $\sqrt{25}$ is a rational number; $\sqrt{8}$ is an irrational number.)

87. True

89. False. The product of a rational number and an irrational number may be a rational number or an irrational number.

91. $3\sqrt{2} + 5\sqrt{2} = 8\sqrt{2}$ **93.** $\sqrt{2} \cdot \sqrt{3} = \sqrt{6}$

95. $\sqrt{2} \neq 1.414$ since $\sqrt{2}$ is irrational and 1.414 is rational.

97. No. π is irrational; therefore, it cannot equal $\frac{22}{7}$ or 3.14, both of which are rational.

99. $\sqrt{4 \cdot 9} = \sqrt{4} \cdot \sqrt{9}$, $\sqrt{36} = 2 \cdot 3$, $6 = 6$

101. a) 10 mph **b)** 20 mph **c)** 40 mph **d)** 80 mph

103. a) Rational. $\sqrt{0.04} = 0.2$, which is a rational number.
   b) Irrational. $\sqrt{0.7} = 0.8366600265\ldots$. Since the decimal number is not a terminating or a repeating decimal number, this number is an irrational number.

105. a) $(44 \div \sqrt{4}) \div \sqrt{4} = 11$ **b)** $(44 \div 4) + \sqrt{4} = 13$
   c) $4 + 4 + 4 + \sqrt{4} = 14$
   d) $\sqrt{4}(4 + 4) + \sqrt{4} = 18$
   Other answers are possible.

SECTION 5.5, PAGE 273

1. The real numbers are the union of the rational numbers and the irrational numbers.

3. If whenever the operation is performed on two elements of a set the result is also an element of the set, then the set is closed under that operation.

5. $a \cdot b = b \cdot a$, the order in which two numbers are multiplied is immaterial. One example is $4(5) = 5(4)$.

7. $(a + b) + c = a + (b + c)$, when adding three numbers, you may place parentheses around any two adjacent numbers. One example is
$(1 + 2) + 3 = 1 + (2 + 3)$.

9. No **11.** Yes **13.** Yes **15.** Yes **17.** Yes

19. No **21.** No **23.** No **25.** Yes **27.** Yes

29. Commutative property of addition. The only difference between the expressions on both sides of the equal sign is the order of 5 and x.

31. $(-3) + (-4) = (-4) + (-3) = -7$

33. No. $4 - 3 \neq 3 - 4$.

35. $[(-2) + (-3)] + (-4) = (-2) + [(-3) + (-4)] = -9$

37. No. $(16 \div 8) \div 2 \neq 16 \div (8 \div 2)$.

39. No. $(81 \div 9) \div 3 \neq 81 \div (9 \div 3)$.

41. Distributive property

43. Associative property of multiplication

45. Associative property of addition

47. Commutative property of multiplication

49. Distributive property

51. Commutative property of multiplication

53. Distributive property

55. Commutative property of multiplication

57. $4z + 4$ **59.** $-\frac{3}{4}x + 9$ **61.** $3x + 4$

63. $2x - 1$ **65.** 2   **67.** $5\sqrt{2} + 5\sqrt{3}$

69. a) Distributive property
b) Associative property of addition

71. a) Distributive property
b) Associative property of addition
c) Commutative property of addition
d) Associative property of addition

73. a) Distributive property
b) Commutative property of addition
c) Associative property of addition
d) Commutative property of addition

75. Yes   **77.** No   **79.** Yes   **81.** No   **83.** Yes

85. Yes   **87.** Yes   **89.** Answers will vary.

91. No. $0 \div a = 0$ (when $a \neq 0$), but $a \div 0$ is undefined.

SECTION 5.6, PAGE 284

1. The 2 is the base and the 3 is the exponent or power.

3. a) To multiply two exponential expressions with the same base, add the exponents and use this sum as the exponent on the common base.
b) $2^3 \cdot 2^4 = 2^{3+4} = 2^7 = 128$

5. a) Any nonzero expression raised to the power of 0 equals 1.
b) $7^0 = 1$

7. a) Any base with an exponent raised to another exponent is equal to the base raised to the product of the exponents.
b) $(3^2)^4 = 3^{2\cdot4} = 3^8 = 6561$

9. a) -1^{500} means $-(1)^{500}$ or $-1 \cdot 1^{500}$. Since 1 raised to any power equals 1, $-1^{500} = -1 \cdot 1^{500} = -1 \cdot 1 = -1$.
b) $(-1)^{500}$ means (-1) multiplied by itself 500 times. Since 500 is even, $(-1)^{500} = 1$.

c) -1^{501} means $-(1)^{501}$ or $-1 \cdot 1^{501}$. Since 1 raised to any power equals 1, $-1^{501} = -1 \cdot 1^{501} = -1 \cdot 1 = -1$.
d) $(-1)^{501}$ means (-1) multiplied by itself 501 times. Since 501 is odd, $(-1)^{501} = -1$.

11. a) If the exponent is positive, move the decimal point in the number to the right the same number of places as the exponent, adding zeros where necessary. If the exponent is negative, move the decimal point in the number to the left the same number of places as the exponent, adding zeros where necessary.
b) 0.0000291   **c)** 7,020,000

13. a) 9   **b)** 8   **15. a)** 25   **b)** -25

17. a) -16 **b)** 16   **19. a)** -64 **b)** -64

21. a) $\frac{1}{64}$ **b)** $\frac{9}{16}$ **23. a)** 1000   **b)** 1

25. a) 243   **b)** -243 **27. a)** 25   **b)** 25

29. a) 1   **b)** -1 **31. a)** 1   **b)** 6

33. a) $\frac{1}{27}$ **b)** $\frac{1}{49}$ **35. a)** $-\frac{1}{81}$ **b)** $\frac{1}{81}$

37. a) 64   **b)** 64   **39. a)** 4   **b)** $\frac{1}{16}$

41. 1.75×10^5 **43.** 2.3×10^{-4} **45.** 5.6×10^{-1}

47. 1.9×10^4 **49.** 1.86×10^{-4} **51.** 4.23×10^{-6}

53. 7.11×10^2 **55.** 1.53×10^{-1} **57.** 170

59. 0.0001097   **61.** 0.0000862   **63.** 0.312

65. 9,000,000   **67.** 231   **69.** 35,000   **71.** 10,000

73. 750,000   **75.** 0.0153   **77.** 250   **79.** 0.0021

81. 20   **83.** 1.0×10^{11} **85.** 4.5×10^{-7}

87. 7.0×10^1 **89.** 2.0×10^{-7} **91.** 3.0×10^8

93. 3.6×10^{-3}; 1.7; 9.8×10^2; 1.03×10^4

95. 8.3×10^{-5}; 0.00079; 4.1×10^3; 40,000

97. 0.046   **99.** 0.168

101. 2.708×10^{51} sec   **103.** $6779

105. 11.95 hours   **107.** 2.9×10^8 cells

109. 8.64×10^9 ft^3

111. a) $720,000,000   **b)** $300,000,000   **c)** $120,000,000
d) $60,000,000

113. 1000   **115.** 333,333 times

117. a) About 5.87×10^{12} (5.87 trillion) mi
b) About 500 sec or 8 min 20 sec

SECTION 5.7, PAGE 294

1. A sequence is a list of numbers that are related to each other by a given rule. One example is 1, 3, 5, 7, 9, ...

3. a) An arithmetic sequence is one in which each term differs from the preceding term by a constant amount. One example is 4, 7, 10, 13, 16, ...
b) A geometric sequence is one in which the ratio of any two successive terms is a constant amount. One example is 3, 6, 12, 24, ...

5. a) a_n is the nth term or the general term.
  b) a_1 is the first term.
  c) d is the common difference.
  d) s_n is the sum of the first n terms.

7. 5, 6, 7, 8, 9 **9.** 12, 10, 8, 6, 4

11. 5, 3, 1, -1, -3 **13.** $\frac{3}{4}$, 1, $\frac{5}{4}$, $\frac{3}{2}$, $\frac{7}{4}$ **15.** 8 **17.** 13

19. $-\frac{91}{5}$ **21.** -8 **23.** $a_n = n$ **25.** $a_n = 2n$

27. $a_n = \frac{3}{4}n - \frac{1}{2}$ **29.** $a_n = \frac{3}{2}n - \frac{9}{2}$ **31.** $s_{50} = 1275$

33. $s_{50} = 2500$ **35.** $s_8 = -52$ **37.** $s_{24} = 60$

39. 1, 5, 25, 125, 625 **41.** 2, -4, 8, -16, 32

43. -3, 3, -3, 3, -3 **45.** 81, -27, 9, -3, 1

47. 160 **49.** $\frac{3}{4}$ **51.** -3645 **53.** $a_{10} = -39{,}366$

55. $a_n = 2^{n-1}$ **57.** $a_n = (-1)(-1)^{n-1}$

59. $a_n = 2 \cdot \left(\frac{1}{2}\right)^{n-1}$ **61.** $a_n = 9 \cdot \left(\frac{1}{3}\right)^{n-1}$

63. 186 **65.** -4095 **67.** $-620{,}011$

69. $-10{,}923$ **71.** 5050 **73.** 10,100

75. a) 63 in. **b)** 954 in.

77. a) $44,800 **b)** $319,200 **79.** 496 pinecones

81. 52.4288 g **83.** $\approx$ $70,088 **85.** $486,000

87. 161.4375 **89.** 267 **91.** 191.3568 ft

SECTION 5.8, PAGE 302

1. The first and second terms are 1. Each term thereafter is the sum of the previous two terms.

3. a) The golden number is $\dfrac{\sqrt{5} + 1}{2}$.

  b) When a line segment AB is divided at a point C, such that the ratio of the whole, AB, to the larger part, AC, is equal to the ratio of the larger part, AC, to the smaller part, CB, then each of the two ratios AB/AC and AC/CB is known as the golden ratio.

  c) The proportion made by using the two golden ratios, $AB/AC = AC/CB$, is known as the golden proportion.

  d) A golden rectangle is one where the ratio of the length to the width is equal to the golden number.

5. Answers will vary.

7. a) 1.618 **b)** 0.618 **c)** 1

9. $\frac{1}{1} = 1$, $\frac{2}{1} = 2$, $\frac{3}{2} = 1.5$, $\frac{5}{3} \approx 1.667$, $\frac{8}{5} = 1.6$, $\frac{13}{8} = 1.625$, $\frac{21}{13} \approx 1.615$, $\frac{34}{21} \approx 1.619$, $\frac{55}{34} \approx 1.6176$, $\frac{89}{55} \approx 1.6182$. The consecutive ratios alternate, increasing and decreasing about the golden ratio.

11. Answers will vary. **13.** Answers will vary.

15. Answers will vary. **17.** Answers will vary.

19. Answers will vary. **21.** Answers will vary.

23. No **25.** Yes; 3, 5 **27.** Yes; 105, 170

29. Yes; -1, -1 **31.** Answers will vary.

33. Answers will vary.

35. a) 1, 3, 4, 7, 11, 18, 29, 47
  b) $8 + 21 = 29$, $13 + 34 = 47$
  c) It is the Fibonacci sequence.

37. Answers will vary.

39. Answers will vary.

REVIEW EXERCISES, PAGE 306

1. 2, 3, 4, 5, 6, 10 **2.** 2, 3, 4, 6, 9 **3.** $2^2 \cdot 3^3 \cdot 5$

4. $3^2 \cdot 7 \cdot 11$ **5.** $2^3 \cdot 3 \cdot 5 \cdot 7$ **6.** $2 \cdot 3^2 \cdot 7^2$

7. $2^2 \cdot 3 \cdot 11^2$ **8.** 15; 210 **9.** 9; 756 **10.** 5; 2250

11. 30; 900 **12.** 4; 480 **13.** 36; 432 **14.** 45 days

15. -3 **16.** 3 **17.** -6 **18.** -4 **19.** -9 **20.** 3

21. 0 **22.** 4 **23.** -15 **24.** 24 **25.** -56 **26.** 5

27. -2 **28.** 6 **29.** 4 **30.** 3 **31.** 0.3 **32.** 0.44

33. 0.375 **34.** 3.25 **35.** $0.\overline{857142}$ **36.** $0.58\overline{3}$

37. 0.375 **38.** 0.6875 **39.** $0.\overline{714285}$ **40.** $\frac{9}{40}$

41. $\frac{14}{10} = \frac{7}{5}$ **42.** $\frac{2}{3}$ **43.** $\frac{51}{99}$ **44.** $\frac{83}{1000}$ **45.** $\frac{73}{10{,}000}$ **46.** $\frac{211}{90}$

47. $\frac{7}{4}$ **48.** $\frac{25}{6}$ **49.** $-\frac{13}{4}$ **50.** $-\frac{283}{8}$ **51.** $2\frac{1}{5}$ **52.** $9\frac{3}{8}$

53. $-1\frac{5}{7}$ **54.** $-27\frac{1}{5}$ **55.** $\frac{13}{12}$ **56.** $\frac{1}{4}$ **57.** $\frac{17}{12}$ **58.** $\frac{1}{4}$

59. $\frac{35}{54}$ **60.** $\frac{53}{28}$ **61.** $\frac{1}{6}$ **62.** $\frac{13}{40}$ **63.** $\frac{8}{15}$ **64.** $2\frac{7}{32}$ tsp

65. $3\sqrt{5}$ **66.** $10\sqrt{2}$ **67.** $8\sqrt{5}$ **68.** $-3\sqrt{2}$

69. $8\sqrt{2}$ **70.** $-20\sqrt{3}$ **71.** $5\sqrt{7}$ **72.** $3\sqrt{2}$

73. $4\sqrt{3}$ **74.** 10 **75.** $2\sqrt{7}$ **76.** $\dfrac{4\sqrt{3}}{3}$ **77.** $\dfrac{\sqrt{35}}{5}$

78. $6 + 3\sqrt{7}$ **79.** $4\sqrt{3} + 3\sqrt{2}$ **80.** $3\sqrt{2} + 3\sqrt{5}$

81. Commutative property of addition

82. Commutative property of multiplication

83. Associative property of addition

84. Distributive property

85. Associative property of addition

86. Commutative property of addition

87. Associative property of multiplication

88. Commutative property of multiplication

89. Distributive property

90. Commutative property of multiplication

91. No **92.** Yes **93.** No **94.** Yes **95.** No **96.** No

97. 25 **98.** $\frac{1}{25}$ **99.** 81 **100.** 125 **101.** 1 **102.** $\frac{1}{64}$

103. 64 **104.** 81 **105.** 8.2×10^9 **106.** 1.58×10^{-5}

107. 2.309×10^{-2} **108.** 4.95×10^6 **109.** 280,000

110. 0.000139 **111.** 0.000175 **112.** 10,000,000

113. 6.0×10^{-5} **114.** 3.75×10^{12} **115.** 2.1×10^1

116. 3.0×10^0 **117.** 1,100,000,000,000 **118.** 0.7

119. 120 **120.** 5 **121.** $\approx$388 times **122.** $\approx$ $5555.56

123. Arithmetic; 17, 21   **124.** Geometric; 8, 16
125. Arithmetic; $-15, -18$ **126.** Geometric; $\frac{1}{32}, \frac{1}{64}$
127. Arithmetic; 16, 19   **128.** Geometric; $\frac{1}{2}, -\frac{1}{2}$
129. 27   **130.** 10   **131.** 25   **132.** 48   **133.** $\frac{1}{4}$
134. -48 **135.** 3825   **136.** -25 **137.** 632
138. 57.5   **139.** 682   **140.** 45   **141.** 33   **142.** -21
143. Arithmetic; $a_n = 3n$
144. Arithmetic; $a_n = 3n - 2$
145. Arithmetic; $a_n = -\frac{3}{2}n + \frac{11}{2}$
146. Geometric; $a_n = 3(2)^{n-1}$
147. Geometric; $a_n = 2(-1)^{n-1}$
148. Geometric; $a_n = 5\left(\frac{1}{3}\right)^{n-1}$ **149.** No
150. Yes; $-8, -13$ **151.** No   **152.** No

CHAPTER TEST, PAGE 309

1. 5   **2.** $2 \cdot 3^2 \cdot 23$ **3.** 8   **4.** -20
5. -175 **6.** $\frac{37}{8}$ **7.** $19\frac{5}{9}$ **8.** 0.625   **9.** $\frac{129}{20}$
10. $\frac{121}{240}$ **11.** $\frac{1}{8}$ **12.** $9\sqrt{3}$ **13.** $\frac{\sqrt{14}}{7}$
14. Yes; the product of any two integers is an integer.
15. Associative property of addition
16. Distributive property   **17.** 64   **18.** 1024   **19.** $\frac{1}{81}$
20. 8.0×10^{11} **21.** $a_n = -4n + 2$ **22.** -187
23. 486   **24.** -510 **25.** $a_n = 3(2)^{n-1}$
26. 1, 1, 2, 3, 5, 8, 13, 21, 34, 55

CHAPTER 6

SECTION 6.1, PAGE 315

1. Letters of the alphabet used to represent numbers are called variables.
3. The solution to an equation is the number or numbers that replace the variable to make the equation a true statement.
5. a) The 4 is the base and the 5 is the exponent.
  b) Answers will vary.
7. a) 12 **b)** 27   **9.** 25   **11.** -4 **13.** 686   **15.** -3
17. 12   **19.** -20 **21.** $-\frac{10}{9}$ **23.** 7   **25.** 9   **27.** 39
29. No   **31.** No   **33.** No   **35.** Yes   **37.** Yes
39. \$62.93   **41.** 96.8 million   **43.** 88.72 min   **45.** 1.71 in.
47. The two expressions are not equal.

SECTION 6.2, PAGE 327

1. The parts that are added or subtracted in an algebraic expression are called terms. In $3x - 2y$, the $3x$ and $-2y$ are terms.

3. The numerical part of a term is called its numerical coefficient. For the term $3x$, 3 is the numerical coefficient.
5. To simplify an expression means to combine like terms by using the commutative, associative, and distributive properties.
7. If $a = b$, then $a \cdot c = b \cdot c$ for all real numbers a, b, and c, where $c \neq 0$. If $\frac{x}{3} = 2$, then $3\left(\frac{x}{3}\right) = 3(2)$.
9. If $a = b$, then $a/c = b/c$ for all real numbers a, b, and c, where $c \neq 0$. If $4x = 8$, then $\frac{4x}{4} = \frac{8}{4}$.
11. A ratio is a quotient of two quantities. An example is $\frac{7}{9}$.
13. Yes. They have the same variable and the same exponent on the variable.
15. $10x$ **17.** $2x + 12$ **19.** $3x + 11y$ **21.** $-8x + 2$
23. $-5x + 3$ **25.** $13.3x - 8.3$ **27.** $-\frac{2}{15}x - 4$
29. $10x - 9y + 3$ **31.** $8s - 17$ **33.** $1.4x - 2.8$
35. $-\frac{5}{12}x + \frac{4}{5}$ **37.** $4.52x - 13.5$ **39.** 17   **41.** -1 **43.** $\frac{24}{7}$
45. $\frac{2}{3}$ **47.** 4   **49.** 3   **51.** 17   **53.** -8 **55.** No solution
57. All real numbers   **59.** 3   **61.** -3 **63.** 5
65. \$209.25   **67.** 4 gallons
69. $\approx 15{,}868{,}800$ households
71. a) 1.6 kph **b)** 56.25 mph   **73.** 0.375 cc
75. a) Answers will vary. **b)** -1
77. a) An equation that has no solution.
  b) You will obtain a false statement.
79. a) $2 : 5$ **b)** $m : m + n$

SECTION 6.3, PAGE 337

1. A formula is an equation that typically has a real-life application.
3. Subscripts are numbers (or letters) placed below and to the right of variables. They are used to help clarify a formula.
5. An exponential equation is of the form $y = a^x$, $a > 0$, $a \neq 1$.
7. 56   **9.** 56   **11.** 10   **13.** 251.2   **15.** 37.1
17. 2   **19.** 3000   **21.** 12   **23.** 25   **25.** 6
27. 200   **29.** 7.2   **31.** 14   **33.** 3240
35. 66.67   **37.** 39   **39.** 7609.81
41. $y = \dfrac{4x - 14}{9}$ or $y = \dfrac{4}{9}x - \dfrac{14}{9}$
43. $y = \dfrac{-8x + 21}{7}$ or $y = -\dfrac{8}{7}x + 3$
45. $y = \dfrac{2x + 6}{3}$ or $y = \dfrac{2}{3}x + 2$

47. $y = \dfrac{2x - z + 15}{3}$ or $y = \dfrac{2}{3}x - \dfrac{1}{3}z + 5$

49. $y = \dfrac{9x + 4z - 7}{8}$ or $y = \dfrac{9}{8}x + \dfrac{1}{2}z - \dfrac{7}{8}$

51. $r = \dfrac{d}{t}$ **53.** $a = p - b - c$

55. $B = \dfrac{3V}{h}$ **57.** $r = \dfrac{C}{2\pi}$

59. $b = y - mx$ **61.** $w = \dfrac{P - 2l}{2}$

63. $c = 3A - a - b$ **65.** $T = \dfrac{PV}{K}$

67. $C = \frac{5}{9}(F - 32)$ **69.** $h = \dfrac{S - 2\pi r^2}{2\pi r}$

71. a) \$112.50   **b)** \$4612.50   **73.** ≈ 4.19 in.^{3}

75. 486,000 bacteria   **77.** $\approx \$4.49 \times 10^{14}$

79. ≈ 1051.47 in.^{3}

SECTION 6.4, PAGE 343

1. A mathematical expression is a collection of variables, numbers, parentheses, and operation symbols. An equation is two algebraic expressions joined by an equal sign.

3. 4 more than x **5.** 2 times x, decreased by 3

7. $8 + x$ **9.** $3 + 2z$ **11.** $6w + 9$ **13.** $4x + 6$

15. $\dfrac{18 - s}{4}$ **17.** $3(x + 7)$ **19.** $x + 5 = 11; 6$

21. $x - 4 = 20; 24$ **23.** $4x - 10 = 42; 13$

25. $4x + 12 = 32; 5$ **27.** $x + 6 = 2x - 3; 9$

29. $x + 10 = 2(x + 3); 4$

31. $150 + 0.42x = 207.54$; 137 miles

33. $x + 0.05x = 42$; \$40 per half hour

35. $0.08x = 250$; 3125 copies

37. $x + 3x = 1000$; \$250 for business, \$750 for liberal arts

39. $2w + 2(w + 3) = 54$; width: 12 ft, length: 15 ft

41. $x + 6x + 2515 = 21{,}730$; Utah: 2745 mustangs; Nevada: 18,985 mustangs

43. $x + 14x + 11 = 341$; December: 22 tornados; May: 319 tornados

45. $3w + 2(2w) = 140$; width: 20 ft, length: 40 ft

47. $70x = 760$; ≈ 11 months

49. $\dfrac{r}{2} + 0.07r = 257$; \$450.88

51. Deduct \$720 from Mr. McAdams's income and \$2920 from Mrs. McAdams's income.

53.
$$x + (x + 1) + (x + 2) = 3(x + 2) - 3$$
$$3x + 3 = 3x + 6 - 3$$
$$3x + 3 = 3x + 3$$

55. $-40°$

SECTION 6.5, PAGE 352

1. Direct variation: As one variable increases, so does the other, and as one variable decreases, so does the other.

3. Joint variation: One quantity varies directly as the product of two or more other quantities.

5. Direct   **7.** Inverse   **9.** Direct   **11.** Inverse

13. Inverse   **15.** Inverse   **17.** Direct   **19.** Direct

21. Answers will vary.   **23. a)** $y = kx$ **b)** 120

25. a) $m = \dfrac{k}{n^2}$ **b)** 0.25   **27. a)** $A = \dfrac{kB}{C}$ **b)** 2.5

29. a) $F = kDE$ **b)** 210   **31. a)** $t = \dfrac{kd^2}{f}$ **b)** 200

33. a) $Z = kWY$ **b)** 100   **35. a)** $H = kL$ **b)** 3

37. a) $A = kB^2$ **b)** 720   **39. a)** $F = \dfrac{kq_1q_2}{d^2}$ **b)** 1200

41. a) $t = kv$ **b)** \$2700   **43. a)** $l = \dfrac{k}{d^2}$ **b)** 80 dB

45. a) $R = \dfrac{kA}{P}$ **b)** 4800 tapes

47. a) $v = \dfrac{k\sqrt{t}}{l}$ **b)** 4 vibrations per second

49. a) $N = \dfrac{kp_1p_2}{d}$ **b)** $\approx 121{,}528$ calls

51. a) Inversely   **b)** Stays 0.3   **53.** \$132.27

SECTION 6.6, PAGE 360

1. $a < b$ means that a is less than b, $a \le b$ means that a is less than or equal to b, $a > b$ means that a is greater than b, $a \ge b$ means that a is greater than or equal to b.

3. When both sides of an inequality are multiplied or divided by a negative number, the direction of the inequality symbol must be reversed.

5. Yes, the inequality symbol points to the -3 in both cases.

7. a) An inequality of the form $a < x < b$ is called a compound inequality.
b) $-5 < x < 3$

9.

11.

13.

15.

17.

19. (number line showing point at 4, ray to right)
  0   4

21. No solution
  0

23.
  −1   3

25.
  10   13

27.
  1 0 1 2 3 4 5 6

29.
  −11 −10 −9 −8 −7 −6 −5 −4

31.
  −1 0 1 2 3 4 5 6

33.
  −9 −8 −7 −6 −5 −4 −3 −2

35.
  −11 −10 −9 −8 −7 −6

37.
  −2 −1 0 1 2 3 4 5

39.
  −6 −5 −4 −3 −2 −1 0 1

41.
  −5 −4 −3 −2 −1 0 1 2

43.
  −1 0 1 2 3 4 5 6

45.
  −3 −2 −1 0 1 2 3 4

47. a) 2009–2015   **b)** 2005–2008
  c) 2006–2015   **d)** 2005–2006

49. 19 videos   **51.** 350 hours

53. 6 hours   **55.** $0.5 < t < 1.5$

57. $94 \le x \le 100$, assuming 100 is the highest grade possible

59. $6.875 \le x \le 11$

61. The student's answer is $x \le -12$, whereas the correct answer is $x \ge -12$. Yes, -12 is in both solution sets.

SECTION 6.7, PAGE 372

1. A graph is an illustration of all the points whose coordinates satisfy an equation.

3. To find the y-intercept, set $x = 0$ and solve the equation for y.

5. a) Answers will vary.   **b)** $-\frac{1}{3}$

7. a) First   **b)** Third

For Exercises 9–15, see the following figure.

For Exercises 17–23, see the following figure.

25. $(0, 2)$ **27.** $(-2, 0)$ **29.** $(-5, -3)$ **31.** $(2, -3)$
33. $(2, 2)$ **35.** $(5, 2), (1, 4)$ **37.** $(8, 2), \left(0, -\frac{10}{3}\right)$
39. $(-3, -2)$ **41.** $(8, 0), (0, 3)$

43.

Slope: undefined

45.

Slope is 0.

47.

49.

51.

53.

55.

57.

59.

61.

63.

65.

67. 2 **69.** $-\frac{5}{4}$ **71.** 0 **73.** Undefined **75.** $-\frac{8}{11}$

77.

79.

81.

83.

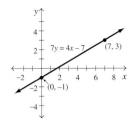

$7y = 4x - 7$ (7, 3)   (0, -1)

85.

$3x - 2y + 6 = 0$ (0, 3)   (-2, 0)

87. $y = -2x + 4$ **89.** $y = 3x + 2$

91. a) $D(3, -2)$ **b)** $A = 20$ square units

93. $(7, 2)$ or $(-1, 2)$ **95.** -3 **97.** 3

99. a)

(500, 190)   (200, 100) $C = 40 + 0.30s$ (0, 40)

Cost ($)   Square feet

b) $130   **c)** 100 square feet

101. a)

(50, 1325)   (20, 575)   (0, 75)

Cost ($)   Number of months

b) $825   **c)** 36 months

103. a) 2   **b)** $y = 2x + 9$ **c)** 15 defects   **d)** 4 workers

105. a) ≈ -0.05 **b)** $y = -0.05x + 2.60$
   c) $\approx$ \$2.50 billion   **d)** 4 years after 2000, or in 2004

107. a) Solve the equations for y to put them in slope–intercept form. Then compare the slopes and y-intercepts. If the slopes are equal but the y-intercepts are different, then the lines are parallel.
   b) The lines are parallel.

SECTION 6.8, PAGE 378

1. (1) Mentally substitute the equal sign for the inequality sign and plot points as if you were graphing the equation. (2) If the inequality is $<$ or $>$, draw a dashed line through the points. If the inequality is $\leq$ or $\geq$, draw a solid line through the points. (3) Select a test point not on the line and substitute the x- and y-coordinates into the inequality. If the substitution results in a true statement, shade in the area on the same side of the line as the test point. If the test point results in a false statement, shade in the area on the opposite side of the line as the test point.

3. A half plane is the set of all the points in a plane on one side of a line.

5. a) No   **b)** Yes   **c)** Yes   **d)** No

7.

$x = 3$

9.

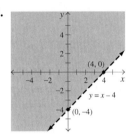

(4, 0) $y = x - 4$ (0, -4)

11.

(3, 0) $y = 2x - 6$ (0, -6)

13.

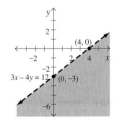

(4, 0) $3x - 4y = 12$ (0, -3)

15.

17.

19.

21.

23.

25.

27.

29. a) $x + y \leq 300$
  b)

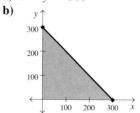

31. a) No, you cannot have a negative number of shirts.
  b)

  c) Answers will vary.

SECTION 6.9, PAGE 388

1. A binomial is an expression that contains two terms in which each exponent that appears on the variable is a whole number. $2x + 3, x - 7, x^2 - 9$

3. Answers will vary.   **5.** $ax^2 + bx + c = 0, a \neq 0$

7. $(x + 5)(x + 3)$ **9.** $(x - 2)(x + 1)$

11. $(x + 6)(x - 4)$ **13.** $(x + 1)(x - 3)$

15. $(x - 7)(x - 3)$ **17.** $(x - 7)(x + 7)$

19. $(x + 7)(x - 4)$ **21.** $(x + 9)(x - 7)$

23. $(2x + 3)(x - 2)$ **25.** $(5x + 1)(x + 3)$

27. $(5x + 2)(x + 2)$ **29.** $(4x + 3)(x + 2)$

31. $(4x - 3)(x - 2)$ **33.** $(4x + 1)(2x - 3)$

35. $3, -6$ **37.** $-\frac{4}{3}, \frac{1}{2}$ **39.** $-5, -2$ **41.** $6, 1$ **43.** $5, -3$

45. $3, 1$ **47.** $9, -9$ **49.** $-9, 4$ **51.** $\frac{2}{3}, -4$ **53.** $-\frac{1}{5}, -2$

55. $\frac{1}{3}, 1$ **57.** $\frac{1}{3}, \frac{3}{2}$ **59.** $3, -5$ **61.** $6, -3$ **63.** $9, -1$

65. No real solution   **67.** $2 \pm \sqrt{2}$

69. $\dfrac{4 \pm \sqrt{13}}{3}$ **71.** $\dfrac{3 \pm 2\sqrt{6}}{3}$ **73.** $-1, -\frac{5}{2}$ **75.** $\frac{7}{3}, 1$

77. No real solution

79. Width $= 12$ m, length $= 22$ m

81. a) The zero-factor property cannot be used.
b) $\approx 8.37, \approx 2.63$

83. $x^2 - 2x - 3 = 0$

SECTION 6.10, PAGE 401

1. A function is a special type of relation in which each value of the independent variable corresponds to a unique value of the dependent variable.

3. The domain of a function is the set of values that can be used for the independent variable.

5. If a vertical line touches more than one point on the graph, then for each value of x there is not a unique value for y and the graph does not represent a function.

7. $x = -\dfrac{b}{2a}$

9. Not a function

11. Function, domain: $\mathbb{R}$; range: $\mathbb{R}$

13. Not a function

15. Function, domain: $\mathbb{R}$; range: $y \geq -4$

17. Not a function

19. Function, domain: $0 \leq x < 12$; range: $y = 1, 2, 3$

21. Not a function

23. Function, domain: $\mathbb{R}$; range: $y > 0$

25. Not a function

27. Yes   **29.** No   **31.** Yes   **33.** 8   **35.** 1

37. -6 **39.** 5   **41.** -23 **43.** 45   **45.** -17

47.

49.

51.

53. a) Upward   **b)** $x = 0$ **c)** $(0, -9)$ **d)** $(0, -9)$
e) $(3, 0), (-3, 0)$
f)

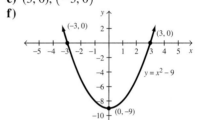

g) Domain: $\mathbb{R}$; range: $y \geq -9$

55. a) Downward   **b)** $x = 0$ **c)** $(0, 4)$ **d)** $(0, 4)$
e) $(-2, 0), (2, 0)$
f)

g) Domain: $\mathbb{R}$; range: $y \leq 4$

57. a) Downward   **b)** $x = 0$ **c)** $(0, -8)$ **d)** $(0, -8)$
e) No x-intercepts
f)

g) Domain: $\mathbb{R}$; range: $y \leq -8$

59. a) Upward   **b)** $x = 0$ **c)** $(0, -3)$ **d)** $(0, -3)$
e) $(-1.22, 0), (1.22, 0)$
f)

(graph: $y = 2x^2 - 3$, points $(-1.22, 0)$, $(1.22, 0)$, $(0, -3)$)

g) Domain: $\mathbb{R}$; range: $y \geq -3$

61. a) Upward   **b)** $x = -1$ **c)** $(-1, 5)$ **d)** $(0, 6)$
e) No x-intercepts
f)

(graph: $f(x) = x^2 + 2x + 6$, points $(-4, 14)$, $(2, 14)$, $(-1, 5)$, $(0, 6)$)

g) Domain: $\mathbb{R}$; range: $y \geq 5$

63. a) Upward   **b)** $x = -\frac{5}{2}$ **c)** $(-2.5, -0.25)$ **d)** $(0, 6)$
e) $(-3, 0), (-2, 0)$
f)

(graph: $y = x^2 + 5x + 6$, points $(-3, 0)$, $(-2, 0)$, $(0, 6)$, $(-2.5, -0.25)$)

g) Domain: $\mathbb{R}$; range: $y \geq -0.25$

65. a) Downward   **b)** $x = 2$ **c)** $(2, -2)$ **d)** $(0, -6)$
e) No x-intercepts

f)

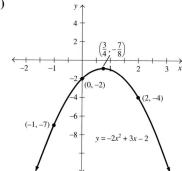

(graph: $y = -x^2 + 4x - 6$, points $(2, -2)$, $(0, -6)$, $(6, -18)$)

g) Domain: $\mathbb{R}$; range: $y \leq -2$

67. a) Downward   **b)** $x = \frac{3}{4}$ **c)** $\left(\frac{3}{4}, -\frac{7}{8}\right)$ **d)** $(0, -2)$
e) No x-intercepts
f)

(graph: $y = -2x^2 + 3x - 2$, points $\left(\frac{3}{4}, -\frac{7}{8}\right)$, $(0, -2)$, $(2, -4)$, $(-1, -7)$)

g) Domain: $\mathbb{R}$; range: $y \leq -\frac{7}{8}$

69. Domain: $\mathbb{R}$; range: $y > 0$

(graph: $y = 3^x$, points $(2, 9)$, $(1, 3)$, $(0, 1)$, $\left(-1, \frac{1}{3}\right)$)

71. Domain: $\mathbb{R}$; range: $y > 0$

(graph: $y = \left(\frac{1}{3}\right)^x$, points $(-2, 9)$, $(-1, 3)$, $(0, 1)$, $\left(1, \frac{1}{3}\right)$)

73. Domain: $\mathbb{R}$; range: $y > 1$

75. Domain: $\mathbb{R}$; range: $y > 1$

77. Domain: $\mathbb{R}$; range: $y > 0$

79. Domain: $\mathbb{R}$; range: $y > 0$

81. $41,000   **83. a)** $\approx 13\%$ **b)** 1970   **c)** $x \approx 36; \approx 5\%$

85. a) 5200 people   **b)** $\approx 14{,}852$ people

87. a) Yes   **b)** $\approx \$59$

89. a) 23.2 cm   **b)** 55.2 cm   **c)** 69.5 cm

91. a) 170 beats per minute
  b) ≈ 162 beats per minute
  c) ≈ 145 beats per minute
  d) 136 beats per minute
  e) 120 years of age

REVIEW EXERCISES, PAGE 407

1. 19   **2.** -8 **3.** 17   **4.** $\frac{1}{4}$ **5.** -13 **6.** 13

7. $4x + 5$ **8.** $15x - 10$ **9.** $7x - 3$ **10.** -10

11. -3 **12.** -13 **13.** -31 **14.** $\frac{128}{5}$ **15.** $\frac{1}{2}$ cup

16. 250 min, or 4 hr 10 min   **17.** 48   **18.** ≈ 173.1

19. 101.5   **20.** 25

21. $y = 2x - 6$

22. $y = \dfrac{-2x + 15}{7}$ or $y = -\dfrac{2}{7}x + \dfrac{15}{7}$

23. $y = \dfrac{2x + 22}{3}$ or $y = \dfrac{2}{3}x + \dfrac{22}{3}$

24. $y = \dfrac{-3x + 5z - 4}{4}$ or $y = -\dfrac{3}{4}x + \dfrac{5}{4}z - 1$

25. $w = \dfrac{A}{l}$ **26.** $w = \dfrac{P - 2l}{2}$

27. $l = \dfrac{L - 2wh}{2h}$ or $l = \dfrac{L}{2h} - w$

28. $d = \dfrac{a_n - a_1}{n - 1}$

29. $7 - 4x$ **30.** $2y + 7$ **31.** $10 + 3r$ **32.** $\dfrac{9}{q} - 15$

33. $3 + 7x = 17; x = 2$ **34.** $3x + 8 = x - 6; x = -7$

35. $5(x - 4) = 45; x = 13$

36. $10x + 14 = 8(x + 12); x = 41$

37. $x + 2x = 15{,}000$; bonds: \$5000, mutual funds: \$10,000

38. $9.50x + 15{,}000 = 95{,}000; \approx 8421$ chairs

39. $x + 6x + 2 = 79$; 11 species of threatened mammals, 68 species of endangered mammals

40. $x + (x + 12{,}000) = 68{,}000$; \$28,000 for B and \$40,000 for A

41. 64   **42.** 2   **43.** 20   **44.** ≈ 426.7

45. a) 150 lb   **b)** 5 bags   **46.** 4 in.

47. \$119.88   **48.** 1.75 in.

49.

50.

51.

52.

53.

54.

55.

56.

For Exercises 57–60, see the following figure.

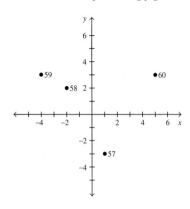

61. $D(-3, -1)$; area $= 20$ square units

62. $D(4, 1)$; area $= 21$ square units

63.

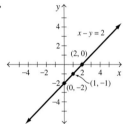

64.

65.

66.

67.

68.

69.

70.

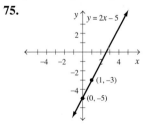

71. $\frac{2}{5}$ **72.** $-\frac{3}{2}$ **73.** $\frac{7}{3}$ **74.** Undefined

75.

76.

77.

78.

79. $y = 2x + 4$ **80.** $y = -x + 1$

81. a)

b) \$600   **c)** 70 cars

82. a)

b) $\approx$ \$6400   **c)** $\approx$ 4120 ft^2

83.

84.

85.

86.

87. $(x + 2)(x + 9)$ **88.** $(x + 6)(x - 5)$

89. $(x - 6)(x - 4)$ **90.** $(x - 5)(x - 4)$

91. $(3x - 1)(2x + 3)$ **92.** $(2x - 1)(x + 7)$

93. $-1, -3$ **94.** $-6, 3$ **95.** $\frac{2}{3}, 5$

96. $-2, -\frac{1}{3}$ **97.** $2 \pm \sqrt{5}$ **98.** $-2, 8$

99. No real solution   **100.** $-1, \frac{3}{2}$

101. Function, domain: $x = -2, -1, 2, 3$; range: $y = -1, 0, 2$

102. Not a function   **103.** Not a function

104. Function, domain: $\mathbb{R}$; range: $\mathbb{R}$

105. 5   **106.** 15   **107.** 39   **108.** -21

109. a) Downward   **b)** $x = -2$ **c)** $(-2, 25)$ **d)** $(0, 21)$

e) $(-7, 0), (3, 0)$

f)

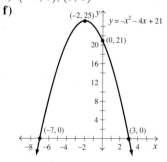

g) Domain: $\mathbb{R}$; range: $y \leq 25$

110. **a)** Upward **b)** $x = -2$ **c)** $(-2, -2)$ **d)** $(0, 6)$
e) $(-3, 0), (-1, 0)$
f)

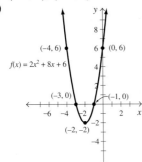

g) Domain: $\mathbb{R}$; range: $y \geq -2$

111. Domain: $\mathbb{R}$; range: $y > 0$

112. Domain: $\mathbb{R}$; range: $y > 0$

113. 22.8 mpg

114. **a)** 4208 **b)** 4250

115. $\approx 68.7\%$

CHAPTER TEST, PAGE 411

1. 9 **2.** $\frac{19}{5}$ **3.** 18 **4.** $3x + 5 = 17; 4$

5. $350 + 0.06x = 710; \$6000$ **6.** 77

7. $y = \dfrac{-3x + 11}{5}$ or $y = -\dfrac{3}{5}x + \dfrac{11}{5}$

8. $3\frac{1}{3}$ **9.** 6.75 ft

10.

11. 2

12.

13.

14.

15. $-1, -4$ **16.** $\frac{4}{3}, -2$

17. It is a function. **18.** 17

19. **a)** Upward **b)** $x = 1$ **c)** $(1, 3)$ **d)** $(0, 4)$
e) No x-intercepts
f)

g) Domain: $\mathbb{R}$; range: $y \geq 3$

CHAPTER 7

SECTION 7.1, PAGE 419

1. Two or more linear equations form a system of linear equations.

3. An inconsistent system of equations is one that has no solution.

5. A dependent system of equations is one that has an infinite number of solutions.

7. The graphs of the equations will be parallel lines.

9. The graphs of the equations will be the same line.

11. (3, 5)

13.

15.

17.

19.

21.

23.

25.

27.

29.

31.

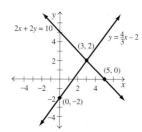

33. a) One unique solution; the lines intersect at one and only one point.

b) No solution; the lines do not intersect.

c) Infinitely many solutions; the lines coincide.

35. An infinite number of solutions

37. One solution   **39.** No solution

41. An infinite number of solutions

43. No solution   **45.** One solution

47. Not perpendicular   **49.** Perpendicular

51. a) Cost to repair: $C = 375x + 250$
   Cost to replace: $C = 225x + 700$

b)

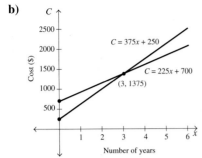

c) 3 years

53. a) $C = 15x + 400$

$R = 25x$

b)

c) 40 backpacks   **d)** $P = 10x - 400$
e) Loss of $100   **f)** 140 backpacks

55. a) $C = 230x + 8400$, $R = 300x$

b)

c) 120 units   **d)** $P = 70x - 8400$
e) Loss of $1400   **f)** 138 units

57. a) Job 1: $s = 0.15x + 500$
Job 2: $s = 650$

b)

c) $1000

59. a) One   **b)** Three   **c)** Six   **d)** Ten
e) To determine the number of points of intersection for n lines, add $n - 1$ to the number of points of intersection for $n - 1$ lines. For example, for 5 lines there were 10 points of intersection. Therefore, for 6 lines ($n = 6$) there are $10 + (6 - 1) = 15$ points of intersection.

SECTION 7.2, PAGE 431

1. Answers will vary.

3. The system is dependent if the same value is obtained on both sides of the equal sign.

5. x, in the first equation

7. $(-2, 6)$ **9.** $(1, -1)$

11. No solution; inconsistent system   **13.** $\left(-2, \frac{8}{3}\right)$

15. An infinite number of solutions; dependent system

17. $(-3, -6)$ **19.** $\left(\frac{11}{5}, -\frac{13}{5}\right)$ **21.** $\left(-\frac{1}{5}, -\frac{8}{5}\right)$

23. No solution; inconsistent system

25. $(3, 0)$ **27.** $(7, 5)$ **29.** $(-2, 0)$

31. $(-4, 5)$ **33.** $(3, 5)$ **35.** $(1, -2)$

37. No solution; inconsistent system   **39.** $(2, 1)$

41. $s = 0.15p + 12,000$

$s = 0.05p + 27,000$

$150,000 annual profit

43. $m + l = 50$

$10.95m + 14.95l = 663.50$

21 medium, 29 large

45. $x + y = 10$

$0.25x + 0.50y = 0.40(10)$

4ℓ of 25%, 6ℓ of 50%

47. $y = 0.02x + 18$

$y = 0.015x + 24$

1200 copies

49. $x + y = 20$

$3x + y = 30$

Mix 5 lb of nuts with 15 lb of pretzels.

51. $x + y = 4600$

$27x + 14y = 104,700$

3100 amphitheatre tickets and 1500 lawn tickets

53. ≈ 11.2 years after 2000 or in 2011

55. $\left(\frac{1}{2}, \frac{1}{3}\right)$ **57.–61.** Answers will vary.

SECTION 7.3, PAGE 441

1. A matrix is a rectangular array of elements.

3. A square matrix contains the same number of rows and columns.

5. 4

7. a) Answers will vary.   **b)** $\begin{bmatrix} 7 & 9 & -7 \\ 2 & 16 & 9 \end{bmatrix}$

9. a) The number of columns of the first matrix must be the same as the number of rows of the second matrix.
b) 2×3

11. a) $I = \begin{bmatrix} 1 & 0 \\ 0 & 1 \end{bmatrix}$ **b)** $I = \begin{bmatrix} 1 & 0 & 0 \\ 0 & 1 & 0 \\ 0 & 0 & 1 \end{bmatrix}$

13. $\begin{bmatrix} -3 & 9 \\ 9 & 9 \end{bmatrix}$ **15.** $\begin{bmatrix} -1 & 4 \\ -5 & 4 \\ 7 & 6 \end{bmatrix}$

17. $\begin{bmatrix} 6 & -7 \\ -12 & 4 \end{bmatrix}$ **19.** $\begin{bmatrix} 1 & 9 \\ 18 & 17 \\ -2 & 2 \end{bmatrix}$

21. $\begin{bmatrix} 6 & 4 \\ 10 & 0 \end{bmatrix}$ **23.** $\begin{bmatrix} -2 & 16 \\ 26 & 0 \end{bmatrix}$

25. $\begin{bmatrix} 16 & 2 \\ 12 & 0 \end{bmatrix}$ **27.** $\begin{bmatrix} 26 & 18 \\ 48 & 24 \end{bmatrix}$

29. $\begin{bmatrix} 15 \\ 22 \end{bmatrix}$ **31.** $\begin{bmatrix} 4 & 7 & 6 \\ -2 & 3 & 1 \\ 5 & 1 & 2 \end{bmatrix}$

33. $A + B = \begin{bmatrix} 9 & 1 & 8 \\ 6 & -1 & 4 \end{bmatrix}$; cannot be multiplied

35. Cannot be added; $A \times B = \begin{bmatrix} 26 & 38 \\ 24 & 24 \end{bmatrix}$

37. Cannot be added; $\begin{bmatrix} 1 \\ -1 \end{bmatrix}$

39. $A + B = B + A = \begin{bmatrix} 7 & 10 \\ 4 & 4 \end{bmatrix}$

41. $A + B = B + A = \begin{bmatrix} 8 & 0 \\ 6 & -8 \end{bmatrix}$

43. $(A + B) + C = A + (B + C) = \begin{bmatrix} 7 & 10 \\ 6 & 13 \end{bmatrix}$

45. $(A + B) + C = A + (B + C) = \begin{bmatrix} 5 & 5 \\ 7 & -37 \end{bmatrix}$

47. No   **49.** No   **51.** Yes

53. $(A \times B) \times C = A \times (B \times C) = \begin{bmatrix} 41 & 13 \\ 56 & 16 \end{bmatrix}$

55. $(A \times B) \times C = A \times (B \times C) = \begin{bmatrix} 16 & -10 \\ -24 & 2 \end{bmatrix}$

57. $(A \times B) \times C = A \times (B \times C) = \begin{bmatrix} 17 & 0 \\ -7 & 0 \end{bmatrix}$

59. Tomatoes Onions Carrots
$\begin{bmatrix} 88 & 58 & 70 \\ 78 & 54 & 71 \end{bmatrix}$ Chase's Farm
Gro-More Farms

61. Total cost
$\begin{bmatrix} 53 \\ 55 \end{bmatrix}$ Java's Coffee Shop
Spot Coffee Shop

63. Large Small
$\begin{bmatrix} 38 & 50 \\ 56 & 72 \\ 17 & 26 \\ 10 & 14 \end{bmatrix}$ Sugar
Flour
Milk
Eggs

65. $[36.04 \quad 47.52]$

67. Answers will vary.

69. Yes   **71.** False

73. a) $28.70   **b)** $60.10   **c)** $\begin{bmatrix} 28.7 & 24.6 \\ 41.3 & 35.7 \\ 69.3 & 60.1 \end{bmatrix}$

75. Yes. Answers will vary. One example is
$A = \begin{bmatrix} 2 & 7 & 6 \\ -3 & 0 & 8 \end{bmatrix}$, $B = \begin{bmatrix} 1 & 2 \\ 3 & 4 \\ 5 & 6 \end{bmatrix}$.

SECTION 7.4, PAGE 452

1. a) An augmented matrix is a matrix formed with the coefficients of the variables and the constants. The coefficients of the variables are separated from the constants by a vertical bar.
b) $\begin{bmatrix} 1 & 3 & | & 7 \\ 2 & -1 & | & 4 \end{bmatrix}$

3. If you obtain an augmented matrix in which a 0 appears across an entire row, the system of equations is dependent.

5. Change the -2 to a 1 by multiplying the numbers in the second row by $-\frac{1}{2}$.

7. $(1, 2)$ **9.** $(3, 2)$

11. An infinite number of solutions; dependent system

13. $\left(\frac{7}{2}, -1\right)$ **15.** $(1, 1)$

17. No solution; inconsistent system

19. $(3, 4)$

21. Fitted caps: 18, stretch-fit caps: 14

23. Truck driver: $7\frac{1}{9}$ hours; laborer: $9\frac{1}{9}$ hours

25. Premium paper: 150 reams, paper for color printers: 50 reams

SECTION 7.5, PAGE 456

1. The solution set of a system of linear inequalities is the set of points that satisfy all inequalities in the system.

3. Yes. A point of intersection satisfies both inequalities and is, therefore, a solution.

5.

$y > x + 4$
$y > 3x$

7.

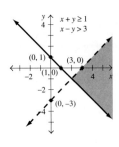

$x + y \geq 1$
$x - y > 3$

19.

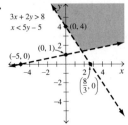

$3x + 2y > 8$
$x < 5y - 5$

21. a) $20x + 30y \leq 600, x \geq 2y, x \geq 10, y \geq 5$

b)

c) One example is 15 small bowls, 7 large bowls

23. $x \leq 0, y \geq 0$

25. Yes. One example is $x \geq 0, y \geq 0, x \leq 0, y \leq 0$ which has solution $(0, 0)$.

27. Answers will vary.

SECTION 7.6, PAGE 461

1. Constraints are restrictions that are represented as linear inequalities.

3. Vertices

5. Answers will vary.

7. Maximum is 30 at $(5, 0)$, minimum is 0 at $(0, 0)$.

9. Maximum is 190 at $(50, 30)$, minimum is 70 at $(20, 10)$.

9.

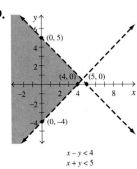

$x - y < 4$
$x + y < 5$

11.

$x + 2y \geq 4$
$3x - y \geq -6$

11. a)

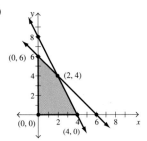

b) Maximum is 30 at $(0, 6)$, minimum is 0 at $(0, 0)$.

13. a)

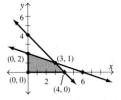

b) Maximum is 28 at $(4, 0)$, minimum is 0 at $(0, 0)$.

13.

$y \leq 3x$
$x \geq 3y$

15.

$x \leq 0$
$y \leq 0$

17.

$4x + 2y > 8$
$x \geq y - 1$

5. If the corresponding sides of two similar figures are the same length, the figures are congruent figures.

7. a) Triangle   **b)** Regular

9. a) Octagon   **b)** Regular

11. a) Rhombus   **b)** Not regular

13. a) Octagon   **b)** Not regular

15. a) Isosceles   **b)** Acute

17. a) Isosceles   **b)** Right

19. a) Scalene   **b)** Acute

21. a) Scalene   **b)** Right

23. Rectangle   **25.** Square   **27.** Rhombus

29. $17°$ **31.** $150°$

33. $m\angle 1 = 90°$, $m\angle 2 = 50°$, $m\angle 3 = 130°$, $m\angle 4 = 50°$, $m\angle 5 = 50°$, $m\angle 6 = 40°$, $m\angle 7 = 90°$, $m\angle 8 = 130°$, $m\angle 9 = 140°$, $m\angle 10 = 40°$, $m\angle 11 = 140°$, $m\angle 12 = 40°$

35. $720°$ **37.** $1080°$ **39.** $3240°$

41. a) $60°$ **b)** $120°$ **43. a)** $108°$ **b)** $72°$

45. a) $144°$ **b)** $36°$ **47.** $x = 2, y = 1.6$

49. $x = 2\frac{2}{5}, y = 7\frac{1}{2}$ **51.** $x = 20, y = 21\frac{1}{4}$

53. 3 **55.** $3\frac{1}{3}$ **57.** 28 **59.** 30 **61.** $28°$ **63.** 9

65. 10 **67.** $70°$ **69.** $55°$ **71.** $35°$ **73.** 70 ft

75. a) ≈ 113.14 mi   **b)** ≈ 75.43 mi

77. $D'E' = 4, E'F' = 5, D'F' = 3$

79. a) $m\angle HMF = m\angle TMB$, $m\angle HFM = m\angle TBM$, $m\angle MHF = m\angle MTB$

   b) 44 ft

SECTION 9.3, PAGE 545

Throughout this section, we used the $\boxed{\pi}$ key on a scientific calculator to determine answers in calculations involving π. If you use 3.14 for π, your answers may vary slightly.

1. a) Answers will vary.
   b) Answers will vary.
   c)

   6
   2

   The area of this rectangle is 12 square units. The perimeter of this rectangle is 16 units.

3. a) To determine the number of square feet, multiply the number of square yards by 9.
   b) To determine the number of square yards, divide the number of square feet by 9.

5. 17.5 in.^{2} **7.** 17.5 cm^2 **9. a)** 210 ft^2 **b)** 62 ft

11. a) 6000 cm^2 **b)** 654 cm

13. a) 288 in.^{2} **b)** 74 in.

15. a) 50.27 m^2 **b)** 25.13 m

17. a) 132.73 ft^2 **b)** 40.84 ft

19. a) 17 yd   **b)** 40 yd   **c)** 60 yd^2

21. a) 12 km   **b)** 30 km   **c)** 30 km^2

23. ≈ 21.46 m^2 **25.** 8 in.^{2} **27.** 90 yd^2

29. ≈ 65.73 in.^{2} **31.** ≈ 307.88 cm^2 **33.** 23 yd^2

35. 132.3 ft^2 **37.** $234{,}000$ cm^2 **39.** 0.8625 m^2

41. a) $\$5973$ **b)** $\$7623$ **43.** $\$2890$

45. $\$2908.80$ **47.** $\$234.21$ **49. a)** 288 ft   **b)** 4700 tiles

51. 21 ft   **53.** ≈ 312 ft

55. a) $A = s^2$ **b)** $A = 4s^2$ **c)** Four times larger

57. 24 cm^2 **59.** Answers will vary.

SECTION 9.4, PAGE 561

Throughout this section, we used the $\boxed{\pi}$ key on a scientific calculator to determine answers in calculations involving π. If you use 3.14 for π, your answers may vary slightly.

1. a) Volume is a measure of the capacity of a figure.
   b) Surface area is the sum of the areas of the surfaces of a three-dimensional figure.

3. A polyhedron is a closed surface formed by the union of polygonal regions. A regular polyhedron is one whose faces are all regular polygons of the same size and shape.

5. Answers will vary.   **7. a)** 8 m^3 **b)** 28 m^2

9. a) 125 ft^3 **b)** 150 ft^2

11. a) 150.80 in.^{3} **b)** 175.93 in.^{2}

13. a) 131.95 cm^3 **b)** 163.22 cm^2

15. a) 381.70 cm^3 **b)** 254.47 cm^2

17. 384 in.^{3} **19.** 524.33 cm^3 **21.** 324 mm^3 **23.** 392 ft^3

25. 284.46 cm^3 **27.** 24 ft^3 **29.** 243 ft^3 **31.** ≈ 5.67 yd^3

33. $3{,}700{,}000$ cm^3 **35.** 7.5 m^3

37. Tubs ≈ 141.37 in.^{3}; boxes $= 125$ in.^{3}

39. $40{,}840$ mm^2 **41.** ≈ 2.50 qt

43. a) $120{,}000$ cm^3 **b)** $120{,}000$ mℓ **c)** 120 ℓ

45. ≈ 283.04 in.^{3}

47. a) ≈ 323.98 in.^{3} **b)** ≈ 0.19 ft^3

49. ≈ 14.14 in.^{3} **51.** Ten edges

53. Six vertices   **55.** Fourteen edges

57. a) $\approx 5.11 \times 10^8$ km^2 **b)** $\approx 3.79 \times 10^7$ km^2
   c) ≈ 13 times larger   **d)** $\approx 1.09 \times 10^{12}$ km^3
   e) $\approx 2.20 \times 10^{10}$ km^3 **f)** ≈ 50 times larger

59. a)–e) Answers will vary.
   f) If we double the length of each edge of a cube, the new volume will be eight times the original volume.

61. a) Answers will vary.
  b) $V_1 = a^3; V_2 = a^2b; V_3 = a^2b; V_4 = ab^2;$
      $V_5 = a^2b; V_6 = ab^2; V_7 = b^3$
  c) ab^2

SECTION 9.5, PAGE 579

1. The act of moving a geometric figure from some starting position to some ending position without altering its shape or size is called rigid motion. The four main rigid motions studied in this section are reflections, translations, rotations, and glide reflections.

3. A reflection is a rigid motion that moves a figure to a new position that is a mirror image of the figure in the starting position.

5. A rotation is a rigid motion performed by rotating a figure in the plane about a specific point.

7. A translation is a rigid motion that moves a figure by sliding it along a straight line segment in the plane.

9. A glide reflection is a rigid motion formed by performing a translation (or glide) followed by a reflection.

11. A geometric figure is said to have reflective symmetry if the positions of a figure before and after a reflection are identical (except for vertex labels).

13. A tessellation is a pattern consisting of the repeated use of the same geometric figures to entirely cover a plane, leaving no gaps.

This figure contains the answers for Exercises 15 and 16.

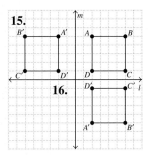

This figure contains the answers for Exercises 17 and 18.

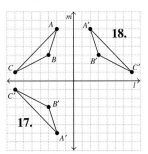

This figure contains the answers for Exercises 19 and 20.

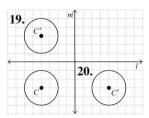

This figure contains the answers for Exercises 21 and 22.

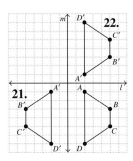

This figure contains the answers to Exercises 23 and 24.

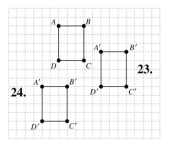

This figure contains the answers to Exercises 25 and 26.

This figure contains the answers to Exercises 27 and 28.

This figure contains the answers to Exercises 35 and 36.

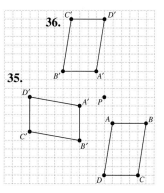

This figure contains the answers to Exercises 31 and 32.

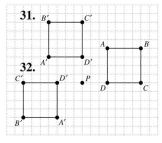

This figure contains the answers to Exercises 37 and 38.

This figure contains the answers to Exercises 33 and 34.

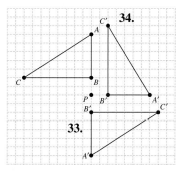

This figure contains the answers to Exercises 39 and 40.

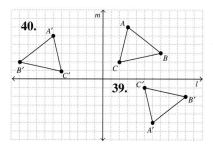

This figure contains the answers to Exercises 41 and 42.

This figure contains the answers to Exercises 43 and 44.

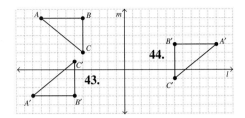

This figure contains the answers to Exercises 45 and 46.

47. a)

b) Yes
c) Yes

49. a)

b) No
c) No

51. a)

b) No
c) No
d)

e) Yes
f) Yes

53. a)–c)

d) No. Any 90° rotation will result in the figure being in a different position than the starting position.

55. a) and **b)**

c) No.
d) The order in which the translation and the reflection are performed is important. The figure obtained in part (a) is the glide reflection.

57. Answers will vary.

59. a) Answers will vary.
   b) A regular pentagon cannot be used as a tessellating shape.

61. Although answers will vary depending on the font, the following capital letters have reflective symmetry about a vertical line drawn through the center of the letter: A, H, I, M, O, T, U, V, W, X, Y.

SECTION 9.6, PAGE 590

1. Topology is sometimes referred to as "rubber sheet geometry" because it deals with bending and stretching of geometric figures.

3. Take a strip of paper, give one end a half twist, and tape the ends together.

5. Four

7. A Jordan curve is a topological object that can be thought of as a circle twisted out of shape.

9. The number of holes that go through the object determines the genus of an object.

11.–19. Answers will vary. **21.** Outside **23.** Inside

25. Outside **27.** Outside **29.** 1 **31.** 5

33. Larger than 5 **35.** 5 **37.** 0 **39.** Larger than 5

41. a)–d) Answers will vary. **43.** One **45.** Two

47. a) No, it has an inside and an outside. **b)** Two
   c) Two **d)** Two strips, one inside the other

49.–51. Answers will vary.

SECTION 9.7, PAGE 600

1.–5. Answers will vary.

7. a) *Euclidean:* Given a line and a point not on the line, one and only one line can be drawn parallel to the given line through the given point.
   b) *Elliptical:* Given a line and a point not on the line, no line can be drawn through the given point parallel to the given line.
   c) *Hyperbolic:* Given a line and a point not on the line, two or more lines can be drawn through the given point parallel to the given line.

9. A plane

11. A pseudosphere

13. Spherical: elliptical geometry; flat: Euclidean geometry; saddle-shaped: hyperbolic geometry

15.

17.

19. a)

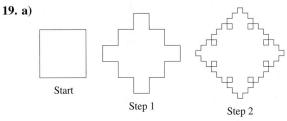

Start   Step 1   Step 2

b) Infinite
c) Finite

REVIEW EXERCISES, PAGE 603

In the Review Exercises and Chapter Test questions, we used the $\boxed{\pi}$ key on a scientific calculator to determine answers in calculations involving π. If you use 3.14 for π, your answers may vary slightly.

1. $\{B\}$ **2.** $\overline{AD}$ **3.** $\triangle BFC$ **4.** $\overleftrightarrow{BH}$ **5.** $\{F\}$

6. $\{\ \}$ **7.** 66.3°   **8.** 55.3°   **9.** 10.2 in.   **10.** 2 in.

11. 58°   **12.** 92°

13. $m\angle 1 = 50°, m\angle 2 = 120°, m\angle 3 = 120°, m\angle 4 = 70°,$
$m\angle 5 = 110°, m\angle 6 = 70°$

14. 540°   **15. a)** 80 cm^2 **b)** 36 cm

16. a) 13 in.^{2} **b)** 19.4 in.

17. a) 84 in.^{2} **b)** 42 in.

18. a) 6 km^2 **b)** 12 km

19. a) 530.93 cm^2 **b)** 81.68 cm

20. 41.20 in.^{2} **21.** 35.43 cm^2 **22.** $1176

23. a) 120 cm^3 **b)** 164 cm^2

24. a) 2035.75 in.^{3} **b)** 904.78 in.^{2}

25. a) 603.19 mm^3 **b)** 435.20 mm^2

26. a) 268.08 ft^3 **b)** 201.06 ft^2

27. 432 m^3 **28.** 28 ft^3 **29.** 75.40 cm^3 **30.** 791.68 cm^3

31. a) $\approx$ 67.88 ft^3 **b)** 4610.7 lb; yes   **c)** $\approx$ 510.3 gal

This figure contains the answers for Exercises 32 and 33.

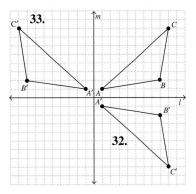

This figure contains the answers for Exercises 34 and 35.

This figure contains the answers for Exercises 36–38.

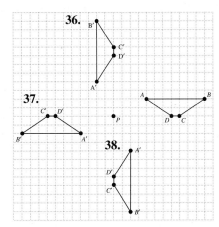

This figure contains the answers for Exercises 39 and 40.

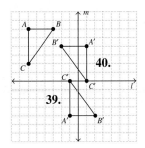

41. Yes   **42.** No   **43.** No   **44.** Yes

45. a)–d) Answers will vary.   **46.** Answers will vary.

47. Outside

48. Euclidean: Given a line and a point not on the line, one and only one line can be drawn parallel to the given line through the given point. Elliptical: Given a line and a point not on the line, no line can be drawn through the given point parallel to the given line. Hyperbolic: Given a line and a point not on the line, two or more lines can be drawn through the given point parallel to the given line.

49.

50.

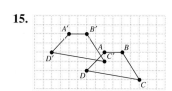

CHAPTER TEST, PAGE 607

1. $\overrightarrow{EF}$ **2.** $\triangle BCD$ **3.** $\{D\}$ **4.** $\overleftrightarrow{AC}$ **5.** $77.6°$

6. $128.3°$ **7.** $64°$ **8.** $1440°$ **9.** ≈ 2.69 cm

10. a) 12 in.   **b)** 30 in.   **c)** 30 in.^{2}

11. a) 1436.76 cm^3 **b)** 615.75 cm^2

12. 59.43 m^3 **13.** 112 ft^3

14.

15.

16.

17.

18. a) No   **b)** Yes

19. A surface with one side and one edge

20. a)–b) Answers will vary.

21. Euclidean: Given a line and a point not on the line, one and only one line can be drawn parallel to the given line through the given point. Elliptical: Given a line and a point not on the line, no line can be drawn through the given point parallel to the given line. Hyperbolic: Given a line and a point not on the line, two or more lines can be drawn through the given point parallel to the given line.

CHAPTER 10

SECTION 10.1, PAGE 617

1. A binary operation is an operation, or rule, that can be performed on two and only two elements of a set. The result is a single element.

3. a) When we add two numbers, the sum is one number: $4 + 5 = 9$.
 b) When we subtract two numbers, the difference is one number: $5 - 4 = 1$.
 c) When we multiply two numbers, the product is one number: $5 \times 4 = 20$.
 d) When we divide two numbers, the quotient is one number: $20 \div 5 = 4$.

5. A mathematical system is a commutative group if all five of the following conditions hold.
 1. The set of elements is closed under the given operation.
 2. An identity element exists for the set.
 3. Every element in the set has an inverse.
 4. The set of elements is associative under the given operation.
 5. The set of elements is commutative under the given operation.

7. An identity element is an element in a set such that when a binary operation is performed on it and any given element in the set, the result is the given element. For the set of integers, the additive identity element is 0 and the multiplicative identity element is 1. Examples: $5 + 0 = 5, 5 \times 1 = 5$.

9. When a binary operation is performed on two elements in a set and the result is the identity element for the binary operation, each element is said to be the *inverse* of the other. For the set of rational numbers, the additive inverse of 2 is -2 since $2 + (-2) = 0$ and the multiplicative inverse of 2 is $\frac{1}{2}$ since $2 \times \frac{1}{2} = 1$.

11. Yes; for a group the commutative property need not apply.

13. (d); the commutative property need not apply.

15. $a + b = b + a$ for any elements a and b; $3 + 4 = 4 + 3$

17. $(a \cdot b) \cdot c = a \cdot (b \cdot c)$ for any elements a, b, and c;
$(2 \cdot 3) \cdot 4 = 2 \cdot (3 \cdot 4)$

19. $4 \div 2 \neq 2 \div 4$

21. $(6 - 4) - 1 \neq 6 - (4 - 1)$
$\qquad 2 - 1 \neq 6 - 3$
$\qquad\qquad 1 \neq 3$

23. Yes; it satisfies the five properties needed.

25. No; there is no identity element.

27. No; the system is not closed.

29. No; there is no identity element.

31. No; not all elements have inverses.

33. Yes; it satisfies the four properties needed.

35. No; the system is not closed. For example, $\frac{1}{0}$ is undefined.

37. No; it does not have an identity element and does not satisfy the associative property.

39. Yes

41. No; the system is not closed. For example, $\sqrt{2} + (-\sqrt{2}) = 0$, which is rational. There is also no identity element.

43. Answers will vary.

SECTION 10.2, PAGE 625

1. Numbers are obtained by starting at the first addend (on a clock face) and then moving clockwise the number of hours equal to the second addend.

3. a) Add $2 + 11$ to get 1, then add $1 + 5$.   **b)** 6

5. a) Add 12 to 2 to get $14 - 6$.   **b)** 8
  c) Since 12 is the identity element, you can add 12 to any number without changing its value.

7. Yes, 12

9. Yes; 1:11, 2:10, 3:9, 4:8, 5:7, 6:6, 7:5, 8:4, 9:3, 10:2, 11:1, and 12:12

11. Yes, $6 + 9 = 9 + 6$ since both equal 3

13. a) 7   **b)** 5. The additive inverse of 2 is 5 since $2 + 5 = 7$.

15. No; the elements are not symmetric about the main diagonal.

17. Yes; C is the identity element since the row next to C is identical to the top row, and the column under C is identical to the left-hand column.

19. a) The inverse of A is B since $A \otimes B = B \otimes A = C$.
  b) The inverse of B is A since $B \otimes A = A \otimes B = C$.
  c) The inverse of C is C since $C \otimes C = C$.

21. 7   **23.** 4   **25.** 4   **27.** 8   **29.** 6   **31.** 6

33. 5   **35.** 11   **37.** 3   **39.** 1   **41.** 12   **43.** 12

45.

+	1	2	3	4	5	6
1	2	3	4	5	6	1
2	3	4	5	6	1	2
3	4	5	6	1	2	3
4	5	6	1	2	3	4
5	6	1	2	3	4	5
6	1	2	3	4	5	6

47. 3   **49.** 3   **51.** 2   **53.** 4

55.

+	1	2	3	4	5	6	7
1	2	3	4	5	6	7	1
2	3	4	5	6	7	1	2
3	4	5	6	7	1	2	3
4	5	6	7	1	2	3	4
5	6	7	1	2	3	4	5
6	7	1	2	3	4	5	6
7	1	2	3	4	5	6	7

57. 4   **59.** 6   **61.** 4   **63.** 7

65. Yes; it satisfies the five required properties.

67. a) $\{0, 2, 4, 6\}$ **b)** £   **c)** Yes
  d) Yes; 0   **e)** Yes; 0–0, 2–6, 4–4, 6–2
  f) $(2 £ 4) £ 6 = 2 £ (4 £ 6)$
  g) Yes; $2 £ 6 = 6 £ 2$ **h)** Yes

69. a) $\{*, 5, L\}$
  b) 🐎
  c) Yes
  d) Yes, L
  e) Yes, $*$–5, 5–$*$, L–L
  f) $\left(* \text{🐎} 5 \right) \text{🐎} 5 = * \text{🐎} \left(5 \text{🐎} 5 \right)$
  g) Yes, $L \text{🐎} * = * \text{🐎} L$
  h) Yes

71. a) $\{f, r, o, m\}$ **b)** 🐎   **c)** Closed
  d) m **e)** f **f)** f **g)** m **h)** r

73. Not closed

75. No inverse for $\odot$ or for $*$, not associative

77. No identity element, no inverses, not associative, not commutative

79. a)

+	E	O
E	E	O
O	O	E

  b) Yes, it is a commutative group; it satisfies the five properties.

81. Answers will vary.

83. a) Is closed; identity element is 6; inverses: 1–5, 2–2, 3–3, 4–4, 5–1, 6–6; is associative; for example,

$$(2 \infty 5) \infty 3 = 2 \infty (5 \infty 3)$$
$$3 \infty 3 = 2 \infty 2$$
$$6 = 6$$

b) $3 \infty 1 \neq 1 \infty 3$
$$2 \neq 4$$

85. a)

*	R	S	T	U	V	I
R	V	T	U	S	I	R
S	U	I	V	R	T	S
T	S	R	I	V	U	T
U	T	V	R	I	S	U
V	I	U	S	T	R	V
I	R	S	T	U	V	I

b) Yes. It is closed; identity element is I; inverses $R-V$, $S-S$, $T-T$, $U-U$, $V-R$, $I-I$; and the associative property will hold.

c) No, it is not commutative. For example, $R*S \neq S*R$

87.

+	0	1	2	3	4
0	0	1	2	3	4
1	1	2	3	4	0
2	2	3	4	0	1
3	3	4	0	1	2
4	4	0	1	2	3

89. Add the number in the top row and the number in the left-hand column and divide the sum by 4. The remainder is placed in the table.

SECTION 10.3, PAGE 638

1. A modulo m system consists of m elements, 0 through $m - 1$, and a binary operation.

3. 3;

0	1	2
0	1	2
3	4	5
6	7	8
.	.	.
.	.	.
.	.	.

5. 16 classes **7.** (b), (c) or (d) **9.** Sunday **11.** Friday
13. Saturday **15.** Friday **17.** February **19.** May
21. September **23.** July **25.** 1 **27.** 2 **29.** 3
31. 2 **33.** 1 **35.** 2 **37.** 0 **39.** 0 **41.** 5 **43.** 2
45. 6 **47.** 9 **49.** 5 **51.** 5 **53.** 3 **55.** 5 **57.** 5
59. { } **61.** 1 and 6 **63.** 5 **65.** 0
67. a) 2020, 2024, 2028, 2032, 2036 **b)** 3004
c) 2552, 2556, 2560, 2564, 2568, 2572
69. a) Resting (for the second of two days)
b) Resting (for the second of two days)
c) Morning and afternoon lessons
d) No

71. a) 5 **b)** No **c)** 54 weeks from this week
73. a) Evening **b)** Day **c)** Day
75. a)

+	0	1	2
0	0	1	2
1	1	2	0
2	2	0	1

b) Yes **c)** Yes, 0 **d)** Yes; 0–0, 1–2, 2–1
e) $(1 + 2) + 2 = 1 + (2 + 2)$
f) Yes; $2 + 1 = 1 + 2$ **g)** Yes **h)** Yes

77. a)

×	0	1	2	3
0	0	0	0	0
1	0	1	2	3
2	0	2	0	2
3	0	3	2	1

b) Yes **c)** Yes, 1
d) No; no inverse for 0 or for 2, inverse of 1 is 1, inverse of 3 is 3
e) $(1 \times 2) \times 3 = 1 \times (2 \times 3)$
f) Yes, $2 \times 3 = 3 \times 2$ **g)** No

79. 7 **81.** 1, 2, 3 **83.** 0 **85.** 2
87. a) Tuesday **b)** Wednesday **89.** 0

REVIEW EXERCISES, PAGE 641

1. A mathematical system consists of a set of elements and at least one binary operation.

2. A binary operation is an operation that can be performed on two and only two elements of a set. The result is a single element.

3. No; for example, $2 - 5 = -3$, and -3 is not a whole number.

4. Yes; the difference of any two real numbers is a real number.

5. 3 **6.** 5 **7.** 9 **8.** 8 **9.** 9 **10.** 11

11. Closure, identity element, inverses, and associative property

12. A commutative group **13.** Yes

14. No; no inverse for any integer except 1 and -1

15. No; no identity element **16.** No; no inverse for 0

17. No identity element; no inverses

18. Not every element has an inverse; not associative. For example, $(P \, ? \, P) \, ? \, 4 \neq P \, ? \, (P \, ? \, 4)$.

19. Not associative. For example, $(! \, \square \, p) \, \square \, ? \neq ! \, \square \, (p \, \square \, ?)$

20. a) $\{ \odot, \bullet, \female, \male \}$ **b)** $\triangle$ **c)** Yes **d)** Yes; $\odot$
e) Yes; $\odot - \odot$, $\bullet - \male$, $\female - \female$, $\male - \bullet$
f) $(\bullet \triangle \female) \triangle \male = \bullet \triangle (\female \triangle \male)$
g) Yes; $\bullet \triangle \male = \male \triangle \bullet$ **h)** Yes

21. 0 **22.** 5 **23.** 1 **24.** 3 **25.** 11 **26.** 2 **27.** 4

28. 12 **29.** 9 **30.** 11 **31.** 1 **32.** 1 **33.** 6 **34.** 1

35. 0, 2, 4, 6 **36.** 4 **37.** 5 **38.** 9 **39.** 7 **40.** 8

41.

+	0	1	2	3	4	5
0	0	1	2	3	4	5
1	1	2	3	4	5	0
2	2	3	4	5	0	1
3	3	4	5	0	1	2
4	4	5	0	1	2	3
5	5	0	1	2	3	4

Yes, it is a commutative group.

42.

×	0	1	2	3
0	0	0	0	0
1	0	1	2	3
2	0	2	0	2
3	0	3	2	1

No; no inverse for 0 or 2

43. a) Yes **b)** Yes

CHAPTER TEST, PAGE 643

1. A set of elements and a binary operation

2. Closure, identity element, inverses, associative property, commutative property

3. Yes

4.

+	1	2	3	4	5
1	2	3	4	5	1
2	3	4	5	1	2
3	4	5	1	2	3
4	5	1	2	3	4
5	1	2	3	4	5

5. Yes, it is a commutative group.

6. 3 **7.** 2

8. a) □ **b)** Yes **c)** Yes, T **d)** S **e)** S

9. No, not closed.

10. Yes, it is a commutative group.

11. Yes, it is a commutative group.

12. 2 **13.** 0 **14.** 6 **15.** 2 **16.** 5

17. 2 **18.** { } **19.** 7

20. a)

×	0	1	2	3	4
0	0	0	0	0	0
1	0	1	2	3	4
2	0	2	4	1	3
3	0	3	1	4	2
4	0	4	3	2	1

b) No; no inverse for 0

CHAPTER 11

SECTION 11.1, PAGE 653

1. A percent is a ratio of some number to 100.

3. Divide the numerator by the denominator, multiply the quotient by 100, and add a percent sign.

5. Percent change $= \dfrac{\left(\begin{array}{c}\text{amount in}\\ \text{latest period}\end{array}\right) - \left(\begin{array}{c}\text{amount in}\\ \text{previous period}\end{array}\right)}{\text{amount in the previous period}} \times 100$

7. 75.0% **9.** 55.0% **11.** 0.8% **13.** 378.0%

15. 0.05 **17.** 0.0515 **19.** 0.0025 **21.** 0.002

23. 0.01 **25.** 25% **27.** 4.7 grams **29.** $208,341.12

31. $651,066.00 **33.** $1550 **35.** $2100 **37.** 41.3%

39. 10.0% **41.** 0.9% decrease

43. a) 98.8% increase **b)** 224.9% increase
c) 145.9% increase **d)** 58.5% increase

45. a) 78.3% increase **b)** 27.4% decrease
c) 28.5% increase **d)** 66.2% increase

47. $6.75 **49.** 25% **51.** 300

53. a) $2.61 **b)** $46.11 **c)** $6.92 **d)** $53.03

55. 12 students **57.** $39,055 **59.** 5.3% decrease

61. 12.5% increase **63.** 18.6% decrease **65.** $3750

67. He will have a loss of $10. **69.** $21.95

SECTION 11.2, PAGE 665

1. Interest is the money the borrower pays for the use of the lender's money.

3. Security or collateral is anything of value pledged by the borrower that the lender may sell or keep if the borrower does not repay the loan.

5. i is the *interest*, p is the *principal*, r is the interest *rate* expressed as a percent, and t is the *time*.

7. The United States rule states that if a partial payment is made on a loan, interest is computed on the principal from the first day of the loan until the date of the partial payment.

9. $15 **11.** $24.06 **13.** $15.85 **15.** $80.06

17. $113.20 **19.** 5% **21.** $600 **23.** 2 years **25.** $6630

27. a) $131.25 **b)** $3631.25

29. a) $182.50 **b)** $3467.50 **c)** $\approx 7.9\%$

31. $23,793.75 **33.** 284 days **35.** 272 days **37.** 264 days

39. June 13 **41.** March 24 **43.** $2017.03 **45.** $1459.33

47. $5278.99 **49.** $850.64 **51.** $6086.82 **53.** $2646.24

55. a) November 3, 2007 **b)** $978.06 **c)** $21.94
d) $\approx 4.44\%$

57. a) $\approx 409.0\%$ **b)** $\approx 204.5\%$ **c)** $\approx 102.3\%$

59. a) 6.663% **b)** $\approx 7.139\%$ **c)** $6663 **d)** $6996.15

SECTION 11.3, PAGE 675

1. An investment is the use of money or capital for income or profit.

3. A variable investment is one in which neither the principal nor the interest is guaranteed.

5. a) The effective annual yield is the simple interest rate that gives the same amount of interest as a compound rate over the same period of time.
   b) Another name for effective annual yield is annual percentage yield.

In the remainder of this exercise set, your answers may vary in the last digit, depending on the calculator used.

7. $A = \$1216.65$; $i = \$216.65$

9. $A = \$2253.98$; $i = \$253.98$

11. $A = \$8252.64$; $i = \$1252.64$

13. $A = \$8666.26$; $i = \$666.26$

15. $30,695.66   17. $85,282.13   19. $54,142.84

21. $1653.36   23. $2341.82

25. a) $4195.14   b) $4214.36

27. $3106.62   29. $7609.45

31. $\approx 3.53\%$ 33. Yes, the APY should be 2.43%.

35. He will earn more interest in the account that pays the 5% simple interest

37. a) $555,000   b) $58,907.61   c) $98.51

39. $212,687.10   41. $12,015.94

43. a) $1040.60, $40.60   b) $1082.43, $82.43
    c) $1169.86, $169.86   d) No

45. $1.53

47. a) 24 years   b) 12 years   c) 9 years   d) 6 years
    e) 3.27%

49. The simple interest is $20,000. The compound interest is $22,138.58. The compound interest is greater by $2,138.58.

SECTION 11.4, PAGE 688

1. An open-end installment loan is one with which you can make different payments each month. A fixed installment loan is one in which you pay a fixed amount each month for a set number of months.

3. The APR is the true rate of interest charged on a loan.

5. The total installment price is the sum of all the monthly payments and the down payment, if any.

7. The unpaid balance method and the average daily balance method

9. a) $1719.55   b) $170.33

11. a) $1834.57   b) $223.85

13. a) $628.40   b) 9.0%

15. a) $1752   b) 9%

17. a) 6%   b) $726.00   c) $7858.00

19. a) $2818.20   b) $689.39   c) $347.90   d) $8614.17

21. a) $26   b) $46.01, which rounds up to $47

23. a) $21.14, which rounds up to $22
    b) $34.39, which rounds up to $35

25. a) $19.76   b) $743.41

27. a) $1.56   b) $133.11

29. a) $512.00   b) $6.66   c) $638.43

31. a) $121.78   b) $1.52   c) $133.07
    d) The interest charged using the average daily balance method is $0.04 less than the interest charged using the unpaid balance method.

33. a) $11.96   b) $886.96

35. a) $25   b) $35.60   c) 8.5%   d) 6.5%

37. $14,077.97

39. a) $6872.25   b) $610.37   c) $2637.42   d) $2501.05

41. $201.48

SECTION 11.5, PAGE 702

1. A mortgage is a long-term loan in which the property is pledged as security for payment of the difference between the down payment and the sale price.

3. The major difference is that the interest rate for a conventional loan is fixed for the duration of the loan, whereas the interest rate for a variable-rate loan may change every period, as specified in the loan agreement.

5. A buyer's adjusted monthly income is found by subtracting any fixed monthly payments with more than 10 months remaining from the gross monthly income.

7. An amortization schedule lists payment dates and payment numbers. For each payment, it lists the amount that goes to pay the interest and the principal. It also gives the balance remaining on the loan after each payment.

9. Equity is the difference between the appraised value of your home and the loan balance.

11. a) $70,000   b) $2438.80

13. a) $262,500   b) $9579.50

15. a) $39,000   b) $156,000   c) $3120

17. a) $802.20   b) $1411.50   c) No

19. a) $187,736.40   b) $112,736.40   c) $38.68

21. a) $17,025   b) $2894.25   c) $1212.40   d) $930.98
    e) $1057.65   f) Yes   g) $127.02

23. Bank B

25. a) $664,704   b) $927,360   c) $1,231,488

CREDITS

Page 1, Couple working on budgets, Digital Vision; Page 2, DisneyWorld finger scanner, Apani Networks; Page 3, ATM eye scan, International Business Machines Corporation; Page 4, Astronaut on moon, NASA; Page 9, Jury, Corbis; Page 9, Buying stamps in post office, Beth Anderson; Page 10, Bushels of grapes in vineyard, PhotoDisc Red; Page 13, Counting birds, Art Wolfe/Stone/Getty Images; Page 14, Repeat detail of birds, Art Wolfe/Stone/Getty Images; Page 15, Picking strawberries, PhotoDisc Blue; Page 18, Bananas, Allen R. Angel; Page 18, Blueberries, PhotoDisc; Page 18, Statue of Liberty, Allen R. Angel; Page 19, Three men and palm tree, Corbis Royalty Free; Page 20, Buying CDs in office supply store, PhotoDisc Red; Page 22, George Polya at blackboard, Palo Alto Weekly; Page 23, Archimedes, The Bridgeman Art Library International; Page 25, Restaurant waitstaff with customers, PhotoDisc Red; Page 26, Vacation in the mountains, Allen R. Angel; Page 27, Spraying weed killer, Corbis; Page 30, Large fishing boat, Digital Vision; Page 33, African safari, PhotoDisc Red; Page 42, Van Gogh's *Sunflowers*, Christie's Images/Corbis; Page 44, Child playing with abacus, Stockbyte Platinum Getty RF; Page 45, Horses, Allen R. Angel; Page 45, Georg Cantor, The Granger Collection; Page 46, Planets in Earth's solar system, Bennett/Donahue/Schneider/Voit, *The Essential Cosmic Perspective, Fourth Edition*, Pearson Addison-Wesley, © 2007; Page 50, Astronauts on the moon, NASA; Page 51, Snow White and the Seven Dwarfs, Walt Disney/The Kobal Collection; Page 52, The Beatles, The Everett Collection; Page 53, J.K. Rowling, Corbis; Page 54, Fireman, PhotoDisc Red; Page 57, Pizza with toppings, Stockbyte Platinum Getty RF; Page 58, Exercise class at gym, Purestock Getty RF; Page 60, Koala and Mrs. Angel, Allen R. Angel; Page 60, Buying a soft drink, Image Source Getty RF; Page 69, San Diego Zoo, PhotoDisc Red; Page 68, Yosemite National Park, PhotoDisc; Page 71, Person with dog, Ingram Publishing Getty RF; Page 71, High school band, Beth Anderson; Page 72, Doctors examining x-rays, PhotoDisc Blue; Page 77, Walmart, Beth Anderson; Page 78, Athlete from summer Olympics, Digital Vision; Page 82, Using treadmill or Stairmaster, PhotoDisc; Page 85, Hawaii, Beth Anderson; Page 87, Movie theatre attendees, PhotoDisc Red; Page 87, Water slide at amusement park, Digital Vision; Page 88, Women at restaurant, Digital Vision; Page 89, Farmer and wheatfield, PhotoDisc; Page 89, Georg Cantor, The Granger Collection; Page 91, Leopold Kronecker, The Granger Collection; Page 95, Tiger, PhotoDisc; Page 97, Actors from *CSI:NY*, Getty Editorial; Page 98, Cotton candy snack, PhotoDisc Red; Page 100, Lawyer arguing case, Digital Vision; Page 101, Morton salt ad, Morton International, Inc.; Page 102, Brooklyn Bridge, Corbis RF; Page 103, Abbott and Costello, Bettmann/Corbis; Page 105, Billie Joe Armstrong/Green Day, AP Wideworld Photos; Page 106, George Boole, The Granger Collection; Page 108, Painting by Beth Anderson, Beth Anderson; Page 109, Martin St. Louis, Getty Images Sport; Page 110, Louis Armstrong playing trumpet, Time & Life Pictures/Getty Images; Page 110, Butterflies and flowers, Allen R. Angel; Page 111, Ken Jennings, Jeopardy champ, Getty Images News; Page 113, Idaho potatoes, Digital Vision; Page 116, Guitarists, PhotoDisc Red; Page 123, Girl with cookies and milk, Corbis Royalty Free; Page 123, Collie, PhotoDisc Blue; Page 123, John Madden, sports announcer, AP Wideworld Photos; Page 125, Airplane travel, Digital Vision; Page 126, Car for A student, Digital Vision; Page 133, Soccer, Corbis Royalty Free; Page 134, Nike shoes, Corbis; Page 136, Sunny beach, Dennis Runde; Page 137, Charles Dodgson, The Granger Collection; Page 137, Alice in Wonderland, The Granger Collection; Page 141, Augustus de Morgan, The Granger Collection; Page 144, Sitar, PhotoDisc; Page 146, Digital photographer, ThinkStock RF; Page 147, Timex watch, Beth Anderson; Page 148, *Jimmy Neutron*, Paramount Pictures/Courtesy: Everett Collection; Page 148, Sailboat, Beth Anderson; Page 149, Yellow lab, Christine Abbott; Page 149, People fishing, PhotoDisc Blue; Page 150, FedEx truck, Beth Anderson; Page 150, Person taking a picture, PhotoDisc Red; Page 151, John Mellencamp, Major League Baseball/Getty Images; Page 153, Alaska glacier, PhotoDisc Blue; Page 156, U.S. Constitution photo, Corbis RF; Page 158, Prices at gas station, Getty Images News; Page 160, Les Paul and guitar, Corbis; Page 160, Graduation, StockDisc; Page 161, Football quarterback about to throw, Stockbyte Platinum/Getty RF; Page 162, Meg Ryan in *In the Land of Women*, Warner Independent/Courtesy: Everett Collection; Page 165, Barcode scanner, PhotoDisc; Page 167, Gingerbread man, National Geographic/Getty Royalty Free; Page 168, Morel mushrooms, PhotoDisc; Page 168, Caterpillar, PhotoDisc Red; Page 169, Reading lamp, Digital Vision; Page 176, Ellen Johnson-Sirleaf, President of Liberia, Jason Szenes/epa/Corbis; Page 177, Bobby Darin and Sandra Dee, Hulton Archive/Getty Images; Page 180, Ontario Science Center logic game, Courtesy Ontario Science Center; Page 181, Kids using math manipulatives, Richard Hutchings/PhotoEdit Inc.; Page 182, Hines Ward and Super Bowl XL logo, Getty Images Sport; Page 182, Denise Schmandt-Besserat, Courtesy Denise Schmandt-Besserat; Page 183, Rhind papyrus, Public domain; Page 184, Clock with Roman numerals, Allen R. Angel; Page 186, Sign showing Hindu-Arabic numerals, Mark Gibson; Page 190, Great pyramid at Khufu, Giza Egypt, Digital Vision (PP); Page 190, Bob Barker game in *Price is Right*, CBS/Photofest; Page 192, Counting board, Taxi/Getty Images; Page 193, Globe, PhotoDisc; Page 194, Photo of Jan Fleck's *Numbers*, Allen R. Angel; Page 197, Boys with iPods, Christine Abbott; Page 198, Students in computer lab, Blend Images/Getty RF; Page 200, Arabic and Chinese numerals, PhotoAlto/Getty RF; Page 204, Middle school kids in computer lab, PhotoDisc; Page 204, Arithmetic teacher and students, Purestock/Getty RF; Page 213, Internet researcher, Stockbyte Platinum/Getty RF; Page 214, Pay check/time clock, PhotoDisc; Page 215, Gradeschool class—new old math, PhotoDisc; Page 216, John Napier, The Granger Collection; Page 223, Teenagers on cell phones. PhotoDisc Red (PP); Page 224, Box of chocolates, Beth Anderson; Page 227, Factoring machine 1914, Jeffrey Shallit; Page 228, Srinivasa Ramanujan, The Granger Collection; Page 230, Bank vault, Digital Vision; Page 233, Barbie dolls, Beth Anderson; Page 233, School band, Merrill Education; Page 235, Playing in snow, PhotoDisc; Page 238, Mount Everest, Corbis RF; Page 243, NYSE floor traders, PhotoDisc; Page 243, Reggie Bush, New Orleans Saints, Getty Editorial; Page 245, Preparing cake mix, Pixland/Corbis Royalty Free; Page 253, Waves of string vibration, Fundamental Photos; Page 257, Measuring boy's height, Corbis RF; Page 258, Farmer and silo, Digital Vision; Page 259, San Antonio, Texas, Allen R. Angel; Page 261, Superbowl champs playing, Getty Editorial; Page 262, Pythagoras of Samos, Corbis; Page 268, Car in motion, Allen R. Angel; Page 269, Elementary school students doing arithmetic, Digital Vision; Page 271, Daniel Tammet, Colin McPherson/Corbis; Page 275, English bulldog, PhotoDisc Red; Page 276, Breaking an egg into a bowl, PhotoDisc; Page 276, Galaxy, NASA headquarters; Page 280, Another galaxy, NASA Media Services; Page 281, Preda Mihailescu, Used with permission of Preda Mihailescu; Page 283, WWI poster, U.S. bonds, Library of Congress; Page 286, Jupiter, U.S. Geological Survey/U.S. Department of the Interior; Page 286, IBM BlueGene/L system computer, Kim Kulish/Corbis; Page 287, Niagara Falls, PhotoDisc; Page 288, U.S. mint/ uncut sheets of currency, Charles O'Rear/Corbis; Page 289, Scientists in lab with microscope, Digital Vision; Page 290, Carl Frederich Gauss, Public domain; Page 292, Sun and planets, PhotoDisc; Page 296, Squirrel, Digital Vision; Page 296, Cuckoo clock, PhotoDisc Red; Page 296, Samurai sword production, Bettmann/Corbis; Page 297, Black Jack game, PhotoDisc; Page 298, Beehive, Digital Vision; Page 298, Leonardo of Pisa, Fibonacci, Corbis;

tower, St. Mark's Basillica, Allen R. Angel; Page 623, Rubik's cube, Silver Burdett Ginn; Page 631, Birthday celebration, PhotoDisc Red; Page 634, Enigma encrypting machine, Allen R. Angel; Page 635, Bus driver, PhotoDisc Red; Page 636, Protons and neutrons, Fermilab/Peter Arnold; Page 637, Amalie Emmy Noether, The Granger Collection; Page 639, Tennis pro, PhotoDisc; Page 639, Movie theater manager, Corbis RF; Page 640, Birthday party, Stockbyte/Getty RF; Page 643, Fire fighters, Corbis RF; Page 645, College graduates, Beth Anderson; Page 646, Woman in the rain, Blend Images/Getty Royalty Free; Page 648, Cherries, PhotoDisc; Page 651, Waitress, Rubberball Productions/Getty Royalty Free; Page 652, Chess game, Corbis Royalty Free; Page 653, Taj Mahal, Photosindia/Getty Royalty Free; Page 656, Charter fishing boat, Digital Vision; Page 656, Man on weight scale, Blend Images/Getty Royalty Free; Page 657, Auto mechanic, PhotoDisc; Page 658, Eye examination, PhotoDisc; Page 659, Pawnshop, Dana White/PhotoEdit; Page 665, Restaurant patrons, PhotoDisc; Page 666, Pawnbroker, Beth Anderson; Page 667, U.S. Treasury Building, MedioImages/Getty Royalty Free; Page 668, Albert Einstein, Library of Congress; Page 669, Benjamin Franklin, The Granger Collection; Page 670, Manhattan, Allen R. Angel; Page 673, Bunker Hill Community College, Beth Anderson; Page 675, Little League game, Courtesy of authors; Page 676, Waitress, PhotoDisc; Page 677, Kieler water tower, Allen R. Angel; Page 677, Larger tractor with cab, Dennis Runde; Page 677, People golfing, Allen R. Angel; Page 678, Woman trying on shoes, PhotoDisc Red; Page 679, Stoves at appliance store, PhotoDisc; Page 680, 12-ft boat, PhotoDisc; Page 681, 1965 Mustang, Corbis/Bettmann; Page 683, Credit cards, Beth Anderson; Page 686, Stereo system, Digital Vision; Page 688, 2006 PT Cruiser GT, AFP/Getty Editorial; Page 690, Paint store shoppers, Somos/Getty Royalty Free; Page 691, Kids watching TV, Digital Vision; Page 693, Young family with their home, Digital Vision; Page 693, Sears catalog page, Public domain; Page 694, Couple with new home, Photographer's Choice RF/Getty Royalty Free; Page 695, Cape-style house, PhotoDisc; Page 699, Cape-style house for sale, PhotoDisc; Page 702, Brooklyn brownstone, PhotoDisc Red; Page 702, Couple in front of house for sale, PhotoDisc; Page 703, Couple applying for mortgage, PhotoDisc; Page 704, House on lake, Digital Vision; Page 706, Active retirement-age couple, PhotoDisc Blue; Page 707, Coke bottle, Beth Anderson; Page 708, Bonds, Corbis Royalty Free; Page 710, Richard playing piano, Dennis Runde; Page 715, Grandchild and grandparents, Stockbyte Silver/Getty Royalty Free; Page 715, Campaign volunteer, Digital Vision; Page 717, Skateboarder, Duomo/Corbis; Page 718, Woman buying new clothes, PhotoDisc Red; Page 720, House for sale (bungalow), PhotoDisc; Page 720, John Deere combine harvesting corn, Digital Vision; Page 721, New touring bicycle, Corbis; Page 723, Lottery winning family, AFP/Getty Editorial; Page 724, Buying music CDs, Corbis Royalty Free; Page 724, Jacob Bernoulli, Corbis/Bettmann; Page 726, Husky with brown and blue eyes, Allen R. Angel; Page 728, Joe Mauer, Minnesota Twins, Getty Images Sport; Page 729, Benjamin Franklin Commemorative Coin, U.S. Mint; Page 729, Bird feeder, PhotoDisc; Page 730, Rabbit, PhotoDisc Red; Page 730, MapQuest web page, © 2007 Mapquest, Inc. All Rights Reserved; Page 733, Casino games, Corbis Royalty Free; Page 734, Northern cardinal, Allen R. Angel; Page 739, Young boy with a basketball, Image Source/Getty Royalty Free; Page 741, Chicago White Sox, 2005 World Series, Major League Baseball/Getty Images; Page 744, NBC game show *Deal or No Deal*, Chris Haston © NBC/Courtesy: Everett Collection; Page 747, NASCAR driver Tony Stewart, Getty Images Sport; Page 747, Bingo board, Allen R. Angel; Page 748, Horse race, Kevin R. Morris/Corbis; Page 749, Rolling die, fStop/Getty Royalty Free; Page 750, JetBlue plane, Beth Anderson; Page 751, Parking ticket, PhotoDisc; Page 752, Hot dog vendor, Digital Vision; Page 756, Candace Parker, University of Tennessee, Corbis; Page 756, Hollywood Bowl, Corbis; Page 757, Student taking a test, PhotoDisc Blue; Page 759, Construction site planning, Digital Vision; Page 761, Parents with three children, PhotoDisc Red; Page 763, Kids eating in restaurant, Stockbyte Platinum/Getty Royalty Free; Page 764, Beyoncé Knowles, Getty Images Entertainment; Page 767, Making paint choices, PhotoDisc Red; Page 769, Couple buying kitchen appliances, Stockbyte Platinum/Getty Royalty Free; Page 770, Pea plant, PhotoDisc; Page 771, Renting a movie, Stockdisc Premium/Getty Royalty Free; Page 774, Family with 9 girls, Corbis/Bettmann; Page 776, 24 kids on steps, PhotoDisc; Page 777, Slot machines, Allen R. Angel; Page 781, Golfers, PhotoDisc; Page 782, Slot machine, Allen R. Angel; Page 784, Basket of apples, PhotoDisc; Page 788, U.S. currency, Beth Anderson; Page 790, Katie Couric, *CBS Evening News*, CBS/Getty Images; Page 791, Display of televisions, Digital Vision; Page 792, Illustration of slot machine, © Tomo Narashima. Used with permission; Page 793, Man shopping for clothes, PhotoDisc Red; Page 794, The Three Stooges, Bettmann/Corbis; Page 794, Seven children, Corbis Royalty Free; Page 796, Ski club, Digital Vision; Page 797, Biking club, Corbis Royalty Free; Page 800, Nurses, Getty Royalty Free; Page 802, Monet's *Water Lily Pond*, Musee d'Orsay Paris/Lauros-Giradon, Paris/SuperStock; Page 803, Student meeting, Ingram Publishing/Getty Royalty Free; Page 805, Bridge hand, Beth Anderson; Page 805, Woman arranging flowers, PhotoDisc Red; Page 806, Hawaii, Getty RF; Page 807, Theatre district, New York City, Beth Anderson; Page 807, Jai alai game, Tom McCarthy/PhotoEdit; Page 808, Tray of hors d'oeuvres, PhotoDisc; Page 809, Woman with playing cards, PhotoDisc Red; Page 810, The Dead Man's Hand, Allen R. Angel; Page 811, Security guard, PhotoAlto/Getty Royalty Free; Page 812, Rare coin collection, PhotoDisc; Page 813, Yellow Lab puppies, Stockbyte/Getty RF; Page 813, Customers at bookstore, Getty Royalty Free; Page 814, Baseball game, Allen R. Angel; Page 816, Waiter and customer, Corbis Royalty Free; Page 820, Flashlights, Beth Anderson; Page 823, Dolphin, Allen R. Angel; Page 825, Person tasting cheese, Stockbyte Platinum/Getty Royalty Free; Page 826, Baseball game, Allen R. Angel; Page 828, Spelling bee, Photo Edit; Page 828, Sled dogs, Corbis Royalty Free; Page 829, Deep sea fishing, PhotoDisc Red; Page 830, Toll booth at Golden Gate bridge Beth Anderson; Page 832, Numbers presentation in a bar graph, Stockbyte/Getty Royalty Free; Page 833, Customer at pharmacy, PhotoDisc; Page 834, Neilsen survey family, Neilsen Media Research; Page 838, iPods on assembly line, PhotoEdit; Page 839, Chewing bubble gum, Westend61/Getty RF; Page 843, Man organizing data, Getty RF; Page 847, Seattle, Digital Vision; Page 848, Eating a hot dog, Digital Vision; Page 849, Mumbai, India, Digital Vision; Page 849, Student graph presentation, Image Source/Getty RF; Page 850, Circus clown, Image Source/Getty RF; Page 854, Bed and Breakfast, courtesy Janet and Rick Wolf; Page 857, Dancers at an audition, Digital Vision; Page 859, Jogging on treadmill, PhotoDisc Red; Page 859, Visiting a museum, Beth Anderson; Page 860, Putting gas in a car, Photographer's Choice RF/Getty RF; Page 862, San Jose, California, Getty; Page 862, Youngstown, Ohio, U.S. Department of Justice; Page 863, Lissy, 2 months old, Bruce Anderson; Page 868, Bookstore, Photographer's Choice RF/Getty RF; Page 868, Airport, Beth Anderson; Page 871, Water park, Allen R. Angel; Page 871, Babe Ruth, Bettmann/Corbis; Page 871, Mickey Mantle, Bettmann/Corbis; Page 872, Airplane engine, Stockbyte/Getty RF; Page 875, Sydney opera house, Corbis Royalty Free; Page 878, Fishermen with poles, Digital Vision; Page 881, Students taking exam, Digital Vision; Page 884, David. H Blackwell, Courtesy David Blackwell, University of California, Berkeley; Page 885, Motorola phones, Beth Anderson; Page 894, Woman with tire, Corbis Royalty Free; Page 896, Cup in vending machine, Allen R. Angel; Page 896, Box of Corn Flakes, Beth Anderson; Page 896, Children in daycare, Corbis Royalty Free; Page 897, Taking a test, PhotoDisc; Page 898, Buying a used car, PhotoDisc Red; Page 909, Fitness class, PhotoDisc; Page 909, Girl studying, PhotoDisc Red; Page 909, Gatorade bottle, Beth Anderson; Page 910, Hotel swimming pool, Allen R. Angel; Page 911, Woman measuring an arm, Stockbyte/PhotoDisc Blue; Page 914, Police car, Corbis Royalty Free.

I N D E X of Applications

INDEX

Note: Page numbers preceded by AA or AB indicate appendices.